ANNUAL REVIEW OF
ECOLOGY AND SYSTEMATICS

EDITORIAL COMMITTEE (1977)

P. W. FRANK
C. S. HOLLING
R. F. JOHNSTON
S. A. LEVIN
D. A. LIVINGSTONE
C. D. MICHENER
L. R. POMEROY
O. T. SOLBRIG

Responsible for the Organization of Volume 8
(Editorial Committee, 1975)

P. W. FRANK
C. S. HOLLING
R. F. JOHNSTON
D. A. LIVINGSTONE
C. D. MICHENER
E. C. PIELOU
L. R. POMEROY
P. H. RAVEN
P. GRANT (Guest)

Production Editor   R. L. BURKE
Indexing Coordinator   M. A. GLASS
Subject Indexer   B. OZAKI

ANNUAL REVIEW OF ECOLOGY AND SYSTEMATICS

RICHARD F. JOHNSTON, Editor
University of Kansas

PETER W. FRANK, Associate Editor
University of Oregon

CHARLES D. MICHENER, Associate Editor
University of Kansas

VOLUME 8

1977

ANNUAL REVIEWS INC.   4139 EL CAMINO WAY   PALO ALTO, CALIFORNIA 94306

ANNUAL REVIEWS INC.
Palo Alto, California USA

COPYRIGHT © 1977 BY ANNUAL REVIEWS INC, PALO ALTO, CALIFORNIA.
ALL RIGHTS RESERVED. No part of this book may be reproduced in any form
or by any means without permission in writing from the publisher.

International Standard Book Number: 08243-1408-5
Library of Congress Catalog Number: 71-135616

Annual Reviews Inc. and the Editors of its publications assume no
responsibility for the statements expressed by the contributors to this *Review*.

REPRINTS

The conspicuous number aligned in the margin with the title of each article in this volume is a key for use in ordering reprints. Available reprints are priced at the uniform rate of $1 each postpaid. The minimum acceptable reprint order is 10 reprints and/or $10.00, prepaid. A quantity discount is available.

PRINTED AND BOUND IN THE UNITED STATES OF AMERICA

PREFACE

The search for an understanding of the world's biota is necessarily undertaken piecemeal. Initial emphasis is put upon one or a few taxa, or on parts of processes or patterns. Understanding itself comes in even smaller pieces.

In many fields of study, finding a simple explanation for a phenomenon is often only the beginning of the sequence. Exceptions to the explanation emerge shortly, and a period follows in which the phenomenon may seem slowly to become intractable to conventional thinking. Each exception demands a detailed explanation; further study shows enormous complexity to have been involved from the start; in retrospect we sometimes appear foolish in having proposed simplicity at all.

Our current review of the evolution of life history traits by S. Stearns suggest why this situation will be with us a great deal longer. Stearns notes that the evolutionary process has been to some extent understood by researchers in both evolutionary ecology and genetics, despite the fact that they have used vastly different assumptions and techniques. The phenotypes with which researchers in both fields work are only fragments of the organism–environment reality; and as each route proceeds from simplicities to complexities, the threat is continual that the latter will overwhelm the former.

What can be gotten from such fragments, however, is frequently impressive, and the reviews in the present volume document this for several subfields in ecology and systematics. R. B. Payne looks at the evolution of brood parasitism in birds, and A. J. Nahmias and D. C. Reanney examine the evolution of viruses. W. Schlesinger details information on carbon contained in terrestrial detritus. P. J. Grubb provides a view of the distribution of tropical montane forests ancillary to that of E. G. Leigh in an earlier volume of the *Annual Review of Ecology and Systematics*. J. S. Jones, B. H. Leith, and P. Rawlings discuss the interplay of factors responsible for the shell polymorphisms in *Cepaea* species. Two approaches to the study of migration are set forth, one by E. Gwinner for birds, and another by W. C. Leggett for fishes.

Natural complexities are rendered more complex when humans modify even a few of the variables under their control. R. M. Baxter reviews environmental effects of dams and water impoundments; O. L. Loucks summarizes the emergence of research on agricultural ecosystems; J. E. Cohen reviews recent developments in the study of schistosome parasites; and A. F. G. Dixon writes on aphid ecology.

Two reviews of studies of the fossil record are included in this volume: J. A. Hopson is concerned with brain size and behavior in archosaurian reptiles, and S. D. Webb deals with the history of savanna vertebrates in North America. Biochemical approaches to systematics are examined by L. H. Throckmorton for a number of drosophilids and by G. B. Johnson for the significance of electrophoretically cryptic genetic variation. Several complex systems of coevolution form the basis of B. L. Bentley's review of plant-nectar–arthropod relationships. Some of the more unlikely convergences between galliform birds and ungulates are summarized by V. Geist.

In short, Volume 8 is again occupied by and preoccupied with heterogeneity. We wish to thank Professor Peter Grant for joining us in planning the contents of this volume. Ms. Susan Futterman and Mr. Richard Burke were critically important to the development of the volume from first to last. We are greatly indebted to our authors for their contributions, and we thank them for their good spirits this past year.

THE EDITORS AND THE EDITORIAL COMMITTEE

CONTENTS

THE ECOLOGY OF BROOD PARASITISM IN BIRDS, *Robert B. Payne*	1
THE EVOLUTION OF VIRUSES, *André J. Nahmias and Darryl C. Reanney*	29
CARBON BALANCE IN TERRESTRIAL DETRITUS, *William H. Schlesinger*	51
CONTROL OF FOREST GROWTH AND DISTRIBUTION ON WET TROPICAL MOUNTAINS, *P. J. Grubb*	83
POLYMORPHISM IN CEPAEA, *J. S. Jones, B. H. Leith, and P. Rawlings*	109
THE EVOLUTION OF LIFE HISTORY TRAITS, *Stephen C. Stearns*	145
EMERGENCE OF RESEARCH ON AGRO-ECOSYSTEMS, *Orie L. Loucks*	173
A COMPARISON OF SOCIAL ADAPTATIONS IN RELATION TO ECOLOGY IN GALLINACEOUS BIRD AND UNGULATE SOCIETIES, *Valerius Geist*	193
MATHEMATICAL MODELS OF SCHISTOSOMIASIS, *Joel E. Cohen*	209
DROSOPHILA SYSTEMATICS AND BIOCHEMICAL EVOLUTION, *Lynn H. Throckmorton*	235
ENVIRONMENTAL EFFECTS OF DAMS AND IMPOUNDMENTS, *R. M. Baxter*	255
THE ECOLOGY OF FISH MIGRATIONS, *William C. Leggett*	285
ASSESSING ELECTROPHORETIC SIMILARITY, *George B. Johnson*	309
APHID ECOLOGY, *A. F. G. Dixon*	329
A HISTORY OF SAVANNA VERTEBRATES IN THE NEW WORLD (Part I), *S. David Webb*	355
CIRCANNUAL RHYTHMS IN BIRD MIGRATION, *Eberhard Gwinner*	381
EXTRAFLORAL NECTARIES AND PROTECTION BY PUGNACIOUS BODYGUARDS, *Barbara L. Bentley*	407
RELATIVE BRAIN SIZE AND BEHAVIOR IN ARCHOSAURIAN REPTILES, *James A. Hopson*	429
INDEXES	
AUTHOR INDEX	449
SUBJECT INDEX	463
CUMULATIVE INDEX OF CONTRIBUTING AUTHORS, VOLUMES 4–8	487
CUMULATIVE INDEX OF CHAPTER TITLES, VOLUMES 4–8	488

ANNUAL REVIEWS INC. is a nonprofit corporation established to promote the advancement of the sciences. Beginning in 1932 with the *Annual Review of Biochemistry,* the Company has pursued as its principal function the publication of high quality, reasonably priced Annual Review volumes. The volumes are organized by Editors and Editorial Committees who invite qualified authors to contribute critical articles reviewing significant developments within each major discipline.

Annual Reviews Inc. is administered by a Board of Directors whose members serve without compensation.

BOARD OF DIRECTORS
1977

Dr. J. Murray Luck
Founder Emeritus, Annual Reviews Inc.
Department of Chemistry
Stanford University

Dr. Joshua Lederberg
President, Annual Reviews Inc.
Department of Genetics
Stanford University Medical School

Dr. James E. Howell
Vice President, Annual Reviews Inc.
Graduate School of Business
Stanford University

Dr. William O. Baker
President
Bell Telephone Laboratories

Dr. Sidney D. Drell
Deputy Director
Stanford Linear Accelerator Center

Dr. Eugene Garfield
President
Institute for Scientific Information

Dr. William D. McElroy
Chancellor
University of California, San Diego

Dr. William F. Miller
Vice President and Provost
Stanford University

Dr. John Pappenheimer
Department of Physiology
Harvard Medical School

Dr. Colin S. Pittendrigh
Director
Hopkins Marine Station

Dr. Esmond E. Snell
Department of Microbiology
University of Texas at Austin

Dr. Harriet Zuckerman
Department of Sociology
Columbia University

Annual Reviews are published in the following sciences: Anthropology, Astronomy and Astrophysics, Biochemistry, Biophysics and Bioengineering, Earth and Planetary Sciences, Ecology and Systematics, Energy, Entomology, Fluid Mechanics, Genetics, Materials Science, Medicine, Microbiology, Nuclear Science, Pharmacology and Toxicology, Physical Chemistry, Physiology, Phytopathology, Plant Physiology, Psychology, and Sociology. The *Annual Review of Neuroscience* will begin publication in 1978. In addition, two special volumes have been published by Annual Reviews Inc.: *History of Entomology* (1973) and *The Excitement and Fascination of Science* (1965).

THE ECOLOGY OF BROOD PARASITISM IN BIRDS

❖4115

Robert B. Payne

Museum of Zoology and Division of Biological Sciences, University of Michigan,
Ann Arbor, Michigan 48109

INTRODUCTION

Birds as brood parasites lay their eggs in the nests of other kinds of birds; these "hosts" incubate and rear the young. In the Old World, cuckoos have long been known as brood parasites. The early Vedic writers of India as well as Aristotle mentioned as common knowledge the fact that cuckoos are reared by other species (42). About 1% of all bird species are brood parasites, including the honey guides (Indicatoridae), nearly half of the 130 species of cuckoos (Cuculidae), two genera of finches (*Vidua* and *Anomalospiza*, Ploceidae), five cowbirds (Icteridae), and a duck (*Heteronetta atricapilla*, Anatidae) (37–42, 91–93, 114, 132, 136, 199). No other vertebrate groups with parental care so consistently parasitize other species. [A few freshwater fish do occasionally (14).] Counterparts of the cuckoos are known among insects, of which several groups are specialized for interactions with social insects, ranging from facultative commensalism to an inquilinism close to the cuckoo nexus. Some adult insects as well as immature depend completely on their hosts (4, 202). Most parasitic birds are altricial; their nestlings depend on the host for food. The parasitic duck, however, obtains only protection and warmth from the host and feeds and cares for itself shortly after hatching (199).

HOST SELECTION AND SPECIALIZATION

Parasitic birds exploit mainly the parental care of altricial birds. Young altricial birds, particularly young passerines, grow rapidly after hatching, and their parents deliver them large amounts of high protein food. Cuckoos, honeyguides, and cowbirds parasitize altricial insectivorous birds (38, 39, 41, 45, 81, 82, 91–93, 182, 184). Cuckoos occasionally parasitize finches that feed seeds to their young; cuckoo young often starve in these nests (81). Seed-eating carduelines are scarcely parasitized, whereas emberizine finches in the same area feed insects to their young and are parasitized (81, 92–94, 196). In the New World, few carduelines are parasitized by cowbirds; some can rear a cowbird to fledging (41). Fruit eaters may be generally avoided; but nearly 8% of waxwing nests found in northern Michigan contained cowbird eggs or young, and a higher proportion may have been parasitized but had the egg ejected by the host before the nests were observed (157). Tropical fruit-eating passerines are avoided by the cuckoos except for birds that feed insects to their young (38, 43, 45, 73, 92, 140). In contrast, the viduine finches parasitize the estrildid grassfinches which feed their young almost exclusively on grass seeds, and young viduines have been reared successfully in captivity on a seed diet (40, 103, 114, 137). Nutritional constraints apparently are not further significant in host selection, as more than 100 species of fosterers are known to have reared young cowbirds *Molothrus ater* to fledging (41, 46, 47), and the growth rates of nestling cowbirds are similar when fed by different host species (122). In some cuckoos, more than 20 species of fosterers in several families have reared the young (45, 73, 92).

Compatible parental behavior of the foster parents is necessary for successful parasitism. Most passerine young beg by opening the gape widely; the parent thrusts food deep into the nestling's throat. Young cowbirds and cuckoos also beg and are fed in this way. Young estrildid finches twist the head upsidedown and wave it from side to side; their parents feed them by pumping regurgitated seeds into their mouths. The viduine young beg in the same way and viduine parasitism is restricted to the estrildids (114, 132).

Another constraint upon host selection is egg size. Eggs are incubated by contact with the parent's brood patch, and contact might be poor if the parasites' eggs were smaller than the eggs of the hosts. Common passerine species with eggs larger than the local parasite generally are not parasitized (6, 39, 47, 95).

The availability of birds as hosts also depends on their abundance. In England, the main hosts of *Cuculus canorus* are among the most abundant species, and in North America the largest number of host records for the cowbird *M. ater* are for the commonest vireos, warblers, and emberizine finches (41, 46, 47, 81). Apparently these species are used in proportion to their numbers. Birds of the open woodland are more often parasitized than are birds in dense forest or birds in treeless grassland or desert, perhaps because the parasites more readily find nests when perched in the open (37–39, 41, 95, 114, 132).

Studies of banded females are needed to determine the degree of host specialization of individual birds. Some cuckoos parasitize only one or a few species in one locality, and different hosts in another (10, 45, 72, 91, 92). The cowbird *M. ater* is an extreme generalist as a species; both generalist and specialist females have been reported (41, 99, 192, 194).

Sympatric cuckoos usually parasitize different species of hosts (6, 38, 44, 45, 73, 101, 140). This selectivity may help to avoid competition (44, 45, 101, 172). Some overlap occurs; most host species in India are shared by two or more cuckoos (6); six cuckoo species are said to parasitize one Australian host (45); both a honeyguide and a cuckoo parasitize *Spreo bicolor* in South Africa (39, 43). [The Indian cuckoo data are questionable; most of Baker's eggs were collected by others, and identifications were often made much later (6).] The difference in host size of the European *Cuculus c. canorus* (most hosts are smaller than 20 g), the only cuckoo in its range, and the African *C. (c.) gularis* (all hosts are larger than 20 g), which coexists with other smaller cuckoo species (44, 81), suggests an effect of other parasite species in host selection. A similar trend is suggested in the differences in sizes of the hosts of *C. canorus* in Europe and in Japan, where other cuckoos breed (44).

The parasitic specificity of individual birds in species that parasitize several host species may be related to learning of the host during parental care, as suggested for cowbirds and cuckoos (95, 172). Host selection can be switched at least within a few generations. Songbirds introduced into New Zealand within the past 120 years are now parasitized by local cuckoos (45). In Sri Lanka the cuckoo *Eudynamis scolopacea* parasitized *Corvus macrorhynchus* prior to 1880 and added *C. splendens* within 60 years (142). An imprinting-like attachment to a new host might be established within a single generation if a cuckoo is reared successfully from an egg laid in the nest of an accidental host species.

EFFECTS OF BROOD PARASITISM ON BREEDING SUCCESS OF THE HOST

Parasitism usually depresses the nesting success of the host. The means by which nesting success is depressed varies with the tactics of the parasite. Female cuckoos may remove one of the host's eggs (18, 19, 32, 56, 72, 203) perhaps to adjust the clutch size according to the hosts' incubation behavior. They may simply eat the eggs to nourish themselves. They may also cause the host to desert and nest again by eating all the eggs if they find the nest after incubation has begun. The nonremoval of eggs in many fresh clutches, the eating of eggs nearly ready to hatch (32, 75, 136, 175, 203), and the murder of nestlings without eating them (75, 86, 104, 187, 203) suggest the importance of the last two hypotheses as the adaptive bases of nest robbing.

Cuckoos sometimes lay from a perch above the nest, and a thick-shelled cuckoo egg may dent or crack a host egg on impact (51, 75, 105). At hatching, some young cuckoos evict all eggs or other nestlings from the nest (45, 71, 73, 75, 87, 92). Some young parasites kill the host young or crowd them until they starve (39, 41, 43, 51, 72, 88, 103). After fledging, the young parasites may affect the survival of host young by monopolizing parental care. The host's own young may starve or be eaten (109). If a parasite is the only fledgling, it may delay another nesting attempt by the host; and if a parasite exhausts its host then the host may not breed again (51, 132).

The impact of brood parasitism on hosts can be calculated from the proportion of nests parasitized locally and the difference in host nesting success between parasitized and unparasitized nests (Table 1). Individual host nests may fail when parasitized, and the net effect on the breeding success of host populations is marked in most host-parasite systems. Brood parasites that do not depress their host's success greatly are the parasitic duck *Heteronetta atricapilla*, which has a negligible effect on the breeding success of nesting waterbirds, and the giant cowbird *Scaphidura oryzivora*, where the young may reduce the mortality of host young by removing botfly larvae from them. In this last species the overall effect of brood parasitism would depend on the degree of botfly infestation in local host populations.

Table 1 Effect of parasitism on fledging success of some foster species

Parasite	Host	Nests parasitized (%)	Survival, egg to fledging (%)[a] nonparasitized	Survival, egg to fledging (%)[a] parasitized	Calculated reduction in recruitment (%)[b]	Refs.
Heteronetta atricapilla	*Fulica rufifrons*	54	{87}	{81}	3.2	(199)
Molothrus ater	*Sayornis phoebe*	24	46	9	8.9[c]	(78)
	Empidonax virescens	67	61	54	4.7	(193)
	Vireo olivaceus	72	81	22	42.5	(173)
	Dendroica kirtlandii	55	32	7	13.8	(95)
	Dendroica kirtlandii	38	38[d]	—	—	(194)
	Dendroica kirtlandii	3	63[d,e]	—	—	(195)
	Melospiza melodia	44	[3.4]	[2.4]	12.9	(113)
	Chondestes grammacus	45	55	20	15.8	(111)
M. bonariensis	*Agelaius xanthomus*	74	38 {50}	18 {25}	14.8	(145)
	Zonotrichia capensis	66	44	12	21.1	(77)
	Zonotrichia capensis	61	30	19	6.7	(165)
Scaphidura oryzivora	*Zarhynchus wagleri* & *Cacicus cela*	28	{46}	{25}	5.9	(170)
	Zarhynchus wagleri & *Cacicus cela*	72	{18}	{45}	−19.4[g,h]	(170)
Vidua chalybeata	*Lagonosticta senegala*	36	28	30	−2.5[h]	(103)
Clamator jacobinus	*Turdoides striatus*	53	59.5	31.6	14.8	(51)
	T. caudatus	30	77	33	13.2	(51)

[a] Figures in brackets are mean number of host young fledged per nest where at least one host young is fledged. Figures in braces are percentages of all nests producing at least one host fledging.
[b] Calculated reduction = percentage of nests parasitized × percentage reduction (difference) in survival of host young in parasitized nests. Calculated reduction in recruitment is lower than estimated reduction in some of the papers cited, where all failures of parasitized nests were attributed directly to the parasites. Estimates in Table 1 are derived from local populations, and studies were sometimes made because the brood parasites were unusually abundant, so the estimates may not be representative of the total impact of parasitism on all host populations.
[c] Loss from parasitism is higher if eggs removed by a female from nests not later parasitized are included.
[d] Nearly all cowbird eggs were removed from nests by the observer.
[e] Cowbirds trapped and removed from breeding area.
[f] Colonies with no botflies.
[g] Colonies with botflies present.
[h] Gain, not reduction, in success.

The overall effect of brood parasites on their host populations is not well known. It appears that cuckoos have little effect on host populations over broad areas. Lack (81) tabulated samples of nest records of small passerines in England and found that fewer than 3% of songbird nests are parasitized by *Cuculus canorus*. These data led him to conclude that "the Cuckoo is an almost negligible cause of egg and nestling losses among English breeding birds." The frequency of parasitism is also low among hosts in southern Africa, where nine cuckoo species are common (140). There, rates of cuckoo parasitism average less than 3%. The incidence of cuckoo parasitism is higher in certain locales (31, 73, 81, 88, 148, 203). In England some populations of Robins *Erithacus rubecula* are unparasitized, whereas in others 35% of the nests are parasitized (81).

Local depression and extinction of host bird populations by a parasite may occur, but the effect is not well documented. A population of Sedge Warblers *Acrocephalus schoenobaenus* was parasitized by *C. canorus* with increasing frequency over several years and the host became scarce and then disappeared (125). Another warbler population decreased when cuckoo parasitism increased (158) and has been cited

BROOD PARASITISM IN BIRDS 5

(155, 172) as evidence of the effect of parasitism on a host species. However, these studies were much too local to sample adequately the population effects of parasitism. One would expect immigration of new warblers from year to year; the warblers may have declined because of changes in local habitats.

The impact of cuckoo parasitism on host populations is best known in the case of the Indian *Clamator jacobinus*, which parasitizes *Turdoides* babblers (51). Host eggs may be cracked when the cuckoo eggs are laid, and the young cuckoos crowd and starve the host young. Babblers remain in family groups until the second year, and yearling babblers defend the group territory and feed the young of their older relatives. Parasitism may decrease the group size of the babblers. Because each territorial group nests by itself, and the number of babblers (not of groups) appears to be constant in an area, parasitism may have the long-term effect of increasing the number of nests available for the cuckoos (51).

Hosts of the cowbird *Molothrus ater* in North America and *M. bonariensis* in South America generally fledge fewer young than unparasitized parents. The presence of a parasite egg apparently does not inhibit the host from ovulating and laying its usual number of eggs in the clutch. Reduced success results from (*a*) removal of a host egg by the female cowbird, usually shortly before she lays her own, (*b*) puncture of the host's eggs with bill or claws by the cowbird, (*c*) lower hatching success of the host eggs, because of desertion by parasitized parents or earlier hatching of the parasite eggs, and (*d*) lower survival due to crowding and competition for parental care (41). Some small flycatchers often fail to fledge any young if a cowbird hatches in the nest (78, 193). Regional and local differences in the proportion of parasitized nests occur (34, 41, 47), and for widespread species the overall impact of cowbird parasitism may be less than suggested in Table 1.

Expansions in the range or number of parasitic cowbirds sometimes have been followed by decrease in host populations, and these observations are the best evidence available of the impact of the brood parasites on the host populations. The cowbird *M. bonariensis* extended its breeding range through the Antilles from 1950 to 1970. Its establishment and increase in Puerto Rico coincided with the decline in numbers and restriction in range of the endemic blackbird *Agelaius xanthomus*. In lowland Puerto Rico the blackbirds may be the only species parasitized and the only areas where the blackbirds now breed successfully are the small cays of the southwest, where the cowbirds are less common (143). An expansion of North American *M. ater* into the range of the warbler *Dendroica kirtlandii* in the late 1800s may have had a similar effect (95). The frequency of parasitism increased in the 1960s (194), while the total population of the warbler decreased (97). From 1966–1971, cowbird eggs were removed from nearly all study area warbler nests, and the local breeding success of warblers was then as great as in the unparasitized nests reported earlier (95, 194). More convincing is the difference in warbler success after cowbirds were removed from this host's breeding habitat (Table 1). The proportion of parasitized nests was 66% in undisturbed areas and 21% in the areas of cowbird removal (195).

In a study of finches, Morel (103) found little effect of parasitism on host young survival. Indigobirds *Vidua chalybeata* are species-specific parasites of firefinches

Lagonosticta senegala (103, 132). The parasites usually do not remove host eggs, and the young of the two species are reared together. The host eggs in parasitized nests have a higher hatching success than in unparasitized nests; the difference may result from closer parental attendance at nests with more eggs (103). Survival of host young from hatching to fledging is lower in unparasitized nests, and more young hosts are fledged from nests with no parasite young than from nests with young parasites. Parasitism seems to have a negligible net result in this species (103).

Facultative brood parasitic ducks depress the breeding success of nesting ducks. Redheads (*Aythya americana*) and ruddy ducks (*Oxyura jamaicensis*) usually incubate their own eggs but sometimes (redheads especially) lay in the nests of other ducks. Cinnamon teal *Anas cyanoptera* parasitized by redheads and ruddy ducks have smaller clutches of eggs than do unparasitized teal because the host eggs are displaced from the nest. Fewer host eggs hatch in nests with more eggs of the redheads or ruddy ducks. The "parasites" often lay after the teal clutch is complete, and so the number of eggs laid by the teal is not significantly affected by the presence of alien eggs. Nest abandonment and predation are no higher in parasitized nests than in unparasitized nests (76). The effect of these facultatively parasitic ducks on their hosts is greater than in the obligate parasitic black-headed duck (199).

BEHAVIOR AND COEVOLUTIONARY ASPECTS OF BROOD PARASITISM

That brood parasites decrease the nesting success of hosts has exerted selective pressure for host behavior that reduces parasitism. Among the possible responses of the foster species to parasitism are shifts of breeding season, attack against the parasite, warning calls, nest concealment, and nest defenses. More directly related to brood parasitism is the behavioral discrimination of the eggs or young in a mixed clutch or brood and the removal of the parasite by the nesting birds.

Adaptive responses by the hosts to parasitism have been followed and accompanied by adaptation in the brood parasites. In some a shorter incubation period has evolved. Some species kill their host's eggs or young. Most striking of the specific adaptations are the mimicry by the eggs of many cuckoos of the eggs of their hosts and the mimicry of their host's nestlings' mouth colors and patterns by the host-specific parasitic finches.

Breeding Seasons

Seasonal timing of breeding in birds generally correlates with the seasonality of food and nesting sites. In temperate regions it is timed proximally by photoperiodism (82, 89, 129). In California the seasonal gonadal development of the cowbirds *Molothrus ater* matches that of common host species; synchrony is mediated by a shared photoperiodic response (128, 134). Cowbirds begin laying within a few days of the onset of nestbuilding by the earliest phoebes *Sayornis phoebe* in Kansas (78). The close similarity in time suggests that ovarian development in the cowbirds is mediated by photoperiodism and the sight of nestbuilding hosts is needed only to complete its final stages and bring about ovulation (128, 134). South American

BROOD PARASITISM IN BIRDS 7

cowbirds *M. bonariensis* closely match the breeding time of their local hosts (77). In Britain, cuckoos *Cuculus canorus* begin to lay weeks after their hosts begin, perhaps because cuckoos cannot form eggs until the caterpillars on which they feed become large and numerous (81). The breeding season of each species-specific *Vidua* finch matches that of its host, rather than that of other local *Vidua* species (9, 102, 103, 132), suggesting that the parasites respond to the same seasonal changes as their hosts.

Behavioral Interactions Between Adult Host and Parasite

Brood parasites usually find the nest of their hosts by watching the nesters build (18, 19, 37, 41, 88, 121, 132, 134, 201). They also search in microhabitats where hosts often nest (121, 132, 172), and they may read locational clues in the intensity of a host's alarm reactions (154, 163). The stealth of nesting birds and the concealment of their nests may not be specific adaptations to avoid parasitism; avoidance of predators is probably more important. Calls given by hosts that see cuckoos are similar among different species (163); it is unknown whether these are identical to predator-alarm calls. Alarm calls may be effective in alerting a mate against going to the nest, in attracting potential mobbers to the parasite, or in attracting predators to the parasite (15, 33, 41, 113, 171).

Nesting birds are sometimes effective in driving the parasite away from the nest (6, 41, 113, 153, 171). Nesting hosts of the cuckoo *Cuculus canorus* typically fly at a cuckoo and peck and beat it vigorously (33, 171). Responses of hosts to models of female cowbirds *Molothrus ater* are generally most aggressive in the most frequently parasitized species. Within some host species, the aggressive response is more pronounced in western populations that have presumably had a longer experience of cowbird parasitism. Hosts at parasitized nests are more aggressive toward cowbird models on the average than hosts at unparasitized nests, suggesting a learned response to a nest intruder (154). The greater aggressiveness of hosts with parasitized nests suggests that the aggressive response is not always advantageous to the host; response may be more harmful than simply ignoring the cowbird, since the parasite might find the nest because of the behavior of the host (154). It would be of interest to test pairs early in the season, before cowbirds lay, to find whether the parasitism of aggressive or of nonresponsive host pairs is more likely.

Brood parasites often ignore the host reactions. When they are larger, cuckoos can withstand attacks during the few seconds taken to lay into the host nest (18). Also, females usually lay when the host is off the nest. In most species, the female parasite finds nests without the aid of a male (18, 19, 37, 41, 132, 134, 147, 180, 203). Observations by Chance (18, 19) and Wyllie (203) do not support the suggestion that male cuckoos *C. canorus* accompany females and flush the host, allowing the female access to the nest (80, 92, 119, 188). However, a male and female work together in *Clamator jacobinus* (51, 88).

The resemblance in plumage of several *Cuculus* cuckoos to that of hawks suggests mimicry directed towards the host. Some populations of *C. canorus* and *C. saturatus* are polymorphic, the females gray (like *Accipiter* hawks) or rufous (like *Falco* falcons), and the proportions of the color phases may vary geographically with the proportions of the hawk "models" (80, 188). Both juvenile and adult drongo cuckoos *Surniculus lugubris* are very similar in plumage to the juvenile and adult drongos *Dicrurus* spp. The significance of this resemblance is unknown—drongo cuckoos only infrequently parasitize the drongos (92).

The behavior and coloration of parasitic birds in general, especially of the females, are adaptations that may reduce their conspicuousness and that may increase their chances of finding an appropriate nest to parasitize. Most female brood parasites are dull, whereas some male cuckoos (especially the glossy cuckoos *Chrysococcyx*) and finches are brightly colored. The plumages of parasitic cuckoos on the average are duller and more cryptic than are the plumages of nesting cuckoos (127). They lack the agonistic releasers that may be present in nesting cuckoos.

Host Discrimination and Egg Mimicry

Nesting birds often incubate a parasite's eggs along with their own. Sometimes, however, they remove the parasite's egg from the nest, build a new nest over the parasite egg or the entire clutch, or nest anew at a different site. Removal of the parasite egg may be a host response specifically selected as a counter to parasitism (155). Many birds eject their own cracked eggs, eggs that fail to hatch, dead young, eggshells, and feces (65, 155). Some cowbird hosts remove other kinds of experimentally introduced eggs sooner than cowbird eggs (155). Some cuckoo hosts remove the cuckoo egg from the nest (6, 75, 186). Experiments beginning as early as 1775, when cuckoo eggs or other alien eggs were introduced into the nests of songbirds, show differences among species in the frequency with which they reject the alien eggs (1, 72, 75, 149, 179). Some hosts discriminate. Others, such as southern African bulbuls (*Pycnonotus* spp.), appear never to reject either natural or experimental parasitism by *Clamator jacobinus*, which lay plain white eggs unlike their own heavily spotted eggs (88, 179). Geographic variation in the color pattern of the eggs of some host-specific cuckoos (*Eudynamis scolopacea*) parallels that of the host eggs (5). Where such resemblance occurs it seems likely that the resemblance between cuckoo and host eggs results from a greater survival of cuckoo eggs that match the hosts' eggs (5–8, 75, 112, 172), and thus the gradual elimination of genotypes in the population that were associated with contrasting eggs. Among European passerines, rejection of cuckoo eggs seems most pronounced among species that are only infrequently parasitized (75).

Several Old World cuckoo populations are polymorphic or variable in egg color and markings, and in some, particularly *Cuculus canorus,* the egg morph often resembles the eggs of the host even when the host eggs themselves are polymorphic (7, 8, 17, 18–20, 70, 75, 150, 151, 172, 177). A female cuckoo apparently seeks out nests with eggs like her own. Mismatches may arise if a nest of the usual host is deserted or destroyed and the parasite deposits her formed eggs in whatever similar nest she can find (18, 75). Over broad geographic regions where cuckoos are egg-polymorphic more than 75% of cuckoo eggs found in the field match those of their host (5–8, 75, 177, 191). *Chrysococcyx caprius* in southern Africa usually parasitizes hosts of the same color and pattern (127).

Eggs of the North American generalist parasite cowbird *Molothrus ater* resemble in color and markings the eggs of many passerines that build open nests. Species of passerines in the same area were tested by Rothstein for their response to artificial cowbird eggs. The species tested were either "acceptors" (did not remove the egg) or "rejectors" (removed the egg) (155–156a). A few individual "acceptors" rejected eggs and a few "rejectors" accepted eggs, but most species were consistent in their responses to this experimental parasitism. Whether these birds accept or reject is independent of whether or not they are parasitized in nature (155, 156). The frequency of rejection within a host species is the same over broad geographic areas (156). The ability of some "rejector" species to discriminate cowbird eggs from their own eggs may depend on whether the cowbird eggs outnumber their own in the nest (156a). To find whether rejectors may learn to discriminate between their own eggs and cowbird eggs, Rothstein removed the first eggs laid by wild gray catbirds *Dumetella carolinensis,* replacing them with artificial cowbird eggs. When the first eggs allowed to remain in the nest were "cowbird" eggs, the catbirds were more likely to eject subsequent catbird eggs than cowbird eggs. The results suggest that birds may learn to recognize their own eggs by seeing them after they are laid (154a).

Some discriminating hosts build over the parasite egg and lay again (37, 41, 75); these nests may be reparasitized, and the process repeated. Parasitized nests may have several layers, each sandwiching one or more parasite eggs. More commonly, birds desert a parasitized clutch and nest elsewhere, sometimes avoiding the parasite. Desertion rates of nests containing cuckoo eggs for four European songbirds (*Phoenicurus phoenicurus, Motacilla alba, Lanius collurio,* and *Erithacus rubecula*) are about 10%, the same as parasite egg ejection rates; a fifth species (*Phylloscopus sibilatrix*) deserted 77% of the parasitized nests (17, 75). Comparable figures are unavailable for desertion of unparasitized nests of these cuckoo hosts.

Most parasite eggs are similar in size to host eggs. Often they are somewhat larger, but seldom are they smaller (e.g. *Clamator glandarius* eggs and *Eudynamis scolopacea* eggs in nests of their crow hosts) (5, 6, 75, 92). Experiments with eggs of one species of cowbird host indicate that eggs ½ normal size are likely to be abandoned, eggs 1½ × normal size are not incubated longer than normal eggs, and larger eggs are abandoned (64). The sizes of cuckoo *C. canorus* eggs in the nests of several European songbird species overlap considerably within each kind, though in a few cases the differences may be statistically significant (85). Minor differences in egg size of parasite and host do not appear important in host acceptance.

Host Discrimination and Mimicry of the Young

Many brood parasites are adapted to produce the signals of the host young that elicit parental care. Old World cuckoos in the subfamily Cuculinae lack the stiff bristle-like down of other groups of cuckoos and also lack the brightly colored spots found inside the nestlings' mouths in these groups [Phaenicophaeinae (63, 144, 146); Crotophaginae (28, 29, 169); Neomorphinae (106); Couinae (3); Centropodinae (164)]. The parasitic New World neomorphine *Tapera naevia* resemble the Old World parasitic cuckoos in lacking the peculiar down and mouth spots (62, 104).

Young passerine hosts lack the bristle-like down and usually have unicolored mouths. Resemblance to young hosts may enhance the survival of young cuckoos by eliciting brooding and foster-parental feeding.

Young brood parasites which upon hatching evict host eggs or young do not closely resemble the host young. Some host-specific parasites which are reared together with the host young resemble the hosts as nestlings. Viduine young differ only slightly from their estrildid nestmates (167); their mouth colors and markings match those of the species or species group of host estrildid (103, 114, 116, 132, 137, 183). Nestling begging calls and behavior also are like those of the host (103, 114, 116, 117, 132). Experimental studies on the responses of the estrildid parent to young with different mouth patterns are highly desirable. Young cuckoos that are reared with the young of host crows sometimes vaguely resemble them (43, 75, 178). Nevertheless, crows accept dissimilar eggs and young experimentally placed in their nests (55).

Begging calls of some young cuckoos and of one honey guide are said to resemble those of their hosts (22, 49, 52, 107, 110, 123, 127, 148; contra 108). However, no tape recordings have been made to document divergence in calls among the parasitic species or convergence between a parasite and its host except in the viduine finches.

Adaptations Promoting the Success of the Young Parasites

Brood parasites usually develop faster than do their hosts, and early hatching gives the parasite a head start on the host young in the nest. Incubation periods of parasitic cuckoos are usually 11–14 days, often 2–4 days shorter than their hosts. Incubation periods of parasitic cuckoos tend to be shorter than those of nesting cuckoo species, but the difference is not quite significant (Mann-Whitney $U = 27.5$, $p > .05$) (Table 2). The nesting cuckoos with the longer incubation periods are larger birds than are the parasitic cuckoos in Table 2. Incubation of eggs of parasitic cuckoos may require less time than those of hosts because cuckoo eggs are laid with partly developed embryos. The eggs of parasitic cuckoos are laid on alternate days. Each egg is retained in the oviduct, where the embryo develops for a day before the egg is laid (72, 87, 135, 136). Some nesting cuckoos also lay at intervals of two or three days, and it is not known whether their eggs are likewise developed at laying time. The honey guide *Indicator minor* may have an incubation period as short as 12 days, short for a piciform bird, but honey guides in the nest have not been extensively studied (39, 147). The cowbird *Molothrus ater* hatches in 11–13 days, often a day or two before its hosts (37, 41, 95). The finch *Vidua chalybeata* hatches in 10–11 days, usually 1–3 days before its host (103).

Growth of the parasite young is no more rapid than in nesting relatives of the same adult size (Table 2). In cuckoos that evict their nestmates, growth periods are long (20–21 days in six species of *Cuculus* and *Chrysococcyx*). In cuckoos that are reared with their nestmates, growth rates are like those of their hosts. Cuckoos the same size as the host at fledging grow at the same rate or slightly faster thereafter (51, 83, 176). When smaller than the host, other cuckoos gain less weight per day but fledge earlier (109, 189, 190). Fledging periods of nesting cuckoos range from

BROOD PARASITISM IN BIRDS   11

Table 2 Incubation and nestling periods in parasitic cuckoos and nesting cuckoos

Species	Parasitic (P) or nesting (N)	Incubation (days)	Fledging (days)	Refs.
Clamator jacobinus	P	11	15	(88)
Clamator levaillantii	P	11	12-17	(73)
Clamator glandarius	P	13.5-15	26-28	(105, 109, 181, 189)
Eudynamys scolopacea	P	13-14	19	(36, 54, 83)
Cuculus canorus	P	12-13	20-23	(19, 119)
Cuculus micropterus	P	12	21	(110)
Cuculus solitarius	P	11.5	20	(87, 88)
Cuculus clamosus	P	13-14	20-21	(72)
Chrysococcyx caprius	P	11-12	21	(148)
Chrysococcyx klaas	P	12	20	(72)
Coccyzus americanus	N	10	6-8	(58, 144)
Coccyzus erythropthalmus	N	10	6-7	(174)
Coccyzus pumilus	N	13	10+	(146)
Centropus senegalensis	N	18-19	15	(48)
Centropus superciliosus	N	14	18-20	(48)
Centropus toulou	N	14-16	19	(48)
Crotophaga ani	N	14	6-7	(79)
Crotophaga sulcirostris	N	12-13	9	(169)
Geococcyx californianus	N	17-18	18	(124)

15–20 days for large terrestrial species (*Geococcyx* and *Centropus*) to 6–10 days for smaller arboreal or terrestrial ones (*Coccyzus* and *Crotophaga*). Both parasitic and nesting cuckoos fledge upon reaching 50–60% of adult body weight (88, 124, 174), whereas passerines fledge when near their adult weight. Passerines feeding cuckoos may occasionally care for the nestling and dependent fledgling parasites longer than their own young would take to reach independence (18, 19, 51, 72, 119). The huge mouth, brightly colored gape, and persistent begging of the fledged cuckoo may elicit prolonged feeding by the foster parent and may also induce other species to feed the young cuckoo (2, 45, 52, 60, 125). Growth of parasitic cowbirds and finches that are reared in the nest with their foster siblings is faster only because of their larger size at hatching, and is not noticeably different from that of related nesting species (41, 77, 103, 113, 122).

The demands for parental care and the larger size of young parasites such as the cowbird *M. ater* often cause death by starvation or by crowding and trampling of the host's smaller young (41). Some cuckoos (*Cuculus, Chrysococcyx*) kill their nestmates. Upon hatching, the young cuckoos use an "evicting response:" They maneuver under a nearby egg or nestling, lift it onto their concave backs, and rear back to the rim of the nest until the egg or young bird plummets from the nest (71–73, 75, 88, 92, 203). Some *Indicator* honey guides have sharp hooks on the ends of their bills at hatching; the blind young bite and stab, killing their nestmates (39, 147).

Vocal Mimicry in Vidua Finches

Most parasitic finches mimic not only the begging calls of the host young, but also as adults mimic the songs of their foster species (114–118, 132, 133, 138, 141). Functional aspects of viduine song are best known for indigobirds *Vidua chalybeata* (132, 141; R. B. Payne, unpublished observations). Young *Vidua* learn the song of the foster species (114, 118, 132) and breeding males subsequently sing the song at special singing sites (116, 132, 133, 141). Females in the field and in experimental studies are attracted to songs of the males that mimic their own foster species (132, 133). Occasional wild males sing the song of another estrildid, suggesting that they were reared in the nest of that species (132).

The ecological significance of song mimicry by adult males is intraspecific, functioning in mate selection of males by females. Hosts are sometimes atracted to the vocalizing male *Vidua* but in nearly 2000 hours of observation of marked singing males, neither male or female parasites obviously responded to the visiting host (132, 141). By mimicking the song of his foster species a singing male advertises his success at being reared by this species, and the female is thereby attracted to a male with early experience similar to her own (132, 133). Modification of mimetic song by early experience with the fosterer allows not only intergeneration tradition of song and host-specific parasitism but also occasional switches of host species and exploitation of previously unparasitized hosts (132, 137).

BREEDING BIOLOGY, SOCIAL BEHAVIOR, AND POPULATION STRUCTURE

Brood parasites may be viewed as natural experiments in the effect of parental care on other variables of breeding ecology. Since they do not rear their own young, their reproduction is not limited by the number of young the parent can rear (82, 126, 136) and they would be expected to lay more eggs than nesting birds. Because the young brood parasites usually do not grow up together or have social interactions with their parents, kin may not form any social bonds. Because the parents do not feed their young, there is no social circumstance that favors an exclusive social organization or monogamy. Parasites with adaptations for parasitizing local populations of hosts might be expected to show a nondispersive population structure. On the contrary, all known parasites are dispersive and they acquire their local behavioral adaptations after they disperse.

Reproductive Strategies

Egg size, clutch size, the number of eggs laid in a season, and the dispersion of eggs among the nests are different in some brood parasites from their nesting relatives. Egg size in relation to body size is small in *Cuculus canorus* (26, 71). Payne (136) had compared egg size and body weight in 65 species of cuckoos. Eggs are smaller both in Old World and New World parasitic cuckoos than in any group of nesting cuckoos. There is no overlap in relative egg size between parasitic and nonparasitic

BROOD PARASITISM IN BIRDS 13

cuckoos. Eggs are relatively larger in parasitic cuckoo species that parasitize hosts larger than themselves (51,136). In finches, the parasitic *Vidua* and *Anomalospiza* have eggs smaller in proportion to body size than do the African Passerinae and Ploceinae (their probable closest relatives), but viduine eggs are the same proportional size as in the estrildids (139). In cowbirds, the eggs are similar in size to those of other icterids (11, 37). Egg size is similar in the parasitic honey guides and in other African piciform birds of similar size (162). Finally, eggs of the parasitic duck are relatively large, like their closest relatives (82, 162, 199).

The small eggs of the parasitic cuckoos may be due to size mimicry of the host eggs (6), to a shorter incubation period for smaller eggs, or to the allocation of energy stores to a large number of small eggs (136). Cuckoo eggs are often similar in size to the eggs of their hosts, but not identical (5, 6, 18, 72, 75, 91, 162, 203) and within a species egg size is not known to vary with that of the host, either among different hosts in one area or in different geographic areas (6, 85, 93, 136). The availability of energy not allocated to the maintenance of a parental behavior may permit parasitic cuckoos to lay more eggs than nesting cuckoos. The eggs of parasitic viduines are larger then the eggs of their hosts; the difference allows the parasites to be larger at hatching and to outcompete the host young for parental care (103, 139). The large egg of the parasitic duck may enable the precocial young to care for itself as soon as two days after hatching (199).

The number of parasite eggs laid in a season has been estimated after searching for nests every day in an area used by a female. Identifying the eggs of each female *Cuculus canorus* by their color and pattern, some observers found that a cuckoo lays 12 eggs or more in a season; other have reported smaller numbers (125, 135, 136, 161, 203). The difficulties in finding nests before any loss and in identifying a female by her eggs make these estimates of questionable accuracy. The location of host nests and the timing of nesting may affect cuckoo laying (13, 17, 150, 203). In field experiments, host nests were collected as the eggs were laid, causing the hosts to nest again and to be parasitized repeatedly (18, 19, 158, 161). Under these circumstances, a cuckoo laid 18–25 eggs in a season. Laying histories have been determined for several species of African cuckoos by examination of the ovaries (135, 136). These cuckoos lay 16–25 eggs in a season. The cuckoos ovulate on alternate days, and 3–6 eggs usually make up a series or clutch (18, 19, 51, 125, 135, 203), as found in both field and morphological studies.

The number of eggs laid by the cowbird *Molothrus ater* has been estimated in the field after searching for parasitized nests. Counts range from 7 to 18 in a season (41, 99, 113, 192). No studies of marked, laying females have been made, and since cowbird eggs are not as variable or polymorphic as are the eggs of cuckoos *C. canorus,* egg identification may have been inaccurate. Captive cowbirds lay more eggs in a series; Friedmann (37) reported a female's laying 13 eggs in 14 days in nests with candy eggs. Ovary histories show that wild females lay in a series of 2–5 eggs; mean clutch size in Oklahoma, California, and Michigan is between 3.9–4.1. Between 11 and 25 eggs are usually laid in a season, with more in areas with a longer breeding season (126, 138). The ovaries of South American cowbirds bore evidence of similar clutch sizes (29).

Viduine finches almost always have a clutch of three (139). Many nests have two or more eggs of the viduine parasite, and several parasite young grow up together in a nest (40, 103, 116, 139, 167, 183). More multiply-parasitized nests occur than could be expected from a random binomial distribution of the parasite eggs (139). The timing and similarity in size of eggs in nests in the field and the laying behavior of captive *V. chalybeata* indicate that an individual female often lays two or three eggs in a nest. A young *V. chalybeata*'s chances of success from egg to fledging are independent of the number of parasite eggs up to the full size of the *Vidua* clutch (103, 139). The nearly constant clutch size may be related to an advantage of multiple parasitism by a female *Vidua*. In contrast, cowbird *M. ater* eggs are laid apparently at random with respect to the number of cowbird eggs already in the nest, and multiple nest parasitism often results from layings by different females (41, 96). Cowbirds are larger than most hosts, and only one cowbird fledges from many of the small hosts' nests. However, in some areas where the common hosts are large, multiple parasitism is common and is often successful (34). Nests parasitized by *C. canorus* almost always have only one cuckoo egg (13, 17, 53, 75, 125, 203), and in the rare instances of two cuckoo eggs, multiple parasitism is due to two females. Interference among young cuckoos would prevent the survival of more than one, and this explains why a female lays only once in a nest.

Clutch sizes in the parasitic cuckoos average slightly larger than they do in nesting cuckoos (136). Clutch size in the viduine finches is larger than in ploceine finches, but usually smaller than in the estrildid finches or the Passerinae sparrows. The other parasitic finch *Anomalospiza imberbis* has a clutch size like that of related nesting ploceines (139). Clutch size in cowbirds is like that of nesting icterids (11, 30, 69, 74, 138, 168). The overall tendency of clutch size in parasitic birds to be like that in related nesting species suggests that constraints other than limitation of family care (82) are generally important in determining clutch size. Repeated renestings and second nestings are uncommon in temperate North American icterids, whereas cowbirds in the same area lay over 20 eggs. Total numbers of eggs in a season in parasitic birds are generally higher than in their nesting relatives (136, 138, 139).

Mating Systems and Social Behavior

Few studies of individually marked parasitic birds have been completed. Cuckoos in the genera *Cuculus* and *Chrysococcyx* appear to be territorial, with persistent singing by dispersed males (19, 38, 45, 66, 148, 203), though some *Clamator* cuckoos are not (43, 51). Individual cowbirds tend to use nonexclusive areas in the breeding season (37, 84, 134). In the finches, *Vidua chalybeata* and related species defend sites where they sing and mate. When males are removed, males without call-sites or males with poorer sites replace them. The traditional call-sites are often held by a male through a breeding season and in successive years. Mating success among males results from their fighting over the better call-sites and also from female mate choice based more on male song than on any habitat differences among the local males (132, 141). The honey guides *Indicator indicator* and *I. minor* have highly traditional, dispersed call-sites where males chase others and where females

BROOD PARASITISM IN BIRDS 15

visit and mate (39, 147). *I. xanthonotus* males spend nearly all their time perched by a bee nest and defend the site against other males, allowing access to the bee nest (these honey guides eat wax) to females with whom they mate (23).

Most brood parasites have no evident pair bond and each male may mate with several females. No studies have been made of the mating systems of parasitic cuckoos. Howard (66) asserted that *Cuculus canorus* is polyandrous, but field observations of several male cuckoos associating with each other or with single females (38, 45,) suggest leks as much as they do polyandry. Females do not remain in the territory of a single male, and several females may lay in the same area (13, 17, 18, 203). In cowbirds, males may mate with one or more females (24, 25, 37, 84, 113). The honey guides *I. xanthonotus* are polygynous; one marked male mated with 18 different marked females (23). Individual male indigobirds *V. chalybeata* also are polygynous, sometimes mating with three or four females in a day. Both the frequency distribution and the individual variation in male mating success in indigobird populations are similar to those of such polygynous but nonparasitic birds as grouse and manakins in which males form leks and display to females, who visit only to mate. In each of two local populations, each characterized by its distinctive song dialect, one male completed more than half of all matings. Females sometimes mate with more than one male in a season (141). Although few species have been studied, the observed mating systems in marked populations of parasitic birds generally show polygynous mating and lack of a prolonged pair bond.

The loose social organization and the lack of social relationships between parent and offspring may explain the apparent rarity of alarm calls in parasitic birds. Viduine finches sometimes call when taking flight, usually in a flock that may provide cover (57). A call that might have been an alarm was heard when a mixed-species flock mobbed a snake (132). The rarity of alarm calls in parasitic birds is consistent with the hypothesis that such calls evolve only among species that remain in kinship groups (15).

Population Structure

Age structure in parasite populations does not differ obviously from that in nesting birds. First-year females breed in cowbirds (24, 25, 126, 134), in most or all parasitic finches (132, 141), and in some cuckoos (135, 136). Males may not breed until two years in cowbirds (134) and in some viduines (39). In the indigobirds, males have breeding plumage in the first year, but all aged and marked males that were seen to mate were at least two years of age (132, 141).

Schedules of natality and mortality are not well known. About 23% of *Cuculus canorus* eggs survive to fledging—fewer than in most European birds (13, 17, 119, 203). In the cowbird *Molothrus ater* the percentage is higher—43% survive in the nests of warblers that show no defensive responses to parasitism (95), 25% in the nests of frequently used and widespread hosts, and 16% in the nests of infrequently used hosts (200, 205). In adults, annual survival is higher in males than in females, as in other sexually dimorphic icterids (25, 35).

Dispersal studies of parasites are based on the recovery of banded cowbirds and cuckoos and on observation of the local movements in color-marked indigobirds.

Many cowbirds *M. ater* return to the natal area. More than 90% of marked birds recovered during the breeding season in a later year were found within 10 km of their birthplace (138). The greater dispersal of the other 10% indicated that cowbird populations undergo considerable mixing of breeding individuals in a year. In the finch *Vidua chalybeata*, ringing studies show no long-term population units at the local level. The birds have local song dialects, but they acquire their characteristic dialect songs after they disperse, and adults may change their dialect songs when they move into another local population [(132, 141); R. B. Payne, unpublished observations].

POPULATION STRUCTURE AND THE GENS CONCEPT IN CUCKOOS Baldamus' observations (7,8) of variation in the color and markings of cuckoo *Cuculus canorus* eggs that matched the eggs of their host led Newton (112) in 1898 to explain egg mimicry in terms of natural selection. Presumably each female lays eggs of a single color and pattern. When she parasitizes a nest containing similar eggs, her own are likely to be accepted. A cuckoo reared in such a nest may then return to the same kind of host when she is adult. Because her eggs probably resemble those of her mother, she may have better success with this host than will a cuckoo whose eggs are unlike the host's. Thus over several generations, clans of cuckoos would be selected whose eggs match the eggs of the host. Every race, or gens, might be reproductively isolated from other local races (6, 172).

The distribution of cuckoo egg-types in central and northern Europe suggests an origin of egg races in populations geographically isolated from one another, where egg races may coincide with major habitats. A gens may be very local or may occur over hundreds of kilometers in Sweden and Germany (17, 91, 93, 94, 196). In central Europe, two to four races or gentes may coexist, each tending to parasitize a single species. Occasionally a gens may parasitize a different host species successfully, and two or more gentes may parasitize the same species locally (17, 75). In Britain, fewer cuckoo eggs resemble the eggs of their host. Some host eggs that are mimicked on the continent are not mimicked in Britain (61). Southern (172) suggests that British cuckoo gentes may have interbred because agriculture has broken up the forest habitat, and genetic recombination may have led to a breakdown of egg mimicry. Local egg races, or local host-specific populations without egg variation, occur in some other cuckoos [*Cuculus solitarius* (38), *C. micropterus* (110), *Chrysococcyx klaas* and *C. caprius* (45, 70, 72, 73, 148)]. Newton (112) and Baker (6) suggested that different cuckoo gentes may live in an area and not interbreed, the resulting mating isolation thereby maintaining the egg polymorphism. Other genetic mechanisms have been suggested (36), but nothing in fact is known about the genetics of egg polymorphism in cuckoos.

Population structure in cuckoos may be examined by analysis of all recoveries of birds ringed as young and found in a subsequent breeding season in Europe. Bird ringing is carried out by thousands of amateurs in the range of *C. canorus*, mainly in Britain and Germany. Of 15 cuckoos ringed as local young birds in Switzerland and Germany, four were recovered within 6 km of their birth site, all were within

BROOD PARASITISM IN BIRDS 17

105 km, and the mean effective dispersal was 26 km (68, 159, 160). Data for British cuckoos (Table 3) show all recoveries in Britain and a mean effective dispersal of 99 km. More than half of the effective dispersal distances recorded for *C. canorus* are greater than 30 km. The recoveries suggest a population structure compatible with regional genotypes for egg color, but not on a very local level. An Australian parasitic cuckoo *Cacomantis pyrrhophanus* recovered after six years was within 2 km of its birth place. For migratory nesting cuckoos, recoveries of a North American *Coccyzus americanus* (10 km) and three *C. erythropthalmus* (16, 174, and 304 km) occurred at about the same distances as for *Cuculus canorus*. These records suggest no marked tendency for parasitic cuckoos to return to their natal areas with any great precision. Because of their apparently promiscuous mating, the absence of obvious morphological differences or behavioral mating differences in local populations where more than one egg morph occurs, and the nonzero dispersal distances of the cuckoos, it is unlikely that cuckoo gentes are noninterbreeding, sympatric races. Perhaps because of an early experience similar to imprinting, females tend to parasitize the species that reared them. This probably explains host specialization in the cuckoos (172).

HOST SPECIFICITY AND SPECIATION IN THE VIDUINE FINCHES   Host-species-specific parasitism in the viduine finches (114–118, 131–133, 137) and the occurrence of complexes of similar parasite species suggest the possible past differentiation of populations to the species level without geographic isolation. Host-specific speciation such as proposed for some animals specialized in certain food plant or animal host species (16) might occur in the viduines by the following steps (132). A female viduine may occasionally parasitize an alternate foster estrildid with success. The young may learn the songs of this host by a process like imprinting. The females

Table 3 Effective dispersal distances of *Cuculus canorus* ringed as nestlings or newly fledged young in Britain and recovered in a subsequent breeding season (May–July)[a]

Ring no.	Date ringed/recovered	Locality ringed/recovered	Distance (km)
251430	17.07.60/23.05.64	Spurn Point/Wetherby, Yorkshire	104
251328	14.08.66/26.06.67	Spurn Point/Ulceby, Lincolns.	32
258901	30.07.58/13.06.60	Isle of May/Cluanie, Glenmoriston	185
287224	5.08.57/(27.06.61)	Beddington, Croydon/Stockport, Cheshire	249
210751	5.06.40/1.07.41	Henley on Thames/Toddington, Dunatable, Bedfords.	53
262445	11.06.54/29.05.57	Boxhill, Dorking, Surrey/Luton, Bedfords.	64
237921	29.06.50/?.05.51	Bishops, Stortford, Hertfords./Sutton Bridge, Lincolns.	101
Bournemouth-1527	—.—.54/13.05.57	St. Mary's, Scilly/St. Agnes, Scilly	<3

[a]Source: unpublished records of British Trust for Ornithology (R. Spencer, J. Parslow).

may mate with a male viduine with similar songs and parasitize the foster species that reared them. If host switches are rare, then chance effects and selection for mouth pattern mimicry may result in genetic differentiation between the populations that parasitize the original foster species and those that follow successfully the new tradition of the second fosterer, regardless of any geographic isolation.

The possible process of host-imprinting leading to speciation in the parasite, suggested by Nicolai (114), has been analyzed by Payne (132) in the species complex showing the most geographic overlap among similar species: the indigobirds. The predictions of two models (allopatric speciation and "cultural" speciation) were tested. The finches are host-specific; the young learn the songs of their foster parent species, and both in the field and in aviary experiments females are more strongly attracted to males who mimic the songs of their foster species than to males who mimic sibling species. Occasional females lay in the nests of the alternate host and their young are sometimes successful in fledging. A few wild males sing the songs of the fosterer of the sibling species rather than their usual host (132, 133, 137). The population complex of the pale-winged indigobirds in West Africa involves three male morphs, each mimicking the songs of three or four different species of *Lagonosticta* in different areas (132, 137). Females of at least three mimics are morphologically indistinguishable (132, 137). Either male plumage colors have evolved independently in four distinct species of indigobirds, or populations are not isolated at the species level and parasite populations have switched hosts repeatedly (132, 137). The observed geographic distribution and variation of indigobirds can be explained by postulating a series of isolations and extensive areas of secondary contact as readily as by postulating a series of host changes and sympatric speciation (132). Each of the two speciation models predicts most of the observed host-parasite behavior and morphologic variation in the species complex. Thus, current understanding of the viduine finches does not let us reject the model of allopatric speciation. It implicates the behavior tradition of host specificity and song in maintaining much of the pattern of mating isolation among viduine populations, if not in the origin of species.

THE EVOLUTION OF BROOD PARASITISM IN BIRDS

Parasitic birds have evolved from birds with nesting and parental care, but the possible intermediate stages of brood parasitism and the selective forces involved are not well known. No populations are known where some individuals always rear their own young and the others are always parasites. Darwin's (26) description of South American cowbirds laying in the nests of conspecifics which then reared the young, was based on a misidentification. Two species were involved (42). Redhead ducks (*Aythya americana*) and ruddy ducks (*Oxyura jamaicensis*) usually incubate their own eggs, but sometimes lay in the nests of other ducks (76, 197). Some individuals that "parasitize" the nesting ducks incubate their own clutch later in the season (197). The prolonged season of parasitism (197) and the high proportion of redhead eggs found in the nests of other duck species (76) suggest that some individuals may be completely parasitic, and others may always incubate their own nests. Individuals have not been followed between seasons, and there is no evidence of a genetic or developmental polymorphism related to nesting behavior. Nor are bird species known with some local populations parasitic and others nonparasitic. The following speculations are made from comparisons of breeding biologies in birds related to the brood parasites and from consideration of the ecology of brood parasitism.

1. Females may occasionally lay a few eggs to be reared in the nests of conspecific neighbors, but more often rear their own young. This may lead to extinction of "parental" conspecifics and dispersal of parasites to other conspecific populations. If population structure were not inbred or viscous (57) then intraspecific parasitism might spread quickly through several foci within a species. Such parasites would rapidly lead to population extinction in nondispersive population systems, or they would be behaviorally counterselected by their more sociable relatives.

Intraspecific parasitism is uncommon in birds (65, 204). Communal nesting babblers *Turdoides squamiceps* have "helpers" at the nest that may remove an egg and lay their own (206). Captive weaver finches *Ploceus cucullatus* sometimes lay in the nests of others. In this bird, a nesting female discriminates and removes eggs unlike those of her own (185). Females in a weaver colony often have eggs of different color and pattern; perhaps this variation may facilitate detection and removal of intraspecific parasites. Some colonial birds discriminate alien eggs as well [references in (65)], suggesting selection against "cheaters." Facultative intraspecific parasitism— "dump nesting" by several females into a nest which sometimes is incubated by one female—is common in ducks. "Dump nests" are usually less successful than the nests of single females, but when they hatch young that then feed themselves (21, 197) such communal clutches may succeed (30). In the altricial species with prolonged parental care after hatching, selection against "cheating" is evidently more severe, and "cheating" behavior is rare. The potential for "cheating" in communal birds has led some to suggest that communal nesting may lead to parasitism (27, 28, 79, 197), but there are no closely related (i.e. congeneric) species pairs of communal nesters and brood parasites to support this idea.

Dispersal and a shift to parasitism of nonconspecifics might occur either early or late in the spread of intraspecific parasitism through conspecific populations. Geographic isolation and differentiation in mating preferences (speciation) followed by reinvasion of the sister species might provide the conditions necessary for maintaining parasitism as well as care for the young. Such a process has been described for the sporadic parasitism of one sister species of humble-bee on another (4, 57). Perhaps this sequence applies to the one known instance in birds (37) where a species (*Molothrus rufoaxillaris*) parasitizes a closely related one (*M. badius*).

2. Birds may lay in the nests of other species and there rear their own young. In Africa alone more than 100 species of birds frequently use the old nests of other birds but rear their own young. Contrary to orthogenetic interpretations (37, 41, 75, 120, 197), a lack of nestbuilding implies neither a "loss of maternal drive resulting in the development of obligate parasitism" nor a lack of synchrony in nestbuilding activities (59). Most birds that rear their young in the nests built by others do so in old

nests when the nestbuilders are absent and could not care for the young "protoparasites." Two widespread species of nonparasitic North American cuckoos occasionally lay in other birds' nests as well as in each other's (26, 120). Darwin suggested that the occasional success enjoyed by a cuckoo whose egg was reared by another species in a mixed clutch might lead to selection for brood parasitism, but he did not discuss the probable failure of mislaid eggs. Present data are insufficient to show whether one cuckoo disproportionately parasitizes the other in their wide range of overlap, but few nests are interspecifically parasitized. Waterfowl also may parasitize other species in mixed nesting aggregations (76, 145, 197).

3. Active takeover of the nest of one species by another, particularly at the time of nestbuilding and egg laying in mixed-species colonies, may have been a precursor of brood parasitism. Chestnut sparrows *Passer eminibey* sometimes build a nest (12, 90) but usually use the nests of colonial nesting finches. Male sparrows display for and attract females who mate with them at the nearly completed nests. The nestbuilders vigorously chase them, but the sparrows may disrupt the nestbuilders sufficiently to force their desertion. The sparrows then rear their own young in the usurped nests, sometimes adding a lining (130; R. B. Payne, unpublished observations). It would be economical for a sparrow to lay quickly in the finch nest, and then to repeat at another nest; laying synchrony is high in weaver colonies and several eggs might be laid in a few days. However, no instances are recorded of a sparrow laying in a nest before desertion and then having the nestbuilder rear the young. Some sparrow populations use the old abandoned nests of weavers (50); one has been reported in which the sparrows use the nests of another species and also build their own (100).

The apparently intense selection against intraspecific parasites makes it more likely that genetic tendencies for brood parasitism would spread through a population and become established in a species when parasitism is interspecific than when it is intraspecific.

The evolution of brood parasitism has sometimes been explained in terms of "preadaptation" (59). Certain traits of ancestral parasites supposedly predisposed selection for parasitism in those groups. Preadaptation is not a rigorous concept. The supposed evolutionary preadaptations may have been involved but not sufficient, and in the examples cited the characters are not unique. The reasoning is *a posteriori*, and different unique attributes are proposed for each group. Preadaptation is an untestable explanation for the evolution of current organisms (98, 166).

A phylogenetic model has been proposed for cowbirds (37), the only group with closely related species of nesters and parasites. As described by Friedmann, *Molothrus badius* sometimes builds a nest but usually uses nests built by other kinds of birds. It either takes over old abandoned nests or drives off the nestbuilders, and then lays and rears its own young. *M. rufoaxillaris* lives within the same range in South America and lays its eggs in nests used by *M. badius*. The young are reared by adult *badius*; the eggs and young of the two species are similar. Friedman regards the parental, non-nestbuilding *M. badius* as ancestral and the specialist *M. rufoaxillaris* as a primitive parasite. He conceives other cowbirds as derived generalist

BROOD PARASITISM IN BIRDS 21

parasites. Parasitism, he believes, evolved because there were preexisting tendencies toward the loss of behavioral "drives." This model shares untestable assumptions with the preadaptation model.

The increased availability of parental care that it assures, and the diminished risk of loss to nest predation are advantages that have led to selection for brood parasitism. That it relieves the burden of parental care is suggested by the prevalence of brood parasitism among altricial birds. Selective advantage must accrue to a female that lays in the nests of other birds if the several fosterers thereby engaged find more food than she could and thus rear more of her young. Facultative parasitism—intraspecific or interspecific—may allow a female to have some young reared by others while she rears more herself. She can thus outproduce any conspecific female whose family is limited by the amount of parental care she can provide. Obligate parasitism might then be favored if a female is likely to leave more young in a season by seeking out new nests later in the season rather than rearing any young of her own. Four of the five families containing obligate parasites are altricial, and parasitism is thought to have evolved independently twice each in the cuckoos and in the finches (40, 136). The number of parasitic species is greater among altricial birds (about 86) than in the precocials (one duck). Among all birds, only a fifth of the families and a tenth of the species are precocial, so altricial parasites are not disproportionately common. A parental care explanation may be appropriate for most brood parasites but cannot be applied to all parasitic birds; the young parasitic duck leaves the nest after two days and feeds unaided (199).

Brood parasitism has a second selective advantage: it improves the chances that at least some offspring will escape predation. Most parasites lay only a single egg in a host nest. This is true of honey guides (39, 182, 184), cuckoos (6, 45, 72, 73), the parasitic duck (198, 199) and perhaps the cowbirds, (41, 96), but not the finches (103, 139). Instances of multiple eggs laid in a host nest by cowbirds and cuckoos are often due to two or more females, who lay sometimes when their other nests have been suddenly lost (41, 67, 203). In all parasites studied, several eggs develop in a series or clutch, but in most species they are divided among several nests. This dispersion of eggs may provide that some eggs of the parasite will escape nest predation. Since most nest predators take all the eggs or nestlings (152), the chances of loss due to predation may be independent of the number of eggs in a nest. A binomial probability distribution of success and failure shows a higher chance that at least one egg in a three-egg clutch will survive if each egg is laid in a different nest (scatter laying) (139). The difference in the probability that at least one egg in a three-egg clutch will survive in (a) scattered laying and (b) multiple parasitism depends on the frequency of nest predation; a higher difference accompanies a higher predation (Figure 1). With high nest survival, the increased likelihood of survival of at least one egg with scatter-laying is only 20%. This decrease in risk of total loss may be offset by the increased costs associated with laying in more than one nest. With low nest survival however, the likelihood of survival of at least one egg in a dispersed clutch is more than twice that when all the eggs are in one nest (Figure 1). Nest loss to predation is high in the tropics (152) where we find most

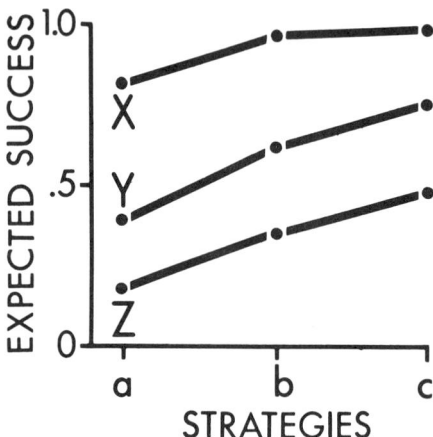

Figure 1 Probabilities of success in escaping nest predation of at least one egg of a hypothetical three-egg clutch, with strategies (*a*) all eggs laid in one nest, (*b*) two laid in one nest and one in a second nest, and (*c*) each egg laid in a separate nest, calculated from a binomial distribution of nesting success and failure. Probabilities of nest survival shown are $(x) = .828$ [*Troglodytes aedon* in Ohio (152)], $(y) = .378$ [*Lagonosticta senegala* in Senegal (103, 139)], and $(z) = .194$ [*Manacus manacus* in Trinidad (152)]. Birds (x) and (z) are small passerines with extremes in nest survival (152), and (y) is a host of the parasite *Vidua chalybeata*.

brood parasitic species. The origins of brood parasitism in the Old World and of brood parasitism in at least the cuckoos in the New World are tropical.

The prevalence of single-egg nest parasitism suggests that brood parasitism in general may be an evolutionary "strategy" to spread the risks of predation. Constraints on high egg dispersion in the brood parasites include the economics of leaving more than one egg in the nest of those hosts than can rear adequately more than one parasite (41, 51, 139, 170). Evicting behavior and other severe interference by young parasites probably evolved later, given the absence of siblings in the same nests.

The strategy of minimizing total loss in the brood parasites is that of Cervantes' Sancho Panza, who tells Don Quixote that "a wise man...[does] not venture all his eggs in one basket."

ACKNOWLEDGMENTS

For unpublished data on banding recoveries I thank D. Purchase (Australian Bird-Banding Scheme, CSIRO), R. Spencer and J. Parslow (British Trust for Ornithology), and J. M. Sheppard (US Bird-Banding Laboratory). I thank H. Friedmann, A. J. Gaston, K. Payne, S. I. Rothstein, P. W. Sherman, and R. W. Storer for comments on an earlier manuscript. My work has been supported by the National Science Foundation.

Literature Cited

1. Ali, S. A. 1931. The origin of mimicry in cuckoo's eggs. *J. Bombay Nat. Hist. Soc.* 34:1067–70
2. Ali, S., Ripley, S. D. 1969. *Handbook of the Birds of India and Pakistan*, Vol. 3. Bombay: Oxford Univ. Press. 325 pp.
3. Appert, O. 1970. Zur Biologie einiger Kua-Arten Madagaskars (Aves, Cuculi). *Zool. Jahrb.* 97:424–53
4. Askew, R. R. 1971. *Parasitic Insects.* New York: Elsevier. 316 pp.
5. Baker, E. C. S. 1923. Cuckoos' eggs and evolution. *Proc. Zool. Soc. London* 1923:277–94
6. Baker, E. C. S. 1942. *Cuckoo Problems.* London: Witherby. 207 pp.
7. Baldamus, E. 1854. Neue Beiträge zur Fortpflanzungsgeschichte des Europäischen Kuckkuks. English transl. in *Zoologist* 1868: 1146–66
8. Baldamus, E. 1892. *Das Leben der Europäischen Kuckucke, nebst Beiträgen zur Lebenskunde der übrigen parasitischen Kuckucke und Stärlinge.* Berlin: Parey. 224 pp.
9. Benson, C. W., Brooke, R. K., Vernon, C. J. 1964. Bird breeding data for the Rhodesias and Nyasaland. *Occas. Pap. Natl. Mus. S. Rhodesia* 27B:30–105
10. Benson, C. W., Irwin, M. P. S. 1972. The thick-billed cuckoo *Pachycoccyx audeberti* (Schlegel) (Aves: Cuculidae). *Arnoldia (Rhodesia)* 5(33):1–24
11. Bent, A. C. 1958. Life histories of North American blackbirds, orioles, tanagers, and allies. *US Natl. Mus. Bull. 211.* 531 pp.
12. Betts, F. N. 1966. Notes on some resident breeding birds of southwest Kenya. *Ibis* 108:513–30
13. Blaise, M. 1965. Contribution à l'étude de la reproduction du Coucou gris *Cuculus canorus* dans le nord-est de la France. *L'Oiseau et R. F. O.* 35:87–116
14. Breder, C. M., Rosen, D. E. 1966. *Modes of Reproduction in Fishes.* New York: Nat. Hist. Press. 941 pp.
15. Brown, J. L. 1975. *The Evolution of Behavior.* New York: Norton. 761 pp.
16. Bush, G. L. 1975. Modes of animal speciation. *Ann. Rev. Ecol. Syst.* 6: 339–64
17. Capek, V. 1896. Beiträge zur Fortpflanzungsgeschichte des Kuckucks. *Ornithol. Jahrb.* 7:41–72, 102–17, 146–57, 165–83
18. Chance, E. 1922. *The Cuckoo's Secret.* London: Sedgwick & Jackson. 239 pp.
19. Chance, E. 1940. *The Truth about the Cuckoo.* London: Country Life. 207 pp.
20. de Chavigny, J., Le Dû, R. 1938. Note sur l'adaptation des oeufs de Coucou de l'Afrique du Nord, *Cuculus canorus bangsi* Oberholser. *Alauda* 10:91–115
21. Cooke, F., Mirsky, P. J. 1972. A genetic analysis of lesser snow goose families. *Auk* 89:863–71
22. Courtney, J. 1967. The juvenile food-begging call of some fledgling cuckoos —vocal mimicry or vocal duplication by natural selection? *Emu* 67:154–57
23. Cronin, E. W., Sherman, P. W. 1977. A resource-based mating system: the orange-rumped honeyguide, *Indicator xanthonotus. Living Bird* 15:In press
24. Darley, J. A. 1968. The social organization of breeding brown-headed cowbirds. PhD thesis. Univ. Western Ontario, London, Ontario
25. Darley, J. A. 1971. Sex ratio and mortality in the brown-headed cowbird. *Auk* 88:560–66
26. Darwin, C. 1872. *The Origin of Species.* New York: Modern Library. 1000 pp. 6th ed. Reprinted
27. Davis, D. E. 1940. Social nesting habits of *Guira guira. Auk* 57:472–84
28. Davis, D. E. 1942. The phylogeny of social nesting habits in the crotophaginae. *Quart. Rev. Biol.* 17:115–34
29. Davis, D. E. 1942. The number of eggs laid by cowbirds. *Condor* 44(1):10–12
30. Delnicki, D., Bolen, E. G., Cottam, C. 1976. An unusual clutch size of the black-bellied whistling duck. *Wilson Bull.* 88:347–48
31. Diesselhorst, G. 1955. Eizahl des Kuckucks. *Vogelwelt* 76:53–58
32. Dow, D. D. 1972. The New Zealand long-tailed cuckoo: nest parasite or predator? *Emu* 72:179–80
33. Edwards, G., Hosking, E., Smith, S. 1949–1950. Reactions of some passerine birds to a stuffed cuckoo. *Br. Birds* 42:13–19, 43:144–50
34. Elliott, P. F. 1977. Cowbird parasitism in the tallgrass prairie of northeastern Kansas. *Auk:* In press
35. Fankhauser, D. P. 1971. Annual adult survival rates of blackbirds and starlings. *Bird-Banding* 42:36–42
36. Ford, E. B. 1964. *Ecological Genetics.* London: Methuen. 335 pp.
37. Friedmann, H. 1929. *The Cowbirds: A Study in the Biology of Social Parasitism.* Springfield, Ill: Thomas. 421 pp.

38. Friedmann, H. 1948. The parasitic cuckoos of Africa. *Wash. Acad. Sci. Monogr. 1.* 204 pp.
39. Friedmann, H. 1955. The honey-guides. *US Natl. Mus. Bull. 208.* 292 pp.
40. Friedmann, H. 1960. The parasitic weaverbirds. *US Natl. Mus. Bull. 223.* 196 pp.
41. Friedmann, H. 1963. Host relations of the parasitic cowbirds. *US Natl. Mus. Bull. 233.* 273 pp.
42. Friedmann, H. 1964. The history of our knowledge of avian brood parasitism. *Centaurus* 10:282–304
43. Friedmann, H. 1964. Evolutionary trends in the avian genus *Clamator. Smithson. Misc. Collect.* 146(4):1–127
44. Friedmann, H. 1967. Alloxenia in three sympatric African species of *Cuculus. Proc. US Natl. Mus.* 124(3633):1–14
45. Friedmann, H. 1968. The evolutionary history of the avian genus *Chrysococcyx. US Natl. Mus. Bull. 265.* 137 pp.
46. Friedmann, H. 1971. Further information on the host relations of the parasitic cowbirds. *Auk* 88:239–55
47. Friedmann, H., Kiff, L. F., Rothstein, S. I. 1977. A further contribution to knowledge of host relations of the cowbirds. *Smithson. Contrib. Zool. 235.* 75 pp.
48. Frith, C. B. 1975. Field observations on *Centropus toulou insularis* on Aldabra atoll. *Ostrich* 46:251–57
49. Fry, C. H. 1974. Vocal mimesis in nestling greater honey-guides. *Bull. Br. Ornithol. Club* 94:58–59
50. Fuggles-Couchman, N. R., Elliot, H. F. I. 1946. Some records and field notes from North-Eastern Tanganyika Territory. *Ibis* 1946:327–47
51. Gaston, A. J. 1976. Brood parasitism by the pied crested cuckoo *Clamator jacobinus. J. Anim. Ecol.* 45:331–48
52. Goodwin, D. 1974. In *Birds of the Harold Hall Australian Expeditions, 1962–70,* ed. B. P. Hall. London: Br. Mus. (Nat. Hist.) Publ. No. 745. 396 pp.
53. Gosnell, H. T. 1932. Two cuckoos laying in the same nest without rivalry. *Br. Birds* 26:226
54. Gosper, D. 1964. Observations on the breeding of the koel. *Emu* 64:39–41
55. Gramet, P. 1970. Le parasitisme des Corvides par le Coucou-Geai (*Clamator glandarius*). *Rev. Comp. Anim.* 4:17–26
56. Gurney, J. H. 1897. Cuckoos sucking eggs. *Zoology* 1897:568
57. Hamilton, W. D. 1971. Selfish and altruistic behavior in some extreme models. In *Man and Beast: Comparative Social Behavior,* ed. J. F. Eisenberg, W. S. Dillon, pp. 59–91. Washington DC: Smithson. Inst. Press
58. Hamilton, W. J. III, Hamilton, M. E. 1965. Breeding characteristics of yellow-billed cuckoos in Arizona. *Proc. Calif. Acad. Sci., 4th Ser.,* 32(14): 405–32
59. Hamilton, W. J., Orians, G. H. 1965. Evolution of brood parasitism in altricial birds. *Condor* 67:361–82
60. Hardcastle, A. 1925. Young cuckoo fed by several birds. *Br. Birds* 19:100
61. Harrison, C. J. O. 1968. Egg mimicry in British cuckoos. *Bird Study* 15(1): 22–28
62. Haverschmidt, F. 1961. Der Kuckuck *Tapera naevia* und seine Wirte in Surinam. *J. Ornithol.* 102:353–59
63. Herrick, F. O. 1910. Life and behavior of the cuckoo. *J. Exp. Zool.* 9:169–233
64. Holcomb, L. S. 1970. Prolonged incubation behavior of red-winged blackbird incubating several egg sizes. *Behaviour* 36:74–83
65. Hoogland, J. L., Sherman, P. W. 1976. Advantages and disadvantages of bank swallow (*Riparia riparia*) coloniality. *Ecol. Monogr.* 46:33–58
66. Howard, H. E. 1913. Cuckoos and the struggle for breeding-territory. *Br. Birds* 7:83–85
67. Hoy, G., Ottow, H. 1964. Biological and oological studies of the molothrine cowbirds (icteridae) of Argentina. *Auk* 81:186–203
68. Hückler, U. 1968. Ringfunde der Kuckucks (*Cuculus canorus*). *Auspicium* 2:338–43
69. Hudson, W. H. 1920. *Birds of La Plata,* Vol. 1. London: Dent. 244 pp.
70. Hunter, H. C. 1961. Parasitism of the masked weaver *Ploceus velatus arundinaceus. Ostrich* 32:55–63
71. Jenner, E. 1788. Observations on the natural history of the cuckoo. *Philos. Trans. R. Soc. London* 78:219–35
72. Jensen, R. A. C., Clinning, C. F. 1975. Breeding biology of two cuckoos and their hosts in South West Africa. *Living Bird* 13:5–50
73. Jensen, R. A. C., Jensen, M. K. 1969. On the breeding biology of southern African cuckoos. *Ostrich* 40:237–46
74. Johnson, A. W. 1967. *The Birds of Chile,* Vol. 2. Buenos Aires: Johnson. 447 pp.
75. Jourdain, F. C. R. 1925. A study of parasitism in the cuckoos. *Proc. Zool. Soc. London* 1925:639–67

76. Joyner, D. E. 1976. Effects of interspecific nest parasitism by redheads and ruddy ducks. *J. Wildl. Manage.* 40: 33–38
77. King, J. R. 1973. Reproductive relationships of the rufous-collared sparrow and the shiny cowbird. *Auk* 90:19–34
78. Klaas, E. E. 1975. Cowbird parasitism and nesting success in the eastern phoebe. *Occas. Pap. Mus. Nat. Hist. Univ. Kans.* 41:1–18
79. Köster, F. 1971. Zum Nistverhalten des Ani, *Crotophaga ani. Bonn. Zool. Beitr.* 22:4–27
80. Kuroda, N. 1966. A note on the problem of hawk-mimicry in cuckoos. *Jpn. J. Zool.* 1966:173–81
81. Lack, D. 1963. Cuckoo hosts in England. With an appendix on the cuckoo hosts in Japan, by T. Royama. *Bird Study* 10:185–203
82. Lack, D. 1968. *Ecological Adaptations for Breeding in Birds.* London: Methuen. 409 pp.
83. Lamba, B. S. 1963. The nidification of some common Indian birds—Part I. *J. Bombay Nat. Hist. Soc.* 60:121–33
84. Laskey, A. R. 1950. Cowbird behavior. *Wilson Bull.* 62:157–74
85. Latter, O. H. 1905. The egg of *Cuculus canorus:* an attempt to ascertain from the dimensions of the cuckoo's egg if the species is tending to break up into subspecies, each exhibiting a preference for some one foster-parent. *Biometrika* 4:363–73
86. Link, J. A. 1890. Ob der Kuckuck Eier und junge Vögel frisst? *Ornithol. Monatsschr.* 1890:25
87. Liversidge, R. 1955. Observations on a Piet-My-Vrou (*Cuculus solitarius*) and its host the cape robin (*Cossypha caffra*). *Ostrich* 26:18–27
88. Liversidge, R. 1971. The biology of the jacobin cuckoo *Clamator jacobinus. Proc. 3rd Pan-Afr. Ornithol. Congr., Ostrich Suppl.* 8:117–37
89. Lofts, B., Murton, R. K. 1968. Photoperiodic and physiological adaptations regulating avian breeding cycles and their ecological significance. *J. Zool.* 155:327–94
90. Lynes, H. 1924. On the birds of North and Central Darfur, with notes on the west-central Kordofan and north Nuba Provinces of British Sudan. *Ibis, 11th Ser.* 6(3):399–446, (4):648–719
91. Makatsch, W. 1937. *Der Brutparasitismus der Kukucksvögel.* Leipzig: Quelle & Meyer. 152 pp.
92. Makatsch, W. 1955. *Der Brutparasitismus in der Vogelwelt.* Radebeul: Neumann. 236 pp.
93. Makatsch, W. 1971. Einige Bemerkungen über die parasitären Kuckucke. *Zool. Abh. Staat. Mus. Tierk. Dresden* 30(20):247–83
94. Malchevsky, A. S. 1960. On the biological races of the common cuckoo, *Cuculus canorus* L. in the territory of the European part of the USSR. *XII Int. Ornithol. Congr., Helsinki,* 1958: 464–70
95. Mayfield, H. 1960. The Kirtland's warbler. *Cranbrook Inst. Sci. Bull.* 40. 242 pp.
96. Mayfield, H. 1965. Chance distribution of cowbird eggs. *Condor* 67:257–63
97. Mayfield, H. 1972. Third decennial census of Kirtland's warbler. *Auk* 89:263–68
98. Mayr, E. 1963. *Animal Species and Evolution.* Cambridge: Harvard. 797 pp.
99. McGeen, D. S., McGeen, J. J. 1968. The cowbirds of Otter Lake. *Wilson Bull.* 80:84–93
100. McInnes, D. 1933. Nesting habits of some East African birds. *J. East Afr. Uganda Nat. Hist. Soc.* 47–48:128–35
101. Moreau, R. E. 1949. Special Review. Friedmann on African cuckoos. *Ibis* 91:529–37
102. Morel, G., Morel, M.-Y. 1962. La reproduction des Oiseaux dans une région semi-aride: la Vallée du Sénégal. *Alauda* 30:161–203, 241–69
103. Morel, M.-Y. 1973. Contribution à l'étude dynamique de la population de *Lagonosticta senegala* L. (estrildides) à Richard-Toll (Sénégal). Interrelations avec le parasite *Hypochera chalybeata* (Müller) (viduines). *Mem. Mus. Nat. d'Hist. Nat. Ser. A Zool.* 78:1–156
104. Morgensen, J. 1927. Nota sobre el parasitismo del "Crespin" (*Tapera naevia*). *Hornero* 4:68–70
105. Mountfort, G., Ferguson-Lees, I. J. 1961. The birds of the Coto Doñana. *Ibis* 103a:86–109
106. Muller, K. A. 1971. Physical and behavioral development of a roadrunner raised at the National Zoological Park. *Wilson Bull.* 83:186–93
107. Mundy, P. J. 1973. Vocal mimicry of their hosts by nestlings of the great spotted cuckoo and striped crested cuckoo. *Ibis* 115:602–4
108. Mundy, P. J., Cook, A. W. 1971. Sokoto province. (I) Sokoto town & environs. *Nigerian Ornithol. Soc. Bull.* 8(30):21–24

109. Mundy, P. J., Cook, A. W. 1974. The birds of Sokoto, part 3: breeding data. *Nigerian Ornithol. Soc. Bull.* 10(37): 1–28
110. Neufeldt, I. 1966. Life history of the Indian cuckoo, *Cuculus micropterus micropterus* Gould, in the Soviet Union. *J. Bombay Nat. Hist. Soc.* 63:399–419
111. Newman, G. A. 1970. Cowbird parasitism and nesting success of lark sparrows in southern Oklahoma. *Wilson Bull.* 82:304–9
112. Newton, A. 1896. *A Dictionary of Birds.* London: Adam & Charles Black. 1088 pp.
113. Nice, M. M. 1937, 1943. Studies in the life history of the song sparrow. *Trans. Linn. Soc. NY*, Vols. 2, 6. 246 pp., 328 pp.
114. Nicolai, J. 1964. Der Brutparasitismus der Viduinae als ethologisches Problem. *Z. Tierpsychol.* 21(2):129–204
115. Nicolai, J. 1967. Rassen- und Artbildung in der Viduinengattung *Hypochera. J. Ornithol.* 108(3):309–19
116. Nicolai, J. 1969. Beobachtungen an Paradieswitwen (*Steganura paradisaea* L., *Steganura obtusa* Chapin) und der Strohwitwe (*Tetraenura fischeri* Reichenow) in Ostafrika. *J. Ornithol.* 110:421–47
117. Nicolai, J. 1972. Zwei neue *Hypochera*-Arten aus West Afrika (Ploceidae, Viduinae). *J. Ornithol.* 113:229–40
118. Nicolai, J. 1973. Das Lernprogramm in der Gesangsausbildung der Strohwitwe *Tetraenura fischeri* Reichenow. *Z. Tierpsychol.* 32:113–38
119. Niethammer, G. 1938. *Handbuch der Deutschen Vogelkunde.* Band II. Leipzig: Akad. Verlagsges. 545 pp.
120. Nolan, V., Thompson, C. F. 1975. The occurrence and significance of anomalous reproductive activities in two North American non-parasitic cuckoos. *Ibis* 117:496–503
121. Norman, R. F., Robertson, R. J. 1975. Nest-searching behavior in the brown-headed cowbird. *Auk* 92:610–11
122. Norris, R. T. 1947. The cowbirds of Preston Frith. *Wilson Bull.* 59:83–103
123. O'Connor, R. J. 1962. Juvenile cuckoo apparently imitating meadow pipit's call. *Br. Birds* 55:481
124. Ohmart, R. D. 1973. Observations on the breeding adaptations of the roadrunner. *Condor* 75:140–49
125. Owen, J. H. 1933. The cuckoo in the Felsted district. *Rep. Felsted School Sci. Soc.* 33:25–39
126. Payne, R. B. 1965. Clutch size and numbers of eggs laid by brown-headed cowbirds. *Condor* 67:44–60
127. Payne, R. B. 1967. Interspecific communication signals in parasitic birds. *Am. Nat.* 101:363–76
128. Payne, R. B. 1967. Gonadal responses of brown-headed cowbirds to long daylengths. *Condor* 69:289–97
129. Payne, R. B. 1969. The breeding seasons and reproductive physiology of tricolored blackbirds and redwinged blackbirds. *Univ. Calif. Publ. Zool.* 90:1–137
130. Payne, R. B. 1969. Nest parasitism and display of chestnut sparrows in a colony of grey-capped social weavers. *Ibis* 113(3):123–34
131. Payne, R. B. 1971. Paradise whydahs *Vidua paradisaea* and *V. obtusa* of southern and eastern Africa, with notes on differentiation of the females. *Bull. Br. Ornithol. Club* 91:66–76
132. Payne, R. B. 1973. Behavior, mimetic songs and song dialects, and relationships of the parasitic indigobirds (*Vidua*) of Africa. *Ornithol. Monogr. 11.* 333 pp.
133. Payne, R. B. 1973. Vocal mimicry of the paradise whydahs (*Vidua*) and response of female whydahs to the songs of their hosts and their mimics. *Anim. Behav.* 21:762–71
134. Payne, R. B. 1973. The breeding season of a parasitic bird, the brown-headed cowbird, in central California. *Condor* 75:80–99
135. Payne, R. B. 1973. Individual laying histories and the clutch size and numbers of eggs of parasitic cuckoos. *Condor* 75:414–38
136. Payne, R. B. 1974. The evolution of clutch size and reproductive rates in parasitic cuckoos. *Evolution* 28:169–81
137. Payne, R. B. 1976. Song mimicry and species relationships among the west African pale-winged indigobirds. *Auk* 93:25–38
138. Payne, R. B. 1976. The clutch size and numbers of eggs of brown-headed cowbirds: effects of latitude and breeding season. *Condor* 78:337–42
139. Payne, R. B. 1977. Clutch size, egg size, and the consequences of single vs multiple parasitism in parasitic finches. *Ecology.* In press
140. Payne, R. B., Payne, K. 1967. Cuckoo hosts in southern Africa. *Ostrich* 38:135–43
141. Payne, R. B., Payne, K. 1978. Social organization and mating success in lo-

142. Phillips, W. W. A. 1948. Cuckoo problems of Ceylon. *Spolia Zeylanica* 25:45–60
143. Post, W., Wiley, J. W. 1976. The yellow-shouldered blackbird—present and future. *Am. Birds* 30:13–20
144. Preble, N. A. 1957. Nesting habits of the yellow-billed cuckoo. *Am. Midl. Nat.* 57:474–82
145. Prevett, J. P., Lieff, B. C., MacInnes, C. D. 1972. Nest parasitism at McConnell River, N.W.T. *Can. Field-Nat.* 86:369–72
146. Ralph, C. P. 1975. Life style of *Coccyzus pumilus*, a tropical cuckoo. *Condor* 77:73–83
147. Ranger, G. A. 1955. On three species of honey-guide: the greater (*Indicator indicator*), the lesser (*Indicator minor*) and the scaly-throated (*Indicator variegatus*). *Ostrich* 36:70–87
148. Reed, R. A. 1968. Studies of the diederik cuckoo *Chrysococcyx caprius* in the Transvaal. *Ibis* 110:321–31
149. Rensch, B. 1924. Zur Entstehung der Mimikry der Kuckuckseier. *J. Ornithol.* 72:461–72
150. Rey, E. 1892. *Altes und Neues aus dem Haushalte des Kuckucks.* Leipzig: Freese. 108 pp.
151. Rey, E. 1894. Beobachtungen über den Kuckuk bei Leipzig aus dem Jahre 1893. *Ornithol. Monatsschr. Dtsch. Vereins Schutze Vogelwelt* 19:159–68
152. Ricklefs, R. E. 1969. An analysis of nesting mortality in birds. *Smithson. Contrib. Zool.* 9:1–48
153. Robertson, R. J., Norman, R. F. 1976. Behavioral defenses to brood parasitism by potential hosts of the brown-headed cowbird. *Condor* 78:166–73
154. Robertson, R. J., Norman, R. F. 1977. The function and evolution of aggressive host behavior towards the brown-headed cowbird (*Molothrus ater*). *Can. J. Zool.* In press
154a. Rothstein, S. I. 1974. Mechanisms of avian egg recognition: possible learned and innate factors. *Auk* 91:796–807
155. Rothstein, S. I. 1975. An experimental and teleonomic investigation of avian brood parasitism. *Condor* 77(3):250–71
156. Rothstein, S. I. 1975. Evolutionary rates and host defenses against avian brood parasitism. *Am. Nat.* 109:161–76
156a. Rothstein, S. I. 1975. Mechanisms of avian egg-recognition: do birds know their own eggs? *Anim. Behav.* 23:268–78
157. Rothstein, S. I. 1976. Cowbird parasitism of the cedar waxwing and its evolutionary implications. *Auk* 93:498–509
158. Schiermann, G. 1926. Beitrag zur Schädigung der Wirtsvögel durch *Cuculus canorus*. *Beitr. Fortpflanz. Biol. Vögel* 2:28–30
159. Schifferli, A. 1951. Bericht der Schweiz. Vogelwarte Sempach für die Jahre 1949 und 1950. *Ornithol. Beob.* 48:181–211
160. Schifferli, A. 1961. Schweizerische Ringfundmeldung für 1959 und 1960. *Ornithol. Beob.* 58:166–96
161. Scholey, G. J. 1927. Abnormal laying of cuckoo. *Br. Birds* 21:179
162. Schönwetter, M. 1960–1971. In *Handbuch der Öologie*, Band I, Nonpasseres, ed. W. Meise, Lief 8, 9, pp. 449–512, 513–76. Berlin: Akademie
163. Seppä, J. 1969. The cuckoo's ability to find a nest where it can lay an egg. *Ornis Fennica* 46:78–79
164. Shelford, R. 1900. On the pterylosis of the embryos and nestlings of *Centropus sinensis*. *Ibis* 6:654–67
165. Sick, H., Ottow, J. 1958. Vom brasilianischen Kuhvogel, *Molothrus bonariensis*, und sienen Wirten, besonders dem Ammerfinken, *Zonotrichia capensis*. *Bonn. Zool. Beitr.* 1:40–62
166. Simpson, G. G. 1953. *The Major Features of Evolution.* New York: Columbia Univ. Press. 434 pp.
167. Skead, D. M. 1975. Ecological studies of four estrildines in the Central Transvaal. *Ostrich Suppl. 11.* 55 pp.
168. Skutch, A. F. 1954. Life Histories of Central American Birds. *Pacific Coast Avifauna 31.* 448 pp.
169. Skutch, A. F. 1959. Life history of the groove-billed ani. *Auk* 76:281–317
170. Smith, N. G. 1968. The advantage of being parasitized. *Nature* 219:690–94
171. Smith, S., Hosking, E. 1955. *Birds fighting. Experimental studies of the aggressive displays of some birds.* London: Faber & Faber. 128 pp.
172. Southern, H. N. 1954. Mimicry in cuckoos' eggs. In *Evolution as a Process*, ed. J. Huxley, A. C. Hardy, E. B. Ford, pp. 219–32. London: George Allen & Unwin
173. Southern, W. E. 1958. Nesting of the red-eyed vireo in the Douglas Lake region, Michigan, Pt. 2. *Jack-Pine Warbler* 36:185–207
174. Spencer, O. R. 1943. Nesting habits of the black-billed cuckoo. *Wilson Bull.* 55:11–22

175. Steyn, P. 1973. Some notes on the breeding biology of the striped cuckoo. *Ostrich* 44:163–69
176. Steyn, P., Howells, W. W. 1975. Supplementary notes on the breeding biology of the striped cuckoo. *Ostrich* 46:258–60
177. Stimming, R. 1927. Meine Beobachtungen über das Legegeschäft des Kucksweibchens in der Mark Brandenburg und in der Provinz Sachsen. *Beitr. Fortpflanz. Biol. Vögel* 3:122–26
178. Stresemann, E. 1927–1934. In *Handbuch der Zoologie*, ed. W. Kükenthal, T. Krumbach, Vol. 7, Pt. 2. Berlin & Leipzig: de Gruyter. 899 pp.
179. Swynnerton, C. F. M. 1918. Rejection by birds of eggs unlike their own: with remarks on some of the cuckoo problems. *Ibis* 1918:27–54
180. Tutt, H. R. 1955. Deposition of eggs by the female cuckoo (*Cuculus canorus*). *XI Int. Ornithol. Congr., Exp. Suppl.* 3:630–31
181. Valverde, J. A. 1971. Notas sobre la biologia de reproduccion del Crialo *Clamator glandarius* (L.). *Aredola* (numero especial), pp. 591–647
182. van Someren, G. R. C. 1970. On *Prodotiscus insignis* (Cassin) parasitizing *Zosterops abyssinica* Guérin. *Bull. Br. Ornithol. Club* 90:129–31
183. van Someren, G. R. C. 1973. *Vidua fischeri* (Reichenow) parasitic on *Granatina ianthinogaster* (Reichenow). *J. East. Afr. Nat. Hist. Soc. Nat. Mus.* 139. 4 pp.
184. Vernon, C. J. 1974. *Prodotiscus regulus* parasitizing *Camaroptera brevicaudata*. *Ostrich* 45:262
185. Victoria, J. K. 1972. Clutch characteristics and egg discriminative ability of the African village weaverbird *Ploceus cucullatus*. *Ibis* 114:367–76
186. Vilks, K. 1972. [Study of the breeding biology of cuckoos (*Cuculus canorus* L.) in Latvia.] *Zoologijas Muzeja Raksti* 9:7–14 (In Latvian with German summary)
187. Vincent, J. 1933. Cuckoo killing nestling meadow-pipits. *Br. Birds* 27:51
188. Voipio, P. 1953. The hepaticus variety and the juvenile plumage of the cuckoo. *Ornis Fenn.* 30:97–117
189. von Frisch, O. 1969. Die Entwicklung des Häherkuckucks (*Clamator glandarius*) im Nest der Wirtsvögel und seine Nachtzucht in Gefangenschaft. *Z. Tierpsychol.* 26:641–50
190. von Frisch, O., von Frisch, H. 1967. Beobachtungen zur Brutbiologie und Jugendentwicklung des Häherkuckucks (*Clamator glandarius*). *Z. Tierpsychol.* 24(2):129–36
191. von Lucanus, F. 1921. Zur Frage der Mimikry der Kuckuckseier. *J. Ornithol.* 69:239–58
192. Walkinshaw, L. H. 1949. Twenty-five eggs apparently laid by a cowbird. *Wilson Bull.* 61:82–85
193. Walkinshaw, L. H. 1961. The effect of parasitism by the brown-headed cowbird on *Empidonax* flycatchers in Michigan. *Auk* 78:266–68
194. Walkinshaw, L. H. 1972. Kirtland's warbler—endangered. *Am. Birds* 26(1):3–9
195. Walkinshaw, L. H., Faust, W. R. 1974. Some aspects of Kirtland's warbler breeding biology. *Jack-Pine Warbler* 52:64–75
196. Wasenius, E. 1936. Om de i Finland funna typerna av gökägg och deras geografiska utbredning. *Ornis Fenn.* 13:147–53
197. Weller, M. W. 1959. Parasitic egg laying in the redhead (*Aythya americana*) and other North American anatidae. *Ecol. Monogr.* 29:333–65
198. Weller, M. W. 1967. Notes on some marsh birds of Cape San Antonio, Argentina. *Ibis* 109:391–411
199. Weller, M. W. 1968. The breeding biology of the parasitic black-headed duck. *Living Bird* 7:169–208
200. Wiens, J. A. 1963. Aspects of cowbird parasitism in southern Oklahoma. *Wilson Bull.* 75:130–39
201. Wilson, A. 1810. *American Ornithology,* 1:201–19. Philadelphia: Bradford & Inskeep
202. Wilson, E. O. 1971. *The Insect Societies.* Cambridge: Harvard Univ. Press. 548 pp.
203. Wyllie, I. 1975. Study of cuckoos and reed warblers. *Br. Birds* 68:369–78
204. Yom-Tov, Y., Dunnet, G. M., Anderson, A. 1974. Intraspecific nest parasitism in the starling *Sturnus vulgaris*. *Ibis* 116(1):87–90
205. Young, H. 1963. Breeding success of the cowbird. *Wilson Bull.* 75:115–22
206. Zahavi, A. 1974. Communal nesting by the Arabian babbler. A case of individual selection. *Ibis* 116(1):84–87

THE EVOLUTION OF VIRUSES ❖4116

André J. Nahmias
Department of Pediatrics, Emory University School of Medicine, Atlanta, Georgia 30303

Darryl C. Reanney
Department of Biochemistry, Lincoln College, Canterbury, New Zealand

INTRODUCTION

"Since Darwin, much of the 'fun' of biological research has been to interpret what directly interests one in evolutionary terms" (9). Applying this aphorism to virology, one arrives at a somewhat paradoxical situation. Where do these smallest of self-replicating units—2.6×10^3 to 10×10^6 nm^3 (35)—fit in our attempts to systematize the biosphere? Viruses have occasionally been considered as a separate kingdom (56). Yet, viruses (with DNA or RNA genomes, a protein coat with or without lipids and carbohydrates) have been identified in large numbers of host species within each of the five major kingdoms of organisms (Table 1).

For many years, viruses have been grouped according to the host in which they have been detected. Among the major problems with this approach are the identification of viruses with similar characteristics in two or even three kingdoms (Table 1), and the well-appreciated recurrent problem that the host from which a virus is initially isolated may not be its "natural" host. Viruses of vertebrates, invertebrates, and bacteria have recently been classified into families, genera, and species on the basis of common characters, while the taxon of plant viruses is still the group (17); viruses of protista and fungi have not as yet been officially classified.[1] Such classifications, while of great practical usefulness, are only a first step toward the construction of a satisfying viral phylogeny which would enable the position of viruses in the biosphere to be clarified. We believe that this question can best be approached by attempting to resolve the issues related to the evolutionary origins of viruses.

The lack of fossils has not deterred the application of evolutionary principles to such subjects as genes and proteins. Not only have the molecular aspects relevant to these issues been unravelled in large part by studies performed with viruses, but

[1] Rickettsiae remain among the major groups of organisms from which viruses have not been identified, perhaps owing to their own incapacity for extracellular replication.

Table 1 Current knowledge of viruses found in the various kingdoms

Nucleic acid — type and strandedness	No. families or groups of viruses	Animalia		Plantae	Fungi	Protista	Monera
		Vertebrates	Invertebrates	Plants	Fungi	Amoebae	Bacteria
DNA, double stranded	13 [3][b]	5 [3]	4 [3]	1 [0]	1 [1]	2 [0]	4 [0]
DNA, single stranded	3 [1]	1 [1]	1 [1]				2 [0]
RNA, double stranded	3 [1]	1 [1]	1 [1]	1 [1]	1 [0]		1 [0]
RNA, single stranded	30 [4]	9[c] [4]	4 [4]	20 [1]	1 [0]		1 [0]
TOTAL	49 [9]	16 [9]	10 [9]	22 [2]	3 [1]	2 [0]	8 [0]

[a] According to the 1969 classification of Whittaker (56).
[b] Brackets indicate viruses of same family or group which have been found in more than one kingdom or subkingdom.
[c] Numbers in italics indicate viruses most commonly found in a particular kingdom.

such evolutionary perspectives have already been applied to virology (3, 8, 16, 22, 28, 31, 58). Indeed, the study of evolutionary questions related to viruses has been tentatively termed evovirology (37).

A large number of methods can now be applied to a range of evolutionary problems, from the definitive description of the nucleic acid sequences in the viral genome (MS-2 and ΦX 174 bacteriophages) to a determination of the wider ecological relations. The information obtained with these approaches, available at various levels of detail in the different virus groups, is applied here to evolutionary questions about the known viruses. In particular, we are guided by two principles which are crucial to the continued survival of virus populations. One relates to the *genetic constitution* of viruses, with its adaptation potential for change and its controlled expression; the other, to mechanisms of *viral transmission,* without which viruses would be biological dead ends.

DIVERSITY AMONG VIRUSES

The versatility of the viral genome can be readily appreciated from the information summarized in Figure 1. This figure is not meant to represent a taxonomic classification; rather, it should serve to provide a base for the delineation of any phylogenetic order and to assist in detecting possible correlations and regularities. When the DNA and RNA genomes and host ranges of the various virus families or groups are thus set out in a symmetric form, all the "slots" are not filled. This can be explained in either of two ways: (*a*) The relevant viruses in particular hosts exist, but have not been discovered to date, owing to the anthropocentricity of interest in disease-producing viruses in humans, animals or plants, or in bacterial viruses which can serve to further basic knowledge. (*b*) Selective pressures, whether due to problems of viral genomic adaptation or of viral transmission, either have prevented the slots from being filled or have not permitted the continued survival of viruses within the slots. The absence of viruses from a particular slot could thus be as informative as their presence.

Two important features of viruses, not included in the figure because of space limitations, are the molecular weight (mol wt) of nucleic acid, and the particle size or hydrated volume (35). With the reservation that new data may change the picture, certain regularities can be noted from Figure 1: (a) There is a generalized correlation between the nucleic acid character of the genome and the host range (see also Table 1). Thus, the vast majority of viruses that occur in monera and protista contain duplex DNA. Almost all known plant viruses contain RNA, as do most fungal viruses described to date. Similarly, many more RNA than DNA viruses affect humans. (b) No physically segmented genomes are known among the DNA viruses, whereas a large number of RNA viruses possess segmented nucleic acids. (c) Although some of the virus groups possess circular DNA, none appears to contain circular RNA. (d) Almost all of the continuous negative or the segmented monoparticulate single-stranded RNA viruses are enveloped.

Rhabdoviruses are the only enveloped viruses found in plants. Plasma membrane budding, which occurs in many animal enveloped viruses, would not be expected to develop within plant cells that have a rigid cellulose wall. Indeed, unlike the vertebrate and invertebrate rhabdoviruses, the plant rhabdoviruses become enveloped at the nuclear or endoplasmic reticulum membranes (26). One might expect from this premise that plant herpesviruses might eventually be discovered, since herpesviruses become enveloped primarily at the nuclear membrane.

Although not noted in the figure, it is also of interest that a tRNA-like structure, which is able to be aminoacylated, has so far been demonstrated only in some of the positive, single-stranded nonenveloped RNA viruses.

STRATEGIES OF VIRAL GENETICS

Figure 1 sets out some of the successful strategies which have resulted from the evolution of viral polynucleotides. In this section we explore the molecular mechanisms by which such genomes may be generated.

In DNA viruses, as in other biological units, evolution appears to have proceeded through mutation, recombination, and selection. The evolution of those RNA viruses which lack a DNA intermediate could have been accelerated if recombination occurred among RNA molecules, perhaps by a copy choice mechanism. However, evidence for RNA recombination, as distinct from reassortment, has been obtained so far only with some picornaviruses (29).

Mutations in viruses occur with varying frequencies. That the mutation rate in phage λ (Styloviridae) is 100 times higher in the lytic state than when the virus is integrated into host DNA as prophage (13) suggests that the mutation rate is, at least in part, a function of generation time. Thus, because of their relatively short generation times, the evolution of viruses may be accelerated with respect to that of [multi]cellular forms of life.

The adaptive role of point mutations has been clearly shown by studies on the $Q\beta$ ribophage (Leviviridae), indicating that an abbreviated mutant differs in three residues from its wild-type parent (27). However, it has become clear over recent years that large-scale recombinational rearrangements, such as duplications and translocations, can provide a more plastic and rapid source of new genotypes than the gradual accumulation of point mutations. The role of recombination in generat-

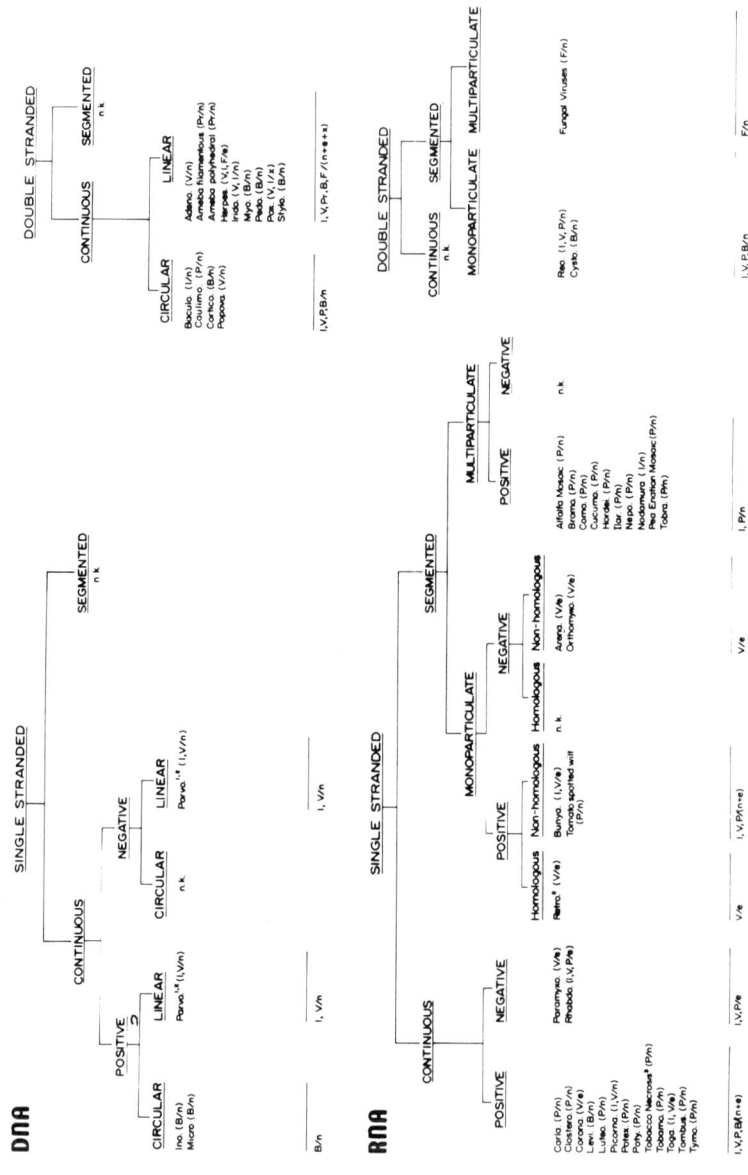

ing novel R plasmids and the evolutionary potential of insertion sequences are well appreciated (41). If the papovavirus, SV40, is passaged at high multiplicity through host cells, a number of novel variants may be generated. Large segments of the original viral genome are replaced by host DNA; the conserved viral DNA, which contains the initiation sequence for DNA replication, is triplicated in one of the variants (25). In this example of rapid adaptation, selective pressures have favored a smaller genome which contains multiple initiation points for DNA synthesis.

In the above examples, viral evolution has exploited the same mechanisms for producing genetic change as those used by cells. Viruses, however, also provide examples of processes which have, as yet, not been convincingly demonstrated in cells. Most striking is the fact that the genome of so many viruses is RNA rather than DNA (Table 1). Retroviruses, which include the RNA tumor viruses, possess an enzyme—reverse transcriptase—able to synthesize DNA on an RNA template (reviewed in 57). Temin (49) has emphasized the versatility introduced into evolution by this reversal of the "normal" information flow; in particular, this enzyme enables the duplication of genes without DNA excision. Nonetheless, recombination has also been demonstrated in the RNA tumor viruses, and it is possible that this occurs via the integrated DNA form (provirus). It thus appears that some RNA viruses can, under appropriate pressures, be driven into DNA forms which could then recombine with cell genes or other integrated viruses.

One striking feature, noted in Figure 1, is the large number of RNA viruses with segmented genomes. The genomes of the retroviruses are polyploid, while RNA genomes of other virus groups occur in discrete segments of varying size or number. One genetic advantage of this segmentation may be an enhanced potential for genetic reassortment. For example, Webster & Laver (54) have presented strong evidence supporting the view that new pandemics of human influenza arise from reassortment between human influenza viruses and those of other mammals or birds.

The existence of segmental RNA genomes may also relate to their evolutionary origins. Novel viral species could have appeared in a "quantum jump" if a single

◄─────────

Figure 1 Versatility of the viral genome–host range. The nomenclature of the International Committee on Taxonomy of Viruses (1977) has been used, and this figure has been prepared with the reservation that newer data may place some of the viruses in different slots. Monoparticulate = segmented genome encapsidated within single virion; multiparticulate = segmented genomes encapsidated in different virus particles.

Positive = RNA genome serves as mRNA: negative = mRNA transcribed from complementary strand of genomic RNA.

Homologous = genome divided into equal size and information; Nonhomologous = genome divided into different sizes and/or information.

nk = not known; e = enveloped; n = nonenveloped; x = envelope not a classical bilayer membrane.

Hosts: B = bacteria, mycoplasma, and blue-green algae; F = fungi; P = plants, Pr = protista; I = invertebrates; V = vertebrates.

[1]Both positive and negative DNA strands may occur in the same population.
[2]May require another "helper" virus for replication.
[3]Has satellite virus (RNA = 4×10^5 daltons) requiring tobacco necrosis virus to replicate.

cell was mixedly infected with two different replicons, each of which *separately* contained genes *jointly* needed for survival in a changed environment (essentially as contemporary bacteria acquire expanded genetic contents in the form of drug-resistant plasmids). An origin of this type could account for the distinction between monoparticulate and multiparticulate genomes. Figure 1 indicates that segmented monoparticulate genomes are characteristic either of enveloped virions or of reoviruses which contain duplex RNA genomes. Such particles, because of their relatively large particle volumes (35), could possibly accommodate "new" coding units within one capsid. By contrast, if the inferred originally independent viruses had small particle volumes, selection may have forced physically separate particles to be retained for each of the new vital coding units. On this basis, multiparticulate viruses would represent a group of ancestrally independent replicons whose *coupled* replication became obligatory in a given environmental context. The advantages offered by such a genetic mechanism must have offset the problem of transmission of separate particles required for infection with many of the plant viruses. There are also examples among both RNA and DNA viruses (Figure 1) of very small virus particles that have been found to replicate only when a cell is simultaneously infected with a large virus; the latter, however, does not need the smaller virus to replicate.

Some features of the eukaryotic RNA viral genomes may reflect the difficulties faced by a polycistronic RNA in a biochemical environment tailored to processing monocistronic mRNAs only (21). Thus, the eukaryotic cell appears to treat the genomes of polioviruses, other picornaviruses, and alphaviruses (Togaviridae), as monocistronic, and translates the entire genome into one or a few large polyproteins which are subsequently cleaved to yield the functional proteins. In case of negative-strand RNA viruses, such as paramyxoviruses (Figure 1), a similar problem in translation does not arise because the mRNAs are transcribed from complementary strands.

STRATEGIES OF VIRAL TRANSMISSION

The strategies used by viruses to spread from one host to another have been at least as numerous as those associated with genetic adaptations. Important among the genetic adaptations influenced by the selective pressures of the micro- or macroenvironment must have been attenuation of the propensity of a virus to kill its host or its vector (if needed for transmission). In multicellular organisms, the distribution of the virus within its host would have had to be such that the virus would be accessible in body sites from which it could be transmitted to other hosts. As long as the host was the same, or in cases of gradual coevolutionary change with its host, there would be no problems with adsorption to, or ability to replicate within, a foreign cell; nor would there be any confrontation with the host's immune or genetic resistance mechanisms. These problems would, however, be particularly pronounced in the case of transmission across taxonomic boundaries and subsequent stabilization of the virus in the "foreign" taxon.

Transmission of a virus from the host parent to its progeny seems particularly favorable to viral survival, especially in the case of small host populations. However, the death of the progeny before they could pass the virus on to their own descendants would favor the development of mechanisms for horizontal host-to-host spread. Intimately associated with these issues is the capacity of the virus to persist within its host. The mechanisms of viral persistence are many: in some cases, the viral genome is integrated in a latent form within the chromosomes of somatic or germ cells; in other cases, the virus, in a noninfectious or infectious form, can persist in somatic cells for prolonged periods.

Vertical parent-to-progeny transmission in asexual forms, associated with persistence of the integrated viral genome in host cell chromosomal DNA, is best exemplified in lysogenic bacteria. This theme recurs in sexually reproducing organisms, such as vertebrates, with the retroviruses. Although similar mechanisms might be expected to occur in other kingdoms, they have not yet been demonstrated.

Other modes of parent-to-progeny transmission have been adopted by many viruses. Because the likelihood of the progeny becoming infected is much less secure when the virus genome is *not* chromosomally integrated, we prefer to group such modes of transmission as "paravertical." Examples of paravertical transmission can be found for viruses in all kingdoms: via spores in certain bacteria and fungi, pollen and seeds in plants, and eggs in invertebrates. Several forms of paravertical transmission can occur in vertebrates: transplacental, intrapartum (around the time of delivery), or via colostrum or milk.

A variation of paravertical transmission, whereby the virus could be transmitted not only from the parent to its progeny, but also from host to host, can be appreciated. Viruses which can persist in an infectious or noninfectious form in the parent (or other) host can be passed on to new progeny after birth. This mechanism, best exemplified by herpesviruses (37), would be particularly valuable in small host populations.

A large number of mechanisms can be found in nature for horizontal spread from one host to another, within or between species. These can be grouped as follows: (*a*) direct physical contact; (*b*) indirect contact via multiple sources, e.g. water; (*c*) by a vector, in which the virus could be spread mechanically or actually replicate. The variations here are immense. They concern characteristics of the virus and its host or vector, and their related evolutionary, geographical, and ecological aspects.

It is tempting, for instance, to speculate that the sparsity of viruses in plants of older origin (34) is related to their relative lack of vectors to transmit archeoviruses. The multiple systems developed by angiosperms to attract insects might well have had the secondary effect of increasing the potential for the spread and survival of viruses in these plants.

Another example of an adaptive method of viral spread is the sexual route which occurs in several animals, including man. Adaptation to this new host microenvironment is correlated with minor to major genetic changes in some viruses, in comparison to their nonvenereal counterparts.

The range of stability of various viruses in extracellular conditions is wide. Thus, unlike most plant viruses, tobacco mosaic virus (TMV) usually spreads in the absence of vectors (34). The capacity of TMV to spread by nonvector routes is closely associated with the greater extracellular stability of this virus. Similarly among the Picornaviridae, the enteroviruses, which spread by oral-fecal contact and can survive in mechanical vectors such as molluscs, have a greater extracellular stability than the related rhinoviruses, which spread by the respiratory route. Temperature must have been an important selective factor also, as exemplified by those viruses which can grow in both poikilothermic insects and homeothermic mammals. Temperature effects can be noted in the types of vectors for alpha- and flaviviruses (Togaviridae). Most of these viruses, transmitted by mosquitoes, are found in the tropics and subtropics, whereas some of the flaviviruses from areas with cold winters are transmitted by ticks.

Highly diverse ecological conditions, some of which have since been altered by human intervention (see below), would facilitate the dissemination of certain adaptable virus groups throughout a diversity of novel host species. Such events might explain the discovery of viruses with many common characters, e.g. rhabdoviruses, within different kingdoms (Figure 1). On the other hand, there seem to be effective barriers against the trans-kingdom spread of some virus groups. Thus, tailed phages are only found in bacteria despite the frequent contact of their bacterial hosts with eukaryotic forms of life for more than a billion years.

When viruses do spread under natural conditions from one host species to another (if infection can be established at all), various outcomes may ensue. In some cases, the virus causes severe, if not fatal, disease in a foreign host, e.g. the B herpesvirus of macaques in man. In other cases, e.g. rabbit myxomatosis viruses, a balance between virus and new host appears to have occurred in a relatively short time (15).

The feline leukemia retrovirus demonstrates a dual adaptation: vertical transmission with germ cell chromosomal integration without apparent disease, and horizontal viral transmission resulting in leukemia (14). Based on nucleotide homology studies, the cat virus and a rodent virus appear to be very similar, suggestive of another probably important mode of viral transmission, that of feeding on virus-containing hosts or vectors. This would be a particularly effective means of crossing taxonomic barriers.

ALTERNATIVES IN VIRUS-HOST INTERACTIONS

The evolutionary implications of two other types of virus-host interactions, besides propagative and persistent infection, deserve special comment.

Viral Oncogenesis

Several DNA and RNA viruses with the common property of persisting in their hosts, have been shown to be oncogenic (14, 28). One way of viewing these viruses is to separate them into two groups: (*a*) those with an oncogenic potential, in that they are carcinogenic only in foreign hosts; (*b*) those with established oncogenicity, inducing cancers in their natural hosts. Nonetheless, it is apparent that viruses in both groups do contain genetic information associated with carcinogenesis. The differences between them most likely reflect host factors, genetic and/or immunological. Recent efforts have been directed at determining the specific base sequence areas associated with oncogenesis.

No readily apparent survival advantage seems to ensue from the ability of a virus to cause cancer in a host. Possible advantages related to transmission are unlikely. In the case of the DNA papovaviruses, adenoviruses, or herpesviruses, very little if any transmissible infectious virus can be detected in the tumors themselves. Even among some of the RNA tumor viruses, infectivity and oncogenicity are not necessarily coupled.

We are left with two interpretations, one of *chance* and one of *necessity*. The numbers of separate virus groups and the numbers of different hosts, from amphibians to primates, in which the phenomenon can be observed repeatedly would not favor a recent chance event, nor could chance alone explain the apparently persistent specificity of both viral and host genetic characters. The more likely alternative is that the genetic information for carcinogenesis is a secondary event and that such genes were, at least initially, responsible for some adaptive survival mechanism important for viral replication or persistence, or possibly even for host functions. In papovaviruses, the genes coding for the T (tumor-associated) antigens appear to be required for viral replication. In case of some of the herpesviruses, such genes might be important for persistence mechanisms. Thus, tumors develop in latent sites of chickens infected with Marek's disease virus, but not in the feather follicles in which the virus propagates. More information is obviously required to define the evolutionary enigma which oncogenic viruses currently present.

Viruses as Agents of Accelerated Cell Evolution

The idea that viruses and other extrachromosomal genetic elements may act as agents of accelerated evolution has been reviewed elsewhere (40, 41, 58). While the concept seems soundly established in bacteria, a convincing factual base is as yet lacking for theories which propose a functional role for retroviral sequences in the ontogeny and evolution of higher organisms (18, 50, 51, 55).

ORIGINS AND PHYLOGENY OF VIRUSES

Hypotheses of Viral Origins

Nowadays, only two hypotheses concerning viral origins are seriously entertained by virologists. The first is that viruses represent highly degenerate bacteria (8). The reader is referred to a detailed schema set out by Matthews (34). His proposed classification of viruses (35) shows that when the log molecular weight of viral nucleic acids is plotted against log particle mass (or volume), a straight line can be obtained. Since it places *E. coli,* mycoplasma, psittacosis, poxvirus, herpesvirus, and the various enveloped RNA viruses in a suggestive order of decreasing complexity, this graph could be used to support a "retrogressive" theory for the origin of at least the enveloped viruses from prokaryotes.

Much of the appeal of the "retrograde evolution" hypothesis comes from examples of apparently analogous processes in parasitology. In particular, if the characteristic organelles of the eukaryocytic cells indeed represent endosymbiotically simplified prokaryocytes (33),[2] it is not difficult to continue the process of simplification to the level of a poxvirus. However, the differences in the particle architecture and genetic strategies between a poxvirus, a herpesvirus, a retrovirus, and a DNA or RNA phage are very great. It is not easy to envisage the types of selective pressure or mechanisms which might have transformed the ancestor of one into the ancestor of the others. The nucleocapsids of the enveloped RNA viruses are not unlike the ribonucleoproteins of normal nucleated cells; however, no such entities are found in bacteria. If viruses are derived from bacteria, then one must note that, at least in eukaryocytic viruses, none of the distinctive features of the prokaryotic cell have been retained despite the evolutionary conservatism displayed by bacterial viruses (1). The polycistronic character of the RNA genome of some eukaryocytic viruses is not a compelling argument for a prokaryocytic ancestry because RNAs of high molecular weight are characteristic of the eukaryocytic nucleus.

The other major alternative hypothesis, that viruses have evolved from cellular polynucleotides, has been put forward in general terms by others (22, 31). With the greater molecular understanding of cell biology that has emerged over recent years, none of the individual steps from cell polynucleotide to self-replicating virus appears implausibly large. Worth noting is that the processes that might be involved could occur with greater probability during embryogenesis than postnatally. The rate of cell division during ontogeny is such that an emerging virus could pass through millions of cellular generations, with immense potential for variation and selection. The problem of transmission would not exist for an archeovirus arising in early ontogeny, as the unit would be spread into a large number of cells by the repeated division of the progenitor clone. Further, in higher metazoans, an origin for an archeovirus during embryogenesis would ensure that the proteins of the replicating unit were recognized as self by the immune system of the host.

Thus, although it is possible to regard the large poxviruses as having originated by retrograde evolution (22, 31), we favor the concept that most viruses represent "escaped" cellular polynucleotides. We consider it significant that sequences homologous to those of various retroviruses have been detected in the normal cells of many vertebrates (18, 50, 51). For example, sequences related to the *src* gene of the avian sarcoma virus have been found in the DNA of normal cells from a variety of vertebrates, from fish to primates, bearing relationships proportional to the evolutionary distance of the species tested from chickens (52).

Differences between the two origin hypotheses can be brought into focus by briefly considering viroids—infectious small replicative units of single-stranded RNA which, unlike true viruses, do not contain protein (12). The number of speculative

[2]The finding of prokaryotic viruses within the mitochondria of animal cells or chloroplasts of plants would offer strong support to this thesis. However, to date, only viruses similar, but not identical, to those affecting other parts of the cell have been observed within chloroplasts (45).

intermediate steps needed to reduce a bacterium to a viroid is enormous. By contrast, if one regards viroids as aberrant cellular RNAs, only a few genetic changes would be required to uncouple such polynucleotides from cell regulatory control. The demonstration that at least 60% of potato spindle viroid RNA is complementary to the DNA of normal solanaceous plants (19) provides strong support for the latter alternative. Any serious comparison of the two origin hypotheses must, however, be made with two qualifications in mind: (a) The two hypotheses are not mutually exclusive;[3] (b) given the genetic flexibility exhibited by extrachromosomal genetic elements (41), one must also accept the possibility that a multiplicity of phenotypically different virions could develop from one ancestral biounit, i.e. many viruses may have evolved not from bacteria or from eukaryotic cell polynucleotides, but from other viruses. (See below.)

Single versus Multiple Origins

A unitary origin for all viruses and viroids appears unlikely (22). A more tenable assumption is that viruses have arisen repeatedly during the course of evolution, and that the characteristic strategies of viral genetics and transmission exemplified in successful groups reflect the relative stages of evolutionary development of the progenitor organisms and the potential for effective spread provided by the prevailing environment(s). The growing body of data relating to the structure and properties of viral RNA polymerases has made it possible to place the hypothesis of multiple origins for viruses on a less speculative basis.

The replicase of ribophages consists of four polypeptides. The presence of three components of the host's translational machinery [the elongation factors EF-Tu and EF-Ts and a ribosomal protein, S1 (53)] in the viral replicase is suggestive, when it is recalled that tRNA-like structures appear to exist in ribophage RNA (39). All three proteins have the ability to recognize tRNA, although the functions characteristically performed by EF-Tu and EF-Ts in translation are not apparently required for viral RNA synthesis (7). This association between a tRNA-like structure in the viral genome and RNA-recognizing proteins in the polymerizing enzyme seems unlikely to be due to chance or to the subsequent addition of such specialized proteins to an independently evolved viral polymerase. We tentatively conclude, therefore, that this particular mode of viral RNA recognition, synthesis, or regulation had its genesis, at least in part, in the translational apparatus of the host cell.

tRNA-like structures have been documented in several plant viral RNAs (21) and in at least one animal virus RNA (mengovirus) (43). Whether the replicases of these viruses resemble the ribophage system has yet to be ascertained. However, to date all viruses which contain tRNA-like structures have positive-strand genomes of relatively low molecular weight. In this regard, the position of the Retroviridae is interesting, in that they possess a sequence complimentary to a tRNA "primer" located about 100 bases from the 5' end of the viral genome.

[3]Indeed, the data of Subak-Sharpe's group (48), obtained with the use of doublet analyses, support the view that the small animal viruses may have originated from host cell polynucleotides, while the larger mammalian viruses could have had different origins.

If RNA viruses arise from cells (by whatever mechanism), the most likely sources for their RNA polymerase enzymes are the DNA or RNA polymerases of their progenitor cells. All cells contain DNA-dependent DNA polymerases and DNA-dependent RNA polymerases, one of the latter in prokaryocytes, several in eukaryocytes. Many vertebrate cells appear to contain endogenous genes of RNA tumor viruses; one of these genes codes for an RNA-dependent DNA polymerase. This may be considered a "viral" or a "cellular" enzyme, depending on point of view (49). Interconversion between the two types of polymerizing activity may be easy: if Mn^{2+} is substituted for Mg^{2+}, the *E. coli* enzyme DNA *pol-1* will incorporate ribonucleotides, rather than deoxyribonucleotides (30). It may be significant that certain RNA polymerases prefer Mn^{2+} to Mg^{2+} (38). Mitzutani & Temin (36) have shown that the DNA polymerase of spleen necrosis virus—SNV (a retrovirus)—can initiate the synthesis of RNA on RNA templates. On the basis of this and other evidence, Temin (50) suggests that many enveloped RNA viral genera evolved from retroviruses. Reciprocally, Biebricher & Orgel (4) have stressed the ease with which an enzyme, tailored to copy RNA from DNA, could acquire the ability to use RNA as template.

Given these multiple possibilities, it is difficult to believe that the multi-unit enzyme exploited by ribophages to replicate their positive-strand genomes could have evolved via the same ancestral source as the enzyme used to copy the negative strand of a paramyxovirus or the duplex RNA of a reovirus. In the face of the diversity of viral replicative and translational strategies, the concept of a unitary origin for all known mechanisms becomes almost impossible to sustain. The hypothesis of multiple reticulate viral origins is reinforced by recent data which support the possibility that some DNA viruses evolved from RNA viruses. The finding that unintegrated DNA from SNV increases in concentration for several days after infection has suggested to Temin (50) that such DNA could represent a precursor to a small animal DNA virus. Similarly, proviral DNA, which could represent intermediate steps on the road to an independently replicating DNA virus, has been demonstrated in another retrovirus—visna (18a).

Viral Systematics

Nahmias (37) superimposed more than 50 herpesviruses isolated from a diversity of hosts (from fungi to man) on the phylogenetic tree of the organisms from which they were originally isolated. This particular representation highlights key questions in viral phylogeny. In theory there are two extreme, albeit not mutually exclusive, views which can explain the observed virus-host distribution:

1. The ancestral herpesvirus arose prior to the separation of the various taxa; in view of the extremely wide divergence between molluscs, fungi, and vertebrates, the earliest common node could be as remote as 1.2×10^9 years ago. Coevolution of the virus with its initial host might help to explain why viruses in all the varied hosts in which they have been demonstrated have maintained so many common characters, while the progenitor host population has split into widely different taxa.

2. The ancestral herpesvirus(es) arose at one or several points during evolution and have spread from their progenitor host(s) across species to higher (and perhaps lower) taxonomic groups. The problem now becomes one of explaining how the virus could adapt itself to such different hosts.

The possibility of convergent evolution is unlikely, as it would require more than 50 separate genetic sources to generate particles of almost indistinguishable morphology and with very similar strategies of viral genetics and transmission, including persistence. Approaches to possible resolution of this problem in systematics are being provided by molecular analyses of the DNAs of various herpesviruses. It has recently been shown (20, 46) that the DNA molecule of herpes simplex viruses (HSV) is a physically continuous, but organizationally bipartite, structure comprised of an L (long) stretch containing 82% of the sequences and an S (small) stretch containing 18%. There are terminal repetitive sequences at both ends of the DNA molecule, as well as internal repetitive sequences. Four kinds of HSV DNAs occur in the virions in roughly equal proportions, which differ in the relative orientation of the L and S subregions. Based on the molecular structure of the DNAs of various other herpesviruses, Sheldrick (personal communication) proposes that the original herpesvirus might have been in the form of the L subregion, common so far in all herpesviruses examined, which would form a circular concatamer with varying S fragments. This in turn would give rise to the various DNA structures found in the different herpesviruses.

Similar patterns have been found between the DNAs of HSV-1 and HSV-2, between the DNAs of equine abortion and porcine pseudorabies herpesviruses, and between the DNAs of two herpesviruses of New World monkeys (*H. saimiri* and *H. ateles*). The DNA configuration of the catfish herpesvirus is so far unique. When this type of technology is applied to herpesviruses of other species, an improved method of classifying viruses in this family should become possible.

Some evolutionary relationships among herpesviruses can also be deduced from viruses affecting the same or different species. Many illustrations can be noted of greater similarities between the herpesvirus of one species to that of another species, rather than to a second herpesvirus affecting the same species (37). For example, the human herpes simplex virus has greater similarities to the macaque monkey's B herpesvirus than to the human cytomegalovirus. Also, the chicken Marek disease virus is closer to the turkey herpesvirus than is another chicken herpesvirus (infectious laryngotracheitis).

A great deal of similar evolutionary information relevant to systematics can be derived from host range data, although it may not be easy to uncouple the relationship between a virus and its primary (or progenitor) host(s) and secondary (or acquired) hosts. The situation is clearest in prokaryotes, where "vectors" as such do not exist. Within a genus, such as *Bacillus,* a given phage can often cross many species lines (42); however, few phages are able to cross *generic* barriers. Of 26,681 lytic reactions analyzed (23), only 0.3% crossed generic divisions in bacteria. Thus, at least at the genus level, the host ranges of phages with distinctive morphology may be a meaningful index of evolutionary relatedness among both host cells and viruses (1).

Among plant viruses it is more difficult to differentiate between possibilities 1. and 2., presented earlier regarding herpesviruses, on the basis of host range. Some plant viruses have limited host ranges: cucumber viruses 3 and 4 are mainly restricted to the Cucurbitaceae (24). By contrast, a large number of plant viruses [e.g. bromo mosaic virus (BMV)] have broad host ranges.

Several explanations may be given for these extraordinarily broad host ranges at the species level. One results from grouping susceptible plants into higher taxonomic categories. While BMV infects 150 species in 8 families, 75% of susceptible plants belong to the Graminae (monocots), the remaining families comprising dicots. Does this reflect ecology—i.e. monocot grasses inhabit an ecosystem which favors viral dissemination—or is a conserved evolutionary relationship still traceable beneath the broader host range? It is also tempting to correlate the host ranges of plant viruses with the feeding propensities of their insect vectors. For example, the aphid *Myzus persicae* is responsible for the transmission of over 50 different viral diseases and will feed upon a wide range of plants.

Once introduced into the plant, the virus may be indifferent to the genus or even family of the host. In our view this is because (*a*) higher plants are rather tolerant to large-scale changes in DNA content, e.g. ploidy, and hence (*b*) the phenotypic and karyotypic differences often used in plant taxonomy may be achieved by minor genetic adjustments. In view of their quite recent evolutionary origin, we suspect that many angiosperms present viral nucleic acids with relatively similar intracellular biochemistry and with similar potentialities for spread through plasmadesmata, phloem, etc. Thus, the differences which seem so important to the plant taxonomist are not reflected in corresponding differences in the molecular and physiological mechanisms which sustain viral reproduction and spread. Set in this context, and recalling that many "unrelated" plant families grow in common ecosystems, the broad ranges of so many plant viruses are predictable rather than unexpected.

Insects provide clear examples of our inability to separate "original" hosts from hosts secondarily acquired through transmission. Arthropod vectors occupy a bridging position between members of several kingdoms. At least several of the vertebrate or plant virus groups have some (tentative) members which multiply endogenously in invertebrates, including the reo-, irido-, bunya-, pox-, parvo-, herpes-, rhabdo-, picorna-, and togaviruses. Although no single rhabdovirus can simultaneously infect a plant, an insect, and a vertebrate, several can infect plants and insects on the one hand, and vertebrates and insects on the other. Knudson (26) suggests that the plant rhabdoviruses originated prior to the insect viruses. However, because of their central position in the persistence and survival of so many animal viruses, insects have been regarded as the likely progenitor and natural hosts for many of the arthropod-borne animal viruses (3). While this hypothesis of a primary association with insects is appealing, the evidence is still primarily inferential.

Several molecular techniques, in addition to those used for DNA mapping, have become available to measure genetic relationship among viruses and hence assist in viral systematics. For instance, nucleic acid hybridization methods have enabled the evolutionary relationship among many of the retroviruses to be traced (18, 51, 52).

However, the usefulness of such techniques is sometimes limited when one notes the absence of nucleic acid homology among viruses with many common characters, e.g. herpesviruses (37). Attempts to resolve this problem, as well as problems raised in earlier sections concerning viral change and conservatism, which are central to the issues of viral systematics, have led us to the following principle: The strongly conserved aspects of a virus genome is its basic genetic *organization*—the grouping, spacing, and ordering of genes—rather than the encoded sequences as such. The following points are relevant: (*a*) No coded information is needed to conserve the physical length of a (linear) genome and the relative order and spacing of genes. (*b*) The folded topologies of specific proteins can be identical, even when 70% of the underlying base sequences differ. This is a consequence of the degeneracy of the genetic code and the ability of amino acids at a variety of positions to be altered by mutation without impairing the performance of the protein (11). (*c*) Proteins of quite different amino acid sequences can fulfill very similar functions. Perhaps the most striking example of this in virology is the ability of the capsomeres of various viruses, totally lacking in antigenic cross-reactivity, to pack together into geometrically similar arrays. In particular, viruses with icosadeltahedral symmetry are found in all five kingdoms and apparently arise from divergent genetic sources; evidently a large number of different polypeptides can be aligned according to the same symmetry principles (10).

Set in this context, the lack of any obvious homology between the genomes of (some) related viruses does not seem so surprising. It seems that wherever the selective pressures of the micro- and macroenvironments remain roughly the same, the general and characteristic structure of viral genomes may be strongly conserved, even though more and more mutations may accumulate over the course of time in individual genes. Certain advantages arising from specific gene orderings are evident; thus, for any DNA virus above a certain level of complexity, genes coding for viral DNA synthesis will be transcribed early. In the tailed phages, genes for structural proteins are transcribed last in all known examples. On a finer scale, one notes that the genes for various head proteins occur before those for various tail proteins in a phage like λ; this is probably a common theme of viral morphogenesis among tailed phages.

This evolutionary conservatism can easily obscure the enormous adaptive flexibility of viral genomes. The examples (6) of the phages λ/p22 show that a virus can acquire a different particle morphology without surrendering the gene order and sequence homologies that make adaptive DNA interchanges with related viruses possible. In the field of microbial genetics, much well-documented evidence shows that the evolution of bacteria, phages, and plasmids is *coupled.* It seems very likely that such processes of "natural genetic engineering" (41) occur among all viruses. For instance, it has been reported that novel C-type retroviruses can arise by picking up fresh sequences from new host cells (2). As noted in the viral genetics section above, this type of recombination-based evolution may be extremely rapid.

In making the above comments, we do not wish to give the impression that all viral evolution has been channelled into one format. Like the evolution of cells, the evolution of viruses is a dynamic, changing, continuing process. Viral speciation may occur (a) whenever a virus group loses by mutation the shared recombination targets which give it access to the coupled evolution of other extrachromosomal elements and cells, or (b) when a virus gains or loses specific key genes by whatever mechanism. Any alteration in a basic viral strategy is a macro-change, which could either be the end product of a very gradual evolutionary process, or represent a sudden quantum jump made possible by opportunistic selection.

HUMAN INTERVENTION IN THE EVOLUTION OF VIRUSES

The term virus (L., venom), betrays the traditional emphasis on its disease-producing characteristics. Yet the potential beneficial uses of viruses were appreciated soon after their discovery at the turn of this century. In this section, we discuss the possible role of human intervention in the evolution of viruses, with particular emphasis on attempts to control viral diseases, as well as the use of viruses for beneficial purposes. Many historical examples of the effects of viruses, e.g. smallpox, on the sociocultural, and even political, aspects of human civilization are well documented (32).

The increasing urbanization, density, and movement of populations and various agricultural, social, and cultural human habits, as well as the increasing contact between people and domestic animals, appear to have played an important role in changes of human–virus interactions. Fenner et al (16) recently emphasized similar correlations between the time scale of culture changes in humans, and the number of generations and size of human communities. Several examples can be cited. It is very likely that viruses with persistent mechanisms, such as herpesviruses, would have greater survival advantage in small populations. In contrast, respiratory viruses, within various families, would require large populations for their continued spread. That this may well be the case is supported by serological studies in primitive isolated human societies in various parts of the modern world (5). With some of the respiratory viruses, particularly influenza A viruses, strong evidence supports the involvement of domestic animals in virus survival adaptations. Interspecies spread of other viruses, such as human measles, canine distemper, and bovine rinderpest, may also have resulted from greater contact between humans and animals. Changes in agricultural practices, such as grafting, which enables transmission of some plant viruses, as well as the interest in certain crops, may be reflected by the larger number of viruses in agricultural plants.

The increase of the human population may have presented a problem in adaptation as well, which those viruses unable to replicate in immune hosts would have to overcome. Variations in antigenic proteins would be likely to have resulted in the many serotypes within some virus groups. Two groups of respiratory viruses convey extremes—the rhinoviruses with 100 or more serotypes, and respiratory syncytial virus with only one.

It has been suggested on the basis of historical records that the virulence of polioviruses may have increased as improved social hygienic conditions brought about a shift in the age of acquisition of these picornaviruses (3). Similar changes in the environment may well explain the increasing disease-producing character with age of other viruses, such as the hepatitis viruses or the Epstein-Barr herpesvirus.

The grouping of newborns in nurseries and of patients in hospitals has created further possibilities for the spread of many human viruses. Furthermore, the use of blood products in medicine has increased the transmission versatility of some viruses, such as hepatitis B and cytomegalovirus, which can also be transmitted by other routes. (It should have been apparent, on evolutionary grounds, that hepatitis B virus, unless a very recent virus, could only have been spread from the blood by a vector and that other modes of transmission, recently demonstrated, would be more likely.) The advances in social hygiene and medicine have enabled more people with cancer or with immunological defects to remain alive; as immunotherapy, such as that used for organ transplants, has become more widely used, the balance of the virus-host relationship has been disturbed. The change may be viewed as an attempt by viruses to increase their transmissibility from poor hosts to healthier susceptible ones. Interestingly, the viruses involved are primarily those with persistent mechanisms, e.g. herpesviruses.

Recent techniques for the treatment and prevention of viral infections of particular severity have affected, in varying degrees, the evolution of viruses. Complete eradication of a virus can be attempted by denying further susceptible hosts to horizontally transmitted viruses without reservoirs or vectors, or by breaking a vital chain in transmission. The case of smallpox is the first example of a conscious and successful effort by humans to eliminate a biological unit from the biosphere. The inability of the smallpox virus to develop successfully either a genetic or a transmission mechanism to ensure its survival contrasts with the adaptability of yellow fever virus, which has an established jungle cycle in monkeys. It is of interest that a virus of probably similar evolutionary origin (cowpox) would have been useful in eradicating smallpox. Similarly, the turkey herpesvirus can prevent the Marek disease herpesvirus from causing cancer in chickens and the human measles virus can prevent infection by the closely related canine distemper virus.

The selection in the laboratory of attenuated variants for vaccine use has been successful with several viruses. The infrequent reversion to a virulent form bespeaks both the genetic stability of many of these man-made virus strains and the great caution required with such approaches. Forced adaptation may also be induced by the accidental introduction into humans of viruses of closely related species, such as occurred when the simian papovavirus (SV40) was inadvertently administered together with the polio vaccines (44).

Yet another recent example of the effect of human intervention, this time at the ecological level, is that of the Lassa fever virus. This virus is carried by a species of small rat whose numbers had been curbed by the brown rat. After human extermination of the brown rat, the small rat frequented human habitations and spread the virus to man.

Even the truly successful antiviral topical therapy for ocular herpes simplex virus infections with iododeoxyuridine (IDU), and the doubtful results of attempts at systemic therapy with this drug, have evoked the specter of two problems of antiviral chemotherapy, other than possible toxic effects. The first is the appearance of IDU-resistant mutants of clinical significance; the second, more hypothetical, is that IDU in vitro has been shown to reactivate latent retroviruses and to increase the yield of infective Epstein-Barr herpesviruses. Could the administration of halogenated pyrimidines in humans have the same effect?

The use of viruses for the benefit of mankind is of increasing interest at two levels—biological control of pests and genetic engineering. After the unsuccessful attempts with bacteriophage therapy in the 1920s and 1930s, a relatively successful application of biological control was made in the 1950s when the rabbit poxvirus (myxomatosis) was introduced in Australia and in other countries to control rabbits detrimental to agriculture (15). The other attempts at biological control have been made with the baculoviruses of certain insects, a particularly attractive approach in view of the increasing toxicity and resistance problems related to the use of chemical insecticides. (Such insecticides most likely also altered the propagation of vector-transmitted viruses and influenced their biogeography.) Another potentially beneficial application of viruses is their use for genetic engineering, i.e. attempting to introduce by means of "innocuous" virus, genetic information lacking or defective in the host.

A related ingenious means of exploiting evolutionary strategies for practical ends has been outlined by Spiegelman (47), based on the development of abbreviated mutants of phage Qβ, some of which contain as little as 17% of the original parental RNA. Since such variants possess a high affinity for the viral replicase and grow at a relatively rapid rate, they can compete most effectively with the wild-type virus for the replicating enzyme. The multiplication of a parent virus could thus be effectively halted, without any side effects harmful to host cells, by derivatives which, because of their greatly reduced size, lack pathogenic ability. The deliberately engineered construction of such variants offers a new dimension to the field of antiviral chemotherapy.

Certain antievolutionary strategies that have already been employed in a more empirical way in the past, could be exploited from the molecular to the epidemiological level. For example, a biomolecular unravelling of key evolutionary survival advantage mechanisms involved in persistence of herpes simplex–like viruses in nerve tissues, might lead to the prevention of clinical recurrences and to the further intransmissibility of the viruses. A better appreciation of the microenvironment of the genital tract, which has long provided a selective advantage to sexually transmitted viruses and multiple microbes, could reveal an effective and desperately needed strategy for the control of sexually transmitted diseases.

In conclusion, a concerted study of the evolutionary aspects of viruses—increasingly feasible as new methodologies are applied —should bring this large and diverse group of agents within the larger evolutionary framework of biology. Ultimately, meaningful systematics can only be developed on the basis of evolutionary relationships. Further, the application of the evolutionary perspective to pragmatic problems may well guide the design of more effective preventive and therapeutic strategies in medicine, agriculture, and horticulture in coming years.

Acknowledgments

This paper was written while A.N. was the recipient of a Macy Faculty Award at the University of Western Australia. We are grateful to our many colleagues who offered suggestions and criticisms, particularly W. Stanley, R. E. F. Matthews, E. Meyr, J. MacKenzie, A. Gibbs, and F. Murphy, as well as to participants in the first workshop on "Evolution of Viruses", 3rd International Congress of Virology, Madrid, September 1975.

Literature Cited[4]

1. Ackermann, H. W. 1975. La classification des bactériophages des cocci Gram-positifs *Micrococcus, Staphylococcus et Streptococcus. Pathol. Biol.* 23:247–53
2. Anderson, G. R., Robbins, K. C. 1976. Rat sequences of the Kirsten and Harvey murine sarcoma virus genomes; nature, origin and expression in rat tumor RNA. *J. Virol.* 17:335–51
3. Andrewes, C. H. 1967. *The Natural History of Viruses.* New York: Norton. 233 pp.
4. Biebricher, C. K., Orgel, L. E. 1973. An RNA that multiplies indefinitely with DNA-dependent RNA-polymerase; selection from a random copolymer. *Proc. Natl. Acad. Sci. USA* 70:934–38
5. Black, F. L. 1975. Infectious diseases in primitive societies. *Science* 187:515–18
6. Botstein, D., Herskowitz, I. 1974. Properties of hybrids between *Salmonella* phage-P22 and coliphage lambda. *Nature* 251:584–89
7. Brown, S., Blumenthal, T. 1976. Reconstitution of Q Beta RNA replicase from a covalently bonded elongation factor Tu-Ts complex. *Proc. Natl. Acad. Sci. USA* 73:1131–35
8. Burnet, F. M. 1946. *Virus as Organism.* Cambridge, Mass.: Harvard Univ. Press. 134 pp.
9. Burnet, F. M. 1974. Invertebrate precursors to immune responses. *Contemp. Top. Immunobiol.* 4:13–35
10. Caspar, D. L. D., Klug, A. 1962. Physical principles in the construction of regular viruses. *Cold Spring Harbor Symp. Quant. Biol.* 27:1–24
11. Dayhoff, M. O. 1972. *Atlas of Protein Sequence and Structure.* Silver Spring, Maryland: Natl. Biochem. Res. Found. 246 pp.
12. Diener, T. O. 1974. Viroids as prototypes or degeneration products of viruses. See Ref. 28, pp. 757–83
13. Dove, W. F. 1968. The genetics of the lambdoid phages. *Ann. Rev. Genet.* 2:305–40
14. Essex, M. 1974. The immune response to oncornavirus infections. See Ref. 28, pp. 514–38
15. Fenner, F., Ratcliffe, F. N. 1965. *Myxomatosis.* London: Cambridge Univ. Press
16. Fenner, F., McAuslan, B. R., Mims, C. A., Sambrook, J., White, D. O. 1974. *The Biology of Animal Viruses.* New York: Academic. 834 pp.
17. Fenner, F. 1976. Classification and nomenclature of viruses. Second report of the International Committee on Taxonomy of Viruses. *Intervirology* 7(1/2):1–115
18. Gillespie, D., Gallo, R. C. 1975. RNA processing and RNA tumor virus origin and evolution. *Science* 188:802–11
18a. Haase, A. T., Stowring, L., Narayan, O., Griffin, D., Price, D. 1977. Slow persistent infection caused by visna virus-role of host restriction. *Science* 195:175–77
19. Hadidi, A., Jones, D. M., Gillespie, D. H., Wong-Staal, F., Diener, T. O. 1976. Hybridization of potato spindle tuber viroid to cellular DNA of normal plants. *Proc. Natl. Acad. Sci. USA* 73:2435–57
20. Hayward, G. S., Jacob, R. J., Wadsworth, S. C., Roizman, B. 1975. Anatomy of herpes simplex virus DNA: evidence for four populations of molecules that differ in the relative orientations of their long and short components. *Proc. Natl. Acad. Sci. USA* 72:4243–47

[4]Owing to space limitations, we could only include a fraction of supporting references in the very large relevant literature on viruses in all the kingdoms.

21. Jaspers, E. M. J. 1974. Plant viruses with a multipartite genome. *Adv. Virus Res.* 19:37-149
22. Joklik, W. K. 1974. Evolution in viruses. *Soc. Gen. Microbiol. Symp.* 24:293-320
23. Jones, D., Sneath, P. H. A. 1970. Genetic transfer and bacterial taxonomy. *Bacteriol. Rev.* 34:40-81
24. Kado, C. I., Knight, C. A. 1970. Host specificity of plant viruses. I: cucumber virus 4. *Virology* 40:997-1007
25. Khoury, G., Fareed, G. C., Berry, K., Martin, M. A., Lee, T. N. H., Nathans, D. 1974. Characterization of a rearrangement in viral DNA: mapping of the circular simian virus 40-like DNA containing a triplication of a specific one-third of the viral genome. *J. Mol. Biol.* 87:289-301
26. Knudson, D. L. 1973. Rhabdoviruses. *J. Gen. Virol.* 20 (Suppl):105-30
27. Kramer, F. R., Mills, D. R., Cole, P. E., Nishihara, T., Spiegelman, S. 1974. Evolution *in vitro:* sequence and phenotype of a mutant RNA resistant to ethidium bromide. *J. Mol. Biol.* 89:719-36
28. Kurstak, E., Maramorosch, K., eds. 1974. *Viruses, Evolution and Cancer.* New York: Academic. 813 pp.
29. Lake, J. R., Priston, R. A. J., Slade, W. R. 1975. A genetic recombination map of foot-and-mouth disease virus. *J. Gen. Virol.* 27:355-67
30. Loeb, L. A., Tartof, K. D., Travaglini, E. C. 1973. Copying natural RNAs with *E. coli* DNA polymerase I. *Nature New Biol.* 242:66-69
31. Luria, S. E., Darnell, J. E. 1967. *General Virology.* New York: Wiley. 512 pp.
32. McNeill, W. H. 1976. *Plagues and Peoples,* Garden City, NY: Anchor. 368 pp.
33. Margulis, L. 1970. *Origin of Eukaryotic Cells.* New Haven, Conn.: Yale Univ. Press. 349 pp.
34. Matthews, R. E. F. 1970. *Plant Virology.* New York: Academic. 778 pp.
35. Matthews, R. E. F. 1975. A classification of virus groups based on the size of the particle in relation to genome size. *J. Gen. Virol.* 27:135-49
36. Mitzutani, S., Temin, H. M. 1976. Incorporation of noncomplementary nucleotides at high frequencies by ribodeoxyvirus DNA polymerases and *E. coli* DNA polymerase I. *Biochemistry* 15:1510-16
37. Nahmias, A. J. 1974. The evolution (evovirology) of herpesviruses. See Ref. 28, pp. 605-24
38. Penhoet, E., Miller, H., Doyle, M., Blatti, S. 1971. RNA-dependent RNA polymerase activity in influenza virions. *Proc. Natl. Acad. Sci. USA* 68:1369-71
39. Prochiantz, A., Benicourt, C., Carre, D., Haenni, A. L. 1975. tRNA nucleotidyltransferase-catalyzed incorporation of CMP and AMP into RNA-bacteriophage genome fragments. *Eur. J. Biochem.* 52:17-23
40. Reanney, D. C. 1974. Viruses and evolution. *Int. Rev. Cytol.* 32:21-52
41. Reanney, D. C. 1976. Extrachromosomal elements as possible agents of adaptation and development. *Bacteriol. Rev.* 40:552-90
42. Reanney, D. C., Teh, C. K. 1976. Mapping pathways of possible phage-mediated genetic interchange among soil bacilli. *Soil Biol. Biochem.* 8:305-11
43. Salomon, R., Littauer, U. Z. 1974. Enzymatic acylation of histidine to mengovirus RNA. *Nature* 249:32-34
44. Shah, K., Nathanson, N. 1976. Human exposure to SV40:review and comment. *Am. J. Epidemiol.* 103:1-12
45. Shalla, T. A., Petersen, L. J., Giunchedi, L. 1975. Partial characterization of virus-like particles in chloroplasts of plants infected with the U5 strain of TMV. *Virology* 66:94-105
46. Sheldrick, P., Berthelot, N. 1974. Inverted repetitions in the chromosome of herpes simplex virus. *Cold Spring Harbor Symp. Quant. Biol.* 39:667-68
47. Spiegelman, S. 1970. Extracellular evolution of replicating molecules. In *The Neurosciences; Second Study Programme,* ed. F. O. Schmitt, pp. 927-45. New York: Rockefeller Univ. Press
48. Subak-Sharpe, J. H., Elton, R. A., Russell, G. J. 1974. Evolutionary implications of doublet analysis. *Soc. Gen. Microbiol. Symp.* 24:131-50
49. Temin, H. M. 1974. On the origin of RNA tumor viruses. *Ann. Rev. Genet.* 8:155-77
50. Temin, H. M. 1976. The DNA provirus hypothesis. *Science* 192:1075-80
51. Todaro, G. J. 1975. Evolution and modes of transmission of RNA tumor viruses. *Am. J. Pathol.* 81:590-604
52. Varmus, H. E., Spector, D. H., Stehelin, D., Deng, C., Padgett, T., Stubblefield, E., Bishop, J. M. 1977. The function and origin of the transforming gene of avian sarcoma virus. *Cold Spring Har-*

bor *Conference on Cell Proliferation*, Vol. 4. In press
53. Wahba, A. J., Miller, M. J., Niveleau, A., Landers, T. A., Carmichael, G. G., Weber, K., Hawley, D. A., Slobin, L. I. 1974. Subunit I of Qβ replicase and 30S ribosomal protein S I of *E. coli. J. Biol. Chem.* 249:3314–16
54. Webster, R. G., Laver, W. G. 1975. Antigenic variations of influenza viruses. In *The Influenza Viruses*, ed. E. D. Kilbourne, pp. 269–314. New York: Academic
55. Weiss, R. 1976. Molecular analysis of the oncogene. *Nature* 260:93
56. Whittaker, R. 1969. New concepts of kingdoms of organisms. *Science* 163:150–60
57. Wu, A. M., Gallo, R. C. 1975. Reverse transcriptase. *Crit. Rev. Biochem.* 3: 289–347
58. Zhdanov, V. M., Tikhonenko, T. I. 1974. Viruses as a factor of evolution: exchange of genetic information in the biosphere. *Adv. Virus Res.* 19:361–95

CARBON BALANCE IN ❖4117
TERRESTRIAL DETRITUS

William H. Schlesinger
Department of Biological Sciences, University of California,
Santa Barbara, California 93106

> Before the leaves can mount again
> To fill the trees with another shade,
> They must go down past things coming up.
> They must go down into the dark decayed.[1]

INTRODUCTION

The fixation of carbon by plants and the subsequent movement of carbon through ecological systems may be viewed as small-scale geochemical processes which are linked to the larger, geological circulation of carbon as a chemical element. On land plant debris often accumulates as dead organic matter, humus, and peat, which exist in and on the mineral soil to the depth of unweathered rock. Models of the worldwide carbon cycle suggest that (*a*) the amount of carbon contained in world detritus is large, but poorly estimated; and (*b*) the amount of CO_2 released from this detritus to the atmosphere is great, but remains poorly evaluated in terms of atmospheric CO_2 balance (214). In this review, I estimate the amount of terrestrial world detritus, paying special attention to detritus which exists as deep and dispersed organic matter in the soil profile. I also review studies of soil respiration—the output of carbon as CO_2 released to the atmosphere. A large literature of soil respiration is reviewed to synthesize worldwide trends in the data.

CARBON ACCUMULATION IN DETRITUS

In forest ecosystems detrital carbon represents the total carbon in dead organic matter in the forest floor and in the underlying mineral soil layers. Only a small

---

[1]From "In Hardwood Groves" from THE POETRY OF ROBERT FROST edited by Edward Connery Lathem. Copyright 1934, © 1969 by Holt, Rinehart and Winston. Copyright © 1962 by Robert Frost. Reprinted by permission of Holt, Rinehart and Winston, Publishers.

amount of carbon is contained in above-ground detritus such as standing dead trees. Ecologists have sampled the forest floor, but in many ecosystems a large amount of carbon is also contained in relatively inert organic compounds in the lower soil layers. As early as 1932, Romell (153) calculated the total humus content in the upper meter of several soil profiles. In each case the humus content dispersed through the lower layers exceeded that in the surface litter layers. Current estimates of detritus have not always considered the dispersed organic matter in the mineral soil. Agronomists have measured organic matter in the various horizons of most soil groups, but their data are almost invariably expressed as percentages and lack the values for bulk density which are necessary to calculate total quantity on an areal basis. Ecologists have only recently looked at the deeper layers. Rodin & Bazilevich (152) gave values derived from the forest floor only for the standing crop of detritus in Soviet forests. Until recently most ecologists have also neglected to include fallen logs in their forest floor collections. In old-growth forests the amount of carbon in fallen logs can be quite large (e.g. 63).

Although detritus has not been carefully evaluated on a worldwide basis, available estimates suggest that the amount of detrital carbon is at least equal to the carbon in living vegetation. Bolin (17) estimated worldwide detritus on land as 700 $\times$ 10^9 mtC, only slightly less than Whittaker & Likens' (200) estimate of world biomass carbon of 830 $\times$ 10^9 mt. The United Nations has compiled values for soil carbon in South American soils (184). Bohn (16) recently used these data to estimate the carbon content of all soils of the world. His value, 3000 $\times$ 10^9 mtC, is over four times greater than Bolin's. Using a slightly different approach, Reiners (145) calculated an indirect estimate of 9120 $\times$ 10^9 mt for detrital carbon from the nitrogen data of Delwiche (36). All of these estimates are much larger than Whittaker's (197) estimate of 55 $\times$ 10^9 mt for world detrital carbon in surface litter. For an accurate understanding of the geochemical cycle of carbon, a better estimate of this major pool and its dynamics is essential.

For my estimate of world detrital carbon I have used studies which include the lower layers of the soil profile. Only studies that gave bulk density values in combination with measures of percent organic matter in soils could be used. With few exceptions (e.g. some tundra sites), the values in Table 1 include detritus to a depth of at least 50 cm as well as surface litter. Many are carbon estimates for the entire upper 1 m of soil. It would be helpful to have a standard sampling depth (a depth unaffected by biotic processes), but different studies must accommodate the greatly varying conditions among ecosystem types. Deep sampling is important in regions of older, more mature soil profiles (i.e. the tropics) and in areas where plants are exceptionally deep-rooted (e.g. grasslands); it would be grossly impractical in the tundra or boreal forests. I tried to select a depth criterion for the inclusion of a value in Table 1. This proved difficult because, for example, in my largest collection of data from a single ecosystem type, temperate forests, there is no apparent correlation between depth of sampling and detritus content. After I had compiled a preliminary list, I eliminated low values from tropical forests and temperate grasslands when it was obvious that the sampling had not been deep enough.

CARBON BALANCE IN DETRITUS 53

Table 1 Total detritus in soil profiles of ecosystems of the world

Ecosystem type	Location	Description and number of stands	Sampling depth (cm)	Mean detritus $kgC\ m^{-2}$	References
Tropical forest	Brazil	rainforest	100	13.8	(101) and H. Klinge, personal communication, 1976
	Colombia	rainforests (2)	100 (mean)	20.5	(80)
	Nigeria	rainforests (4)	50-150	12.0	(150)[a]
		rainforest	178	7.9	(86)[a]
	Suriname	rainforests (5)	100	11.5	(163)[a]
	Thailand	rainforest	100	7.6	(215)
		monsoon forests (2)	100	6.5	(215)
		seasonal forests along mountain gradient (5)	50	9.9 (mean) 5.8-13.7 (range)	(215)
		broadleaf evergreen	65	3.7	(177)
		TROPICAL FOREST MEAN		10.4	
		coefficient of variation[b]		44%	
Temperate forest	Belgium	mixed hardwoods (4)	80	7.6	(41, 42)
	England	oak-ash	?[c]	12.3	(160)
	Germany	beech and spruce (2)	50	11.3	(183)
	Japan	cool temperate (4)	65	10.5	(177)
		evergreen oak forests along mountain gradient (3)	60	10.0 (mean) 8.9-11.5 (range)	(126)
	Sweden	beech and spruce (2)	65	11.0	(127, 128)
	Thailand	evergreen	20[d]	12.5	(131)
	USA	northern hardwood	45	6.1	(61)
		oak-hickory	70	13.8	G. E. Lang, and R. T. T. Forman, personal communication, 1976
		northern hardwoods (6)	100	24.0	(153)
		birch and spruce (3)	60+	17.7	(121)
		Tulip-poplar	75	12.5	(143)
		Ponderosa pine	60	10.6	(100)
		oak-hickory	100	9.3	W. H. Schlesinger, unpublished
		pine	114	5.6	(7)
		pine-fir	91	11.8	(38)
		Douglas fir (2)	100	7.8	(63)
		coniferous forests along mountain gradient (3)	60	19.5 (mean) 17.7-22.6 (range)	(68)
		oak-pine (2)	127	7.2	(80)
		pine	107	10.2	(49)
		fir	60	14.9	(178)
	USSR	mean for forests on slightly podzolized soils	100	11.8	(102)
		subtropical on krasnozem soil	100	14.1	(102)
		Gray forest soils (5)	100	11.3	(156)
		TEMPERATE FOREST MEAN		11.8	
		coefficient of variation		35%	

Table 1 *(Continued)*

Ecosystem type	Location	Description and number of stands	Sampling depth (cm)	Mean detritus kgC m^{-2}		References
Boreal forest	Canada	aspen, birch and spruce (9)	61-91	17.5		(65)[a]
	Sweden	spruce (2)	100	15.0		(153)
	USA (Alaska)	spruce	100	31.8		Van Cleve's data cited in (43)
	USSR	mean for forests on podzolic soils	100	5.0		(102)
		spruce	120	7.9		(175)
		spruce and cedar (3)	150-230	13.9		(133)[a]
		coniferous forests (5)	130-185	13.5		(216)[a]
	BOREAL FOREST MEAN			14.9		
		coefficient of variation			53%	
Woodland and shrubland	England	*Calluna-Erica* heath	40	4.4		(25, 26)
	France	oak-scrub	30	7.1		(111)
	Sweden	wet *Calluna-Erica* heath	20	13.5		(181)
	USA	oak-pine	?	3.2		(213)
		chaparral scrub (5)	60	10.5		(68)
		oak woodland	45	2.4[e]		(85) and F. Johnson, personal communication, 1976
	WOODLAND MEAN			6.9		
		coefficient of variation			59%	
Tropical grassland and savanna	Africa (east)	mean of a large literature review	30	5.7		(12)[a]
	(west)	mean of a large literature review	30	0.3		(87)[a]
	Suriname	savannas (3)	100	8.7		(71)[a]
		savannas (24)	100	7.4		(39, 71)[a]
	Thailand	savanna forest	100	2.6		(215)
		savanna forests (2)	20[d]	0.9		(131)
	Venezuela	Llanos	?	0.5		(18)
	TROPICAL SAVANNA MEAN			3.7		
		coefficient of variation			87%	
Temperate grassland	Japan	mixed grasslands (3)	100	26.2		(93)
	USA	prairie (2)	168	17.1		(196)
		short-grass prairie	228	17.6		(125)
	USSR	mean for chernozem soils (5)	100	25.3		(102)
		ordinary chernozems (7)	100	15.4		(57)
		Solonchak soil	106	13.3		(134)
	TEMPERATE GRASSLAND MEAN			19.2		
		coefficient of variation			25%	
Tundra and alpine	Canada	Devon Island (2)	AL[f]	10.3		(15)
	Sweden	Stordalen (3)	30	12.9		(155)
	USA	Alaska tundra (4)	75-175	49.8		(48)
		Alaska tundra (4)	AL[f]	41.8		(40)
		Alaska tundra (2)	35	35.8		(55)
		Wash. alpine heath (3)	35 or rock	27.0		(62)
		N.H. alpine (2)	to rock	16.3		(14)
		N.H. alpine tundra (5)	to rock	4.0		(147)

Table 1 *(Continued)*

Ecosystem type	Location	Description and number of stands	Sampling depth (cm)	Mean detritus kgC m^{-2}		References
	USSR	tundra	100	3.7		(103, 105)
		mountain meadow	100	22.2		(103, 105)
		Agape IBP Station (4)	50	14.2		(187)
	TUNDRA AND ALPINE MEAN			21.6		
		coefficient of variation			68%	
Desert scrub and	USA	Bajada (3)	60	4.5		(68)
semidesert		sagebrush-grassland (6)	33–168	3.4		(56)
		Nevada deserts (8)	97–173	7.2		(169)
	USSR	Sierozem soil mean	100	4.1		(102)
		Steppe deserts (3)	40	4.8		(189)
		dry grasslands	100	9.6		(102)
	DESERT SCRUB MEAN			5.6		
		coefficient of variation			38%	
Swamp and	Canada	*Alnus* and meadow	60	38.5		(31)
marsh		flood plains				
		peatlands (3)	?^{c}	73.4		(142)
	England	heath bog	60–100	26.0		(59)
	USA	*Thuja* swamp	70^d	92.0		(148)
		Taxodium swamp	291	149.8		(162)
	USSR	flood plains (3)	100+	31.7		(64)
	SWAMP AND MARSH MEAN			68.6		
		coefficient of variation			63%	

^{a}Calculated by estimating bulk densities following Jeffrey (75).
^{b}Coefficient of variation = SD/mean × 100.
^{c}Not known, but believed to include all horizons containing organic matter.
^{d}Actual sampling depth, but estimation is an extrapolation for the whole profile.
^{e}Minimum estimate, does not include soil detritus > 0.5 mm diameter.
^{f}Active layer.

For some ecosystem types (e.g. boreal forest and tropical savanna) good data are extremely limited. To increase the number of values in Table 1 for these systems, I estimated bulk densities for some studies which failed to provide these values, using the formula of Jeffrey (75). If the original data were given as organic matter or loss-on-ignition, they were divided by 2 to obtain carbon content. This mean factor is probably sufficient for global estimates (1).

Coefficients of variation for the mean profile value for each ecosystem type are generally large (Table 1). Variation is greatest in boreal forest, woodland, savanna, and tundra regions where there have been few studies and where it is often necessary to group communities of greatly differing structure and productivity. Only the mean temperate forest value would contain a standard error of less than 10% of the mean. Some of the values in Table 1 are derived from large compilations of data (e.g. 102). When a single study included values for several stands of a particular vegetation type, I averaged the values before including them in the appropriate category. This technique reduces sample size, but it avoids giving undue weight to any one study or area. The data have been arranged by ecosystem and mean ecosystem values multiplied by estimated world areas for total detritus calculation (Table 2).

Table 2 Distribution of detritus and biomass by ecosystem types

Ecosystem type	Mean total profile detritus (kgC m^{-2})	CV[a] (%)	World area[b] (ha × 10^8)	Total world detritus (mtC × 10^9)	Total world surface detritus[b] (mtC × 10^9)	Total world biomass[b] (mtC × 10^9)
Tropical forest	10.4	44	24.5	255	3.6	460
Temperate forest	11.8	35	12	142	14.5	175
Boreal forest	14.9	53	12	179	24.0	108
Woodland and shrubland	6.9	59	8.5	59	2.4	22
Tropical savanna	3.7	87	15	56	1.5	27
Temperate grassland	19.2	25	9	173	1.8	6.3
Tundra and alpine	21.6	68	8	173	4.0	2.4
Desert scrub	5.6	38	18	101	0.2	5.4
Extreme desert, rock and ice	0.1[c]		24	3	0.02	0.2
Cultivated	12.7[d]		14	178	0.7	7.0
Swamp and marsh	68.6	63	2	137	2.5	13.6
Totals			147	1456	55.2	826.9

[a]CV = Coefficient of variation = SD/mean × 100.
[b]Areas, biomass and surface detritus from Whittaker (197).
[c]Based on ratio of detritus to biomass in desert scrub.
[d]Based on the depletion of nitrogen and humus in cultivated soils over time (66, 156); 40% in surface soil, 20% in subsoil.

Important world patterns of detrital accumulation are apparent (Table 2). The average amount of detrital carbon per unit area of soil surface increases from tropical regions poleward to boreal forests. Frozen tundra soils and waterlogged swamp and marsh soils contain the greatest accumulations. My mean boreal forest value is higher than that of Kononova (102, 105), probably because it includes some values from northern, forested peatlands. Soils in temperate grasslands contain very large amounts of detritus, confirming earlier observations for soils of the USSR (102, 105). Low values in tropical regions presumably reflect high rates of decomposition which compensate for the high productivity and litterfall in these regions. The mean value for tropical savanna, 3.7 kgC m^{-2}, is particularly low. This may reflect either the small sample size or the effect of fire in reducing the accumulation of organic matter in these areas. Most of the published tropical data do not confirm Jenny's (80, 83) observations and predictions of large accumulations in these regions.

The distribution of detrital carbon in the soil profile differs among ecosystem types. The values given by Whittaker (197) are included in Table 2 to show the potential for underestimation of detritus in considering only the upper soil layers. On a worldwide basis, carbon in the lower profiles of soils exceeds that in surface soils by a factor of 25. The lower soil layers contain a large percentage of the soil carbon in tropical forests and in both tropical and temperate grasslands. Most of the soil carbon in tropical forests occurs as dispersed organic matter: light-colored fulvic acids in the deep, lower profile [(102); M. Cline, personal communication,

1973]. In high latitudes and at high elevations soil profiles are shallower, and mixing and decomposition are retarded by frozen conditions. Litter on the soil surface as a percentage of total profile detritus increases from 1% in tropical forests to 13% in boreal forests. Recently G. E. Lang and R. T. T. Forman (personal communication, 1976) have compared the mass of the forest floor in oak-hickory forests of the eastern US. Even within this temperate forest region there is a distinct tendency toward greater surface accumulations in the northernmost stands.

While a large percentage of the detritus in temperate forests is exposed to a period of rapid mineralization on the forest floor, a major portion of grassland detritus is derived deep within the soil from the death of roots. This difference in origin may partially explain why there is generally more carbon in the soils of temperate grasslands than in those of temperate forests. Within the temperate forest soils, there is no consistent tendency for either deciduous or coniferous stands to have higher total amounts of detritus (Table 1; see also 153). This suggests that during decomposition, differences in the rate of fragmentation and mixing rather than in remineralization account for the differences in appearance between mull and mor soil profiles (84, 153, 154).

My estimate of the total carbon in detritus exceeds world biomass carbon as estimated by Whittaker & Likens (200). In areas of inhibited decomposition, such as the boreal forest, tundra, and swamps, detritus greatly exceeds biomass. Tropical forests are unique in containing far less detritus than biomass. In contrast to tropical savannas, detritus in temperate grasslands worldwide vastly exceeds total grassland biomass. This may be partially due to lower decomposition rates and fewer fires.

The world value for total detritus, 1456×10^9 mtC, lies between Bolin's (17) estimate of 700×10^9 mtC and Bohn's (16) estimate of 3000×10^9 mtC. My value suggests that Reiners' (145) use of a mean world C/N ratio to estimate detritus from Delwiche's (36) nitrogen data results in an exorbitant world estimate which should be discarded. My estimate should be regarded as preliminary; it is based upon a small data base and upon many assumptions. The estimate is conservative in that deeper field sampling might increase mean profile values, but the estimate is probably too large for present conditions in that much of the natural land area has been disturbed by man.

Identity and Turnover of Soil Carbon Compounds

A brief consideration of the identity of carbon-containing compounds and their position in the soil profile seems desirable. Kononova's (102, 105) terminology separates the various forms of soil organic matter into fresh, incompletely decomposed material, and humus. Humus is in turn subdivided into strictly humus substances (e.g. humic and fulvic acids) and other products synthesized by soil microbes. Within these categories various compounds may be identified. In fresh litter, for instance, compounds present in living plants are most common—cellulose, hemicellulose, lignins, etc. Soil organisms themselves form a major portion of soil organic matter (1). Up to 10% of the weight of the fermentation layer (Fo) may be microbial mass (209).

Both macro- and microorganisms are important in the formation of humus through decay and synthesis. Humus substances are high molecular weight compounds, secondarily derived during microbial processes which are not wholly understood but which involve the formation of aromatic units containing nitrogen, phenol, and -COOH groups (104, 136). Humic substances are often classed by their relative solubilities in acid and alkaline solutions, although soil scientists avoid these harsh analytical treatments in recent studies of the structure of humic acids. Some humus substances are very stable; but masked cellulose, condensed polyphenols (123), and lignin are among the most stable of all organic compounds, particularly if lignin is complexed with amino acids (1).

Since these compounds decay and accumulate at different rates, a general geochemical summary of detrital carbon obscures subtle differences. The turnover time of carbon in world detritus varies greatly depending upon its form and geography. It would be instructive to divide the estimate for total world detritus into fractions turning over in 10, 100, and 1000+ years, but at present we do not know how. Ganzhara (53) discussed this problem for the upper profile of Russian chernozems. To his suggested figures for pool sizes and annual transfers I have added approximations from other studies to obtain a rough model for carbon dynamics in the surface layers of this profile type (Figure 1). Noteworthy is the large amount of carbon in

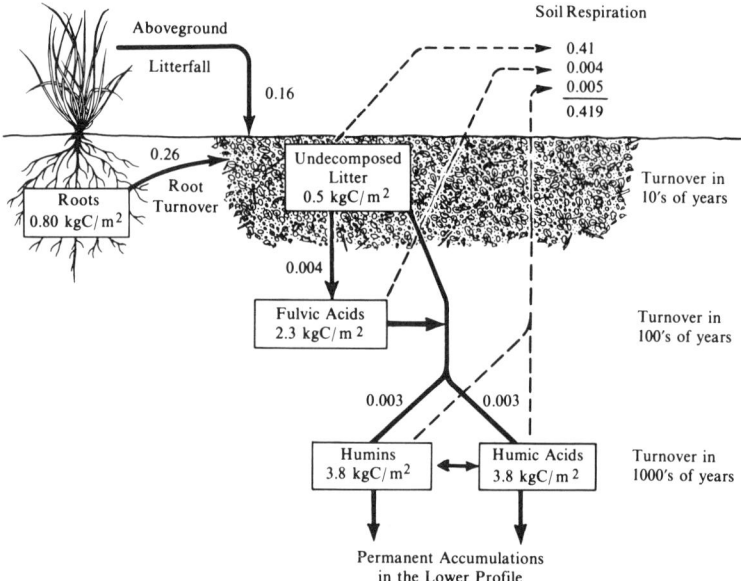

Figure 1 Detrital carbon dynamics for the 0–20 cm layer of a chernozem grassland soil. Carbon pools (kgC m^{-2}) and annual transfers (kgC m^{-2} yr^{-1}) are indicated. Total profile content to 20 cm is 10.4 kgC m^{-2}. Values are derived and/or adapted from this paper and from (24, 30, 53, 102, 105, 136, 145, 197).

pools which turn over very slowly due to small annual additions and great stability of humic substances. Carbon 14 has been used to date organic matter in grassland soils. For soil humins of the USSR, this technique has indicated ages of up to 9000 yr (157). Future investigators of carbon in detritus might apply improved versions of this model to various ecosystems.

Carbon content in soil organic matter changes during decomposition and varies in the different soil horizons. In forests, carbon in living plants and fresh residues ranges between 45–50% by dry weight (143). A standard value used for carbon in soil organic matter has been 58%, resulting in the widespread use of 1.724 as a factor for calculating the dry weight of organic matter. This factor has been criticized (5, 20, 72, 116, 141, 191). A factor of 1.9 probably more accurately represents the situation in surface soils, and carbon in subsoils is often depleted to the extent that 2.5 should be used for conversion (20). The point, however, is that carbon is relatively depleted in the refractory compounds in deep, dispersed humus.

Steady-state Models of Detritus

The decomposition of organic matter (1, 37, 102), and its importance in nutrient cycling are reflected by the research emphasis of the International Biological Program. Specific parameters affecting decomposition rates are discussed later in the context of soil respiration. For a general discussion of carbon geochemistry and the pattern of detritus accumulation (Table 2), more general factors affecting both plant production and decomposition on a global scale must be considered.

The production of forest litterfall is related to a worldwide gradient of net primary productivity (19, 88). Decomposition is affected by worldwide gradients in temperature, moisture, and microbial biomass. However, to explain worldwide detrital carbon accumulations, we must be as concerned with the resynthesis and occurrence of stable humus substances as with the breakdown and mineralization of plant residues. Most ecological studies have stressed the latter processes, with the result that detritus accumulation and nutrient turnover are often considered merely as a result of litter deposition and disappearance.

Several theoretical attempts have treated surface litter accumulations in terms of mass balance (59, 82, 123, 132; 102, p. 230). Litterfall is modeled either as a pulse or as a continuous annual input, while decay is modeled as exponential or stepwise exponential to resemble seasonal fluctuations. The simplest model allows the calculation of decay constants, which are most useful for comparative purposes (Table 3; see also 112). Here (132),

$$dX/dT = L - kX,$$

where X is the mass of the forest floor, L is the amount of annual litterfall, and k is the decay constant representing the instantaneous fractional loss rate. Generalized decay equations fail to fit actual decay curves because after only a few years the organic accumulation is largely composed of the relatively resistant compounds and not of material resembling fresh litterfall (123). Initial organic leachates from litterfall are lost too rapidly to be approximated by decay equations for litter (60). Calculated decay constants may change greatly during the first month of decomposi-

Table 3 Decay constants and time required for the loss of one half and 95% of the original leaf dry weight for three northern hardwood forest species [Adapted from Gosz et al (60)]

Species	Decay parameters[a]			Time parameters (years)	
	Based on original dry weight (k)	Based on dry weight following leaching[b] (k)	Specific loss rate (k')	Halflife $0.693/k$	95% ($3/k$)
Yellow birch	0.85	0.61	0.43	0.8	3.5
Sugar maple	0.51	0.33	0.26	1.4	5.9
Beech	0.37	0.27	0.22	1.9	8.1

[a] k calculated from $x/x_0 = e^{-kt}$; $k' = 1 - e^{kt}$.
[b] One month.

tion (Table 3). Thus, during long-term studies, or where components of the forest floor differ markedly in their decay patterns (i.e. wood vs leaves), k may vary both with time and between materials (107, 122, 123, 132, 138; 102, pp. 230–231). Jenny et al (82) expressed specific loss, k', as a fraction of the original weight (Table 3):

$$k' = 1 - X/X_0 = 1 - e^{-kt}.$$

Elaborations of these studies allow calculation of half-life for litter, and turnover times for the forest floor (132).

Carbon-14 studies have been used in conjunction with steady-state modeling as an independent means of checking predictions of turnover and accumulation times. These studies have shown that the age of soil organic matter tends to increase with depth (161). Ages exceeding 10,000 years have been reported from ancient, buried soil profiles, but a weighted mean age for the organic matter in an entire profile is normally far less. Most of the detritus entering the soil near the surface is probably mineralized within a few years (e.g. Figure 1; see also 61, 112). While several centuries are required for the carbon in the lower soil profile to achieve a steady state (30, 80, 82, 103, 132), prehistoric conditions are generally not important in determining current accumulations of detrital carbon (83).

The stability of organic matter in the mineral soil has been confirmed in subtropical Florida (151). One hundred years after cutting, carbon isotope ratios in the soil organic matter still reflect the C–3 plant metabolism of the original vegetation, rather than an isotopic ratio characteristic of the C–4 plant communities that now predominate. In California long periods are needed to change grassland soils to forest soils by planting pines (83). For tropical soils, the large amount of deep, dispersed humus results in accumulation times far longer than those calculated by ecologists who have studied only the forest floor (80). In higher latitudes, the

CARBON BALANCE IN DETRITUS 61

majority of soil organic matter lies near the surface, and traditional sampling schemes and turnover calculations have been more realistic.

The long-term accumulation of a large amount of deep, dispersed organic matter in soils has important implications. The annual accumulation of organic matter, both living and dead, has been defined as net ecosystem production (NEP) (129, p. 46; 198, 201). A NEP of zero (i.e. community production = community respiration) has been used as an index of climax status for plant communities. Communities developing on disturbed sites may inherit undisturbed and refractory detritus accumulations from earlier communities on the same site (e.g. 26). Nevertheless, the time needed for detritus to achieve a steady state with existing vegetation may exceed the normal life span of most communities. Thus, zero NEP is probably a rare occurrence. Theoretically, the mass-balance approach to total organic matter in the soil is sound, but it is a mistake to apply this thinking only to litter dynamics in the upper soil layers or the forest floor. Ecologists should further investigate the accumulation of living and dead organic matter in various above- and below-ground compartments.

Russian workers first examined the major environmental factors which explain detritus accumulation (102, pp. 230–34). Moisture and temperature were combined into a "hydro-factor" which predicted the observed distribution of organic matter in Russian soils—highest in chernozems and decreasing in the northern tundra soils and in the southern sierozems. More recently, Kononova (102, pp. 241–43) has reviewed microbe biomass and activity data along this gradient. Within limits, microbes respond directly to moisture and temperature. Less to be expected is evidence for high potential microbiological activity in the dry sierozem soils. Jenny (78, 79) suggested a similar explanatory correlation between the nitrogen and humus in prairie soils and annual indexes of temperature and moisture in the United States. At constant temperature, nitrogen was related to a moisture factor (m) by

$$N = 0.320 \ (1 - e^{-0.0034m}).$$

At constant moisture, soil nitrogen decreased exponentially with increasing temperature. These equations predicted very small detritus accumulations for ecosystems in tropical latitudes. Later, trying to apply these equations to field data from tropical regions, Jenny and others (12, 80–82, 168) reported disproportionately high organic matter accumulations. They suggested that the year-round growing conditions provide a greater enhancement of production over decomposition in the humid tropics and that while decomposition was rapid on the forest floor, humic compounds transported to lower soil layers had great stability. Jenny's original equations (78, 79) are still useful for the regions where they were developed, but their usefulness is otherwise limited.

The interaction of abiotic factors with both the decay and resynthesis activity of microbes must be considered to explain the asymptotic level of detrital accumulation in a region (102). Optimum abiotic conditions for microbes result in maximum decay and synthesis activity, and rapid humus formation, but certainly not in maximum organic accumulation. In marginal environments (e.g. the tundra) accumulation can be large despite low vegetative production if conditions for decay are usually poor. In temperate montane regions, production is usually lower at higher elevations, but the rate of decomposition is inhibited to a greater extent, so that the greatest accumulations may occur near ridgetops (50, 68, 79, 126, 147, 199). These situations contrast with desert communities, which have minimal organic accumulations because their productivity is low and conditions for litter disappearance are periodically very favorable.

Future Investigation

The complexity of variables affecting both vegetation and soil microbe communities has undoubtedly been responsible for the failure of general, analytical equations based on only a few factors to explain detrital accumulations worldwide. In particular, both the Russian work and Jenny's treatment recognized plant production as a possible parameter but did not include it in their equations. Thus, the value of these equations lies not with their current empirical level of sophistication but with their suggestive potential. Illustrating a possible new approach, Jenny et al (83) used principal component analysis to relate soil organic matter and nitrogen to precipitation, elevation, and other variables in 97 California soils. A similar approach has been applied to East African soils (74, 104). For many biomes data are too few to produce models with widespread applicability. As the data accumulate and computer modeling of ecological phenomena matures, predictive equations for both broad and limited areas could represent new levels of sophistication that will allow valuable insight to the role of explanatory variables.

CARBON RELEASE

On land most net primary production is delivered to and decomposed on the forest floor (117, p. 170) or in the humus layers (153, 154). Early studies of the release of carbon compounds during the decomposition and oxidation of organic matter were especially concerned with the use of soil respiration as a fertility index for agriculture (32, 52, 114, 115, 159, 166, 176). Waksman & Starkey (190) demonstrated an increased bacterial response when soils were fertilized. In Europe, and later in the United States, a rich literature of soil respiration work was more perceptive than is generally acknowledged. While lacking sophisticated equipment, early investigators framed many questions that are still unanswered, including that of the importance of diffusion as a control of CO_2 release (167).

With the advent of the modern community and ecosystem concepts, soil respiration studies were again undertaken in natural communities. Lieth & Ouellette (110) compared soil respiration in communities of differing net primary productivity and observed respiration rates substantially lower than those in an earlier study in Europe (153). In theory, since global patterns of litterfall and litter accumulation are recognized (19, 88, 132), it should be possible to balance the level of soil CO_2 release in steady-state communities with carbon input by litter and the CO_2 produced by root respiration. In practice, it is often difficult to show good correlations between productivity and soil respiration in specific areas. Measurements of the loss of carbon from forest floors are confounded by numerous difficulties of method.

Complexities of the Soil System

The inherent complexity of the soil system presents many problems. Fungi, bacteria, and higher organisms are all actively involved in the decomposition process. Methods are lacking which can partition the energy flow through these largely microscopic populations. The populations themselves are transitory. Methods for separating respiratory contributions of roots and mycorrhizal fungi are similarly primitive (69). To avoid these complexities, investigators have employed the "blackbox" approach (135) by monitoring only inputs and outputs of arbitrarily defined units of the forest floor.

More than any other factor, root respiration has interfered with studies of the microbial release of CO_2 from the soil surface. Root respiration causes most measurements of CO_2 evolution to be much greater than the actual loss of carbon from detritus. By subtracting carbon delivered to the soil in litterfall from carbon released from the soil in CO_2, investigators have attributed 30–70% of soil CO_2 evolution in forests to roots (138, 202). These studies have not separated true root respiration from microbial respiration of detritus derived below ground (138) or from the potentially important contribution of mycorrhizal respiration (69, 115). A recent study which included actual root respiration measurements and partitioning of forest floor layers suggested that 35% of the CO_2 may originate from roots (46). In the beech forest at Wytham Woods, Phillipson et al (138) attributed only 3–4% of the soil respiration to roots, 1% to mycorrhizae, and 27% to the decomposition of organic matter contributed by roots. There is a clear need for additional studies which separate the contributions of CO_2 released from various below-ground processes.

Carbon dioxide measurements will underestimate carbon loss from soils if there are losses of volatile carbon compounds that are not completely oxidized and losses of both these compounds and dissolved CO_2 to groundwater. These carbon losses have been insufficiently studied; they are probably minor in most communities. For example, C^{13}/C^{12} isotope fractionation data (151) suggest that current dissolved carbon in groundwater in Florida does not reflect the recent change from C–3 to C–4 plants in the dominant vegetation. Thus, despite the importance of subsurface drainage in this region, subsurface carbon losses are probably slight. In the Hubbard Brook Experimental Forest in New Hampshire, various types of carbon losses in groundwater are extremely minor relative to the annual circulation of carbon in the ecosystem (61; see also 118). In an Illinois old-field, maximum soil CO_2 concentrations near the water table suggested that groundwater as a major sink for dissolved CO_2 may not be important (165).

Soil Respiration Methodology

Many of the functional intricacies of the soil system complicate the development of accurate monitoring techniques for soil respiration. Diffusion is the dominant process of gaseous movement in many soils, especially those with fine texture (167). Temperature and diffusion rate are closely dependent as seen in the equation for equilibrium diffusion potential:

$$q = \frac{kt^2}{(273)^2} \frac{dc}{dz} = -D\frac{dc}{dz},$$

where q is the rate of flux in g cm^{-2}sec^{-1}, k is a constant, t is the temperature in °K, c is the concentration in g cm^{-3}, and z is the depth in cm (34). Many studies correlating temperature and CO_2 evolution really illustrate a relationship between temperature and diffusion. Reiners (145) has outlined the significance of this problem for respiration studies which attempt to calculate a total yearly flux from annual temperature records by using regressions developed from short, intensive studies of the CO_2 evolution rate and temperature. For shallow soils the problem is probably not serious, but the sizeable CO_2 storage capacity of deep forest soils means that convection or diffusion rate rather than current levels of microbial activity is what is often measured. (145).

Apparatus for measuring CO_2 flux should not interfere with the diffusion process. Many early studies used the inverted box (static, or Haber) method in which an airtight container is placed over the soil surface and the evolved CO_2 absorbed in a vial of KOH or NaOH (193). The CO_2 flux is calculated after titration with acid. Since the inverted box is obviously an artificial environment, there are several problems inherent in its operation. Increased upward CO_2 diffusion is caused by depletion of CO_2 in the chamber as it is absorbed by the hydroxide. Due to changes in efficiency of KOH absorption with temperature, the method may measure only 75% of the CO_2 flux (67), but this correction has seldom been applied (e.g. 110, 194, 195). Kucera & Kirkham (106) found the absorption method resulted in values 39% below those obtained by infrared gas analysis, while Ino & Monsi (73) and Anderson (4) found very close agreement, which indicated efficient absorption. Clearly, the efficiency of the Haber method depends on the conditions of the particular system being used (46). Other methodological comparisons have been studied in some detail (96, 97, 99, 124).

In more recent studies air has been pumped through a chamber and CO_2 release calculated from concentration changes in the circulating air. Some investigators have employed such dynamic measurement systems using absorption techniques, but most now prefer infrared gas analyzers. Flow rate and pressure must be carefully monitored since they directly affect diffusion and are functions of the experimental system employed (6, 46, 95, 130, 144). In a pine forest in Tennessee, Witkamp & Frank (210) obtained relatively low values using a very low flow rate and attributed this result to minimal interference with the normal diffusion process by such factors as air turbulence and suction. Maintaining a slight positive pressure in the measurement chamber is also essential to avoid drawing abnormally great amounts of CO_2 from the soil air (106, 144). Positive pressures resulted in CO_2 evolution values nearly ten times lower than those obtained under conditions of suction in a study of an agricultural soil (91).

CARBON BALANCE IN DETRITUS   65

Soil respiration studies have gradually employed more sophisticated techniques: gas chromatography (22), diffusion theory applied to measured CO_2 profiles (34), and moving chambers mounted over partitioned forest floor layers (44–46). The latter are good because they allow the humus to undergo natural drying between measurements. Because they avoid disruption of the normal soil system, it is likely that most in situ methods are preferable to the extraction of soil cores and their transport to the laboratory. Nevertheless, careful partitioning or removal of layers of the soil profile has allowed increased perception of ecosystem function (28, 33, 45, 46, 210).

Regardless of the selection of technique, measurements of soil respiration must be made over complete 24-hour periods and at different seasons. Large seasonal variations in the rate of CO_2 evolution occur. Witkamp (207) documented the importance of diurnal changes and explained their occurrence by diffusion theory. Generally, a predawn minimum and an afternoon maximum occur in response to the daily temperature cycle. A second peak of CO_2 evolution may occur during the night when soil temperatures exceed air temperatures and create a thermal convection of soil air to the surface. In such cases of an inverse relation of CO_2 evolution rate and temperature, doubt is cast upon studies predicting yearly flux from yearly temperature records.

Seasonal trends of CO_2 evolution are equally apparent for temperate forests (45, 51, 144) and grasslands (34, 106, 204). Summer CO_2 evolution rates can be up to an order of magnitude greater than winter rates. However, significant amounts of decomposition can occur during long winters even in particularly cold climates if the soil is protected by the insulative properties of a deep snowpack (13, 45, 60, 70). Seasonal changes in soil temperature and moisture and particular seasonal events affecting the soil environment are important regulators of carbon release and are discussed in further detail below.

Patterns of Soil Respiration

Despite the number of problems encountered in the measurement of soil respiration and the diversity of techniques applied to estimate CO_2 evolution, general patterns in the data exist. In 1932 Romell (153) compiled data for hourly summertime carbon release from soils in Europe. He noted only a weak tendency for higher mean hourly CO_2 evolution rates as one moved from north to south:

Rate of evolution $= 1.70 - 0.02$ (Latitude); $R^2 = 0.14$.

For comparisons of annual detritus turnover, hourly rates measured during the growing season were inadequate because the season for decomposition is so much shorter in northern climates. Most of the data compiled by Romell (153) also ignored diurnal variations in CO_2 evolution by measuring only short periods. Below I compare trends in more recent data.

Daily rates for tropical forests range from minimum estimates of 0.3–0.8 gC m^{-2} day^{-1} as CO_2 in Puerto Rico (130), to extremely high values of 2.4–16.7 gC m^{-2} day^{-1} in Costa Rican forests (164). Edwards & Sollins (46) reported an annual range varying seasonally from 0.76 to 7.2 gC m^{-2} day^{-1} in a temperate forest in

Tennessee, while Reiners (144) presented data which suggest means rates of 3.9–6.5 gC m^{-2} day^{-1} during the summer in Minnesota forests. For high latitude forests Repnevskaja (149) gave a summer range of 1.6–2.5 gC m^{-2} day^{-1} and Van Cleve & Sprague (185) determined maximum rates of 3.7 gC m^{-2} day^{-1}. Since temperate and high latitude communities are subject to marked seasonality, for comparisons between forest types summer or maximum evolution rates are probably better measures of the potential for CO_2 evolution on a daily basis. Clearly, microbes in cooler climates are able to respond to seasonal warm periods and concentrate decomposition in short periods of relatively intensive respiration (29, 137). Temperate grassland studies have reported rates ranging up to 2.94 gC m^{-2} day^{-1} (106) and from 0.31 to 2.28 gC m^{-2} day^{-1} (27).

Latitudinal trends in values for annual CO_2 release have been examined (106). I have attempted to compile the available studies reporting total yearly carbon release (Table 4). For temperate forests and grasslands, I have included only those studies which presented data for total yearly release of carbon (or gave data from which this value could be calculated with few assumptions). For tropical and tundra ecosystems, a scarcity of studies necessitated my calculating yearly release from daily or sometimes from hourly measurements. These values are subject to large errors and they show great variability. The tropical values of Schulze (164) are probably too high while those of Wanner and his associates (194, 195) are from dry season measurements only. These values were omitted during the calculation of the regression (Figure 2) for the rate of evolution of carbon as a function of latitude in forest ecosystems.

The linear regression for carbon release is uniformly higher than a similar relation for carbon input (19) based upon mean values for above-ground litterfall. The area between the lines represents the contribution of detritus derived below ground, and the root and mycorrhizal respiration. In forests, carbon loss in soil respiration averages about 2.5 times the carbon input in above-ground litterfall. There are few data from temperate coniferous stands, but at steady state, similar CO_2 output from coniferous and deciduous forests of similar environment and productivity (206) confirms Jensen's (84) suggestion that remineralization rates in these forest soils are similar and that physical processes in decay are likely to account for differences in profile appearance (cf 153, 154). If plotted, grassland and desert values (Table 4) would generally lie below the latitudinal regression (Figure 2).

Soil respiration in montane forests is low (188). Along an elevational gradient in Japan, Nakane (126) found the lowest rates of litterfall and soil respiration at the highest site. Since carbon accumulations were greatest at high elevations, decomposition must be retarded to a greater degree than productivity.

Long-term measurements of CO_2 evolution from soils in steady-state communities should reflect carbon losses that balance productivity and root respiration. A latitudinal gradient in plant production is at least partially responsible for the latitudinal trend of soil respiration values in Figure 2. This gross analysis eliminates the confusion caused by examination of the specific controlling parameters within each community. Yet it is instructive to examine the factors which mediate the rate of soil CO_2 loss over short periods.

Table 4 Total annual release of carbon as CO_2 evolved from soils in various natural ecosystems of the world

Ecosystem type	Reference	Latitude (°N unless indicated)	Soil carbon output $gC\ m^{-2}\ yr^{-1}$	Notes on location and methods
Tropical forests	(119)	1°	1260 $gC\ m^{-2}\ yr^{-1}$	Mean for four rainforests. Field, static, absorption techniques. Hourly rates extrapolated to year.
	(215)	9°	1610	Rainforest at Khao Chong, virgin growth. Field, static, absorption techniques. Hourly rates.
	(194) (195)	10° S-8°	405-655	Range for nine mountain rainforests. Field, static, absorption techniques. Hourly measures in the dry season.
	(164)	10°	890-6100	Range for four forests. Field, static, absorption techniques. Hourly rates.
	(130)	18°	2117	Rainforest, virgin. Field, dynamic infrared analysis. Estimate is 5.8 $gC\ m^{-2}\ day^{-1}$ extrapolated for year.
Temperate forests	(98)	33°	1054	Evergreen forest, field, static, absorption techniques. Mean of five plots. Yearly values are the sum of observed daily rates for one year.
	(98)	35°	1414	Evergreen oak forest, field, static, absorption techniques. Mean for two years; yearly values are the sum of observed daily rates.
	(126)	35°	1098	Mean for three evergreen oak forests on an elevational gradient at the same site as Kirita (98).
	(45), (46)	36°	1069	*Liriodendron* forest, 40-50 years old. Field, dynamic, infrared analysis.
	(73)	36°	598	Mean for four *Pinus* forests, less than 15 years old. Laboratory, static, absorption techniques.
	(206)	36°	414	Mean for three forests (*Quercus, Acer,* and *Pinus*). Field, static, absorption techniques. Data converted: 1 liter CO_2 = 0.54 gC.

Table 4 *(Continued)*

Ecosystem type	Reference	Latitude (°N unless indicated)	Soil carbon output	Notes on location and methods
	(210)	36°	303	*Pinus* forest, 40 years old. Field, dynamic, infrared analysis.
	(54)	39°	1010	*Quercus-Carya* forest, 50 years old. Field, dynamic, infrared analysis.
	(188)	41°	427	Mean for montane *Fagus* and *Fagus-Abies* forests, 60–90 years old. Field, static, absorption techniques.
	(144)	45°	751	Mean of *Quercus* forest, 35–45 years old and fen forest. Field, dynamic, infrared analysis.
	(51)	50°	175	Mixed *Quercus-Fagus* forest, young. Field, static, absorption techniques. Calculated as total of biweekly rates for 2 years.
	(138)	52°	171	*Fagus* forest, 150 years old. Field, static, absorption techniques.
	(4)	52°	554	Mean of *Castanea* and *Fagus* forests, 40–60 years old. Field, static, absorption techniques.
	(33)	53°	222	*Quercus-Fagus* forest, 135 years old. Laboratory, oxygen uptake in a Gilson respirometer converted following Witkamp (206).
Boreal forests	(70)	66°	232	*Picea* forest, 250 years old. Laboratory and field, dynamic, infrared analysis.
	(149)	77°	147	*Pinus* forest, young. Field methods unknown. Data converted using rate of 60 kg ha^{-1} day^{-1} × 3 mo. decomposition season.
Woodland and shrubland	(212; cf 197, p. 196)	41°	653	*Quercus-Pinus* woodland, 45 years old. Determined by total community respiration measurements under atmospheric inversions. Value is saprobe respiration + one-half of total plant community respiration.

Table 4 (Continued)

	(21)	56°	399	*Calluna* heath. Field, static, absorption techniques. Calculated from a yearly rate based on daily measurements using 4.6 Kcal = 0.5 gC.
	(10)	48°	513	Mean of two *Quercus ilex* woodlands, 150 years old. Field, static, absorption techniques.
Tropical savanna	(195)	7° S	515	Field, static, absorption techniques.
	(164)	10°	785	Field, static, absorption techniques.
Temperate grassland	(27)	33°	357 421	Field, static, absorption techniques, 2 years.
	(73)	38°	74	Laboratory, static, absorption techniques.
	(106)	39°	452	Field, dynamic, infrared analysis.
	(204)	46°	165	Field, static, absorption techniques.
	(34), (35)	50°	150 447	Special diffusion theory calculations, 2 years.
Tundra	(40)	71°	37	Mean value. Field, dynamic absorption techniques.
	(137)	71°	38	Field, dynamic, infrared analysis. Estimate is for 2 mo. frost-free season from hourly rates, assuming 60 per cent of total flux is from underground.
	(29)	71°	210	Total community respiration based on assumption of a steady-state ecosystem. Special aerodynamic exchange method.
Desert	(23)	42°	22	Field, static, absorption techniques.
Swamp and marsh	(144)	45°	739	*Thuja* swamp, 100 years old. Field, dynamic, infrared analysis.
	(58)	18°	730	Mangrove swamp. Field, dynamic, infrared analysis. Daily rates.
	(113)	26°	1350	Mangrove swamp. Field, dynamic, infrared analysis. Daily rates.

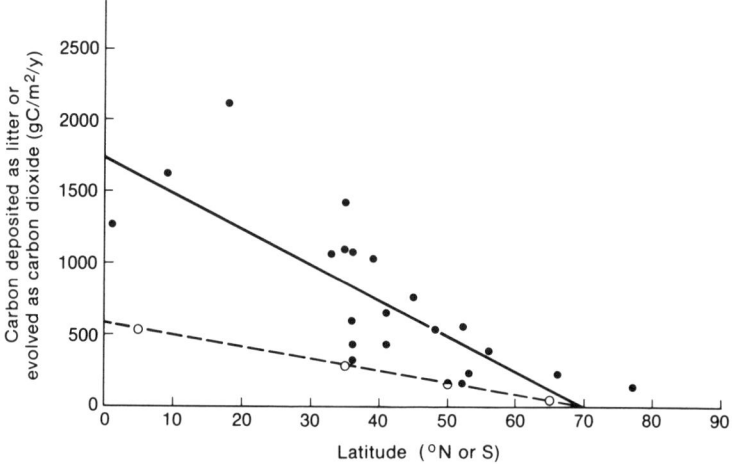

Figure 2 Latitudinal trends for carbon dynamics in forest and woodland soils of the world. The dashed line shows the mean annual carbon input to soils in litterfall; the solid line shows the pattern of carbon loss as CO_2 evolved from soils. Litterfall data are mean values for world regions derived from (19); the linear regression for CO_2 evolution is calculated from the data in Table 4. The equation is EVOL = -24.2 (LAT) + 1721.5; R^2 = 0.60, F = 30.05.

Environmental Factors Affecting Soil Respiration

Because temperature has a direct influence on diffusion rate, research techniques must account for but not disrupt the normal temperature and diffusion regime of the soil. The significance of moisture and chemical concentrations in the control of CO_2 evolution has also been studied. Investigators have used multiple linear regression techniques to deduce the importance of certain variables (35, 73, 144, 185, 204–206). Usually soil temperature has an overriding influence, perhaps since its effect is more strictly monotonic than that of other variables. Studies have shown exponential relations between soil temperature and CO_2 evolution (4, 45, 98, 106, 138, 144, 203); others have calculated Q_{10} values which have ranged from 1.6 (106) to 3.2 (4), tending to be higher at lower temperatures (102, pp. 244–46). Havas & Mäenpää (70) described measurable respiration under snow even when ambient temperatures fell below 0°C.

Since soil moisture and temperature are often inversely correlated, investigators have sometimes found it difficult to separate their effects (35, 144, 185, 204). Over a range of normal environmental temperatures, respiration and moisture are often related in a parabolic fashion (40, 45, 73). In general, it is likely that moisture is seldom limiting except during periods of marked drought in summer months (4, 10, 33, 45, 47, 51, 185, 204) and in tropical dry seasons (194, 195, 215). De Boois (33) noted that moisture affected only the litter layer, while temperature was of importance in the fermentation and humus layers where the majority of CO_2 was released.

CARBON BALANCE IN DETRITUS 71

Carbon dioxide evolution and microbial activity are related to particular environmental events and disturbances (4, 144). Heavy precipitation may displace CO_2 from soil pore spaces (4). McColl (120) used principal component analysis to study the controls of ion release from forest floors. While ion release is only indirectly related to carbon mineralization, he noted a positive effect of large rainstorms. These flushes of decomposition in remoistened soils may be due to microbial growth stimulated by nutrients released during a previous dry phase (11, 170–172, 186). Similarly, freeze-thaw cycles may stimulate the release of CO_2 (171, 208) and the mobilization of nutrients (9). Agricultural workers have long known that the addition of fresh plant residues despite their wide C/N ratios often stimulates a flush of microbial activity (76, 77, 109, 190). This has been called "the priming effect." Laura (108, 109) has suggested that protons are added or made available by the addition of fresh materials or salts (e.g. KNO_3, NH_4Cl), and that the decomposition of relatively stable soil organic matter is stimulated by the increased supply of protons. He noted, however, that a net decrease in soil organic matter is unlikely since the added materials lose protons and become relatively stable, remaining residues. In forests with large detritus accumulations, seasonal litterfall in autumn appears to have little effect on CO_2 evolution rates (4, 144); but in a forest with relatively little humus, CO_2 evolution was stimulated by fresh litterfall (45).

Sites with a highly variable soil microclimate may show an increased annual carbon release because microbial populations may remain perpetually in an exponential phase of growth (144). De Jong and his co-workers (35) concluded that the number of disturbances in the environment of microorganisms was the most important determinant of annual respiration based on a three-year comparison at a grassland site. In contrast, the environmental stability of waterlogged soil environments and their anaerobic conditions inhibit microbial metabolism and allow deep organic accumulations (e.g. 148). Contrary to all of these views, Kononova (102, p. 240) has suggested that rhythmical conditions in the soil environment can explain the very high organic content in chernozem soils of the USSR. Cycles stimulate microbial activity and the formation of humic substances during favorable periods but allow organic matter to form more stable complexes with the mineral soil during depression periods.

PHYSICO-CHEMISTRY Due to their high percentage of protein, soil microbes demand large amounts of nitrogen. Various workers have ascribed great importance to nitrogen in regulating decay rate and organic accumulation. In general, forest leaves and crop residues high in nutrient content (including N) tend to disappear at a faster rate (60, 192, 205). Allison et al (2) have graphically illustrated a relation between cumulative CO_2 evolved from the soil and the nitrogen immobilized (presumably in microbial biomass). Adequate nitrogen for reasonable development of the decomposer biomass is certainly an important determinant of decay rate.

On the other hand "the percentage of added carbon left in the soil after an *extended* period of decomposition is determined primarily by the composition of the materials added and not by their carbon-nitrogen ratios" [(139), italics added; cf (3)]. Reviewing the results of many C^{14} studies, Jenkinson (77) summarized that "the proportion of the added plant carbon retained in the soil under different climatic conditions, using different plant materials and soils, is remarkably similar." Thus, he stressed the development of relatively stable, more refractory compounds in soil organic matter. The rapid disappearance of litter high in nutrient elements does not necessarily mean that the carbon it contains is rapidly mineralized; instead, this carbon may be transformed to a more stable, humic form. Since soil scientists have shown the great control that the C/N ratio exerts on nitrogen mineralization, it may be logical to believe in concurrent, equivalent carbon mineralization; but these results suggest that nitrogen supply should not be used to explain long-term carbon release or accumulation (cf 83).

While the importance of the C/N ratio to the rate of disappearance of leaves and to nitrification in humus has been well documented for many years, studies specifically investigating a relation between soil chemistry and soil respiration are less numerous. Little or no correlation between respiration and carbon, nitrogen, and phosphorus was reported by Van Cleve & Sprague (185) and Ino & Monsi (73). Carbon dioxide evolution increased with experimental NPK fertilization in studies by Waksman & Starkey (190) and Repnevskaja (149), but DeJong et al (35) reported the opposite result. Similarly, trace metal additions generally decrease microbiological activity (8, 89, 179, 180); but in one study, increased available NO_3^- resulted from the addition of Cd and Pb (182). Decomposition was inhibited by decreasing pH in laboratory soil cultures (92); but in arctic soils Van Cleve & Sprague (185) noted no significant soil respiration differences between two sites of differing pH. Perhaps the most comprehensive study of the importance of soil chemical properties is of the nutrient-depleted soils of the Piedmont region (90). Using multiple linear regression to explain variance in soil respiration rates, Jorgensen & Wells found that the important, correlated variables were Cu, Ca, and Fe in organic soil layers, C and N in mineral soil layers. Phosphorus was only a weakly correlated variable in one instance. Moreover, since calcium influences the availability of many other elements and since trace metals may be chelated with the remaining organic matter, the causal significance of any element in a regression may be misjudged.

The physical properties of a soil are most likely to affect respiration through their influences on diffusion rate and aeration (173). Like calcium, however, clay minerals may complex organic matter in an unavailable state and retard microbial action (102, pp. 251–56; 174).

Conclusions

In general, the physicochemical properties of litter determine the rate of its disappearance (60, 192, 205, 206, 209, 211), but temperature and moisture are the variables of overshadowing importance to soil respiration rate. Thus, even if they differ in species compositions, natural communities with similar net primary productivities in the same general environment will tend to have similar soil CO_2 evolution rates (90, 144, 185, 206). Current field methodology is still too crude to pick out fine differences and the effects of minor variables. Understanding the significance of specific chemical elements would enhance our knowledge of microbiological function as well as of the effects of pollutant elements in natural systems (89, 158, 179, 180).

CARBON CYCLE SYNTHESIS

The values assigned to the same compartments and to annual transfers in the carbon cycle have differed by several orders of magnitude in various treatments. This variation emphasizes both the primitive nature and the problems of global geochemical estimates. A synthesized diagrammatic model (146) perhaps best represents our current understanding of the carbon cycle. My world detritus value of 1456 $\times$ 10^9 mtC is about twice as high as the 700 $\times$ 10^9 mtC value of the Brookhaven model, presumably due to my inclusion of dispersed humus. The world detritus input estimate of 37.5 $\times$ 10^9 mtC yr^{-1} of Reiners (145) is probably the best approximation from literature values ranging from 17 $\times$ 10^9 mtC yr^{-1} (102, p. 206; 140) to 64 $\times$ 10^9 mtC yr^{-1} (145). As a whole, then, soil organic matter turns over in about 40 years, but certainly specific fractions are far more or far less refractory.

In forests, we might estimate that total soil respiration averages about twice the total annual detritus input of carbon from above- and below-ground sources (2.5 times the input from the above-ground sources only). In other ecosystems such as grasslands, soil output is perhaps slightly less than twice the annual input. If we use a factor of two, we may calculate a world output of carbon of 75 $\times$ 10^9 mt yr^{-1} from soil respiration in terrestrial ecosystems. Above-ground plant and heterotrophic respiration and carbon release in fires must be added to this value to obtain total world terrestrial carbon output to the atmosphere. The Brookhaven model (146) balances total terrestrial outputs against input by gross photosynthesis estimated at 100 $\times$ 10^9 mtC yr^{-1} (200). Thus, the input and output calculated separately are balanced within 25%. This is encouraging, given the crude nature of these estimates. More important is the large size of these transfers compared to fossil fuel combustion [3.6 $\times$ 10^9 mtC yr^{-1}; (94, 146)]. A minor percentage shift or unbalance in the mass balance of the biomass and/or detritus pools could result in a large release of CO_2 to the atmosphere. Given man's current activities, the potential and effects of decreasing the detritus pool should not be underrated.

ACKNOWLEDGMENTS

I thank B. E. Mahall, P. L. Marks, R. K. Peet, W. A. Reiners, T. R. Wentworth, and R. H. Whittaker for stimulating discussions, encouragement, and careful, critical reviews of the manuscript.

Literature Cited

1. Allison, F. E. 1973. *Soil Organic Matter and Its Role in Crop Production.* New York: Elsevier. 637 pp.
2. Allison, F. E., Murphy, R. M., Klein, C. J. 1963. Nitrogen requirements for the decomposition of various kinds of finely ground woods in soil. *Soil Sci.* 96:187–90
3. Allison, F. E., Sherman, M. S., Pinck, L. A. 1949. Maintenance of soil organic matter. I. Inorganic soil colloid as a factor in retention of carbon during formation of humus. *Soil Sci.* 68:463–78
4. Anderson, J. M. 1973. Carbon dioxide evolution from two temperate, deciduous woodland soils. *J. Appl. Ecol.* 10:361–78
5. Ball, D. F. 1964. Loss-on-ignition as an estimate of organic matter and organic carbon in non-calcareous soils. *J. Soil Sci.* 15:84–92
6. Ballard, T. M. 1970. Gaseous diffusion evaluation in forest humus. *Soil Sci. Soc. Am. Proc.* 34:532–33
7. Barnette, R. M., Hester, J. B. 1930. Effect of burning upon the accumulation of organic matter in forest soils. *Soil Sci.* 29:281–84
8. Bhuiya, M. R. H., Cornfield, A. H. 1972. Effects of addition of 1000 ppm Cu, Ni, Pb and Zn on carbon dioxide release during incubation of soil alone and after treatment with straw. *Environ. Pollut.* 3:173–77
9. Biederbeck, V. O., Campbell, C. A. 1973. Soil microbial activity as influenced by temperature trends and fluctuations. *Can. J. Soil Sci.* 53:363–75
10. Billes, G., Cortez, J., Lossaint, P. 1971. L'activité biologique des sols dans les écosystèmes méditerranéens. I. Minéralisation du carbone. *Rev. Ecol. Biol. Sol* 8:375–95
11. Birch, H. F. 1958. The effect of soil drying on humus decomposition and nitrogen availability. *Plant Soil* 10:9–31
12. Birch, H. F., Friend, M. T. 1956. The organic-matter and nitrogen status of east African soils. *J. Soil Sci.* 7:156–67
13. Bleak, A. T. 1970. Disappearance of plant material under a winter snow cover. *Ecology* 51:915–17
14. Bliss, L. C. 1966. Plant productivity in alpine microenvironments on Mt. Washington, New Hampshire. *Ecol. Monogr.* 36:125–55
15. Bliss, L. C. 1975. Devon Island, Canada. In *Structure and Function of Tundra Ecosystems, Ecol. Bull. 20,* ed. T. Rosswall, O. W. Heal, pp. 17–60. Stockholm: Swedish Nat. Sci. Res. Coun.
16. Bohn, H. L. 1976. Estimate of organic carbon in world soils. *Soil Sci. Soc. Am. J.* 40:468–70
17. Bolin, B. 1970. The carbon cycle. *Sci. Am.* 223(3):124–32
18. Bourlière, F., Hadley, M. 1970. The ecology of tropical savannas. *Ann. Rev. Ecol. Syst.* 1:125–52
19. Bray, J. R., Gorham, E. 1964. Litter production in forests of the world. *Adv. Ecol. Res.* 2:101–57
20. Broadbent, F. E. 1953. The soil organic fraction. *Adv. Agron.* 5:153–83
21. Brown, A., MacFadyen, A. 1969. Soil carbon dioxide output and small-scale vegetation pattern in a *Calluna* heath. *Oikos* 20:8–15
22. Burford, J. R., Bremner, J. M. 1972. Gas chromatographic determination of carbon dioxide evolved from soils in closed systems. *Soil. Biol. Biochem.* 4:191–97
23. Caldwell, M. M., DePuit, E. J., Fernandez, O. A., Wiebe, H. H., Camp, L. B. 1974. Gas exchange, translocation, root growth, and soil respiration of Great Basin plants. *US/IBP Desert Biome Res. Memo 74–9.* 32 pp.
24. Campbell, C. A., Paul, E. A., Rennie, D. A., McCallum, K. J. 1967. Applicability of the carbon-dating method of analysis to soil humus studies. *Soil Sci.* 104:217–24
25. Chapman, S. B. 1967. Nutrient budgets for a dry heath ecosystem in the South of England. *J. Ecol.* 55:677–89
26. Chapman, S. B. 1970. The nutrient content of the soil and root system of a dry heath ecosystem. *J. Ecol.* 58:445–52
27. Coleman, D. C. 1973. Soil carbon balance in a successional grassland. *Oikos* 24:195–99
28. Coleman, D. C. 1973. Compartmental analysis of "total soil respiration": An exploratory study. *Oikos* 24:361–66
29. Coyne, P. I., Kelley, J. J. 1975. CO_2 exchange over the Alaskan arctic tundra: Meteorological assessment by an aerodynamic method. *J. Appl. Ecol.* 12:587–611
30. Dahlman, R. C., Kucera, C. L. 1965. Root productivity and turnover in native prairie. *Ecology* 46:84–89
31. Daly, G. T. 1966. Nitrogen fixation by nodulated *Alnus rugosa. Can. J. Bot.* 44:1607–21

32. Darbishire, F. V., Russell, E. J. 1907. Oxidation in soils and its relation to productiveness. Part II. The influence of partial sterilisation. *J. Agric. Sci.* 2:305–26
33. De Boois, H. M. 1974. Measurement of seasonal variations in the oxygen uptake of various litter layers of an oak forest. *Plant Soil* 40:545–55
34. De Jong, E., Schappert, H. J. V. 1972. Calculation of soil respiration and activity from CO_2 profiles in the soil. *Soil Sci.* 113:328–33
35. De Jong, E., Schappert, H. J. V., MacDonald, K. B. 1974. Carbon dioxide evolution from virgin and cultivated soil as affected by management practices and climate. *Can. J. Soil Sci.* 54:299–307
36. Delwiche, C. C. 1970. The nitrogen cycle. *Sci. Am.* 223(3):136–46
37. Dickinson, C. H., Pugh, G. J. F., eds. 1974. *Biology of Plant Litter Decomposition*, Vols. 1 and 2. New York: Academic
38. Dickson, B. A., Crocker, R. L. 1953. A chronosequence of soils and vegetation near Mt. Shasta, California. II. The development of the forest floors and the carbon and nitrogen profiles of the soils. *J. Soil Sci.* 4:142–54
39. Donselaar, J. Van. 1965. *The Vegetation of Suriname. IV. An Ecological and Phytogeographic Study of Northern Surinam Savannas*. Amsterdam: Van Eedenfronds. 163 pp.
40. Douglas, L. A., Tedrow, J. C. F. 1959. Organic matter decomposition rates in arctic soils. *Soil Sci.* 88:305–12
41. Duvigneaud, P., Denaeyer-DeSmet, S. 1970. Biological cycling of minerals in temperate deciduous forests. In *Analysis of Temperate Forest Ecosystems*, ed. D. E. Reichle, pp. 199–225. New York: Springer
42. Duvigneaud, P., Kestemont, P., Ambroes, P. 1971. Productivité primaire des forêts tempérées d'essences feuillues caducifoliées in Europe occidentale. pp. 259–270. In *Productivity of Forest Ecosystems, Proc. Brussels Symp., 1969.* ed. P. Duvigneaud. Paris: UNESCO
43. Edmonds, R. L., ed. 1974. An initial synthesis of results in the coniferous forest biome, 1970–1973. *US/IBP Coniferous Forest Biome Bull. No. 7.* Seattle: Univ. Washington
44. Edwards, N. T. 1974. A moving chamber design for measuring soil respiration rates. *Oikos* 25:97–101
45. Edwards, N. T. 1975. Effects of temperature and moisture on carbon dioxide evolution in a mixed deciduous forest floor. *Soil Sci. Soc. Am. Proc.* 39:361–65
46. Edwards, N. T., Sollins, P. 1973. Continuous measurement of carbon dioxide evolution from partitioned forest floor components. *Ecology* 54:406–12
47. Ellis, R. C. 1969. The respiration of the soil beneath some *Eucalyptus* forest stands as related to the productivity of the stands. *Aust. J. Soil Res.* 7:349–57
48. Everett, K. R. 1971. Composition and genesis of the organic soils of Amchitka Island, Aleutian Islands, Alaska. *Arctic Alp. Res.* 3:1–16
49. Foster, N. W., Morrison, I. K. 1976. Distribution and cycling of nutrients in a natural *Pinus banksiana* ecosystem. *Ecology* 57:110–20
50. Franz, G. 1976. Der Einfluss von Niederschlag, Höhenlage und Jahresdurchschnittstemperatur im Untersuchungsgebiet auf Humusgehalt und mikrobielle Aktivität in Bodenproben aus Nepal. *Pedobiologia* 16:136–50
51. Froment, A. 1972. Soil respiration in a mixed oak forest. *Oikos* 23:273–77
52. Gainey, P. L. 1919. Parallel formation of carbon dioxide, ammonia and nitrate in soil. *Soil Sci.* 7:293–311
53. Ganzhara, N. F. 1974. Humus formation in chernozem soils. *Sov. Soil Sci.* 1974:400–7
54. Garrett, H. E., Cox, G. S. 1973. Carbon dioxide evolution from the floor of an oak-hickory forest. *Soil Sci. Soc. Am. Proc.* 37:641–44
55. Gersper, P. L. 1972. Chemical and physical soil properties and their seasonal dynamics at the Barrow intensive site. In *Proc. 1972 Tundra Biome Symp.,* ed. S. Bowen, pp. 87–93. Seattle: Univ. Washington
56. Gilkeson, R. A., Marshall, W. G., Smith, H. W. 1957. Characteristics of some soils of the arid region of Washington. *Soil Sci. Soc. Am. Proc.* 21:644–49
57. Godlin, M. M., Son'ko, M. P. 1970. Humus of ordinary steppe chernozems in the Ukraine. *Sov. Soil Sci.* 1970:8–18
58. Golley, F. B., Odum, H. T., Wilson, R. F. 1962. The structure and metabolism of a Puerto Rican red mangrove forest in May. *Ecology* 43:9–19
59. Gore, A. J. P., Olson, J. S. 1967. Preliminary models for accumulation of organic matter in an *Eriophorum/Calluna* ecosystem. *Aquilo Ser. Bot.* 6:297–313

60. Gosz, J. R., Likens, G. E., Bormann, F. H. 1973. Nutrient release from decomposing leaf and branch litter in the Hubbard Brook forest, New Hampshire. *Ecol. Monogr.* 43:173–91
61. Gosz, J. R., Likens, G. E., Bormann, F. H. 1976. Organic matter and nutrient dynamics of the forest and forest floor in the Hubbard Brook forest. *Oecologia* 22:305–20
62. Grier, C. C. 1973. Organic matter and nitrogen distribution in some mountain heath communities of the source lake basin, Washington. *Arctic Alp. Res.* 5:261–67
63. Grier, C. C., Cole, D. W., Dyrness, C. T., Fredriksen, R. L. 1974. Nutrient cycling in 37- and 450-year-old Douglas-fir ecosystems. In *Integrated Research in the Coniferous Forest Biome, Coniferous Forest Biome Bull. No. 5,* ed. R. H. Waring and R. L. Edmonds, pp. 21–34. Seattle: Univ. Washington
64. Grishina, L. A., Srebnova, L. V. 1973. The humus and nitrogen of some soils of the Tambov lowland. *Sov. Soil Sci.* 1973:315–19
65. Gross, R. A. 1946. The composition and classification of forest floors and related soil profiles in Saskatchewan. *Sci. Agric.* 26:603–21
66. Haas, H. J., Evans, C. E., Miles, E. F. 1957. Nitrogen and carbon changes in Great Plains soils as influenced by cropping and soil treatments. *USDA Tech. Bull. No. 1164.* 111 pp.
67. Haber, W. 1958. Ökologische Untersuchung der Bodenatmung. *Flora* 146:109–57
68. Hanawalt, R. B., Whittaker, R. H. 1976. Altitudinally coordinated patterns of soils and vegetation in the San Jacinto Mountains, California. *Soil Sci.* 121:114–24
69. Harley, J. L. 1971. Fungi in ecosystems. *J. Ecol.* 59:653–68
70. Havas, P., Mäenpää, E. 1972. Evolution of carbon dioxide at the floor of a *Hylocamium-Myrtillus*-type spruce forest. *Aquilo Ser. Bot.* 11:4–22
71. Heyligers, P. C. 1963. *Vegetation and Soil of a White-sand Savanna in Suriname.* Amsterdam: North-Holland. 148 pp.
72. Howard, P. J. A. 1965. The carbon-organic matter factor in various soil types. *Oikos* 15:229–36
73. Ino, Y., Monsi, M. 1969. An experimental approach to the calculation of CO_2 amount evolved from several soils. *Jpn. J. Bot.* 20:153–88
74. Jagnow, G. 1971. The effect of altitude, precipitation and cultivation on the organic carbon and nitrogen content of east African soils. *Trans. Int. Symp. "Humus and Plants V,"* Prague
75. Jeffrey, D. W. 1970. A note on the use of ignition loss as a means for the approximate estimation of soil bulk density. *J. Ecol.* 58:297–99
76. Jenkinson, D. S. 1966. The priming action. In *The Use of Isotopes in Soil Organic Matter Studies,* pp. 199–208. Oxford: Pergamon.
77. Jenkinson, D. S. 1971. Studies on the decomposition of C^{14} labelled organic matter in soil. *Soil Sci.* 111:64–70
78. Jenny, H. 1930. A study on the influence of climate upon the nitrogen and organic matter content of the soil. *Univ. Missouri Agric. Exp. Stn. Res. Bull. 152.* 66 pp.
79. Jenny, H. 1941. *Factors of Soil Formation, A System of Quantitative Pedology.* New York: McGraw-Hill. 281 pp.
80. Jenny, H. 1950. Causes of the high nitrogen and organic matter content of certain tropical forest soils. *Soil Sci.* 69:63–69
81. Jenny, H., Bingham, F., Padilla-Saravia, B. 1948. Nitrogen and organic matter contents of equatorial soils of Colombia, South America. *Soil Sci.* 66:173–86
82. Jenny, H., Gessel, S. P., Bingham, F. T. 1949. Comparative study of decomposition rates of organic matter in temperate and tropical regions. *Soil Sci.* 68:419–32
83. Jenny, H., Salem, A. E., Wallis, J. R. 1968. Interplay of soil organic matter and soil fertility with state factors and soil properties. In *Study Week on Organic Matter and Soil Fertility,* pp. 6–37. Amsterdam: North-Holland
84. Jensen, V. 1974. Decomposition of angiosperm tree leaf litter. See Ref. 37, pp. 69–104
85. Johnson, F. L., Risser, R. G. 1974. Biomass, annual net primary production, and dynamics of six mineral elements in a post oak-blackjack oak forest. *Ecology* 55:1246–58
86. Jones, E. W. 1955. Ecological studies on the rainforest of southern Nigeria. IV. The plateau forest of the Okomu forest reserve. *J. Ecol.* 43:564–94
87. Jones, M. J. 1973. The organic matter content of the savanna soils of west Africa. *J. Soil Sci.* 24:42–53
88. Jordan, C. F. 1971. A world pattern in plant energetics. *Am. Sci.* 59:425–33

89. Jordan, M. J., Lechevalier, M. P. 1975. Effects of zinc-smelter emissions on forest soil microflora. *Can. J. Microbiol.* 21:1855–65
90. Jorgensen, J. R., Wells, C. G. 1973. The relationship of respiration in organic and mineral soil layers to soil chemical properties. *Plant Soil* 39:373–87
91. Kanemasu, E. T., Powers, W. L., Sij, J. W. 1974. Field chamber measurements of CO_2 flux from soil surface. *Soil Sci.* 118:233–37
92. Katznelson, B. H., Stevenson, I. L. 1956. Observations on metabolic activity of the soil microflora. *Can. J. Microbiol.* 2:611–22
93. Kayama, R. 1975. Soil factor influencing biological production. In *Ecological Studies in Japanese Grasslands*, ed. M. Numata, pp. 93–102. Tokyo: Univ. Tokyo
94. Keeling, C. D. 1973. Industrial production of carbon dioxide from fossil fuels and limestone. *Tellus* 25:174–98
95. Kimball, B. A., Lemon, E. R. 1971. Air turbulence effects upon soil gas exchange. *Soil Sci. Soc. Am. Proc.* 35:16–21
96. Kirita, H. 1971. Re-examination of the absorption method of measuring soil respiration under field conditions. II. Effect of the size of the apparatus on CO_2-absorption rates. *Jpn. J. Ecol.* 21:37–42
97. Kirita, H. 1971. Re-examination of the absorption method of measuring soil respiration under field conditions. III. Combined effect of the covered ground area and the surface area of KOH solution on CO_2-absorption rates. *Jpn. J. Ecol.* 21:43–47
98. Kirita, H. 1971. Studies of soil respiration in warm-temperate evergreen broadleaf forests of southwestern Japan. *Jpn. J. Ecol.* 21:230–44
99. Kirita, H., Hozumi, K. 1966. Re-examination of the absorption method of measuring soil respiration under field conditions. I. Effect of the amount of KOH on observed values. *Physiol. Ecol.* 14:23–31
100. Klemmedson, J. O. 1975. Nitrogen and carbon regimes in an ecosystem of young dense ponderosa pine in Arizona. *For. Sci.* 21:163–68
101. Klinge, H., Rodrigues, W. A., Bruning, E., Fittkau, E. J. 1975. Biomass and structure in a central Amazonia rain forest. In *Tropical Forest Ecosystems*, ed. F. B. Golley, E. Medina, pp. 115–22. New York: Springer
102. Kononova, M. M. 1966. *Soil Organic Matter: Its Nature, Its Role in Soil Formation and in Soil Fertility.* Oxford: Pergamon. 554 pp. 2nd ed.
103. Kononova, M. M. 1968. Humus of the main soil types and soil fertility. See Ref. 83, pp. 361–79
104. Kononova, M. M. 1972. Current problems in the study of soil organic matter. *Sov. Soil Sci.* 1972:420–28
105. Kononova, M. M. 1975. Humus of virgin and cultivated soils. In *Soil Components.* Vol. 1, *Organic Components*, ed. J. E. Gieseking, pp. 475–526. New York: Springer
106. Kucera, C. L., Kirkham, D. R. 1971. Soil respiration studies in tallgrass prairie in Missouri. *Ecology* 52:912–15
107. Lang, G. E. 1974. Litter dynamics in a mixed oak forest on the New Jersey piedmont. *Bull. Torrey Bot. Club* 101:277–86
108. Laura, R. D. 1974. Effects of neutral salts on carbon and nitrogen mineralisation of organic matter in soil. *Plant Soil* 41:113–27
109. Laura, R. D. 1975. On the "priming effect" of organic materials. *Soil Sci. Soc. Am. Proc.* 39:807–8
110. Lieth, H., Ouellette, R. 1962. Studies on the vegetation of the Gaspé peninsula. II. The soil respiration of some plant communities. *Can. J. Bot.* 40:127–40
111. Lossaint, P., Rapp, M. 1971. Répartition de la matière organique, productivité et cycles des éléments minéraux dans des écosystèmes climat méditerranéen. See Ref. 42, pp. 597–617
112. Lousier, J. D., Parkinson, D. 1976. Litter decomposition in a cool temperate deciduous forest. *Can. J. Bot.* 54:419–36
113. Lugo, A. E., Snedaker, S. C. 1974. The ecology of mangroves. *Ann. Rev. Ecol. Syst.* 5:39–64
114. Lundegårdh, H. G. 1924. *Der Kreislauf der Kohlensäure in der Natur.* Jena: Fischer. 308 pp.
115. Lundegårdh, H. 1927. Carbon dioxide evolution of soil and crop growth. *Soil Sci.* 23:417–53
116. Lunt, H. A. 1931. The carbon-organic matter factor in forest soil humus. *Soil Sci.* 22:27–33
117. Lutz, H. J., Chandler, R. F. Jr. 1946. *Forest Soils.* New York: Wiley. 514 pp.
118. Maier, W. J., Gast, R. C., Anderson, C. T., Nelson, W. W. 1975. Carbon contents of surface and underground waters in south-central Minnesota. *J. Environ. Qual.* 5:124–28

119. Maldague, M. E., Hilger, F. 1963. Observations faunistiques et microbiologiques dans quelques biotopes forestiers equatoriaux. In *Soil Organisms,* ed. J. Doeksen, J. Van Der Drift, pp. 368–74. Amsterdam: North-Holland
120. McColl, J. G. 1973. A model of ion transport during moisture flow from a Douglas-fir forest floor. *Ecology* 54: 181–87
121. McFee, W. W., Stone, E. L. 1965. Quantity, distribution, and variability of organic matter and nutrients in a forest podzol in New York. *Soil Sci. Soc. Am. Proc.* 29:432–36
122. McFee, W. W., Stone, E. L. 1966. The persistence of decaying wood in the humus layers of northern forests. *Soil Sci. Soc. Am. Proc.* 30:513–16
123. Minderman, G. 1968. Addition, decomposition and accumulation of organic matter in forests. *J. Ecol.* 56:355–62
124. Minderman, G., Vulto, J. C. 1973. Comparison of techniques for the measurement of carbon dioxide evolution from soil. *Pedobiologia* 13:73–80
125. Mogen, C. A., McClelland, J. E., Allen, J. S., Schroer, F. W. 1959. Chestnut, chernozem and associated soils of western North Dakota. *Soil Sci. Soc. Am. Proc.* 23:56–60
126. Nakane, K. 1975. Dynamics of soil organic matter in different parts on a slope under evergreen oak forest. *Jpn. J. Ecol.* 25:204–16
127. Nihlgård, B. 1971. Pedological influence of spruce planted on former beech forest soils in Scania, South Sweden. *Oikos* 22:302–14
128. Nihlgård, B. 1972. Plant biomass, primary production and distribution of chemical elements in a beech and a planted spruce forest in South Sweden. *Oikos* 23:69–81
129. Odum, E. P. 1971. *Fundamentals of Ecology.* Philadelphia: Saunders. 574 pp. 3rd ed.
130. Odum, H. T., Lugo, A., Cintrón, G., Jordan, C. F. 1970. Metabolism and evapotranspiration of some rain forest plants and soil. In *A Tropical Rainforest,* ed. H. T. Odum, pp. I:103–64. Springfield, Va: US AEC, Div. Tech. Info.
131. Ogawa, H., Yoda, K., Kira, T. 1961. A preliminary survey on the vegetation of Thailand. *Nature and Life in S. E. Asia* 1:21–157
132. Olson, J. S. 1963. Energy storage and the balance of producers and decomposers in ecological systems. *Ecology* 44:322–31
133. Ovchinnikov, S. M., Sokolova, T. A., Targul'yan, V. O. 1973. Clay minerals in loamy soils in the taiga and forest tundra of west Siberia. *Sov. Soil Sci.* 1973:709–22
134. Pachikina, L. I., Sharoshkina, N. B. 1970. Composition, properties, and formation of compact meadow soils in the ancient Ural River delta. *Sov. Soil Sci.* 1970:385–96
135. Patten, B. C., Witkamp, M. 1967. Systems analysis of 134 cesium kinetics in terrestrial microcosms. *Ecology* 48: 813–24
136. Paul, E. A. 1970. Plant components and soil organic matter. *Rec. Adv. Phytochem.* 3:59–104
137. Peterson, K. M., Billings, W. D. 1975. Carbon dioxide flux from tundra soils and vegetation as related to temperature at Barrow, Alaska. *Am. Midl. Natur.* 94:88–98
138. Phillipson, J., Putman, R. J., Steel, J., Woodell, S. R. J. 1975. Litter input, litter decomposition and the evolution of carbon dioxide in a beech woodland—Wytham Woods, Oxford. *Oecologia* 20:203–17
139. Pinck, L. A., Allison, F. E., Sherman, M. S. 1950. Maintenance of soil organic matter. II. Losses of carbon and nitrogen from young and mature plant materials during decomposition in soil. *Soil Sci.* 69:391–401
140. Plass, G. N. 1959. Carbon dioxide and climate. *Sci. Am.* 201(1):41–47
141. Read, J. W., Ridgell, R. H. 1922. On the use of the conventional carbon factor in estimating soil organic matter. *Soil Sci.* 8:1–6
142. Reader, R. J., Stewart, J. M. 1972. The relationship between net primary production and accumulation for a peatland in southeastern Manitoba. *Ecology* 53:1024–37
143. Reichle, D. E., Dinger, B. E., Edwards, N. T., Harris, W. F., Sollins, P. 1973. Carbon flow and storage in a forest ecosystem. *Brookhaven Symp. Biol.* 24: 345–65
144. Reiners, W. A. 1968. Carbon dioxide evolution from the floor of three Minnesota forests. *Ecology* 49:471–83
145. Reiners, W. A. 1973. Terrestrial detritus and the carbon cycle. *Brookhaven Symp. Biol.* 24:303–27
146. Reiners, W. A. 1973. A summary of the world carbon cycle and recommenda-

tions for critical research. *Brookhaven Symp. Biol.* 24:368–82
147. Reiners, W. A., Marks, R. H., Vitousek, P. M. 1975. Heavy metals in subalpine and alpine soils of New Hampshire. *Oikos* 26:264–75
148. Reiners, W. A., Reiners, N. M. 1970. Energy and nutrient dynamics of forest floors in three Minnesota forests. *J. Ecol.* 58:497–519
149. Repnevskaja, M. A. 1967. Liberation of CO_2 from the soil in the pine stands of the Kola peninsula. *Sov. Soil Sci.* 1967:1067–72
150. Richards, P. W. 1939. Ecological studies on the rain forest of southern Nigeria. I. The structure and floristic composition of the primary forest. *J. Ecol.* 27:1–61
151. Rightmire, C. T., Hanshaw, B. B. 1973. Relationship between the carbon isotope composition of soil CO_2 and dissolved carbonate species in groundwater. *Water Resour. Res.* 9:958–67
152. Rodin, L. E., Bazilevich, N. I. 1967. *Production and Mineral Cycling in Terrestrial Vegetation.* London: Oliver & Boyd. 288 pp.
153. Romell, L. G. 1932. Mull and duff as biotic equilibria. *Soil Sci.* 34:161–88
154. Romell, L. G. 1935. Ecological problems of the humus layer in the forest. *Cornell Univ. Agric. Exp. Stn. Mem. 170.* 28 pp.
155. Rosswall, T., Flower-Ellis, J. G. K., Johansson, L. G., Jonsson, S., Rydén, B. E., Sonesson, M. 1975. Stordalen (Abisko), Sweden. See Ref. 15, pp. 265–94
156. Rubilin, Y. V., Dolotov, V. A. 1967. Effect of cultivation on the amounts and composition of humus in gray forest soils. *Sov. Soil Sci.* 1967:733–38
157. Rubilin, Y. V., Kozyreva, M. G. 1974. The age of Russian chernozem. *Sov. Soil Sci.* 1974:383–92
158. Rühling, A., Tyler, G. 1968. An ecological approach to the lead problem. *Bot. Notiser.* 121:321–42
159. Russell, E. J., Appleyard, A. 1915. The atmosphere of the soil: Its composition and the causes of variation. *J. Agric. Sci.* 7:1–48
160. Satchell, J. E. 1971. Feasibility study of an energy budget for Meathop Woods. See Ref. 42, pp. 619–30
161. Scharpenseel, H. W. 1975. Natural radiocarbon measurements on humic substances in the light of carbon cycle estimates. In *Humic Substances: Their Structure and Function in the Biosphere*, ed. D. Povoledo, H. L. Golterman, pp. 281–92. Wageningen, The Netherlands: Cent. Agric. Publ. Docum.
162. Schlesinger, W. H. 1976. *Biogeochemical Limits of Plant Community Organization in the Cypress Forest of Okefenokee Swamp.* PhD thesis. Cornell Univ., Ithaca, NY. 142 pp.
163. Schulz, J. P. 1960. *Ecological Studies on Rain Forest in Northern Suriname.* Amsterdam: Van Eedenfonds. 267 pp.
164. Schulze, E. D. 1967. Soil respiration of tropical vegetation types. *Ecology* 48:652–53
165. Schwartz, D. M., Bazzaz, F. A. 1973. *In situ* measurements of carbon dioxide gradients in a soil-plant-atmosphere system. *Oecologia* 12:161–67
166. Smith, F. B., Brown, P. E. 1931. Soil respiration. *J. Am. Soc. Agron.* 23:909–16
167. Smith, F. B., Brown, P. E. 1933. The diffusion of carbon dioxide through soils. *Soil Sci.* 35:413–23
168. Smith, R. M., Samuels, G., Cernuda, C. F. 1951. Organic matter and nitrogen build-ups in some Puerto Rican soil profiles. *Soil Sci.* 72:409–27
169. Soil Conservation Service. 1970. *Soil Survey Laboratory Data and Descriptions for Some Soils of Nevada.* 219 pp.
170. Sørensen, L. H. 1974. Rate of decomposition of organic matter in soil as influenced by repeated air drying-rewetting and repeated additions of organic material. *Soil Biol. Biochem.* 6:287–92
171. Soulides, D. A., Allison, F. E. 1961. Effect of drying and freezing soils on carbon dioxide production, available mineral nutrients, aggregation, and bacterial population. *Soil Sci.* 91:291–98
172. Stevenson, I. L. 1956. Some observations on the microbial activity of remoistened air-dried soils. *Plant Soil* 8:170–82
173. Stolzy, L. H., Van Gundy, S. D. 1968. The soil as an environment for microflora and microfauna. *Phytopathology* 58:889–99
174. Stotzky, G. 1967. Clay minerals and microbial ecology. *Trans. NY Acad. Sci. Ser. II,* 30:11–21
175. Surorov, A. K. 1974. Characteristics of migration of organic and mineral substances in plowed sod-podzolic soils. *Sov. Soil Sci.* 6:18–25
176. Truog, E. 1915. Methods for the determination of carbon dioxide and a new form of absorption tower adapted to the titrimetric method. *J. Ind. Eng. Chem.* 7:1045–49

177. Tsutsumi, T. 1971. Accumulation and circulation of nutrient elements in forest ecosystems. See Ref. 42, pp. 543–52
178. Turner, J., Singer, M. J. 1976. Nutrient distribution and cycling in a sub-alpine coniferous forest ecosystem. *J. Appl. Ecol.* 13:295–301
179. Tyler, G. 1974. Heavy metal pollution and soil enzymatic activity. *Plant Soil* 41:303–11
180. Tyler, G. 1976. Heavy metal pollution, phosphatase activity, and mineralization of organic phosphorus in forest soils. *Soil Biol. Biochem.* 8:327–32
181. Tyler, G., Gullstrand, C., Holmquist, K. A., Kjellstrand, A. M. 1973. Primary production and distribution of organic matter and metal elements in two heath ecosystems. *J. Ecol.* 61:251–68
182. Tyler, G., Mörnsjö, B., Nilsson, B. 1974. Effects of cadmium, lead, and sodium salts on nitrification in a mull soil. *Plant Soil* 40:237–42
183. Ulrich, B., Ahrens, E., Ulrich, M. 1971. Soil chemical differences between beech and spruce sites—an example of the methods used. In *Integrated Experimental Ecology, Methods and Results of Ecosystem Research in the German Solling Project*, ed. H. Ellenberg, pp. 171–90. Berlin: Springer
184. United Nations. 1971. *Soil Map of the World*. Paris: Food & Agric. Org., UNESCO
185. Van Cleve, K., Sprague, D. 1971. Respiration rates in the forest floor of birch and aspen stands in interior Alaska. *Arctic Alp. Res.* 3:17–26
186. Van Schreven, D. A. 1967. Effect of soil atmosphere on bacterial populations and activity. *Plant Soil* 36:561–69
187. Vassiljevskaja, V. D., Ivanov, V. V., Bogatyrev, L. G., Pospelova, E. B., Shalaeva, N. M., Grishina, L. A. 1975. Agapa, USSR. See Ref. 15, pp. 141–58
188. Virzo de Santo, A., Alfani, A., Sapio, S., 1976. Soil metabolism in beech forests of Monte Taburno (Campania Apennines). *Oikos* 27:144–52
189. Volkovintsev, V. I. 1969. Soil formation in the steppe basins of southern Siberia. *Sov. Soil Sci.* 1969:383–91
190. Waksman, S. A., Starkey, R. L. 1924. Microbiological analysis of soil as an index of soil fertility. VII. Carbon dioxide evolution. *Soil Sci.* 17:141–61
191. Waksman, S. A., Stevens, K. R. 1930. A critical study of the methods for determining the nature and abundance of soil organic matter. *Soil Sci.* 30:97–116

192. Waksman, S. A., Tenney, F. G. 1928. Composition of natural organic materials and their decomposition in the soil. III. The influence of nature of plant upon the rapidity of its decomposition. *Soil Sci.* 26:155–71
193. Walter, H., Haber, W. 1957. Über die Intensität der Bodenatmung mit Bemerkungen zu den Lundegårdhschen Werten. *Ber. Dtsch. Bot. Ges.* 70:275–82
194. Wanner, H. 1970. Soil respiration, litter fall and productivity of tropical rain forest. *J. Ecol.* 58:543–47
195. Wanner, H., Soerohaldoko, S., Santosa, P. D. N., Panggabean, G., Yingchoi, P., and Nguyen, T. T. H. 1973. Die Bodenatmung in tropischen Regenwäldern Südost-Asiens. *Oecologia* 12:289–302
196. Weaver, J. E., Hougen, V. H., Weldon, M. D. 1935. Relation of root distribution to organic matter in prairie soil. *Bot. Gaz.* 96:389–420
197. Whittaker, R. H. 1975. *Communities and Ecosystems*. New York: MacMillan. 387 pp. 2nd ed.
198. Whittaker, R. H., Bormann, F. H., Likens, G. E., Siccama, T. G. 1974. The Hubbard Brook ecosystem study: Forest biomass and production. *Ecol. Monogr.* 44:233–54
199. Whittaker, R. H., Buol, S. W., Niering, W. A., Havens, Y. H. 1968. A soil and vegetation pattern in the Santa Catalina Mountains, Arizona. *Soil Sci.* 105:440–50
200. Whittaker, R. H., Likens, G. E. 1973. Carbon in the biota. *Brookhaven Symp. Biol.* 24:281–302
201. Whittaker, R. H., Woodwell, G. M. 1969. Structure, production and diversity of the oak-pine forest at Brookhaven, New York. *J. Ecol.* 57:155–74
202. Wiant, H. V. Jr. 1967. Has the contribution of litter decay to forest "soil respiration" been overestimated? *J. Forest.* 65:408–9
203. Wiant, H. V. Jr. 1967. Influence of temperature on the rate of soil respiration. *J. Forest.* 65:489–90
204. Wildung, R. E., Garland, T. R., Buschbom, R. L. 1975. The interdependent effects of soil temperature and water content on soil respiration rate and plant root decomposition in arid grassland soils. *Soil Biol. Biochem.* 7:373–78
205. Witkamp, M. 1966. Decomposition of leaf litter in relation to environment, microflora, and microbial respiration. *Ecology* 47:194–201
206. Witkamp, M. 1966. Rates of carbon di-

oxide evolution from the forest floor. *Ecology* 47:492–94
207. Witkamp, M. 1969. Cycles of temperature and carbon dioxide evolution from litter and soil. *Ecology* 50:922–24
208. Witkamp, M. 1971. Forest soil microflora and mineral cycling. See Ref. 42, pp. 413–24
209. Witkamp, M. 1974. Direct and indirect counts of fungi and bacteria as indexes of microbial mass and productivity. *Soil Sci.* 118:150–55
210. Witkamp, M., Frank, M. L. 1969. Evolution of CO_2 from litter, humus, and subsoil of a pine stand. *Pedobiologia* 9:358–65
211. Witkamp, M., Van Der Drift, J. 1961. Breakdown of forest litter in relation to environmental factors. *Plant Soil* 15:295–311
212. Woodwell, G. M., Dykeman, W. R. 1966. Respiration of a forest measured by carbon dioxide accumulation during temperature inversions. *Science* 154:1031–34
213. Woodwell, G. M., Marples, T. G. 1968. The influence of chronic gamma irradiation on production and decay of litter and humus in an oak-pine forest. *Ecology* 49:456–65
214. Woodwell, G. M., Pecan, E. V. 1973. *Carbon and the Biosphere, CONF-720510.* Springfield, Va: US AEC, Div. Tech. Info. 392 pp.
215. Yoda, K., Kira, T. 1969. Comparative ecological studies on three main types of forest vegetation in Thailand. V. Accumulation and turnover of soil organic matter, with notes on the altitudinal soil sequence on Khao (Mt.) Luang, peninsular Thailand. *Nature and Life in S. E. Asia* 6:83–109
216. Zaydel'man, F. R., Narokova, R. P. 1975. Genesis of brown, podzolic, and bog-podzolic soils on coarse-textured parent materials. *Sov. Soil Sci.* 1975:11–25

Ann. Rev. Ecol. Syst. 1977. 8:83–107
Copyright © 1977 by Annual Reviews Inc. All rights reserved.

CONTROL OF FOREST ❖4118
GROWTH AND DISTRIBUTION
ON WET TROPICAL MOUNTAINS:
with Special Reference
to Mineral Nutrition

P. J. Grubb

Botany School, University of Cambridge, Cambridge CB2 3EA, England

INTRODUCTION

On the wetter slopes of the highest tropical mountains, wherever man's depradations are not in evidence, a continuous forest cover may be found from sea level to an altitude of about 4000 m. With increase in altitude the forest becomes markedly changed in character. Most obviously it decreases in height from about 45 m to about 2 m, but there are also many changes in the representation of life forms and in the physiognomy of the plants. Very frequently these changes are relatively sharp over narrow altitudinal bands so that it is practicable to recognize four major formation types with relatively narrow ecotones between them. The formation types have been given many different names (8), but the system adopted in this review is set out in Table 1. This system has been shown to be workable in the Andes (40), in the Caribbean (42), and in Malesia, i.e. that area stretching from northern Malaya to the eastern end of New Guinea (41, 105). There is no tradition of separating lower montane and upper montane forests in Africa, but examination of the literature reveals that the trends shown in Table 1 are also found on that continent.

The several formations in a given area may be constituted from almost completely different suites of species, e.g. in Malaya (105), or show very considerable overlap in species composition, e.g. in New Guinea (41). In the latter case fully grown plants of particular species differ greatly in height in adjacent formations; some also differ significantly in leaf size (41). There is virtually always a marked fall in species richness with increase in altitude: In an exhaustive study of one area in New Guinea, 152 species of trees and shrubs were found in the lower montane rain forest (LMRF), 69 in the upper montane rain forest (UMRF), and 20 in the subalpine rain forest (SARF) (41). Dominance is typically shared by several to many species at all altitudes, although some exceptions are known (41).

83

Table 1 The formation types recognized in this review and the characters used in their definition (the most useful characters in italics)

Formation type	Lowland rain forest	Lower montane rain forest	Upper montane rain forest	Subalpine rain forest
Abbreviation	LRF	LMRF	UMRF	SARF
Height of forest[a]	25–45(–67) m	15–33(–45) m	1.5–18(–26) m	1.5–9(–15) m
Dominant leaf-size class[b] of trees and shrubs	*mesophyll*	*notophyll* or *mesophyll*	*microphyll*	*nanophyll*
Buttresses on trees	*usually frequent and large*	*infrequent or small or both*	usually none	none
Cauliflorous tree species	frequent	rare	none	none
Compound leaves on trees	*abundant*	occasional	few	none
Drip tips on leaves	*abundant*	frequent or occasional	few or none	none
Climbers	*thick-stemmed woody species frequent*	*thick-stemmed woody species usually none; other species often frequent*	usually very few	very few
Vascular epiphytes	frequent	*abundant*	frequent	occasional

[a] Emergents in parentheses.
[b] According to the systems of Raunkiaer (82) and Webb (104).

The altitudinal limits of the formation types vary greatly. They are higher on taller, more massive mountains than on smaller mountains and outlying ridges. This effect, which parallels an effect first recorded in the European Alps, is known as the *Massenerhebung* effect (83). Examples of it in all parts of the tropics have been summarized elsewhere (43). On the mainland of New Guinea the upper limit of LMRF can be as high as 3050 m and as low as 1800 m and all intermediates have been documented (41). On many small islands in the area the limit is at 700 m or lower (41, 105). The upper limits of LRF have been less closely studied, but appear to range from 1500 to 700 m; in extreme cases LRF abuts on UMRF and no LMRF is to be found (43, 105). The lower limit of SARF varies much less, e.g. only in the range 3650–3200 m in New Guinea (41).

Most montane and subalpine forests are characterized by a high incidence of low level cloud (fog). The frequency and duration of fog vary greatly within each formation type, but wherever the forest limits are markedly lower, these limits can be correlated with persistent cloud (43).

GROWTH OF TROPICAL MOUNTAIN FORESTS   85

Only 15,000 years ago, at the end of the last glaciation, many forest limits were at least 500 m lower, and there is evidence that some forests were different in kind (49).

In this review I deal with two basic sets of questions, which may be summarized as follows:

1. What environmental factors lead to the changes in forest form found with increased altitude, and how are the plants in the various forests adapted to these factors?
2. What environmental factors determine the altitudinal limits of any given forest type, and in particular what are the mechanisms involved in the *Massenerhebung* effect in the tropics?

It is necessary first to review the results of recent work on the changes with altitude in biomass, productivity, leaf life, leaf area index, leaf form, and disappearance of litter. Next I give a brief account of the parts played by factors other than mineral nutrition; the rest of the review is concerned with the involvement of mineral supply.

BIOMASS, PRODUCTIVITY, AND DECAY

Biomass and Productivity

Many difficulties are encountered when making estimates of any of the quantities discussed in this section; these difficulties are discussed elsewhere (22, 23, 105). The major trends to appear so far are as follows:

1. Biomass decreases less than stature on passing from LRF to UMRF (Tables 1 and 2).
2. Production of woody parts declines on passing from LRF to UMRF, but there are too few data for us to be sure of the magnitude of the effect; the best data for primary stands of LRF have been summarized elsewhere (51) and the rates are 3–6 t ha^{-1} yr^{-1} (dry matter), while the rate in LMRF in Puerto Rico (12, 71) was found to be only ~1.4 t ha^{-1} yr^{-1} and in UMRF (101) even less. This trend is supported by other data for increments in basal area in LRF, LMRF, and UMRF in the Philippines (13, 14). There is circumstantial evidence for extremely slow growth in subalpine forest (52).
3. Production of litter, particularly leaf litter, declines much less than biomass or, so far as we can tell, production of woody parts (Table 2).
4. The standing crop of leaves declines much less than the overall biomass (Table 2).
5. The mean leaf life lengthens only slightly, being about 12–14 months in the lowlands (83), 14–16 months in LMRF in New Guinea (22), and 14–18 months in UMRF in Jamaica (98). Values of 14–18 months probably reflect mixtures of species in the canopy, some living one year and others two years: In particular the leaves of montane conifers may commonly last at least two years; there is

Table 2 The best available estimates of total biomass, leaf biomass, leaf area index, production and disappearance of litter, and organic matter content in various types of rain forest [unless otherwise indicated, data from Table 10 in (22) and Table 8 in (23)]

Forest type, country and reference	Approximate altitude (m)	Above-ground biomass (t ha^{-1})	Below-ground biomass (t ha^{-1})	Leaf biomass (t ha^{-1})	Leaf area index (m^2 m^{-2})	Annual litter fall[a] (t ha^{-1})	Annual leaf litter fall (t ha^{-1})	Rate of disappearance of litter (% per year)	Organic matter in soil[b] (t ha^{-1})
Lowland rain forests and semideciduous forests									
Brazil	50	406[c]	67	9.3	—	7.3[d]	6.0[d]	—	248[e]
Ghana	150	233[f]	54	—	—	10.5	7.0	465	86
Ivory Coast	50–100	435[c,g]	49	7.9–8.7	10–12	8.3–13.4[h]	6.4–10.4[h]	330[i]–420[j]	100–170
Malaya	100	431[c]	—	7.8	6.9	—	7.5m	—	136
Panama	100	316[k]	11[l]	9.3	10.6	—	5.8–7.5m	—	—
Thailand	150–200	365[k]	32	7.8	10.7[n]	—	—	—	97–112
Trinidad	200	—	—	—	—	—	6.8	164	—
Lower montane rain forests									
India (9)	2100	—	—	—	—	5.6	3.9	—	—
Jamaica (98)	1500	(153)[o]	—	—	—	6.5	5.7	71	160
New Guinea	2500	310[c]	40	8.9	~5.5	7.4[i]–7.7[j]	6.5[i]–6.2[j]	104[i]–155[j]	1200
Puerto Rico	500	148–194[p]	72	7.9–8.1	6.4	—	4.8	94	146
Upper montane rain forests									
Jamaica (98)	1500								
mor ridge		216[c]	54	7.5	5.3	6.6	5.2	47	500
mull ridge		(219)[c,o]	—	(7.6)[o]	—	5.5	5.3	45	180
wet slope		(152)[c,o]	—	(5.3)[o]	—	5.5	4.4	43	60

[a]In some cases measured in same type of forest as used for biomass study but at some distance from it.
[b]All estimated as 2 times organic carbon content; sampling has been to variable depth and all estimates are underestimates, but probably not by > 10–20%.
[c]Old forests apparently little disturbed by man.
[d]From (54).
[e]This relatively high value is associated with the development of "moder" humus.
[f]Young forest, 50 (?) years old.
[g]Estimate revised by Edwards & Grubb (23).
[h]From (5).
[i]Plateau or ridge site.
[j]Valley site.
[k]Old secondary forest, less than 400 years old.
[l]Value admitted by authors to be a gross underestimate.
[m]From (60).
[n]Probably overestimated through overrepresentation of shade leaves in sample used to determine specific leaf area.
[o]Values in parentheses are only preliminary, based on regressions using an extensive sample of trees in mor ridge forest plus a few trees in mull ridge and gap forest.
[p]These figures span the range of published values; for details see Edwards & Grubb (23). All the values exclude tree ferns, epiphytes, and individuals of diameter < 2.5 cm, none of which would be likely to raise the highest estimate much above 200 t ha^{-1}.

evidence that the leaves of *Podocarpus urbanii* in Jamaica may persist for seven years (98). Leaves in the undergrowth persist longer than those in the upper canopy (50).
6. The leaf area index falls much more than the leaf standing crop (Table 2) because the leaves become thicker and more coriaceous. The mean specific leaf area declines from ~90–130 cm^2 g^{-1} in LRF to ~80 cm^2 g^{-1} in LMRF and ~70 cm^2 g^{-1} in UMRF [Table 2; (60)]. The mean values for trees (and shrubs exposed in the upper part of the canopy) are lower: 61 and 58 cm^2 g^{-1} respectively in LMRF in Puerto Rico and New Guinea (23), 56 cm^2 g^{-1} in mor-forming UMRF in Jamaica, and 57 and 45 cm^2 g^{-1} in UMRF and SARF in New Guinea (P. J. Grubb and E. A. A. Grubb, unpublished).

Tropical montane plants appear to resemble their temperate counterparts in investing relatively more of their production in leaves when total production falls [as at the cold and dry limits of temperate deciduous forest (24, 84)], but certainly differ markedly in their leaf form and leaf life. They almost certainly differ in having a lower ratio of total production to leaf production, even at favorable sites. Temperate deciduous and coniferous trees have been shown to have ratios in the range 2.7:1 to 6.3:1 (see 11), while the data for LMRF in Puerto Rico suggest a ratio of <1:1. The higher ratios found for temperate trees may partly reflect the fact that in the tropics relatively undisturbed stands have been studied, where production is severely limited by mutual competition between trees (19), while stands disturbed or planted by man have been most frequently studied in the temperate zone. However, the major explanations of the higher ratios in temperate plants are (*a*) the inexpensive nature of the leaves in deciduous species [they are thin and have a high specific leaf area—about 200 cm^2 g^{-1} (See 60)—but are at least as effective in photosynthesis on a unit area basis as evergreen species (see 59, 88)] and (*b*) the long leaf life (2–5 years) in the conifers. It is probably highly significant in this context that the great majority of the tallest and largest-girth trees in montane and subalpine forests in New Guinea are conifers with leaves that almost certainly last at least two years or more (*Dacrycarpus, Papuacedrus, Phyllocladus,* and *Podocarpus*), although it remains to be proved that they grow faster than the dicotyledons. It is not the case that fully grown dicotyledonous trees never have leaves lasting more than one year; many—perhaps most—tree species in warm temperate rain forests and mediterranean sclerophyll communities have leaves lasting two years [(81); P. J. Grubb, unpublished]. Production of most woody plants in tropical montane forests may be seen as severely limited by the expensive nature of their leaves and the short leaf life. We are bound to ask what advantages accrue from having thick, expensive leaves and from relatively short leaf life, although it is important to remember that high productivity is not necessarily the key to success (see below).

Decay

In rain forests disappearance of litter may be equated effectively with decay, which is therefore seen to be much slower in montane forests (Table 2). The rate of decay is not to be related simply to environmental factors such as temperature because it is strongly dependent on species, and the tough leaves of montane plants are almost certainly inherently slow to decay (22). At a given altitude decay is more rapid in valley sites than on ridges or plateaus (Table 2), but this difference is not controlled by water relations. Edwards (22) and Tanner (98) have shown that samples of a given type of litter decay at the same rate in the two situations. The greater rate for naturally fallen litter in valley sites must reflect either a difference in the tree species present (certainly true in Jamaica) or chemical differences in the leaves of species which grow in the two situations (possibly the explanation in New Guinea).

There is a decline with altitude in the rate of "decay," i.e. mineralization, of humus in the soil; and the ratio of organic matter in the soil to that in the plants rises from between 0.2 and 0.5:1 in LRF, to between 0.7 and 4.0:1 in LMRF and UMRF [Table 2; (see also 23)]. This trend is complicated by certain soil minerals, e.g. allophane, which complex the humus and prevent its mineralization (21, 23).

LEAF STRUCTURE

I have presented elsewhere (36) details of the changes found on passing from LRF to SARF in New Guinea, together with data for a few related forests (Table 3). It is plain that the increase in mean thickness is accounted for chiefly by increases in the mean thickness of the palisade layer and in the incidence of hypodermis. I have called the typical leaves of tropical montane and subalpine forests "pachyphylls" (36), defining them as normally having (*a*) a lamina >300 μm thick, (*b*) a well-developed palisade ($>\frac{1}{2}$ but $<\frac{3}{4}$ of the mesophyll thickness), (*c*) the outer walls of the epidermides markedly thickened, and (*d*) a hypodermis frequently present. Such leaves generally differ from the classical sclerophylls of mediterranean climates in their greater thickness, their greater incidence of hypodermis, and the lesser relative thickness of the palisade (Table 3). Although it is not evident from Table 3, they also differ in possessing a spongy mesophyll that contains relatively large cavities rather than being compact and often palisadiform. There is no consistent difference in stomatal density; the highest stomatal densities tend to occur in tall trees of lowland rain forest rather than in the shorter trees of either montane forests or dry climate forests (38).

PARTS PLAYED BY FACTORS OTHER THAN MINERAL NUTRITION

I have reviewed this issue in some detail elsewhere (36) and I will reiterate only the main points here.

Limitations on the Distribution of Forest Formations

It appears that some function of temperature sets the uppermost limit for each formation or for any individual species. The first order explanation of the *Massenerhebung* effect is through the incidence of cloud. A lowering of forest limits on outlying ridges almost always accompanies a lowering of the normal level of cloud formation (43). I have suggested (36) that the most important effect of cloud is that

GROWTH OF TROPICAL MOUNTAIN FORESTS   89

Table 3   A summary of selected features of leaf structure in the trees and shrubs of lowland (L), lower montane (LM), upper montane (UM), and subalpine (SA) rain forests in New Guinea (NG), Malaya (MAL), and Puerto Rico (PR), and in mediterranean-climate sclerophylls (MCS) of California (CAL) and Europe (EUR) [data chiefly from Table 7 of (36); the rest are original]

	NG				MAL	PR	CAL	EUR
	L	LM	UM	SA	UM	UM	MCS	MCS
Number of species studied	40	55	61	19	38	24	25	17
Estimated mean lamina thickness, rounded off (μm)	230	300	380	430	(490)	380	290	330
Percent with hypodermis	25	46	56	61	32	46	12	6
Percent with multiseriate palisade	51	68	70	79	73	71	98	88
Estimated ratio of non-palisade to palisade (thickness)	2.5	–	1.9	1.6	–	2.0	1.0	1.5
Estimated mean thickness of outer walls of upper epidermis (μm)	3.8	5.5	7.4	10.5	(10.2)	–	–	9.3
The same, lower epidermis	2.2	2.6	3.5	4.6	(6.6)	–	–	7.3
Mean stomatal density (number per mm^2)[a]	408	255	–	271	–	406[b]	–	305[c]

[a] In >95% of species stomata occur only on underside, but densities have been calculated for one side including any stomata on the upper surface.
[b] The mean for 20 species (excluding 1 *Eugenia* and 3 Melastomaceae) is 238.
[c] The mean for 35 sclerophylls in South Australia is 250.

it prevents bright sunlight from raising the leaf temperature toward the optimal level, which is above air temperature. The amount of radiation available for photosynthesis is also reduced by the cloud, and there is probably an interaction between temperature and light factors unfavorable to growth.

The lower limit for each formation or species, as on mountains in the temperate zone (108), is probably determined primarily by competition (35, 41) rather than by excessively high temperatures, although temperatures may set the ultimate limits upon extremely high altitude species.

The lower water vapor pressure deficit under cloudy conditions may be responsible for depressing the growth of some species, but not through any effect on the uptake of mineral nutrients (see below). The excessive wetness of the soil in the most persistently clouded conditions may be extremely important in depressing some forest limits (41, 101), but experimental evidence is lacking.

The lessened water stress under cloudy conditions is clearly what assures the greater abundance of vascular and nonvascular epiphytes in LMRF as compared to LRF (43), and the same factor, plus particularly the abundant direct supply of liquid water droplets, accounts for the greater development of epiphytic bryophytes in some upper montane and lower montane forests than in others (36).

It appears that favorable water relations are not a major factor in confining nonepiphytic species to LMRF or UMRF. Simple experiments on cut shoots of trees and shrubs have shown that under standard conditions the mean percentage water loss in 24 hours is not greater, but a little less, in the more xeromorphic species of SARF and UMRF than in the less xeromorphic species of LMRF and LRF (36). The ability of the protoplasts to resist desiccation has been investigated much less fully but seems to be similar in the few species of LRF and UMRF tested in Malaya (R. A. Buckley and R. T. Corlett, unpublished). Favorable water relations may be important in confining to montane forests certain delicate, shade-tolerant herbs (36) and possibly some of the montane species that are confined to waterlogged sites when they occur in the lowlands (105). Wind does not affect forest limits on most tropical mountains.

Limitations on Forest Growth

The decrease in stature, biomass, productivity, number of life forms, and number of species can be attributed primarily to the decrease in both temperature and amount of photosynthetically active radiation. The lapse rate for mean temperature is 0.5–0.6°C per 100 m on most tropical mountains and the importance of leaf-warming effects in direct sunshine has already been emphasized. The mean number of hours of bright sunshine per day falls from between 5 and 7 in the lowlands (17, 83) to between 3 and 6 in LMRF (41, 44), and to about 3 in the UMRF-SARF (48). Baynton (3) showed that the amount of radiation reaching UMRF in Puerto Rico was only 60% of that received in the lowlands nearby.

The low air temperatures and infrequency of direct radiation affect not only leaf metabolism but the development of flowers (66) and probably of fruits too. They also lead to a reduction in most forms of animal life. There is no evidence for a decrease in insect damage to leaves on passing from LRF to LMRF [levels of 11% being reported for LRF in Panama (60) and between 7 and 10% for LMRF in Puerto Rico and New Guinea (72, 22)], but casual observations suggest a marked decrease in SARF (41). There is some evidence for a decrease in dependence on insects for pollination and on birds for dispersal on passing from LMRF to SARF in New Guinea (41).

There is no reason why excessive soil moisture should reduce forest growth. Many of the most productive communities in the world occur on waterlogged ground (36).

Wind effects are rare on equatorial mountains. They occur somewhat more frequently in the Trade Wind belt, but even there wind-clipping is quite restricted in occurrence (41, 42).

Leaf Life

It is suggested here that the primary determinant of leaf life is the high humidity, which facilitates invasion by fungi and bacteria, and by epiphyllous lichens and bryophytes. Certainly the black hyphae of fungi are prominent on the mature leaves of many species in UMRF and SARF. Small bryophytes and lichens are common on twigs and seem frequently to be in an early stage of colonization of the leaf lamina when the leaves fall. Epiphyllous hepatics not only shade the leaves but, by puncturing the cuticle, destroy the plant's protection against water loss under the drying conditions (6) which occur not infrequently during sunny spells in montane and subalpine conditions. Fungi, too, may interfere with control of water loss, perhaps by upsetting stomatal control and by increasing cuticular permeability.

If infection by epiphyllous organisms is seen as a major hazard to life under montane conditions and if drip tips are supposed to facilitate runoff of surface water and thus to minimize such invasion (83), why then are drip tips less common in montane forests than in lowland forests? The answer seems to be that drip tips are effective under conditions of alternating heavy storms and sunny spells, but ineffective where the moisture is supplied by persistent fog and drizzle. The very thick outer walls of pachyphylls (Table 2) may well be a major adaptation to minimize penetration of the leaf by epiphylls, particularly fungi.

Leaf Form

It seems unlikely that the xeromorphism represents an adaptation to drought stress. Montane and subalpine forests are subjected to severe drought at most only once a decade, and it seems that most mature trees on reasonably deep soils are not seriously affected at such times. This was the impression gained during the notable drought in New Guinea in 1972 (P. F. Stevens, personal communication) and such an impression parallels experience with many trees of the temperate deciduous forest in England during the great drought of 1976. As in England in 1976, so in New Guinea in 1972, the greatest hazard to forest was man-borne fire. It must also be emphasized that the sclerophylls of mediterranean climates, which certainly are adapted to withstand drought, are consistently different in structure from pachyphylls; they rarely have a hypodermis, and they always have a compact, small-celled mesophyll. Cut shoots of European sclerophylls lose less water in 24 hours under standard drying conditions than do pachyphylls (P. J. Grubb, unpublished). Sclerophylls also have protoplasts that can tolerate much more severe drying out than those of pachyphylls. European sclerophylls can recover to 90% of the original saturated water content after drying out to 40–70%, whereas pachyphylls from UMRF in Malaya and Jamaica can only tolerate drying to 65–80% [(73); R. A. Buckley, R. T. Corlett, P. J. Grubb, and E. V. J. Tanner, unpublished].

Parkhurst & Loucks (77) have argued that if the ratio of CO_2 taken up to H_2O lost is the most important criterion for optimal leaf size, then we should expect the smaller leaves that are found in the cooler, damper forests. For plants growing in an environment that is generally wet, it seems remarkable that the ratio between CO_2 uptake and H_2O loss should matter, and yet the internal features of the pachyphyll reviewed above are likely to maximize this ratio (36). The thickness of the lamina will tend to increase the ratio of internal to external surface, and so reduce the ratio of the carboxylation resistance [sensu Monteith (64)] to the diffusive resistances for uptake of CO_2. Other things being equal, this change in the ratio of resistances must increase the ratio of CO_2 in to H_2O out (cf 67). The increased thickness of the palisade may also fit into this scheme (palisade cells are generally smaller than spongy mesophyll cells and therefore offer a greater area for CO_2 absorption per unit volume of cells), but the functional significance of the division between palisade and spongy mesophyll is in general obscure (37). The hypodermis may act as a water reservoir to be exploited by the mesophyll when it comes under moderate stress.

It is envisaged that, if the $CO_2:H_2O$ ratio is maximized in the way suggested, the rate of transpiration per unit area under standard well-lit conditions should be about the same in pachyphylls as in other rain forest leaves (actually a little higher if the stomatal densities are the same because of the smaller leaf area and thinner layer of still air), while the rate of photosynthesis per unit area should be higher in the pachyphylls. There is some evidence that the rates of transpiration are indeed similar in the two leaf types (103). However, the postulated difference in photosynthetic rate may well not materialize. In mediterranean sclerophylls, which we might reasonably expect to maximize the $CO_2:H_2O$ ratio, the maximum photosynthetic rates tend to be lower than in tropical lowland forest trees (59). This latter finding is in accord with the mounting evidence that plants of many unfavorable habitats have inherently slow growth rates (34)—surviving stress is more important than productivity. If pachyphylls turn out to have low photosynthetic rates and low $CO_2:H_2O$ ratios under standard conditions, some new rationale of their structure is called for.

Wardle (102) suggested that the xeromorphic features of many subalpine plants in New Zealand could be explained in terms of their causing leaf temperatures to rise during the periodic sunny spells. The chief difficulty in applying this idea to the tropics is that most subalpine species have tiny leaves. These are likely to have thin still-air layers, and to lose heat readily by conduction-convection. The extent to which the effect of small leaf size is offset by close packing of leaves is in need of experimental investigation, especially since the dark color and compact crown structure of some tiny-leaved species are likely to encourage tissue warming. Wardle's hypothesis almost certainly applies to the minority type in subalpine forest, i.e. to the hairy mesophyll or macrophyll [sensu Raunkiaer (82)] exemplified by many *Senecio* species in New Guinea, Africa, and South America.

It is unlikely that the xeromorphy represents primarily an adaptation to reduce insect damage. The pachyphylls are soft when young, which is when they are most susceptible to insect damage, and many of the toughest leaves in UMRF in New Guinea were found to have lost large portions of lamina at this stage (41).

THE ROLE OF MINERAL NUTRIENT SUPPLY

Reasons for Considering Mineral Supply

The reasons that have been suggested may be summarized as follows: (*a*) The reduced rate of transpiration in cloudy conditions is thought by some to reduce the uptake of mineral ions (60, 70, 103). (*b*) The xeromorphy of the leaves has been interpreted as a response to oligotrophy (43, 83). (*c*) The crooked and gnarled growth of many montane trees has been likened to the type of growth seen on infertile lowland soils and in bonsai plants deliberately grown in nutrient-poor soil (60). (*d*) The slower rate of decay of litter and slower rate of mineralization of humus have been supposed to leave much of the nitrogen and phosphorus in the soil-plant system locked up in unavailable forms (23, 35). (*e*) Forest limits and forest growth are often most depressed on sites we would expect to be leached of mobile nutrients, i.e. ridges and especially knolls on ridges (95). (*f*) The most dwarfed forests are often found on mor or peat of pH ≤ 3.5, which we should expect to be infertile for most plants (42). (*g*) Some of the plants frequent in UMRF on peat in Malaya and Borneo are widespread on highly acidic infertile soils in the lowlands (105). (*h*) Insectivorous plants (*Nepenthes*) are frequently abundant in UMRF on peat in Malesia (105). Reasons *c* through *g* merge with each other and are considered later along with reason *b*. That reason *a* may be dismissed is shown in the following section.

Transpiration Rate and Mineral Supply

There is much evidence that may be marshalled against Odum's theory (70) that reduction in the amount of transpiration is likely to depress the supply of mineral nutrients.

First, there are the results of many experiments on the uptake of ions by whole plants from simple salt solutions as a function of the amount of flow of water through the plants. Although the results have seemed hopelessly contradictory to some, a clear picture emerges if the concentrations of the salts used are noted. Generally, if the ions are at a high concentration in the external solution (say 10–100 meq per liter) water flow has a marked effect on throughput of ions; sometimes there is an almost linear relation. In contrast, if the ions are at a low concentration (10–100 μeq per liter) there is a negligible effect of water flow on the amount of ion throughput. This important effect of concentration was shown explicitly by Kylin & Hylmö (57) and by Russell & Shorrocks (87). There can be little doubt that for most ions in most natural soils the concentrations in the soil solution are in the "low" range (<100 μeq per liter). Levels of Ca^{2+}, Mg^{2+}, NO_3^- and SO_4^{2-} can exceed 1 meq per liter in the soil solution in fertilized, alkaline soils (2), but even in such soils K^+ and $H_2PO_4^-$ are at low levels. No measurements of soil solution levels are available for montane forest soils, but it is virtually certain that the levels of K^+ and $H_2PO_4^-$ are always in the low range. The few data available for concentrations of NO_3^- and NH_4^+ in montane forest soils (99) show them to be low, as we would expect by analogy with the better-known soils of forests in the temperate zone. Thus we should expect that the key macronutrients (N, P, and K) will all be at low levels in natural soils and that, by analogy with laboratory results, transpiration will have a negligible effect on ion uptake by the whole plant. So long as there is a little transpiration, which there must be in all montane forests most of the time (29), ions will move to the aerial parts. The general lack of effect of transpiration is a result, of course, of the fact that the primary limiting factor is the rate of metabolically controlled uptake in the root.

So far we have considered uptake in an uninfected plant; but most, and possibly all, rain forest trees are mycorrhizal. In plants growing on poor soils and having either ecto- or endomycorrhizae, uptake of phosphorus is not only mediated but greatly facilitated by the mycorrhiza (46). There is no reason to suppose that passage of phosphorus through the mycorrhizal root system is increased by an increase in transpiration. There are some grounds for believing that in certain soils uptake of cations is facilitated by mycorrhizae (46) and that nitrogen-fixing bacteria increase in number and activity in the perimycorrhizal region (92). Whether they have this effect in montane forest soils is unknown.

Diffusion up to the root, a most important limiting step in the movement of ions from soil to shoots, is effective even earlier than the primary uptake by phanerogam root cells or mycorrhizal fungal cells. Nye (69) and Tinker (100) have considered this problem and have produced evidence that under summer conditions in southern England a strong increase in the rate of mass flow of water through the plant-soil system has a very small effect ($<10\%$) on the rate of movement of potassium up to the root surface. Potassium movement, in other words, is largely by diffusion and not by mass flow. Two key macronutrients, K^+ and especially $H_2PO_4^-$, are known to diffuse only slowly through soils generally, and this would certainly be true in highly organic montane forest soils with abundant cation-exchange sites and a high potential for phosphate fixation.

A process acting even earlier in overall uptake is the microbial breakdown of organic matter which releases N and P in forms that the plant can take up (46). Possibly mycorrhizal fungi, and possibly wall-bound phosphatases of the root itself (109), are directly involved in this. Certainly free-living bacteria and fungi are involved. Whatever the agency, the rate of breakdown could well be the major limiting factor in a cool, very wet soil where microbial metabolism is reduced.

Within the shoot only the primary upward supply is affected by transpiration; this supply occurs through the xylem and is received only by the organs that transpire [see (96), plate II]. However, the actively growing parts (shoot apices, young leaves, etc) do not transpire very much, and their supply is effectively limited to the secondary redistribution carried out by the phloem. A proportion of the nutrients that enter the actively transpiring organs is redistributed to the younger organs in the phloem (7). There is no evidence that this redistribution is increased by increased transpiration in the very young organs.

The various nutrient levels in the several parts of the plant are certainly under a subtle system of controls, at least some of them hormonal (18, 65, 78, 79). The lack of any simple net effect of transpiration is nicely shown by the remarkably constant nitrogen or phosphorus concentration (% dry weight basis) found within several species for sun leaves (which transpire a lot) and shade leaves (which transpire little) and by the generally higher levels of potassium in shade leaves (which transpire less but have thinner walls). For details see (1, 10, 15, 89–91).

In summary, it is most unlikely that the rate of transpiration will significantly affect the rate of supply of nutrients to young growing shoots because (a) the soil solution concentrations are likely to be very low, so that active processes of uptake are most likely to be limiting within the plant (including its mycorrhizae); (b) the rate of diffusion up to the root, or the rate of release from insoluble material in the soil, is likely to limit as much as the rate of active uptake; (c) the direct supply to young growing regions is via the phloem and is not increased by transpiration; and (d) some transpiration must occur in the most persistently clouded forests, and just a very little will suffice for the primary upward movement in the xylem.

GROWTH OF TROPICAL MOUNTAIN FORESTS   95

Odum (70) quotes Winneberger (107) as finding poor growth of sunflowers at very high humidities, and he attributes this to lack of throughput of nutrients. It is true that many crop plants seem to grow less well at very high humidities, e.g. broad beans, dwarf beans, peas, strawberries, and many others (76). Useful summaries of the work done on this topic have been published (28, 56). It is important to realize, however, that the reduction in growth at high humidities may have nothing to do with the rate of supply of mineral nutrients to young growing organs. An alternative explanation concerns a reduction in the rate of photosynthesis in fully turgid leaves compared with slightly desiccated leaves. This remarkable phenomenon was noted by Larcher (58) in the Mediterranean xerophyte *Quercus ilex,* and was later reported for cotton (75) and soybeans (4). It may well be that high humidity has an inhibitory effect on the photosynthesis (and thereby on the growth) of lowland but not montane plants. However, no feature evolved by the plant as an adaptation that maximizes transpiration could significantly reduce this inhibition in foggy periods. In fact it is most unlikely that the xeromorphic features of the leaves of UMRF trees (other than small size) actually facilitate transpiration, as supposed by Odum (70). As pointed out earlier, there is no evidence for higher stomatal densities in trees and shrubs of upper montane as opposed to lowland forest; in fact the position tends to be the reverse (Table 3).

If mineral uptake is inhibited by cloudy conditions, it is much more likely to be due to the low light intensities and reduced supply of metabolites to the roots (47) than to the slow rate of transpiration.

Kinds of Evidence for Mineral Shortage

EFFECTS OF ADDITIONS OF FERTILIZER   If a factor is limiting a process, alleviation of that factor must increase the rate of the process. Only tests on the rate of growth with and without added N, P, K, etc can ultimately determine which mineral deficiencies, if any, are limiting growth. So far no appropriate experiments have been reported for tropical montane forests.

Tanner (98) added N, P, and K to mull-forming UMRF on a ridge site in Jamaica and monitored the effects on shoot extension in saplings. He found no effect in a year, but perhaps we should not expect a result until the second year—there is often such a delay in experiments with trees in plantations.

A less satisfactory approach is to grow test species on the forest soils in pots and add various fertilizers. Experiments with *Amaranthus viridis* and *Brassica pekinensis* on soil from UMRF and LMRF in Jamaica (99) and with *Holcus lanatus* on LMRF soils in Tanzania (J. A. C. Smith, D. N. S. Allen, and A. F. U. Powell, unpublished) have shown massive improvement with addition of P, but no effect of N or K. Whether young tree seedlings, established naturally in the forest, suffer the same phosphorus stress is not known. If they do, it is possible that they develop more effective mycorrhizae with age and phosphorus becomes relatively less limiting (26, 39). An alternative possibility is that pot experiments may overestimate the importance of $H_2PO_4^-$ because that ion is so immobile in the soil while NO_3^- is very mobile. A seedling in a pot may have available to it all the NO_3^- mineralized in the pot but only the $H_2PO_4^-$ from a tiny volume surrounding the roots (R. J. Allanach, unpublished). In nature the soil is permeated by roots and the volume of soil supplying NO_3^- to a seedling is much smaller. In experiments with various seedlings in pots of acidic montane soils in Britain, additions of a nitrogen source alone have been found to promote growth considerably (62). Therefore the results obtained in Jamaica and Tanzania do suggest that nitrogen may be relatively less limiting in the montane soils of the tropics than in those of the temperate zones. This is supported by Tanner's finding an excessively high concentration of N in seedlings of *Brassica pekinensis* grown on UMRF soil in Jamaica, compared with larger, healthier plants grown on fertile garden soil (99). However, there is evidence from nutrient-cycling studies that N is in short supply to mature trees (see below).

In all such studies it should be remembered that different mineral elements may limit the growth of different species on a given soil. This has been nicely demonstrated for crops grown on highly acidic peat in lowland Malaya [summarized in (42)].

SOIL ANALYSES Many conventional soil analyses are now available for tropical montane soils, but they tell us very little about the availability of mineral nutrients to plants. Very few incubations to study release of nitrogen have been reported (99). What is needed is a combination of soil analyses, soil incubations, bioassays, and fertilizer treatments, with foliar analyses and studies on mineral cycling. In the context of tropical montane forests the only attempt at such a combined study has evidently been that of Tanner (98, 99).

FOLIAR ANALYSES For at least 40 years foliar analysis has been a routine part of studies on mineral nutrition of all sorts of plants in the temperate zone and for crops in the tropics (31, 80, 94, 97). The operative principle is that when the growth of a plant has been reduced by the paucity of a given mineral nutrient, the plant's tissues show a lowered concentration of that nutrient. Various difficulties are encountered in diagnostic foliar analysis—especially for very mobile elements (see below), and when there is an interaction between elements, e.g. when shortage of P inhibits uptake of K (32).

Foliar analysis concentrates on particular species at different sites, and the parts used for analysis are standardized very closely. The work done on rubber *(Hevea brasiliensis)* in Malaya illustrates the technique very well (45, 89–91). The chief variable is leaf age. Work with *Hevea, Theobroma cacao* (1), and the evergreen but nontropical *Citrus aurantium* (15, 55), *Nothofagus truncata* (63), and *Quercus ilex* (81) shows (*a*) that after the completion of the initial leaf expansion there are generally only rather small changes in the concentrations of P, K, Mg, Cu, and Zn; (*b*) that the level of N often decreases considerably; and (*c*) that the levels of Ca, Mn, Fe, Na, and Si almost always increase appreciably with age. The next most important variable is shading: Shade leaves and sun leaves generally contain similar concentrations of N, P, S, Fe, and Cu; but shade leaves usually have appreciably higher concentrations of K, Mg, Zn, and Mo, and lower concentrations of Mn and

B (1, 15, 89-91). The published results show no clear trend for Ca. A third variable is the leaf's height on the tree; but it appears from studies on *Hevea* (90) and *Citrus* (15, 55) that there are very small differences in concentrations in sun-leaves at different heights in the crown.

Normal foliar analysis has been attempted for species growing in a series of four montane forests in Jamaica (42, 99) and has yielded consistent results, which are mentioned later.

In an attempt to provide some standards by which foliar analyses of rain forest trees may be judged in various biogeographic regions, I have amassed data for numerous species in various situations. These are summarized in Table 4, together with some relevant data from other authors. There is a general decline in the concentration of N on passing from LRF to UMRF and SARF. As argued in another context (61), such a decline almost certainly reflects an increasing ratio of cell wall material (low in N) to cytoplasm (rich in N). The dependence of foliar mineral concentrations on leaf anatomy means that we can conclude nothing directly regarding the N supply from the results in Table 4, and that we must consider adjusting the results for other elements, e.g. by studying the P/N and K/N ratios as a means of estimating the availability of P and K per unit cytoplasm. The P/N and K/N ratios are indeed suggestive. The soils of the LRF in New Britain with high values (Table 4) are some of the best in Papua New Guinea for growth of coconuts and cocoa, while those of the LRF in the Brazilian Amazon with low values are decidedly infertile for most crops. The P/N ratios are commonly low in UMRF, particularly in the forests on peat in Malaya, and especially so in the most dwarfed forest studied (the second in Table 4), that on Gunong Ulu Kali, where the canopy was only 1.5–2.0 m tall. The same range of K/N values (0.3 to >0.8) is to be found within each formation type and there is no regular association of a low ratio with forests of low stature. This situation accords with Edwards' finding (21) that although in LMRF on a 45° slope in New Guinea much less K was cycling in the litter and throughfall than at ridge-top and valley sites nearby, the slope forests did not appear to differ from the others in stature. In Jamaica the much lower K/N ratio in mor-forming forest than in mull-forming forest is associated with a marked decline in stature. (This ratio is supported by normal foliar analyses of several species and by much smaller amounts of K cycling in the litter and throughfall.) Tanner (99), however, has tentatively interpreted the K paucity as a secondary effect, regarding the extreme acidity of the soil (pH 2.8–3.5) as the primary factor inhibiting growth. The apparent paucity of K in markedly dwarfed UMRF on muck soil in Puerto Rico, suggested by low K/N values (Table 4), is supported by the very few analyses available for the same species there and in LMRF nearby (36). The pH is ~4.5 and not likely to be harmful as such; in the absence of data on the cycling of N and P (see below) it is not possible to assess the likely importance to the plants of K shortage. The high foliar values of Fe may indicate a potential toxicity of Fe^{2+} in the very wet, partly anaerobic soil.

Inspection of the mean foliar concentrations for different forests in Table 4 (and of data for individual species not given there) shows that the nature of the soil's

Table 4 Mean foliar concentrations of mineral elements in relatively well-lit leaves[a] of trees and shrubs in various tropical rain forests; values as percent dry weight for N to Mn, ppm for Fe to Cu [data mostly original, but many preliminary values are given in (36)]

Country (ref.)	Soil parent material (ref.)	Humus type	Number of trees or shrubs	Number of species	N	P	K	Ca	Mg	Mn	Fe	Zn	Cu	P/N	K/N
Lowland rain forest															
Brazil (53)	sandy alluvia	moder	?	?	1.84	0.066	0.33	0.21	0.16	–	–	–	–	0.036	0.18
Malaya	granite	mull	30	10	1.55	0.071	1.00	0.58	0.18	0.024	33	15	5.5	0.046	0.65
New Britain	volcanic ash	mull	42	14	2.08	0.148	1.67	2.04	0.30	0.005	83	48	5.3	0.071	0.80
Lower montane rain forest															
Jamaica (99)	mudstone/sandstone	mull	16	9	1.76	0.100	1.49	1.26	0.39	0.029	–	–	–	0.057	0.85
Malaya	granite	mull	14	14	1.38	0.067	0.77	0.21	0.20	0.020	49	20	5.5	0.049	0.56
New Guinea	gabbritic alluvium	mull	69	23	1.52	0.090	1.11	0.95	0.25	0.022	66	24	4.9	0.059	0.73
	gabbritic alluvium (21)	mull	12	12	1.32	0.083	0.91	1.19	0.29	–	–	–	–	0.063	0.69
Puerto Rico	andesite	mull	75	28	1.62	0.078	1.00	0.98	0.36	–	–	–	–	0.048	0.62
Upper montane rain forest															
Australia	granite	podzolic soil	26	26	1.11	0.048	0.62	0.65	0.36	0.02	–	–	–	0.043	0.56
Jamaica (99)	"volcanics"	mull	32	14	1.61	0.074	1.23	0.93	0.43	0.024	–	–	–	0.046	0.77
	mudstone/sandstone	mor	26	13	1.05	0.054	0.55	0.62	0.33	0.018	–	–	–	0.051	0.53
	mudstone/sandstone (42)	mor	22	11	1.12	0.048	0.35	0.64	0.28	0.016	85	27	4.2	0.043	0.31
Malaya	granite	peat	30	10	1.00	0.042	0.59	0.22	0.19	0.011	43	17	~3	0.042	0.59
	(3 different mountains)	peat	39	13	1.15	0.030	0.66	0.35	0.22	0.021	34	13	~3	0.026	0.57
		peat	27	9	0.81	0.027	0.66	0.48	0.20	0.025	30	13	3.5	0.035	0.82
New Guinea	basalt	alpine humus soil	30	15	1.42	0.080	1.01	1.09	0.28	0.051	–	–	–	0.056	0.71
Puerto Rico	andesite	muck	40	15	0.97	0.053	0.39	0.76	0.19	0.019	138	27	6.6	0.055	0.40
Subalpine rain forest															
New Guinea	basalt	alpine humus soil	28	14	0.97	0.054	0.64	0.97	0.13	0.205	–	–	–	0.056	0.66

[a] The values for the Brazilian LRF and Puerto Rican LMRF and the second set of analyses for the New Guinean LMRF are overall values for leaves of all ages.

GROWTH OF TROPICAL MOUNTAIN FORESTS   99

parent materials and the selective uptake system in the plant's roots have more effect than leaf anatomy on the concentrations of Ca, Mg, Mn, Fe, and Zn. Similarly, proximity to the sea has an enormous effect on the level of Na (not shown in Table 4). Ca levels can be higher in leaves of persistently clouded UMRF on mor and peat of pH <3.5, or on acidic muck, than in leaves at some LRF and LMRF sites. This is further evidence against the notion that uptake of bases is depressed in persistently clouded forest (cf 70, 103).

The foliar concentration of Cu does fall more or less in parallel with that of N and is very low in the upper montane forests on peat in Malaya. Copper shortage is known to be important for some crops on acidic peat in lowland Malaya and Indonesia [references in (42)], and may well be important in UMRF, especially for potential invaders from LMRF. Very few data are available for S (36), but they suggest there is no shortage in UMRF on peat in Malaya. No data are available on B or Mo levels in montane forest plants; B can be limiting for some crops on peat (42).

In conclusion, foliar analyses suggest (a) that the P supply varies appreciably within each formation type but may be particularly deficient in UMRF on peat; (b) that the supplies of K and Ca vary even more within each formation type; (c) that the supplies of Mg, Mn, Fe, and Zn vary less than that of Ca and are unlikely to limit growth; and (d) that the Cu supply varies even less but may be deficient in UMRF on peat. All these suggestions assume that foliar levels are more dependent on the soil supply than on the activities of the selective uptake systems in the roots. The finding regarding P gives support to f through h in the section above cataloging the reasons for considering mineral supply.

The analyses for N have an indirect value in offering some support for the traditional interpretation of oligotrophic xeromorphism—that xeromorphic plants can produce more per unit N taken up. If the soils, on the average, supply less N on passing from LRF to UMRF (d in our catalog of reasons above), then xeromorphic plants appear to be preadapted to their situation. When there is a successive fall in the mean foliar values of N (and P and K) on passing from gully forest to mull-forming forest on a ridge and then to mor-forming forest on knolls, as there is in Jamaica (Table 4), the idea of preadaptation (i.e. that only those plants with successively lower foliar levels of macronutrients can survive) seems plausible, and reasons e through f in our catalog above appear justified. However, the relevance of this idea seems to be ruled out by the results of the normal analysis of single species across this series of forests. These data provide evidence not of any fall in the supply of N or P, but only of the fall in K (42, 99). Nevertheless one is left with a nagging feeling that preadaptation may be relevant, and there are many examples in the Far East to provide support. The dwarfed UMRF forests there are often dominated by Ericaceae and Myrtaceae, with Theaceae quite common. Plants of these families tend to have low foliar levels of N, P, and K wherever they occur. Some families abundant in LMRF, but usually sparsely represented on peat, almost always have high foliar levels of N and P, e.g. the Lauraceae.

If the xeromorphism of UMRF plants has anything to do with oligotrophy, it is unlikely that the functional significance will prove to be the same as in the north temperate zone. There the prime adaptation to nutrient stress is seen to be the longevity of the leaves, and therefore the greater photosynthetic production per N taken up (93). In the temperate zone the xeromorphic characteristics provide a guard against water stress during the winter.

ANALYSES OF THE WHOLE BIOMASS   Data are available for only a few forests (Table 5). Between them the data for the Brazilian and Panamanian forests span the whole range of lowland soils, from highly acidic podzolized sand to calcareous margalitic cracking clay. Between them there is an order of magnitude difference in the amounts of Ca in circulation. The soil of the Ghanaian forest is intermediate in texture and pH (5.2). There is a very good parallel between the foliar levels of the five elements listed and the concentrations of these elements in the whole aboveground biomass. Concentration rankings agree perfectly for N and K and show only one reversal for P, Ca, and Mg. These results greatly strengthen our confidence in the value of the foliar analyses in Table 4.

The concentration of N in the whole biomass is consistently lower in the montane forests and the concentrations of P, K, and Ca are lower than in the two lowland forests on moderately fertile soils. Clearly the data are consistent with the idea that montane plants are adapted to poor N and P supply (reason d in our catalog above), but analyses from many more forests are needed. Since so much of the biomass in primary forest consists of trunks and branches, we are chiefly concerned with the concentrations of N and P in these parts. Is there in montane plants a more effective redistribution of N and P out of the wood and bark as these tissues age?

MINERAL CYCLING STUDIES   The chief question to be asked is whether the amounts of mineral nutrients in the "external cycle," i.e. in litter fall and throughfall, are relatively smaller in montane forests than those in the "internal cycle," i.e. those withdrawn into the wood and bark from the leaves before these fall. In order to make an accurate estimate of the amounts withdrawn from senescent leaves it

Table 5 The weights of the above-ground standing crops of living plants (dry t ha^{-1}) and the concentrations of mineral elements (percent dry weight) in the leaves (L) and the whole above-ground standing crops (W) in six rain forests (original data for New Guinea)

Country (Ref.)	Above-ground biomass	N		P		K		Ca		Mg	
		L	W	L	W	L	W	L	W	L	W
Upper montane rain forest											
Jamaica mor forming (98)	263[a]	1.05	0.20	0.054	0.012	0.55	0.12	0.62	0.20	0.33	0.074
Lower montane rain forests											
New Guinea	310	1.32	0.24	0.083	0.015	0.91	0.22	1.19	0.23	0.29	0.062
Puerto Rico (74)	197	1.62	0.41	0.078	0.022	1.00	0.26	0.98	0.45	0.36	0.17
Lowland rain forests											
Brazil (53)	406	1.84	0.60	0.066	0.015	0.33	0.11	0.21	0.10	0.16	0.05
Ghana (33)	233	–	0.62	–	0.040	–	0.31	–	0.79	–	0.12
Panama (30)	316	–	–	0.06–0.24	0.046	1.12–1.95	0.94	2.06–2.71	1.13	0.21–0.29	0.12

[a]Data given here for a plot 10 × 10 m. Biomass on a 0.1 ha plot equivalent to 216 t ha^{-1} (see Table 2).

GROWTH OF TROPICAL MOUNTAIN FORESTS 101

is necessary to know the concentrations in mature leaves and in fresh litter, the amount of organic material (dry weight CHO) translocated out of the leaves during senescence, and the amounts leached into the throughfall. Edwards (21) has provided the necessary concentration data for leaves, litter, and throughfall, and a preliminary mean value of ~10% for withdrawal of dry weight, but complete data are not available for any tropical forest. We may nevertheless gain a useful preliminary picture by comparing the concentrations in mature leaves and freshly fallen leaves. First analyses of two tree species (*Alchornea latifolia* and *Clethra occidentalis*) in four forest types in Jamaica have given evidence for withdrawal of N, P, and K, but not of Ca, Cu, Fe, Mg, Mn, Na, or Zn (98). These results are in general compatible with those obtained with crop plants and deciduous trees (106) and with the mediterranean sclerophyll *Quercus ilex* (81). The few data available for crop plants suggest that withdrawal of S and Mo is minimal. For the elements that are only minimally withdrawn, or are not withdrawn at all, we may expect that any deficiency in the soil will be reflected in the concentration in the mature leaves; but this is not necessarily true of the highly mobile elements N, P, and K, which can be cycled quickly from older leaves to newer ones. Foresters have appreciated for some time the importance of analyzing leaves of several ages to investigate this point when comparing plantations (27).

In most forests as much K reaches the forest floor in throughfall as in litter, and the changes in concentration between mature leaves and freshly fallen leaves are likely to be affected more by leaching than the comparable changes in N and P. Thus differences between forests in the amounts of internal cycling of N and P are our chief concern here. Nye (68) was the first to draw attention to the high N content of fresh litter in LRF and the apparent lack of withdrawal from old leaves, which contrasts strongly with the situation in most temperate deciduous forests. Subsequent studies of LRF in the Ivory Coast (5), Brazil (54), and Guatemala (25) have confirmed the high levels of N in the litter and, by implication, the lack of withdrawal. Such a finding fits with the high overall concentration in the biomass of very different lowland forests (Table 5) and also with the abundant mineralization of N in LRF soils (20). In contrast the levels of P in LRF litter (5, 25, 54, 68) are not particularly high. A comparison of the leaf concentration given by Klinge for the central Amazon (53) with the leaf litter concentration in similar forest (54) suggests a withdrawal on the order of 50%. However, Cornforth's data (16) for *Mora excelsa* on infertile soil in Trinidad show a reversed picture with ~50% withdrawal of N but none of P. The level of P in the mature leaves is very low (0.04%)—even lower than in the litter of some other lowland forests. Perhaps it is impracticable to withdraw P from such leaves. Presumably further studies will emphasize the variability in amounts of withdrawal of N and P in the LRF formation type.

Certainly there are marked differences between the two areas of montane forest that have been critically investigated: the 33-m tall LMRF at 2500 m in New Guinea and the 12-m tall UMRF in Jamaica. The leaf litter concentrations of N are 86% and 52% of the mature leaf concentrations in the two forests respectively, and the values for P are 71% and 34% (21, 98). It is thus virtually certain that the low-stature forest is cycling much more N and P internally than the high-stature forest;

and it is suggested that despite the lack of any indication from the analyses of mature leaves (Table 4), the difference in forest stature is largely the result of differences in the supply of N and P. This interpretation is supported by the fact that the levels of N and P in the throughfall in the Jamaican forest proved to be too low for accurate measurement (98). In contrast, in the New Guinean forest a great enrichment in N of the rain was found to occur as it passed through the canopy, perhaps as a result of N-fixing bacteria on the leaves and twigs (85, 86). This enrichment in N has been seen by Edwards (21) as the main source of the great quantities of N locked up in the organic matter of the soil in many New Guinean lower montane forests. The activity of N-fixing bacteria may be dependent on an adequate supply of P leaking from the leaves, and possibly too little is leaked from the leaves of the Jamaican forest.

When the 6-m tall mor-forming UMRF in Jamaica is compared with the adjacent 12-m tall mull-forming UMRF, the litter concentrations of N and P are about the same percentage of the mature leaf concentrations in the two forests. Again there is no evidence of differences in N and P supply for the plants that can grow in both situations. This finding supports the view set out above that the primary cause of poor plant growth in the mor-forming forest is the extreme acidity of the soil. It would be particularly interesting to see a comparison in this respect between the dwarfed forest on mor in Jamaica, and the dwarfed forests on muck soil in Puerto Rico and on peat in Malaya.

In summary, the results available suggest that variation in the amount of internal cycling of N and P may be an important guide to mineral shortages affecting some forests. There are clearly differences between forests of a given formation type and there are too few data to be sure whether, in general, there is more internal cycling in montane forests.

CONCLUSIONS AND OUTLOOK

The factors primarily responsible for limiting the growth and distribution of tropical montane forests are air temperature and the radiation climate. However, the high humidities in montane forests make invasion of leaves by fungi and other epiphylls easy, and most of the dicotyledonous trees there have leaves that last only one year. These leaves are characteristically very thick (pachyphylls) and are therefore expensive in terms of dry weight per unit area. The leaf thickness and short leaf life mean that relatively much less production can be invested in woody parts in montane forest than in temperate deciduous forest or conifer forest; and therefore growth—already slowed down by air temperatures and lack of bright sunlight—is further decreased. The decline in mean leaf size with altitude has been interpreted by some as an adaptation maximizing the ratio of carbon dioxide absorbed to water lost. It is suggested here that the peculiar structural features of the pachyphyll may have the same effect, although the thick outer walls may also serve to minimize invasion by fungi. Measurements of photosynthesis and transpiration in nature are now urgently needed. The possibility exists that in pachyphylls the maximum photosynthetic rates per unit area (and inherent growth rates) will prove to be low. If this is true, productivity cannot be the most important reason for success in the montane environment, and a new rationale for leaf size and leaf form will have to be found. Mineral supply varies greatly within each formation type recognized. However, there is some evidence that the plants of montane forests in general are adapted to a relatively poor supply of N and P, not because a low transpiration rate inhibits uptake of ions, but because ion uptake is reduced when photosynthates are in short supply and because the plants have to contend with large proportions of the potentially available N and P being locked up in undecayed litter and unmineralized humus. Montane plants seem to have relatively low concentrations of N and P not only in the leaves (where they result from the thick outer walls) but also in their woody parts.

Lower montane, upper montane, and subalpine forests often come to their lowest limits on ridges. This effect may sometimes reflect greater incidence of fog on the ridge-tops; in other cases it may result from reduced competition being provided by the plants of the next lower formation, which suffer an unfavorable interaction between climate and a depleted mineral supply on leached soils. The marked dwarfing of upper montane forest on certain ridge and plateau soils may result primarily from the extreme acidity of the soil (as on mor in Jamaica) or from extreme acidity plus a shortage of P and Cu (as on peat in Malaya). In the case of dwarfed forest on muck soil in Puerto Rico the only indication of mineral shortage is for K. Here excessive waterlogging and high levels of Fe^{2+} may be important in inhibiting invasion by most species of nearby lower montane forest. We have no information on N supply in the forests on peat and muck. Forest growth may also be limited severely by mineral supply on slopes and in gullies.

A comparison of montane forests in Jamaica and New Guinea suggests that it is not sufficient to assess forests by analyses of mature leaves in the case of N and P, but that a study of the amounts cycled internally is very important. Internal cycling of these elements is seen as an important adaptation to shortage.

Much research needs to be done to confirm or refute the ideas set out tentatively in this review and to extend our understanding of the mineral relations of all rain forests.

ACKNOWLEDGMENTS

I thank all the many people who have assisted me in visiting tropical mountains and who have discussed the problems of montane forests with me, particularly Drs. P. F. Stevens and T. C. Whitmore. I owe a special debt to Drs. P. J. Edwards and E. V. J. Tanner, who have shared my fascination for the relevance of mineral cycling studies and who have allowed me to quote freely from their studies in press or still unpublished. I am also very grateful to Messrs. M. J. Smith and G. J. Timmins for analyzing innumerable rain forest samples, and to my wife and Messrs. R. H. Whybrow and R. W. Hill for much help with anatomical work.

Literature Cited

1. Acquaye, D. K. 1964. Foliar analysis as a diagnostic technique in cocoa nutrition. I. Sampling procedure and analytical methods. *J. Sci. Food Agric.* 15:855–63
2. Barber, S. A., Walker, J. M., Vasey, E. H. 1962. Principles of ion movement through the soil to the plant root. *Trans. Int. Conf. Soil Sci. Com. II & IV*, pp. 121–24
3. Baynton, H. W. 1968. Ecology of an elfin forest in Puerto Rico, 2. The microclimate of Pico del Oeste. *J. Arnold Arbor.* 49:419–30
4. Beardsell, M. F., Mitchell, K. J., Thomas, R. G. 1973. Transpiration and photosynthesis in soybean. Effects of temperature and vapor pressure deficit. *J. Exp. Bot.* 24:587–95
5. Bernhard, F. 1970. Étude de la litière et sa contribution au cycle des éléments minéraux en forêt ombrophile de Côte-d'Ivoire. *Oecol. Plant.* 5:247–66
6. Berrie, G. K., Eze, J. M. O. 1975. The relationship between an epiphyllous liverwort and host leaves. *Ann. Bot. London* 39:955–63
7. Biddulph, O. 1959. Translocation of inorganic solutes. In *Plant Physiology—A Treatise*, ed. F. C. Steward, pp. 553–603. New York: Academic
8. Blasco, F. 1971. *Montagnes du Sud de l'Inde. Forêts, Savanes, Ecologie.* Madras: Inst. Français de Pondichery. 436 pp.
9. Blasco, F., Tassy, B. 1975. Étude d'un écosystème forestier montagnard du Sud de l'Inde. *Bull. Ecol.* 6:525–39
10. Bötticher, R., Behling, L. 1939. Licht, Transpiration, Salzaufnahme und Blattstruktur. *Flora Jena* (NF) 34:1–44
11. Bray, J. R., Gorham, E. 1964. Litter production in forests of the world. *Adv. Ecol. Res.* 2:101–57
12. Briscoe, C. B., Wadsworth, F. H. 1970. Stand structure and yield in the Tabunoco forest of Puerto Rico. In *A Tropical Rain Forest*, ed. H. T. Odum, R. F. Pigeon, pp. B79–89. Springfield, Va: US AEC
13. Brown, W. H. 1917. The rate of growth of *Podocarpus imbricatus* at the top of Mount Banahao, Luzon, Philippine Islands. *Philipp. J. Sci. Sect. C* 12:317–28
14. Brown, W. H. 1919. *The Vegetation of Philippine Mountains.* Manila:Bur. Sci. 433 pp.
15. Chapman, H. D., Brown, S. M. 1950. Analysis of orange leaves for diagnosing nutrient status with reference to potassium. *Hilgardia* 19:501–39
16. Cornforth, I. S. 1970. Leaf fall in a tropical rain forest. *J. Appl. Ecol.* 7:603–8
17. Dale, W. L. 1964. Sunshine in Malaya. *J. Trop. Geogr.* 19:20–26
18. Davies, C. R., Wareing, P. F. 1965. Auxin-directed transport of radiophosphorus in stems. *Planta* 65:139–56
19. Dawkins, H. C. 1959. The volume increment of natural tropical high forest and limitations on its improvement. *Emp. For. Rev.* 38:175–80
20. de Rham, P. 1970. L'azote dans quelques forêts, savanes et terrains de culture d'Afrique tropicale humide (Cote-d'Ivoire). *Veröff. Geobot. Inst. Zürich*, Vol. 45. 124 pp.
21. Edwards, P. J. 1973. *Nutrient cycling in New Guinea montane forest.* PhD thesis. Univ. Cambridge, Engl. 198 pp.
22. Edwards, P. J. 1977. Aspects of mineral cycling in a New Guinea montane forest. II. The production and disappearance of litter. *J. Ecol.* 65:In press
23. Edwards, P. J., Grubb, P. J. 1977. I. The distribution of organic matter in the vegetation and the soil. *J. Ecol.* 65:In press
24. Elkington, T. T., Jones, B. M. G. 1974. The biomass and primary productivity of birch (*Betula pubescens* s-lat) in southwest Greenland. *J. Ecol.* 62:821–30
25. Ewel, J. J. 1976. Litter fall and leaf decomposition in a tropical forest succession in eastern Guatemala. *J. Ecol.* 64:293–308
26. Fenner, M. 1975. *Factors limiting the distribution of strict calcicoles.* PhD thesis. Univ. Cambridge, Engl. 241 pp.
27. Florence, R. G., Chuong, P. H. 1974. The influence of foliar nutrients in *Pinus radiata* plantations. *Aust. For. Res.* 6:1–8
28. Ford, M. A., Thorne, G. N. 1974. Effects of atmospheric humidity on plant growth. *Ann. Bot. London* 38:441–52
29. Gates, D. M. 1969. The ecology of elfin woodland in Puerto Rico, 4. Transpiration rates and temperatures of leaves in cool humid environment. *J. Arnold Arbor.* 50:93–98
30. Golley, F. B., McGinnis, J. T., Clements, R. G., Child, G. I., Deuver, M. J. 1975. *Mineral Cycling in a Tropical*

Moist Forest System. Athens, Ga: Univ. Georgia Press. 248 pp.
31. Goodall, D. W., Gregory, F. G. 1947. Chemical composition of plants as an index of their nutritional status. *Tech. Commun. Imp. Bur. Hort. Plantation Crops 17.* 167 pp.
32. Greenham, C. G., Randall, P. J., Ward, M. M. 1972. Impedance parameters in relation to phosphorus and calcium deficiencies in subterranean clover (*Trifolium subterraneum* L.). *J. Exp. Bot.* 23:197–209
33. Greenland, D. J., Kowal, J. M. L. 1960. Nutrient content of moist tropical forest in Ghana. *Plant Soil* 12:154–74
34. Grime, J. P., Hunt, R. 1975. Relative growth-rate: its range and adaptive significance in a local flora. *J. Ecol.* 63: 393–422
35. Grubb, P. J. 1971. Interpretation of the "Massenerhebung" effect on tropical mountains. *Nature* 229:44–45
36. Grubb, P. J. 1974. Factors controlling the distribution of forest types on tropical mountains: new facts and a new perspective. In *Altitudinal Zonation in Malesia,* ed. J. R. Flenley, pp. 1–25. Hull, Engl: Univ. Hull Geogr. Dep. Misc. Ser. 16
37. Grubb, P. J. 1977. Leaf structure and function. *An Encyclopaedia of Ignorance, Vol. 2, Biological Sciences.* Oxford: Pergamon. In press
38. Grubb, P. J., Grubb, E. A. A., Miyata, I. 1975. Leaf structure and function in evergreen trees and shrubs of Japanese Warm Temperate Rain-forest. I. The structure of the lamina. *Bot. Mag. Tokyo* 88:197–211
39. Grubb, P. J., Key, B. A. 1975. Clearance of scrub and reestablishment of chalk grassland on the Devil's Dyke. *Nat. Cambridgeshire* 18:18–22
40. Grubb, P. J., Lloyd, J. R., Pennington, T. D., Whitmore, T. C. 1963. A comparison of montane and lowland rain forest in Ecuador. I. The forest structure, physiognomy, and floristics. *J. Ecol.* 51:567–601
41. Grubb, P. J., Stevens, P. F. 1977. *The forests of the Fatima basin and Mt. Kerigomna and a review of Montane and Subalpine forests elsewhere in Papua New Guinea. Dep. Biogeogr. Geomorph., Publ. BG/5.* Canberra: Aust. Natl. Univ. Press. 228 pp.
42. Grubb, P. J., Tanner, E. V. J. 1976. The montane forests and soils of Jamaica: a reassessment. *J. Arnold Arbor.* 57: 313–68
43. Grubb, P. J., Whitmore, T. C. 1966. A comparison of montane and lowland rain forest in Ecuador. II. The climate and its effects on the distribution and physiognomy of the forests. *J. Ecol.* 54:303–33
44. Grubb, P. J., Whitmore, T. C. 1967. A comparison of montane and lowland rain forest in Ecuador. III. The light reaching ground vegetation. *J. Ecol.* 55:33–57
45. Guha, M. M., Narayanan, R. 1969. Variation in leaf nutrient content in *Hevea* with clone and age of leaf. *J. Rubb. Res. Inst. Malaya* 21:225–39
46. Harley, J. L. 1972. *The Biology of Mycorrhiza.* London: Hill. xxii + 334 pp. 2nd ed.
47. Hatrick, A. A., Bowling, D. J. F. 1973. A study of the relationship between root and shoot metabolism. *J. Exp. Bot.* 24:607–13
48. Hnatiuk, R. J., Smith, J. M. B., McVean, D. N. 1976. *Mt. Wilhelm Studies. 2. The Climate of Mt. Wilhelm. Dep. Biogeogr. Geomorphol. Publ. BG/4.* Canberra: Aust. Natl. Univ. Press. xiv + 76 pp.
49. Hope, G. S. 1976. The vegetational history of Mt. Wilhelm, Papua New Guinea. *J. Ecol.* 64:627–63
50. Howard, R. A. 1969. The ecology of an elfin forest in Puerto Rico, 8. Studies on stem growth and form and of leaf structure. *J. Arnold Arbor.* 50:225–67
51. Huttel, C., Bernhard-Reversat, F. 1975. Recherches sur l' écosystème de la forêt subéquatoriale de Basse Côte d'Ivoire. V. Biomasse végétal et productivité primaire cycle de la matière organique. *Terre Vie* 29:203–28
52. Janzen, D. H. 1973. Rate of regeneration after a tropical high elevation fire. *Biotropica* 5:117–22
53. Klinge, H. 1975. Bilanzierung von Hauptnährstoffen im Ökosystem Tropischer Regenwald (Manaus)—vorläufige Daten. *Biogeographica* 7:59–76
54. Klinge, H., Rodrigues, W. A. 1968. Litter production in an area of Amazonian terra firme forest. *Amazonia* 1:287–310
55. Koo, R. C. J., Sites, J. W. 1956. Mineral composition of Citrus leaves and fruit as associated with position on the tree. *Proc. Am. Soc. Hort. Sci.* 68:245–52
56. Krizek, D. T., Bailey, W. A., Klueter, H. H. 1971. Effects of relative humidity and type of container on the growth of F_1 hybrid annuals in controlled environments. *Am. J. Bot.* 58:544–51

57. Kylin, A., Hylmö, B. 1957. Uptake and transport of sulfate in wheat. Active and passive components. *Physiol. Plant.* 10:467–84
58. Larcher, W. 1960. Transpiration and photosynthesis of detached leaves and shoots of *Quercus pubescens* and *Quercus ilex* during desiccation under standard conditions. *Bull. Res. Counc. Isr. Sect. D* 80:213–24
59. Larcher, W. 1975. *Physiological Plant Ecology.* Berlin: Springer. 252 pp.
60. Leigh, E. G. Jr. 1975. Structure and climate in tropical rain forest. *Ann. Rev. Ecol. Syst.* 6:67–86
61. Loveless, A. R. 1962. Further evidence to support a nutritional interpretation of sclerophylly. *Ann. Bot. London* (NS) 26:551–61
62. Miles, J. 1974. Experimental establishment of new species from seed in Callunetum in north-east Scotland. *J. Ecol.* 62:527–52
63. Miller, R. B. 1963. Plant nutrients in hard beech. II. Seasonal variation in leaf composition. *NZ J. Sci.* 6:378–87
64. Montieth, J. L. 1963. Gas exchange in plant communities. In *Environmental Control of Plant Growth,* ed. L. T. Evans, pp. 95–112. New York:Academic
65. Mullins, M. G. 1970. Hormone-directed transport of assimilates in decapitated internodes of *Phaseolus vulgaris* L. *Ann. Bot. London* (NS) 34:897–909
66. Nevling, L. I. 1969. The ecology of an elfin forest in Puerto Rico, 5. Chromosome numbers of some flowering plants. *J. Arnold Arbor.* 50:99–103
67. Nobel, P. S., Zaragoza, L. J., Smith, W. K. 1975. Relation between mesophyll surface area, photosynthetic rate, and illumination level during development for leaves of *Plectranthus parviflorus* Henckel. *Plant Physiol.* 55:1067–70
68. Nye, P. H. 1961. Organic matter and nutrient cycles under moist tropical forest. *Plant Soil* 13:333–46
69. Nye, P. H. 1969. The soil model and its application to plant nutrition. In *Ecological Aspects of the Mineral Nutrition of Plants, Symp. Br. Ecol. Soc. 9,* ed. I. H. Rorison, pp. 108–14. Oxford: Blackwell
70. Odum, H. T. 1970. Rain forest structure and mineral-cycling homeostasis. See Ref. 12, pp. H3–52
71. Odum, H. T. 1970. Summary: an emerging view of the ecological system at El Verde. See Ref. 12, pp. I191–289
72. Odum, H. T., Ruiz-Reyes, J. 1970. Holes in leaves and the grazing control mechanism. See Ref. 12, pp. I69–80
73. Oppenheimer, H. R. 1963. Zur Kenntnis kritischer Wasser-Sättigungsdefizite in Blättern und ihrer Bestimmung. *Planta* 60:51–69
74. Ovington, J. D., Olson, J. S. 1970. Biomass and chemical content of El Verde Lower Montane Rain forest plants. See Ref. 12, pp. H53–77
75. Pallas, J. E., Michel, B. E., Harris, D. G. 1967. Photosynthesis, transpiration, leaf temperature, and stomatal activity of cotton plants under varying water potentials. *Plant Physiol.* 42:76–88
76. Pareek, O. P., Sivanayagam, T., Heydecker, W. 1969. Relative humidity: a major factor in crop plant growth. *Rep. Sch. Agric. Univ. Nottingham,* 1968–1969, pp. 92–95
77. Parkhurst, D. F., Loucks, O. L. 1972. Optimal leaf size in relation to environment. *J. Ecol.* 60:505–38
78. Pitman, M. G., Cram, W. J. 1977. Regulation of ion content in whole plants. In *Integration of Activity in the Higher Plant, Symp. Soc. Exp. Biol. 31,* ed. D. H. Jennings, pp. 391–424. London: Cambridge
79. Pozsár, B. I., Király, Z. 1966. Phloem-transport in rust infected plants and the cytokinin-directed long-distance movement of nutrients. *Phytopathol. Z.* 56:297–309
80. Prevot, P., Ollagnier, M. 1957. Directions for use of foliar analysis. *Fertilité* 2:3–12
81. Rapp, M. 1969. Production de litière et apport au sol d'éléments minéraux dans deux écosystèmes Mediterranéens: le forêt de *Quercus ilex* L. et la garigue de *Quercus coccifera* L. *Oecol. Plant.* 4:377–410
82. Raunkiaer, C. 1934. *The Life-Forms of Plants and Statistical Plant Geography.* Oxford: Clarendon. xvi + 632 pp.
83. Richards, P. W. 1952. *The Tropical Rain Forest.* London: Cambridge. 450 pp.
84. Rochow, J. J. 1974. Estimates of aboveground biomass and primary productivity in a Missouri forest. *J. Ecol.* 62:567–77
85. Ruinen, J. 1965. The phyllosphere. II. Nitrogen fixation in the phyllosphere. *Plant Soil* 22:375–94
86. Ruinen, J. 1975. Nitrogen fixation in the phyllosphere. In *Nitrogen Fixation by Free-living Micro-organisms,* ed. W.

D. P. Stewart, pp. 85–100. London: Cambridge
87. Russell, R. S., Shorrocks, V. M. 1957. The effect of transpiration on the absorption of inorganic ions by intact plants. *Proc. 1st UNESCO Int. Conf. on Radioisotopes in Sci. Res., Paris,* Rep. No. 178
88. Sestak, Z., Catsky, J., Jarvis, P. 1971. *Plant Photosynthetic Production. Manual of Methods.* The Hague: Junk. 818 pp.
89. Shorrocks, V. M. 1961. Leaf analysis as a guide to the nutrition of *Hevea brasiliensis.* I. Sampling technique with mature trees: principles and preliminary observations on the variations in leaf nutrient composition with position on the tree. *J. Rubb. Res. Inst. Malaya* 17:1–13
90. Shorrocks, V. M. 1962. Leaf analysis as a guide to the nutrition of *Hevea brasiliensis.* II. Sampling technique with mature trees: variations in nutrient composition of the leaves with position on the tree. *J. Rubb. Res. Inst. Malaya* 17:91–101
91. Shorrocks, V. M. 1965. Leaf analysis as a guide to the nutrition of *Hevea brasiliensis.* VI. Variations in leaf nutrient composition with age and with time. *J. Rubb. Res. Inst. Malaya* 19:1–8
92. Silvester, W. B., Bennett, K. J. 1973. Acetylene reduction by roots and associated soil of New Zealand conifers. *Soil Biol. Biochem.* 5:171–79
93. Small, E. 1972. Photosynthetic rates in relation to nitrogen recycling as an adaptation to nutrient deficiency in peat bog plants. *Can. J. Bot.* 50:2227–33
94. Smith, P. F. 1962. Mineral analysis of plant tissues. *Ann. Rev. Plant Physiol.* 13:81–108
95. Steenis, C. G. G. J. van, Hamzah, A., Toha, M. 1972. *The Mountain Flora of Java.* Leiden: Brill. 90 pp.
96. Sutcliffe, J. F. 1962. *Mineral Salts Absorption.* Oxford: Pergamon. 194 pp.
97. Tamm, C. O. 1964. Determination of nutrient requirements of forest stands. *Int. Rev. For. Res.* 1:115–70
98. Tanner, E. V. J. 1977. *Mineral cycling studies in montane forest in Jamaica.* PhD thesis. Univ. Cambridge, Engl. 296 pp.
99. Tanner, E. V. J. 1977. Four montane rain forests of Jamaica: a quantitative characterization of the floristics, the soils and the foliar mineral levels, and a discussion of the interrelations. *J. Ecol.* 65:In press
100. Tinker, P. B. 1969. The transport of ions in the soil around plant roots. See Ref. 69, pp. 135–47
101. Wadsworth, F. H., Bonnet, J. A. 1951. Soil as a factor in the occurrence of two types of montane forest in Puerto Rico. *Caribb. For.* 12:67–70
102. Wardle, P. 1965. Significance of xeromorphic features in humid subalpine environments in New Zealand. *N.Z. J. Bot.* 3:342–43
103. Weaver, P. L., Byer, M. D., Bruck, D. L. 1973. Transpiration rates in the Luquillo Mountains of Puerto Rico. *Biotropica* 5:123–33
104. Webb, L. J. 1959. A physiognomic classification of Australian rain forests. *J. Ecol.* 47:551–70
105. Whitmore, T. C. 1975. *Tropical Rain Forests of the Far East.* Oxford: Clarendon. xiii + 282 pp.
106. Williams, R. F. 1955. Redistribution of mineral elements during development. *Ann. Rev. Plant Physiol.* 6:25–42
107. Winneberger, J. H. 1958. Transpiration as a requirement for growth of land plants. *Physiol. Plant.* 11:56–61
108. Woodward, F. I., Pigott, C. D. 1975. The climatic control of the altitudinal distribution of *Sedum rosea* (L.) Scop. and *Sedum telephium* L. 1. Field observations. *New Phytol.* 74:323–34
109. Woolhouse, H. W. 1969. Differences in the properties of the acid phosphatases of plant roots and their significance in the evolution of edaphic ecotypes. See Ref. 69, pp. 357–80

POLYMORPHISM IN *CEPAEA:* ♦4119
A Problem with Too Many Solutions?

J. S. Jones, B. H. Leith, and P. Rawlings

Department of Genetics, Royal Free Hospital School of Medicine,
University of London, London WC1N 1BP

INTRODUCTION

When faced with a problem or a series of related problems, scientists tend to look for a single unifying solution. When alternative solutions present themselves, it is usually accepted that only one is likely to be true. Further research often justifies this assumption. The successes of this type of reasoning have tended to obscure the fact that not all biological observations must necessarily be explicable in a simple and unitary way and that not all hypotheses to be tested need be mutually exclusive. We hope to illustrate this by considering some of the evolutionary processes affecting one well known genetic polymorphism, that of shell pattern in the land snail *Cepaea*. The development of techniques for detecting molecular polymorphisms has led to considerable disagreement between those who believe that such genetic variation is actively maintained by natural selection and those who suggest that random processes are the only significant factor in its control. The most important lesson to be gained from an intensive study of the *Cepaea* polymorphism is that many types of evolutionary force act upon it and that their relative importance varies between different polymorphic loci, or even when the same locus is studied in different populations. It is largely meaningless to ask whether selection or drift explains the observed variation in gene frequency, or indeed to attempt to identify the single selective mechanism acting on the polymorphism. The nature of the evolutionary process means that the genetic structure of each *Cepaea* population usually requires a complex and perhaps a unique explanation. This is a point which has not been sufficiently emphasized by students of other polymorphic systems.

THE GENUS *CEPAEA*

Cepaea is a cross-fertilizing hermaphrodite. There are two pairs of sibling species in the genus. *C. nemoralis* (L.) and *C. hortensis* Mull. have a haploid number of 22, *C. vindobonensis* Pf. and *C. sylvatica* one of 24 (154). There are considerable differences in the anatomy of the species (7, 154). Hybrids between them are very rare although a few have been produced in the laboratory (139). There is a hierarchy of isolating mechanisms. The species are isolated to some extent by habitat preferences and by differences in time of activity (39, 41–43). Courtship is complex and mixed courtships are usually abandoned before mating takes place (78). Interspecific differences in genital structure ensure that any attempts at mating are unlikely to succeed (139, 154). Finally, such matings rarely lead to fertilization, and only infertile offspring are produced (154).

C. nemoralis is a western European species. *C. hortensis* is also widespread in western Europe, but has a more northerly distribution. It is also common in eastern North America, and has probably been there for far longer than its recently introduced sibling species (161, 180). *C. vindobonensis* occurs in southeastern Europe and western Russia, while *C. sylvatica* is found only in the western Alps.

Cepaea tolerates a very wide range of habitats. *C. nemoralis,* for example, is found on dunes, in cultivated areas, and as high as 2100 meters in the Pyrenees (3). The microgeographical distribution of the various species often reflects their large-scale climatic preferences. Where *nemoralis* and *hortensis* occur together the latter is frequently found in damper and cooler places, while *nemoralis* prefers warmer microclimates (2, 18, 29, 39, 43, 49, 108, 109). Subfossil shells show that in southern England the range of *hortensis* has expanded at the expense of that of *nemoralis* since the warmer conditions which prevailed several thousand years ago (27, 72). In the laboratory, *C. hortensis* is more active at low temperatures, while *C. nemoralis* survives better and conserves water more efficiently in hot dry conditions (41, 42). In regions where *nemoralis* and *vindobonensis* overlap, there is some evidence that *vindobonensis* prefers the drier and more exposed places (120). As might be expected from their geographical ranges, in the Alps *nemoralis* reaches its upper limits first, followed by *hortensis* and then *sylvatica* (86). In many places there are differences in the microgeographical distribution of each of the four *Cepaea* which cannot easily be related to the environment. This suggests that the snails are detecting environmental discontinuities not apparent to us.

Where the geographical ranges of *C. nemoralis* and *C. hortensis* overlap, approximately 20% of the colonies for which we have information contain both species. There is some evidence of competition between them. South of its northern limit in Britain, *C. nemoralis* excludes *hortensis* from dunes. North of this limit, however, *hortensis* is abundant in such habitats (28, 121). Competitive exclusion on a smaller scale has also been suggested for other types of habitat (39, 43, 108), although no experiments have been carried out involving the removal of one species and subsequent investigation of niche expansion by the other. In mixed colonies there is an increase in the incidence of shell breakage in the rarer species, which suggests that competitive interactions may lead indirectly to physical damage (182). There may be a stochastic element in competition; in some regions where *nemoralis* is abundant, *hortensis* is confined to woods, while in other places the opposite pattern is found (29).

Various aspects of the ecology of *C. nemoralis* have been studied. Information is available on actual and effective population size (32, 97, 131, 145), temporal fluctuations in population size (187), breeding activity (32, 166, 187, 188), generation interval and length of life (32, 187), fecundity (188, 189, 191, 199), metabolic cost of reproduction (183), juvenile viability and developmental time (32, 184, 195, 196), density dependent and independent mortality (32, 187, 197, 199), predators (36, 167), field activity in relation to climate (32, 41, 42, 183, 190), habitat choice (18, 32, 39, 108), movement (32, 46, 93, 131), biomass (166, 187), feeding rates, food preference and nutritional value of foods (88, 98, 99, 165, 185, 188, 200), assimilation efficiency (165, 183), and metabolic rates (183).

THE POLYMORPHISM

Since the time of Linnaeus *C. nemoralis* has been known as one of the most polymorphic members of the European fauna. Nineteenth century biologists coined scores of varietal names to describe the various forms, and its shell polymorphism was the subject of one of the earliest investigations into the genetics of natural populations (75). The other species of *Cepaea* also show genetic variation, although less so than does *C. nemoralis*.

The shell of *C. nemoralis* is polymorphic for color and for the presence, number, and appearance of up to five dark bands (Figure 1). The major color classes are yellow, pink, and brown. Several shades of each of these can be distinguished. Many shells have five bands (pattern 12345), but unbanded individuals (00000) and those with a single central band (00300) are often found. Other less common banding patterns are also known. There is much variation in the expression of the bands. Fully pigmented bands are common, but there is often interruption or even complete loss of the band pigment. Fusions of adjacent bands and spreading of band pigment over the whole shell are also frequently found in some populations. Changes in the color of the band pigment to orange or to light yellow are less common. *C. nemoralis* is polymorphic for the color of the lip of the shell and for a range of body colors. Fewer than 20 of 3,000 British colonies are monomorphic, and many populations segregate simultaneously for up to 6 loci controlling the shell polymorphism.

C. hortensis shares nearly all the polymorphisms found in *C. nemoralis*. Approximately 90% of British *hortensis* populations are polymorphic for the major shell characters. One polymorphism is almost diagnostic (8, 103): Nearly all *C. nemoralis* shells have dark lips, while the lips of the great majority of *C. hortensis* are white (Figure 1). *C. vindobonensis* and *C. sylvatica* are much less polymorphic. They have no clear-cut color variation, and banding polymorphism is restricted to a few variations in band number (12345 being the most common, with 10345, 00345, and other minor variants at lower frequencies), and in degree and intensity of band pigmentation. In *C. vindobonensis,* many individuals have their band pigment reduced from the normal black to a faint yellow, and in some populations of *C. sylvatica* a high proportion of the snails have bands reduced to a series of flecks. Occasional individuals with completely unpigmented bands are found in both species.

Genetically determined variation in shell shape and size is known (65, 68, 76, 147) as is continuous variation in a number of body structures (7). *C. nemoralis* is more

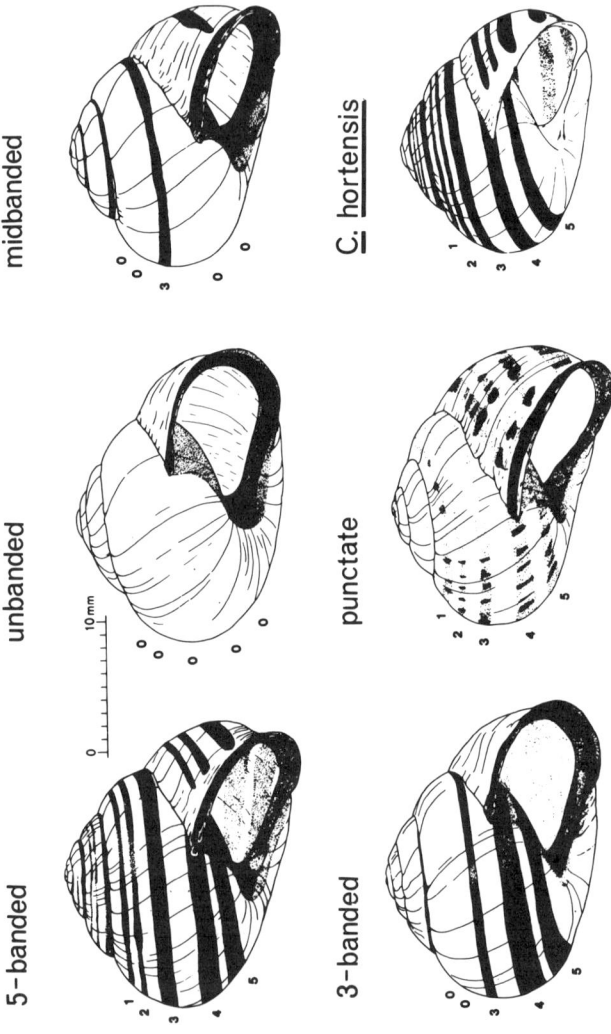

Figure 1 Banding polymorphism in *C. nemoralis* and five-banded *C. hortensis*. From (2) with permission.

variable in this respect than is *C. hortensis.* Genetic variation also exists for fecundity, viability, developmental time, and some aspects of behavior (190, 195–197). Cell surface polymorphism is manifested by differences in the ability of various *Cepaea* to agglutinate mammalian blood (129). Little intraspecific variation in chromosome number has been found (11) [although closely related snails do possess B chromosomes (85)]. There is a single large chromosome pair in *C. nemoralis* and *hortensis,* the other 21 pairs being very small. There can be up to 6 chiasmata on this chromosome (11) and there is some variation in chiasma frequency and position between populations (156–59). Several protein polymorphisms have been described in *C. nemoralis* and *hortensis* (21–23, 118, 140, 142, 150–53, 178, 203).

The genetics of most of the *C. nemoralis* shell polymorphisms is well understood (33, 37, 65, 131, 147, 194). Genes controlling the major polymorphisms for shell color, presence or absence of bands, lip color, and type of band pigmentation are borne together as a supergene. Other unlinked loci modify the number of bands on a banded shell. The degree of band fusion, minor variation in band number, and the degree of dominance of some alleles are polygenically controlled. There are also epistatic interactions between some of the polymorphic loci. Figure 2 is a summary of the genetic architecture of the visible polymorphism of *C. nemoralis.*

Recombination between members of the supergene has occasionally been recorded (33, 77, 147). There may be more than one linkage state for color and banding; in most experimental lineages there are no crossovers, but occasional families show recombination frequencies of up to 20% (87). Suggestions that this effect is an artifact due to the use of previously mated snails in breeding experiments (131) are probably not correct (66).

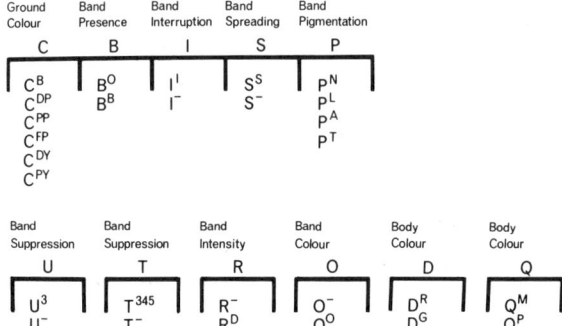

Figure 2 Genetic architecture of the *C. nemoralis* polymorphism. Alleles (in decreasing order of dominance): C: Dark Brown, Dark Pink, Pale Pink, Faint Pink, Dark Yellow, Pale Yellow; B: Unbanded, Banded; I: Punctate Bands, Continuous Bands; S: Spread Bands, Unspread Bands; P: Fully-pigmented Bands and Lip, Light Brown Bands and Lip, White Lip and Dark Bands, White Lip and Unpigmented Bands; U: Suppression of Bands 1, 2, 4, and 5, Unsuppressed; T; Suppression of Bands 1 and 2, Unsuppressed; R: Band Complete, Band "Darkens" from Apex to Lip; O: Black Bands, Orange Bands; D: Red Body, Grey Body; Q: Dark Body, Pale Body. The order of loci in the supergene is not known.

Shell size is under polygenic control (63, 65). Its heritability is surprisingly high, although this may be an overestimate due to the use of parents from different populations with very different shell sizes. Fecundity and developmental time are also polygenic characters (188, 189, 195, 199). The very rare sinistral (left-coiling) *Cepaea* probably result from developmental accidents (17). The genetics of most polymorphisms detected by electrophoresis can be inferred from the pattern of stained bands in the gel. However, the esterase polymorphism is complex, involving several loci producing many zones of activity on a gel (150-152). Its interpretation is further complicated by the induction of additional active zones by certain food plants (153).

The inheritance of shell characters in *C. hortensis* is almost identical to that in *C. nemoralis*, to the extent that dominance is preserved in interspecies hybrids (67, 101, 102, 139, 147). Their shared polymorphism presumably antedates speciation. No information is available on the formal genetics of *C. vindobonensis* or *C. sylvatica*.

PATTERNS OF GENE FREQUENCY DISTRIBUTION

C. nemoralis has long been notable for the great variation in morph frequency from place to place. It is not unknown for populations only a few hundred meters apart to be fixed for alternative alleles at a particular locus. It is possible to identify three general patterns of morph frequency distribution. In most low-lying populations from dunes and marshlands there are considerable fluctuations in morph frequency over very short distances (Figure 3*a*). In cultivated areas at slightly greater altitudes adjacent regions a few kilometers apart may show characteristic general differences in gene frequency, but within each region there remains considerable differentiation between local populations (Figure 3*b*). Finally, in high and exposed country, large areas containing a variety of habitats are characterized by very uniform gene frequencies, while adjacent and superficially similar areas have quite different frequencies of alleles at these loci (Figure 3*c*). These "area effects" (29), which are separated by sharp steps in allele frequency over apparently uniform habitats, have excited particular interest because they may represent, on a small and easily studied scale, patterns of population differentiation found in several other organisms (119). *C. hortensis* shows patterns of gene frequency distribution similar to those found in *C. nemoralis*. Although less information is available for *C. vindobonensis* and *C. sylvatica*, the populations that have been studied show area effects rather similar to those of *C. nemoralis* (122, 123).

EVOLUTION OF THE SHELL POLYMORPHISM

The visible polymorphism of *Cepaea* provides an excellent opportunity to study the interactions between ecology and genetics. The polymorphism is extensive and easy to score, and *Cepaea* is common in a variety of habitats. Much information is available on the genetic structure of natural populations. We have data on over a million specimens from ten thousand populations, representing a body of information on populations in their natural habitats greater than that which exists for any other animal except man. In the remainder of this review we show that this mass

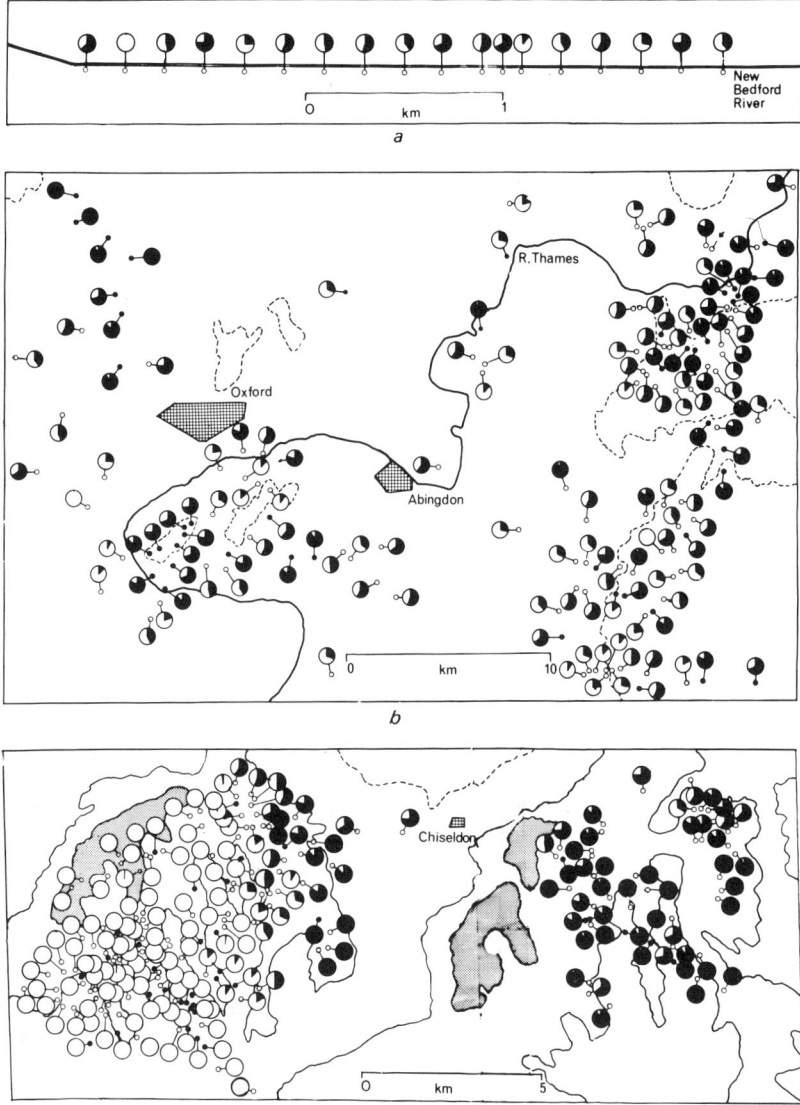

Figure 3 Patterns of morph frequency distribution in *C. nemoralis*. Large circles: morph frequency; attached small circles: habitat type (open: open habitats, shaded: woodlands). Contours at 400 feet (discontinuous line) and 600 feet (continuous line). Land over 800 feet shaded. (*a*) Frequency of yellow (white sector) along a river bank (92). (*b*) Frequency of yellow near Oxford (29, 36, 49). Note generally lower frequency of this allele in woodlands. (*c*) Frequency of non-five-banded (white sector) on a chalk upland (29, 49). Note "area effects" and lack of association of morph frequency with topography or background vegetation.

of information has led to an increase in the complexity of our explanations of genetic variation, rather than to the emergence of a single unifying explanation similar to that sought by workers on ecogenetic problems for which far fewer data exist. The processes acting on the polymorphism are considered in turn.

Visual Selection

Selection by predators is well known to exert a directional effect on morph frequencies in *Cepaea*. The song thrush *Turdus ericetorum* has the convenient habit of breaking open the shells on stones. Because the snails are rarely carried very far (73, 144) it is in principle possible to compare morph frequencies in the selected population with those in the population which has escaped predation. The thrushes discriminate against conspicuous shells, so that there is, for example, selection against yellow shells in populations living on dark backgrounds, such as those found in beech woods. Mammals may select against conspicuous shells on the basis of tone rather than color (25). Remarkably enough, glowworms *Lampyris noctiluca* also select *C. nemoralis* on the basis of their appearance. When given a choice of morphs in the laboratory, they avoid yellow five-banded shells at the expense of brown and yellow unbanded (148). Other predators include various birds, beetles, and the predatory snail *Oxychilus cellarius*.

Snails are not the favorite food of the thrushes. They appear to be eaten primarily during periods of food shortage (73, 90, 167). The ecological efficiency of the snail-thrush trophic link is rather low. In a sand-dune population, less than 1% of the food energy consumed in the previous year by the snails was passed on to the birds (167). There are great seasonal and regional differences in the amount of predation. In some localities, *Cepaea* is eaten only in the winter (167), in others in the summer (187), and in still others throughout the year (40, 73). Thrush predation may account for more than 90% of total adult deaths over a period of several weeks (187). On sand dunes, predation by thrushes is the major cause of winter mortality (167). Similarly, in an experimental colony of *C. nemoralis* at least half (and probably many more) of the original snails were killed by thrushes in 2½ years (188). The degree of predation varies greatly from year to year both on chalk uplands and on sand dunes.

Attempts to estimate selective intensities by comparing morph frequencies in predated and unpredated shells may be inaccurate because of uncertainty as to the exact location of (and hence the gene frequencies in) the population being attacked by the thrushes. One estimate is that selection against yellow shells in woodlands is of the order of 5% (29). Even stronger visual selection has been recorded on short grass (32). Release of marked snails in a woodland and subsequent examination of predated shells showed that the proportion of yellow snails killed decreased from April to May as the colour of the background changed from brown to green (176). Similar experiments later in the year show a reversal of this trend (49, 188). Twelve years after the original perturbation the frequency of yellow shells in one of these experimental populations had returned from 0.25 to the original 0.08 (48) suggesting that visual selection had indeed exerted an effect on gene frequencies.

The effectiveness of visual selection in controlling morph frequencies in *C. nemoralis* was shown by comparing the structure of a series of adjacent sand-dune populations in 1926 and 1960 (60). During this period the dunes became more stable and the background upon which the snails were living altered. There was a corresponding change in morph frequency, with a decrease in dark brown and an increase in mid-banded. The intensity of selection against the less cryptic morphs was approximately 10% when averaged over the whole period (58, 61). Because of the episodic nature of thrush predation, selection must occasionally be even stronger than this. Similar changes in gene frequency associated with an increase in the degree of thrush predation have also been recorded in an upland colony (32). Other cases in which morph frequencies have apparently changed over a period of years (many of which have been ascribed to selective predation) are of uncertain value since it is not clear whether the same populations were sampled at the beginning and end of the period of observation (58, 131, 173).

Selective predation has an important effect on the spatial as well as the temporal distribution of genes in many populations of *Cepaea*. In some areas of arable land in the south of England (where thrushes are common) there is a clear association between morph frequencies in *C. nemoralis* and the background upon which the snails are living (34–36, 71, 96). Woodland populations, which live on a brown and uniform background of dead leaves, have a higher frequency of the more cryptic brown, pink, and "effectively unbanded" shells (which appear unbanded to a predator because the uppermost bands are missing). In green and variegated backgrounds such as short grass, yellow and banded shells predominate. Intermediate habitats such as hedgerows have intermediate morph frequencies (Figures 3*b* and 4*a*). *C. hortensis* in central southern England achieves a similar visual effect using a different genetic strategy (Figure 4*b*). Woodland populations have a higher frequency of shells with all their bands fused together (so that they appear uniformly dark) than do those from open habitats (51). Some *C. vindobonensis* populations from woods also have a higher frequency of band fusions than do nearby nonwoodland populations (168).

Climatic Selection

Differences in the absorption of solar energy between dark and light colored individuals have an important effect on the thermal relations of many animals (113). In some polymorphic species [such as *Colias* butterflies and the ladybird *Adalia* (69, 181)] they appear to cause local selective responses to microclimate. Differences among the shell morphs of *Cepaea* lead to considerable differences in their rate of heating in sunshine. Temperature probes placed inside living *C. nemoralis* (or inside mercury-filled shells) exposed to the sun show that banded shells attain a higher equilibrium temperature than do unbanded, and brown shells a higher temperature than yellows (83, 113).

In *C. vindobonensis*, fully pigmented shells heat up more in the sun than do shells with reduced pigmentation (122). The temperature difference between morphs in each of these experiments is about 1°C. In this species at least, cool and resting dark snails when placed in sunshine become active before more faintly colored snails. There are no differences in the thermal relations of different colored snails in the shade. The wavelength of radiation from a surface depends on its temperature. Animals lose heat by radiation at 8–10 μm. The Kirchoff Law relates the degree

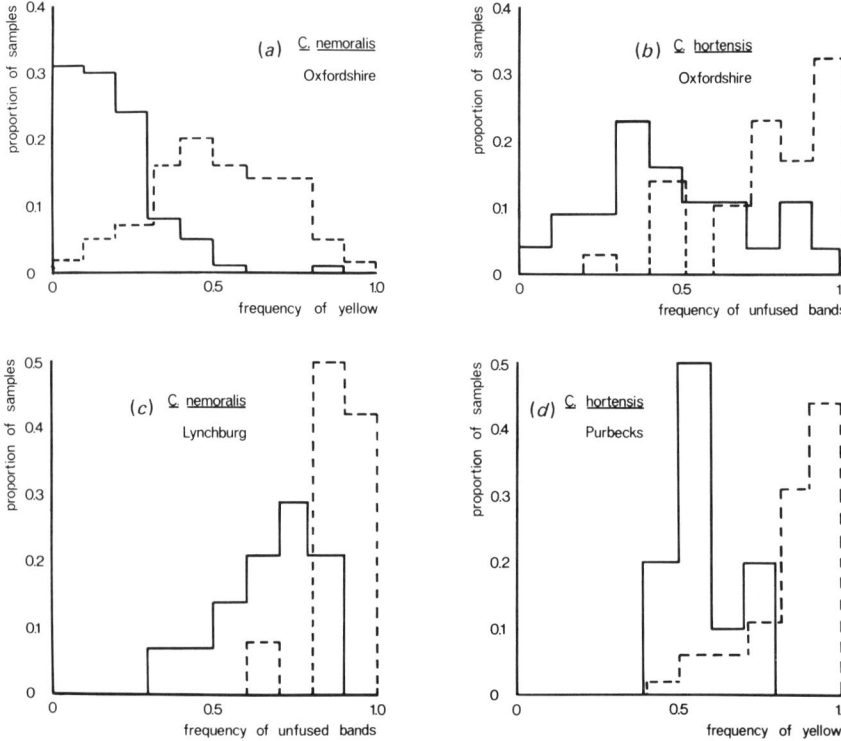

Figure 4 Strategies of adaptation of *C. nemoralis* and *C. hortensis* to visual selection by predators. Solid line: woodland habitats. Broken line: open habitats (36, 49, 51, 163).

of radiation from an object to its absorption of radiation at the same wavelength. A body's loss of heat by long wave radiation therefore bears no relation to its absorption of the much shorter wave visible light which determines its color. Dark-colored animals will not cool more than light-colored ones when out of the sun.

The effects of phenotype on thermal input are likely to be of selective importance. Live *C. nemoralis* frequently find themselves at a temperature near their lethal limit. In a sand-dune population of this species (164) almost any sunny day produced nearly lethal temperatures at the grass surface. The frequency of pink and banded shells was significantly higher among snails dying from heat shock than among survivors, as would be expected from the effects of their darker coloration in increasing the uptake of solar energy. Several laboratory experiments have claimed to demonstrate similar differences in survival rates between morphs in extremes of temperature (20, 132, 174); but these are all open to severe criticism, either on statistical grounds or because the experimental animals came from different populations which may show physiological differences which are independent of shell phenotype (58, 119). In population cages of *C. nemoralis* containing a variety of morphs placed in the open in a region known to have a cold microclimate, there was a significant tendency for brown-shelled individuals to survive more often than pink or yellow, and for single-banded shells to survive more often than five-banded (12). This experiment, however, may have confounded the physiological differences within a population with those between populations because the snails were collected from several donor colonies. Experimental transfers of 7000 marked *C. vindobonensis*, faint and dark, from warmer to colder sites did not demonstrate any differences in survival rate after one year (127). However, the scale of the experiment would not have detected small selective differences; and only one component of fitness, the survival of adults, was measured. Further investigation of the effects of thermal physiology on fitness in *Cepaea* is needed.

Climate is strongly implicated in causing gene frequency changes in natural populations of *Cepaea* over time, over large distances, and between local populations. One problem in invoking climatic selection as a general cause of geographical variation in *Cepaea* is that it is always possible (and perhaps sometimes justified) to claim that genetic differences between populations living in different places are due to subtle and undetected differences in their climate. It is therefore impossible to falsify the hypothesis of climatic selection simply by examining the distribution of genes. This has led to a certain lack of consistency among authors in their interpretation of climatic selection (Table 1). Climate can be accepted as a selective agent if there is some insight into the physiological mechanisms of selection. Alternatively there should be independent evidence from several geographically separated sets of populations that associations of gene frequency with climate are consistent and not merely spurious local coincidences. There are several cases in which these criteria have been met.

In *C. nemoralis* the effect of shell color on thermal relations is reflected in the large scale geographical distribution of alleles at this locus (Figures 5 and 6). There is a highly significant tendency for yellow shells to be commoner in the hottest parts

Table 1 Apparent local associations of morph frequency with climate in *C. nemoralis*[a]

Region	Hot	Cold	Dry	Wet	Harsh	Mild	Ref.
Southeastern France	Y, 0		P, 5				(2)
Touraine	3, 0						(5)
South Downs						P, 5	(6)
Steep Holm		B, 0					(13)
Somerset	P, B				3	5	(18)
Northern Lancashire				5	3	5	(45)
Brittany				Y			(100)
Cornwall, Pembrokeshire	None obvious						(105)
Southwestern France, Northern Spain				Y			(106)
Yorkshire Wolds	No consistent associations						(110)
France	0	Y					(132)

[a] Y = yellow, P = pink, B = brown, 5 = five-banded, 3 = mid-banded, 0 = unbanded.

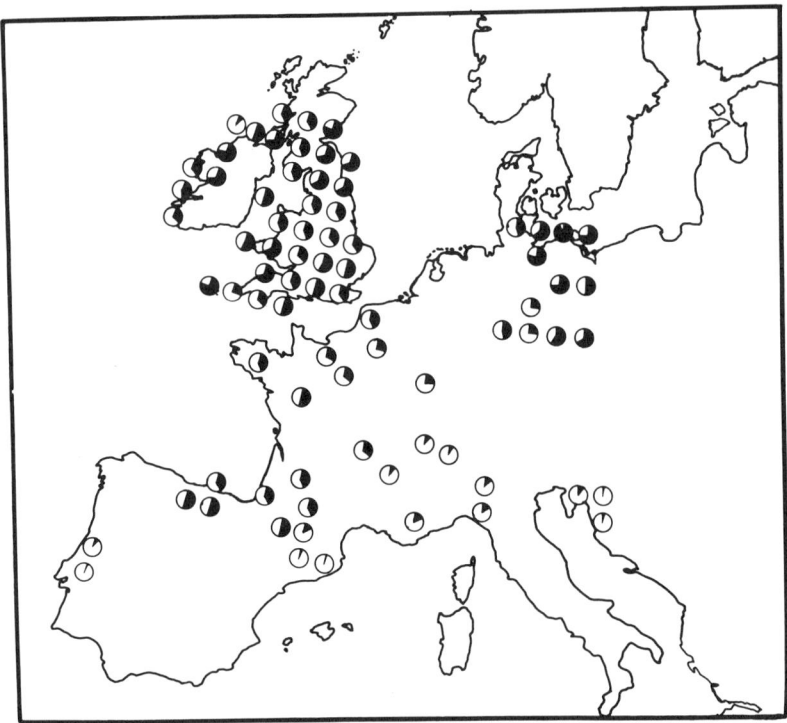

Figure 5 Frequency of yellow shells (white sector) in European populations of *C. nemoralis* mapped on the basis of 100 km grid squares. Only grid squares containing more than 100 snails in more than 10 samples included. (3-6, 13, 18, 25-27, 29, 31, 34-36, 39, 44, 45, 49, 53, 59, 60, 71, 72, 74, 89, 91, 92, 96, 100, 105-107, 110, 119-129, 126, 131-135, 137, 146, 170-172, 177, 192, plus unpublished data).

of this species' range. This is not simply a geographical effect which might be due to the spread of a selectively neutral allele from its centre of origin; in the Iberian peninsula, populations living in relatively cold conditions have a lower frequency of yellow shells than do populations from warmer places further north. Neither is it an artifact of visual selection resulting from southern populations originating from more open habitats than do those from the north. Omission of all woodland samples has little effect in any region on the frequency of yellow shells. In the south of England, for example, it increases their frequency from 50% to 53% (119). The association of yellow shells with high temperature is a causal one; this morph is more resistant to overheating during periods of thermal stress, while pink and brown shells have a higher efficiency of absorption of solar energy and in a cold climate are able to attain a temperature suitable for activity more often than can yellow snails. The effect is probably most marked in the early morning, since *C. nemoralis* is usually crepuscular (41).

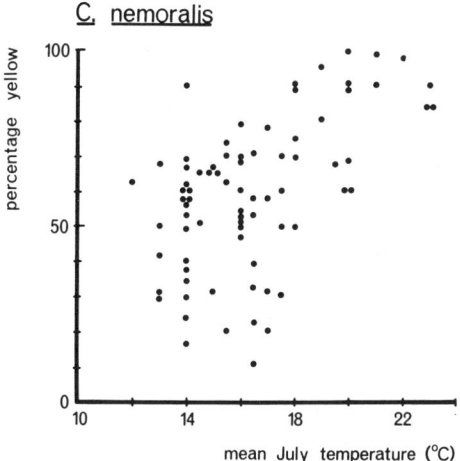

Figure 6 Association between the frequency of yellow shells and mean July temperature in *C. nemoralis* ($p < 0.001$).

It is unlikely that microclimatic selection can provide a complete explanation of local (rather than the large-scale) distribution of genes in populations of *C. nemoralis* in which visual selection is not acting. Repeated and geographically independent local associations between microclimate and allele frequency have been found at two loci in this species and at one in each of *C. hortensis* and *C. vindobonensis*.

Consistent associations of dark brown shells with topography and microclimate are found in sand-dune populations of *C. nemoralis* (26). Two types of dune can be distinguished. Simple dunes consist of a single line of sand hills (or a few lines parallel to each other), while complex dunes include numbers of large completely enclosed hollows up to about 100 m across. The brown allele was present on every one of 9 complex dunes in the British Isles, but on only 2 of 24 simple dunes. Furthermore, detailed mapping of the distribution of brown shells on one complex dune in North Wales showed a clear association with topography. This morph was more common within the enclosed hollows than elsewhere. The association is probably due to the accumulation of cold air in the hollows during clear nights when heat is lost from the ground by radiation, and to the greater efficiency of absorption of solar energy (and hence the earlier onset of activity) of brown shells in the early morning. The selective advantage of brown shells in cold climates is further demonstrated by examining subfossil shells (27). Occasionally brown shells buried in prehistoric earthworks (whose age is known from archaeological evidence) retain their color. The frequency of this allele can therefore be calculated at various times over the past several thousand years. Information on climate can be gained from studies of pollen. In the south of England brown shells were most common during periods of cold climate. This allele occurred much less frequently in the hot and dry conditions of the Neolithic than in the colder and wetter Iron Age. It is still common today. The distribution of dark brown shells in modern populations from southern English downlands also appears to coincide to some extent with valley bottoms in which cold air accumulates at night (29, 49). However, not all frost hollows contain brown-shelled snails (39) and this allele occurs in places where there is no ponding of cold air (31).

Microclimate also affects the microgeographic distribution of faint and dark *C. vindobonensis* (122, 123). This species is widespread in a limestone area of North Yugoslavia which has an unusual topography. Many large and very level basins are surrounded by steep mountains up to 2000 m high. There is a close association between morph frequency and topography. Faint banded individuals are found almost exclusively on the hillsides. This is not due simply to height above sea level. The floors of some of the basins are at a greater altitude than are some of the mountain slopes, but still contain no faint banded snails. The association is consistent over 14 separate basins. Microclimate is strongly implicated in this pattern of gene distribution. The basins are well known to act as frost hollows because they accumulate the cold air shed from the hillsides on clear nights. The night minimum temperature in the basins is sometimes as much as 18°C below that of the mountain slopes, and such temperature inversions may persist over several days. One hillside locality only 100 m above a nearby basin had an average night minimum temperature 2.3°C above that of the basin, and temperature inversions developed on 27 of 30 nights. Temperature probes show that fully pigmented *C. vindobonensis* heat up more rapidly and attain a higher equilibrium temperature in sunshine than do faint banded. Cooled dark snails, when placed in the sun, become active several minutes sooner than faint banded. Selection presumably controls the geographic distribution of alleles through the medium of the differences in the thermal relations of the two phenotypes.

We have preliminary and unpublished information of another type of climatic selection acting on the pigmentation of the shell lip in *C. nemoralis* and *C. hortensis*. This is of particular interest because the same selective factor appears to act locally, on the larger scale, and over time in each of the two sibling species. In *C. nemoralis* the lip is usually dark, but may occasionally be white. In *C. hortensis* the opposite pattern is found. However, in some regions there are high frequencies of white lip in *C. nemoralis* and in others an equally high frequency of dark lip in *hortensis*. A local survey of *C. nemoralis* along a gradient of rainfall in west Cornwall shows that white lipped shells are found preferentially in the wetter areas. The same is true on the European scale: this allele is found almost exclusively in the west of Scotland, western Ireland, high limestone in the north of England, west Cornwall, the Pyrenees, and the Cantabrians—all regions of high rainfall. Similarly, subfossil samples from the south of England (27) where this allele is now extremely rare, show that white lipped *C. nemoralis* were common during periods of wet climate. In *C. hortensis*, dark lipped shells are found only in the drier parts of this species' range. This is an unusual case of a parallel response to selection of a shared polymorphism in two sibling species. The mechanism of selection is not yet known.

Although it is clear that climatic selection exerts an effect on the frequency of alleles at some polymorphic loci in *Cepaea*, it is less certain which aspects of fitness are involved. Climate is known to have an effect on egg production and developmen-

POLYMORPHISM IN CEPAEA   123

tal time in *C. nemoralis* (189, 191, 199). In the closely related *Helix aspersa*, weather is the prime controller of natality. Other processes (such as predation) are much more important in affecting adult mortality (155). Differences in metabolism associated with differences in the environmental physiology of the various shell phenotypes could affect these components of fitness, and in one experimental population of *C. nemoralis* yellow shells reproduced more successfully in hot weather than did pink (191). Differences in thermal physiology could also affect mating activity, which is known to be an important fitness component in *Drosophila* (24).

Because at least two generation-transitions are needed to assess the total selection acting on a polymorphic locus (160), and because *Cepaea* has a four year generation interval in the wild (32), experimental estimation of the intensity of climatic selection will not be easy. Inconsistencies in patterns of distribution of gene frequency between places with apparently similar climates make it unlikely that climate can be used as a general predictor of morph frequencies. No climatic association is absolute. For example, some areas of complex dune have no brown *C. nemoralis*, some of the wettest parts of Europe have no white lipped members of this species, and some steep hillsides in Yugoslavia have no faint banded *C. vindobonensis*. There is also genetic differentiation between adjacent *Cepaea* populations living in topographically uniform places where it is hard to see how microclimate could be selectively important (74, 92, 125, 192). These problems are discussed below. Nevertheless, climate is a major factor in determining gene frequency distribution in many *Cepaea* populations.

Both climatic and visual selection act *against* the polymorphism: Neither of them alone can explain how it is maintained. However, morph frequencies in *Cepaea* populations are often stable for many generations. Subfossil samples show that in a number of populations the frequency of some morphs has not changed for thousands of years (27, 72, 75). Several instances of genetic stability over shorter periods have been described (29, 59, 89, 91, 172, 193). As is the case for many other widely studied polymorphisms, much more information is available on the evolutionary processes acting to remove alleles from a population than on those which act to retain them. Balancing mechanisms which may help to explain why so few *Cepaea* populations are monomorphic for the shell characters include frequency- and density-dependent selection, disruptive selection in a complex environment, heterozygous advantage, and a balance between random gain and loss of alleles.

Frequency-Dependent Selection

A selective advantage of rare morphs over more common ones is often important in maintaining visible polymorphism in prey species whose predators hunt by sight (9). The predators form a "searching image" for the first prey morph encountered and subsequently hunt for it to the exclusion of other morphs. An allele will be favored when it is rare because predators are unlikely to encounter it when they first sample the prey population. It will therefore become more common until a frequency-dependent equilibrium is reached. "Apostatic" selection of this kind (52, 82) is known to be important in maintaining polymorphism in, for example, mimetic butterflies. It tends to promote complex polymorphisms involving many strikingly different morphs. In *C. nemoralis* and *C. hortensis* almost every variant is distinct and stands out to some extent. Some of the shell phenotypes differ quite emphatically from others in the same population; thus, *C. nemoralis* with a single central band often have an unpigmented region on either side of the band which accentuates their difference from unbanded or five banded snails. The *Cepaea* polymorphism is also notable for the variety of visually different types within a single population. Recent experiments show that thrushes, the major predators of *Cepaea*, exert very strong frequency-dependent selection when presented with colored artificial baits at different frequencies (1). It is not surprising that apostatic selection has been put forward as a mechanism of maintaining polymorphism in this genus.

The initial evidence for apostatic selection in *Cepaea* rested largely on comparing the frequencies of shell phenotypes in mixed colonies of *C. nemoralis* and *C. hortensis*. In some mixed colonies around Oxford and in parts of France there is a negative association between morph frequencies in these two species (53, 131). The effect is sometimes striking. In some French mixed colonies one shell color allele is fixed in one species and an alternative allele in the other. Thrushes do not appear to discriminate between the sibling species when hunting. For example, in an experimental population containing both *nemoralis* and *hortensis*, snails of intermediate size were taken preferentially, without respect to the species to which they belonged (14). The negative associations between the genetic structure of the two species in mixed colonies might therefore be due to apostatic selection by the thrushes. In such colonies an allele in one species which was not present in the other would be favored because it would be rare in the *Cepaea* population as a whole.

There has been disagreement about the interpretation of data from mixed colonies. The initial analysis has been criticized on statistical grounds (56) and also because it was biased by including populations in which visual selection did not affect morph frequencies (47, 56, 64). An extensive survey of *Cepaea* populations in southern England (47), in which mixed colonies were compared with nearby colonies containing only one species, showed that *nemoralis* and *hortensis* were often *more* similar to each other when present together than when apart. This conflicts with the results expected from the hypothesis of apostatic selection. Nevertheless, it remains true that in some places there is a negative association between morph frequencies in the two species, which can be interpreted as indirect evidence for frequency-dependent selection.

There is now experimental evidence that thrushes can form a searching image for *Cepaea* morphs. They select brown bread-filled shells more often in a polymorphic population after having been "trained" on a bait population containing only brown individuals (111). In one artificial colony of *C. nemoralis* and *C. hortensis*, thrushes apparently produced a frequency-dependent equilibrium of morph frequencies (15). However, because the equilibrium was not reached until predation had ceased, its stability is in doubt. Other experiments using marked live snails in a natural population (R. W. Arnold, unpublished) provide some evidence that rare morphs are indeed at an advantage, but the results are not statistically significant. Further experiments on apostatic selection are needed.

We have preliminary data suggesting that among the helicid molluscs the species which climb most actively are the most polymorphic. In a series of laboratory experiments, the mean distance climbed by *C. nemoralis* was 242 cm, by *C. hortensis* 216 cm, by *C. sylvatica* 89 cm, and by *Arianta arbustorum* 29 cm. There is therefore a positive association between visible polymorphism and habitat selection, as expected from the hypothesis that conspicuousness to predators promotes visible polymorphism. The most active and conspicuous species, *C. nemoralis* has a complex system of variation involving many strikingly different morphs, while the more retiring species (such as *A. arbustorum,* a close relative of *Cepaea*) are much less variable. This evidence supports the suggestion that apostatic selection helps to maintain the shell polymorphism. However there has been as yet no direct demonstration of apostatic selection in natural populations of *Cepaea*.

Negative assortative mating, which maintains polymorphisms in a frequency-dependent fashion in *Drosophila* and in the Lepidoptera (9) does not appear to affect the *Cepaea* shell polymorphism. Examination of copulating pairs in the field (131) shows no deviation from random mating. As courtship in the helicid molluscs does not depend on visual stimuli (117) this is not surprising.

Disruptive Selection

It has frequently been suggested that a diverse environment will tend to promote polymorphism as a result of a balance between selection acting in different directions in different parts of a population and gene flow between the subpopulations. *Cepaea* has a low mobility, and usually experiences a "coarse-grained" environment. The fitness set model predicts that in these circumstances many populations will contain alleles favored in only one of the microhabitats. Disruptive selection and gene flow might therefore be important in maintaining shell polymorphism in *Cepaea*.

Cepaea is rather sluggish. Estimates of mobility (measured as the standard deviation of dispersal over one year) for *C. nemoralis* range from about 5 m to 10 m per year (32, 46, 97). Movement of this order often leads to significant emigration from local populations. About half the surviving snails emigrated from 20 m × 20 m plots on duneland and on short grass during the course of a year (46, 187). The panmictic unit is between 20 and 30 m for a continuous, and between 30 and 50 m for a linear population (92, 131, 192). Gene frequencies often vary significantly over distances shorter than this. In a linear *C. nemoralis* population living on a Dutch roadside there was a change in the frequency of pink shells from 0.15 to 0.88 within 35 m, and in another locality nearby there was a significant change in the frequency of yellow unbanded shells within 15 m (192). The latter cline was sampled on 22 occasions over 12 years, and remained stable (193). Similar changes have been recorded over distances as short as 5 m in the south of England (58). In one small area there were significant differences in morph frequency between a patch of nettles and the surrounding short grass despite measurable gene flow between the microhabitats (32). The exchange of genes will inevitably lead to an increase in the frequency of the inappropriate allele within both the nettle patch and the short grass populations and will thus promote polymorphism in each of them.

In some cases it is possible to identify the opposing forces of selection which, when coupled with gene flow, increase heterozygosity in certain *Cepaea* populations. Thus, some populations of *C. vindobonensis* living near the edge of a basin contain low frequencies of faint banded individuals, presumably as a result of gene flow from nearby hillsides where climatic selection favors this allele (122). Similarly, woodland and grassland populations of *C. nemoralis* living near the edge of a wood sometimes have a relatively high frequency of visually inappropriate morphs which enter the population by gene flow from the adjacent habitat (36). The efficacy of disruptive selection in maintaining polymorphism is shown by comparing gene frequencies in *C. nemoralis* where this species is confined to woods with those in regions where it lives both in woods and open habitats (36, 39, 71, 97). The mean frequency of pink shells in woods among narrow niche (woodland only) populations is higher than in regions where *C. nemoralis* has a broad niche (woods plus open habitats). In the latter case the joint action of selection and gene flow promotes polymorphism in a patchy environment. Where this species has a narrow niche and is confined to woods there can be no gene flow from alternative habitats, and simple directional selection reduced polymorphism. Disruptive selection of this kind is further illustrated in Somerset, where *C. nemoralis* shows gradually increasing niche restriction to woodland over a distance of 10 km (18). In the woodland populations there is a parallel decrease in the frequency of yellow shells and in the degree of shell color polymorphism which is again presumably due to restriction of gene flow from nearby open habitats with different optimum gene frequencies.

Cepaea can respond to changes in the direction of selection over very short distances. Disruptive selection and gene flow in populations encompassing several environmental patches may be important in explaining why so few populations are monomorphic.

Density-dependent Selection

The ultimate goal of ecological genetics must be to understand the control of population density in terms of the genetic variation within a population. Several models suggest that in density-regulated populations changes in density might lead to genetic changes (57) but this has not yet been convincingly shown in nature. Population ecology and population genetics have developed as almost separate disciplines.

There is much variation in density in *Cepaea*—from $0.04/m^2$ for adult *C. nemoralis* on chalk grassland, to $20/m^2$ in long grass (97). Food is unlikely to be a limiting factor because snails consume only a small part of the total production of their food plants (99, 165, 185, 200). The accumulation of slime-trail pheromones may limit the number of snails which can live in a particular area and hence lead to density-dependent control of population size (186). Interactions among snails in dense populations reduce activity, slow juvenile growth rate, and reduce the size of adults. In 9 separate study areas on chalk downland (186) and in a Dutch population (198) there was a reduction of up to 31% in body weight as density increased from 0.5 to 5.5 adults/m^2. Since small snails are less fecund, have a lower growth rate, and have a lower overwinter survival rate than do large (68, 184, 196, 199), this could lead to density-dependent control of population size.

Sparse *C. nemoralis* populations have more chiasmata on the largest bivalent chromosome than do dense populations (156–159). (This is apparently not true for *C. hortensis.*) This may be an adaptation to increase the diversity of gametic types within families and hence facilitate the release of variability in small and relatively inbred populations.

Although there may be density-dependent limitation of population size and of recombination, it has not yet been possible to relate this to the shell polymorphism. Climatic selection presumably acts in a density-independent fashion, at least insofar as it affects adult survival. In several populations of *Cepaea* in which climatic selection is acting there are no associations between shell polymorphism and population density (124). Visual selection could have a density-dependent component, since predators are known to change their hunting behavior when satiated by a particularly abundant species of prey (115). Directional selection by predators might therefore be less effective in very dense *Cepaea* populations. Density is also known to exert an effect on frequency-dependent hunting behaviour. When birds are presented with very dense populations of polymorphic prey they sometimes cease to exert apostatic selection, and instead attack the rarer morphs preferentially (112). This might occur in dense populations of *Cepaea* (95). In mixed colonies of *C. nemoralis* and *C. hortensis* where apostatic selection is acting and in which the birds hunt without discriminating between the species, the relative density of each species is bound to affect the fitness of alleles within them. The less dense species will appear to the predator to contain rarer morphs, which are to be avoided at the expense of alleles found only in the more dense species.

In some organisms, increased density leads to an increase in migration (130). This expands certain niche dimensions experienced by an average member of the species and may lead to disruptive selection in a heterogeneous environment. This does not seem to occur in *C. nemoralis*. A fivefold increase in population density does not increase the amount of emigration from a population (46).

There has been little interaction between genetics and demography in *Cepaea*. Experimental investigation of density-dependent selection would be worthwhile.

Heterozygote Advantage and Stabilizing Selection

Heterozygote advantage has been put forward as an important stabilizing influence on the *Cepaea* polymorphism (29, 36). There is, however, no direct evidence of its action. Unfortunately, no reliable method exists for detecting heterozygotes for any of the shell polymorphisms. Some populations show delayed penetrance of the dominant color allele in heterozygotes, giving, for example, pink shells with a yellow apex (33). This effect is not absolute and cannot be used to calculate the frequency of heterozygotes. Attempts to distinguish color heterozygotes from homozygotes using spectrophotometry and chromatography have been unsuccessful (120). From the observed variation of gene frequencies between habitats in the south of England and from estimates of the strength of visual selection any advantage of color heterozygotes over homozygotes has been calculated to be less than 5% (29). An apparent case of heterozygote advantage at the leucine amino-peptidase locus in this species (21) is probably due to the sampling of mixed populations combined with a statistical artifact (58). The role of heterozygote advantage in maintaining polymorphism in *Cepaea* remains an object of speculation (as indeed does its general importance as a mechanism of maintaining genetic variation in natural populations).

Stabilizing selection on shell size in *Arianta arbustorum* was demonstrated by a reduction in variance between young and adults (50). This type of selection also acts in *Cepaea*, as the genetic architecture of shell size (and of growth rate) suggests that it has been subject to a history of stabilizing selection (65, 68, 128, 195).

Random Processes

Genetic differentiation among populations of *C. nemoralis* was quoted for many years as an example of the importance of random processes in evolution (78, 79, 116, 143). It was assumed that shell color and banding had no biological significance and that allele frequencies were therefore determined by a balance between their origin by mutation and their random loss. This assumption resulted in a lengthy period during which data on *Cepaea* were collected without any attempt to relate gene frequencies to the environment. The history of attitudes toward the *Cepaea* polymorphism parallels that of attitudes toward many other polymorphic systems.

There is now abundant evidence that selection of various kinds affects gene frequencies in *Cepaea*. This does not, of course, indicate that random processes can have no importance. *Cepaea*'s hermaphroditism increases effective population size in comparison to that of monoecious species. Sperm is stored after insemination and is released for fertilization over a period of time, further increasing the effective size of the population (145). Multiple insemination and sperm storage means that even a colony founded by a single individual will contain on the average six haploid genomes. In spite of these effects (which may themselves have evolved to reduce inbreeding and loss of alleles through drift), many populations of *Cepaea* are very small and are isolated in patches of suitable habitat. The liability to interpopulation divergence by random drift is increased by the population bottlenecks caused by the frequent failure of whole year-classes to reproduce. For example, in a population of *C. nemoralis* living on a chalk hillside, annual recruitment of adults varied from 30 to 1400 over a six-year period (187).

Evidence that random processes could be an important cause of divergence among *Cepaea* populations came from a survey of heterozygosity at the shell-banding locus in relation to population size in French colonies of *C. nemoralis* (131–133). The variance in allele frequencies among small populations was estimated to be considerably greater than the variance among large populations. By assuming that the mutation rate at this locus was approximately 10^{-4} (a remarkably high figure) it was possible to fit the observed distribution of phenotype frequencies to those expected to arise from a balance between the random gain and loss of alleles.

Aspects of this analysis have been severely criticized (34, 36, 58, 177). Arbitrary adjustments of estimates of mutation rate, migration rate, and population size make it possible for almost any distribution of gene frequencies to accord with that predicted on the hypothesis of genetic drift. Models of drift show that divergence between populations is greatly reduced by even small amounts of long-distance migration (70). Such migration is now known to be important in other organisms. Snails and their eggs are carried for surprising distances affixed to birds' legs, and even adult molluscs are liable to transport by high winds (162). A helicid mollusc has been found at the top of St. Peter's Basilica in Rome, whence it was presumably

POLYMORPHISM IN *CEPAEA*   129

carried by a bird or a gust of wind. Long distance migration of this kind was not considered when estimating gene flow among the French populations. Also, heterogeneous data were combined. It is now known that some of the populations respond to visual selection (5). Finally, no information was given on the size of the sampling areas. It is possible that the large populations came from larger and hence more varied habitats than did the small. The decreased variance in allele frequencies among large colonies could then be explained as being due to natural selection. Their habitats would be the average of several habitat types and would hence be more similar to each other than would those of the small colonies, each of which came from a single type of habitat (177).

The analysis of French *C. nemoralis* populations cannot be accepted as firm evidence that drift has an important general effect on the polymorphism. However, it remains true that random loss of alleles must occur in any population which is sufficiently small and isolated. Without a more accurate analysis of population sizes in relation to heterozygosity the importance of the action of random processes acting in each generation in small *Cepaea* populations cannot be estimated.

There is however one random process which is clearly important in some populations. This is the founder effect. *Cepaea* is a very opportunistic snail, which rapidly invades suitable habitats. *C. nemoralis* in particular often lives in disturbed and transient habitats in which there must be frequent population extinctions and recolonizations. It has been aptly described as a "molluscan weed." As in other weeds (10), sampling errors in the founding colony may have an important influence on the genetic structure of at least the early generations of a new population. This will of course tend to promote divergence among populations. The founder effect can easily be identified in the *C. nemoralis* populations of North America. The first known introduction of this species was in the mid-nineteenth century, and it has since become widespread in eastern North America with isolated colonies as far west as Vancouver (80, 81, 161). Local colonizations and extinctions are common (161). In a few cases, the origin of a local colony can be traced, and it is sometimes found that an allele present in the parental population is absent in the new colony because of sampling effects (163). Occasionally it has been possible to locate the part of Europe in which an American colony originated. Allozyme frequencies show that the genetic structure of the American population is usually a function of its history of origin rather than of any new selective forces which may be acting (22).

Historical accidents have also played a part in determining gene frequencies in European populations of *Cepaea*. Isolated colonies of *C. nemoralis* north of the limit of general distribution of this species in Scotland have much lower levels of heterozygosity for the shell polymorphism than do most populations in this species' main range. In some cases these colonies are monomorphic (119). This is probably a result of the founder effect. In parts of the low lying fens of East Anglia, populations of *Cepaea* were wiped out by extensive flooding in 1948. Snails rapidly recolonized the area after the floods had subsided. In 1952 a survey of *C. nemoralis* populations living along a roadside verge which had been submerged for several months showed that there were great fluctuations in allele frequency between adjacent colonies (92). There were no apparent associations with the environment, which was in any case remarkably uniform along the length of the bank (Figure 3a). Reexamination of the populations 16 years later (94) showed that gene frequencies had not changed significantly at any locality. Since this period represents only 3 or 4 snail generations, the result is not surprising. Similar extreme microgeographical differences in morph frequency without apparent relation to the environment are found in *C. nemoralis* populations which have colonized the newly drained Dutch polders (192). Once again, the founder effect is presumably the primary agent responsible for genetic differentiation among populations.

Random processes are strongly implicated in several of the few *Cepaea* populations for which historical information is available. It may be that the founder effect is also important in determining the present-day distribution of genes in other *Cepaea* populations with unknown histories. In unstable habitats such as sand dunes or roadside verges population bottlenecks must occur quite frequently. This may explain why there is often extremely localized genetic differentiation among such populations.

Suggestions that sampling errors in founding populations which later expanded are responsible for the areas of homogeneous allele frequency found in many upland populations (Figure 3c) are unlikely to be correct (30, 93). The boundaries of area effects at different loci are often not contiguous, so that some of the steepest gene frequency steps for shell banding occur within areas of constant frequency of the shell color genes. It is difficult to see how this pattern could result from the expansion of a founding population. The stability of some area effects over thousands of years is also hard to explain unless they are actively maintained.

The hypothesis that the distribution of genes is determined by historical or modern sampling accidents is difficult to disprove. Uncritical use of this hypothesis without attempting to test alternative explanations has led many biologists to question its value. Nevertheless the evidence for the action of random processes in *Cepaea* is strong.

Linkage Disequilibrium and Coadaptation

The close linkage of several of the loci which control shell polymorphism is itself evidence that their interaction is subject to selection. Even loci which are not part of the supergene (such as U in Figure 2) are associated with it without the necessity of linkage, as they are expressed only as modifiers of its effects. There are usually strong linkage disequilibria within the supergene in both *C. nemoralis* and *C. hortensis* (Figure 7). The mechanism responsible for these can sometimes be identified. In regions where *C. nemoralis* responds to visual selection, pink and unbanded shells are favored in woods, yellow and five banded in open habitats. The joint action of selection on these two loci often leads to disequilibrium between them. In other cases, the selective mechanisms producing disequilibria are harder to identify (136). For example, the direction of disequilibrium between color and banding in *C. nemoralis* is reversed in two Pyrenean valleys with no apparent relationship to the ecological differences between them (Figure 7a). Similar geographic changes in disequilibrium exist in *C. hortensis* (Figure 7b). In *C. nemoralis* there is an almost absolute disequilibrium between the allele for dark brown shell color and that for

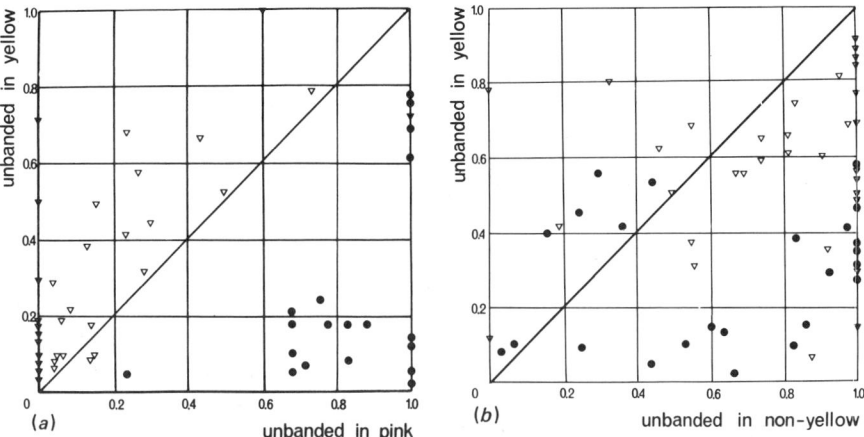

Figure 7 Linkage disequilibria between color and banding in *C. nemoralis* (7a) and *C. hortensis* (7b). (a) ▽ Garonne Valley; ● Segre Valley of Pyrenees (3) (b) ▽ South Germany (172); ● South Central England.

the absence of shell bands (38). How this is maintained is not known. Some linkage disequilibria are shared by both species. Thus, in both *C. nemoralis* and *C. hortensis* there is usually an excess of white lips among yellow shells in populations segregating at both loci. This disequilibrium therefore probably arose before speciation and its conservation over many generations within each species may be evidence that it is maintained by selection.

Disequilibria can also arise from sampling errors in small populations (114). In the absence of selection the linked loci will return to equilibrium at a rate depending on their rate of recombination (70). An example of this was provided by the *C. nemoralis* population which recolonized a Cambridgeshire river bank after flooding. Soon after the recolonization there was at one locality a very significant excess of unbanded shells among the pink. This excess bore no apparent relation to any local peculiarity of habitat. Sixteen years later this disequilibrium had disappeared, although there had been no changes in allele frequencies at either of the loci (94). The original disequilibrium probably resulted from the population bottleneck which occurred after the flooding, but was not maintained in the absence of selection.

It has often been suggested that it is meaningless to study polymorphisms at single loci without reference to the genetic context in which they exist. Linkage disequilibria between the locus actually studied and undetected loci which respond to selection may often be important in controlling the distribution of polymorphisms (141). The hypothesis that "coadaptations" of this kind can affect the fitness of the shell polymorphisms in *Cepaea* is attractive. It might explain how differences in gene frequency arise without the need to invoke subtle and often undetectable environmental selection. This is particularly true for area effects (Figure 3c), between which sharp steps in gene frequency may occur without any obvious relationship to the environment. Sampling errors in founding populations followed by local coadaptation of the isolated populations might explain such mosaic patterns of population structure. When the isolated populations expand and meet, hybrids between them are relatively unfit, so that stepped clines develop (93).

As gene frequency steps at different loci are rarely concordant, this simple model of genetic divergence between *Cepaea* populations is unlikely to be correct (30). However, several models (54, 84, 201) now suggest that different coadaptation could arise without geographic isolation. Accumulation of different modifying genes in adjacent populations could lead to the steepening of an originally gradual cline at a locus controlling shell polymorphism. The adjacent populations would then attain very different allele frequencies at this locus. Each area effect is a separately evolved population which represents on a very small scale a geographic race or even a subspecies. In principle such divergence could lead to reproductive isolation and hence to the evolution of new species by "parapatric" speciation (8, 55).

The pattern of distribution of some species of low mobility does suggest that they arose by parapatric speciation (8). The Polynesian snail *Partula* (which has area effects similar to those of *Cepaea*) is found on many small and isolated islands. Most islands have a number of endemic species. It is hard to identify geographic patterns of speciation that could be explained on the allopatric model. There are zones of hybridization at species' borders that are easily explained if speciation resulted from the evolutionary divergence of adjacent populations (62). If *Cepaea* area effects do indeed represent an early stage of speciation, this adds new interest to their study. Several attempts have been made to test whether the regions of uniform gene frequency found in many *Cepaea* populations are more than simple one-locus selective responses to the environment.

Subspecies often differ from each other at a significant proportion of the loci detected by electrophoresis. Several populations of *C. nemoralis* have been examined for possible coincidences of molecular and visible population differentiation. The results are equivocal. In populations on the Berkshire Downs which are strongly differentiated from each other at the U and S loci (Figure 2) there are parallel differences at two allozyme loci (118). This effect is far from absolute and the geographic patterns of other protein polymorphisms do not coincide with area effects for shell characters. In North Wales and Cumbria where there are strong area effects there are no clear relationships between the patterns of visible and molecular population differentiation (Jones, Selander & Hudson, unpublished). There are topological problems associated with measuring the geographic covariance of overlapping patterns, especially when only a few samples are available. However, the present evidence is that geographic differences in allele frequencies for the shell polymorphism are not accompanied by marked differences at a random sample of other loci. Overall population differentiation at the molecular level is much less evident between area effects in *Cepaea* than between, for example, subspecies of mice (175).

There is nevertheless some evidence that the fitness of alleles controlling the shell polymorphism can be influenced by their genetic background. The identical mode of inheritance of the shared polymorphisms of *C. nemoralis* and *C. hortensis* suggests that the same genes are involved and that the polymorphism arose before speciation. However, there are often great differences between *C. nemoralis* and *C. hortensis* in overall morph frequencies, which reflect different responses to selection of the shell polymorphism within each species. Selective responses are affected by interactions with other polymorphisms which have accumulated since the species diverged.

Figure 8 shows the distribution of yellow shells in European populations of *C. hortensis*. In *C. nemoralis* (Figures 5 and 6) the frequency of this allele is closely associated with temperature. This results from a selective response at this locus to the relatively low uptake of solar energy by yellow shells. In *C. hortensis*, however, there are no obvious trends in the distribution of yellow shell color, and no clear associations with temperature (Figure 9). (A barely significant negative association of the frequency of yellow with mean July temperature is due to uneven sampling.) The selective value of this allele is therefore affected by the "genetic environment" of the species in which it is placed.

Differences in genetic environment between populations of the same species may also affect the shell polymorphism. In *C. nemoralis* the degree of dominance of some

Figure 8 Frequency of yellow shells (white sector) in European populations of *C. hortensis* mapped on the basis of 100 km grid squares. Only grid squares containing more than 100 snails in more than 10 samples included (16, 19, 28, 29, 31, 39, 47, 49, 51, 53, 90, 104, 119, 138, 149, 172, 179, plus unpublished data).

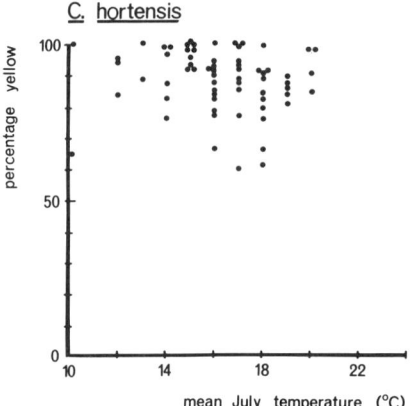

Figure 9 Association between the frequency of yellow shells and mean July temperature in *C. hortensis.*

alleles varies from population to population (33, 38, 194), probably as a result of differences in their interaction with local modifying genes. Such interactions may lead to differences in the response to similar selective forces in geographically separated populations of the same species. In Oxfordshire, woodland *C. nemoralis* populations have a relatively high frequency of pink shells, while woodland *C. hortensis* in this region have a high frequency of shells with their bands fused together (Figure 4a and 4b). Elsewhere, however, these evolutionary strategies are reversed (49, 163); *C. hortensis* in woodland responds to visual selection by an increase in the frequency of pinks, while *C. nemoralis* increases the frequency of fusions (Figure 4c and 4d). The frequency of alleles at these loci therefore results from their response to both the ecologic and the genetic environment.

Responses to microclimate may also depend on the genetic background of local populations. In some Pyrenean valleys unbanded *C. nemoralis* are much more common at the highest and lowest altitudes than in intermediate zones. This has been interpreted as resulting from the greater resistance of unbanded shells to climatic extremes (3, 4). In other Pyrenean valleys however, this pattern is not found (44, 126, 134, 135) and in some places is even reversed (Figure 10). This effect might again reflect interactions within local populations between the shell banding locus and other loci which modify its response to selection.

Local "coadaptations" of this kind could explain much of the population differentiation in *Cepaea* without having to depend on apparently ad hoc arguments based on cryptic environmental selection. However, the hypothesis of coadaptation is difficult to test and easy to abuse. It has sometimes been used, without sufficient consideration of natural selection's complexity and subtlety, to dismiss the possibility that selection acts directly on the shell polymorphism. However, there is good evidence that genetic interactions can in some circumstances modify responses at the loci controlling shell polymorphism.

POLYMORPHISM IN *CEPAEA* 135

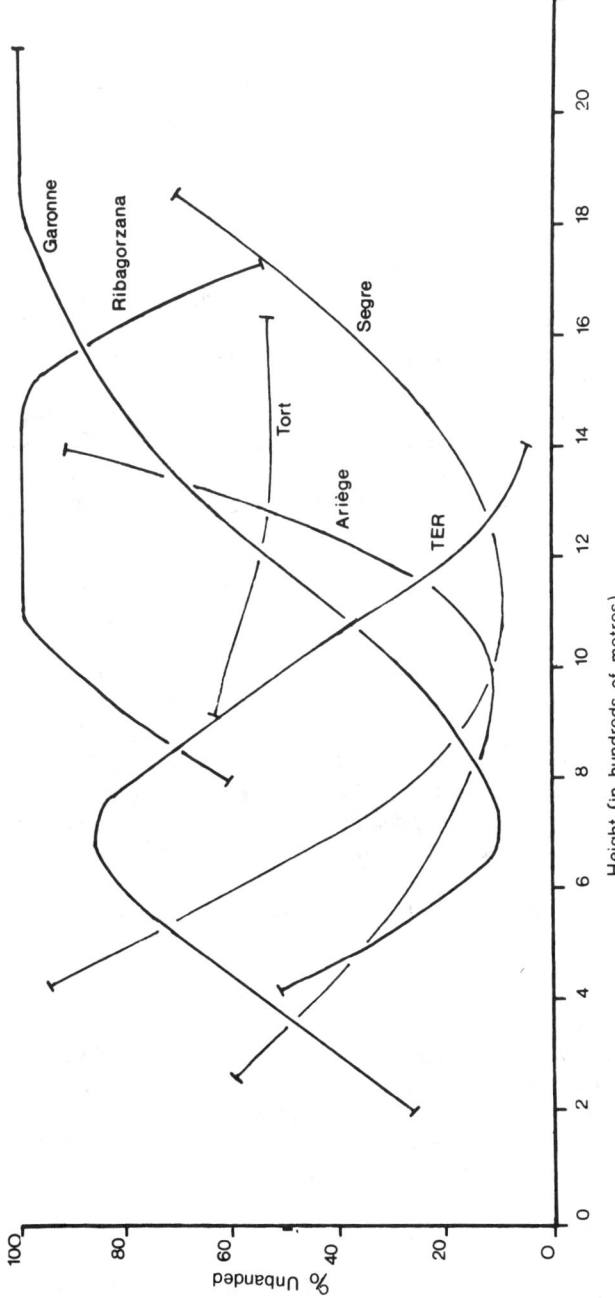

Figure 10 Patterns of association of the frequency of unbanded *C. nemoralis* with altitude in various Pyrenean valley systems (126).

CONCLUSIONS

Cepaea has a relatively simple system of variation. Nevertheless, at least eight evolutionary forces are now known to affect its shell polymorphism. In most populations, several of these act simultaneously to control morph frequency. It is possible to imagine, for example, a grassland *C. nemoralis* population in which the frequency of yellow shells is determined first by the cool climate of its location in northern Europe, and also by the fact that it is in a valley in which cold air accumulates. Predators act to improve the camouflage of the population against its green background by increasing the frequency of yellow shells. The selective advantage of this allele under predation is limited by the predators' frequency-dependent searching behavior, which favors the rarer alternative color alleles. This itself is modulated by the density of the *C. nemoralis* population and by the presence of pink *C. hortensis* in the area. Gene flow from an adjacent woodland population increases the frequency of nonyellow shells. Finally, there is a fortuitous local linkage disequilibrium between the allele for yellow shell and that for the presence of shell bands resulting from a recent population bottleneck. This increases the selective advantage of the yellow allele in this population to a level greater than that in a nearby population in which the disequilibrium is in the opposite direction.

The relative importance of the various mechanisms will vary from population to population, but almost nowhere will it be possible to explain the observed pattern of genetic variation by invoking only one of them. Complex and perhaps unique explanations are needed for almost every *Cepaea* population. It is not possible to produce general rules from detailed studies of a few populations.

Authors writing at a time when relatively little information was available on *Cepaea* emphasized the importance of random processes in controlling the distribution of genes. In 1932 Haldane estimated that the selective differences between alleles were unlikely to exceed 10^{-5} (104a). Ten years later Dobzhansky suggested that the observed gene-frequency distributions were a good example of "haphazard differences ... which are expected to arise from the random fixation or loss of genes in colonies with limited effective population size" (79). Similarly Huxley (1942) claimed that in *Cepaea* "the distribution of types appears to be wholly random" and Mayr (1942) that such variation was "obviously of very insignificant selective value" (116, 143). Opinion has now moved to the opposite extreme, to such an extent that Lewontin (1974) suggests that "the case of *Cepaea* is regarded as a paradigm by selectionists" (141).

We hope to have shown that this polymorphism cannot be explained by simplistic models which depend entirely on random processes, entirely on one type of selection, or even on a few types of selection. Acceptance of one evolutionary factor need not necessarily involve the rejection of others. The moral for the currently fashionable topic of protein polymorphism should be clear.

ACKNOWLEDGMENTS

We are most grateful to Drs. R. W. Arnold, C. R. Bantock, A. J. Cain, M. A. Carter, B. C. Clarke, L. M. Cook, C. B. Goodhart, J. J. D. Greenwood, P. H. Harvey, M.

H. Williamson, P. Williamson, and H. Wolda for providing us with unpublished data on *Cepaea*; to C. R. Bantock, R. J. Berry, and R. Probert for their comments on the manuscript; and to Ms. M. Roper for her help in preparing the manuscript. Our research on *Cepaea* is supported by the Natural Environment Research Council and the Royal Society.

Literature Cited

1. Allen, J. A. 1976. Further evidence for apostatic selection by wild passerine birds—9:1 experiments. *Heredity* 36: 173–80
2. André, J. 1973. Recherches écologiques sur les populations de *Cepaea nemoralis* du Languedoc et du Rousillon. PhD thesis. Univ. Paris. 117 pp.
3. Arnold, R. W. 1968. Studies on *Cepaea*. VII. Climatic selection in *Cepaea nemoralis* (L.) in the Pyrenees. *Philos. Trans. R. Soc. London Ser. B* 253: 549–93
4. Arnold, R. W. 1969. The effects of selection by climate on the land snail *Cepaea nemoralis* (L.). *Evolution* 23: 370–78
5. Arnold, R. W. 1970. A comparison of populations of the polymorphic land snail *Cepaea nemoralis* (L.) living in a lowland district in France with those in a similar district in England. *Genetics* 64:589–604
6. Arnold, R. W. 1971. *Cepaea nemoralis* on the East Sussex South Downs, and the nature of area effects. *Heredity* 26:277–98
7. Aubertin, D. 1927. On the anatomy of the land snails (Helicidae) *Cepaea hortensis* (Mull) and *Cepaea nemoralis* (L.). *Proc. Zool. Soc. London* 1927: 553–82
8. Ayala, F. J. 1975. Genetic differentiation during the speciation process. In *Evolutionary Biology*, ed. Th. Dobzhansky, M. K. Hecht, W. C. Steere, 8:1–78. New York: Plenum
9. Ayala, F. J., Campbell, C. A. 1974. Frequency-dependent selection. *Ann. Rev. Ecol. Syst.* 5:115–38
10. Baker, H. G. 1974. The evolution of weeds. *Ann. Rev. Ecol. Syst.* 5:1–24
11. Bantock, C. R. 1972. Localisation of chiasmata in *Cepaea nemoralis* L. *Heredity* 29:213–21
12. Bantock, C. R. 1974. Experimental evidence for non-visual selection in *Cepaea nemoralis*. *Heredity* 33:409–37
13. Bantock, C. R. 1974. *Cepaea nemoralis* (L.) on Steep Holm. *Proc. Malacol. Soc. London* 41:223–32
14. Bantock, C. R., Bayley, J. A. 1973. Visual selection for shell size in *Cepaea*. *J. Anim. Ecol.* 42:247–61
15. Bantock, C. R., Bayley, J. A., Harvey, P. H. 1975. Simultaneous selective predation on two features of a mixed sibling species population. *Evolution* 29: 636–49
16. Bantock, C. R., Noble, K. 1973. Variation with altitude and habitat in *Cepaea hortensis* (Mull). *J. Linn. Soc. London Zool.* 53:237–52
17. Bantock, C. R., Noble, K., Ratsey, M. 1973. Sinistrality in *Cepaea hortensis*. *Heredity* 30:397–98
18. Bantock, C. R., Price, D. J. 1975. Marginal populations of *Cepaea nemoralis* (L.) on the Brendon Hills, England. I. Ecology and ecogenetics. *Evolution* 29:267–77
19. Bengtson, S.-A., Nilsson, A., Nordstrom, S., Rundgren, S. 1976. Polymorphism in relation to habitat in the snail *Cepaea hortensis* in Iceland. *J. Zool.* 178:173–88
20. Boettger, C. R. 1954. Zur Frage der Verteilung bestimmter Varianten bei der Landschneckengattung *Cepaea* Held. *Biol. Zentralbl.* 73:318–33
21. Brussard, P. F. 1974. Population size and natural selection in the land snail *Cepaea nemoralis*. *Nature* 251:713–15
22. Brussard, P. F. 1975. Geographic variation in North American colonies of *Cepaea nemoralis*. *Evolution* 29:402–10
23. Brussard, P. F., McCracken, G. F. 1974. Allozymic variation in a North American colony of *Cepaea nemoralis*. *Heredity* 33:98–101
24. Bundgaard, J., Christiansen, F. B. 1972. Dynamics of polymorphisms. I. Selection components in an experimental population of *Drosophila melanogaster*. *Genetics* 71:439–60
25. Cain, A. J. 1953. Visual selection by tone of *Cepaea nemoralis*. *J. Conchol.* 23:333–36
26. Cain, A. J. 1968. Studies on *Cepaea*. V. Sand dune populations of *Cepaea nemoralis* (L.). *Philos. Trans. R. Soc. London Ser. B* 253:499–517

27. Cain, A. J. 1971. Colour and banding morphs in subfossil samples of the snail *Cepaea*. In *Ecological Genetics and Evolution*, ed. R. Creed, pp. 65–92. Oxford: Blackwell
28. Cain, A. J., Cameron, R. A. D., Parkin, D. T. 1969. Ecology and variation of some helicid snails in Northern Scotland. *Proc. Malacol. Soc. London* 38:269–99
29. Cain, A. J., Currey, J. D. 1963. Area effects in *Cepaea*. *Philos. Trans. R. Soc. London Ser. B* 246:1–81
30. Cain, A. J., Currey, J. D. 1963. The causes of area effects. *Heredity* 18:467–71
31. Cain, A. J., Currey, J. D. 1963. Area effects in *Cepaea* on the Larkhill Artillery ranges, Salisbury Plain. *J. Linn. Soc. London Zool.* 45:1–15
32. Cain, A. J., Currey, J. D. 1968. Studies on *Cepaea*. III. Ecogenetics of a population of *Cepaea nemoralis* (L.) subject to strong area effects. *Philos. Trans. R. Soc. London Ser. B* 253:447–82
33. Cain, A. J., King, J. M. B., Sheppard, P. M. 1960. New data on the genetics of polymorphism in the snail *Cepaea nemoralis* L. *Genetics* 45:393–411
34. Cain, A. J., Sheppard, P. M. 1950. Selection in the polymorphic land snail *Cepaea nemoralis*. *Heredity* 4:275–94
35. Cain, A. J., Sheppard, P. M. 1952. The effects of natural selection on body colour in the land snail *Cepaea nemoralis*. *Heredity* 6:217–23
36. Cain, A. J., Sheppard, P. M. 1954. Natural selection in *Cepaea*. *Genetics* 39:89–116
37. Cain, A. J., Sheppard, P. M. 1957. Some breeding experiments with *Cepaea nemoralis*. *J. Genet.* 55:195–99
38. Cain, A. J., Sheppard, P. M., King, J. M. B. 1968. Studies on *Cepaea*. I. The genetics of some morphs and varieties of *Cepaea nemoralis* (L.). *Philos. Trans. R. Soc. London Ser. B* 253:383–96
39. Cameron, R. A. D. 1969. The distribution and variation of three species of land snail near Rickmansworth, Hertfordshire. *J. Linn. Soc. London Zool.* 48:83–111
40. Cameron, R. A. D. 1969. Predation by song thrushes *Turdus ericetorum* (Turton) on the snails *Cepaea hortensis* (Mull) and *Arianta arbustorum* (L.) near Rickmansworth. *J. Anim. Ecol.* 38:547–53
41. Cameron, R. A. D. 1970. The effect of temperature on the activity of three species of helicid snail (Mollusca: Gastropoda). *J. Zool.* 162:303–15
42. Cameron, R. A. D. 1970. The survival, weight-loss and behaviour of three species of land snail in conditions of low humidity. *J. Zool.* 160:143–57
43. Cameron, R. A. D. 1970. Differences in the distributions of three species of helicid snail in the limestone district of Derbyshire. *Proc. R. Soc. London Ser. B* 176:131–59
44. Cameron, R. A. D., Carter, M. A., Haynes, F. N. 1973. The variation of *Cepaea nemoralis* in three Pyrenean valleys. *Heredity* 31:43–74
45. Cameron, R. A. D., Cook, L. M. 1971. *Cepaea nemoralis* on Whitbarrow Scar, Lancashire. *Proc. Malacol. Soc. London* 39:399–408
46. Cameron, R. A. D., Williamson, P. 1977. Estimating migration and the effects of disturbance in mark-recapture studies on the snail *Cepaea nemoralis* L. *J. Anim. Ecol.* 46:173–80
47. Carter, M. A. 1967. Selection in mixed colonies of *Cepaea nemoralis* and *Cepaea hortensis*. *Heredity* 22:117–39
48. Carter, M. A. 1968. Thrush predation of an experimental population of the snail *Cepaea nemoralis* (L.). *Proc. Linn. Soc. London* 179:241–49
49. Carter, M. A. 1968. Studies on *Cepaea*. II. Area effects and visual selection in *Cepaea nemoralis* (L.) and *Cepaea hortensis*. *Philos. Trans. R. Soc. London Ser. B* 253:397–446
50. Cesnola, A. P. D. 1907. A first study of natural selection of *Helix arbustorum*. *Biometrika* 5:387–99
51. Clarke, B. C. 1960. Divergent effects of natural selection on two closely related polymorphic snails. *Heredity* 14:423–43
52. Clarke, B. C. 1962. Balanced polymorphism and the diversity of sympatric species. *Syst. Assoc. Publ.* 4:47–70
53. Clarke, B. C. 1962. Natural selection in mixed populations of two polymorphic snails. *Heredity* 17:319–45
54. Clarke, B. C. 1966. The evolution of morph-ratio clines. *Am. Nat.* 100:389–402
55. Clarke, B. C. 1968. Balanced polymorphism and regional differentiation in land snails. In *Evolution and Environment*, ed. E. T. Drake, pp. 351–68. New Haven: Yale Univ. Press
56. Clarke, B. C. 1969. The evidence for apostatic selection. *Heredity* 24:347–52
57. Clarke, B. C. 1972. Density-dependent selection. *Am. Nat.* 106:1–13

58. Clarke, B. C., Arthur, W., Horsley, D. T., Parkin, D. T. 1978. Genetic variation and natural selection in pulmonate molluscs. In *The Pulmonates*, ed. J. Peake. London: Academic. In press
59. Clarke, B. C., Diver, C., Murray, J. J. 1968. Studies on *Cepaea*. VI. The spatial and temporal distributions of phenotypes in a colony of *Cepaea nemoralis* (L.). *Philos. Trans. R. Soc. London Ser. B* 253:519–48
60. Clarke, B. C., Murray, J. J. 1962. Changes in gene frequency in *Cepaea nemoralis* (L.). *Heredity* 17:445–65
61. Clarke, B. C., Murray, J. J. 1962. Changes in gene frequency in *Cepaea nemoralis* (L.); the estimation of selective values. *Heredity* 17:467–76
62. Clarke, B. C., Murray, J. J. 1969. Ecological genetics and speciation in land snails of the genus *Partula*. *J. Linn. Soc. London Biol.* 1:31–42
63. Cook, L. M. 1964. Inheritance of shell size in the snail *Arianta arbustorum*. *Evolution* 19:86–94
64. Cook, L. M. 1965. A note on apostasy. *Heredity* 20:631–36
65. Cook, L. M. 1967. The genetics of *Cepaea nemoralis*. *Heredity* 22:397–410
66. Cook, L. M. 1969. Results of breeding experiments of Diver and Stelfox on *Helix aspersa*. *Proc. Malacol. Soc. London* 38:351–58
67. Cook, L. M., Murray, J. J. 1966. New information on the inheritance of polymorphic characters in *Cepaea hortensis*. *J. Hered.* 57:245–47
68. Cook, L. M., O'Donald, P. 1971. Shell size and natural selection in *Cepaea nemoralis*. In *Ecological Genetics and Evolution*, ed. E. R. Creed, pp. 93–108. Oxford: Blackwell
69. Creed, E. R. 1975. Melanism in the two spot ladybird: the nature and intensity of selection. *Proc. R. Soc. London Ser. B* 190:135–48
70. Crow, J. F., Kimura, M. 1970. *An Introduction to Population Genetics Theory*. New York: Harper & Row. 591 pp.
71. Currey, J. D., Arnold, R. W., Carter, M. A. 1964. Further examples of variation of populations of *Cepaea nemoralis* with habitat. *Evolution* 18:111–17
72. Currey, J. D., Cain, A. J. 1968. Studies on *Cepaea*. IV. Climate and selection of banding morphs in *Cepaea* from the climatic optimum to the present day. *Philos. Trans. R. Soc. London Ser. B* 253:483–98
73. Davies, P. W., Snow, D. W. 1965. Territory and food of the song thrush. *Br. Birds* 58:161–75
74. Day, J. C. L., Dowdeswell, W. H. 1968. Natural selection in *Cepaea* on Portland Bill. *Heredity* 23:169–88
75. Diver, C. 1929. Fossil records of mendelian mutants. *Nature* 124:183
76. Diver, C. 1931. A method of determining the number of whorls on a shell and its application to *Cepaea hortensis* (Mull) and *C. nemoralis* (L.) *Proc. Malacol. Soc. London* 19:234–39
77. Diver, C. 1932. Mollusc genetics. *Proc. 6th Int. Congr. Genet., Ithaca* 2:236–39
78. Diver, C. 1940. The problem of closely related species living in the same area. In *The New Systematics*, ed. J. Huxley, pp. 303–28. London: Oxford Univ. Press
79. Dobzhansky, T. 1941. *Genetics and the Origin of Species*. New York: Columbia Univ. Press. 505 pp. Rev. ed.
80. Draycot, W. M. 1961. Mollusks introduced into British Columbia. *Can. Field Nat.* 75:164
81. Dundee, D. S. 1974. Catalog of introduced molluscs of eastern North America (north of Mexico). *Sterkiana* 55:1–37
82. Elton, R. A., Greenwood, J. J. D. 1970. Exploring apostatic selection. *Heredity* 25:629–33
83. Emberton, L. R. B., Bradbury, S. 1963. Transmission of light through shells of *Cepaea nemoralis* (L.). *Proc. Malacol. Soc. London* 35:211–19
84. Endler, J. A. 1973. Gene flow and population differentiation. *Science* 179:243–50
85. Evans, H. J. 1960. Supernumary chromosomes in wild populations of the snail *Helix pomatia*. *Heredity* 15:129–38
86. Favre, J. 1927. *Les Mollusques postglaciares et actuels du bassin de Geneve*. *Mem. Soc. Phys. Hist. Nat. Geneve* 40:171–434
87. Fisher, R. A., Diver, C. 1934. Crossing-over in the land snail *Cepaea nemoralis* (L.). *Nature* 133:834–35
88. Fromming, E. 1954. *Biologie der mitteleuropaischen Landgastropoden*. Berlin: Duncker & Humboldt. 389 pp.
89. Goodhart, C. B. 1956. Genetic stability in populations of the polymorphic snail *Cepaea nemoralis* (L.). *Proc. Linn. Soc. London* 167:50 67
90. Goodhart, C. B. 1958. Thrush predation in the snail *Cepaea hortensis* (L.). *J. Anim. Ecol.* 27:47–58

91. Goodhart, C. B. 1962. Genetic stability in the snail *Cepaea nemoralis* (L.): a further example. *Proc. Linn. Soc. London* 169:163-67
92. Goodhart, C. B. 1962. Variation in a colony of the snail *Cepaea nemoralis* (L.). *J. Anim. Ecol.* 31:207-37
93. Goodhart, C. B. 1963. "Area effects" and non-adaptive variation between populations of *Cepaea* (Mollusca). *Heredity* 18:459-65
94. Goodhart, C. B. 1973. A sixteen-year survey of *Cepaea* on the Hundred-Foot Bank. *Malacologia* 14:327-31
95. Greenwood, J. J. D. 1969. Apostatic selection and population density. *Heredity* 24:157-61
96. Greenwood, J. J. D. 1974. Visual and other selection in *Cepaea:* a further example. *Heredity* 33:17-31
97. Greenwood, J. J. D. 1974. Effective population numbers in the snail *Cepaea nemoralis*. *Evolution* 28:513-26
98. Grime, J. P., Blythe, G. M. 1969. An investigation of the relationships between snails and vegetation at the Winnats Pass. *J. Ecol.* 57:45-66
99. Grime, J. P., MacPherson-Stewart, S. F., Dearman, R. S. 1968. An investigation of leaf palatability using the snail *Cepaea nemoralis* (L.) *J. Ecol.* 56: 405-20
100. Guerrucci-Henrion, M.-A. 1966. Recherches sur les populations naturelles de *Cepaea nemoralis* en Bretagne. *Arch. Zool. Exp. Gen.* 107:369-417
101. Guerrucci-Henrion, M.-A. 1971. Etude de la transmission de quelques caracteres de la pigmentation chez *Cepaea hortensis*. *Arch. Zool. Exp. Gen.* 112: 211-19
102. Guerrucci-Henrion, M.-A. 1973. Etude de la transmission du caractere peristome colore chez *Cepaea hortensis*. *Arch. Zool. Exp. Gen.* 114:313-6
103. Guerrucci-Henrion, M.-A. 1973. Aspects generaux du polymorphisme de la couleur du peristome chez *Cepaea hortensis* en France. *Malacologia* 14: 333-38
104. Guerrucci-Henrion, M.-A. 1974. Le polymorphisme de la coquille chez *Cepaea hortensis* Muller (Mollusques pulmones) en France. *Mem. Soc. Zool. Fr.* 37:103-27
104a. Haldane, J. B. S. 1932. *The Causes of Evolution*. London: Longmans, Green. 235 pp.
105. Harvey, P. H. 1971. *Cepaea nemoralis* in Brittany, Cornwall and Pembrokeshire. *Heredity* 26:365-72
106. Harvey, P. H. 1971. Populations of *Cepaea nemoralis* from south-western France and northern Spain. *Heredity* 27:353-63
107. Harvey, P. H. 1972. *Cepaea nemoralis* on cliff-tops in south-west England. *Proc. R. Soc. London Ser. B* 181:375-93
108. Harvey, P. H. 1973. The distribution of three species of helicid snail in East Yorkshire. I. General Survey. *Proc. Malacol. Soc. London* 40:523-30
109. Harvey, P. H. 1974. The distribution of three species of helicid snail in East Yorkshire. II. Intensive Survey. *Proc. Malacol. Soc. London* 41:57-64
110. Harvey, P. H. 1976. Factors influencing the shell pattern of *Cepaea nemoralis* (L.) in east Yorkshire: a test case. *Heredity* 36:1-10
111. Harvey, P. H., Birley, N., Blackstock, T. H. 1975. The effect of experience on the selective behaviour of song thrushes feeding on artificial populations of *Cepaea* (Held.). *Genetica* 45:211-16
112. Harvey, P. H., Jordan, C. A., Allen, J. A. 1974. Selection behaviour of wild blackbirds at high prey densities. *Heredity* 32:401-4
113. Heath, D. J. 1975. Colour, sunlight and internal temperatures in the land snail *Cepaea nemoralis* (L.). *Oecologia* 19: 29-38
114. Hill, W. G., Robertson, A. 1968. Linkage disequilibrium in finite populations. *Theor. Appl. Genet.* 38:226-31
115. Holling, C. S. 1966. The functional response of invertebrate predators to prey density. *Mem. Entomol. Soc. Can.* 48: 1-85
116. Huxley, J. 1942. *Evolution, the Modern Synthesis*. London: George Allen & Unwin. 645 pp.
117. Jeppesen, L. L. 1976. The control of mating behaviour in *Helix pomatia* (L.) (Gastropoda: Pulmonata). *Anim. Behav.* 24:275-90
118. Johnson, M. S. 1976. Allozymes and area effects in *Cepaea nemoralis* on the western Berkshire Downs. *Heredity* 36:105-21
119. Jones, J. S. 1971. Studies on the ecological genetics of *Cepaea*. PhD thesis. Univ. Edinburgh. 342 pp.
120. Jones, J. S. 1973. The genetic structure of a southern peripheral population of the snail *Cepaea nemoralis*. *Proc. R. Soc. London Ser. B* 183:371-84
121. Jones, J. S. 1973. Ecological genetics of a population of the snail *Cepaea nemoralis* at the northern limit of its range. *Heredity* 31:201-11

122. Jones, J. S. 1973. Ecological genetics and natural selection in molluscs. *Science* 182:546–52
123. Jones, J. S. 1974. Area effects in the snail *Cepaea vindobonensis* in the Lika region of Yugoslavia. *Heredity* 32:165–70
124. Jones, J. S. 1974. Comment on letter by D. F. Owen. *Science* 185:376–77
125. Jones, J. S. 1975. The genetic structure of some steppe populations of *Cepaea vindobonensis. Genetica* 45:217–25
126. Jones, J. S., Irving, A. J. 1975. Gene frequencies, genetic background and environment in Pyrenean *Cepaea* populations. *J. Linn. Soc. London Biol.* 7:262–71
127. Jones, J. S., Parkin, D. T. 1977. Experimental manipulation of some snail populations subject to climatic selection. *Am. Nat.* In press
128. Kearsey, M. J., Kojima, K. I. 1967. The genetic architecture of body weight and egg hatchability in *Drosophila melanogaster. Genetics* 56:23–37
129. Kothbauer, H., Schnitzler, S. 1972. Haemagglutinins in *Cepaea hortensis, C. nemoralis* and *C. vindobonensis* (Helicidae, Gastropoda); their importance in systematics. *Z. Zool. Syst. Evolutionsforsch.* 10:133–37
130. Krebs, C. J., Myers, J. H. 1974. Population cycles in small mammals. *Adv. Ecol. Res.* 8:267–399
131. Lamotte, M. 1951. Recherches sur la structure genetique des populations naturelles de *Cepaea nemoralis* (L.). *Bull. Biol. Fr. Belg. Suppl.* 35:1–239
132. Lamotte, M. 1959. Polymorphism of natural populations of *Cepaea nemoralis. Cold Spring Harbor Symp. Quant. Biol.* 24:65–80
133. Lamotte, M. 1966. Les facteurs de la diversité du polymorphisme dans les populations naturelles de *Cepaea nemoralis* (L.). *Lav. Soc. Malacol. Ital.* 3:33–73
134. Lamotte, M. 1968. Les traits generaux du polymorphisme de la coquille dans les populations naturelles de *Cepaea nemoralis* (Mollusques Helicidae) des Pyrénées françaises. *C. R. Acad. Sci. Ser. D* 267:1318–21
135. Lamotte, M. 1968. Influence de l'altitude sur la frequence du caracter "absence de bandes" dans les populations de *Cepaea nemoralis* (Mollusques, Helicidae) des Pyrénées françaises. *C. R. Acad. Sci. Ser. D* 267:1649–52
136. Lamotte, M. 1969. Relations entre deux couples de caracteres dependant de deux locus etraitement lies: les caracteres jaune/rose et avec bandes/sans bandes dans les populations naturelles de *Cepaea nemoralis* (Mollusques Helicides) du Sud de l'Aquitaine. *C. R. Acad. Sci. Ser. D* 268:2476–79
137. Lamotte, M. 1972. The "white peristome" characteristic in the populations of *C. nemoralis* (L.) (pulmonate moll.) of the Ariège Valley. *C. R. Acad. Sci. Ser. D* 274:1558–61
138. Lamotte, M., Guerrucci, M.-A. 1970. Traits generaux du polymorphisme du systeme de bandes chez *Cepaea hortensis* (Mollusque Helicidae) en France. *Arch. Zool. Exp. Gen.* 111:393–409
139. Lang, A. 1908. Über die Bastarde von *Helix hortensis* Mueller und *H. nemoralis* L. *Festschr. Univ. Jena* 1908:1–120
140. Levan, G., Fredga, K. 1972. Isozyme polymorphism in three species of land snails. *Hereditas* 71:245–52
141. Lewontin, R. C. 1974. *The Genetic Basis of Evolutionary Change*. New York: Columbia Univ. Press. 346 pp.
142. Manwell, C., Baker, C. M. A. 1968. Genetic variation of isocitrate, malate, and 6-phosphogluconate dehydrogenases in snails of the genus *Cepaea*— introgressive hybridization, polymorphism and pollution? *Comp. Biochem. Physiol.* 26:195–209
143. Mayr, E. 1942. *Systematics and the Origin of Species*. New York: Columbia Univ. Press. 334 pp.
144. Morris, D. 1954. The snail-eating behaviour of thrushes and blackbirds. *Br. Birds* 47:33–49
145. Murray, J. J. 1964. Multiple mating and effective population size in *Cepaea nemoralis. Evolution.* 18:283–91
146. Murray, J. J. 1966. *Cepaea nemoralis* in the Isles of Scilly. *Proc. Malacol. Soc. London* 37:167–81
147. Murray, J. J. 1975. The genetics of the mollusca. In *Handbook of Genetics,* ed. R. C. King, 3:3–31. New York: Plenum
148. O'Donald, P. 1968. Natural selection by glow-worms in a population of *Cepaea nemoralis. Nature* 217:194
149. Owen, D. F., Bengtson, S.-A. 1972. Polymorphism in the land snail *Cepaea hortensis* in Iceland. *Oikos* 23:218–25
150. Oxford, G. S. 1973. The biochemical properties of esterases in *Cepaea* (Mollusca: Helicidae). *Comp. Biochem. Physiol.* 45:529–38
151. Oxford, G. S. 1973. The genetics of *Cepaea* esterases. I. *Cepaea nemoralis. Heredity* 30:127–39

152. Oxford, G. S. 1973. Molecular weight relationships of the esterases in *Cepaea nemoralis* and *Cepaea hortensis* and their genetic implications. *Biochem. Genet.* 8:365–82
153. Oxford, G. S. 1975. Food induced esterase phenocopies in the snail *Cepaea nemoralis*. *Heredity* 35:361–70
154. Perrot, J.-L., Perrot, M. 1938. Monographie des *Helix* du groupe *Cepaea*. Contribution a la notion d'espece. *Bull. Biol. Fr. Belg.* 72:232–59
155. Potts, D. C. 1975. Persistence and extinction of local populations of the garden snail *Helix aspersa* in unfavourable environments. *Oecologia* 21:313–34
156. Price, D. J. 1974. Variation in chiasma frequency in *Cepaea nemoralis*. *Heredity* 32:211–18
157. Price, D. J. 1975. Chiasma frequency variation with altitude in *Cepaea hortensis* (Mull). *Heredity* 35:221–29
158. Price, D. J. 1975. Position and frequency distribution of chiasmata in *Cepaea nemoralis* (L.). *Caryologia* 28:261–68
159. Price, D. J., Bantock, C. R. 1975. Marginal populations of *Cepaea nemoralis* (L.) on the Brendon Hills, England. II. Variation in chiasma frequency. *Evolution* 29:278–86
160. Prout, T. 1965. The estimation of fitnesses from genotypic frequencies. *Evolution* 19:546–51
161. Reed, C. F. 1964. *Cepaea nemoralis* in eastern North America. *Sterkiana* 16:11–8
162. Rees, W. 1965. The aerial dispersal of Mollusca. *Proc. Malacol. Soc. London* 36:269–82
163. Richards, A. V., Murray, J. J. 1975. The relation of phenotype to habitat in an introduced colony of *Cepaea nemoralis*. *Heredity* 34:128–31
164. Richardson, A. M. M. 1974. Differential climatic selection in natural populations of the land snail *Cepaea nemoralis*. *Nature* 247:572–73
165. Richardson, A. M. M. 1975. Food, feeding rates and assimilation in the land snail *Cepaea nemoralis* (L.). *Oecologia* 19:59–70
166. Richardson, A. M. M. 1975. Energy flux in a natural population of the land snail *Cepaea nemoralis* (L.). *Oecologia* 19:141–64
167. Richardson, A. M. M. 1975. Winter predation by thrushes, *Turdus ericetorum* (Turton) on a sand dune population of *Cepaea nemoralis* (L.). *Proc. Malacol. Soc. London* 41:481–88
168. Rotarides, M. 1926. Über die Bändervariation von *Cepaea vindobonensis* Fer. *Zool. Anz.* 67:28–44
169. Ruiter, L. De. 1958. Natural selection in *Cepaea nemoralis Arch. Neer. Zool.* 12:571–73
170. Sacchi, C. F., Valli, G. 1975. Recherches sur l'ecologie des populations naturelles de *Cepaea nemoralis* (L.) (Gastr., Pulmonata) en Lombardie meridionale. *Arch. Zool. Exp. Gen.* 116:549–78
171. Schilder, F. A. 1950. Die Ursachen der Variabilität bei *Cepaea*. *Biol. Zentralbl.* 69:79–103
172. Schilder, F. A., Schilder, M. 1953. *Die Bänderschnecken. Eine Studie zur Evolution der Tiere*. Jena: Fischer. 92 pp.
173. Schnetter, M. 1950. Veränderungen in genetischen Konstitution in natürlichen Populationen der polymorphen Bänderschnecken. *Verh. Dtsch. Zool. Ges. Marburg* 1950:192–206
174. Sedlmair, H. 1956. Verhaltens-, Resistenz-, und Gehäuseunterschiede bei den polymorphen Bänderschnecken *Cepaea hortensis* (Mull) und *Cepaea nemoralis* (L.). *Biol. Zentralbl.* 75:281–313
175. Selander, R. K., Hunt, W. G., Yang, S. Y. 1969. Protein polymorphism and genic heterozygosity in two European subspecies of the house mouse. *Evolution* 23:379–90
176. Sheppard, P. M. 1951. Fluctuations in the selective value of certain phenotypes in the polymorphic land snail *Cepaea nemoralis* (L.) *Heredity* 5:125–34
177. Sheppard, P. M. 1952. Natural selection in two colonies of the polymorphic land snail *Cepaea nemoralis*. *Heredity* 6:233–38
178. Tegelstrom, H., Haggstrom, A., Kvassman, S. 1975. Esterases of the snails *Helix pomatia* and *Cepaea hortensis*: variation and characteristics of different molecular forms. *Hereditas* 79:117–24
179. Valovirta, I., Halkka, O. 1976. Color polymorphism in northern peripheral populations of *Cepaea hortensis*. *Hereditas* 83:123–26
180. Walden, H. W. 1963. Historical and taxonomical aspects of the land Gastropoda in the North Atlantic Region. In *North Atlantic Biota and Their History*, ed. A. Love, D. Love, pp. 153–71. New York: Macmillan
181. Watt, W. B. 1968. Adaptive significance of pigment polymorphisms in *Colias* butterflies. I. Variation of melanin pigment in relation to thermoregulation. *Evolution* 22:437–58

182. Williamson, M. H. 1959. Differential damage in a mixed colony of the land snails *Cepaea nemoralis* and *C. hortensis. Heredity* 13:261–63
183. Williamson, P. 1975. Use of ^{65}Zn to determine the field metabolism of the snail *Cepaea nemoralis* (L.). *Ecology* 56: 1185–92
184. Williamson, P. 1976. Size–weight relationships and field growth rates of the land snail *Cepaea nemoralis* (L.). *J. Anim. Ecol.* 45:875–86
185. Williamson, P., Cameron, R. A. D. 1976. Natural diet of the land snail *Cepaea nemoralis. Oikos* 27:493–500
186. Williamson, P., Cameron, R. A. D., Carter, M. A. 1976. Population density affecting adult shell size of snail *Cepaea nemoralis* (L.). *Nature* 263:496–97
187. Williamson, P., Cameron, R. A. D., Carter, M. A. 1977. Population dynamics of the land snail *Cepaea nemoralis* (L.): a six-year study. *J. Anim. Ecol.* 46:181–94
188. Wolda, H. 1963. Natural populations of the polymorphic land snail *Cepaea nemoralis* (L.). *Arch. Neer. Zool.* 15:381–471
189. Wolda, H. 1965. The effect of drought on egg production in *Cepaea nemoralis* (L.). *Arch. Neer. Zool.* 16:387–99
190. Wolda, H. 1965. Some preliminary observations on the distribution of the various morphs within natural populations of the polymorphic land snail *Cepaea nemoralis* (L.). *Arch. Neer. Zool.* 16: 280–92
191. Wolda, H. 1967. The effect of temperature on reproduction in some morphs of the land snail *Cepaea nemoralis* (L.). *Evolution* 21:117–29
192. Wolda, H. 1969. Fine distribution of morph frequencies in the snail *Cepaea nemoralis* near Groningen. *J. Anim. Ecol.* 38:305–27
193. Wolda, H. 1969. Stability of a steep cline in morph frequencies of the snail *Cepaea nemoralis* (L.). *J. Anim. Ecol.* 38:623–35
194. Wolda, H. 1969. Genetics of polymorphism in the land snail, *Cepaea nemoralis. Genetica* 40:475–502
195. Wolda, H. 1970. Variation in growth rate in the land snail *Cepaea nemoralis. Res. Pop. Ecol.* 12:185–204
196. Wolda, H. 1970. Ecological variation and its implications for the dynamics of populations of the land snail *Cepaea nemoralis.* In *Dynamics of Populations,* ed. P. J. den Boer, G. R. Gradwell, pp. 98–108. Wageningen: Pudoc
197. Wolda, H. 1972. Ecology of some experimental populations of the land snail *Cepaea nemoralis* (L.). I. Adult numbers and adult mortality. *Neth. J. Zool.* 22:428–55
198. Wolda, H. 1972. Changes in shell size in some experimental populations of the land snail *Cepaea nemoralis* (Linnaeus). *Argamon, Isr. J. Malacol.* 3:63–71
199. Wolda, H., Kreulen, D. A. 1973. Ecology of some experimental populations of the land snail *Cepaea nemoralis* (L.). II. Production and survival of eggs and juveniles. *Neth. J. Zool.* 23:168–88
200. Wolda, H., Zweep, A., Schuitema, K. A. 1971. The role of food in the dynamics of populations of the land snail *Cepaea nemoralis. Oecologia* 7:361–81
201. Wright, S. 1965. Factor interaction and linkage in evolution. *Proc. R. Soc. London Ser. B* 162:80–104

THE EVOLUTION OF LIFE HISTORY TRAITS: A Critique of the Theory and a Review of the Data

♦4120

Stephen C. Stearns
Department of Zoology, University of California, Berkeley, CA 94720

INTRODUCTION

In 1976, I reviewed (84) two models which give alternative explanations for the adaptation of life history traits to stable and fluctuating environments. Deterministic models (r- and K-selection) predict that organisms exposed to high levels of density-independent mortality, wide fluctuations in population density, or repeated episodes of colonization will evolve towards a combination of earlier maturity, larger broods, higher reproductive effort, and shorter lifespans than will organisms exposed to density-dependent mortality or constant population density (48, 65). Stochastic models (58, 77) predict the evolution of the same combinations of life history traits, but for different reasons: when fluctuations in the environment result in highly variable juvenile mortality, then a syndrome of delayed maturity, smaller reproductive effort, and greater longevity should evolve.

Several years ago I set out to test these predictions by measuring the reproductive traits of two species of small fish that had been introduced to Hawaiian reservoirs in 1907 and 1922. Ambiguities appeared in the interpretation of the results, some of them inherent in the theory, others in the observations. I could not decide which of several possible causal systems had produced the pattern I observed. To determine what my results meant, I first tried to understand what life history data could mean in general, given the present state of our knowledge. In brief, the theory is not yet refined enough to be tested by crucial experiments that can pinpoint flaws. Under these circumstances, observation and experiment cannot falsify predictions definitively, but they can profitably arbitrate among the various simplifying assumptions that theorists may want to try out in their pursuit of unambiguous predictions that cleanly touch reality.

In this paper, I hope to demonstrate that the interpretation of data is ambiguous because the theory is incomplete. Theory can form the empiricists' search image, which then contains just as many unarticulated assumptions as does the theory. That is the subject of the first section of this essay. I then briefly list some of the obstacles to empirical work in the second section. In the third section, I review a representative sample of life history data, for two reasons. First, although there are difficulties relating theory and observation, the data show clearly that the number of types of life histories is limited. Thus hope for a general explanation of life history diversity is justified. Second, a review of life history diversity can itself challenge theorists by revealing the complex nature of the phenomena.

SOURCES OF AMBIGUITY IN THE THEORY

Students of life history evolution seek to explain variation in age at maturity, number of young, reproductive effort, size of young, and interbrood interval. Empiricists take one of two general approaches. On the one hand, they compare existing forms, assume that the conditions under which these forms are living represent the conditions under which they evolved, and test possible explanations against field observations. This, the comparative approach, often makes use of those interspecific or intergeneric comparisons that are subject to the pitfalls ably summarized by Lack (42). Empiricists taking a direct approach, on the other hand, measure the selective difference between habitats and predict how life history traits should change if a population is introduced into one from the other. If the traits do not change as predicted, new theories are required. Placing evolutionary predictions at risk has not been popular. Evolution moves slowly and biologists are impatient. I believe that the logic of comparison is weaker than the logic of prediction, which should be used wherever feasible. The foundations of both will be strengthened by making explicit seven sources of ambiguity in the theory.

These ambiguities all share a general form. Each represents an unanalyzed complexity or subtlety, and for each we do not know whether explicit consideration of the problem would make any difference to our predictions. So long as we can accumulate confirmations of predictions that take real risks, we can continue on the assumption that these ambiguities make no difference. However, when falsifications force us to reconsider the assumptions, a skeptic could argue that the model failed to fit reality because it ignored or misrepresented one or more of these sources of ambiguity, and not because of the other features in the model with which we are usually concerned. Without relaxing his assumptions and examining the behavior of models that incorporate none of these complexities, without, in other words, showing that these ambiguities had no influence upon his predictions, one could not answer this objection. I call this the Dilemma of the Faustian Empiricist, who pursues the basis of his knowledge perhaps a bit too far for his own comfort.

These considerations should not inhibit theorists or empiricists, precisely because we do not yet know if they make any difference, and the only way we are going to find out is by testing energetically the interaction of theory and experiment.

1 Diploid Genetics and Ontogeny

What are the consequences of ignoring the complexities of diploid genetics and ontogeny? I will begin with one classical model of genetic fitness. My purpose is not to criticize this particular model, but to use the discussion of the ambiguities associated with it to illustrate a more general problem. In population genetics, fitness is sometimes measured as the malthusian parameter, m, defined as the fitness of a *genotype*. Lotka's (46) equation defines m:

$$1 = \int_A^W e^{-mx} l_x b_x \, dx,$$

where m is the rate at which that genotype is increasing, x is age, l_x is the probability of survival to age x, b_x is age-specific fecundity, A is age at maturity, and W is age at last reproduction. Whereas m is a property of a genotype, l_x and b_x can only be measured on a population. That population is made up of a single, stable genotype only if it is an asexually produced clone (27). As Kempthorne & Pollak (37) state, "... there is deep obscurity in the malthusian formulation. It is fundamental that a diploid individual contributes genes and not individuals to the next generation." By making the assumption that the complexities of sex and ontogeny do not matter, life history theorists are ignoring the essence of selection in a mendelian population: the coadaptation of the gene complex (51). The usual *Gedanken* experiment consists of endowing a series of clones with certain reproductive traits, putting them in competition, and seeing which one wins (e.g. 58).

You may object that some models have successfully tied genetics to life history evolution. In fact, commendable efforts have been made, but no one has avoided these objections. Murphy (58), MacArthur (47), and Roughgarden (74) attempted to account for sex. They assumed that age at maturity and fecundity, or r and K, are so tightly associated genetically that they can be productively modelled as alleles at a single locus. Implicit in that approach is the necessity of assuming what one is trying to explain: the association of early maturity and high fecundity, or low r's and high K's. Charlesworth (13) generalized the one-locus-two-allele case to an age-structured population, thus considerably extending the realism attained by population genetics. The remaining difficulties appear intractable. For three reasons, we cannot generalize the one locus case to life history traits, most of which are influenced by many genes. First, as Wright (102) showed for pelage color of guinea pigs, the interaction of two or more loci can be wildly nonlinear. Second, the selective value of an allele at one locus can depend on the frequency of alleles at other loci, and always depends on whether the trait influenced by that allele is currently above or below the optimum value in the population at large (102). Third, evidence of two sorts indicates that changes in a few regulatory genes with large effects, rather than many structural genes with small, additive effects, and rearrangements of large blocks of genes determine large differences between species: (a) chimpanzees differ strongly from man in morphology, behavior, ecology, and life history traits, but are so nearly identical at electrophoretically detectable loci that the difference between chimpanzee and man for structural loci is less than the equivalent difference between pairs of sibling species of fruit flies or mammals (39); (b) differences in rates of anatomical, molecular, and chromosomal evolution between frogs and placental mammals indicate that there are two kinds of evolution: serum albumins in frogs and mammals have evolved at about the same rate, but chromosome number has changed 20 times faster in mammals than in frogs, paralleling the much greater anatomical diversity of placental mammals (100). These kinds of evidence convince me that detailed genetical models of life history evolution will have a strong flavor of unreality for some time to come, perhaps forever. If regulatory genes are preeminently important, if loci in general interact nonlinearly, and if linkage rearrangements result in more rapid evolution than changes in structural loci, then the assumptions of classical populations genetics are profoundly violated, and its generality is limited to the single locus case. Traits determined by single loci are rarely important to ecologists.

Another approach should clarify the point. Consider the way a population geneticist, who tries to deal with the complexities of diploid genetics, and a life history theorist, who usually does not, view the structure of evolutionary theory. One population geneticist (44) has posed as the general problem of evolution the understanding of four transformations within and between genotype space and phenotype space (Figure 1). The initial distribution of genotypes in the population, G_1, is transformed by the (as yet unknown) laws of ontogeny, T_1, into the initial distribution of phenotypes, P_1. These phenotypes are not all equally fit, and T_2 consists of the laws of ecology (as yet unknown) that determine the relative survival of phenotypes, resulting in the set of selected phenotypes, P_2. T_3, inverse epigenetic laws (as yet unknown), permit the inference of genotypes from phenotypes, giving us G_2, the distribution of genotypes that underlies the P_2 set of phenotypes. The partially understood rules of Mendel and Morgan, T_4, produce the next generation of genotypes: the subject of theoretical population genetics.

Lewontin's model poses the problem of evolutionary ecology with striking clarity: T_1, T_2, and T_3 are not understood, and without them, knowing a bit about T_4 is of little help. Life history theorists (and most other evolutionary ecologists) approach the problem a different way. They see associated with each transformation a surface structure of observables and a deep structure (the relationships embodied in the transformation rules) in terms of which they seek to explain the surface structure (85). In life history work (Figure 2), the surface structure consists of individual organisms, their demographic and physiological characteristics, and a set of environmental measures that describe the conditions they encounter. Statistical inference connects the surface structure to an intermediate structure consisting of estimates of age at maturity, age-specific survivorship and fertility, growth rates, size of young, the time course of resource availability and weather, and so forth. The deep structure relates parameters that measure fitness, such as r, K, or the probability of leaving no young at all; these connect to the intermediate structure by such models as Lotka's demographic equation, the Lotka-Volterra equations, or analogous difference equations.

EVOLUTION OF LIFE HISTORY TRAITS   149

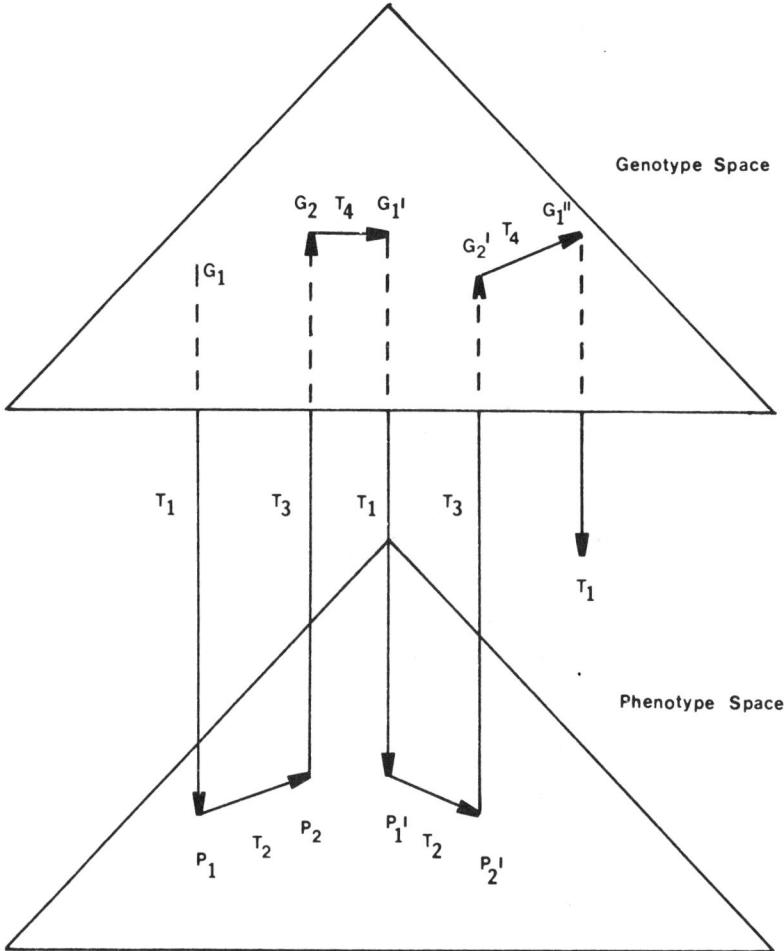

Figure 1 A population geneticist's view of evolution. A generational cycle starts with an initial distribution of genotypes, G_1, proceeds through a series of epigenetic and selective transformations, and finishes with a new distribution of genotypes, G_1'. This initial conceptualization implies that both dynamics and statics are part of the problem statement. After (44).

Life history theorists do not isolate any of Lewontin's transformation sets cleanly. Life history theory is a set of optimality models; theoretical population genetics is a set of mechanistic models. They approach the problem of evolution in profoundly different ways. To make predictions about the relative survival of phenotypes, life history theorists ignore ontogeny (T_1) and genetics (T_4). This method is attractive because genotypes are not being continuously destroyed and reshuffled by sex, and because the complications of developmental plasticity and canalization can be ig-

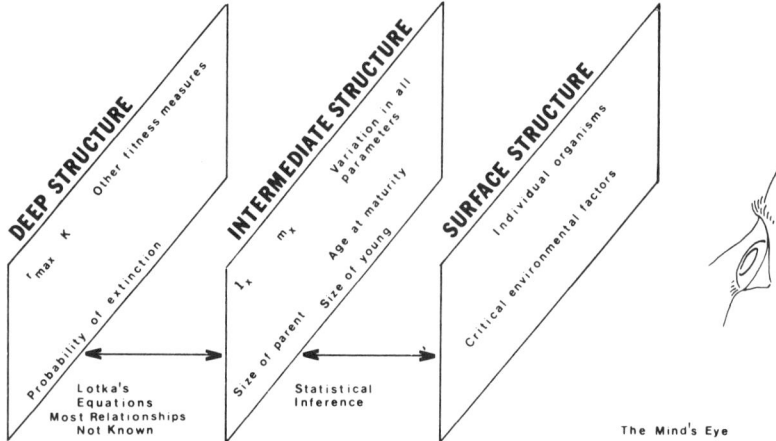

Figure 2 An evolutionary ecologist's view of evolution. The surface structure of observables is related by statistical inference to an intermediate structure of demographic parameters. These are in turn shaped by the deep structure of relationships implied by fitness and optimization assumptions. Statics are emphasized; dynamics are not. Compare with Figure 1.

nored. But they buy their ability to make statements about the relative fitness of phenotypes at a considerable cost in realism, and by compressing Lewontin's sequence,

$$G_1 \xrightarrow{T_1} P_1 \xrightarrow{T_2} P_2 \xrightarrow{T_3} G_2 \xrightarrow{T_4} G_1 \xrightarrow{T_1} P_1',$$

into the much shorter sequence,

$$P_1 \xrightarrow{T_{LH}} P_1',$$

where T_{LH} are the transformation rules defining the relative fitness of life history phenotypes.

Thus life history theory may suffer from a lack of realism, whereas the theory of population genetics founders on too much realism as soon as the mechanisms are pursued beyond one locus. In order to drive that point home, I mention a problem explored by Rocklin & Oster (73): The dimensionality of the phenotype space in Figure 1 is much lower than the dimensionality of the genotype space. For each generation the possible genotypes greatly exceed the realized genotypes; and to the extent that canalization is important, the realized genotypes greatly exceed the phenotypes produced. This difference in dimensionality seems to make impossible genetical models that predict phenotypic traits, such as life histories.

Some have suggested that the analysis of evolutionarily stable strategies (ESS's) is practically assumption-free and gets around all these difficulties. Smith & Price

(79) defined an ESS as "... a strategy such that, if most of the members of a population adopt it, there is no 'mutant' strategy that would give higher reproductive fitness." At least two difficulties make the analysis of ESS's, promising as it is, less than the solution to all our problems. First, nothing guarantees that the analyst can conceive of all the adaptive options; second, nothing guarantees that the options conceived will be genetically attainable, particularly if the options consist of combinations of complex traits, like life histories.

2 Design Constraints

Design constraints may keep populations from reaching predicted optima. To visualize this source of ambiguity, consider Figure 3a, which represents in three dimensions the relationships among the elements of the deep structure. The independent variables, $P_1, P_2, \ldots P_n$ are the elements of a life history that can evolve. Their interaction determines the values of the dependent variable, fitness, which in this three-dimensional representation is a surface with peaks and valleys. A and B represent local optima on the fitness surface determined by P_1 and P_2 (e.g. age at maturity and clutch size) in an unspecified environment and with other life history traits held constant in some particular combination. These points are located either through the analysis of the maxima of equations relating fitness measures to life history traits, or through graphical analysis where the assumptions are embodied as the shapes of curves rather than as equations. Then the theorist asserts that because those are the combinations of traits that are maximally fit in the model, those should be the combinations found in nature. Here a problem characteristic of all optimality models surfaces.

Nothing guarantees that the optimum point will be located in an accessible portion of phase space. For example, first, the chitinous exoskeleton of insects restricts growth and limits the number of eggs that can be carried at one time. Second, the water vascular system of echinoderms prevents colonization of land and, through osmotic constraints, of fresh water. Third, the complex interrelationships and multiple functions of mammalian hormones make the evolution of reversed sex roles in mammals difficult. None of these examples represent trade-offs between opposing selection forces. They are design barriers, limits beyond which organisms cannot operate. Design constraints should not be confused with trade-offs or costs, all of which share the characteristic that if the opposing selection force were removed, the phenotype would be free to move beyond the point already attained.

In the language of our model, this means that the fitness surface is dissected by barriers implicit in the design of particular groups of organisms. These barriers will be located in different places, and have different shapes, for different groups. What are the consequences? Consider Figure 3b. Here B is the optimum point, but the population started at A. The population cannot get from A to B because it runs into the barrier implied by design constraints. As the population evolves, it should travel upward along the fitness surface, following the high ground, as far as it can go, stopping at the barrier. If our theory were flawless, we could predict that a population should be at B; if we observed not-B, we could then state that its evolution had encountered a design constraint, and we could check for that. But our

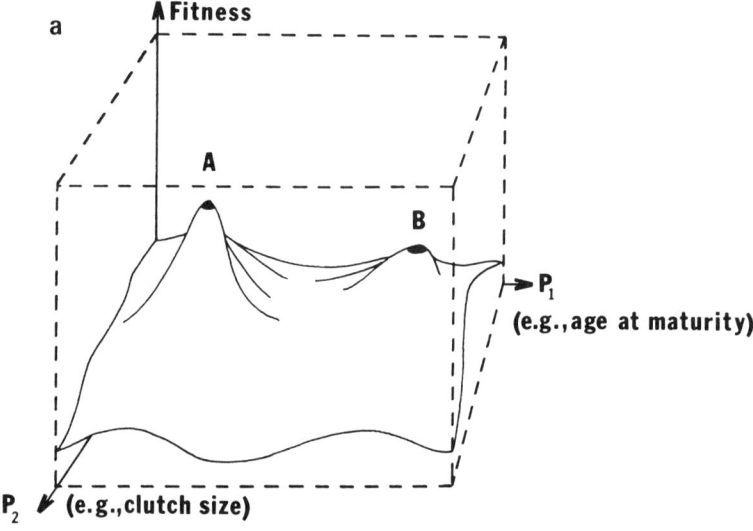

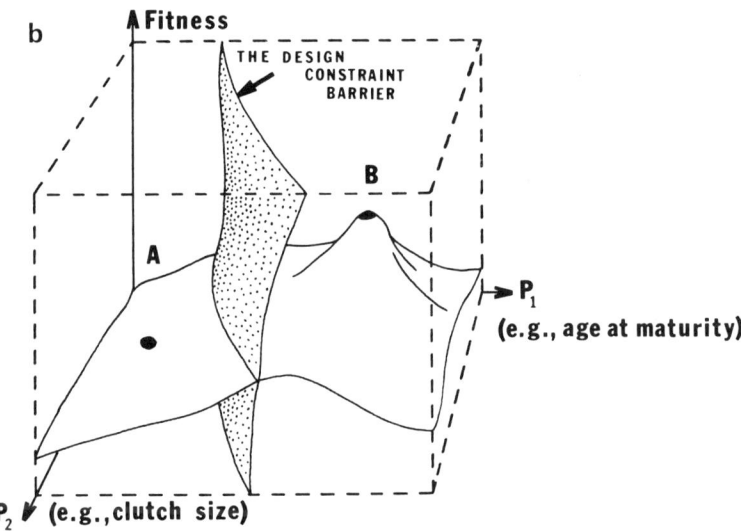

Figure 3 The optimization-design constraint ambiguity. (*a*) Life history theory may predict optimum combinations of traits, e.g. points *A* (a global optimum) and *B* (a local optimum); but (*b*) a design constraint may keep the population from attaining the optimum (point *B* in this case) if it started on the wrong side of the design constraint barrier. See text for the epistemological consequences.

EVOLUTION OF LIFE HISTORY TRAITS   153

theory cannot be trusted. We do not know where the design barriers are located; and when we observe not-B, we can only say that (a) the population is up against the limits of an unknown design constraint, or (b) not enough time has elapsed for the optimum to be attained (i.e. the constraint is dynamic, not static), or (c) there is something wrong with the theory.

I should insert a *caveat* here. Figures 3a and 3b seriously misrepresent the actual state of affairs. We are accustomed to thinking of smooth Wrightean (101) fitness surfaces. In fact, the fitness surfaces generated by simple models are non-picturable—so convoluted and discontinuous that, like four dimensions, the human imagination cannot grasp them (Oster, personal communication). We cannot actually trace the possible paths between two points in fitness space to determine if access is possible. This limits further the conclusions that may be drawn from optimality models.

3 Multiple Causation

Alternative explanations of life history diversity, all dealing with evolutionary rather than proximate causes, are possible and are rarely mutually exclusive. Multiple causation frequently operates on life histories.

Two *leitmotifs* dominate the development of life history theory: the deterministic view (19, 26, 43, 48, 83) and the stochastic view (17, 55, 56, 58, 77). The consequences of the deterministic view are relatively well understood, at least for r-selection. Although not as thoroughly explored as the deterministic approach, the stochastic viewpoint offers a plausible alternative explanation for the trends "explained" by r- and K-selection. Murphy (58) showed, with a simulation model of an age-structured population, that where juvenile mortality fluctuates, organisms with delayed reproduction, small reproductive efforts, and a few young are favored. Thus he predicted that where fluctuations in population densities result primarily from fluctuations in recruitment, a syndrome of traits should evolve that would appear to be K-selected, not r-selected. Schaffer (77) confirmed Murphy's conclusion from a simple analytical model of a population without age structure. In considering the contrasting case where adult mortality fluctuates, Schaffer predicted the evolution of a syndrome of traits similar to those predicted by r-selection: early maturation and large reproductive efforts. Table 1 compares the predictions of the two viewpoints.

One could argue that the deterministic and stochastic approaches make the same predictions for the same environments. In a stationary population where resources are limiting and competition is fierce, variation in juvenile survival may be greater than variation in adult survival (32). Similarly, in a population moving through a series of colonizing episodes, adult survival must vary considerably, perhaps much more than juvenile survival. One can suggest that the two approaches are dealing with the same phenomena at two different levels, as though there were two levels of deep structure in Figure 2. This is essentially the point made by (99): "Because neither the carrying capacity nor the mechanism of population regulation is known for most natural populations, data on life history parameters are often consistent with more than a single hypothesis." I add that even if such knowledge were available, the data could still be consistent with more than a single hypothesis.

Table 1 A Summary of the theory

Model	Assumptions	Predictions
Deterministic		
r-selection	exponential population growth stable age distributions repeated colonizations or fluctuations in population density	earlier maturity more, smaller young larger reproductive effort shorter life
K-selection	environment stable population near equilibrium density logistic population growth competition important	later maturity fewer, larger young smaller reproductive effort longer life
Stochastic	environment fluctuates population near equilibrium	
	(a) juvenile mortality or birth rate fluctuates, adult mortality does not	later maturity smaller reproductive effort fewer young
	(b) adult mortality fluctuates, juvenile mortality or birth rate does not	earlier maturity larger reproductive effort more young

A priori criteria can aid the choice among models that are empirically indistinguishable. Models couched in terms of the means and variances of adult and juvenile mortality rates would make more readily falsifiable predictions than models couched in terms of population regulation because their assumptions can be checked. [It may not be possible to distinguish density-dependent mortality from density-independent mortality; see (72)]. This approach would avoid the indefinable relationship between K and life history traits, and by making explicit the variability in juvenile and adult mortality, would make possible the examination of other fitness measures, such as the minimization of the probability of leaving no young at all. I find these advantages persuasive. Such a theory does not exist; its development should challenge us all.

The first three theoretical ambiguities afflict all models in evolutionary ecology. The remaining four are peculiar to life history theory.

4 *The Assumption of a Stable Age Distribution*

One must assume fixed age schedules of survivorship and fecundity to write an equation relating the elements of a life history to one measure of fitness, r. Because this is equivalent to assuming a temporally constant, spatially homogeneous

EVOLUTION OF LIFE HISTORY TRAITS   155

environment, most graphical models make an analogous assumption. Although the stability of an age distribution is rarely checked (12), it is reasonable to assume that few, if any, natural populations have achieved it. Again, the dilemma is that of the Faustian Empiricist: A skeptic could always claim that the failure of the model to fit reality lay in its assumption of a stable age distribution, and not in any of its other features.

This criticism could be blunted by showing that the departures from stable age distribution were small [e.g. perhaps (97)], but where the mortality schedule has been shown to vary wildly (e.g. 57), the criticism retains considerable force. The only detailed theoretical analysis of populations not in stable age distribution (62) supports this point. In that model, population densities varied in a very surprising way, leading to unanticipated selection forces on life histories.

5   K as a Function of Life History Traits

Unlike r, K cannot be realistically expressed as a function of life history traits. Some of the most stimulating life history work has taken the following pattern. Let r be the measure of fitness, and examine its sensitivity to changes in age at maturity, fecundity, etc. Then predict that the trait to which r is most sensitive should be under the strongest selection pressure, should be found closest to its theoretical optimum (or against a design barrier), and should exhibit the least additive genetic variance of any of the traits (e.g. age at eclosion in *Drosophila melanogaster*). This process reduces life history traits to a common currency, units of r, and permits direct comparisons. For example, Lewontin (43) could say that at certain values of other life history traits, the increase in lifetime fertility necessary to increase r from 0.30 to 0.33 was a change from 780 to 1350 eggs, and that a decrease in age at maturity from 12 to 9.8 days, with no change in fecundity, made the same impact on r. Thus an additional 570 eggs were equivalent to 2.2 fewer days maturation time; both added 0.03 units of r.

This analysis is possible only because we can express r as a function of A, l_x, and b_x. No one could possibly write an equivalent expression for K because the general relationship between the sensitivity of a species to changes in its population density on the one hand, and its age at maturity, reproductive effort, and so forth on the other, is not clear. K is not a population parameter, but a composite of a population, its resources, and their interaction. Calling K a population trait is an artifact of logistic thinking, an example of Whitehead's Fallacy of Misplaced Concreteness. Thus r and K cannot be reduced to units of common currency. If they do trade off, so that higher r's imply lower K's, then the mechanisms by which that trade-off is accomplished are not demographic, but are bound up in physiology and social behavior, and as such could be expected to change from taxon to taxon.

In short, the theory of r- and K-selection contains a serious, and to my way of thinking fatal, flaw: A population that has a life history thought to result from r-selection is called "r-selected"; a population with the opposite traits is called "K-selected" in the absence of either evidence or deductive logic indicating that such traits have been molded by density-dependent effects. Such traits may eventually be shown to result from density-dependence, but the connection has yet to be demonstrated.

6 Post-Reproductive Survival: Alternative Equilibria

Certain classes of trade-offs between reproductive effort and subsequent adult survival can lead to any of several stable equilibria, depending on initial conditions.

Few have tried to model explicitly the effect that the act of reproduction has in decreasing subsequent survival [but see (67)]. Schaffer (75, 76) and Schaffer & Rosenzweig (78) have used graphs to analyze such trade-offs. When the relationships are purely concave, iteroparity, or repeated reproduction, is favored; when they are convex, semelparity, or "big-bang" reproduction is favored. The ambiguities enter when more complex, but biologically plausible, curves are analyzed. Sigmoid trade-offs lead to multiple stable equilibrium points, even in a simple, three-stage life history, and the number of possible equilibrium points can be expected to rise rapidly with the number of stages in the life history. Even simple stochastic models also show history dependence, as Cohen (18) has pointed out. The results obtained from any particular experiment that appears to settle down to "good" behavior may stem from a series of random events converging on a unique endpoint. Each repetition of the experiment will give a different result. In such cases, order and generality only emerge when one deals with the distribution of results from a large number of replicates.

These results lend plausibility to the idea that patterns of life history diversity have been influenced by chance historical events. In comparative work, the existence of either multiple stable endpoints or a broad distribution of possible outcomes results in terrible ambiguities. To make a precise and testable prediction in the first case, we would need to know not only the nature of the trade-off, which could be measured, but also the initial conditions, which can only be specified in manipulative experiments. In the second case, we would need a large number of well controlled replicates to discern the possibility of a strong stochastic effect.

7 Choice of Time Scale

Should we use a relative (generation time) or absolute (solar time) scale? This question becomes pressing as we try to extend the generality of the field to accommodate comparisons of higher taxa in a productive way. Choosing one time scale or the other leads to differing interpretations of just what we are trying to explain. Consider the life histories of two barnacles on the California coast (31).

The life histories of *Chthamalus fissus* and *Tetraclita squamosa* differ in absolute time (Figure 4a). *Chthamalus* is smaller, shorter-lived, matures earlier, and may be making a larger reproductive effort than *Tetraclita* (it appears to spawn as frequently as it can acquire the energy to do so). Because it lives in the relatively unstable upper intertidal where physical factors are limiting, *Chthamalus* could be called *r*-selected. In contrast, *Tetraclita,* living in the relatively stable subtidal zone, where biotic factors are limiting, could be called *K*-selected.

On the other hand, the life histories of the two barnacles recalculated in generation time appear quite similar (Figure 4b). Our problem is to explain the similarity, rather than the difference. By reproducing many times within a season, *Chthamalus* appears to be dealing with intraseasonal unpredictability, which could be more

EVOLUTION OF LIFE HISTORY TRAITS 157

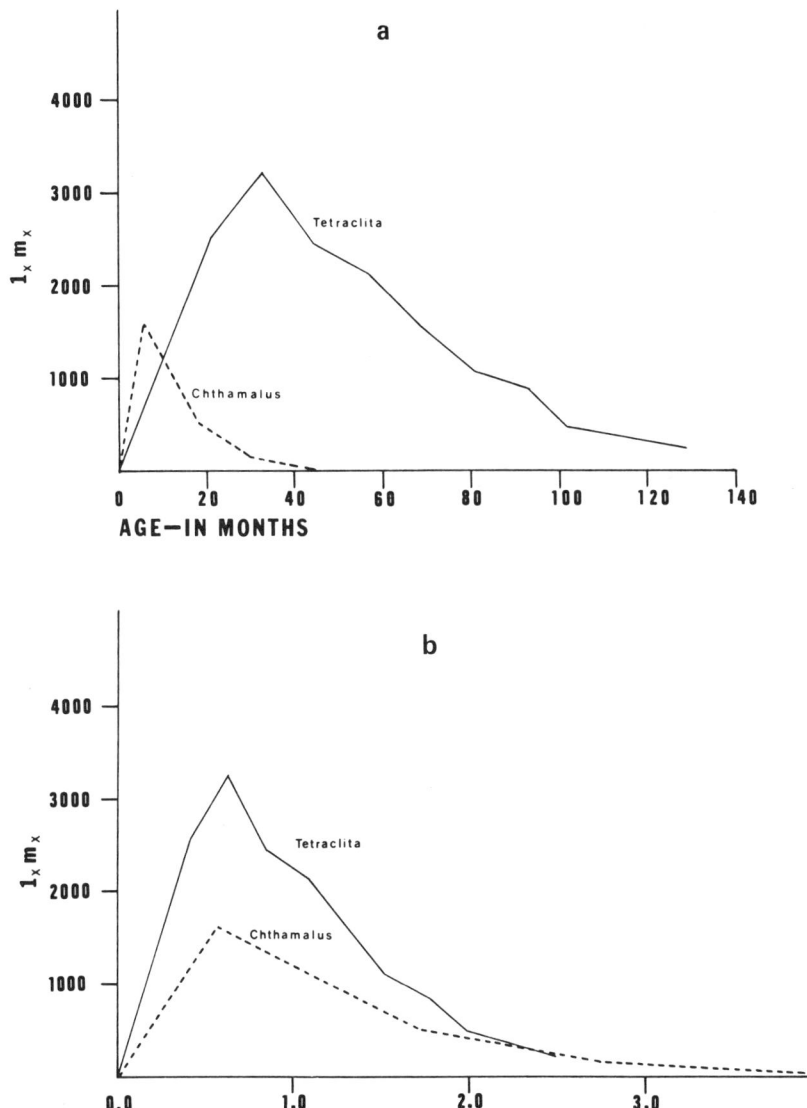

Figure 4 The ambiguity arising from the choice of time scales. Should we use generation time or absolute time? (*a*) The life histories of two intertidal barnacles graphed in absolute time [From (31)]. (*b*) The same life histories graphed in generation time. Figure 4*a* emphasizes the differences between the two species, Figure 4*b*, the similarities. Which should we be trying to explain?

important in the upper intertidal. By spreading its reproduction out over many years, *Tetraclita* appears to be dealing with inter-year variability, which could be more important in the subtidal. In both cases, the variable factor is probably the survival and settling of planktonic larvae.

Which of the two measures should we choose? One could argue either way on a priori grounds. The relativistic measure, generation time, has the strong advantage of placing the scientist in the organism's frame of reference. The absolute measure, solar time, corresponds much more generally to the natural periodicities of the environment. Note that posing the question as a dichotomy may itself be misleading, because we may not have to choose between them. The issue is whether population growth shall be modelled as a function of age, $n(a)$, as a function of time, $n(t)$, or as a function of both, $n(a,t)$ (Oster, personal communication). Leslie matrix models use $n(a)$, Lotka-Volterra models use $n(t)$, and the von Foerster equations use $n(a,t)$ (61).

SOURCES OF DIFFICULTY IN OBSERVATION AND EXPERIMENT

In the foregoing I want to have suggested enough of the difficulties that afflict life-historical empiricists to convince you that published life history data are not reliable enough to justify any great faith in the generality of conclusions drawn from them. The review of published data in the next section will extend this point. Consider first, however, five problems which further compound our difficulties.

1. We often cannot identify the unit of evolution in the field. The spatial structure of most populations is almost impossible to observe, making it hard to measure the rate and direction of gene flow in nature.

For life history traits, on which selection is almost certainly operating, the degree of correspondence between the genetic component of the trait observed in the field and local environmental conditions will depend on the magnitude of local selection pressures, the rate of gene flow, and the nature of the average global selection pressure encountered by the whole interbreeding unit (Spieth, personal communication). When migration rate exceeds selection pressure, the local group must frequently be adapted to the entire spatiotemporal mosaic encountered by the panmictic unit, and any correlation between local environmental conditions and reproductive traits could be spurious.

2. There are severe technical problems with getting reliable life tables (12).

3. The criteria used for selecting organisms for meaningful measures of reproductive effort necessarily conflict.

Energy budgets require two measurements, the time spent by an organism in various activities, measured in the field, and the energy expended on those activities per unit time, measured in the lab. Time budgets are most easily measured on large, long-lived species, sessile species, or species living in open habitats. Energy expenditures are most easily measured on small, short-lived organisms that will grow, mate, and reproduce under laboratory conditions. Clearly, time budgets cannot usually be estimated with the same accuracy as energy budgets for the same species. [See (32) for a thoughtful discussion of the second and third empirical difficulties.]

EVOLUTION OF LIFE HISTORY TRAITS 159

4. Laboratory work requires organisms with short generations (thus ruling out comparisons with long-lived species) for two reasons: First, differences observed in the field may contain a strong element of developmental plasticity. To check this, one must rear the organisms through two generations in the laboratory to eliminate, in the first generation, first-order developmental plasticity, and in the second generation, maternal effects. Second, to carry out a program that tests evolutionary predictions with selection experiments, one needs organisms with short generation times. Because most animals are immediately excluded, rigorous empirical generalizations are impossible. Charlesworth (personal communication) has suggested an additional difficulty inherent in any laboratory approach: "A species which is simply better adapted to standard lab conditions would have a greater longevity *and* fecundity than one which is not so well adapted."

5. To measure the reduction in survival rates caused by reproduction, one must keep the organisms from reproducing and monitor subsequent survival, preferably in the field. But in any organism with no parental care, one cannot stop the commitment of essentially full reproductive effort by withholding mates.

SUMMARY OF SOURCES OF AMBIGUITY

Table 2 summarizes the theoretical and empirical sources of ambiguity just discussed. Each of the twelve items presents a worthy challenge. Overcoming or clarifying any would be significant. In certain cases, it would be profound.

The ideal organism for a complete life history study would have the following attributes. It should be possible (*a*) to define the boundaries of the field populations and to measure emigration and immigration rates; (*b*) to construct, by marking or aging, cohort life tables from field data with confidence limits on the mortality and fertility columns, and to measure the temporal component of the coefficient of variation in adult and juvenile mortalities, exclusive of measurement error; (*c*) to measure activity budgets in the field and energy budgets in the laboratory; (*d*) to achieve a relatively short ($\leqslant$ 8 weeks) generation time in the laboratory; and (*e*) to make the organism forego reproduction (preferably in the field) and then to follow its subsequent survival.

These criteria, taken together, are quite restrictive, and adhering to them rigidly would preclude most of the interesting work that could be done. I do recommend against working on species for which none of the criteria could be satisfied. Investigators should focus on those questions for which most of the criteria can be met in a tractable species.

PUBLISHED EVIDENCE ON LIFE HISTORY DIVERSITY

This section evaluates present knowledge of diversity in life history traits for many taxa. In judging these papers by the criteria developed here and in the previous section, I am not implying that their authors have in some way failed; rather, I am trying to establish, on a uniform basis, what we do and do not know. Much of the work examined here was done before the field attained its current state of sophistication; and it would be unreasonable to criticize someone for not checking a point

Table 2 Summary of sources of ambiguity

Ambiguity	Difficulty it leads to
Theoretical	
1. Experimenter assumes that the complexities of diploid genetics and ontogeny are irrelevant.	epistemological and general
2. Design constraints restrict optimality arguments.	epistemological and general
3. Several alternative explanations which are not necessarily mutually exclusive are usually possible for any adaptation.	epistemological and general
4. Experimenter assumes a stable age distribution.	epistemological and peculiar to life history work
5. K cannot be expressed as a function of life history traits.	technical and peculiar to life history work
6. Sigmoid trade-offs between the cost of reproduction and subsequent adult survival lead to multiple stable equilibria.	epistemological and peculiar to life history work
7. What is the appropriate time scale?	epistemological and peculiar to life history work
Empirical	
1. What is the unit of evolution?	technical and general
2. Reliable life tables are hard to get.	technical and particular
3. Measurement of reproductive effort restricts the choice of organisms and thus of comparisons.	technical and particular
4. Need for short generation times restricts choice of organisms.	technical and particular
5. Measurement of cost of reproduction restricts choice of organisms.	technical and particular

that had not yet been raised, or to criticize data gathered as a byproduct of a study with other objectives. I have used papers frequently cited and readily accessible. While I have not exhausted the literature, I believe my review conveys an accurate impression of the reliability of the better papers. I relied most heavily on papers that discuss intra- or interspecific comparisons.

Both r- and K-selection and the stochastic models can be applied in three ways: as *systems of classification,* as *explanations*, and as *predictions*. Thus many workers have used the r- and K-selection dichotomy as a convenient shorthand, referring to the combination of early maturity, large clutches, large reproductive effort, small young, and a short life as r-selected, perhaps without meaning to imply the evolutionary origin of that suite of traits by their use of the term. The stochastic models predict the association of the same traits for different reasons. I refer to that syndrome as fitting the *accepted scheme,* because as yet no predictions of alternative associations of traits have been made. Reproductive traits may covary according to the accepted scheme without distinguishing between deterministic and stochastic

EVOLUTION OF LIFE HISTORY TRAITS 161

models as *explanations,* or even guaranteeing that either model is involved. But if reproductive traits do not covary according to the accepted scheme, we can be sure that neither model provides a sufficient explanation. Therefore, if the data were adequate, I determined whether each comparison fit the accepted scheme.

I used six criteria of reliability: (*a*) Did the author rear the organisms under constant conditions to isolate the genetic component of the variability observed in the field? [Where enough field data are available, laboratory work may not be necessary, e.g. the analysis of heritability of clutch size in the Great Tit (64).] (*b*) Did the author attempt to measure the environmental factors later invoked to explain differences in reproductive traits? (*c*) Did the author attempt to measure the degree of density-dependent or density-independent regulation? (*d*) Did the author attempt to measure the year-to-year variability in the mortality schedule? (*e*) Were the statistics convincing? For intraspecific comparisons, were analyses of variance or covariance done? For interspecific comparisons, did the author demonstrate that interspecific differences were still significant when intraspecific variation was taken into account? (*f*) Was an attempt made to measure reproductive effort? If so, at what level of sophistication? Was the rate calculated as the calories diverted to reproduction per unit time as a proportion of total calories passing through the organism? For each of the six criteria, one could make a commendable effort with less than convincing results. Therefore, in the Tables I have indicated satisfaction of the criteria at three levels: "no" indicates that no attempt was made; "maybe" signals a good try whose data remain unconvincing; "yes" signifies a criterion completely satisfied. The review is organized by taxa.

Salamanders

Table 3 summarizes the quality of data on salamander life history adaptation. This literature is strong in large field samples (criterion *e*) that demonstrate intra- and interspecific differences, but weak in measuring the factors on which explanations of life history diversity have been based (criteria *b, c,* and *d*) or on documenting the genetic basis of the diversity (criterion *a*). Only one of five salamander comparisons fits the accepted scheme. Five species of *Desmognathus* show unusual resistance to change, while *Batrachoseps attenuatus* has unusual developmental flexibility.

Table 3 Reliability of life history data on salamanders

			Criterion						Fit	Number of
Ref.	Comparison[a]	Taxon	a	b	c	d	e	f	scheme?	criteria met
(60)	inter	*Desmognathus*	no	no	no	no	yes	maybe	no	2
(87)	inter	*Desmognathus*	no	no	no	yes	yes	no	no	2
(11)	intra	*Gyrinophilus*	no	no	no	no	yes	no	yes	1
(88)	intra	*Desmognathus*	no	no	no	no	yes	no	no	1
(29)	intra	*Notophthalmus*	no	no	no	no	yes	no	?	1
(50)	inter	*Batrachoseps*	no	maybe	no	maybe	maybe	no	no	3

[a] Inter = interspecific; intra = intraspecific.

Lizards

Table 4 summarizes lizard life history data. Reproductive effort is more frequently measured on lizards than on most taxa; the field statistics are good and there is much comparative life table information. Broad, intergeneric comparisons of many species reveal a set of syndromes that resemble the accepted scheme (95), but detailed studies break the trend (3, 92–94, 97). Although no study satisfied all criteria, the best work published in 1974–75 represented a considerable advance over 1969–70. The lizard data emphasize that rigorous definitions of reproductive effort can be made in the presence of year-to-year variability in mortality schedules and should be included in future studies.

In a very thorough study, Tinkle & Hadley (94) measured the calories in reproductive effort per season for ten species, estimated the annual energy budgets for three species, and examined correlations of the reproductive data with age at maturity, with adult and juvenile survival rates, and with generation time. Only one correlation was significant: the ratio of clutch calories to body calories (a rough measure of reproductive effort) was negatively correlated with mean annual adult survivorship. The authors recognized that, had they been able to measure annual energy budgets for all ten species, there might have been less noise in their correlations. Of the three species for which they had estimates of the annual energy budgets, the one with delayed maturity and long life, *Sceloporus graciosus,* had the highest per season reproductive effort (0.23), while an essentially annual species, *Uta stansburiana,* made an intermediate effort (0.19). Sample sizes were small, and from too few populations to make the interspecific comparisons airtight. Nevertheless, these data, which are among the best available, force the conclusion that lizards do not neatly fit the accepted scheme.

All long-term studies of lizard populations have shown year-to-year variation in reproductive success (9, 22, 23, 89, 90, 96, 97, 103), and certainly not all lizards that encounter such variability are late-maturing, with one clutch per season as Murphy's model would predict, e.g. *Uta stansburiana* (90, 96). Some of the variation in reproductive success stems from changes in fecundity, some from changes in mortality, and the relative contribution of each is hard to measure because of the

Table 4 Reliability of life history data on lizards

Ref.	Comparison	Taxon	Criterion a	b	c	d	e	f	Fit scheme?	Number of criteria met
(91)	inter	112 species	no	no	no	no	yes	no	yes	1
(95)	inter	37 species	no	no	no	no	yes	no	yes	1
(66)	intra	*Cnemidophorus*	no	yes	no	no	maybe	no	?	2
(92, 93)	intra	*Sceloporus*	no	maybe	no	maybe	yes	wt	no	4
(5)	inter	*Sceloporus*	no	no	no	no	yes	no	yes	1
(33)	inter	*Typhlosaurus*	no	no	no	no	maybe	no	yes	1
(3)	inter	*Anolis*	no	no	no	no	maybe	wt	no	2
(94)	inter	10 species	no	no	no	yes	maybe	cal/rate	no	3
(97)	inter	*Sceloporus*	no	yes	no	yes	yes	wt	no	4

difficulty in counting the number of clutches per season. It is possible, by judicious selection, to find series of both intra- and interspecific comparisons of lizards that fit the accepted scheme, but any attempt to fit all possible comparisons into the accepted scheme founders when the best-studied populations are examined.

Birds

Table 5 summarizes the quality of avian data. I have relied heavily on review articles (4, 16, 42) and included Hussell's (34) study as an example of one of the better pieces of detailed work. No one has reared large numbers of wild birds under constant conditions; therefore the genetic basis of observed diversity is unknown (criterion a). Hussell (34), Ashmole (4), and Lack (42) all tried to measure relevant environmental factors and, while failing to exclude alternative explanations, did render more plausible the hypothesis that food places the proximate limit on clutch size in birds. Criteria c through f were rarely satisfied, which indicates that many types of data are hard to obtain for birds, not that ornithologists have not tried to get them. For example, Ashmole (4) reported estimates of adult and juvenile mortality in sea birds based on banding returns, and pointed out that the assumptions of band survival and bird movement required for making such calculations are so frequently violated that the estimates are not reliable. Acquiring such data takes great effort. Although the data on avian life histories are not reliable, they have triggered much productive speculation, particularly because the patterns emerging from broad comparisons among species (16) and subfamilies (42) offer alluring glimpses of generality.

Mammals

The data on mammalian life histories (Table 6) are less reliable than the avian data. Convincing statistics on age at maturity are entirely missing, except for extremely broad comparisons, and virtually no comparative data exist on reproductive effort. Information is generally limited to litter size. The work done on the snowshoe hare stands out (e.g 36). It demonstrated a genetic component to geographical variation in litter size. Other than that, we know almost nothing about the coadaptation of age at maturity, litter size, longevity, and reproductive efforts in mammals. When such information is gathered, I suspect social systems and behavioral peculiarities will be shown to interact strongly with reproductive traits.

Table 5 Reliability of life history data on birds

Ref.	Comparison	Taxon	Criterion a	b	c	d	e	f	Fit scheme?	Number of criteria met
(42)	subfamilies	most birds	no	yes	no	no	no	no	?	1
(4)	inter	many sea birds	no	yes	no	yes	no	no	yes	2
(16)	inter	many birds	no	no	no	no	no	no	yes	0
(34)	intra	Calcarius, Plectrophenax	no	yes	no	no	yes	wt	?	3

164 STEARNS

Table 6 Reliability of life history data on mammals

Ref.	Comparison	Taxon	a	b	c	d	e	f	Fit scheme?	Number of criteria met
(21)	inter	Peromyscus	no	no	no	no	no	no	?	0
(45)	inter	many mammals	no	no	no	no	no	no	?	0
(54)	inter	many squirrels	no	no	no	no	maybe	no	?	1
(6)	intra	Sylvilagus	no	no	no	no	yes	no	?	1
(35)	intra	Rattus	no	no	no	no	no	no	?	0
(36)	intra	Lepus	yes	no	no	no	yes	no	?	2
(80)	inter	Peromyscus	no	no	no	no	no	no	?	0
(20)	intra	Sylvilagus	no	no	no	no	yes	no	?	1

Both technical problems and the confounding effects of complex behavior make it difficult to draw reliable conclusions about evolutionary causes from the avian and mammalian data. This largely descriptive work illustrates the strengths and weaknesses of the comparative method. Its strength lies in suggesting hypotheses; its weakness lies in testing them. It cannot exclude alternative explanations, nor can it evaluate the relative impact of multiple causes. When one makes broad comparisons across either class, the accepted scheme emerges: Delayed maturity, small clutches, and long lives seem to come as a unit, as do early maturity, large clutches, and short lives. Where there are data available on mortality patterns, as for some seabirds (4), they are consistent with both deterministic and stochastic models. When intraspecific and intrageneric comparisons are examined, generalizations seem to vanish.

Fish

Table 7 summarizes data on fish life histories. The fish literature is strong in field data with large sample sizes, but weak in attempts to measure explanatory environmental factors, degree of density-dependence, mortality rates, or reproductive efforts. It has suggested three points of broader significance. First, Murphy (58) arrived at his hypothesis (that the combination of delayed reproduction, low repro-

Table 7 Reliability of life history data on fishes

Ref.	Comparison	Taxon	a	b	c	d	e	f	Fit scheme?	Number of criteria met
(40)	intra	Gambusia	no	no	no	no	yes	no	?	1
(2)	intra	Salmo, others	yes	no	no	no	yes	no	?	2
(41)	intra	Gambusia	no	no	no	no	yes	no	yes	1
(58)	inter	herring-like fish	no	no	no	maybe	maybe	no	yes	2
(84a)	intra	Neoheterandria	no	no	no	no	yes	wt	yes	2
Stearns, unpublished data	intra	Gambusia, Poecilia	yes	yes	no	no	yes	wt	no	4

ductive effort, and increased longevity adapts the organism to fluctuations in recruitment) through a consideration of fisheries data. That was the first alternative hypothesis suggested that deals with the same reproductive trend "explained" by r- and K-selection. Second, nowhere is the impact of developmental plasticity more obvious than in fish. Growth rate, age at maturity, and fecundity are all very sensitive to temperature and food. This makes interpretation of field data difficult, and laboratory work necessary. Third, because of their broad range in fecundity and egg size, it was in fish that the trade-off between a few large young and many small young was first noted (86).

Insects

The outstanding characteristics of insect life history data (Table 8) are the carefully controlled experimental studies of maturation, longevity, and fecundity, with fairly good attempts to measure the factors that might account for the differences in these traits, such as adult and juvenile mortality. But there are no attempts to measure reproductive effort. While none of the studies reviewed here involved field work, the model laboratory systems have considerable relevance to natural conditions, particularly for the grain beetles.

Mertz (53) performed an elegant experimental analysis of the coevolution of fecundity and senescence in *Tribolium castaneum,* a flour beetle. He subjected beetles to three treatments: (a) certain death after ten days as an adult, (b) certain death after 20 days as an adult, and (c) no imposed mortality, in which case some beetles lived more than 400 days. In treatments a and b, 11 to 12 generations elapsed during the selection phase of the experiment; in treatment c some of the founders probably survived to the end. *Tribolium* begin reproduction at 3 to 4 days of adult age, and by the fifth day of adult life females are producing eggs at a rate of 10 to 18 per day, a level which they maintain until senescent decline sets in at about 70 to 80 days of adult age.

One would predict that the age distribution of fecundity should shift to an earlier peak in beetles from treatment a and c, with b intermediate, and that beetles from

Table 8 Reliability of life history data on insects

Ref.	Comparison	Taxon	Criterion						Fit scheme?	Number of criteria met
			a	b	c	d	e	f		
(7)	inter	Calandra, Rhizopertha	yes	no	no	yes	yes	no	no	3
(63)	inter	Tribolium	yes	no	no	no	yes	no	yes	2
(8)	intra	Drosophila	yes	no	no	no	yes	no	no	2
(52)	intra	Tribolium	yes	yes	no	yes	yes	no	?	4
(38)	intra	Tribolium	no	yes	no	no	yes	no	no	2
(68)	inter	248 species of Ichneumonids	no	yes	no	no	yes	no	?	2
(69)	inter	10 species of Ichneumonids	no	yes	no	no	yes	no	?	2
(53)	intra	Tribolium	yes	yes	no	yes	yes	no	?	4

c should live longer than beetles from a, with b again intermediate. In fact, when he assayed for fertility and longevity after rearing them under constant conditions for two generations to eliminate maternal effects, Mertz observed a statistically significant shift only in fecundity: Beetles from treatment a had fecundities about 15% higher than beetles from treatments b or c during the first ten days of their adult lives. Later in life, the fecundity relationships reversed, so that beetles from all treatments gave birth to about the same number of young during their lifespans. Mertz's experiment provides rare, direct evidence that a change in adult mortality can shift the age distribution of fecundity.

At present, $T.$ confusum is probably the only species for which one can answer the question, "Are r and K negatively correlated as most theory assumes?" The answer is, "Sometimes they are, and sometimes they aren't" (38). King & Dawson measured population growth rates and equilibrium densities of $T.$ confusum exposed to all 16 possible combinations of four temperatures and four food treatments. Developmental plasticity resulted in quite different growth rates and equilibrium densities. The correlation between r and K was negative but not significant (−0.148) in one type of rearing chamber, positive and significant (+0.561, $p < 0.001$) in another. Such experiments cannot establish the genetic basis of a trade-off between r and K, but they can show that where differences are based on phenotypic plasticity, that trade-off need not exist.

In summary, the insect data do not support the generality of the accepted scheme. Of the well-studied cases, only the comparison between $T.$ castaneum and $T.$ confusum (63) fits the accepted scheme, and it is complicated by the developmental response to experimental conditions. Neither comparisons of geographical races of the fly, Drosophila serrata (8), nor comparisons of the grain beetles Calandra oryzae and Rhizopertha dominica (7), fit the scheme. Field evidence on ichneumonid wasps (68, 69) and the results of selection experiments on Tribolium (53) suggest that patterns of adult mortality can dominate life history evolution.

Herbaceous flowering plants

Table 9 summarizes data on herbaceous flowering plants. The consistently high quality of work on this group probably stems from the influence of Clausen, Keck & Hiesey (15), who emphasized the necessity of separating genetically based variation from developmental plasticity; and of Harper (28), who championed the analysis of reproductive adaptations in plants. Many plant ecologists raise individuals under equivalent conditions (six of ten studies reviewed) and attempt to measure reproductive effort (six of ten studies reviewed). However, there were few attempts to measure the relevant selective forces (four of ten studies), to measure age- or stage-specific mortality rates, or to measure degree of density-dependence (one of ten studies), even though the latter was frequently invoked as an explanation (1, 24, 25).

Solbrig's experiments on dandelions (24, 81, 82) equal the best on any organism. By growing four biotypes of parthenogenetic dandelions under constant conditions, he found that he could neatly arrange them into a series ranging from small leaves and high reproductive output (type A) to large leaves and low reproductive output

EVOLUTION OF LIFE HISTORY TRAITS 167

Table 9 Reliability of life history data on plants

Ref.	Comparison	Taxon	Criterion						Fit scheme?	Number of criteria met
			a	b	c	d	e	f		
(70)	intra	Echinochloa	yes	no	no	no	yes	no	no	2
(71)	intra	Euphorbia	yes	no	no	no	yes	no	no	2
(14)	inter	Lemna, Salvinia	yes	no	yes	no	yes	no	no	3
(24)	inter	Solidago	no	no	no	no	yes	wt	yes	1
	intra	Taraxacum	yes	maybe	no	no	yes	wt	yes	4
(1)	inter	Solidago	no	no	no	no	yes	wt	yes	2
(25)	inter	Helianthus	no	no	no	no	yes	wt	yes	2
(30)	intra	Polygonum	yes	yes	no	no	yes	wt	?	4
(49)	inter	Typha	yes	yes	no	no	yes	wt	yes	4
(98)	inter	Asclepias	no	yes	no	no	yes	no	no	2
(59)	both	Oryza	yes	no	no	yes	yes	no	yes	3

(type D). In three plots varying from highly disturbed (frequently mowed and trampled) to relatively undisturbed (mowed once a year to a height 20 cm above the ground), the biotypes fell into a convincing cline, type D dominating the undisturbed site, type A the disturbed site, and types B and C sharing the intermediate site with type A. Type D always outcompeted type A. Solbrig's data fit neatly into the accepted scheme, but patterns emerging from recent work on dandelions from Montana (Solbrig, personal communication) break that trend.

McNaughton's (49) data on cattails (*Typha*) are also reliable. He raised them under constant conditions, tried to measure relevant environmental factors, and measured reproductive effort. Cattails also neatly fit the accepted scheme. Populations from North Dakota (*T. latifolia* and *T. angustifolia*) matured a month faster than populations from Texas (*T. latifolia* and *T. dominguensis*), produced more and smaller fruits per clone, more and smaller rhizomes per clone, and grew to a lower height.

In a provocative paper, Oka (59) examined inter- and intraspecific variability in the reproductive traits and phenotypic plasticity of wild (*Oryza perennis*) and domestic (*O. sativa*) rice. Under seminatural conditions annual forms had higher seed production, more soil-buried seeds, strong seed dormancy, less vegetative reproduction, and higher juvenile mortality than perennial forms. All annuals, of course, died after their first season of reproduction, but during that first reproductive period they experienced lower adult mortality rates than the perennials. These results roughly fit the accepted scheme, but go beyond it in emphasizing the broad range of reproductive adaptations in plants: Seed dormancy, vegetative reproduction, and developmental plasticity have to be considered.

In herbaceous plants the accepted scheme is not general, but does fit some cases. Sunflowers, goldenrods, dandelions (so far), and cattails seem to fit; grasses, milkweeds, and duckweeds do not. Wilbur (98) demonstrated that a more complex set of selective factors, accounting for predation, competition, and mortality, could explain more types of reproductive variability than r- and K-selection. Hickman (30) showed that all the reproductive variability observed among populations in the field could be due to developmental plasticity.

As a whole, the published life history data are moderately reliable as description, but are not reliable as tools for uncovering evolutionary causation. To what degree does this conclusion depend on the criteria used and the papers selected? I tried to select the best papers, which would make the conclusion conservative. The criteria arise naturally from the models being tested: criteria a to d were included to assay attempts to check assumptions and avoid circular reasoning; criterion e, to establish the statistical reliability of the data; and criterion f, to probe the degree of realism achieved. Additional criteria could have been used, but I would oppose the elimination of any of the six employed, with the possible exception of the requirement that degree of density-dependence be measured. That seems to have been asking too much. I doubt that adding criteria would change the conclusions.

How well do the data fit the accepted scheme? In about half the studies containing sufficient information, the organisms fit the accepted scheme ($n = 18$); in the other half, they did not ($n = 17$). Authors of a broad view (16, 95) were slightly more likely to perceive the accepted scheme than authors of a detailed study on intraspecific variability, but there are a number of counterexamples (24, 41). Studies fitting the scheme satisfied 1.5 of the 6 criteria, on average; those that did not satisfied 1.9. That slight difference in reliability is not significant. I am satisfied that neither the deterministic nor the stochastic models are empirically sufficient. Their predictions are not consistent with much of the evidence. We do not yet have a general and reliable theory of life history evolution, and the crux of the problem is: What will be an empirically sufficient set of parameters in which to couch the theory?

ACKNOWLEDGMENTS

My thanks to Mike Soulé, Bill Lidicker, Ray Huey, Paul Sherman, George Oster, Dave Wake, Tuck Hines, George Williams, Mike Hirshfield, Brian Charlesworth, and Ric Charnov, whose constructive criticisms greatly improved early drafts. In all cases, the interpretations and conclusions are my own, for I have rejected some substantive suggestions. I was supported by a Miller Postdoctoral Fellowship at the University of California, Berkeley.

Literature Cited

1. Abrahamson, W. G., Gadgil, M. 1973. Growth form and reproductive effort in goldenrods (*Solidago,* Compositae). *Am. Nat.* 107:651–61
2. Alm, G. 1959. Connection between maturity, size, and age in fishes. *Rep. Inst. Freshw. Res., Drottingholm* 40:5–145
3. Andrews, R., Rand, A. S. 1974. Reproductive effort in anoline lizards. *Ecology* 55:1317–27
4. Ashmole, N. P. 1971. Sea bird ecology and the marine environment. In *Avian Biology,* Vol. I, Ch. 6, ed. Farner & King. New York: Academic
5. Ballinger, R. E. 1973. Comparative demography of two viviparous iguanid lizards (*Sceloporus jarrovi* and *Sceloporus poinsetti*). *Ecology* 54:269–83
6. Barkalow, F. S. Jr. 1962. Latitude related to reproduction in the cottontail rabbit. *J. Wildl. Mgmt.* 26:32–37
7. Birch, L. C. 1953. Experimental background to the study of distribution and abundance of insects. I. The influence of temperature, moisture, and food on innate capacity for increase of grain beetles. *Ecology* 34:698–711
8. Birch, L. C., Dobzhansky, Th., Elliott, P. O., Lewontin, R. C. 1963. Relative

fitness of geographic races of *Drosophila serrata*. *Evolution* 17:72–83
9. Blair, W. F. 1960. The rusty lizard. A population study. Austin: Univ. Texas Press. 185 pp.
10. Bradshaw, A. D. 1965. Evolutionary significance of phenotypic plasticity in plants. *Adv. Genet.* 13:115–55
11. Bruce, R. C. 1972. Variation in the life cycle of the salamander *Gyrinophilus porphyriticus*. *Herpetologica* 28:230–45
12. Caughley, G. 1966. Mortality patterns in mammals. *Ecology* 47:906–18
13. Charlesworth, B. 1976. Natural selection in age-structured populations. *Lect. Math. Life Sciences* 8:69–87
14. Clatworthy, J. N., Harper, J. L. 1962. The comparative biology of closely related species living in the same area. V. Inter- and intraspecific interference within cultures of *Lemna* spp. and *Salvinia natans*. *J. Exp. Bot.* 13:307–24
15. Clausen, J., Keck, D. D., Hiesey, W. M. 1940. Experimental studies on the nature of species. I. Effect of varied environments on Western North American plants. *Carnegie Inst. Washington Publ.* 520. 452 pp.
16. Cody, M. L. 1971. Ecological aspects of reproduction. In *Avian Biology*, Vol. I, Ch. 10, ed. Farner & King. New York: Academic
17. Cohen, D. 1966. Optimizing reproduction in a randomly varying environment. *J. Theor. Biol.* 12:119–129
18. Cohen, J. E. 1976. Irreproducible results and the breeding of pigs. *Bioscience* 26:391–94
19. Cole, L. C. 1954. The population consequences of life history phenomena. *Quart. Rev. Biol.* 29:103–37
20. Conaway, C. H., Sadler, K. C., Hazelwood, D. H. 1974. Geographic variation in litter size and onset of breeding in cottontails. *J. Wildl. Mgmt.* 38:473–81
21. Dunmire, W. W. 1960. An altitudinal survey of reproduction in *Peromyscus maniculatus*. *Ecology* 41:174–82
22. Fitch, H. S. 1956. An ecological study of the collared lizard (*Crotaphytus collaris*). *Univ. Kans. Publ. Mus. Nat. Hist.* 8:213–74
23. Fitch, H. S. 1958. Natural history of the sixlined racerunner (*Cnemidophorus sexlineatus*). *Univ. Kans. Publ. Mus. Nat. Hist.* 11:11–62
24. Gadgil, M., Solbrig, O. T. 1972. The concept of *r*- and *K*-selection: evidence from wild flowers and some theoretical considerations. *Am. Nat.* 106:14–31
25. Gaines, M. S., Vogt, K. J., Hamrick, J. L., Caldwell, J. 1974. Reproductive strategies and growth-patterns in sunflowers (*Helianthus*). *Am. Nat.* 108:889–94
26. Green, R., Painter, P. R. 1975. Selection for fertility and development time. *Am. Nat.* 109:1–10
27. Hairston, N. G., Tinkle, D. W., Wilbur, H. M. 1970. Natural selection and the parameters of population growth. *J. Wildl. Mgmt.* 34:681–90
28. Harper, J. L. 1967. A Darwinian approach to plant ecology. *J. Ecol.* 55:242–70
29. Healy, W. R. 1974. Population consequences of alternative life histories in *Notophthalmus v. viridescens*. *Copeia* 1974:221–29
30. Hickman, J. C. 1975. Environmental unpredictability and plastic energy allocation strategies in the annual *Polygonum cascadense* (Polygonaceae). *J. Ecol.* 63:689–701
31. Hines, A. H. Jr. 1976. Comparative reproductive biology of three species of intertidal barnacles. PhD Thesis. Univ. California, Berkeley. 259 pp.
32. Hirshfield, M. F., Tinkle, D. W. 1975. Natural selection and the evolution of reproductive effort. *Proc. Nat. Acad. Sci. USA* 72:2227–31
33. Huey, R. B., Pianka, E. R., Egan, M. E., Coons, L. W. 1974. Ecological shifts in sympatry: Kalahari fossorial lizards (*Typhlosaurus*). *Ecology* 55:304–16
34. Hussell, D. J. T. 1972. Factors affecting clutch size in arctic passerines. *Ecol. Monogr.* 42:317–64
35. Jackson, W. B. 1965. Litter size in relation to latitude in two murid rodents. *Am. Midl. Nat.* 73:245–47
36. Keith, L. B., Rongstad, O. J., Meslow, E. C. 1966. Regional differences in reproductive traits of the snowshoe hare. *Can. J. Zool.* 44:953–61
37. Kempthorne, O., Pollak, E. 1970. Concepts of fitness in mendelian populations. *Genetics* 64:125–45
38. King, C. E., Dawson, P. S. 1972. Population biology and the *Tribolium* model. In *Evolutionary Biology* 5:133–228, ed. Dobzhansky, Hecht, & Steere
39. King, M.-C., Wilson, A. C. 1975. Evolution at two levels in humans and chimpanzees. *Science* 188:107–16
40. Krumholz, L. A. 1948. Reproduction in the western mosquitofish, *Gambusia affinis affinis* (Baird and Girard), and its use in mosquito control. *Ecol. Monogr.* 18:1–44

41. Krumholz, L. A. 1963. Relationships between fertility, sex ratio, and exposure to predation in populations of the mosquitofish. *Gambusia manni* Hubbs at Bimini, Bahamas. *Int. Rev. Gesamten Hydrobiol.* 48:201–56
42. Lack, D. 1968. *Ecological Adaptations for Breeding in Birds.* London: Methuen. 409 pp.
43. Lewontin, R. C. 1965. Selection for colonizing ability. In *The Genetics of Colonizing Species,* ed. Baker & Stebbins, pp. 77–91. New York: Academic
44. Lewontin, R. C. 1974. *The Genetic Basis of Evolutionary Change.* New York: Columbia Univ. Press. 346 pp.
45. Lord, R. D. Jr. 1960. Litter size and latitude in North American mammals. *Am. Midl. Nat.* 64:488–99
46. Lotka, A. J. 1907. Studies on the mode of growth of material aggregates. *Am. J. Sci.* 24:199–216
47. MacArthur, R. H. 1962. Some generalized theorems of natural selection. *Proc. Nat. Acad. Sci. USA* 48:1893–97
48. MacArthur, R. H., Wilson, E. O. 1967. *Theory of Island Biogeography.* Princeton: Princeton Univ. Press. 203 pp.
49. McNaughton, S. J. 1975. *r*- and *K*-selection in Typha. *Am. Nat.* 109:251–61
50. Maiorana, V. C. 1976. Size and environmental predictability for salamanders. *Evolution* 30:599–613
51. Mayr, E. 1963. *Animal Species and Evolution.* Cambridge: Belknap. 797 pp.
52. Mertz, D. B. 1971. Life history phenomena in increasing and decreasing populations. In *Statistical Ecology,* 2:361–392, ed. Patil, Pielou, & Waters. University Park, Pa: Penns. State Univ. Press
53. Mertz, D. B. 1975. Senescent decline in flour beetle strains selected for early adult fitness. *Physiol. Zool.* 48:1–23
54. Moore, J. C. 1961. Geographic variation in some reproductive characteristics of diurnal squirrels. *Bull. Am. Mus. Nat. Hist.* 122:1–32
55. Mountford, M. D. 1968. The significance of litter size. *J. Anim. Ecol.* 37:363–67
56. Mountford, M. D. 1971. Population survival in a variable environment. *J. Theor. Biol.* 32:75–79
57. Murphy, G. I. 1966. Population biology of the Pacific sardine (*Sardinops caerulea*). *Proc. Cal. Acad. Sci.* 34:1–84
58. Murphy, G. I. 1968. Pattern in life history and the environment. *Am. Nat.* 102:391–403
59. Oka, H.-I. 1976. Mortality and adaptive mechanisms of *Oryza perennis* strains. *Evolution* 30:380–92
60. Organ, J. A. 1961. Studies on the local distribution, life history, and population dynamics of the salamander genus *Desmognathus* in Virginia. *Ecol. Monogr.* 31:189–220
61. Oster, G. F., Auslander, D. M., Allen, T. T. 1976. Deterministic and stochastic effects in population dynamics. *J. Dyn. Syst., Meas., Control.* No. 76-Aut-F:1–5
62. Oster, G. F., Takahashi, Y. 1974. Models for age-specific interactions in a periodic environment. *Ecol. Monogr.* 44:483–501
63. Park, T., Mertz, D. B., Petrusewicz, K. 1961. Genetic strains of *Tribolium:* their primary characteristics. *Physiol. Zool.* 34:62–80
64. Perrins, C. M., Jones, P. J. 1974. The inheritance of clutch size in the Great Tit (*Parus major* L.). *Condor* 76:225–28
65. Pianka, E. R. 1970. On "r" and "K" selection. *Am. Nat.* 104:592–97
66. Pianka, E. R. 1970. Comparative autecology of the lizard *Cnemidophorus tigris* in different parts of its geographic range. *Ecology* 51:703–20
67. Pianka, E. R., Parker, W. S. 1975. Age-specific reproductive tactics. *Am. Nat.* 109:453–64
68. Price, P. W. 1973. Reproductive strategies in parasitoid wasps. *Am. Nat.* 107:684–93
69. Price, P. W. 1974. Strategies for egg production. *Evolution* 28:76–84
70. Ramakrishnan, P. S. 1960. Ecology of *Echinochloa colonum* Linn. *Proc. Ind. Acad. Sci.* 52:73–90
71. Ramakrishnan, P. S. 1960. Studies in the autecology of *Euphorbia hirta* Linn. *J. Ind. Bot. Soc.* 39:455–73
72. Reddingius, J. 1971. Gambling for existence. *Acta Biotheoretica XX, Suppl.* 1:1–208
73. Rocklin, S., Oster, G. F. 1976. Competition between phenotypes. *J. Math. Biol.* 3:225–61
74. Roughgarden, J. 1971. Density-dependent natural selection. *Ecology* 52:453–68
75. Schaffer, W. M. 1972. Evolution of optimal reproductive strategies. PhD Thesis. Princeton University, Princeton, N.J. 129 pp.
76. Schaffer, W. M. 1974. Selection for optimal life histories: the effects of age structure. *Ecology* 55:291–303

77. Schaffer, W. M. 1974. Optimal reproductive effort in fluctuating environments. *Am. Nat.* 108:783-90
78. Schaffer, W. M., Rosenzweig, M. L. 1977. Selection for life histories. II. Multiple equilibria and the evolution of alternative reproductive strategies. *Ecology:* 58:60-72
79. Smith, J. M., Price, G. R. 1973. The logic of animal conflict. *Nature* 246:15-18
80. Smith, M. H., MacGinnis, J. T. 1968. Relationships of latitude, altitude, and body size to litter size and mean annual production of offspring in *Peromyscus*. *Res. Popul. Ecol.* 10:115-26
81. Solbrig, O. T. 1971. The population biology of dandelions. *Am. Sci.* 59:686-94
82. Solbrig, O. T., Simpson, B. B. 1974. Components of regulation of a population of dandelions in Michigan. *J. Ecol.* 63:473-86
83. Southwood, T. R. E., May, R. M., Hassell, M. P., Conway, G. R. 1974. Ecological strategies and population parameters. *Am. Nat.* 108:791-804
84. Stearns, S. C. 1976. Life history tactics: a review of the ideas. *Quart. Rev. Biol.* 51:3-47
84a. Stearns, S. C. 1977. Interpopulation differences in reproductive traits of *Neoheterandria tridentiger* (Pisces: Poeciliidae) in Panamá. *Copeia.* In press
84b. Stearns, S. C. 1977. Reproductive variation in two poeciliid fish in stable and fluctuating Hawaiian reservoirs. Manuscript
85. Stent, G. S. 1975. Limits to the scientific understanding of man. *Science* 187:1052-57
86. Svardson, G. 1949. Natural selection and egg number in fish. *Ann. Rep. Inst. Freshw. Res., Drottingholm* 29:115-22
87. Tilley, S. G. 1968. Size-fecundity relationships and their evolutionary implications in five desmognathine salamanders. *Evolution* 22:806-16
88. Tilley, S. G. 1973. Life histories and natural selection in populations of the salamander. *Desmognathus ochrophaeus. Ecology* 54:3-17
89. Tinkle, D. W. 1965. Effects of radiation on the natality, density, and breeding structure of a natural population of lizard, *Uta stansburiana. Health Phys.* 11:1595-99
90. Tinkle, D. W. 1967. The life and demography of the side-blotched lizard, *Uta stansburiana. Misc. Publ. Mus. Zool. Univ. Mich.* 132:1-182
91. Tinkle, D. W. 1969. The concept of reproductive effort and its relation to the evolution of life histories in lizards. *Am. Nat.* 103:501-16
92. Tinkle, D. W. 1972. The dynamics of a Utah population of *Sceloporus undulatus. Herpetologica* 28:351-59
93. Tinkle, D. W., Ballinger, R. E. 1972. *Sceloporus undulatus:* a study of the intraspecific comparative demography of a lizard. *Ecology* 53:570-84
94. Tinkle, D. W., Hadley, N. F. 1975. Lizard reproductive effort: caloric estimates and comments on its evolution. *Ecology* 56:427-34
95. Tinkle, D. W., Wilbur, H. M., Tilley, S. G. 1970. Evolutionary strategies in lizard reproduction. *Evolution* 24:55-74
96. Turner, F. B., Hoddenbach, G. A., Medica, P. A., Lannom, J. R. 1970. The demography of the lizard, *Uta stansburiana* Baird and Girard, in Southern Nevada. *J. Anim. Ecol.* 39:505-19
97. Vinegar, M. B. 1975. Demography of the striped plateau lizard, *Sceloporus virgatus. Ecology* 56:172-82
98. Wilbur, H. M. 1976. Life history evolution in seven milkweeds of the genus *Asclepias. J. Ecol.* 64:223-40
99. Wilbur, H. M. Tinkle, D. W., Collins, J. P. 1974. Environmental certainty, trophic level, and resource availability in life history evolution. *Am. Nat.* 108:805-17
100. Wilson, A. C., Sarich, V. M., Maxson, L. R. 1974. The importance of gene rearrangement in evolution: evidence from studies on rates of chromosomal, protein, and anatomical evolution. *Proc. Nat. Acad. Sci. USA* 71:3028-30
101. Wright, S. 1932. The roles of mutation, inbreeding, crossbreeding, and selection in evolution. *Proc. Sixth Int. Congr. Genet.* 1:356-66
102. Wright, S. 1968. *Evolution and the Genetics of Populations*. Chicago: Univ. Chicago Press. 469 pp. Vol. I
103. Zweifel, R. G., Lowe, C. H. 1966. The ecology of a population of *Xantusia vigilis*, the desert night lizard. *Am. Mus. Novitates* 2247:1-57

EMERGENCE OF RESEARCH ON AGRO-ECOSYSTEMS

❖4121

Orie L. Loucks

Departments of Botany and Forestry, University of Wisconsin, Madison, Wisconsin 53706

INTRODUCTION

Ecosystem research using a systems approach has been under way in ecology for about fifteen years, mostly as studies of energy flow and nutrient dynamics. During this time the methods of ecosystem analysis have been applied to a few lightly managed natural resource systems such as fisheries, forestry, and rangelands (30). Systems methods have been applied in agriculture for about the same length of time, but usually in relation to reducing costs, improving marketability, or broadening income options.

Agricultural ecosystems are more complex than other natural resource systems in many ways (11), and it is understandable that an ecosystem viewpoint has not developed. In addition to the cycling of energy and materials in the agricultural system, there are many man-manipulated processes going on, mostly modifying inputs and exports, but also affecting rate relationships within the system. These interventions are largely the result of economic and market processes, and, ultimately, they control the dominant characteristics of the systems. Although humans can be viewed as a natural constituent of the system, the study of agricultural ecosystems cannot focus on marketing and decision-making processes alone, nor on energy exchange and related ecosystem processes. The economic system governing the intensity of inputs and exports, and the economic viability (survival) of the farm operator determining the inputs are essential, integral components of the agricultural ecosystem. Thus, students of agricultural ecosystems must recognize that we are only beginning the long process of molding diverse viewpoints together as a coherent field of inquiry. For convenience, the generic term for the object of study has become "agro-ecosystems." This paper reviews the recent development of an interfacing among the sciences involved and the emergence of an art and science of agro-ecosystem analysis.

For the purposes of the review, ecosystems are defined as functioning units of the biosphere, usually self-maintaining (often with perturbations), and deriving distinctive properties from their structural components as well as from interactions among those components. For studies of the functioning of agro-ecosystems, the concept must be elaborated to include the agents of the expanded inputs and exports, and to provide for quantifiable boundary conditions such that exchange of materials with adjacent systems is minimized, and, in any case, is measurable.

This paper examines studies that view culturally dominated agricultural systems as ecosystems, amenable to intensive, holistic analysis. Intensive efforts to increase agricultural production, due perhaps to the prospect of world food shortages (7, 22), have created an urgent need for stability and control analysis of agro-ecosystems on a world-wide scale. To increase agricultural production, the United States and other countries will be forced to use more fertilizers, more fossil energy, and more marginal land (37), all at the risk of serious degradation of at least some agro-ecosystems or of nonsustainable dependencies on others. It is therefore essential to develop methods for analyzing and forecasting the consequences of increased stress on agro-ecosystems, and of considering new ecosystem designs that minimize the risks.

It is important to recognize several properties which distinguish agro-ecosystems from other terrestrial ecosystems. Maximum harvestable (i.e. economic) productivity is a dominant goal. It is achieved through the use of monocultures of opportunistic species, whose ecology, genetics, and physiology make up much of agronomy. The system is supported by external perturbations, and additions of minerals and water. The residual detritus of the system, including nutrients, is reduced by harvest and export, as well as by exposure and the continual disturbance of the soil. Materials not converted to plant productivity are in excess of the needs of the system, and are readily leached into the groundwater or to nearby streams where they contribute to increased productivity in aquatic ecosystems (26). The conversion of natural ecosystems to agricultural systems and the expansion of intervention in established agro-ecosystems lead to a variety of changes in nutrient cycling and material flow. In this review I first survey the autecology of the domesticated plant and animal species that make up intensively managed agro-ecosystems. Second, I review the use of systems methods or ecosystem modeling to examine the functioning of ecosystems in general, or agro-ecosystems in particular. This inevitably leads to a focus on the energy and materials that flow or cycle in agro-ecosystems, independent of specific management or harvest practices. Finally, I include studies of the need and/or techniques for evaluating control mechanisms within the systems, and for assessing the balances of inputs, plant products, animal responses, and total exports.

THE CONTRIBUTION OF AGRICULTURAL ECOLOGY

The study of agro-ecosystems necessarily draws on the existing broad foundations of agronomy and crop ecology. Azzi (4) traces the early development of agricultural ecology to its origin within the field of agriculture. He cites agricultural meteorology and soil science as subjects that grew from a common root within agriculture. However, between agricultural meteorology, soil science, and agronomy (crop management), Azzi suggests there is a deep rift which agricultural ecology must fill. What interests the agronomist is not so much the relationships among soils, insect pests, and fertilization, but the effect of each of these upon the desired crop.

While meteorology, soil science, and entomology are distinct components according to Azzi, their study in relation to the potential responses of crop plants converges on the single branch of science he identifies as agricultural ecology. He recognizes four important goals: to determine climatically equivalent regions, to explore the ecological characteristics of the species, to develop geographical race trials for all commonly cultivated species, and to analyze yields differentially to discover what controls the complex relationships among plants, environment, and yield.

Agricultural ecology was further elaborated in the 1960s by Tischler (49) and is an integral part of the Agricultural College curriculum through a variety of courses on the ecology of various agronomic groups. Because it provides a precise statement of the biophysical relations of the crop species, it is a subject practically applicable to farm management. The courses teach the bases for an ecological point of view on crop adaptation, pest management, and land husbandry. Along with work on insect and plant pathogen ecology (5), the broad topic of agricultural ecology seems to have been the basis for recent moves toward an ecosystem approach to soil management (50), pest control (8), and crop management (41).

THE CONTRIBUTION OF ECOSYSTEM ANALYSIS

Origins of the Approach

For years, scientists have sought to understand the temporal, spatial, and structural complexities of dynamic, renewable natural resource systems. To some, the landscape may appear as populations of organisms influenced by a complex of climatic, edaphic, genetic, and other interactive factors. To others, it appears as a mosaic of interdependent units, ecosystems, the product of evolutionary processes mediated by climate, geology, and geography (40). Ultimately, resource management requires an understanding of the systems at both levels of resolution. Such a view of the landscape can include the various types of cultivated systems as well as the forests, grasslands, lakes, deserts, and oceans which comprise the biosphere.

Differentiation of the functional and dynamic components of ecosystems is commonly credited to Lindeman (23). A great impetus to understanding ecological processes at the landscape level came from Odum (31), who showed that ecosystem ecology has an intrinsic tie to our understanding of nature and the environment. The flow of materials within the ecosystem became identified with the overall metabolism of the ecosystem; in turn, system metabolism was recognized as a measure of the collective processes which served to maintain the integrity and stability of the system (40).

The analogy to cellular and organism metabolism provided a basis for the application of similar methodologies to ecosystem analysis. The principles, techniques, and analytical power of radiotracer methods for quantifying trophic dynamics in ecosystems were demonstrated by Auerbach (2) and others. Application of transfer coefficient matrices in ecosystem modeling by Olson (33) illustrated the potential for using advances in systems analysis and computers to provide answers to questions about interactions within ecosystems.

By the early 1960s, a number of studies had demonstrated the advantages of using data processing for the solution of long-standing problems such as predator-prey interactions (17), populations and their exploitation (52), and ecosystem dynamics (33). But the full potential of the computer for large-scale ecosystem studies could not be realized until a new level of formal theory was achieved. Once the new concepts of ecosystem level research had been introduced, and a parallel understanding of mathematical and numerical analytic techniques had been developed, several investigators (32, 36, 43) began to apply systems analysis procedures to the study of ecosystems with convincing results. The Biome programs of the US International Biological Program (IBP) and other similar studies further stimulated the application of systems methods to the study of natural resource systems (30).

The earliest work on ecosystem analysis grew out of descriptive research on the flow of energy in food webs. This work uses the "compartmental" approach to the definition of functional parts within ecosystems, viewing the whole ecological system as a series of parallel compartments or "pools" of energy, carbon, and nutrients. The energy and nutrients in a pool are utilized or "transferred" to other pools by a variety of processes within the ecosystem. Descriptive models of the systems take the form of "box and arrow" diagrams which indicate the defined pools (primary producers, herbivores, decomposers etc) as boxes and show the relationships among them with arrows. The relationships pictured in the diagrams are then often expressed by systems of simple linear differential equations which allow a preliminary study of the dynamics and interactions among the components of the system.

The US contribution to the International Biological Program provided an additional impetus to the study of ecosystem dynamics and in particular to quantitative modeling of ecosystem processes (30). This "Analysis of Ecosystems" program set out to develop realistic mathematical constructs for the basic biological, physical, and chemical processes within ecosystems (the arrows of the box and arrow diagrams), and to aggregate these into ecosystem level models where possible. Emphasis was placed on clarifying ecosystems and the functional relationships within them. Some lightly managed ecosystems such as forests and rangelands were considered in the studies, but treatment of the magnified perturbations as well as the human inputs proved too difficult at the time.

Definition of Universal Components of Ecosystems

To achieve the goals of basic ecosystem analysis, whether of natural or intensively managed systems, consensus on the definition of components is essential. The ecosystem functions in the exchange of materials and energy. It has feedback dependencies and therefore self-regulatory attributes; it has identifiable boundaries and recurring, recognizable relationships among subcomponents. The properties of individual ecosystems arise from the magnitude of their dominant compartments, from the interaction, feedback, and synergism among their components; and from their exchange of materials with other systems. The variables (compartments) in the system, the processes operating on them, the parameters regulating the processes, the inputs and losses, and the external environmental influences operating on the system, all have to be defined, quantified, and evaluated with respect to their significance in each system.

The *variables* (compartments) are the system's individual species or functional groups of species, its dead organic matter, water, nutrient elements, and other significant materials. The *processes* are biological and physical activities or reactions that move or transform the materials of the system. Some processes operate within variables (e.g. respiration within an organism or functional group of related species), or they may be links between variables (e.g. water and nutrient uptake from soil). The processes can be expressed as a continuing rate of transfer, or they can be represented as complex functions, dependent on temperature, age, population density, and other factors. The relevant parameters of the system are coefficients that operate in the expressions for process relationships. Some, such as Q_{10} values, are constant for a variable and a process, while others vary continuously with the environment. Environmental influences (and in agro-ecosystems these include cultivation and other human interventions) operate to accelerate or limit the system by controlling process rates.

Some of the ecosystem processes under study by the US/IBP (30) and other research programs include the following:

1. *Plant Processes*
   Uptake: Net carbon fixation, water uptake, nutrient uptake
   Growth: Translocation, vegetation growth
   Life process: Respiration, flowering, nonherbivorous mortality
   Losses: Transpiration
2. *Animal Processes*
   Uptake: Food consumption, predation
   Growth: Assimilation, individual growth and development
   Life process: Respiration, reproduction, non-predatory mortality
   Losses: Excretion
3. *Water-Related Process*
   Input: Rainfall, interception
   Flows: Runoff, runon, infiltration
   Losses: Evaporation, percolation to groundwater
4. *Nutrient Process*
   Inputs: Deposition, nutrient transport, nitrogen fixation
   Transformations: Humification, volatization, ammonification
   Losses: Denitrification, decomposition, nitrification

Results from the studies of these processes, viewed in their context as ecosystem processes, are being reported in the various papers and books of the US/IBP. Preliminary results have been reported by Innis (18), O'Neill (34), Overton (35), Goodall (14), and Miller et al (29), and other papers are in press. Generally similar information is available for some agricultural ecosystems, but virtually no effort has been made to create a comprehensive model of the properties and functions of agro-ecosystems.

EMERGENCE OF A FOCUS ON AGRO-ECOSYSTEMS

Attempts to Define the Field

Agricultural practice has been understood as a system for decades, but relatively little has been done to formalize the body of relationships implicit in the system. Spedding (45) has reviewed some of the diversity of agricultural "systems," summarized in Figure 1. These systems are like the "natural" resource systems of the type studied as ecosystems during the IBP. Spedding notes that description and study of these systems can be more or less detailed. For a classification, the minimum description requires information about the dominant features that characterize a group. But for the purposes of analysis and extension of results to other areas, much more information is required about the functioning of the system.

To illustrate the conceptualization of agricultural systems by agriculturalists, it is useful to consider Spedding's example of a lowland sheep production system (Figure 2). The breed, size, age, weight, and longevity of the livestock are clearly important, as are the dates of mating and lambing. Spedding notes the need for detailed consideration of ram management, including the ratio of ewes to rams, and how many can share a paddock. Ultimately, some framework must be used to relate the components to each other and to the purpose of the system.

Diagrams of agricultural systems are likely to be oversimplifications. In any case they are static. However, as models they have many similarities to (but also some fundamental differences from) ecosystem models. According to Spedding the questions to be answered are: What kinds of models will be useful and to what extent will they serve their purpose? He especially notes that the value of a mathematical model in eliminating ambiguity and allowing system manipulation should not obscure the fact that it usually originates with a picture model.

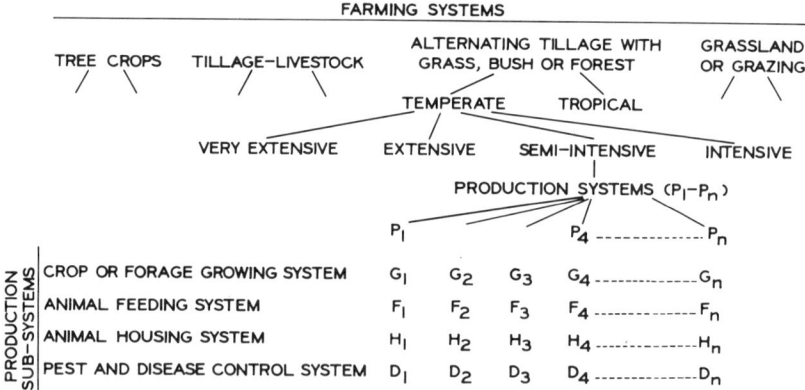

Figure 1 Relationships among agricultural systems (45). The subsystems are describable in terms of major components, different combinations of crop, cultivation method, fertilizer input, sowing date, etc, these being represented by $G_1 \ldots, G_n$.

AGRO-ECOSYSTEMS ANALYSIS 179

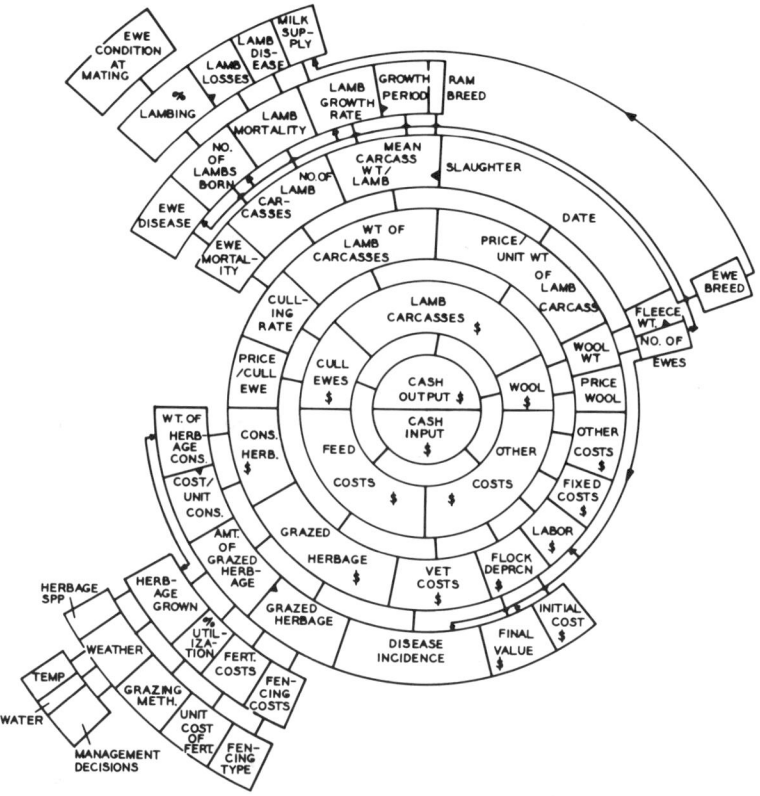

Figure 2 Components of a sheep production system showing the relationships among components. The system objective is at the center and successive circles contain those factors that directly affect components internal to them, indicated by radial lines. [From (45).]

Spedding's views, and the use of the term "agricultural system," reflect the past several decades in agricultural science. During that period a principal objective was the development of forecasting models to analyze agricultural production alternatives and marketing priorities, both in relation to cost and profit of the operation. However, agricultural policymaking also grew in scope and began to interrelate variables such as land use, farm management, agricultural residuals (pollution), price fluctuations, and taxation. Since then a new generation of agricultural models has evolved, consisting of optimization programs that include the detail necessary for modern agricultural policy analysis (1). Production processes for certain crops have been incorporated in the simulation for the purpose of determining crop spacing, water use efficiency, and interactions between water availability and fertilizer requirements. To date, however, there has been little if any application of agricultural simulation models for analysis of such environmental consequences of modern agricultural practice as nutrient leakage. A few models have been developed to address such problems as agricultural nitrate pollution of streams and groundwater (6, 19), but many of these are viewed as hydrologic or water quality simulation systems, rather than as components of an agricultural system.

In introducing the first number of the new journal *Agro-Ecosystems,* Harper (15) notes that even before the ecosystem concept was introduced into ecology both agriculture and forestry had been concerned with the practical management of ecosystems. The rotation of crops, control of weeds and pests, manipulation of stocking rates, drainage, manuring, and fertilization were acknowledgments of interactions of the type known in ecosystems. They simply had not been studied quantitatively. Harper suggests that at all levels of sophistication, farming, as well as forestry, involves extracting materials from an ecosystem and directing them away from the cycles that characterize the natural system. Agriculture deliberately channels energy and mineral resources out of one area into another, and in addition commonly creates "leakages" in otherwise self-maintaining systems. In the most intensive forms of agriculture, mineral resources may be lost by overcropping, particularly when the biomass holding much of the nutrient pool in a living cycle is suddenly reduced and the regulating mechanisms are lost. The science of agro-ecosystems will study the changes that follow the conversion of natural vegetation to agriculture.

Harper goes on to discuss the question of whether agricultural or forest systems can maintain productivity when they require a continual replenishment of the nutrients removed in cropping, or lost through "leaks" in the ecosystem. However organic or inorganic fertilizers are applied, whether simply at a level sufficient to make up losses, or at rates designed to maximize crop production, the intervention does not correspond with ecosystem functioning. Such use of fertilizers can create changes in the system itself, as well as in neighboring ecosystems into which excess nutrient leaches (26). Some of the most dramatic deteriorations of habitats have come as the result of irrigation—by the development of creeping salt deserts. Harper notes that a strong science is a predictive science, and the strength of the science of agro-ecosystems must be proved by its ability to predict complex interactions of ecosystems. Progress toward this science and its use in decision-making by the agricultural sector have been outlined in a recent report by the Scientific Committee on Problems of the Environment (16).

The contrasting views of Spedding (45), Harper (15), and classical ecosystem researchers (40) seem to suggest that the definition of agro-ecosystem science is as incomplete as the field is young. The absence of consensus appears to be a normal first step in the development of a discipline. A central issue is the comparative weakness of the applied sphere of ecosystem research: analysis of manipulations of production as well as harvesting and marketing of the yield from a natural resource system. The adoption of system methods for research in production aspects of agriculture has been late in developing, perhaps because factors controlling the intensity of inputs and extraction make the system much more complex than other natural resource ecosystems studied to date. The continual variation in the nature

AGRO-ECOSYSTEMS ANALYSIS 181

and intensity of human interventions must be included, as well as the very survival (economically) of the individual agents of intervention, the farm operators. Thus, development of an ecosystem approach in agriculture seems to have depended on emergence of relatively sophisticated models for the physical and biological processes of agricultural systems, and an understanding of how these process models could be utilized in an overall system. Progress required major innovations in techniques for coupling the physical and biological models with existing models for decision-making processes, optimization, and marketing systems. The latter, of course, govern the levels of human intervention in the biological and mineral cycling systems. Without them the dominant feedback is missing and the cycle cannot be closed. The papers in the *Study of Agricultural Systems* volume edited by Dalton (11), and to a lesser extent the papers in the journal *Agro-Ecosystems,* indicate the time is ripe for an assimilation of all of these elements. Although this integration has not moved far yet, the important first steps have been taken. A unique need and the specific questions have been recognized, and new research now seems likely to develop.

Agro-Ecosystem Research as Integrative Science

Despite the recent origin of agro-ecosystem analysis as a field of serious inquiry, several significant innovations are already recognizable. The following comments concern properties of entire agro-ecosystems, or the linkages operating among subsystems within these systems, and necessarily pass over the essential related work on subsystems (e.g. irrigation hydrology) or system processes (e.g. water use or photosynthesis). The INTECOL Working Group on Agro-ecosystems (19) has summarized some of this core research as it has involved intensive temperate and tropical agricultural systems.

PRODUCTION AND CYCLING IN THE TEMPERATE ZONES   In spite of the extensive analyses of energy flow and nutrient cycling in forest, grassland, and water ecosystems during the IBP, few studies were concerned with the functioning of whole agro-ecosystems. Recently, Markov (28) has outlined in qualitative terms a holistic approach to agro-ecosystem analysis. He distinguishes the structural components of agro-ecosystems, both biotic and abiotic, and emphasizes the significance of the pattern of cultivated fields in modern landscape management. Markov also distinguishes three broad types of interrelations among organisms in agro-ecosystems: those concerned with trophic relationships; those concerned with environmental conditioning (e. g. allelopathy, mechanical influences, etc); and those concerned with the influence of external factors, including human activities. In his view, these interrelations form a complex network and ultimately determine the functioning of agro-ecosystems.

An agro-ecosystem study concerned with energy flow and matter cycling has been under way in Poland since 1971. It is being carried out by a research team organized around the topic "Ecological Effects of Intensive Agriculture" (41). Long-term goals of the project include measurement of the heat balance of cultivated fields; estimates of total primary production (above and below ground) for various crop species; study of the mineralization of organic matter, with a consideration of the role played by microorganisms and soil animals; measurement of the cycling of water, nitrogen, phosphorus, and other elements, with emphasis on leaching processes; and estimation of the importance to the agro-ecosystem of animal components, both vertebrate and invertebrate. The team is committed to a synthesis of results from all the component studies, but no analysis of dynamic models appears to be under way.

Other studies of the net primary production of various agro-ecosystems and an analysis of the efficiency of solar energy utilization have been carried out in Japan (20). When the net primary production of these agro-ecosystems was compared with that of nearby forests it was found that similar or slightly higher values were obtained from the cultivated fields. The efficiencies of solar radiation utilization varied from 0.69 to 1.43%.

Results on utilization of solar energy and water by various ecosystems have been published by Rauner (39) in the USSR. The studies were carried out in an oak forest and a forb steppe, and can be compared with results from a barley field having similar climatic characteristics (41). The forest intercepted the highest amount of photosynthetically active radiation, and the cultivated field the least (Table 1). The ratio of transpired water to the total rainfall also decreased in this sequence. However, in efficiency of photosynthesis, calculated as the ratio of net production to the radiant energy intercepted by the ecosystem, the cultivated system surpassed the forest. The ratio of net production to the energetic cost of transpiration is also higher in cultivated plants.

If these results can be confirmed in other areas, an important characteristic of agro-ecosystems will emerge: that cultivated field ecosystems use intercepted solar energy and available water more efficiently than do natural ecosystems. Further testing of this principle will require large national research programs, international cooperation, and uniformity in definition of the system components.

Despite the numerous studies on nutrient and yield relationships in agriculture, the patterns of carbon and nutrient cycling within agro-ecosystems are still insufficiently understood (19). Compared to natural soil ecosystems (3, 53), agro-ecosystems have open cycles of mineral transfer, and are prone to losses (leakage) due to the inactivity of plant roots, to soil arthropods, or to other factors at certain times of the year under some cropping systems. A serious problem in studies of nutrient cycling in ecosystems has been evaluation of microbial activity under field conditions, its contribution to the food chain supporting the soil fauna, and the importance of both in intercepting and holding nutrients (3). The role played by soil fauna in the mineral cycling of agro-ecosystems appears to have been seriously neglected.

PRODUCTION AND CYCLING IN TROPICAL AGRO-ECOSYSTEMS   In contrast to temperate agro-ecosystems, tropical agro-ecosystems exhibit a wide range of culturally based structure and a wide range of cycling patterns. Some regions are characterized by a highly developed system of material recycling, and others have almost none (44). The rural ecology of Indonesia as described by Soemarwoto is an example of a traditional, intensive recycling agro-ecosystem in which man is an

AGRO-ECOSYSTEMS ANALYSIS 183

Table 1 Photosynthetic and water energy economy of three ecosystems (39, 41)

Energetic parameters	Ecosystem		
	oak forest	forb steppe	barley field
Photosynthetically active radiation Q (cal cm^{-2} yr^{-1})	45,000	45,000	45,000
Intercepted radiation Q_i (cal cm^{-2} yr^{-1})	25,000	18,000	11,000
Q_i/Q	0.56	0.40	0.24
Precipitation r (mm yr^{-1})	750	680	680
Transpiration E (mm yr^{-1})	500	300	170
$E{:}r$	0.67	0.44	0.25
Net production P (cal cm^2 yr^{-1})	640	520	440
$(P/Q_i) \times 100$	2.6	2.9	4.0
$(P/EL^a) \times 100$	2.2	2.9	4.4

$^a L$ = latent heat of water vaporization.

integral part of the food web and contributes to the mechanisms of cycling by returning wastes to the producing system. Soemarwoto points out that rural development programs are likely to modify the traditional cycling of these systems.

Less intensive agro-ecosystems are maintained in other tropical regions such as Central Africa and South America. Talbot (47) has presented data on the domestic livestock grazing systems of east Africa, and in other papers has discussed the mechanisms for self-regulation there. Once again, economic and regional development programs seem likely to change the input and harvest relationships of these ecosystems before regulatory mechanisms can be developed. The basis for the productivity of tropical systems has not been documented. This productivity is now being threatened by cultural changes, and production by the agro-ecosystems these changes have instituted there is uncertain. A rich variety of species is available for the development of new combinations of crop systems in these regions, systems which might well bring about increased yields and long-term stability at the same time.

THE PROBLEM OF BYPRODUCT WASTES FROM AGRICULTURAL TECHNOLOGY   In a stable ecosystem, inputs balance outputs over some reasonable period. Agro-ecosystems are managed to provide a substantial export of crops or forageable products. This increased output demands increased input. The periodic interventions in the system by cultivation and the addition of step-inputs that exceed the immediate capacity of the system tend to generate significant excesses in some compartments. These become "wastes" and are added to other systems in the landscape. Much of our understanding of waste releases has come from studies of excess nitrogen leaching into the groundwater of many regions (9, 27), and from work on natural cycling processes (3, 53). Few studies have looked for alternative agro-ecosystem structures and cycling patterns which could maintain adequate levels of production and still prevent unnecessary losses. The crop rotations described by Ryszkowski (41) are an outstanding example of a system that should minimize leakage, but the results are unquantified.

Researchers are similarly concerned about persistent use of biocides in agriculture (25, 27) and about salinity problems resulting from extensive additions of wastes in arid regions (22). A strategy for balancing the inputs and outputs of agro-ecosystems will minimize the wastes generated by agriculture, and conserve the water, soil, fertilizers, and other amendments used as inputs to the system (37).

Simulation Models in the Analysis of Agro-Ecosystems

Models as a form of synthesis are commonly used in some of the literature on agricultural systems (11). Van Dyne (51) has written an extensive review on agricultural systems models. He notes that when representing an agro-ecosystem mathematically, one must consider great diversity in the time and space responses of agro-ecosystems, the existence of threshold processes, limiting values, and discontinuities. Social, economic, and market considerations further complicate the problem. Although mathematical models have been used in some aspects of agricultural systems, there are few detailed analyses of such models.

Van Dyne provides several lengthy tables comparing the models he examined. The topics covered by these models include

Dairying (all aspects of feeding, herd size, pastures, weather, and capital investment)
Pest control (aspects of methodologies for finding optimal pest control policies, estimating the impact in crop and land use)
Irrigation (decision-making in the operation of irrigation systems)
Grazing (aspects of grazing systems management, live weight changes, and population variations)
Plant Growth (influences of moisture extraction, nitrogen, and light)
Forestry (aspects of optimal continuous yield cutting, and estimation of maximum allowable yields)
Agricultural Economics (including economic replacement problems, determining optimal animal combinations, farm growth, etc)
Fisheries (including population dynamics of salmon spawning in a large river system, and other exploited fish populations)
Wildlife (including herd management over time, consequences of pollution, and land use for big game).

Van Dyne notes that by and large the reporting of the models he examined was inadequate. He found it difficult to assess the attributes of the models or to find complete specifications for them. Almost all of the models he examined were deterministic; most authors seemed not to have recognized that their parameters had variances and covariances. One exception (10) varied one parameter at a time to obtain information about the covariances among parameters. According to Van Dyne's analysis of the simulation models used in agriculture, there are now increasing numbers of multiple-material flow models where once energy was the sole quantity modelled. He considers this to be an improvement. In addition, recent simulation models show an increasing ratio of flows to state variables, which indicates an increase in model complexity and in the coupling among variables within the model.

Van Dyne also comments on the recent abundance of optimization models utilizing agricultural economics rather than ecological relationships. He suggests that the closer the field is to practical economic problems, the more likely it is that optimization models will be used. The closer the field is to basic science, on the other hand, the more likely the models will use differential equations or use simulations to examine population or system phenomena. There is a need for models that have both simulation and optimization components.

Large-scale systems studies of the agriculture enterprise have appeared since Van Dyne's review. These show an increasingly comprehensive view of the system, but no improvement in the precision with which physical and biological processes are represented. To examine the energy balance of Israel's agriculture, Stanhill (46) uses an energy analysis for the year 1969/1970, together with data on changing trends in inputs and production. The overall energy balance of crop production in 1969/1970 was 66 kcal m^{-2} yr^{-1} and is increasing at the rate of 22 kcal m^{-2} yr^{-1}. Irrigated crop production had a negative balance (-143 kcal m^{-2} yr^{-1}) despite the fact that the productivity of irrigated land was twice that of agriculture as a whole. Data for the various branches of crop production suggest an inverse relationship between the energy balance and the intensity of production.

Another comprehensive agro-ecosystem study focused on dynamic simulation of a grazing ecosystem in Colorado (21). The model was developed by the Grassland Biome study group of the US/IBP, but it was used in this analysis to examine alternate grazing rotations (30). The goal was to study the long-term outcome of options that allowed more livestock yield and better quality herbaceous production. The general scheme of relationships used in the model is shown in Figure 3.

The most comprehensive systems studies continue to focus on the management paradigm of the system, rather than the energy and material flows. Up to a point, such a view seems to be essential for agro-ecosystems because of the dominance of human interventions (i. e. markets and prices, which in turn govern optimal fertilizer amendments). Arnold & Bennett (1) used an advanced set of crop growth, animal growth, and water balance submodels to simulate a mixed farming operation. They concluded that the consequences of a variety of tactics in managing a complex system with many feedbacks can be studied effectively. Optimization gives a single "best" strategy for a given set of conditions. It is of value in deciding short-term issues and in guiding long-term decisions. Simulation allows the decision-maker to see the consequences of choices other than the optimal one. The optimal solution to a problem depends on the income that the farmer is willing to accept, on what he is willing to risk, on labor availability, and on other factors.

Another study extended this approach to entire agricultural regions (12). These investigators defined inputs and outputs at the level of regional imports and exports, and examined the apparent optimal dynamics of the regional agro-ecosystem under

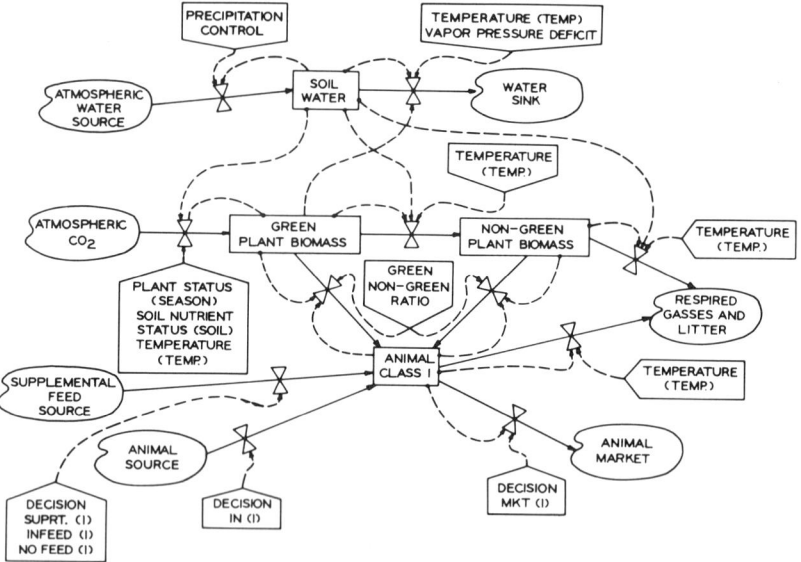

Figure 3 A simulation model of state variables (*boxes*) in a grazing ecosystem, utilizing biological and physical process submodels (*arrows*). [From (30).]

various control strategies. The study concludes that optimization models greatly help regional agricultural administrators to make their decisions.

THE PRESENT OPPORTUNITY FOR AGRO-ECOSYSTEMS

The research reviewed up to this point leads to several conclusions: much work on the functioning and dynamics of agricultural systems is under way, largely from disciplinary or subsystem viewpoints; serious gaps exist for some parts of the systems (such as below-ground organic production and mineral cycling); whole-system studies are dominated by operations-oriented work which pays little attention to questions of basic materials balances or long-term stability. If operations models are to be made more reliable, both for ecologists and managers, comprehensive work at the ecosystem level is needed.

One view of this necessity was expressed by Brown et al (7) in a report entitled *International Conference on Crop Productivity—Research Imperatives.* The conference focused on food crops and what might be done with fundamental biological processes to enhance their productivity. Preservation and improvement of the land base was recognized as essential. Arable land, water, and fossil fuel energy were recognized as finite resources. More land can be put into production, but much of that land is marginal and may require substantial inputs of water and fossil fuel energy to be productive. Little attention has been directed toward increasing the efficiency of water usage by crops. That efficiency is 30–40% worldwide, but Israel has achieved a remarkable efficiency of water utilization that exceeds 80%.

AGRO-ECOSYSTEMS ANALYSIS 187

Soil erosion control was recognized as another research imperative. Despite erosion control efforts, losses are estimated at 3.6 billion MT of topsoil annually, equivalent to 31 MT per hectare of tilled land. At the same time, improved efficiency of fertilizer uptake is needed. Crops in temperate zones are acknowledged to utilize 50% or less of the applied nitrogen; less than 35% is utilized in the tropics. Other issues recognized as global concerns included unemployment, inflation, and inadequate food supplies.

The conference report suggests that the operational approach to finding maximum agricultural productivity will have to use more than a simple modification of traditional ad hoc techniques; it will have to be "predicated on a scheme that integrates the entire system." Cropping systems with a high yield but sparing of nonrenewable resource inputs can and must be developed. Interacting factors such as crop cultivars, rotations, row spacing, soil nutrients, structure and moisture, temperature, sunlight, plant protection from pests, harvesting procedures, environmental concerns, and public health are recognized as having led to a complexity that cannot be accommodated without sophisticated management schemes. Further breakthroughs in agricultural productivity will be acceptable only if the total agricultural system is considered by new research strategies. Interdisciplinary work aimed at effective management of the total food production system will be required.

Many observers of agricultural research are less than sanguine about the prospect of achieving this goal. A National Academy of Sciences report (38) questions how adequately the present agricultural research community has responded to new imperatives. Commenting on the NAS report, Levins (24) observes that "agriculture research in the United States, whether federally supported or carried out by the agricultural chemicals industry, seed companies, or by the faculties of agricultural colleges, has suffered from a narrow pragmatism that is reinforced by the general indifference of researchers in basic science to problems of agriculture." Within agricultural science, he says, America's tradional anti-intellectualism survives as an antitheoretical bias, seen in the reluctance to take intellectual detours, and a preference for accumulating direct experience with a crop as a guide to practice. Even consideration of mixed plantings is inhibited by the present design of farm machinery, and research into the ecology of a mixed sowing only makes sense as part of a broader program that must include redesign of the machines.

Levins suggests that the long-term improvement of agricultural research requires, first of all, an intellectual rapprochement between pragmatic research and research into the theoretical underpinnings of applied science. "Fundamental" researchers must be convinced that there is a rich intellectual content in the study of agriculturally practical systems. The cultivated field and its associated fauna should be viewed as an historically new system, ideal for the study of dynamic aspects of evolutionary ecology. As the major interest in population biology shifts from equilibrium to changing systems, the environments, communities, and organisms of agricultural significance will acquire increasing fundamental significance. Levins believes that the goal of better *management* would tend to break down disciplinary boundaries, and would create an integrated population biology in which community structure, physiological responses, migration–extinction patterns, bioclimatology, and natural selection could be studied together.

At the same time, Levins says, people working in agricultural science must be convinced that theory is practical. As research horizons broaden, we deal with increasingly complex systems. Commonsense intuition, derived from the behavior of single components studied in isolation, becomes less reliable. We inevitably reach the point where a basic science approach to understanding the interactions among variables is the only practical approach. Among the areas of agriculturally relevant fundamental research which are relatively lagging, and which suggest themselves for new work, Levins includes the modeling and analysis of the cultivated field as an ecosystem, including the microbiological and cryptozoan communities.

Initiatives in these directions now seem to be developing among traditionally basic sciences, as well as in agriculture. Fiering & Holling (13) have outlined a comprehensive and conceptually precise approach to setting standards for the perturbation of ecosystems. They introduce constructs for measuring the effectiveness of environmental standards and for formalizing the economic and social costs of meeting them, emphasizing the element of recovery time (or its manipulation) in ecosystem management.

In an analysis of large-scale agriculture policy problems, Pimentel et al (37) examined patterns of arable land loss in the United States, alternatives for the conservation and protection of arable lands, and the economic and energy effects of soil erosion versus conservation. They project the land, water, and energy needed to feed a growing US population and to contribute to world food supplies. They find that even though soil erosion losses are serious, and have a detrimental effect on reservoirs, rivers, and lakes, impediments to soil conservation (such as its cost, though moderate) make substantial containment of soil erosion impossible. They then estimate the increased quantities of fossil energy, in the form of fertilizers and other inputs, that have had to be expended to offset the decline in productivity due to erosion. They estimate that the production potential of US cropland has been reduced by 10–15%. The input of fossil energy for current crop production is about 3 million kcal per acre. About half of this is utilized to increase crop productivity, whereas the other half is used to reduce labor. Thus, the estimated energy input required to offset past soil losses is about 200,000 kcal per acre; for the estimated 400 million acres in production, energy equivalent to 2.1 billion gal of fuel is used annually to offset past soil erosion losses in the United States. This was equal to about 4% of the nation's total oil imports in 1970. These results illustrate the fundamental interdependence among energy resources, land resources, and functioning agro-ecosystems.

Intensive studies also can be made of the tolerance of an agricultural system to stress. Let us assume that pressure greatly to increase agricultural productivity in the United States and other countries will continue to increase.[1] Let us then examine the methods by which we could assess potential long-term responses to this pressure.

The challenge is to assess the stability of the more productive agro-ecosystems, given the very subtle forms in which the new interventions will be introduced (new crop varieties, increased nitrogen fixation, etc). Recent ecosystem level models

[1]The following is based on a paper by the author presented at the Annual Founders Meeting of The Institute of Ecology, Morgantown, West Virginia, 1975.

provide the means for testing the apparent long-term sensitivity and internal control characteristics of natural resource systems (13, 42, 48). The response patterns are of the form expected, but the functional mechanisms and their sensitivity to human intervention can now be investigated.

The effects of introduced stress are apparent in their simplest form in simulation studies using simple stepwise changes in rates or rate-regulating coefficients within models of ecosystems. Unintended responses can be postulated as a result of subtle changes in stress over extended periods of time. Simulation studies of a tundra ecosystem (48) used as a stress a temperature reduction of 2°C, and a decrease in solar input of 10%. The result was a gradual deterioration in primary production over a decade. However, the population of consumers (lemmings) dwindled at a higher rate than primary production—so much so that the predator of the lemmings, the Arctic Jaeger, vanished from the system. The prospective long-term result appeared to be a leveling out of the fluctuations in lemming population, a resultant lower equilibrium in the remineralization cycle, and a stagnation of the system at an artificially reduced rate of remineralization and organic production. Thus, serious degradation of the system appears to be induced by a modest change in the environment, largely due to unanticipated secondary effects of the original stress.

The tundra example also provides an illustration of the importance of fluctuations in a natural system. It is important to look at the frequency and amplitude of input fluctuations (e.g. seasonal cycles of precipitation, nutrient dryfall, or flooding), and to determine the extent to which man-related stresses modify these fluctuations and inadvertently induce or dampen secondary oscillations which influence ecosystem stability.

Studies by Shugart et al (42) at Oak Ridge National Laboratory provide some insight into these response patterns. Control in ecosystems operates through feedback functions. Remineralization of essential nutrients is frequently an agent of such control. Shugart chose to study calcium because this element, for which relevant data were already available, has many of the response characteristics representative of remineralized nutrients. The general structure of Shugart's model provides for a "controllable" input (i.e. an input with a frequency modifiable from days to centuries), and a calcium output (in the soil water) in which oscillation magnitude and phase delay can be observed.

Of the 116 feedback control mechanisms examined in their study, Shugart et al found approximately 20% had a tendency to amplify fluctuations at some level. Figure 4 shows that most of these amplifications were two-fold, but others were as much as six-fold. On the whole, the model systems showed a high degree of stability in the face of small-scale modifications of calcium input periodicities. In addition, the results showed that nutrients in the soil water are one of the most sensitive monitors of developing imbalances. The study also showed that the most significant perturbation frequencies are those between 20 and 200 years—a time span that does not now concern most agricultural managers.

These results, and those cited in the previous pages, show that an exceptional opportunity has developed for new, integrative research on agro-ecosystems. New kinds of questions are being asked about short- and long-term maintainance of

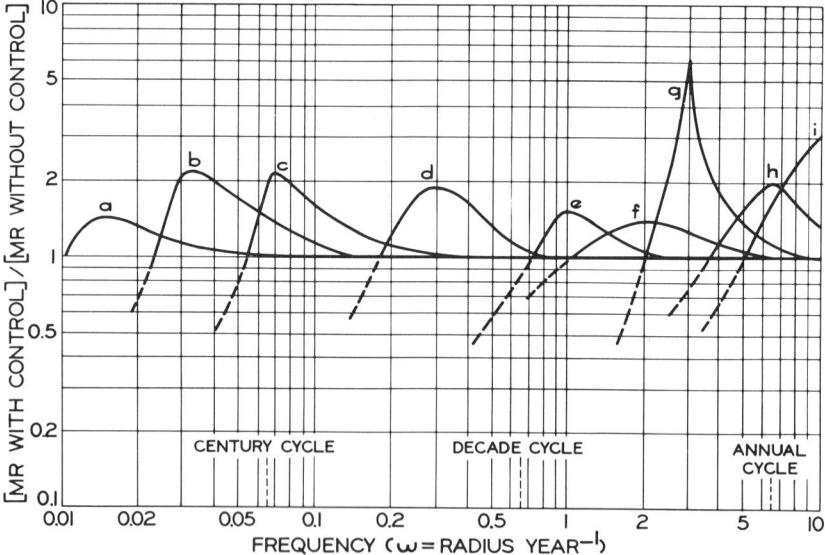

Figure 4 Deviation curves for negative feedbacks that amplify input variations in 100-1 yr frequencies. The deviation curve plots the ratio of [MR with feedback]/[MR without feedback] vs frequency (w = rad yr^{-1}). MR is the ratio of system component amplitude to input amplitude. Feedback control systems plotted are (*a*) monitoring of Ca content in canopy boles, altering input with a gain of –10; (*b*) monitoring Ca content in canopy central root, input with a gain of –100; (*c*) canopy central root, gain = –1000; (*d*) standing dead wood, gain = –1000; (*e*) O_1 wood, gain = –1000; (*f*) O_2, gain = –10; (*g*) O_2, gain = –100; (*h*) small dead root, gain = –100; and (*i*) O_2 leaf, gain = –100.

agro-ecosystems, and about their effects on neighboring ecosystems. Needed improvements in systems approaches are developing, but the definitive studies still lie ahead.

CONCLUSIONS

In the face of inadequate food supplies and the pressure upon agro-ecosystems to increase harvests worldwide, new cropping systems can and will be developed. The methods of ecosystem analysis are beginning to figure in the development of these new systems. Obtaining maximum agricultural productivity over long periods requires more than a simple modification of traditional ad hoc techniques; it will have to build on analysis that integrates the entire system. Management methods will be developed to accommodate interacting components such as crop cultivars, rotations, row spacing, soil nutrients, structure and moisture, protection from pests, harvesting procedures, and public health.

Preliminary results from studies of the responses of agro-ecosystems to subtle changes in stress indicate an exceptional opportunity for integrative research on agro-ecosystems. In the near future this research will be directed toward answering major questions about the long-term stability of agro-ecosystems and their effects on neighboring ecosystems in the agricultural landscape.

ACKNOWLEDGMENT

Permission from Applied Science Publishers (London) and C. R. W. Spedding, author, to reproduce Figures 1 and 2 is gratefully acknowledged.

Literature Cited

1. Arnold, G. W., Bennett, D. 1975. The problem of finding an optimum solution. See Ref. 11, pp. 129–73
2. Auerbach, S. I. 1958. The soil ecosystem and radioactive waste disposal to the ground. *Ecology* 39:522–29
3. Ausmus, B. S., Witkamp, M. 1974. Litter and soil microbial dynamics in a deciduous forest stand. *Oak Ridge Natl. Lab. EDFB Rep.* 73–10. 183 pp.
4. Azzi, G. 1956. *Agricultural Ecology.* London: Constable. 424 pp.
5. Baker, K. F., Snyder, W. C., eds. 1965. *Ecology of Soilborne Plant Pathogens.* Berkeley: Univ. Calif. Press. 571 pp.
6. Bigger, J. W., Corey, R. B. 1969. Agricultural drainage and eutrophication. In *Eutrophication: Causes, Consequences, Correctives,* ed. G. A. Rohlich, pp. 404–45. Washington DC: Natl. Acad. Sci.
7. Brown, A. W. A., Byerly, T. C., Gibbs, M., San Pietro, A., eds. 1975. *Crop Productivity—Research Imperatives.* East Lansing, Mich: Michigan Agric. Exp. Stn.; Yellow Springs, Ohio: Kettering Found. 399 pp.
8. Council on Environmental Quality. 1972. *Integrated Pest Management.* Washington DC: GPO. 41 pp.
9. Commoner, B. 1970. Threats to the integrity of the nitrogen cycle: Nitrogen components in soil, water atmosphere and precipitation. In *Global Effects of Environmental Pollution,* ed. S. S. Singer. New York: Springer. 218 pp.
10. Dalton, G. E. 1971. Simulation models for the specification of farm investment plans. *J. Agric. Econ.* 22:131–42
11. Dalton, G. E., ed. 1975. *Study of Agricultural Systems.* London: Applied Science. 441 pp.
12. Egbert, A. C., Estacio, F. 1975. Regional agricultural planning. See Ref. 11, pp. 317–60
13. Fiering, M. B., Holling, C. S. 1974. Management and standards for perturbed ecosystems. *Agro-Ecosyst.* 1(4): 301–21
14. Goodall, D. W. 1975. Ecosystem modeling in the Desert Biome. See Ref. 36, 3:73–93
15. Harper, T. L. 1974. Agricultural ecosystems. *Agro-Ecosyst.* 1:1–6
16. Holcomb Research Institute. 1976. *Environmental Modeling and Decision Making, The United States Experience.* New York: Praeger. 152 pp.
17. Holling, C. S. 1966. The functional response of invertebrate predators to prey density. *Mem. Entomol. Soc. Can.* 48: 1–85
18. Innis, G. S. 1975. Role of total systems models in the Grassland Biome. See Ref. 36, 3:14–47
19. International Association for Ecology. 1976. *Report on an International Programme for Analysis of Agro-Ecosystems.* Madison, Wisconsin: Intecol. 14 pp.
20. Iwaki, H. 1974. Comparative productivity of terrestrial ecosystems in Japan with emphasis on the comparison between natural and agricultural systems. In *Proc. First Int. Congr. Ecol.,* pp. 40–43. Wageningen: Cent. Agric. Publ. Doc.
21. Jameson, D. A. 1974. Management of ecosystems: Information supplied by simulation models. See Ref. 20, pp. 233–36
22. Kovda, V. A. 1975. Biosphere, soils and their utilization. *Pochvovedenie* 1:3–16
23. Lindeman, R. L. 1942. The trophic-dynamic aspect of ecology. *Ecology* 23:399–418
24. Levins, R. 1973. Fundamental and applied research in agriculture. *Science* 181:523–24
25. Loucks, O. L. 1972. Contaminants and recycling in relation to biogeochemical cycles. In *Challenging Biological Problems. Directions Toward their Solution,* ed. J. Behnke, pp. 297–312. New York: Oxford Univ. Press

26. Loucks, O. L. 1975. Models linking land-water interactions around Lake Wingra, Wisconsin. In *Coupling of Land and Water Systems,* ed. A. D. Hasler, pp. 53–63. New York: Springer
27. Manners, I. R. 1974. The environmental impact of modern agricultural technologies. In *Perspectives on Environment,* ed. I. R. Manners, M. W. Mikesell. Washington DC: Assoc. Geogr. 395 pp.
28. Markov, M. V. 1974. Agrobiogeocenoses. In *Program and Methods of Biogeocenological Investigations,* ed. N. Dylis. Moscow: Nauka. 401 pp.
29. Miller, P. C., Collier, B. D., Bunnell, F. L. 1975. Development of ecosystem modeling in the Tundra Biome. See Ref. 36, 3:95–116
30. National Academy of Science. 1974. *US Participation in the International Biological Program. Rep. No. 6 of the US Natl. Comm. for the IBP.* Washington DC: Natl. Acad. Sci. 166 pp.
31. Odum, E. P. 1969. The strategy of ecosystem development. *Science* 164: 262–70
32. Olson, J. S. 1961. Analog computer models for movement of isotopes through ecosystems. In *Radioecology,* ed. V. Schultz, A. K. Klement, Jr., pp. 121–25. New York: Reinhold
33. Olson, J. S. 1965. Equations for cesium transfer in a *Liriodendron* forest. *Health Phys.* 11:1385–95
34. O'Neill, R. V. 1975. Modeling in the Eastern Deciduous Forest Biome. See Ref. 36, 3:49–72
35. Overton, W. S. 1975. The ecosystem modeling approach in the Coniferous Forest Biome. See Ref. 36, 3:117–38
36. Patten, B. C., ed. 1971. *Systems Analysis and Simulation in Ecology.* New York: Academic. 607 pp.
37. Pimentel, D., Terhune, E. C., Dyson-Hudson, R., Rochereau, S., Samis, R., Smith, E. A., Denman, D., Reifschneider, D., Shepard, M. 1976. Land degradation: Effects on food and energy resources. *Science* 194:149–55
38. Pound, G. S. 1973. Report of the Committee Advisory to the US Department of Agriculture. Washington DC: Natl. Acad. Sci. 91 pp.
39. Rauner, Y. L. 1972. Teplovi balans rastitelnogo pokrova. Leningrad: Gidrometeizdat. 210 pp.
40. Reichle, D. E., Auerbach, S. I. 1972. Analysis of ecosystems. In *Challenging Biological Problems,* ed. J. A. Behnke, pp. 260–80. New York: Oxford Univ. Press
41. Ryszkowski, L., ed. 1974. *Ecological Effects of Intensive Agriculture.* Warsaw: Polish Scientific. 84 pp.
42. Shugart, H. H. Jr., Reichle, D. E., Edwards, N. T., Kercher, J. R. 1976. A model of calcium-cycling in an east Tennessee *Liriodendron* forest: Model structure, parameters and frequency response analysis. *Ecology* 57(1):99–109
43. Smith, F. E. 1970. Analysis of ecosystems. In *Analysis of Temperate Forest Ecosystems,* ed. D. E. Reichle, pp. 7–18. New York: Springer
44. Soemarwoto, O. 1974. Rural ecology in development. See Ref. 20
45. Spedding, C. R. W. 1975. The study of agricultural systems. See Ref. 11, pp. 3–19
46. Stanhill, G. 1974. Energy and agriculture: a national case study. *Agro-Ecosyst.* 1(3):205–17
47. Talbot, L. M. 1965. The meat production potential of wild animals in Africa; a review of biological knowledge. Farnham Royal, Bucks., Eng: Commonwealth Agric. Bur. 42 pp.
48. Timin, M. E., Collier, B. D., Zich, J., Walters, D. 1974. A computer simulation of the arctic tundra ecosystem near Barrow, Alaska. *US Tundra Biome Rep.* 73-1. Fairbanks, Alaska: Univ. Alaska. 82 pp.
49. Tischler, W. 1965. *Agraökologie.* Jena: Gustaw Fisher. 499 pp.
50. University of California, Division of Agricultural Science. 1970. *Statewide Conf. Crises and Conflicts in Agro-Ecosyst., Summ. Pap.* Davis, Calif: Univ. Calif. 77 pp.
51. Van Dyne, G. M., Abramsky, Z. 1975. Agricultural systems models and modelling: An overview. See Ref. 11, pp. 23–106
52. Watt, K. E. F. 1963. Mathematical models for five agricultural crop pests. *Mem. Entomol. Soc. Can.* 32:83
53. Witkamp, M. 1971. Soils as components of ecosystems. *Ann. Rev. Ecol. Syst.* 2:85–110

A COMPARISON OF SOCIAL ADAPTATIONS IN RELATION TO ECOLOGY IN GALLINACEOUS BIRD AND UNGULATE SOCIETIES

❖4122

Valerius Geist
Faculty of Environmental Design, University of Calgary, Calgary, Alberta, Canada

INTRODUCTION

One may wonder what profit there is in comparing the social adaptations of organisms as diverse as birds and ungulates. Yet such a comparison indicates that useful insights into adaptive strategies may be gained not despite but because of their great differences. It shows that old phylogenetic constraints greatly affect social adaptations; that in ungulates as in birds, social and ecological specialization appear to run parallel; that environmental extremes tend to produce "grotesque giants" or monomorphic forms as the terminal product of social and ecological evolution; that social evolution proceeds from stable to unstable habitats, warm to cold, moist to dry, low latitude and altitude to high latitude and altitude; that therefore zoogeographic patterns are closely related to social and ecological specialization within a given family; and that the evolution of sexual dimorphism, weapons, defences, and display organs parallels in birds the patterns seen in ungulates and is probably subject to the same explanations. In short, a comparison of sociobiology in birds and ungulates reveals unifying factors by explaining attributes of birds first explained in ungulates, and vice versa. The present paper is a limited attempt to point out explanations apparently applicable both to bird and to ungulate societies.

PAIRBONDING

We begin our enquiry by asking how it is possible that while some 91% of birds form pair bonds to raise offspring (33), by my estimate no more than 5% of the 187 odd ungulate species (1) are likely to do so. This is despite the fact that the monogamous pair bond is likely to be ancestral and therefore primitive in ungulates because their evolution proceeded from the warm, productive forests to dry, cold, or unstable, unpredictable montane and high latitude environments (15, 21). An explanation of this difference in occurrence of pair bonds emphasizes the importance of phylogenetic constraints on social evolution on the one hand, and the effects of feeding strategies on the other.

It has been widely recognized that the inability of mammalian males to produce milk and to nourish neonatal young severely reduces the potential role of the male in prenatal care (14). It is at present not known why mammalian males failed to evolve lactating organs; among the birds the males in the Columbidae did evolve "crop milk" and do participate in feeding the young with their own body secretions. Nobody has asked or answered the question of what prevents mammalian males from developing lactating organs, probably because we have taken it for granted that only females lactate.

The inability of mammalian males to lactate does not preclude pairbonding, but only makes it less likely than in birds in which the young usually are fed undigested food rather than body secretions. It does preclude however the evolution of polyandry of the types observable in birds (32) in which females distribute eggs to males and the males then incubate and lead the young. Whether or not a mammalian male will enter into a pair bond depends very much on the food resources to be exploited. As has been pointed out by (14, 39), difficulty in procuring food may select for males that capture food for both their young and their mates during the ontogeny of the young. This strategy of pairbonding is found in some carnivores, and is best known in the Canidae, in particular the hunting dog (*Lycaon*) and the wolf (*Canis lupus*) (36).

If the young are capable of finding their own food during and after weaning, the male can still contribute to parental care by guarding in a territory a defendable food resource (6). Since ungulates eat vegetation and the young have no difficulty in procuring that food, ungulate males are necessary only in this role. We find such pair bonds, be they monogamous or polygynous, confined to a few species from low latitudes in which either there is a continuous vegetation season and high productivity, or, as in the vicuna (*Lama vicugna*), the habitat is insular and very limited (19). Monogamous pairbonding in ruminants is confined to several small-bodied forms that feed on highly digestible forage of low bulk and high caloric density (15, 31). The scarcity of such food is the reason pairbonding is rare.

The success of ungulates is based on the fact that they adapted to feeding on coarse, bulky plant matter of very low digestibility. They evolved highly complex digestive systems with grinding teeth that shred plant fibers, and complex alimentary canals that possess one or several chambers for fermentation of plant matter by symbiotic micro-organisms. These not only reduce the undigestible cellulose and hemicellulose to digestible volatile fatty acids, but also produce vitamins and, from inorganic components, synthesize essential amino acids before (29) being in turn digested in the true stomach and supplying the animal with its protein requirements. To maximize the rate of energy and nutrient flow from the ingested forage, ungulates feeding on relatively tough-fibered vegetation must choose the forage containing the least fiber. This, plus the relatively large amount of food they have to ingest daily, imposes long feeding periods on them. Yet ungulates originated in forms that fed on small, diverse pieces of food of low fiber content and relatively high digestibility. These ancestors faced many competitors for this type of food, such as rodents, primates, primitive carnivores and large birds. The presence of competitors may well have precluded ungulates from adopting arboreal habits as found in primates.

Ungulates today are confined—with very few exceptions—to feeding within a short distance of the soil surface, essentially in a two-dimensional habitat. Some of the giants among them such as the giraffe or probably the extinct chalicotheres, homaladotheres, giraffe-camels, indricotheres and giant edentates evolved arboreal browsing. The success of early ungulates was probably a consequence of their ability to digest plant fibers, thus supplementing the energy and nutrients they obtained directly from highly digestible fruits, flowers, and shoots. (Among living ungulates only the Suidae and Tayassuidae penetrate below the soil surface to extract hidden, highly digestible plant and animal food.) The early surface feeders, however, had to exploit large home ranges. This made it difficult to defend their source of food. Nor was there much surplus time and energy to support activities as expensive in energy as fighting or running. Ruminants ingest daily a small surplus of food at best (23). The surface-feeding ungulate can only afford to defend a small but highly productive feeding ground; yet his feeding ground must not be so rich as to attract hordes of competitors. This clearly confines defendable food resources to highly productive warm climates. [There are some exceptions, such as the vicuna (91).] Freedom from competitors for food on the ground is a condition of tropical and subtropical forests with their dense canopies; here food production at the soil surface is low, and only a few ungulates per unit area can be supported. Those that live there and are surface feeders rely on a supplement of fruit and flowers falling from the tree canopy. In addition they feed on diverse seedlings on the ground and on small animals, or exploit the vegetation of small clearings created by fallen trees (3). According to the Jarman-Bell principle (see 23), we expect only small-bodied ungulates to exploit forage of such high digestibility. Therefore pairbonding and defence of a resource territory is in ungulates largely confined to small-bodied ungulates from tropical and subtropical climates.

One must add here that since pigs can penetrate below the soil to reach stored plant and hidden animal foods, they can exploit a greater biomass per unit area than can surface feeders. We would therefore expect to find pair bonds and resource defence among pigs; and indeed both appear to be present in the giant forest hog (*Hylochoerus*) (17, 18).

So efficient are the Suidae and Tayassuidae in extracting much food from little area that the establishment of group territories and the development of inter-deme selection and therefore kin selection may have been favored. The number of adult defenders per unit of defendable resource is thereby maximized, resulting in monomorphism, low reproductive rates, high gregariousness, and severe aggression against members of other demes. Sowls' account (47) of collared peccary behavior suggests that *Dicotyles tajacu* may indeed be a kin-selected species. We have as yet no information on other Suidae, but we are not aware that kin and deme selection could or does operate in any ungulate outside the Suidae and Tayassuidae. Kin selection being rare or weakly developed, we do not expect altruism to have evolved to any noticeable degree in ungulates.

It is evident that the social systems of only a minority of bird species can be similar to those of ungulates, namely those in which males participate little in parental care. Such systems appear to be present among sandpipers (40), and among gallinaceous birds (11, 42, 46, 55). Even a cursory investigation of the latter at once reveals very strong parallels in ecological and social evolution, zoogeographic patterns, the appearance of "grotesque giants" and monomorphic plains dwellers as end points of ecological and social specialization, the distribution of sexual dimorphism, and in patterns of distribution of weapons and display organs in relation to adaptive syndrome. These parallels suggest that the explanations proffered for the social and ecological evolution of ungulates may also apply to birds, in particular gallinaceous birds.

In my discussion I rely primarily on Delacour's (11) and Raethel's (42) accounts of the pheasants. All references are from these authorities unless otherwise indicated.

WEAPONS

I have elsewhere pointed out (25) that among ungulates, weapons that maximize surface damage to the body, and therefore "pain," are associated with the defence of food resources. The teleonomic aim of the defender is to inflict enough pain upon its opponent (see 2) to prevent a retaliatory attack, and cause the opponent to associate the site with the pain. Therefore weapons that maximize damage to the body surface are adaptive, since there is evidence that such damage is more painful than damage to internal organs (28).

Such weapons, however, are not compatible with an anti-predator strategy based on intense gregariousness because the use of damaging weapons reduces group size. An antagonist using damaging weapons in a group is likely to trigger retaliation and be damaged; he may cause his victim to leave, reducing group size and therefore increasing general vulnerability to predators; or he may cause others to shun him, relegating him to the periphery of the group where the dangers of predation are greatest.

The foregoing predicts a maximum use of damaging weapons by animals that defend resources, a decline in their use among animals that do not, and a reduction or absence of such weapons among animals that are dependent on gregariousness as an anti-predator strategy, particularly if they exploit a highly variable habitat. In K-selected species with long life expectancies, damaging weapons are less likely to be selected for than in short-lived, r-selected species (23).

Since birds are generally small of body they can be highly mobile and easily evade strikes by an opponent during fighting. We therefore do not expect morphological defences to have evolved against weapons, as they have among some of the larger ungulates (20, 22). I am not aware of any studies on ratites which examine their anatomy from the viewpoint of offensive and defensive adaptations; the cassowaries may be fruitful subjects for this purpose.

SEXUAL DIMORPHISM OR MONOMORPHISM

In ungulates, sexual monomorphism appears in small, resource-territory-defending forms, and in highly gregarious ones from open plains. The former instance has been explained as a product of the female's mimicry of the male (23). Since two can defend a territory better than one, a female ought not only to engage a male defender of her resources, but she ought to be armed with damaging weapons and to be relatively aggressive. She ought to mimic the male externally in order to convince an intruder that her territory will be defended. The well adapted female presents the same fighting colors as the male; showy monomorphism is the extreme result of this selection process. Also adaptive is a signal denoting full occupation of a resource-territory, such as the duetting now recorded for several monomorphic or territory-defending birds and mammals (35; 56, p. 222). The factors in this adaptive syndrome are monogamy, prolonged pair formation, the use of damaging weapons, and a relatively low level of development of display organs and display behavior.

In tropical and subtropical habitats with a continuous vegetation season, the most primitive members of evolutionary lineages ought to be characterized by this syndrome of pairbonding; sexual monomorphism or weak dimorphism; sharp, damaging weapons and a high level of aggressiveness in both sexes; and relatively less developed display organs and display behaviors. We see there a parallel between the duikers (*Cephalophus*) and chevrotains (Tragulidae) among the ungulates (15, 43), and the primitive peacock pheasants (*Polyplectron*) among the Argusianinae, the green jungle fowl (*Gallus varius*) among jungle fowls, the Malaya fireback (*Houppifer erythrophthalmus*), the ocellate-turkey (*Agriocharis ocellata*) compared to its northern relatives *Meleagris gallopavo*, and probably the razorbilled curassows (*Crax tomentosa, C. salvini, C. mitu*) among the curassows proper. The possession of damaging weapons (other than spurs on the tarsi as in many gallinaceous birds) —namely horny, sharp spurs on the wings, or a specialized stiletto claw as in the cassowaries—is associated with sexual monomorphism or very weak dimorphism, as in the screamers (Anhimidae), the spur-winged goose (*Plectropterus gambienses*), the spur-winged plover (*Hoplopterus spinosus*), and Cayenne lapwing (*Belonopterus cayennensis*). These birds are all from tropical or subtropical climates at low latitudes.

Sexual dimorphism increases in forms which progressively depart from ancestral pairbonding, territorial species adapting to environments requiring increasingly greater opportunism in food exploitation. (This requirement may be due to occupation of areas of higher altitude, greater aridity, lower temperature, or earlier seral stages of plant association.) Females become less conspicuous to predators by becoming more cryptic in coloration and less conspicuous in their behavior. The males become more conspicuous by virtue of larger display organs and more prolonged dominance displays than the males of the parent species. There are a number of explanations for this increase in sexual dimorphism.

The first explanation is based on the ideas of Maynard Smith, Verner, Willson, Pianka, Lack, and Orians (see 39). It states in essence that in habitats with a high productivity it is more profitable to a female to join with at least one other female on the territory of a male defending a very rich resource than to expel all other females. The burden of territorial defence shifts to the male, and sexual selection by the female favors the larger, more conspicuous male, the most competent defender of the richest resource-territory. As Wiley has pointed out (55), this explanation cannot easily be applied to the Tetraonidae. I know of no evidence that it applies to the other gallinaceous birds.

The concept (49) that males freed from parental duties maximize their reproductive fitness by maximizing the number of females inseminated, and consequently, due to sexual competition, evolve to large size with showy display organs, is acceptable but too imprecise. We need to know the circumstances under which the dimorphism takes the form of greater display or greater overt aggressiveness by the male; we must also understand those cases where, despite male freedom from parental duties, it does not occur. The latter appears to be the case for small forest antelopes. These are very nearly monomorphic, yet no care-giving behavior among males has been observed. Protection of the resources required for reproduction may be sufficient to retain monomorphism of sexes even if the male is freed of parental duties.

The second explanation of sexual dimorphism applies to habitats with diffuse food sources. As the male becomes more capable of exploiting the habitat but less capable of defending resources for reproduction, he cannot directly contribute to parental care. Therefore the female can best guard her genes and maximize her reproductive fitness by mating not necessarily with the most dominant, but with the male who is better at exploiting food resources than are other males. Such a male can devote surplus resources to growth and "expensive" activities. He has resources with which not only to maximize body size, but also to grow large display organs ["luxury organs," as Portmann (41) aptly named them]; he can afford to spend more time at complex, showy, and dangerous "advertising" or display activities. By mating with such a male, the female insures that his superior foraging skills will be available to support her offspring. Presumably such a male possesses superior skills in escaping predation, since obviously males which display more have more frequently been noticed by predators. Moreover, in this system the female ought to choose the older of two equally showy males, since the latter has undergone more testing by predators.

The virtue of this explanation is that it predicts a parallel in ecological and social specialization. Females will select not only on the basis of displays, but on age dependent differences between otherwise similar males. The available data from both the ungulates and the gallinaceous birds appear to be compatible with this hypothesis. For instance, the caribou (*Rangifer*) is regarded as a highly opportunistic feeder, and carries relative to body size a far greater set of antlers than any cervid of comparable body size. Among the Argusianinae, we find that the species adapted to either the high montane or the very xeric habitats have the largest bodies, with the most elaborate display feathers and displays, but also with the most reduced weapon systems. These are the ocellate pheasant or crested argus pheasant (*Rheinartia ocellata*), and the great argus pheasant (*Argusianus argus*), the latter adapted to dry, rocky forests.

The third explanation states that as females flock to rich forage sites or to traditional mating grounds, they ought to chose the male in whose company they are least molested by other males. This will be in the company of the most dominant cock. In choosing him, they enhance reproductive success of their male offspring and therefore maximize reproductive fitness. They also reduce the cost of mating since they escape expensive harassment and can invest the energy and nutrient saved in reproduction. This third explanation is really an adjunct to the second explanation.

The fourth explanation derives from the "dispersal theory" that will be developed separately below (21). It predicts not only a parallel evolution of social and ecological adaptations in warm climates, but also a social evolution independent of an ecological one where closely related sympatric species are absent. It predicts as well a divergence of display organs and behavior—not only an increase in these social attributes with increased ecological specialization, but also a diagnostic pattern of zoogeographic dispersal, along with the evolution of "grotesque giants."

At this point we must pause to consider an important distinction. Opportunism may be of two kinds: one in which a species is highly mobile and exploits small pockets of food as such become available and then moves on; and a second in which a species is mobile only until a large food resource is found, adequate not only for maintenance, but also for reproduction. The former we may call small-scale opportunism in contrast to the latter, large-scale opportunism. The former leads directly to extremes in K-selection and the latter to r-selection. The more a species exploits unpredictable food resources that are found by roaming, the less it must be tied down during or for reproduction. Therefore selection favors a very small number of large, well-developed, highly mobile young which can readily follow the parent in their common quest for food. There is also selection for large body size, since endurance in roaming is proportional to body mass (48) and the greater the endurance the greater the home range that may be covered.

Large-scale opportunism, on the other hand, ties down the reproducing individual to a superabundant food source, and therefore to a locality. It can maximize reproduction—hence the selection for a maximum number of young. Its social system is "designed" to disperse young once they reach near-adult size, and the young in turn explore and settle where a superabundant food source becomes available. Among the small-scale, K-selected opportunists, traditions and long-distance migrations of a regular nature can develop; such cannot develop among the large-scale opportunists: the unpredictability of the availability of resources in lumps large enough to permit reproduction precludes both socialization and regularity of migrations.

A good example of the small-scale opportunist among ungulates is the caribou (*Rangifer*), and that of a large-scale opportunist is the moose. In both, males carry large display organs probably for the reasons I have just indicated. An exact parallel is provided by the peacocks, and the peacock pheasants. The terminal species of the peacock pheasants, the ocellate and greater argus pheasant, appear to be K-selected, small-scale opportunists. Peacock pheasants exploit primarily animal food in what are probably climax communities. They produce only two young, which follow the female closely, sticking to her by keeping under her tail. Note that incubation is incompatible with time-consuming, long-distance roaming; therefore it does not surprise one to read that the female of the greater argus pheasant apparently does not feed during incubation. I suspect she may live off a stored fat deposit during that time. Since the resources of a small-scale opportunist are not defendable, the absence of spurs on the tarsi of argus and ocellate pheasants is not surprising.

A third terminal lineage of the peacock pheasants, I suggest, are the peacocks, which I identify as large-scale opportunists. These birds broke away from the climax food to exploit the highly productive ecotone between forest and water. Compared with the peacock pheasants and argus pheasants, they have diversified, omnivorous food habits. The requisites of life for peafowl may be found in a relatively small area; resources are thus more likely to be defendable. Significantly, although the green peacock (*Pavo muticus*) has enormous display organs, its sexual dimorphism is small; both the male and the female carry spurs. The clutch size is larger than that of the peacock pheasants, but the young still crowd under the hen's tail in true ancestral fashion. The Indian peafowl (*Pavo cristatus*) appears to be a greater departure towards opportunism; it is less aggressive than green peafowl, its sexual dimorphism is greater, and it is quite polygynous. All in all, the trend towards increasingly opportunistic food habits in the Argusianinae has given rise to three giants with grotesque display organs: two small-scale opportunists exploiting montane and xeric forests, and a large-scale opportunist exploiting productive ecotones.

When a species moves from a habitat of plentiful cover and adapts to one poor in hiding places, gregariousness becomes a means of reducing predation (56). Under these circumstances sexual dimorphism can be expected to decline for several reasons. First, females can no longer rely on protection from dominant males to reduce harassment by subordinate ones. Male mimicry will evolve in the female; it will protect females against bioenergetically expensive harassment (23). Second, the center of a flock is the place safest from predators, where a maximum of time may be spent feeding. [The latter was demonstrated in wood pigeons by Murton et al (38).] There is thus considerable jostling to get to the center of the flock (38). That males and females may compete for food when foraging in the same flock, and females suffer as a consequence, is indicated in studies of African weaver birds (*Quelea quelea*) (10, 52). In order to compete effectively against males, females clearly ought to assume male form and size; this they achieve by male mimicry. Male-colored and male-shaped females are prevalent among bovids and also among reindeer; in the gallinaceous birds, they are exemplified by the eared pheasant (*Crossoptilon*), the snowcock (*Tetraogallus*), and the Barbary partridge (*Alectoris*).

Under circumstances in which strife is severely selected against (such as extreme predation pressure; or when the food source of a flock cannot be defended, and the individual's best strategy is to feed as speedily as possible), bright, agonistic colors will also be selected against. Colors and display organs associated with aggression increase the cost of existence to their bearer because (*a*) his close presence forces defensive behavior in others, and thus occasions his own frequent arousal, and (*b*) his feeding time is reduced by such distractions. Neutral monomorphic colors thus permit in a flock a maximum of time devoted to feeding; they also permit a closer social distance (37). Plains-dwelling gallinaceous birds, and also parrots (4), tend to be monomorphic, with a plumage resembling juveniles. Of the parrots, as of ungulates or gallinaceous birds, it is true that monomorphic forms of great coloration are found in warm, stable environments; dimorphic forms occupy the ecotone; and plain, neutral, or juvenile-colored forms are found in open, extreme habitats.

It is tempting to suggest that among the curassows (Cracidae), the chachalacas (*Ortalis*) are an expression of this evolutionary trend. Compared to the guans (*Penelope, Penelopina, Aburria, Chamaepetes,* and *Oreophasis*) and the curassows proper (*Nothocrax, Crax*), the chachalacas inhabit dry, shrubby habitats; guans and curassows are found in the tropical forests, the guans being mainly arboreal, the curassows leaning toward a more terrestrial existence (12).

There can, however, be another approach—a temporary one—to reduction in dimorphism. Dimorphic males may cast off their symbols of maleness and mimic females in external appearance, living with these in flocks outside the breeding season. Female mimicry by males can be expected either where males are exhausted after the rut and thus become an easy target for predators (5), or where they are the less abundant sex and are therefore conspicuous to predators. Such we find among ungulates in the pronghorned antelope (*Antilocapra*) (5); we may similarly interpret the casting of antlers in *Rangifer, Capreolus,* and *Odocoileus* among the deer. The phenomenon occurs among the pheasants in the Indian jungle fowl (*Gallus gallus*). A change to "female" monomorphism serves (37) to reduce strife within the flock and permit males, the minority sex, to function as equal flock members.

Another mechanism by which monomorphism may be achieved has been suggested (7) for kin-selected, group-territory-defending jays: paedomorphism, the retention of juvenile plumage, form, and behavior after sexual maturity. The term *neoteny* has been incorrectly used (7) to denote this process. Neoteny implies continued ontogenetic development towards a final adult form; paedomorphism does not. A paedomorphic individual is frozen in a juvenile image for life. Neoteny and paedomorphism are central to the evolution of mountain sheep (22), but there is as yet no example among ungulates or gallinaceous birds to which these concepts are applicable in the form proposed in (7). They may be found, however, to apply to the collared peccary, once the sociobiology of the Tayassuidae has been investigated. Paedomorphism may of course also account for the neutral colors of so many monomorphic plains-dwelling partridges and quails, but for obviously different reasons than those proposed in (7).

An alternative to the coupling of monomorphism with bisexual groups is the retention of sexual dimorphism and the formation of separate unisexual groups. Such are indeed found in bovids, cervids, but also in the gallinaceous birds such as the Himalayan monal pheasant (*Lophophorus impejanus*), and the ptarmigan (*Lagopus*) among the Tetraonidae (54). This separatism is linked to the possibility of a sexual segregation in food habits and thus the exploitation of different habitats or foods by the sexes (9). In the capercallie (*Tetrao urogallus*) we find indeed that males feed from different parts of the spruce trees than do females (see 55). There is some evidence that in the sexually dimorphic red deer (*Cervus elaphus*), males feed on somewhat more fibrous forage than do females (13) and that the males vacate the areas occupied by females (see 26). That sexually dimorphic, social males in ungulates feed in areas different from females is well known for the caprids, various African antelopes, and some cervids (22, 51).

The foregoing explains the existence of monomorphism based on male mimicry in the primitive territorial as well as in the highly advanced social forms—in the former, as adaptations of females against strangers; and in the latter, against males as a whole. It also explains bright and showy, rather than neutral, monomorphism.

DISPERSAL THEORY

Among ungulates we find two zoogeographic patterns in which geographic situation, the development of display organs, and often both body size and ecological specialization are linked in an orderly fashion. First, in the north-temperate and arctic regions we find geographic clines of races or subspecies in which display organs and often body size increase progressively, best illustrated by mountain sheep (*Ovis*) and old world deer (21, 22) but also by bison (*Bison*) (27). Second, species of ungulates radiate geographically from relatively warm to cold or dry habitats and increase concomitantly in body size, size of display organs, and ecological specialization. There are, however, some ecologically specialized species with large bodies and large display organs at the origin of radiation. Thus "grotesque giants" arise both at the periphery of a taxonomic group radiation, and at its center. As I have shown, both patterns of distribution and concurrent social and ecological specialization are explained by the same theory (21). We find, among ungulates, the second type of pattern among old world deer (21) and gazelles (24); and, to my knowledge, it is the only pattern found among birds. It is well illustrated by the argus pheasants and by the pheasants proper with a center of radiation in southeast Asia, as well as by hornbills (Bucerotidae) with a center of radiation in tropical Africa. In the latter we find the small-bodied, relatively insectivorous, small-billed members (*Tockus*) in central Africa, and the largest, most bizarre, fruit-eating forms, the Indian hornbills (*Buceros*) and helmeted hornbills (*Rhinoplax*), in southern Asia (44). There are African and Asian species adapted to diverse niches, but body size, degree of coloration, complexity of display organs, and ecological specialization (as indicated by divergence from forest life and a diet of much animal matter) tend to correlate. In South America we find a similar radiation pattern among the curassows (Cracidae) in the genus *Crax*, with monomorphic, primitive forms of relatively smaller body size found in lowland tropical forests, grotesque large forms in the mountains within the ranges of primitive forms, and dimorphic giants in peripheral areas of distribution, with some leaning toward dryer habitats. Note that in the guans, close relatives of the curassows, we also find the most grotesque members in the montane forests (12).

The phenomenon of the "grotesque giants," highly specialized ecologically and strongly divergent in external appearance from smaller members of a lineage, is well represented in mammals but also in birds. It is exemplified by the mammoth, the end product of elephant evolution; by the woolly rhino (*Coelodonta*), the terminal member of the dicerorhinids; by the Irish elk (*Megaloceros*), terminal product of the old world deer; by the caribou and moose, terminal products of the new world deer; and by *Ovis ammon, Capra ibex, Ovibos moschatus* and *Euceratherium* as different terminal lineages of the rupicaprid radiation. We can include our own species in this collection since, in comparison with other primates, we are not only giants and grotesque, but also the only species adapted to cold and periglacial environments. In the gallinaceous birds the Argusianinae terminate in three grotesque giants, the montane ocellate pheasant (*Rheinartia ocellata*), the dry-forest giant argus pheasant (*Argusianus argus*), and the ecotone-adapted peafowl (*Pavo*). In the pheasants we find a succession of genera that produces grotesque giants both within genera or adaptive radiations, and between genera. Let us look at the pheasant radiations in detail.

The most primitive radiation is that of the firebacks, the Malay fireback (*Houppifer erythrophthalmus*) in particular, a species from humid, lowland forests with male-colored females of low sexual dimorphism, robust bodies, long spurs in both sexes, and a very aggressive disposition. These are apparently monogamous, rather silent, and inactive birds. Within the firebacks, we find a radiation to montane and ecotone habitats. Thus *Houppifer inornatus,* which inhabits the mountain forests of Sumatra, is, as expected, sexually dimorphic. From the Malay fireback in order of grotesqueness or complexity of display organs, yet each distinctly different from the other, we find the crested fireback (*Lophura ignita*), the Siamese fireback (*Diardigallus diardi*), and the Bulwer's pheasant (*Lobiophasis bulweri*). These tend to be larger than the Malay fireback; in the Bulwer's pheasant the spurs are shorter than in the other species. The ecological differences among these species are not well known. Here we deal essentially with a tropical radiation of forest pheasants, at least one of which, the Siamese fireback, ventures into second growth forests. In the same region we find the primitive, monogamous green jungle fowl (*Gallus varius*), the sexually weakly dimorphic and aggressive green peacock (*Pavo muticus*), and the weakly dimorphic, double-spurred peacock pheasants (*Polyplectron*). These forms contrast with their relatives to the north or in dry or mountain habitats; the latter tend to be more dimorphic, polygynous, and apparently less aggressive or less well armed.

More advanced than the fireback radiation is that of the silver pheasants (*Gennaeus*). These are adapted, compared to firebacks, to higher altitudes and latitudes, and live in secondary forest. They are sexually more dimorphic than the firebacks (excepting the Haman silver pheasant from damp, montane forests) since the females of the silver pheasants have relatively fewer feather patterns indicative of social communication. They are polygynous and lean toward a gregarious existence.

A more specialized offshoot of the silver pheasants appears to be the blue pheasants (*Hierophasis*), of whose biology apparently little is known. They are still more sexually dimorphic than the silver pheasants, by virtue of more strikingly colored and wattled males, and more cryptically colored females. Zoogeographically, they lie within or at the edges of silver pheasant distribution.

The next more advanced adaptive radiation appears to be that of the eared pheasants (*Crossoptilon*). These are high mountain birds living in the ecotone between forest and alpine. They spend much of their time in the open. They are highly gregarious, large-bodied, monomorphic but showy or "male-colored," with even larger ornate tail coverts and tail feathers than the silver pheasants. Unlike the silver pheasants, they are monogamous. These are the most grotesque giants among pheasants. In the extreme environments of mountains, "grotesque giants" tend to occur as the end point of evolution. Note the Caprinae as extreme forms of the rupicaprid evolution; the white-lipped deer (*Przewalskium albirostris*) as the end point of the rusa radiation (16); and the mountain nyald (*Tragelaphus buxtoni*), a counterpart to the also large, twisted-horned greater kudu (*T. stresiceros*) from the dry bush, as end products of tragelaphine radiation (virtually a parallel to the radiation of the Argusianinae). Among birds, note the giant snowcock (*Tetraogallus*), or the large, bizarre, mountain rails such as the horned coot (*Fulica cornuta*) and the giant coot (*F. gigantea*) of the Andes.

The conquest of north-temperate habitats by the pheasants can be envisioned as a progression beginning with the monomorphic, aggressive, male-colored, monogamous cheer pheasant (*Catreus wallichii*) of the Himalayas, which shares many features with the silver pheasants, eared-pheasants, and firebacks but has a plumage reminiscent of the ring-necked pheasant. The male cheer pheasant takes part in chick raising. It is somewhat cryptically colored, and has simple social displays and large clutch sizes—both indications of high predation pressure. It inhabits open, steep, rocky habitats with low grasses and shrubs at high elevation; its food habits vary but it tends to eat animal matter. The next step appears to be the polygynous, larger, long-spurred, sexually dimorphic band-tailed pheasants (*Syrmaticus*), which have a distribution at lower elevations south and east of the Himalayas. They appear to be inhabitants of secondary growth and forests in cool climates. Their display appears to be simple, and both sexes are highly aggressive. The final step was taken by the ring-necked pheasants (*Phasianus*). Their sexual dimorphism is extreme. Compared to band-tailed pheasants, the hens are cryptically colored and have no noticeable display organs, while the cocks are at least as colorful as the band-tailed pheasants and have more complex wattles and head ornaments. This appears to be the most opportunistic species, judging from its preference for ecotones and its very wide distribution. Since resource defence declines with increasingly opportunistic ecological strategy, it follows that ring-necked pheasants would need to be less aggressive than the probably less opportunistic band-tailed pheasants. In harmony with this notion, they do have the shorter spurs.

This zoogeographic pattern of ecological and social adaptation can be explained, as in the case of ungulates, by a theory that I termed "dispersal theory" (21, 22). Let us consider the following scenario.

As a consequence of geographic isolation and ecological specialization, a form begins to disperse. Such dispersal may occur over an area inhabited by its parent species. Since the colonizing population is freed of intraspecific competition, and is exploiting a niche different from its parent species, the colonizing individuals can ingest the most nutritious forage in relatively large amounts. This permits them and their offspring to develop phenotypically near their full genetic potential. That gallinaceous birds do respond phenotypically by growing larger in response to superior food has been shown for the red grouse of Scotland (53) and is known for jungle fowl reared in captivity (11). We also know (8) that very small differences in the linear body dimensions of the kittiwake are associated with very great differences in social and reproductive success. We have reason to suspect that in a parent form with sexual dimorphism, and hence intense competition of males, even small differences in superior phenotypic development will translate themselves into significant enhancements of reproductive fitness.

Maximum physical growth is not the only response to the improved forage conditions during dispersal. A constellation of changes warrant terming the new phenotype a "dispersal phenotype." The dispersal phenotype is healthy, socially active, plays and explores much, reproduces successfully, but has a relatively short life expectancy (22, 30, 45). The dispersal phenotype maximizes phenotypically the diagnostic features of its species. In doing so, the environmental variance of these diagnostic features is minimized and the genetic variance is exposed to selection. Therefore, the phenotypic trends maximized become genetically enhanced and the colonizing population begins to diverge genetically from its parent species.

Since social interactions are intense and display organs are maximized phenotypically, display organs are selected to become more effective. However, if the colonizing population is sympatric with its parent species, we can expect that there will be divergent selection of display organs between the two forms. Since the colonizing form is in lower abundance, that form will diverge from the parent species. Once the colonizing population has filled the habitat to capacity, the above selection comes to a halt as food resources become insufficient for the production of dispersal phenotypes. During allopatric dispersal, the dispersal phenotype is likely to meet conditions previously not encountered by its species, and it is likely to alter its behavior to deal with them. Since it tends to be active, explorative, and well fed, it may diverge phenotypically from the parent population. This phenotypic deviation under strong selection will become genetically enhanced, as shown by (50). Therefore the colonizing population may evolve adaptations different from those of the parent stock, and incompatible with it. Upon meeting the parent population, selection against hybridization probably forces distinct displays upon the form in lesser abundance.

Although during colonization display organs will become sophisticated, they are not likely to remain large unless the species adapts to a larger annual pulse of nutrients than its parent species (34). Retention of large body size is thus an indication of environmental instability, a concept that fits well both with the correlation between large body size and the montane, arctic, or xeric habitats so frequently inhabited by the "bizarre giants," and with the highly specialized diet of some of the giants. In the latter case we assume that such species are subject to periodic losses of food and the concomitant necessity to move to new sources of food. It is here that large body size and large home ranges are adaptive.

The fossilization of the large bones of ungulates permits us to follow not only size differences in species, but also the correlation between ecological and social adaptations over geologic time. We do find a correlation between complexity of dentition and hornlike organs, and we find that where a species colonized first it tends to be the largest (21). The proposed relationships between ecology and social adaptations in birds need further study—in particular the ecological specializations, which are less readily discerned than the morphological ones.

Literature Cited

1. Anderson, S., Jones, J. K. 1967. *Recent Mammals of the World.* New York: Ronald Press. 453 pp.
2. Azrin, N. H. 1964. Aggressive responses of paired animals. In *Symposium on Medical Aspects of Stress in the Military Climate*, pp. 329-52. Washington DC: Walter Reed Army Inst. Res.
3. Barrette, C. 1974. *Social Behaviour of Muntjac.* PhD thesis. Univ. of Calgary, Calgary. 234 pp.
4. Brereton, J. L. 1971. A self-regulation to density independent continuum in Australian parrots and its implications for ecological management. In *The Scientific Management of Plant and Animal Communities for Conservation*, ed. E. Duffey, A. S. Watt, 207-21. London: Blackwell
5. Bromley, P. T. 1976. *Aspects of the behavioural ecology and sociobiology of the pronghorn.* PhD thesis. Univ. of Calgary, Calgary, 370 pp.
6. Brown, J. L. 1964. The evolution of diversity in avian territorial systems, *Wilson Bull.* 76:160-69
7. Brown, J. L. 1974. Alternate routes to sociality in jays—with a theory for the evolution of altruism and communal breeding. *Am. Zool.* 14:63-80
8. Coulson, J. C. 1968. Differences in the quality of birds nesting in the centre and on the edges of a colony. *Nature* 217(4127):478-79
9. Crook, J. H. 1966. Gelada baboon herd structure and movement: a comparative report. *Symp. Zool. Soc. London* 18: 237-58
10. Crook, J. H., Butterfield, P. A., 1970. Gender role in the social system of the quelea. In *Social Behaviour in Birds and Mammals: Essays on the social ethology of animals and man*, ed. J. H. Crook, pp. 211-48. New York: Academic
11. Delacour, J. 1951. *The Pheasants of the World.* London: Country Life. 351 pp.
12. Delacour, J., Amadon, D. 1973. *The Curassows and Related Birds*, New York: American Museum of Natural History. 247 pp.
13. Dzieciolowski, R. 1969. *The Quantity, Quality and Seasonal Variation of Food Resources Available to Red Deer in Various Environmental Conditions of Forest Management.* Warsaw, Poland: Forest Res. Inst. 295 pp.
14. Eisenberg, J. F. 1966. The social organization of mammals. *Handbuch der Zoologie* 10(7):1-92
15. Estes, R. D. 1974. Social organization of the African Bovidae. In *The Behaviour of Ungulates and its Relation to Management*, ed. V. Geist, F. Walther, 166-205. Morges, Switzerland: IUCN Publ. NS 24
16. Flerov, K. K. 1952. Musk deer and deer. In *Fauna of the USSR Mammals, Vol. 2.* Moscow: Academy of Sciences of the USSR. English translation, Washington DC: US Dept. Commerce
17. Frädrich, H. 1967. Das Verhalten der Schweine *(Suidae, Tayassuidae)* und Flusspferde *(Hippopotamidae).* Handbuch der Zoologie 10(26):1-44
18. Frädrich, H. 1974. A comparison of behaviour in the Suidae. See Ref. 15, pp. 133-43
19. Franklin, W. L. 1974. The social behaviour of the vicuna. See Ref. 15, pp. 477-87
20. Geist, V. 1966. The evolution of hornlike organs. *Behaviour* 27:175-214
21. Geist, V. 1971. On the relation of the social evolution and dispersal in ungulates during the Pleistocene, with emphasis on the Old World deer and the genus *Bison. Quat. Res.* 1:283-315
22. Geist, V. 1971. *Mountain Sheep: A Study in Behaviour and Evolution*, Chicago: Univ. of Chicago Press. 383 pp.
23. Geist, V. 1974. On the relation of social evolution and ecology in ungulates. *Am. Zool.* 14:205-20
24. Geist, V. 1974. On the relationship of ecology and behaviour in the evolution of ungulates: theoretical considerations. See Ref. 15, pp. 235-46
25. Geist, V. 1977. On weapons, combat and ecology. In *Aggression, Dominance, and Individual Spacing.* Vol. 4. New York: Plenum. In press
26. Geist, V. 1977. Adaptive strategies in the behaviour of elk. In *The Ecology and Management of the North American Elk*, ed. J. W. Thomas. Washington DC: Wildlife Management Institute. In press
27. Geist, V., Karsten, P. 1977. The wood bison in relation to hypotheses on the origin of the American bison. *Z. Säugetierk.* 42:119-22
28. Guyton, A. C. 1971. *Basic Human Physiology*, Philadelphia: Saunders
29. Hanson, H. C., Jones, R. C. 1976. *The Biogeochemistry of Blue, Snow and Ross*

Geese, pp. 247–60. Carbondale and Edwardsville, Ill: Southern Ill. Univ. Press
30. Horejsi, B. L. 1976. *Mother-Young Behaviour in Bighorn Sheep.* PhD thesis. Univ. of Calgary, Calgary. 227 pp.
31. Jarman, P. J. 1974. The social organization in antelope in relation to their ecology. *Behaviour* 58:215–67
32. Jenni, D. A. 1974. Evolution of polyandry in birds. *Am. Zool.* 14:129–44
33. Lack, D. 1968. *Ecological Adaptations to Breeding in Birds.* London: Methuen. 409 pp.
34. Margalef, R. 1963. On certain unifying principles in ecology. *Am. Nat.* 97:357–74
35. Marshall, J. T., Marshall, E. R. 1976. Gibbons and their territorial song. *Science* 193:235–37
36. Mech, L. D. 1970. *The Wolf.* Garden City, NY: Natural History Press. 384 pp.
37. Moynihan, M. 1960. Some adaptations which help to promote gregariousness. *Proc. 12th Int. Ornithol. Congr. Helsinki 1958,* 523–41
38. Murton, R. K., Isaacson, A. J., Westwood, N.J. 1966. The relationship between wood pigeons and their clover food supply and the mechanism of population control. *J. Appl. Ecol.* 3:55–96
39. Orians, G. H. 1969. On the evolution of mating systems in birds and mammals. *Am. Nat.* 103:589–603
40. Pitelka, F., Holmes, R. T., MacLean, S. A. 1974. Ecology and evolution of social organization in arctic sandpipers. *Am. Zool.* 14:185:204
41. Portmann, A. 1953. *Das Tier als Soziales Wesen.* Zurich: Rhein
42. Raethel, S. 1968. Subfamily: argus and pheasants, pp. 26–38; subfamily: Congo peafowl, pp. 38–42; subfamily: pheasants, pp. 49–78. In *Animal Encyclopaedia, Birds II.* Vol 8. New York: Van Nostrand
43. Ralls, K., Barasch, C., Minkowski, K. 1975. Behaviour of captive mouse deer *Trogulus napu, Z. Tierpsychol.* 37:356–78
44. Sanft, K. 1968. Family: hornbills, pp. 49–61, In *Animal Encyclopaedia, Birds III.* Vol 9. New York: Van Nostrand
45. Shackleton, D. M. 1973. *Population Quality and Bighorn Sheep.* PhD thesis. Univ. Calgary, Calgary. 227 pp.
46. Skutch, A. 1968. Subfamily: turkeys, pp. 19–26. In *Animal Encyclopaedia, Birds II.* Vol 8. New York: Van Nostrand
47. Sowls, L. K. 1974. Social behaviour of the collared peccary *Dicotyles tajacu.* See Ref. 15, pp. 144–65
48. Thomsson, D. A. W. 1961. *On Growth and Form,* ed. J. T. Bonner. London: Cambridge. Abridged ed. 345 pp.
49. Trivers, R. L. 1972. Parental investment and sexual selection. In *Sexual Selection and the Descent of Man 1871–1971,* ed. B. Campbell, pp. 136–79. Chicago: Aldine
50. Waddington, C. H. 1975. *The Strategy of the Genes.* London: Allen & Unwin. 261 pp.
51. Walther, F. 1968. *Verhalten der Gazellen.* Wittenberg Lutherstadt, West Germany: A. Ziemsen. 144 pp.
52. Ward, P. 1965. Feeding ecology of the black-faced dioch *Quela quela* in Nigeria. *Ibis* 107:173–214
53. Watson, A., Moss, R. 1972. A current model of population dynamics in red grouse. *Proc. 15th Int. Ornithol. Congr., Brill, Leiden,* pp. 134–49
54. Weeden, R. B. 1964. Spatial segregation of sexes in rock and willow ptarmigan in winter. *Auk* 81:534–41
55. Wiley, R. H. 1974. Evolution of social organization and life history patterns among grouse (*Aves: Tetraonidae*). *Quart. Rev. Biol.* 49:201–27
56. Wilson, E. O. 1975. *Sociobiology.* Cambridge, Mass.: Belknap Press of Harvard Univ. 697 pp.

MATHEMATICAL MODELS OF SCHISTOSOMIASIS

♦4123

Joel E. Cohen

The Rockefeller University, New York, NY 10021

INTRODUCTION

Human schistosomiasis (or synonymously, bilharzia) is a family of diseases caused primarily by three species of the genus *Schistosoma* of flatworms. The adult worms inhabit the blood vessels lining either the bladder or intestine, depending on the species of worm. The worms are also known as blood flukes.

The worldwide prevalence of schistosomal infections has not been measured credibly. A figure conventionally cited is 200 million people, or one of every 20 people on the planet. Except for imported cases, the disease is virtually unknown in the rich countries of the world.

"There is little doubt that all three schistosomes can cause considerable pathological change, sometimes in a comparatively large proportion of the population, but the evidence suggests that only a proportion of those so affected die of the disease" (29, p. 168). The absence of quantitative information from this assessment of the impact of the infection on health fairly reflects the information available.

Jordan & Webbe (29) review human schistosomiasis. Malek (40) and Hairston (24) emphasize the ecological point of view. Warren & Newill (59) cite 10,286 references. Some material here is drawn from Cohen (11) and Fine (18).

After sketching the life cycle of *Schistosoma mansoni*, this chapter reviews mathematical models of schistosomiasis. The bibliography of published works aspires to completeness through 1976.

LIFE CYCLE OF SCHISTOSOMIASIS

The life cycle of the three major human schistosome species (Figure 1) consists of an obligatory alternation of sexual and asexual generations. The sexual generation occurs in man (and sometimes other mammals). The asexual generation must pass through specific snails. The quantitative estimates in the following refer chiefly to *S. mansoni*.

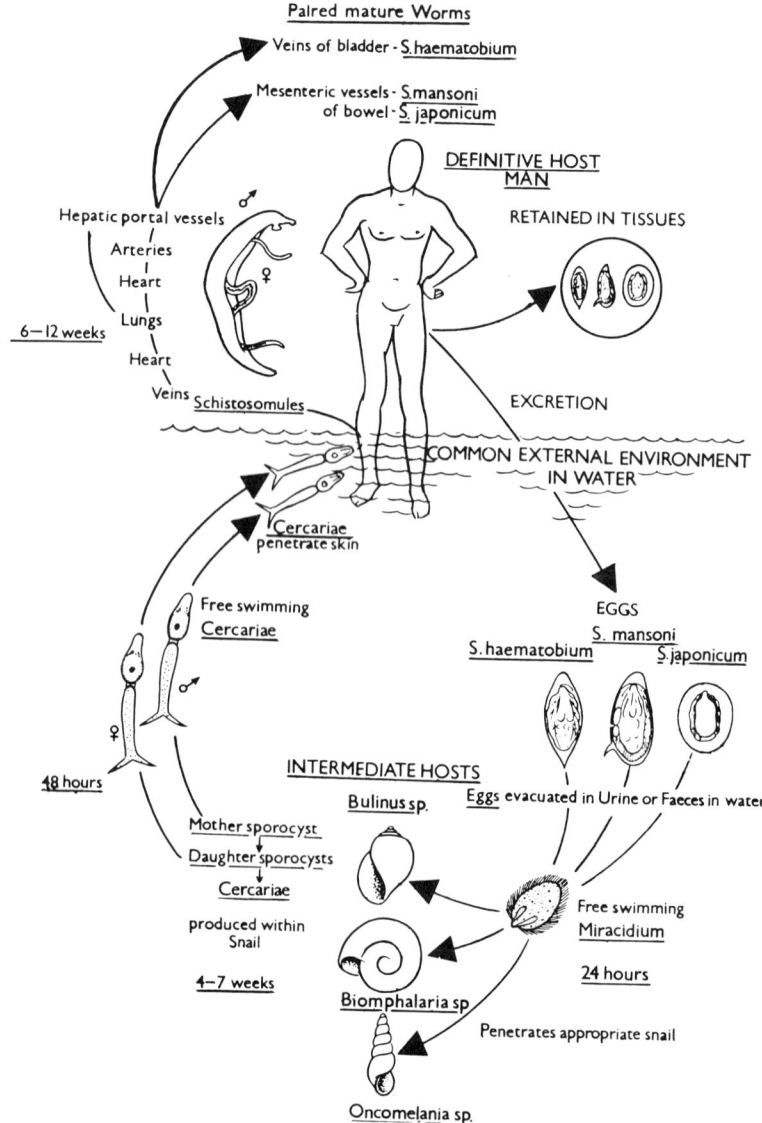

Figure 1 The life cycle. [From (29), p. 7. Courtesy of the authors and Charles C Thomas, publisher.]

MATHEMATICAL MODELS OF SCHISTOSOMIASIS   211

Eggs produced by the sexual stage leave people via urine (in the case of *S. haematobium*) or feces (*S. mansoni* and *S. japonicum*). Eggs that reach water shed their shells and hatch a ciliated free-swimming stage called a miracidium.

A miracidium that locates a snail within approximately one day penetrates it. If the snail is of the appropriate species and genotype, the miracidium multiplies asexually through two larval stages into thousands of cercariae.

Each cercaria that escapes from the snail, starting 4–5 weeks after the initial infection (6 weeks in *S. japonicum*), lives approximately 2 days. It swims until it encounters a skin of suitable warmth and smell. When one of the human schistosomal cercariae enters human skin, it becomes a wormlike "schistosomule."

A schistosomule of *S. mansoni* migrates to the lung, sometimes producing a cough, then appears in the portal system of the liver where it reaches sexual maturity and mates. Worm pairs then migrate to the blood vessels lining the lower small intestine and the large intestine.

At this point the couple of worms resemble a hot dog in a roll. The female, 7–17 mm long, lies in the gynecophoric canal of the male, who is 6–13 mm long and cylindrically shaped to correspond to the walls of their home, a blood vessel. The forward third of the female's body is devoted to the uterus, which contains one to two eggs at a time. The female is estimated to lay from 100 to 300 eggs a day.

Some of these eggs work through the wall of the blood vessel into the lumen of the intestine. Carried by feces, these eggs again begin the life cycle. The interval from the entry of cercariae into human skin to the first detectable passage of eggs in the feces can vary from 4 weeks for *S. japonicum* to 5 or 6 weeks for *S. mansoni* and 13 weeks for *S. haematobium*.

Apart from an occasional aberrant worm that wanders into the wrong organ, such as the brain or eye, most of the disease caused by the infection results from the eggs that do not escape with feces. Some of these get stuck immediately in the tissue near where they are laid, causing fibrosis and granuloma as the host tries to protect itself. Other eggs get washed to the liver and spleen where they may cause similar damage.

The medicines available to kill the schistosomes in people have so many dangerous side effects that they must be administered under medical supervision. They are costly. They do not protect a person in an endemic area against reinfection. Even if enough medical personnel were available to treat all the infected population in a single month, the snail (and sometimes nonhuman mammalian) reservoirs of infection would persist. The control or eradication of schistosomiasis is a truly ecological, as opposed to a purely medical or technological, problem.

PROPORTION EVER INFECTED AS A FUNCTION OF AGE

The mathematical models in this section and the next use cross-sectional data about a population presumed to be in steady state in order to make inferences about the dynamics of infection in a cohort.

Suppose a cohort is entirely susceptible to infection at some initial time, usually taken to be birth. Suppose that this cohort is exposed to a constant force of infection per unit time. "This force is to be measured in effective contacts per unit time, no matter how complex may be the events leading up to these contacts" (42, p. 16). The force of infection a summarizes the contact between cercariae and people and the establishment of a detectable infection.

Let N be the number of individuals in the cohort. Let $x(t)$ be the fraction of the cohort that has never been infected, and $y(t)$ the fraction that has ever been infected, by time t. By definition $x(t) + y(t) = 1$. Assume $x(0) = 1$ and $y(0) = 0$. Then $Nx(t)$ is the number of individuals never infected at time t. These individuals are constantly exposed to a force a of infection. So the change per unit time in the number $Nx(t)$ of people never infected is $d[Nx(t)]/dt = -aNx(t)$, or, cancelling N, assumed constant, $dx/dt = -ax$, $x(0) = 1$. Similarly, for the number ever infected, $dy/dt = ax = a(1-y)$, $y(0) = 0$. The solution is

$$y(t) = 1 - e^{-at}. \qquad 1.$$

Death or emigration will have no effect on the fraction $y(t)$, so long as the loss rate (including death and emigration) is identical for both previously infected and never infected individuals (7).

If past conditions were constant, and all previous infections were detected, then (22) a cross-sectional survey should give a graph of the fraction of people ever infected as a function of age that looks like equation 1.

Figure 2 takes $t = 0$ as 5 years of age. Infections before that age are neglected. The data are the fractions of people in each age group judged ever to have been infected with schistosomes on the basis of a skin test. Particularly for the younger age groups, the fit of equation 1 to the data is reasonable. The discrepancy at the upper ages is explained as due to an insensitivity of the skin test to previous infection if the person has not recently been exposed to female cercariae or has no living female worms.

The numerical value of the parameter $a = 0.12$ used in Figure 2 was not obtained by fitting that curve to those data. The parameter was estimated by fitting another equation (number 5 below) to different data, from stool examinations, on the same population. This finding suggests that an incredibly simple mathematical model can usefully interpret the age distribution of previous infection and provide information about the dynamics of infection which would otherwise be unavailable.

PROPORTION CURRENTLY INFECTED

Female *S. mansoni* worms in human beings live an average of 3–4 years; other species of human schistosomes are comparable (21, p. 52; 29, p. 152). A negative exponential distribution of length of life for female worms is widely assumed. A person in whom all female worms have died no longer discharges eggs. (A person may also no longer discharge eggs because tissue traps the eggs or because living females are unmated. We ignore these complications.) Hence some individuals previously infected may pass from currently discharging eggs to no longer doing so.

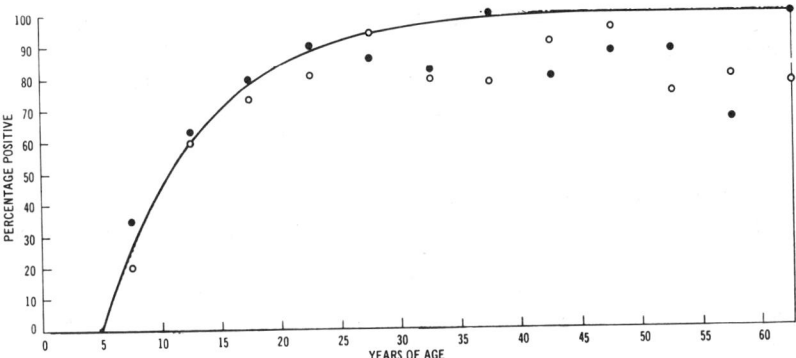

Figure 2 The proportions observed positive in response to *S. japonicum* antigen skin tests as a function of age in the coastal division, Palo, Philippines. Open circles, 1954; solid circles, 1962. Solid line, prediction from equation 1. [Adapted from (22), p. 172. Courtesy of Nelson G. Hairston and the World Health Organization.]

Irreversible Loss of Infection

Let us assume that a previously infected person who is no longer discharging eggs has no risk of reinfection. Let $y(t)$ (not the same as in the previous section) be the proportion of a cohort which is currently giving evidence of infection by excreting eggs. Let $z(t)$ be the proportion that has been previously infected, but is no longer passing eggs and has no risk of reinfection. As before, let $x(t)$ be the proportion which has never been infected. Assuming no death or emigration, $x(t) + y(t) + z(t) = 1$.

If the cohort is subject to a constant force of infection a, and those individuals currently giving evidence of infection are now further subject to a constant risk of loss of infection b, here assumed to be independent of the number of worms or worm pairs in the host, then under constant conditions the proportions x, y, and z are described (42) by:

$dx/dt = -ax,$ $x(0) = 1,$ 2.

$dy/dt = +ax - by,$ $y(0) = 0,$ 3.

$dz/dt = +by,$ $z(0) = 0.$ 4.

All individuals are uninfected initially. The sum of the derivatives is zero, as it must be since the cohort does not change size. Then:

$y(t) = a(e^{-bt} - e^{-at})/(a-b),$ if $a \neq b$;

5.

$y(t) = ate^{-at},$ if $a = b.$

If death and emigration occur at equal rates in all three fractions of the cohort, the same equations hold.

Figure 3 plots $y(t)$ and the observed proportions with *S. japonicum* eggs in their feces by age in the same Philippine population pictured in Figure 2. Hairston (22) fits equation 5 by the method of moments (42). The annual rate $b = 0.02$ of becoming negative is not the annual death rate of individual female worms because (assuming the eggs are not blocked in the person's tissues) all the females in the person have to die, without replacement, for the person to stop passing eggs. Lewis (33) refits the same data by maximum likelihood, with similar results.

The model's assumption that an individual's probability per unit time of losing infection is independent of the individual's age, immune status, duration of infection, and worm burden means that the effects of varying other ecological parameters cannot be calculated. A micro-theory which interprets the model's parameters would be useful.

Snails too pass through the stages of being never infected, being infected and shedding (cercariae, instead of eggs), and (possibly) being no longer infected (53, 55).

Reversible Loss of Infection

Reinfection of previously but no longer infected individuals is observed. At the opposite extreme from the assumption just made that a loss of infection is irreversible is the assumption that a person no longer infected is exposed to a risk of infection identical to that of a person never previously infected.

If, as in the previous model, it is assumed that the instantaneous rates of infection a and of loss of infection b are constant, then the model is identical to one widely used for malaria and other diseases (17). All else being constant, this model predicts

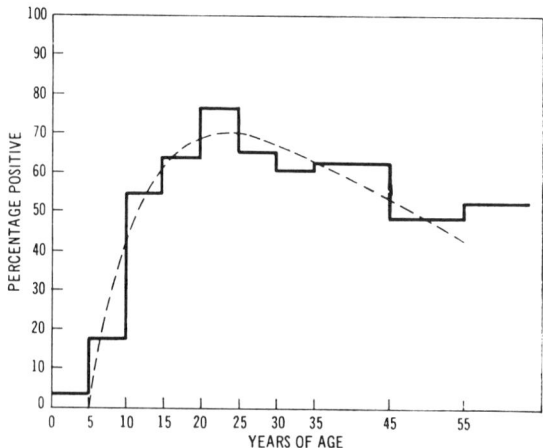

Figure 3 Observed age-specific prevalence rates (solid line) and theoretical age-specific prevalence rates (dashed line) from equation 5 of human infection with *S. japonicum* in the coastal division, Palo, Philippines, neglecting transmission before 5 years of age. [From (22), p. 171. Courtesy of Nelson G. Hairston and the World Health Organization.]

a prevalence rate which increases monotonically to an asymptote, contrary to observation (Figure 3). The assumption of reversible loss of infection is retained in a modified form of this model used for economic evaluation (50, 51).

The assumption of completely reversible loss of infection has appeared in models which view the number of worms in each human host as an immigration-death process (25, 34, 43–47). A risk of infection which is constant in time, or independent of age in a cohort, implies a monotonically increasing prevalence rate of humans who carry at least one mated pair of worms (26). Assuming that the risk of infection decays negative exponentially with increasing time (or age) to some positive lower asymptote predicts an age prevalence curve that fits observations of *S. mansoni* and *S. haematobium* reasonably and gives estimates of the life expectancy of the worms compatible with other findings.

Neither completely irreversible nor completely reversible loss of infections seems likely. Intermediate possibilities are discussed in the section below on immunity.

Differential Mortality Due to Infection

For people, the increment, if it exists, in the probability of death at any age due to infecting schistosomes has never been measured credibly (8, 9). Snails shedding cercariae of *S. mansoni* show an increase in death rate compared with uninfected snails.

If μ is the mortality or emigration rate of individuals not currently shedding eggs (in the case of humans) or cercariae (snails), and $\mu + \epsilon$ is the increased mortality or emigration rate of individuals currently shedding, then suppose, assuming irreversible loss of infection,

$$(1/N)\mathrm{d}(Nx)/\mathrm{d}t = -ax - \mu x, \qquad x(0) = 1, \qquad 6.$$
$$(1/N)\mathrm{d}(Ny)/\mathrm{d}t = +ax - by - (\mu + \epsilon)y, \qquad y(0) = 0, \qquad 7.$$
$$(1/N)\mathrm{d}(Nz)/\mathrm{d}t = by - \mu z. \qquad z(0) = 0. \qquad 8.$$

When $\epsilon = 0$, putting $N(t) = N(0)e^{-\mu t}$ leads back to equation 5.

If $y(t)$ obtained from equations 6–8 is a better approximation to reality than equation 5, but a curve of the form of equation 5 is fitted to data in ignorance of ϵ, then the resulting estimates of the parameters a and b may be biased (8). For humans, the differences are small among the age prevalence curves predicted by assuming that all the bias is absorbed either by a or by b, although the possible bias in the parameter estimates is not. For snails, even the possible bias in the parameter estimates is small (53).

This example illustrates a sensitivity analysis which can profitably accompany the study of ecological models. The model with differential mortality is more realistic than the model without it because differential mortality does occur. The more complicated model is more complicated to study mathematically. It does not cause major alterations in how the age prevalence data are understood. Hence, for rough purposes, one can be more assured of the adequacy of equations 2–4; for finer purposes, one has a more refined tool, equations 6–8.

Latency

The lag or latency of several weeks between the infection of a person with cercariae and the appearance of eggs in feces or excreta is short compared to the 1–5 year age groups used in collecting human prevalence data, and very short compared to the human life span. Hence the assumption of an instantaneous transition from uninfected status to detectably infected status may serve adequately for humans.

With snails, however, the lag of 4–5 weeks exceeds the one week age grouping ordinarily used for age prevalence curves and is a substantial fraction of the snail life span. Nasell (45) distinguishes "exposed" snails infected by miracidia from those shedding cercariae, and derives the age prevalence curve of infective snails. Susceptible, exposed, infective (or shedding), and recovered (or no longer shedding) snails each have a characteristic death rate:

$$dx/dt = -ax - \mu_1 x, \quad x(0) = 1, \quad \text{(susceptible);} \quad 9.$$
$$du/dt = +ax - Au - \mu_2 u, \quad u(0) = 0, \quad \text{(exposed but latent);} \quad 10.$$
$$dy/dt = +Au - by - \mu_3 y, \quad y(0) = 0, \quad \text{(infective, shedding);} \quad 11.$$
$$dz/dt = +by - \mu_4 z, \quad z(0) = 0, \quad \text{(no longer shedding).} \quad 12.$$

The fraction of the cohort infected, $y/(x + u + y + z)$, need not vanish with increasing time if the mortality μ_4 of recovered individuals is large enough.

This model implies that the distribution of the interval from successful infection of a snail by a miracidium to the first shedding of cercariae should be negative exponentially distributed, with the parameter A which appears in equations 10 and 11. The mode of such a distribution is at intervals of length zero, contrary to observation.

In a model which incorporates real latent periods between infection and infectivity, Lee & Lewis (32) estimate the latent period in humans to be 2 months. In snails, the latent period is taken to vary from 5 months in the cool season to 1 month in the warm. The implied age prevalence distribution in humans or snails is not shown.

Immunity

In trying to explain why observed human age prevalence distributions of schistosomiasis initially peak and then decline with increasing age, some medical authorities (3) emphasize the importance of human immunity. Others (58) emphasize declining human contact, for cultural and behavioral reasons, with cercariae-laden water. The fitting of models to age prevalence distributions cannot decide the relative importance of these two explanations. An immigration rate of worms to humans which declines with age may result either from immunity to new infections or from declining water contact (26).

The same qualitative effect is obtained (33) by assuming a constant immigration rate and a temporary immunity following loss of infection in a modified two-stage catalytic model. The modified model yields a substantial and statistically significant improvement in fit to Hairston's (22) data on *S. haematobium* and *S. mansoni,* but describes the *S. japonicum* data no better than equations 2–4. Let $x(t)$ and $y(t)$ be interpreted as in equations 2–4. Let $z(t)$ be the proportion of a cohort that was previously infected, which now no longer shows patent infection, and which is now temporarily immune. Assume that immune individuals are subject to a constant risk c of loss of immunity, after which they are as susceptible to reinfection as individuals never previously infected. Thus:

$$dx/dt = +cz - ax, \quad x(0) = 1, \qquad 13.$$
$$dy/dt = +ax - by, \quad y(0) = 0, \qquad 14.$$
$$dz/dt = +by - cz, \quad z(0) = 0. \qquad 15.$$

For certain parameter values the proportion $y(t)$ of infective individuals initially increases with age, peaks, and then decays exponentially to a positive limit $ac/(ab + ac + bc)$. For other parameter values, $y(t)$ performs damped oscillations in approaching this limit. When $c = 0$, immunity is permanent and this model reverts to equations 2–4.

Lewis (33) extends the model of equations 13–15 by recognizing that an individual never previously infected can shed eggs only if it has been infected by at least one male and at least one female worm. Male and female cercariae are assumed equally likely to enter a host never previously infected, in a Poisson stream with constant parameter. Assuming that worms of the opposite sex survive from a previous infection, previously infected individuals who have lost their immunity require infection only by one more cercaria in order to reestablish infectivity. In this model, permanent immunity can again be represented by taking $c = 0$.

This model describes Hairston's (22) *S. japonicum* data better than equations 13–15, primarily owing to the representation of sexual pairing of worms.

Linhart's (37) predicted age prevalence curves have not been tested against observations, except where they coincide with the two-stage catalytic model.

At each time t in Linhart's three models, every individual is either manifest (showing proof of current worms according to some test) or not. An individual not manifest at t but manifest before t is called cured at t. Every individual is also either immune or not immune at each time t. Immunity is permanent, once achieved. An infection is defined as the attempted entry of cercariae into the individual. An infection is ineffective if the individual is immune at the time, effective otherwise.

Infections arrive as a Poisson stream with parameter α. Let $t_1, t_2, \ldots$ denote the times at which the first, second, ... infections occur.

The first model assumes that an individual becomes manifest at t_1. The assumption would be reasonable if the definition of "manifest" were based on a serological or other assay of the metabolic products of a single schistosomule (30).

The model assumes that the individual becomes immune and cured at $t_1 + c + b$, where c is a nonnegative constant delay and b is a random variable with negative exponential distribution and parameter β. The expected fraction manifest at time t in a cohort not subject to differential mortality or emigration, is

$$y(t) = 1 - e^{-\alpha t}, \qquad t \leq c,$$

$$= e^{-\alpha t}(e^{\alpha c}-1) + \alpha[e^{-\beta(t-c)} - e^{-\alpha(t-c)}]/(\alpha-\beta), \quad \alpha \neq \beta, \; c \leq t. \qquad 16.$$

If $c = 0$, this equation becomes identical to equation 5 when the stochastic rates α and β are replaced by their corresponding deterministic equivalents a and b.

The second model assumes that an individual becomes manifest at t_1 and immune to any further infections at $t_1 + c$, where c is a positive constant. Between t_1 and $t_1 + c$, further effective infections may occur. If the last of these occurs at t_k, then the individual becomes cured at $t_k + d$, where $d \leq c$ is a constant. Under these assumptions,

$$y(t) = 1 - e^{-\alpha t}, \qquad\qquad\qquad\qquad\qquad 0 \leq t \leq d,$$
$$= 1 - e^{-\alpha t} - (1/2)e^{-\alpha c}[e^{\alpha(t-d)} - e^{-\alpha(t-d)}], \quad d \leq t \leq c+d, \qquad 17.$$
$$= e^{-\alpha t}\{[e^{\alpha(c+d)} + e^{-\alpha(c-d)}]/2 - 1\}, \qquad\qquad c+d \leq t.$$

The third model (37) assumes (again) that an individual becomes manifest at t_1. The individual becomes immune and cured when his "infection time" mounts up to a positive constant threshold w. If an individual has received k effective infections by t, then his infection time $I(t)$ at t is $I(t) = \sum_{j=1}^{k} (t - t_j)$, where t_j is the time of the jth effective infection. Every infection is assumed to be effective as long as $I(t) \leq w$. The probability that an individual is manifest is just the probability that $0 < I(t) \leq w$. Hence, for given t, if s is the integer satisfying $st \leq w < (s+1)t$,

$$y(t) = e^{-\alpha t}\{\sum_{k=0}^{\infty} [\alpha t]^k/k! - 1$$

$$+[\; \sum_{k=s+1}^{\infty} \alpha^k/k!] \; \sum_{j=0}^{s} [-1]^j[w-jt]^k/[j!(k-j)!]\}. \qquad 18.$$

Biological Aspects

An important task in the modeling of schistosomiasis is to translate the burgeoning biological information about the immunology of schistosomiasis into mathematically explicit, empirically testable, and epidemiologically useful form.

It would seem useful to develop, and to test against data, a model incorporating: (*a*) a risk of exposure to cercariae which is variable with age, season, and infection status; (*b*) sexual pairing of worms (see below); (*c*) true latency between infection and the first shedding of eggs; (*d*) a risk of loss of apparent infection dependent on worm load, age of worms (since older worms may lay fewer eggs), and age of host (since long-term pathology may interfere with the escape of eggs); (*e*) concomitant immunity, in which the host's response to established infections inhibits superinfection; (*f*) subsequent immunity, in which the host's response to previous infections inhibits reinfection; (*g*) a decay of immunity. Quantitative tests against varied age prevalence data and against direct observations of the component processes assumed in the model might lead more rapidly than the present piecemeal approach to a focus on the features important for a control of prevalence.

SNAIL POPULATION DYNAMICS

The models considered so far are implicitly conditional on the invariance of the half of the schistosome life cycle which is not being modeled. For example, the studies of human prevalence assume the supply of cercariae from snails is steady in time.

Food and Crowding

Biomphalaria glabrata is the snail principally responsible for the transmission of *S. mansoni* in the New World. Jobin & Michelson (28) raised laboratory populations of these snails with varying amounts F of food (measured in grams of watercress), numbers N of snails (each 15 mm in diameter), and volumes V of water (4.5 and 7.6 liters), at 25°C. For each such population they measured the fecundity (E) by the numbers of eggs laid per snail per day (Figure 4):

$$E = kF/NV. \qquad 19.$$

It is plausible that, over a certain range at least, fecundity should increase with food and decrease with the number of snails competing for that food. What is counterintuitive about equation 19 is that a larger volume of water *decreases* fecundity. The reason is that in larger volumes of water the snails have a harder time finding the food.

In very large volumes of water, such as lakes, which are not crowded with snails, the addition of one more snail has no effect on the fecundity of the other snails present. So Jobin & Michelson (28) assume that the inverse dependence on N in

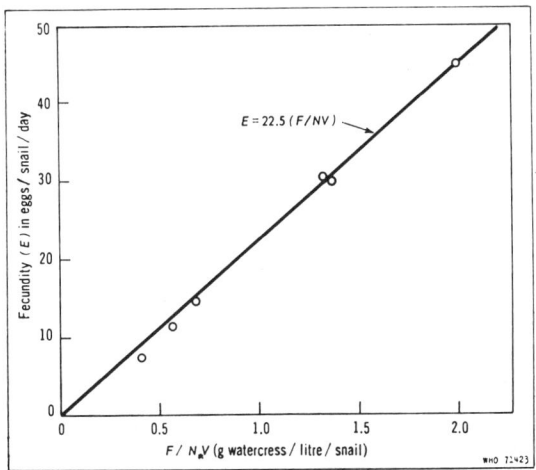

Figure 4 Fecundity of *Biomphalaria glabrata* snails as a function of food F, number N of snails, and water volume V. [From (28), p. 659. Courtesy of E. H. Michelson and the World Health Organization.]

equation 19 disappears whenever the volume of water per snail exceeds some threshold.

Since, moreover, equation 19 predicts that snails in a vanishingly small volume of water have an infinite fecundity, equation 19 should be regarded as a linear approximation to a nonlinear function over the range of variables used in one set of experiments.

In a simulation, the proportionality constant k becomes a function of the species and age of the snails, the nutritional value of food, and water temperature. Mortality is assumed to depend on demographic and ecological factors. The parameters are estimated from observations by another investigator of *Bulinus globosus*, a Rhodesian snail carrying *S. haematobium*. The predicted snail populations agree roughly with censuses. Sensitivity analysis with respect to both parameter values and the form of entire components of the model would be desirable.

Infection and Age Structure

Coutinho & Coutinho [(12); see (13) for errata] explain two generalizations from previous empirical work. The first generalization is that $D = A/(B + p)$, where D is an index of the size of *B. glabrata* snails in a lake and p is the prevalence rate among those snails of infection with *S. mansoni*. A and B are constants.

The explanation offered is that higher prevalence is associated with a higher force of mortality due to infection. A higher force of mortality results in a younger age structure in the population, or a smaller probability that any egg will survive to any given age. Since younger snails are smaller than older snails, a higher prevalence is associated with a smaller snail population. By suitable quantitative assumptions, $D = A/(B + p)$ follows.

The prevalence rates in the 12 lakes reported (12) vary from 0.6% to 8.3%. It would be desirable to represent the life table in these lakes as a mixture of two negative exponential life tables, one with low mortality for the uninfected majority and one with high mortality for the infected minority, rather than assuming a single negative exponential life table as at present.

The explanation offered (12) overlooks the direct inhibitory effect (54) of being infected on snail growth. The inverse relation between snail size and prevalence of infection in different populations very probably results from the effect of infection on both snail growth and mortality.

The second generalization is that "when the infection rate [prevalence rate p] was small, the snails, even though they attained a large size, were less abundant, whereas in sites of greater infection rate which caused smaller shell sizes the abundance per unit area was considerably higher."

The explanation offered assumes that the quantity of food consumed by the snails per unit time and per unit area is constant, and therefore that the supportable biomass of snails is constant from one lake to another. When the mortality due to infection is small, a higher proportion of snails survives to older ages. Being bigger at older ages, those snails consume more food, leaving less for the many small snails. Hence the total number, or abundance, of snails is less than when mortality due to infection is large.

Direct evidence for the constancy of the food supply or the biomass density is not offered. If F is constant, and if the variation in V in equation 19 from one lake to another is negligible, then according to equation 19, the fecundity E is inversely proportional to the abundance N of individuals without regard to their size. Thus lower abundance N should be associated with a higher fecundity E which, all else being equal, should (by standard demographic arguments) be associated with a younger age structure, that is, typically smaller snails. Evidently not all else is equal, because the size index is larger where the snails are less abundant (12). It would be desirable in future work to reconcile the generalization, equation 19, from laboratory work, with the second generalization, based on field work. Measurement of N in terms of biomass, rather than numbers, might suffice.

Coutinho & Coutinho (13) adduce a similar explanation for observations that the maximum diameter of snail shells in a uniform, swiftly flowing channel increases from the input to the outlet of the channel. Along the same axis the number of snails per unit of channel length decreases progressively. Subsequently, Coutinho & Coutinho (14) study the age structure of a snail population resulting from a time-varying, age-independent cause of mortality.

Other Treatments

Nasell & Hirsch (45–47) take the population of snails to be constant, either always or asymptotically. Lewis (34) models the snail population as a stochastic immigration–birth–death process. The linear birth and death rates for susceptible snails are not assumed to be the same as those for infected snails. These models are components of life cycle models (below).

SEX AMONG THE SCHISTOSOMES

Only mated adult worms produce the eggs which are believed to be the primary cause of schistosomal pathology when they do not escape from the human host (36), and which sustain the transmission cycle when they do escape. Mathematical models of mating quantify possibilities within the range of present ignorance.

Monogamy

Assuming monogamous pairing, if a person is infected with an even number n of worms, each of which is male or female with probability 1/2, then (38) the probability that a given larva is matched by one of opposite sex, or the expected proportion of worms that are matched, is $1 - n!/\{[(n/2)!]^2 2^n\} \approx 1 - 0.7979 n^{-1/2}$. When n is odd, the probability that a larva is matched is the same as that for the preceding even number. This probability of matching is not necessarily identical to the probability of mating heterosexual pairs, because the members of each pair have to find each other in the dark, but Hairston (personal communication) knows of no record of mature worms of opposite sexes remaining unmated.

Suppose (16, 20, 21, 37, 38) that the number of larvae per person is Poisson-distributed with a mean, say, of m larvae. Then the proportion of larvae matched by one of the opposite sex, averaged over all people, is $\Psi(m) = 1 - e^{-m}[I_0(m) + I_1(m)]$, where I_n is the modified Bessel function of the first kind of order n (Figure 5). Nasell (43) corrects Macdonald's formula 7 (38, p. 503) and obtains the simple expression for $\Psi(m)$ just given. It is assumed that pairs are strictly monogamous and that a worm that dies is immediately replaced by another worm of the same sex.

The expected proportion of people with at least one potential heterosexual pair of worms (the prevalence, as usually assayed) is $(1 - e^{-m/2})^2$, again assuming monogamy and a Poisson distribution of larvae. This expression, due to Hairston (unpublished) and, independently, Nasell (43), simplifies a result of Macdonald's [(38), p. 504, equation 10].

Suppose (47) that the male and female worms are subject to identical independent processes of immigration and death, where the risk of death is constant for all age and sex classes of worms and hosts. At equilibrium in a life cycle model, the immigration rate of larvae becomes constant, over time and over people. If ρ is the ratio of the equilibrium immigration rate of worms of either sex per host to the constant rate of death per worm, then $m = 2\rho$ and the average number of mated pairs of worms per person in the population is $(m/2)\Psi(m) \sim m^2/4$ as $m \to 0$, and $\sim m/2$ as $m \to \infty$.

Tallis & Leyton (56) study this immigration–linear-death model as well as contagious arrivals and age dependent deaths.

Assuming equilibrium in the input of cercariae, suppose that male and female larvae enter a person according to independent identical Poisson processes with

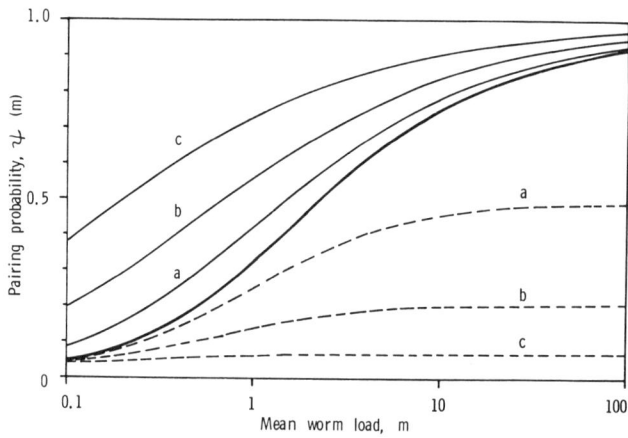

Figure 5 The proportion $\Psi(m)$ of adult worms that can be paired monogamously with an adult worm of opposite sex as a function of the mean worm load m per host, under various assumptions about the distribution of worms. Heavy solid line: Poisson distribution; light solid lines: negative binomial distribution, with both sexes together; light broken lines: negative binomial distribution, with both sexes separate. The coefficient of variation increases from *a* to *b* to *c*. (Courtesy of Robert M. May.)

MATHEMATICAL MODELS OF SCHISTOSOMIASIS   223

constant parameter, and that when a mated worm dies, its former partner does not remate (16). Under the same definition of ρ, the expected number of paired worms per host then approximates ρ^2 as $\rho \to 0$, but $\rho/2$ as $\rho \to \infty$. For large ρ or m the assumption that a mated worm does not remate after the death of its partner leads to half as many expected worm pairs as Macdonald's assumption.

Diversity

Apparently independently of Macdonald (38), Leyton (35) studies several possible modes of reproduction in helminthic infections. He relates the distribution of the number of worms laying eggs to the bivariate distribution of the numbers of sexually mature male and female worms, allowing the worm population's size distribution to be general.

Assuming parthenogenesis, any female can lay eggs. The distribution of egg-layers is the marginal distribution of female worms.

Assuming hermaphroditism, every worm can fertilize itself, though cross fertilization can occur. The distribution of egg-layers is the distribution of the total number of worms.

Assuming monogamous heterosexual mating, and the remating of widowed worms [the case considered by (47)], Leyton (35, p. 418) finds the probability distribution of the number of egg-layers. He also (35, p. 419) writes out explicitly the distribution of egg-layers assuming a (bivariate) negative binomial distribution of worms per host.

Leyton (35) also obtains the general distribution of egg-layers under the assumption that each male can mate up to k females (and that only mated females lay eggs). If a single male renders every female present an egg-layer, the distribution of egg-layers is a simple function of the numbers of males and females.

Overdispersion

Almost certainly in real human populations the variance in the number of infecting worms significantly exceeds the mean (2, 15), even within particular age and sex groups. As analytically tractable approximations to the distribution of adult worms, there are two overdispersed extreme alternatives to a Poisson distribution of worms (4, 41).

First, when worms of both sexes are distributed "together," assume that the total number n of worms in a human host is negative-binomially distributed and that each worm is equally likely to be male or female; that is, conditional on n, the number of male or female worms is binomially distributed with parameters n and 0.5. The negative binomial distribution may be parameterized in terms of its mean m (as for the Poisson distribution above) and an index k which is inversely related to overdispersion. When k is infinite, the negative binomial becomes the Poisson distribution; when $k = 1$, it becomes the geometric; and as k vanishes, the distribution becomes increasingly overdispersed.

The second alternative to a Poisson distribution assumes worms of both sexes are distributed "separately": the number of males in a host is negative binomial with mean $m/2$, and similarly for the number of females, and males and females are independent. The average number m of worms per host is therefore assumed to be the same, whether the males and females are distributed together or separately.

Worms of both sexes are likely to be distributed together where people are exposed to infection from snails with a high prevalence of infection (4). If the prevalence of snail infection is low, however, a person is more likely to be exposed to cercariae of only one sex, and the model of separate infections may be more appropriate.

The expected proportions of worms that are mated under the two assumptions differ from the Poisson expectations (Figure 5) in opposite directions. When both sexes are distributed together, the proportion mated always exceeds the proportion under the Poisson model. The difference is least when the mean worm load is high, which is when the assumption that the sexes are distributed together is most plausible. When the sexes are distributed separately, the proportion mated is always less than the proportion under the Poisson model. The difference is smallest at low worm loads where the model of independent distribution of the sexes is most plausible.

Biological Aspects

These studies point to several needs for additional biological information.

Which of the several models of mating considered by Leyton (35) and others actually holds? Hairston reports (unpublished) that *S. japonicum* is certainly not monogamous. In animals experimentally infected with an excess of females, all the females become mated and carry the full complement of eggs in utero; the same appears to hold for *S. mansoni*.

The public health impact of the difference between monogamy of worm pairs (38, 47) and faithfulness after death (16) depends on the causes of schistosomal pathology. Pathology due to metabolic products or simply the presence of worms may be presumed proportional to the total number of worms [as in (37)]. If the pathology were primarily due to eggs laid, and if only currently mated females could lay eggs, then faithfulness after death would be important. If the pathology were primarily due to eggs laid, but any female once mated could continue to lay eggs, then the relevant variable, namely number of females ever mated, has not been investigated.

Are the assumptions of Nasell & Hirsch (47) regarding the equality of male and female worm death rates close to true? Lewis (34) takes the death rate of worm pairs to be that of the female worms. Some evidence suggests this assumption may be preferable.

Since a female worm must leave the male's gynecophoric canal to lay eggs in the person's small blood vessels, she is more exposed than the male to damage by the host's reactions and suffers a higher death rate (20, p. 51). Human autopsies consistently show an excess of male worms at all levels of infection [(5, p. 45); his Table 3] and in all human age groups [(5, p. 52); his Table 12]. There is no tendency for the fraction of worms that are female to change with host age, however.

The death rate of worms may also be density dependent, rising due to crowding in people with a large number of worms.

How overdispersed are adult schistosomes in human populations? Under what circumstances are the two sexes distributed jointly together or separately? If the sexes are independently distributed at low average worm prevalence and together at high, then the sensitivity analysis of models of Bradley & May (4) offers assurance that deviations from a Poisson distribution are no cause for great concern. If the sexes are distributed otherwise, then it is important to know how before implementing recommendations which presuppose a Poisson distribution of parasites.

LIFE CYCLE MODELS

Some models attempt to comprehend the entire schistosome life cycle (20, 21, 25, 32, 34, 38, 43–47).

A Life Table Model

In Palo in the Philippines (20, 21), the proportion of people infected with *S. japonicum* at each age changed very little from 1945 to 1959. Hence it is plausible that the per capita risk of infection also changed very little over that period. Barring large changes in the human population, the total worm population must have been stationary (constant in total numbers and age structure). The demographic model of a stationary population, a life table, therefore may describe the population of *S. japonicum* there.

In this model, the net rate of reproduction NRR of the worm may be written as the product of four factors reflecting four major stages in the worm's life cycle. The worm's NRR is equal to the larval NRR in the snail, times the cercariae's probability of infecting a mammal, times the adult worm's NRR in the mammal, times the egg's probability of infecting a snail. Each factor decomposes further. For example, the probability of an egg's infecting a snail is the product of the probability that an egg is able to hatch, times the probability that the egg is deposited near snails, times the probability of penetrating a snail, times the probability of establishing an infection in the snail after penetration (21).

The estimate of *S. japonicum*'s NRR is 0.6. Based on less reliable data from Egypt, the estimate for *S. mansoni* is 1.9 and for *S. haematobium* is 2.8. If the model and observations were correct, these estimates should be 1. That the estimates do not differ from 1 by orders of magnitude indicates the coherence of the observations and the approximate correctness of the model, under the given conditions.

Still, some qualifications deserve note. First, the deviation of the worm's NRR from 1 which was necessary to maintain constant prevalence rates in the growing human population was probably much less than the uncertainty of the data. Second, the independence of the elementary events which make up the life cycle is so unlikely that Hairston (20, 21) avoids the assumption in practice by estimating from his data, not the elementary probabilities that appear as factors in some of his formulas, but clusters of these factors representing compound events. Third, since "the parasite population is able to come into equilibrium at different rates of transmission" in different ecological settings, "net reproduction in one or both of the hosts must be curtailed with increasing transmission and enhanced with decreasing transmission." Hence "there is a range of transmission rates over which compensatory mechanisms operate to keep the parasite population in equilibrium" (21, pp. 46–47). This means that if an intervention program reduces one of the four factors that determine the worm's NRR, over at least some range, the product of the other factors will increase to keep the worm's NRR near 1. Hairston (20, p. 52) estimates quantitatively the inhibiting effect of increasing the number of female worms of *S. japonicum* present in a person of specified age on the average daily number of eggs in feces per female worm. Similarly, in human autopsies, not based on a randomly sampled population, among so-called "asymptomatic" cases of *S. mansoni*, the mean number of eggs per gram of feces per worm pair may decline with increasing numbers of worm pairs [(5, p. 45); his Table 3, where worm pairs are defined as the lesser of the number of male and female worms recovered]. Animal experiments show that increasing parasite loads decrease reproductive output per parasite (31).

As a result of such compensatory mechanisms (3), one cannot use Hairston's calculated values of the worm's NRR factors in a simpleminded way when evaluating a control program that affects the values of some of those factors.

Hairston (21, p. 47) guesses that in schistosomiasis, "the most important cause of the failure of compensatory mechanisms at low transmission rates is the increasing probability that single parasites which succeed in entering the definitive host [man] will remain unmated."

Without belaboring the calculations, he explicitly [(20), his Figures 2, 3, 5] uses the chances of being mated, as well as the effects of crowding of worms within the human host, in estimating the mean number of hatchable female eggs per female worm per day as a function of mean worm load and in estimating the NRR of female worms in humans as a function of mean worm acquisition rate.

These calculations were overlooked by Macdonald (38) who cited Hairston's (20) paper. Macdonald is generally, but inaccurately, credited with introducing the role of pair formation in schistosome models.

Dynamic Models

Dynamic models aim to describe what will happen when the life cycle is perturbed. To represent the compensatory mechanisms which regulate population numbers they must be nonlinear. One such model (38) emphasizes the nonlinearity introduced by supposed monogamous mating in the sexual stage of the worms, and borrows other nonlinear bits from existing models of malaria. Numerical analysis of this model suggests the existence of a threshold in m, the mean worm load in people. Once m is below this threshold, transmission of infection disappears in a few years; once above it, infection remains endemic indefinitely.

Macdonald (38, p. 500) claims that a very high level of environmental sanitation, meaning a great reduction in the number of eggs reaching water, has a negligible effect on the mean worm load compared to the combined effects of treating infected people and keeping them out of infected water. This conclusion results from Macdonald's implicit assumption, not generally true, that the water in which snails live is saturated with miracidia and that nearly all snails are infected [(23) and N. G. Hairston, unpublished].

Hybrid Dynamic Models

Nasell & Hirsch (47) assume a fixed number N_1 of humans and fixed number N_2 of snails. The state of the model is specified by the number $M_k(t)$ of male and the number $F_k(t)$ of female worms at time t in each person, $k = 1, 2, \ldots N_1$, and by the number $S(t)$ of infected snails.

Individual people may differ in the number of worms they carry initially, but are otherwise subject to identical Markovian laws. Each worm in a person has a fixed probability intensity μ_1 per unit time of dying, identical to and independent of all other worms. (This μ_1 is not the same as in equation 9 above.) New worms enter a person at a rate $\nu_1 E[S(t)]$, which is proportional to the *expected* number of infected snails at that time. Nasell's use here of the expected, rather than actual, number of infected snails makes this model a hybrid of stochastic and deterministic elements.

Every infected snail has an identical and constant death rate μ_2; every snail that dies is replaced instantaneously by an uninfected snail. Each uninfected snail risks infection at a rate which is proportional to the *expected* number of mated worm pairs in all the human hosts put together, with proportionality constant ν_2.

The model explicitly (47, p. 401) ignores the possible influence of human age and sex on human infection rates; the effect of worm and snail age and population density on worm and snail death rates, respectively; the effect of age on egg-laying by female worms and on cercarial shedding by snails; as well as the development of resistance to infection and of latent periods.

In a closed community, with no infection from without, let $W(t)$ be the expected number of worms invading an individual person since time 0 and still alive at t, let $X(t)$ be the expected number of monogamously mated worm pairs in all people added together, and let $Y(t) = E[S(t)]$ be the expected number of infected snails at time t. Then

$$dW/dt = -\mu_1 W + \nu_1 Y,$$

$$dY/dt = \nu_2 X(N_2 - Y) - \mu_2 Y.$$

20.

Since $X(t)$ is a complicated but explicit function of $W(t)$ and of t, $X(t)$ may be eliminated to give a pair of differential equations in W and Y. To study the asymptotic behavior of the solution(s) of these equations, introduce two "transmission factors" $T_1 = \nu_1 N_2/\mu_1$ and $T_2 = \nu_2 N_1/\mu_2$. T_1 measures the maximum ability of the snail population to deliver live schistosomes to a person, because it is the product of the ability ν_1/μ_1 of the one infected snail to deliver schistosomes times the maximum number N_2 of infected snails. T_2 measures the ability of the human population to deliver live miracidia to an uninfected snail.

Asymptotically equations 20 have one, two, or three critical points (points where the derivatives are zero), depending on a relation between the transmission factors. A human population will ultimately move to a level of infection corresponding to one of these critical points. If the human population initially has a nonzero worm load, and if the transmission factors lie above a certain threshold function, then the critical point reached asymptotically depends on that initial level of infection.

Control of infection is a practical problem only when, in addition to the stable critical point corresponding to the elimination of infection, there are two critical points with positive levels of infection in people and snails. One of these points is stable, the other not. Nasell (46) considers the possibility of controlling the initial conditions of infection in the human and snail populations. Nasell & Hirsch (47, pp. 444 ff.) study changes in the transmission factors.

If costs to diminish a transmission factor are proportional only to the percentage reduction obtained, but are otherwise the same for both transmission factors, then a strategy of reducing T_1 is more efficient than a strategy of reducing T_2. This conclusion has been interpreted as supporting Macdonald's (38) claim that control of human feces by sanitation is a worse strategy than snail control. The applicability of the conclusion depends on the parameter values and costs which must be evaluated in each application (see below).

Nasell observes (unpublished) that if Hairston's condition for stationarity is translated into the notation of his model it becomes $T_1 T_2 \Psi(m) (1 - Y_\infty/N_2) = 2m$. This equation specifies a relation which must hold between the mean worm load m in people and the expected prevalence rate of infection among snails Y_∞/N_2 at equilibrium in Nasell's model.

The introduction of latency in snails (45) does not alter the qualitative conclusions drawn from the model.

Nasell (44) extends the model (47) by assuming that a source of infection external to the community adds to the infection rate per uninfected snail a constant ϵ_2 and adds to the rates of infection of people by male and by female cercariae each a constant $\epsilon_1/2$. Let $\delta_1 = \epsilon_1/(\nu_1 N_2)$ be the increase in the proportion of infected snails that would be equivalent in the effect on people to the external source of cercariae. Let $\delta_2 = \epsilon_2/(\nu_2 N_1)$ be the increase in the mean number of paired female worms per person that would be equivalent in the effect on snails to the external source of miracidia or eggs. T_1 remains a more efficient control of mean worm load m than T_2, assuming equal costs for equal proportional reductions. Further, T_2 is a more efficient control than δ_2, which measures the external input of miracidia or eggs. The transmission factor T_1 is similarly a more efficient control than δ_1.

The asymptotic mean worm load m in humans as a function of the transmission capacity T_2 of eggs from people to snails is shown in Figure 6. Each curve corresponds to a different input δ_2 of eggs. All the curves correspond to the same transmission capacity T_1 of cercariae from snails to people and the same external input δ_1 of cercariae. For the largest value $\delta_{2,III}$ of external input of eggs (the uppermost curve), the only way to reduce m is to lower T_2 (all else held constant), and m varies smoothly as T_2 varies.

For external egg inputs such as $\delta_{2,I}$ and $\delta_{2,II}$ which are smaller than a certain threshold, the initial levels of infection in people and snails determine whether m will fall on the lower curve or on the upper curve. If m falls on an upper curve, then one dramatic possibility of control is to shift the initial conditions by coordi-

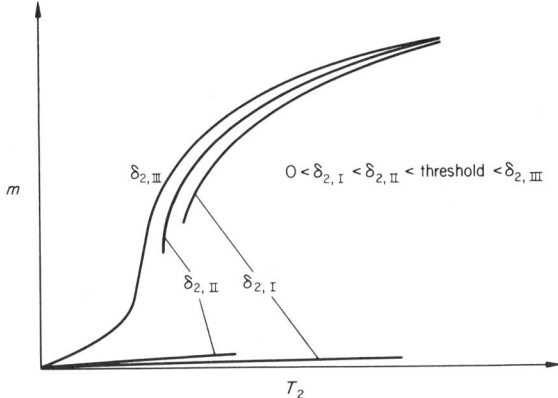

Figure 6 Asymptotic or equilibrium mean worm load m in humans as a function of the transmission capacity T_2 of eggs from people to snails, for various external egg inputs δ_2 and fixed cercarial transmission capacity and fixed external input of cercariae. [Adapted from (44). Courtesy of Ingemar Nasell.]

nated programs of human case finding and treatment so that m falls on the lower line. (When there is no external infection, the lower line is just the abscissa where $m = 0$.) This possibility leaves the community vulnerable to being transferred back to a higher value of m if an uncontrolled influx of infection effectively reverses the change in initial conditions. This possible mode of control arises discontinuously as a function of T_2 and of δ_2.

The curves shift if the parasites are overdispersed rather than Poisson-distributed (4). When worms of both sexes are distributed together, the discontinuities are decreased and the disease becomes harder to control. When worms of each sex are distributed separately, the discontinuities are emphasized, the possibilities of pushing initial conditions below a breakpoint or threshold are enhanced, and the disease becomes easier to control.

Now suppose (34) the human population varies as an immigration–death process, and immigrants are free from infection. Only worm pairs die (at the presumed death rate of females), instead of individuals independently. Instead of having a fixed number N_2 of snails, susceptible snails may immigrate; and both susceptible and infected snails may die and give birth at specific rates. The qualitative behavior of this more realistic hybrid model closely resembles that of the model of Nasell & Hirsch (47).

For one set of parameters, Lewis (34) compares the solution of his hybrid model with the average of ten numerically computed solutions of a fully stochastic model that is identical except for the hybrid simplifications. Since hybrid and stochastic solutions are very close, the hybrid approximations, introduced for mathematical convenience, are quantitatively useful in the range of parameters examined. See (34a) for threshold results.

A Delay Time Model

A deterministic model of Lee & Lewis (32), like that of (21), represents the human and snail populations implicitly through their effects on various stages of the parasite life cycle. All relationships are linear (e.g. eggs produced are directly proportional to the number of mated worm pairs) except for two: Only monogamously mated females are assumed fertile, following (38); and the proportion of miracidia that find snail hosts is assumed to decline nonlinearly with increasing numbers of miracidia per unit volume of water.

Time lags between the infection of snails and the emergence of cercariae, and between the infection of humans and the emergence of eggs, and the two nonlinearities, give the model a breakpoint, which varies seasonally because of the seasonality in the snail latent period. The relative timing of chemotherapy and the application of molluscicide is crucial in determining whether a combined attack using both can eventually eliminate infection.

Caveat Lector

Some models (19, 52; see 57) contain so many internal inconsistencies that they should be perused with extreme caution, if at all.

DECISION MAKING AND ECONOMICS

Sound ecological models of schistosomiasis are a necessary, but not a sufficient, condition for sound public decision making concerning the disease. Many of the models described so far compare strategies for control or eradication. Without reference to the social and economic costs and benefits of these strategies, these comparisons provide no guide for action. Because models which evaluate costs and benefits in relation to schistosomiasis go beyond the scope of ecology, we cite such efforts only briefly.

Using economic and demographic models and data from St. Lucia, a Caribbean island, Weisbrod et al (60) ask: Does infection cause disease? They find it impossible to evaluate the impact of *S. mansoni* on birth and death rates, the achievement of school children, and the labor productivity of adults without considering four intestinal nematodes of man which are also very widespread there.

Rosenfield (50, 51) evaluates the cost effectiveness of molluscicide use, environmental alterations, and chemotherapy, singly and in combination, using detailed data from a *S. haematobium* control project in Iran. Muench's (42) reversible catalytic model is modified to a discrete-time difference equation. The force of infection is estimated by regression as a Cobb-Douglas function of two variables: the meters of snail habitats accessible to the human population, and the number of infected persons. The combined strategy actually used yields a lower prevalence at the end of the project than would any of the three strategies used singly, within the same budgetary constraints.

Other quantitative approaches to evaluating the economic and demographic effects and costs of controlling schistosomiasis are available (1, 9, 10, 27, 39, 48, 49).

MATHEMATICAL MODELS OF SCHISTOSOMIASIS   231

Though sound ecological understanding makes the task easier, a nation's ability to rid itself of schistosomiasis ultimately reflects social and political will (6).

ACKNOWLEDGMENTS

I thank Klaus Dietz, Nelson G. Hairston, Robert M. May, and Ingemar Nasell for criticisms, and the US National Science Foundation for partial support of some of the work described here.

Literature Cited

1. Ansari, N., Junker, J. 1969. Economic aspects of parasitic diseases. *Biotechnol. Bioeng. Symp.* 1:235–52
2. Bradley, D. J. 1972. Regulation of parasite populations. A general theory of the epidemiology and control of parasitic infections. *Trans. R. Soc. Trop. Med. Hyg.* 66(5):697–708
3. Bradley, D. J. 1974. Stability in host–parasite systems. In *Ecological Stability*, ed. M. B. Usher, M. H. Williamson, pp. 71–87. London: Chapman & Hall. 196 pp.
4. Bradley, D. J., May, R. M. 1977. Consequences of helminth aggregation in the dynamics of schistosomiasis. *Trans. R. Soc. Trop. Med. Hyg.*
5. Cheever, A. W. 1968. A quantitative post-mortem study of schistosomiasis mansoni in man. *Am. J. Trop. Med. Hyg.* 17:38–64
6. Cheng, T. 1971. Schistosomiasis in mainland China: A review of research and control programs since 1949. *Am. J. Trop. Med. Hyg.* 20(1):26–53
7. Cohen, J. E. 1972. When does a leaky compartment model appear to have no leaks? *Theor. Popul. Biol.* 3(4):404–5
8. Cohen, J. E. 1973. Selective host mortality in a catalytic model applied to schistosomiasis. *Am. Nat.* 107(954): 199–212
9. Cohen, J. E. 1974. Some potential economic benefits of eliminating mortality due to schistosomiasis in Zanzibar. *Soc. Sci. Med.* 8:383–98
10. Cohen, J. E. 1975. Livelihood benefits of small improvements in the life table. *Health Serv. Res.*, Spring:82–96
11. Cohen, J. E. 1976. Schistosomiasis—a human host–parasite system. In *Theoretical Ecology: Principles and Applications*, ed. R. M. May, pp. 237–56. Oxford: Blackwell. 317 pp.
12. Coutinho, A. B., Coutinho, F. A. B. 1968. The conservation of the biomass density and the age structure of the populations. *Bull. Math. Biophys.* 30: 553–63
13. Coutinho, A. B., Coutinho, F. A. B. 1973. Snail population in running water. *Bull. Math. Biol.* 35:449–58
14. Coutinho, F. A. B., Coutinho, A. B. 1974. Dynamics of populations of *Biomphalaria glabrata* and the von Foerster equation. *Bull. Math. Biol.* 36:29–37
15. Crofton, H. D. 1971. A quantitative approach to parasitism. *Parasitology* 62: 179–93
16. Dietz, K. 1975. A pairing process. *Theor. Popul. Biol.* 8(1):81–6
17. Fine, P. E. M. 1975. Ross's *a priori* pathometry—a perspective. *Proc. R. Soc. Med.* 68:547–51
18. Fine, P. E. M., rapporteur. 1976. *Mathematical Models of Schistosomiasis. Proc. Workshop Bellagio, Italy, 9–14 May 1976.* New York: Edna McConnell Clark Found. 58 pp.
19. Goffman, W., Warren, K. S. 1970. An application of the Kermack-McKendrick theory to the epidemiology of schistosomiasis. *Am. J. Trop. Med. Hyg.* 19(2):278–83
20. Hairston, N. G. 1962. Population ecology and epidemiological problems. In *CIBA Found. Symp. Bilharziasis*, ed. G. E. W. Wolstenholme, M. O'Connor, pp. 36–62. London: Churchill. 433 pp.
21. Hairston, N. G. 1965. On the mathematical analysis of schistosome populations. *Bull. WHO* 33:45–62
22. Hairston, N. G. 1965. An analysis of age-prevalence data by catalytic models. *Bull. WHO* 33:163–75
23. Hairston, N. G. 1971. Book review, *Human Schistosomiasis* by P. Jordan, G. Webbe. *Am. J. Trop. Med. Hyg.* 20(1): 164–5
24. Hairston, N. G. 1973. The dynamics of transmission. In *Epidemiology and Control of Schistosomiasis (Bilharziasis)*, ed.

N. Ansari, pp. 250–336. Baltimore: University Park Press. 752 pp.
25. Hirsch, W. M., Nasell, I. 1975. The transmission and control of schistosome infections. In *Mathematical Analysis of Decision Problems in Ecology*, ed. A. Charnes, W. R. Lynn, pp. 271–98. Berlin; Heidelberg; New York: Springer. 421 pp.
26. Holford, T. R., Hardy, R. J. 1976. A stochastic model for the analysis of age-specific prevalence curves in schistosomiasis. *J. Chronic Dis.* 29:445–58
27. Jobin, W. R. 1972. Computer simulation of costs and benefits for alternate strategies of schistosomiasis control—a dynamic life table model. *WHO/Schisto/WP/72.37*
28. Jobin, W. R., Michelson, E. H. 1967. Mathematical simulation of an aquatic snail population. *Bull. WHO* 37:657–64
29. Jordan, P., Webbe, G. 1969. *Human Schistosomiasis*. Springfield, Ill: Thomas. 212 pp.
30. Kagan, I. G. 1968. Serologic diagnosis of schistosomiasis. *Bull. NY Acad. Med.* 44(3):262–77
31. Kennedy, C. R. 1975. *Ecological Animal Parasitology*. Oxford: Blackwell. 163 pp.
32. Lee, K., Lewis, E. R. 1976. Delay time models of population dynamics with application to schistosomiasis control. *IEEE Trans. Biomed. Eng.* BME-23(3): 225–33
33. Lewis, T. 1975. The loss of immunity in age prevalence models of bilharziasis in man. *Math. Biosci.* 23:205–18
34. Lewis, T. 1975. A model for the parasitic disease bilharziasis. *Adv. Appl. Probab.* 7:673–704
34a. Lewis, T. 1976. Threshold results in the study of schistosomiasis. *Math. Biosci.* 30:205–11
35. Leyton, M. K. 1968. Stochastic models in populations of helminthic parasites in the definitive host, II: Sexual mating functions. *Math. Biosci.* 3:413–19
36. Lichtenberg, F. von. 1975. Schistosomiasis as a worldwide problem: Pathology. *J. Toxicol. Environ. Health* 1: 175–84
37. Linhart, H. 1968. On some bilharzia infection and immunisation models. *S. Afr. Stat. J.* 2:61–66
38. Macdonald, G. 1965. The dynamics of helminth infections, with special reference to schistosomes. *Trans. R. Soc. Trop. Med. Hyg.* 59(5):489–506
39. Macdonald, G., Farooq, M. 1973. The public health and economic importance of schistosomiasis—Its assessment. See Ref. 24, pp. 337–53
40. Malek, E. A. 1961. The ecology of schistosomiasis. In *Studies in Disease Ecology*, ed. J. M. May, pp. 261–327. New York: Hafner
41. May, R. M. 1977. Togetherness among schistosomes: its effect on the dynamics of infection. *Math. Biosci.* In press
42. Muench, H. 1959. *Catalytic Models in Epidemiology*. Cambridge: Harvard Univ. Press. 110 pp.
43. Nasell, I. 1976. On transmission and control of schistosomiasis, with comments on Macdonald's model. Stockholm, Sweden: R. Inst. Technol., Dep. Math., TRITA-MAT-1976-4
44. Nasell, I. 1975. Schistosomiasis in a community with external infection. *Proc. 8th Int. Biometric Conf. 25–30 Aug. 1974, Constanta, Romania*, pp. 123–31
45. Nasell, I. 1976. A hybrid model of schistosomiasis with snail latency. *Theor. Popul. Biol.* 10(1):47–69
46. Nasell, I. 1976. On eradication of schistosomiasis. *Theor. Popul. Biol.* 10: 133–44
47. Nasell, I., Hirsch, W. M. 1973. The transmission dynamics of schistosomiasis. *Commun. Pure Appl. Math.* 26:395–453
48. Paulini, E. 1974. On the Problem of Allocating Funds for Molluscicides and Drugs in Schistosomiasis Control. *WHO/Schisto/74.35*
49. Paulini, E. 1975. Cost-benefit analysis of the use of molluscicides in different endemic areas. *Proc. Int. Conf. Schistosomiasis, Oct. 1975, Cairo, Egypt*
50. Rosenfield, P. 1975. *Development and Verification of a Schistosomiasis Transmission Model*. PhD dissertation. Johns Hopkins Univ., Baltimore; Washington DC: Agency Int. Dev. 162 pp.
51. Rosenfield, P. L., Smith, R. A., Wolman, M. G. 1977. Development and verification of a schistosomiasis transmission model. *Am. J. Trop. Med. Hyg.* In press
52. Stirzacker, D. 1974. A singular perturbation analysis for models of schistosomiasis. *Math. Biosci.* 21:183–205
53. Sturrock, R. F., Cohen, J. E., Webbe, G. 1975. Catalytic curve analysis of schistosomiasis in snails. *Ann. Trop. Med. Parasitol.* 69:133–34
54. Sturrock, R. F., Sturrock, B. M. 1971. Shell abnormalities in *Biomphalaria glabrata* infected with *Schistosoma mansoni* and their significance in field

transmission studies. *J. Helminthol.* 45:201–12
55. Sturrock, R. F., Webbe, G. 1971. The application of catalytic models to schistosomiasis in snails. *J. Helminthol.* 45(2/3):189–200
56. Tallis, G. M., Leyton, M. K. 1969. Stochastic models of populations of helminthic parasites in the definitive host. Part I. *Math. Biosci.* 4:39–48
57. Voors, A. W. 1971. Correspondence re Goffman and Warren 1970. *Am. J. Trop. Med. Hyg.* 20(1):167
58. Warren, K. S. 1973. Regulation of the prevalence and intensity of schistosomiasis in man: Immunology or ecology. *J. Infect. Dis.* 127:595–609
59. Warren, K. S., Newill, V. A. 1967. *Schistosomiasis—A Bibliography of the World's Literature from 1852 to 1962.* Cleveland, Ohio: Western Reserve Univ. Press. 993 pp.
60. Weisbrod, B. A., Andreano, R. L., Baldwin, R. E., Epstein, E. H., Kelley, A. C. 1973. *Disease and Economic Development: the Impact of Parasitic Diseases in St. Lucia.* Madison, Wis: Univ. Wis. Press. 218 pp.

… 4124

DROSOPHILA SYSTEMATICS AND BIOCHEMICAL EVOLUTION

Lynn H. Throckmorton

Department of Biology, The University of Chicago, Chicago, Illinois 60637

INTRODUCTION

The pendulum swings back and forth, and then back again, and the species problem seems always to be with us. This certainly is not an area of biology about which we dare be complacent, though some would have it so, and a few of those concerned with *Drosophila* are not least among them. Fortunately the field advances whether we will or no, and recent work with *Drosophila* does contribute its share to the eventual decline of controversy over speciation and its mechanisms. These contributions derive most directly from studies of protein evolution, and this review concentrates on that. It does not deal with protein variation within populations, which has been much reviewed recently (35, 52, 62). Some topics, particularly rates of molecular evolution and biochemical systematics, are commented upon when the occasion permits.

MAJOR INVESTIGATIONS

Early Protein Studies

These focused on the potentialities of biochemical approaches for shedding light on evolutionary problems. They emphasized the possibilities for measuring evolutionary change by tracing the pathways of molecular evolution and for analyzing the speciation process. In the latter case they sought direct evidence for the extensive reorganization of gene pools so much emphasized by Mayr (44) and questioned by others (74). The earliest studies were expected, as it were, to explore the routes to answers rather than to arrive at the answers themselves. In spite of that they produced suprisingly potent data.

THE VIRILIS GROUP The first of these pathfinders was a small study of soluble proteins from one strain each of nine members of the virilis group (32). These could not be related to specific gene loci, and even the number of loci contributing to the data could be estimated only crudely. Nonetheless, the distribution of proteins of identical mobilities among the species could be determined, and since a chromosome phylogeny was available for these species a partial recontruction of ancestral gene pools could be, and was, attempted. The resulting picture was consistent with established evolutionary theory, but not necessarily with the theories of established evolutionists. Genetic change during evolution from the group ancestor had been gradual, with some gain and some loss of ancestral alleles. Changes in some lineages were more extreme than the changes in others, and the members of one cytological phylad diverged less from the group ancestor than did those of the other. The members of the less conservative phylad showed greater differences among themselves than did the members of the less conservative phylad (a mean similarity of 41% compared to 72%). The average species diverged from its immediate ancestor in about 25% of its proteins (range, 16–41%), and the gene pool of the average species comprised 60% of alleles from the group ancester, 25% of alleles from the ancestor of its cytological phylad, and 15% of alleles unique to itself. The subspecies, *D. a. americana* and *D. a. texana*, appeared to have diverged in the same manner and to the same degree as had full species. The phenetic relationships of these species based on their protein similarites bore only a rough resemblance to the established chromosome phylogeny, although Nei (48, 50) thought otherwise. By no means definitive at the time, these data did suggest the following working hypotheses: (*a*) Biochemical evolution does not occur at a constant rate, either for individual speciation events or averaged over several events; (*b*) there is no evidence for a quantitative difference between speciation and subspeciation; and (*c*) protein "phylogenies" are not necessarily good genealogies.

With respect to item *b,* two matters need to be clarified. First, the "divergence" between species measured in this study, and most others since, confounds at least two types of genetic change: that "necessary" for speciation, and that involved in evolution independent of speciation. Thus, when we look at species differences, we are not concerned with averages. When we ask how much change can be attributed to speciation we look only for minimum difference. This will be a single observation for any particular study. When we have found this minimum we still must treat it as a maximum estimate of the change we are trying to assess, because it incorporates both relevant and irrelevant genetic changes. Hence, in reporting results of the various studies I emphasize the observed minimum divergence. The observation in the study of the virilis group that both species and subspecies diverged in the same manner and to the same degree raised immediate questions. The chief one was, of course, whether speciation necessarily involved a reorganization of the gene pool at all. Might speciation instead involve only a few loci that have special effects on the biology of populations. Might it involve changes only in control systems (70)?

THE SIBLING SPECIES STUDY   The study of general proteins provided challenging data, but it was desirable that these questions be reappraised using methods that permitted more accurate identification of the genetic systems involved. The utility of assay systems for specific enzymes (31) for this kind of work was demonstrated next (33).

Seven assays were used, six for specific enzymes and one for hemolymph proteins. Sibling species representing six species groups and three subgenera were selected, and a nonsibling near relative was chosen to match each pair. Nine triads, a sibling pair and one near relative, were used. Siblings on the average shared proteins of identical mobility 50% of the time (range, 22–85%), while a sibling and a morphologically distinct member of the same group shared proteins of identical mobility only 18% of the time. The approximately 15% minimum divergence compared surprisingly closely with the approximately 16% minimum divergence noted in the virilis group. Even when the apparent defects of this study were duly acknowledged (small sample size, failure to detect cryptic alleles, complications arising from polymorphism, etc), the parallel was too strong to ignore. There seemed to be no sharp difference between intraspecific and interspecific differentiation. It was concluded that an extensive reorganization of gene pools did not appear to be required for speciation, although it was uncertain just how slight the reorganization could be.

The Bandwagon Era

THE OBSCURA GROUP In other laboratories similar work had been going on briskly. The first detailed comparison of closely related species was between *D. pseudoobscura* and *D. persimilis* (53, 55), long studied by Dobzhansky and his coworkers (23) and, after *D. melanogaster* and *D. simulans*, probably the most widely known sibling species pair. The comparison was based on one population of *D. persimilis* and on several of *D. pseudoobscura*. The study was later expanded. Summarizing the evidence to 1974, Lewontin (43) noted that there was slight or no gene frequency differentiation at 88% of the loci, clear gene frequency differentiation at 8%, and something approaching differentiation at 4% of the loci. He concluded that they probably differed completely at certain loci involving just a special part of the genome and that most of the genome was not differentiated.

Until recently the known range of *D. pseudoobscura* was from the Rocky Mountains to the Pacific Ocean and from British Columbia to Guatemala. Then a population was discovered in the Andes near Bogotá, Colombia (27). When analyzed cytologically and genetically it was discovered to differ from the rest of the species, and it is regarded as rather old, probably resulting from passive transport thousands of years ago (21). In one direction crosses between it and the North American populations produce fertile female and sterile male hybrids (54). When compared electrophoretically with other populations it differs from them by about 20% and has been described as a distinct subspecies, *D. p. bogotana* (4). A 20% difference between this and the nominate subspecies contrasts with the 12% difference between *D. p. pseudoobscura* and *D. persimilis*. When full species differ by less than subspecies the importance of giving weight only to minimum differences is underscored.

Lakovaara and his colleagues have used genetic distance from protein study to evaluate evolutionary relationships among 22 species of the obscura group (41, 42). Enzymes at 21 loci were checked, but only the most common allele at each locus was used in calculating genetic distance. Values ranged from about 14% to about

65%. The greatest similarity, 65%, between *D. pseudoobscura* and *D. persimilis*, was rather different from the approximately 88% observed when populations were compared more thoroughly (43). A dendrogram was included and some of the difficulties in equating this with genealogy were considered.

D. athabasca is one of the North American members of the affinis subgroup of the obscura group, and it is made up of a western and two eastern forms sometimes referred to as semispecies (46). The "western" and the "eastern A" forms are sympatric through an area east and north of Minnesota, but they do not overlap outside this region. Strongly isolated ethologically, with only 10% probability of the gene flow between them, there is limited and equivocal evidence from chromosomes for gene flow even though hybrids produced in the laboratory are fully fertile. Protein evidence indicates that 4 of 17 loci (24%) show evidence of significant differentiation between the two forms. Among populations within forms similarities range from 0.93 to 0.99, with an average of 0.97 (Nei's index), and between the two forms the similarity is 0.90. Rogers' index (61) and Nei's index (49) have been used most commonly for comparisons of this type, and both can be equated roughly with percentages, for present purposes at least, without serious distortion. When allozymes are checked for progeny of wild caught females from areas where both forms were collected together, there is no indication of gene flow between the "western" and "eastern" forms. It was decided that these forms are most probably species, and since the minimum divergences between local populations (7%) and between eastern and western forms (10%) were so nearly the same, speciation required very little gene pool reorganization (34).

THE MELANOGASTER GROUP   The best known pair of siblings is *D. melanogaster* and *D. simulans*. Of seventeen loci studied (40), none was fixed for alternative alleles, although one was nearly so, and there was clear gene frequency differentiation at 11% of the loci. The authors made no comment on the significance of these data for the species problem, but the pattern is very similar to that seen for *D. pseudoobscura* and *D. persimilis*.

The bipectinata complex is found mainly in southeast Asia and nearby areas, and its protein variation has been reported by Yang et al (78). They observed allelic distributions at 23 loci over 194 isofemale lines from 30 localities. Four species were involved, two of which were composed of two subspecies each. The subspecies are completely interfertile. Crosses between species usually produce some F_1s, at least in one of the reciprocal crosses, and hybrid males are sterile. Highest similarities, about 92%, were seen between one of the pairs of subspecies, followed closely by the most similar species pair, *D. bipectinata* and *D. malerkotliana*, at 91%. This is, accordingly, another case where speciation and subspeciation show no quantitative differences. Chromosomally the four species share no intraspecific inversions but they have accumulated an average of 9.3 fixed interspecific inversions, and the authors suggest that speciation in this complex was more closely associated with the accumulation of inversion differences than with genetic changes. Not surprisingly, they see no evidence that an extensive reorganization of the gene pool is necessary for speciation. The bipectinata complex provides another case of two species, intercrossable in the laboratory, maintaining themselves as separate species in nature. They are separated by ten fixed inversions and no shared inversions, and when sympatric neither inversions nor proteins are found intermingled.

THE NASUTA COMPLEX   Samples of three species collected from Australia, New Guinea, Malaysia, Hawaii, and points between were assayed for eight enzymes. Two of the species, *D. albomicans* and *D. sulfurigaster,* can be regarded as siblings. *D. pallidifrons* is an only slightly more distant relative. Three subspecies of *D. sulfurigaster* were included. The results, while not presented quantitatively, indicated that the differences between species were few and in close agreement with degrees of relationship established by cytological and hybridization studies. *D. pallidifrons* was most different, diverging from the others at about 40% of its loci. *D. albomicans* differed from its sibling, *D. sulfurigaster,* at about 25% of its loci, and all strains of all three subspecies of *D. sulfurigaster* were indistinguishable. Thus, species were clearly distinct and subspecies were not (37).

THE MESOPHRAGMATICAS   In 1971 a study of six species of the mesophragmatica group was completed. These are primarily Andean in their distribution, from Colombia to Chile. Proteins at 24 loci were assayed, and a cytological phylogeny was available for the group. Using this it was possible to reconstruct the ancestral gene pool in a manner similar to that used for the virilis group. It could then be inferred that 11 out of 24 loci had been polymorphic in the ancestral gene pool as compared, for example, with 12 out of 24 polymorphic in one of the extant species, *D. pavani.* When similarities were calculated using Rogers' index, *D. brncici* and *D. gascici* were closest (0.77); next closest were *D. mesophragmatica* and *D. gasici* (0.67); while a sibling pair, *D. gaucha* and *D. pavani* came in at 0.63. The minimum divergence here, about 30%, is somewhat larger than observed elsewhere. The authors conclude that changes at a large number of loci are not required for speciation, and they suggest that changes in genetic regulatory mechanisms and polygenic systems are the critical factors in the origin of species. The observed genetic variation showed a close relationship to the morphological and cytological variations between species and phylogenetic relationships paralleled the postulated cytotaxonomic relationships. Thus, isozyme variations can be useful diagnostic aids for taxonomy (47).

THE HAWAIIAN DROSOPHILA   A new perspective on the species problem was added with the emergence of investigations of the Hawaiian Drosophilidae. Recent reviews describe the evolutionary biology of these forms and note their diversity and the paucity of sibling species among them (14, 17). Among other unusual features are pairs or groups of species for which no detectible differences are observed in the banding sequences of any of the polytene chromosomes. For continental forms, and even for sibling species, this is rarely the case. Such species are referred to as "homosequential" (15).

An early study of population variation included a small comparison between the very similar *D. mimica* and *D. kambysellisi.* The results were not quantified and the proteins were described only as being very similar (59). In a much larger study the allozymes of about 80 of the Hawaiian species were reported (60). These represented a wide cross section of Hawaiian groups, including most of the major types. Homosequential species were compared in lieu of siblings, and divergence was measured as the proportion of the gene pool that was unique between species. It ranged from 26% to 80% for homosequential species, with a mean divergence of 55% and a variance of 133. A random sample of an equal number of species pairs gave a range of 44% to 87%, a mean divergence of 66% and a variance of 15. Thus, randomly chosen pairs are less similar (34% compared to 45%) than are homosequential species, homosequentials are somewhat less similar (45% compared to 50%) than siblings, and random pairs of Hawaiian species are considerably more similar (34% compared to 18%) than the average nonsiblings reported elsewhere (33). The minimum divergence (26%) is comparable to that seen in the nasutas, the mesophragmaticas, and so on. When substitutions between homologous loci were calculated the homosequential species gave a value of 35%. This contrasts sharply with the value of 12% for complete or near complete substitution between *D. pseudoobuscura* and *D. persimilis,* and of 11% for *D. melanogaster* and *D. simulans* (43). The authors conclude that the large range and high variance of the homosequential sample emphasize the essential lack of parallel between gene sequence stability and genic similarity, and that speciation among Hawaiian *Drosophila* is not accompanied by a major reorganization of the gene pool.

Johnson and his collaborators provide a detailed look at a more compact group of Hawaiian species (36). This is the planitibia subgroup of the Hawaiian picture wing *Drosophila*. These are among the world's largest *Drosophila,* most of them are single island endemics, and usually each is restricted to a single Hawaiian volcano. A cytological phylogeny is available for them, although the direction of evolution within it cannot be specified unambiguously. For sixteen species the similarities ranged from 0.11 to 0.99, with a mean of 0.44 (Roger's index). These values do not differ greatly from those of the sibling species study (33), except that the minimum divergence is much less, here approaching zero. Out of 120 comparisons possible between 16 species, 4 show a divergence of less than 10%, 11 show a divergence of less than 20% and 18 (15% of the comparisons) show a divergence of less than 30%. The authors conclude that the four cases of unusually high similarity, from 0.96 to 0.99, probably resulted from allopatric speciation occurring through fragmentation of a widely distributed population by the appearance of strong isolating barriers (e.g. the rise of sea level to separate formerly continuous land masses, disruption of host ranges by climatic change, etc). They argue that one would expect minimal divergence in that case unless the isolates were subjected to strong differential or directional selection. They regard the most dissimilar species as most likely to have been founded by a few individuals, through interisland colonization for the most part, and hence to have undergone a "flush" and "genetic revolution" as postulated by Carson (10–12). Regardless of how differences are explained, these are the first cases where almost no difference is seen between full species.

Comparing the pattern of genetic relationships shown by the allozyme data with the cytological phylogeny, only a rough congruence is seen. Some subclusters agree, but for the most part the patterns in one set are a scrambled version of those in the other. It is true that in neither case can the direction of evolution be unequivocally specified, more nearly so for the chromosomes than for the enzymes, but the disagreements are not such that they can be reconciled by any simple reorientation of a diagram. The authors conclude that the rate of divergence between species varies greatly depending on the kind of speciation event involved, and similarity values have small phylogentic significance. Whether or not this interpretation is correct, the conflict between allozyme and chromosome results casts doubt on the reliability of protein data for phylogenetic research.

D. setosimentum and *D. ochrobasis* are members of the adiastola subgroup of the picture wing group and are near sibling species, with males distinct but living females not easily separable. They are widely distributed in wet highland forests on the island of Hawaii, show altitudinal and breeding site differences, and both are found together at some localities. Chromosomes and allozymes show significant differences between species and between some of the local populations within species. There is evidence for the differentiation of a low altitude ecotype, and at one locality (Mawae) recent hybridization has been postulated to account for unusual variation observed there (13). At one area (Kahuku Ranch) natural hybridization was detected, with 4 unusual specimens of 180 examined (2.2%), one an F_1 and three backcross hybrids. So it is possible that gene exchange can occur between these species. The gene pools, however, give no evidence of merging. The species pair is regarded as representing a stage that may be rather common for allopatric speciation in general, that is, where two separated populations come back into contact after nearly completing "integration" of "new" gene pools. Reciprocal gene flow is thought to be insignificant in this case because natural selection works against inferior F_2 and backcross combinations (16).

A quantitative assessment of the differentiation, both for chromosomes and for allozymes, has been made (13, 18). *D. setosimentum* could be divided into a set of six highland populations, the "main body" of the species, and one Kona population; *D. ochrobasis* could be divided into two southern populations, the "main body" of the species, and one population from Ohu isolated at the north end of the island. Comparisons within *D. setosimentum* showed similarities of 0.98 (Nei's index) within the main body of the species and 0.92 with the Kona population. Within *D. ochrobasis* the similarity was 0.98 within the main body of the species of 0.87 with the isolated Ohu population. Comparisons between species, however, show that while the main bodies of the two are rather distinct (0.79), the isolated Ohu population is very similar (0.98) to the main body of *D. setosimentum*. Chromosome comparisons give somewhat different results, with a mean intraspecific similarity of 0.96, compared to 0.94 for the allozymes; and for interspecific comparisons a mean value of 0.69 is seen with chromosomes, compared to 0.90 for allozymes. The allozymes are much less precise in distinguishing these species than the chromosomes. This is the reverse of the case for homosequential species, where 100%

similarity of chromosomes contrasts with only 20–74% similarity for allozymes. The authors conclude that for these species, as for *D. silvestris* and *D. heteroneura* (36), speciation has occurred with only a small amount of allozymic reorganization. They suggest this is a result of the recency of speciational events, and that speciation may involve regulatory genes which usually are not revealed by electrophoretic data.

THE WILLISTONI GROUP The biology of these species is reviewed by Dobzhansky and Powell (24). Those of greatest interest here, *D. willistoni* and its siblings, including the semispecies of *D. paulistorum* (26), comprise one of three major clusters among the 25 species that make up the group (73). There are six siblings, three of which consist of a pair of subspecies each and one of which consists of six semispecies. The siblings are nearly indistinguishable morphologically and their distributions overlap broadly. In many areas three or four siblings are collected together, but the range of *D. willistoni*, from southern Florida and central Mexico to northern Argentina, is the greatest of the six. Sexual isolation between these species is quite strong and only occasional sterile hybrids are produced in laboratory crosses. Some fertile hybrids have been reported (76, 77), but they do not seem to appear regularly or generally (25). The subspecies show no sexual isolation in the laboratory, but at least one of the reciprocal crosses between them produces sterile males, and they are largely allopatric (5, 6).

The semispecies of *D. paulistorum* are not distinguishable morphologically (51). With few exceptions, when crossed in the laboratory they produce fertile female and sterile male progenies. A range from little to almost complete sexual isolation is exhibited, depending on the cross (9, 22). They thus show incomplete reproductive isolation from each other. Transitional strains also exist which partially bridge the reproductive gaps, at least in theory, especially between the Centroamerican and the Andean-Brazilian semispecies. In many cases two or three semispecies exist sympatrically without losing their identity. The area of highest sympatry, northern South America, is also the area of highest sympatry for the sibling species themselves (65).

The first protein comparison appeared in 1970 with a report of genetic differentiation at 14 loci among four of the siblings (3). Other reports over the years have filled out the story, and a complete report is given by Ayala and his collaborators (7). More limited discussions are also available (1, 2, 24). Dealing with averages, and using Nei's index as a comparator, it is shown that a graded series of similarities exists among these forms, from local populations to nonsibling species. The local populations range in similarity from 0.91 for transitional strains of *D. paulistorum* to 0.99 for local populations of *D. willistoni*, with a mean for local populations of 0.97. Between subspecies the values were 0.81 and 0.78 with a mean of 0.80, and for semispecies the mean was 0.80 for similarities ranging from 0.81 to 0.96. For sibling species the range was from 0.28 to 0.80, and the mean was 0.56. The nonsibling comparison was made between the siblings and *D. nebulosa* and gave a mean similarity of 0.35. The minimum divergence between species, approximately 20%, is in the usual range for continental species, and this range is overlapped by that for subspecies, which is also about 20%, and for semispecies, which ranges from about 5% to 20%. These data are in nice accord with other results, and they extend the series first described in the sibling species study (33) to the level of the local population. The series of means now stands: local populations, 0.97; subspecies, 0.80; semispecies, 0.80; sibling species, 0.56 (compared to 42% for the sibling species study); nonsibling near relatives, 0.35 (compared to 20% for the sibling species study); and from the sibling species study only, somewhat less than 6% for distantly related species.

The unique thing about this study is, of course, the inclusion of two different subdivisions between local populations and species. The authors regard subspecies and semispecies as different evolutionary levels, using the ranking: geographical populations, subspecies, semispecies, sibling species, morphologically distinct species. Subspecies are regarded as the "first" stage of species formation, representing a type of allopatric differentiation that might give rise to species upon sympatry, according to the model of geographic speciation. Because of their often nearly complete reproductive isolation, the semispecies of *D. paulistorum* are regarded by them as representative of the "second" state of speciation, when reproductive isolation is being completed. Since substantially the same amount of divergence exists between subspecies and between semispecies, on the average, it is concluded that a considerable amount of genetic differentiation occurs during the first stage of speciation but that little is needed during the second stage. In fact, however, the minimum divergence, the operational value for evaluating these results, is only about 5%, and that is for semispecies. The minimum value for subspecies, about 20%, is much higher. Hence, the conclusion of a first stage of considerable divergence, and a second stage of little divergence, is not really supported by these data.

No chromosome phylogeny is available for these species. The chromosomes have been repatterned extensively, with *D. paulistorum,* as the extreme example, exhibiting at least 85 rearrangements of its own (38, 39). Ayala and his coworkers provide a dendrogram constructed from the protein data following one of the "maximum parsimony" approaches (29), and hence having the limitations that result from such procedures. The tree is, nonetheless, a very fair approximation of that derived by Spassky et al (65) from morphology, behavior, and so on. An interesting relationship is observed for the pairs of subspecies, each of which consists of one widespread form and one of narrower range. In each case, the more widely distributed subspecies (the main body of the species, as it were) is more similar to the other siblings than is the narrowly distributed form. It is speculated that the less conservative subspecies is an offshoot of the main body of the species; it has adapted to new habitats and hence has diverged most. Or, they suggest, the less conservative subspecies may have originated through colonization and so may give evidence of a "founder effect" (44) or a "population flush" (10). However that may be, the evidence is thought to indicate an inconstancy of the rate of genetic change among different evolutionary lineages.

THE MULLERI SUBGROUP   Two species pairs have been investigated closely: *D. aldrichi* and *D. mulleri,* and *D. mojavensis* and *D. arizonensis* (79). *D. mojavensis* is divided into two races, and one of these is further subdivided into two subraces.

The biological relationships are somewhat parallel to those among the siblings and semispecies of the willistoni group, and the closeness of the pairs can be appreciated by the fact that *D. mojavensis* was originally described as a subspecies of *D. mulleri*. Crosses between *D. aldrichi* and *D. mulleri* produce either sterile hybrids or none at all. *D. arizonensis* and *D. mojavensis* are fully interfertile. Nevertheless they remain cytologically separate entities, and in the cytological phylogeny the members of pairs are phylogenetically closer to each other than to members of the other species pair (75). The two races of *D. mojavensis* reflect both chromosomal and morphological differences, they are allopatric, and they are completely interfertile. The subraces of *D. mojavensis* are allopatric, one in Arizona and Sonora, the other in Baja California and the islands of the Gulf of California, and they differ in the species of host cactus they exploit. The Arizona and Sonora populations of *D. mojavensis* are sympatric with *D. arizonensis*. The strains of *D. aldrichi* and *D. mulleri* were from Texas, where the two species are sympatric. Similarity means for these forms provide another nicely graded series: populations, 99%; subraces, 97%; races, 88%; sibling species, 84%; and nonsibling species, 70%. The values for siblings and nonsiblings are considerably higher than those from the willistoni group, which were about 56% and 35% respectively. The minimum difference, about 12% for *D. aldrichi* and *D. mulleri*, is in the usual range. It is also less than the 20% minimum difference between *D. arizonensis* and *D. mojavensis*, even though this pair shows much weaker post-mating isolating mechanisms, being fully fertile in crosses, than *D. aldrichi* and *D. mulleri* which produce sterile hybrids or none at all. This is a good example of the failure of reproductive divergence to serve as an index of evolutionary divergence. The author concludes that: (*a*) the hybrid sterility or breakdown observed in species crosses may be the result of changes in a rather small proportion of the genome; (*b*) degree of reproductive isolation is a less reliable index of the genetic change involved during early evolutionary divergence than is the degree of ecological differentiation; and (*c*) while there is a regular accumulation of genic difference as speciation procedes, most genetic change appears only after genetic isolation is insured.

A rather different type of study has been carried out by Richardson and his collaborators (56–58). Stressing that single electromorphs need not represent single alleles, electrophoretic mobilities were analyzed and compared by analysis of variance. A distance measure based on average electrophoretic mobilities was defined for investigating evolutionary relationships. Seven assay systems were employed with species of the mulleri complex whose cytological relationships are specified (75). Variation among species was greater than within species and that among sibling species clusters was greater than that within clusters. Grouping on the basis of average relative mobility yielded results similar to, but not identical with, those given by the chromosomes. The authors conclude that such clusterings can be taxonomically useful in a gross way but that uncritical use of mobility can lead to difficulty (57); modes of evolution are usually different among different loci (56); and the taxonomic diversity of larval substrates and the electrophoretic diversity of *Drosophila* populations using them appear to be associated (58).

DROSOPHILA SYSTEMATICS   245

THE VIRILIS GROUP AGAIN   The virilis group originated most probably in warm temperate Asia somewhat more than 15 million years ago. It subsequently split into two groups, as is indicated by the chromosome phylogeny. The descendents of one, the virilis phylad, are associated so far as is known with willow in riparian communities of the temperate zone, the probable ancestral habitat; those of the other group, the montana phylad, are associated with aspen and alder near lakes and streams and on mountainsides at higher altitudes, higher latitudes, or both. At present the group has a Holarctic distribution, the new world having been colonized by at least one population from each of the phylads, apparently shortly after their founding (73). Thus, by comparing phylads we can check relative rates of evolution, and from biogeographical evidence we can make "ball park" estimates of absolute rates of biochemical evolution.

The first update on proteins came in 1972 with a report of assays covering a wide cross section of laboratory strains from nine species and dealing with 11 loci (71). Genetic changes during speciation averaged 12% in one phylad and 27% in the other, and these amounts were in the same range as those seen for intraspecific differentiation among the same populations. The average speciation event involved a reduction of genetic variability, new species budding off, as it were, from stem populations that retained the reservoirs of genetic variability. The phylads evolved at different rates. The one that exploited novel ecological opportunity evolved at approximately twice the rate of the one that remained in the ancestral habitat (willows). There was a heterogeneous pattern of change at the individual loci, with different loci changing at different rates in different lineages and parts of lineages. It was concluded that: (a) there was no evidence for an extensive reorganization of the gene pool during speciation, (b) there is no constant rate of biochemical evolution, (c) there is evidence for the maintenance of polymorphism through natural selection since polymorphisms had persisted at about 25% of the loci since the founding of the group (72).

The most recent version of these data has not yet been published, although there is some prospect of that (L. H. Throckmorton and J. L. Hubby, in preparation). The number of strains assayed has been increased five-fold, covering large new samples from many new localities for almost every one of the eleven species available. The results substantially agree with and enlarge upon previous ones. The gene pool of the average species consists 62% of ancestral alleles, 31% of phylad specific alleles, and 7% of alleles unique to itself. This compares favorably with the 60%, 25%, and 15% values from the general protein study (32). The proportion of unique alleles per species ranges from 2.2% to 14.4%, giving by that criterion a minimum divergence of about 2%. The similarity between the closest species, 80% for *D. montana* and *D. lacicola,* approximates that of 82% for the most different of local populations, so by that criterion also, interspecific and intraspecific differentiation do not differ markedly.

Having added 1.3 compared to 2.2 alleles per million years of evolution, the ecologically conservative (willow) phylad changed at 60% the rate of species evolving in new habitats and over diverse terrain. The average speciation event still involved a reduction in genetic variation, diverging populations gaining 0.8 times as many alleles and losing 2.4 times as many as the stem population. There was a more than five-fold difference in evolutionary rates among the different loci, with some loci even evolving more rapidly in the conservative phylad, although the total gene pool of those species was changing more slowly. Polymorphism was perpetuated at 56% of the loci from the time of founding of the group (for more than 15 million years), and only partial agreement was obtained between the protein phenogram and the cytological phylogeny. The protein data did partition the group into the same two phylads as did the chromosome data, but the details within the phylads differed greatly.

The obvious and by now tedious conclusions follow: (a) There is no evidence for an extensive reorganization of the gene pool during speciation. (b) There is no evidence for a usable molecular clock operating in these proteins. The rate of molecular evolution is apparently related most closely to the opportunity open to the evolving population. (c) Most speciation events in the group involve a reduction of genetic variability. (d) The dendrogram based on "genetic distance" is not a good phylogeny.

Still More Genetic Variability

REFORMATION AND ENLIGHTMENT   Our first paper on protein differences introduced the topic of cryptic variation; that is, of genetic change not disclosed by electrophoresis of proteins, and specifically of alterations of proteins that did not result in a net charge change for the molecule (32). The topic turned out to have much in common with the weather. Lots of people talked about it, but no one did anything about it. On the basis of early evidence it seemed that about two thirds of the actual variation might be disclosed by electrophoresis. As time passed the estimated fraction grew smaller. The consequences of cryptic variation for comparative studies became more serious.

The problem was finally explored by adding a discriminating technique to the standard procedure for screening alleles. Heat denaturation was used with the *xanthine dehydrogenase* (*XDH*) locus of the virilis group. Almost twice as many alleles were discovered, both within populations and within species, and across the group as a whole the number of alleles increased from 11 to 32. On the basis of small samples, geographically close populations were found to have three of four electromorphs in common but only one of nine "thermoelectromorphs". The average similarity between species dropped to 35% of the value based on electromorphs alone (8).

Suddenly the outlook was bleak if not actually grim. The specter of still more genetic variation had always been with us. Some had wished to ignore it. Now it turned out to be more material than had been expected. It was necessary to take stock at this turn of events.

TWELVE YEARS OF WORK DOWN THE DRAIN?   The news was received with astonishment, some resentment, and some disbelief; but additional studies verified the phenomenon. At the *octanol dehydrogenase* (*ODH*) locus in the virilis group heat denaturation disclosed 2.6 times as many alleles as before observed, and similarities between species dropped accordingly. The major electromorph of this study was one that was shared by all species of the group and usually it was the predominant allele, having a frequency of 85% or higher in eight of the ten species investigated. After heat sensitivity studies there was still one predominant allele with this mobility, but its importance was much reduced. It was present in 17 of 34 populations investigated, and still present in all species. Samples were too small to tell much about frequencies within species (63).

Alleles at the *esterase-6* locus in *D. melanogaster* increased in number from four to seven when heat denaturation was used (19). In this case no allele was unique to any one of the six populations but the genetic similarity between populations decreased. One allele remained most common in different populations, but two different alleles were now second most important in some of them. The sharp differentiation seen for heat sensitive alleles between populations for *XDH* in the virilis group was not seen here.

The *esterase-5* locus was reinvestigated in gels of different concentrations and with different buffer systems. Six different alleles were revealed within the most frequent class and three in the second most frequent one, to make this a highly heterogeneous system (45). Another study of *D. pseudoobscura* employed four different electrophoretic conditions and a heat sensitivity test on *XDH*. Thirty-seven alleles were revealed where only six had been observed before, and *D. p. bogotana* was shown to be different from *D. p. pseudoobscura* at this locus where before they had been thought to share the same most common allele (64). A detailed comparison was also made between *D. p. pseudoobscura* and *D. persimilis* for cryptic variation at the *XDH* locus (20). Sequential electrophoresis was used with different gel concentrations and different gel buffers. Five alleles became twenty-three. Most were unique or nearly so to the populations where they were seen, but the same allele predominated in all populations. Interpopulation genetic similarities (Nei's index) that had ranged from 0.966 to 0.995 for electromorphs now ranged from 0.858 to 0.956, a drop of about 10%. The number of alleles shared at this locus between *D. p. pseudoobscura* and *D. persimilis* dropped from 25% to 6%, and the similarity value fell from 0.55 to 0.03. The most common alleles in the two species are quite different. The author concludes that previous interspecific comparisons now bear reinvestigation. Until now the evidence indicated that marked genetic change need not accompany the appearance of reproductive isolation, but that conclusion may have been premature.

SPECIES AND SPECIATION

The species controversy, in its most recent incarnation, materialized in the early 1960s. In part it was born of reaction to the extreme views of Mayr (44) and particularly to his opinions regarding the cohesion of the gene pool and the universality of the geographic mode of speciation. According to his view, speciation entailed an extensive reorganization of the gene pool, and the only situation in which a gene pool can be completely reconstituted genetically, while all of its elements remain well integrated and coadapted, is geographic isolation.

As I pointed out at the time (74), the evidence supporting this position was meager and equivocal, and it by no means required the conclusions that Mayr insisted upon. Legitimate alternatives existed. His view of speciation was in fact a set of hypotheses resting upon yet other unproven hypotheses. Scientists should not accept opinions uncritically, as many of our contemporaries were doing, but rather should try to show where the truth lies. Biochemical methods were available with which to launch a direct assault upon the problem. After these methods had been exploited to the full we would be nearer to an appreciation of the facts.

At least some of these methods have now been in use for over a decade, and not all of the conclusions from them have been made obsolete by the most recently perceived variability. The genetics of the speciation process remains unsettled, nevertheless. The conclusion of Coyne (20) that previous interspecific comparisons must be reinvestigated is quite proper. Until a thorough reinvestigation is completed we will not *know* whether the situation at the *XDH* locus in *D. pseudoobscura* and *D. persimilis* exposes a general pattern or just a quirk in the data. The heterogeneities among recent results are such as to indicate the latter, but we cannot be certain. The data from the *XDH* locus in the virilis group indicated as much intraspecific divergence as interspecific divergence, suggesting that new information might only shift the mean similarity downward without changing the general pattern. And it also suggested that much of the total variability might be unseen by natural selection. That is, it seemed to have the random distribution that might be expected for neutral alleles or for a mixture of selected and neutral alleles (8). But when Cochrane sought randomness in the distribution of heat sensitive alleles in *D. melanogaster* —randomness of the sort suggested by the data from the virilis group—he did not find it (19).

The data from the *ODH* locus in the virilis group are also pertinent. There one electromorph, common and at high frequency in all species of the group, was discovered to consist of six alleles, one of which was present in all species and apparently predominant in at least one species from each of the phylads. Again, the samples of individual species were small, but with one allele distributed throughout the group it does not seem that speciation necessarily or often completely replaced alleles at this locus. This contrasts with Coyne's data from the obscura group. So, for the present, no common pattern is yet evident, and there is much to indicate that earlier work was not badly misleading.

As documented in the preceeding pages, investigators have been nearly unanimous in agreeing that there is no evidence for extensive reorganization of gene pools during speciation. The basis for that conclusion is easy to see. The concept elicits a vision of speciation that entails adaptive changes at many loci, with these changes precipitating others until a swelling cascade of revision overtakes most loci—a molecular domino effect that leaves little sacred and less untouched. But when the gene pools were assayed, in many different species from all parts of the world, no

DROSOPHILA SYSTEMATICS 249

eivdence of such a cataclysm appeared. True, some siblings may differ at well over 30% of their loci, and that sounds like a lot. But the consequences of high levels of heterozygosity, the observed high levels of intraspecific differentiation in some cases, and the extremely low levels of interspecific differentiation in others, place this in perspective. Against the yardstick of infraspecific differentiation, speciation does not stand out as a special process. And a very strange situation must pertain if it is to appear otherwise when the data from cryptic variation are in. The new variants would have to be predominantly disposed so as to sharpen the differences between species and to minimize those within them, which would imply a peculiar and positive relation between the speciation process and cryptic alleles. Adaptation would then involve chiefly electromorphs while speciation would involve chiefly the cryptic variants within electromorphs. Such a case seems very improbable. Why should we expect extensive reorganization of gene pools to be visible only at the level of variability within electromorphs?

Therefore, while recent work raises uncertainties concerning the details of population genetics and the details of genetic differences between species, it does not negate the conclusions reached to date. These cannot be discarded until strong positive evidence is mustered against them. And while the probability of discovering such evidence is not negligible, it is small. We can expect to discover the dynamics of populations in even more discriminating detail—especially the relationships of selected and nonselected alleles. And we can expect still to be left with no evidence for an extensive reorganization of gene pools and with every indication that changes at a few loci are sufficient for speciation.

MOLECULAR CLOCKS

In principle the discovery of extensive additional variation, variation unassessed in the major studies to date, casts doubt on the reliability of conclusions about molecular clocks and rates of molecular evolution. The doubt is of the same sort raised with respect to the genetics of speciation, and its significance must be questioned on the same grounds. Let us use the data from the virilis group as a case in point. It will be recalled that the two phylads evolved at different rates. The one remaining in its ancestral habitat added fewer alleles than did the one evolving in new environments. If the "within electromorph" alleles are to change this evidence to allow the molecular clock to keep accurate time, they must do it by very nicely countering the differences recorded to date. Most cryptic variants must be added to the conservative phylad's gene pools. But why should populations evolving within ancestral habitats evolve cheifly through "within electromorph" variation while other populations do not? Present evidence increases the uncertainty regarding the accuracy of some estimations of evolutionary rates, but it does not invalidate the general conclusions derived from them. Data from Hawaiian *Drosophila* (36), the willistoni group (7), and the virilis group (32, 71, 72) indicate an inconstancy of evolutionary rates. (The most decisive data are from the virilis group.) New evidence must be abundant and very one-sided if this view is to be changed.

BIOCHEMICAL SYSTEMATICS

The problems of systematics are so diverse that some selection must be made among them. It happens that, aside from speciation, phylogeny is the problem that attracts the interest of most investigators employing protein analysis. Since it most interests me also (66–69), and since protein phylogenies are being promoted quite ostentatiously now, I emphasize phylogeny here. The appropriate questions are, "Can adequate phylogenies be obtained from protein data using present analytical methods?" And, "Why or why not?"

Phylogeny means many things to many people, and I use it in the sense of "genealogy," to refer to the sequence of origin of different taxa during the course of evolution. In studies of the proteins of *Drosophila* two general procedures have been employed to produce "phylogenies." One uses distance measures of some sort combined with one or another of the conventional phenetic clustering methods. The other uses one of the so called "maximum parsimony" methods. The first of these is unsound because, among other things, its validity would depend on the accuracy of molecular clocks. Since molecular clocks do not operate accurately, methods relying on them are unsound. Occasionally, e.g. (6), it is asserted that even though there is some heterogeneity of evolutionary rate, over the long term the average is constant, and hence usable. But even if rates were constant over the long term, so long as they are heterogeneous over the short term they would be unusable for constructing phylogenies. It is a question of what one can and cannot do with averages. One cannot, on the basis of the average height of the United States citizen, predict the height of the next person to enter a room. Neither can one, on the basis of an average evolutionary rate, assert the position, or the length, of any particular internode on a dendrogram. The different sections of a dendrogram are as much individual observations as are the heights of individuals, and they cannot be specified by an average.

The "maximum parsimony" methods are equally unsound, partly because many of them use the inference of constant evolutionary rates as an expedient for "rooting" their trees (30), but mostly because, in fact, they are not parsimonious at all. These undertake to produce trees of minimum length, and they try to arrange the different branches in such a way that the fewest possible character state changes occur over the entire tree or some sections of it (29). The difficulty here, of course, is that evolutionary theory has never proposed that evolution proceeds by a minimum number of steps. At most it requires that changes from one generation to the next be by small steps—and not necessarily by the smallest possible steps. There is nothing whatever in evolutionary theory to suggest that a collection of species should have evolved one from the other, or from annectant ancestors, over the shortest possible pathway. These methods, far from being maximally parsimonious, are completely nonparsimonious. They, and the genetic distance methods, force sets of data to conform to preset bias, very much in the tradition of the Biogenetic Law and Orthogenesis. It is not surprising that while some *Drosophila* studies (32, 36, 57, 71, 72) find protein "phylogenies" incompatible with available information, some (47) find that they match rather well. Protein "phylogenies" may or may not approximate the real genealogy for the species. But they are only as sound as the analyses that produce them. And since there is no guarantee that they must match genealogy, or even that there is good reason to think they will, they have only limited usefulness.

CONCLUSION

Three topics have been dealt with, two with dispatch and one more thoroughly. The *Drosophila* literature on the genetics of speciation, including the most recent disclosures of unanticipated variation, provides no evidence for an extensive reorganization of gene pools during speciation. Hence speciation mechanisms involving only minimal genetic change are not precluded on those grounds. This evidence also contradicts widespread faith in molecular clocks and in the usual analytic methods for deriving "phylogenies" from proteins.

ACKNOWLEDGMENT

This work was supported in part by grants GM 11216 and GM 23007 from the National Institutes of Health.

Literature Cited

1. Avise, J. D. 1976. Genetic differentiation during speciation. In *Molecular Evolution,* ed. F. J. Ayala, pp. 106–22. Sunderland, Mass: Sinauer. 277 pp.
2. Ayala, F. J. 1975. Genetic differentiation during the speciation process. In *Evolutionary Biology,* ed. Th. Dobzhansky, M. K. Hecht, W. C. Steere, 8:1–78. New York: Plenum. 396 pp.
3. Ayala, F. J., Mourão, C. A., Pérez-Salas, S., Richmond, R., Dobzhansky, Th. 1970. Enzyme variability in the *Drosophila willistoni* group, I. Genetic differentiation among sibling species. *Proc. Natl. Acad. Sci. USA* 67:225–32
4. Ayala, F. J., Dobzhansky, Th. 1974. A new subspecies of *Drosophila pseudoobscura* (Diptera: Drosophilidae). *Pan-Pacific Entomol.* 50:211–19
5. Ayala, F. J., Tracey, M. L. 1973. Enzyme variability in the *Drosophila willistoni* group. VIII. Genetic differentiation and reproductive isolation between two subspecies. *J. Hered.* 64:120–24
6. Ayala, F. J., Tracey, M. L., Barr, L. G., Ehrenfeld, J. G. 1974. Genetic and reproductive differentiation of the subspecies, *Drosophila equinoxialis caribbensis. Evolution* 28:24–41
7. Ayala, F. J., Tracey, M. L., Hedgecock, D., Richmond, R. C. 1975. Genetic differentiation during the speciation process in *Drosophila. Evolution* 28:576–92
8. Bernstein, S. C., Throckmorton, L. H., Hubby, J. L. 1973. Still more genetic variability in natural populations. *Proc. Natl. Acad. Sci. USA* 70:3928–31
9. Carmody, G., Collazo, D. A., Dobzhansky, Th., Ehrman, L., Jaffrey, I. S., Kimball, S., Obreebski, S., Silagi, S., Tidwell, T., Ulrich, R. 1962. Mating preferences and sexual isolation within and between the incipient species of *Drosophila paulistorum. Amer. Midl. Nat.* 68:67–82
10. Carson, H. L. 1968. The population flush and its genetic consequences. In *Population Biology and Evolution,* ed. R. C. Lewontin, pp. 123–37. Syracuse, NY: Syracuse Univ. Press. 205 pp.
11. Carson, H. L. 1973. Reorganization of the gene pool during speciation. In *Genetic Structure of Populations,* ed. N. E. Morton, 3:274–80. Honolulu: Univ. Press of Hawaii
12. Carson, H. L. 1975. The genetics of speciation at the diploid level. *Am. Nat.* 109:83–92
13. Carson, H. L., Johnson, W. E. 1975. Genetic variation in Hawaiian *Drosophila.* I. Chromosome and allozyme polymorphism in *D. setosimentum* and *D. ochrobasis* from the island of Hawaii. *Evolution* 29:11–23

14. Carson, H. L., Kaneshiro, K. Y. 1976. *Drosophila* of Hawaii: Systematics and ecological genetics. *Ann. Rev. Ecol. Syst.* 7:311–45
15. Carson, H. L., Clayton, F. E., Stalker, H. D. 1967. Karyotypic stability and speciation in Hawaiian *Drosophila*. *Proc. Natl. Acad. Sci. USA* 57:1280–85
16. Carson, H. L., Nair, P. S., Sene, F. M. 1975. *Drosophila* hybrids in nature: proof of gene exchange between sympatric species. *Science* 189:806–7
17. Carson, H. L., Hardy, D. E., Spieth, H. T., Stone, W. S. 1970. The evolutionary biology of the Hawaiian Drosophilidae. In *Essays in Evolution and Genetics in Honor of Theodosius Dobzhansky*, ed. M. K. Hecht, W. C. Steere, pp. 437–543. New York: Appleton Century Crofts
18. Carson, H. L., Johnson, W. E., Nair, P. S., Sene, F. M. 1975. Allozyme and chromosomal similarity in two *Drosophila* species. *Proc. Natl. Acad. Sci. USA* 72:4521–25
19. Cochrane, B. J. 1976. Heat stability variants of esterase-6 in *Drosophila melanogaster*. *Nature* 263:131–32
20. Coyne, J. A. 1976. Lack of genic similarity between two sibling species of *Drosophila* as revealed by varied techniques. *Genetics* 84:593–607
21. Dobzhansky, Th. 1974. Genetic analysis of hybrid sterility within the species *Drosophila pseudoobscura*. *Hereditas* 77:81–88
22. Dobzhansky, Th., Pavlovsky, O. 1967. Experiments on the incipient species of the *Drosophila paulistorum* complex. *Genetics* 55:141–56
23. Dobzhansky, Th., Powell, J. R. 1975. *Drosophila pseudoobscura* and its American relatives, *Drosophila persimilis* and *Drosophila miranda*. In *Handbook of Genetics*, ed. R. C. King, 3:537–87. New York: Plenum. 874 pp.
24. Dobzhansky, Th., Powell, J. R. 1975. The *willistoni* group of sibling species of *Drosophila*. See Ref. 23, 3:589–622
25. Dobzhansky, Th., Powell, J. R. 1975. See Ref. 23, p. 594
26. Dobzhansky, Th., Spassky, B. 1959. *Drosophila paulistorum*, a cluster of species *in statu nascendi*. *Proc. Natl. Acad. Sci. USA* 45:419–28
27. Dobzhansky, Th., Hunter, A. S., Pavlovsky, O., Spassky, B., Wallace, B. 1963. Genetics of natural populations. XXXI. Genetics of an isolated marginal population of *Drosophila pseudoobscura*. *Genetics* 48:91–103
28. Ehrman, L. 1965. Direct observation of sexual isolation between allopatric and between sympatric strains of different *Drosophila paulistorum* races. *Evolution* 19:459–64
29. Farris, J. S. 1972. Estimating phylogenetic trees from distance matrices. *Am. Nat.* 106:645–68
30. Farris, J. S. 1972. See Ref. 29, p. 657
31. Hubby, J. L., Lewontin, R. C. 1966. A molecular approach to the study of genic heterozygosity in natural populations. I. The number of alleles at different loci in *Drosophila pseudoobscura*. *Genetics* 54:577–94
32. Hubby, J. L., Throckmorton, L. H. 1965. Protein differences in *Drosophila*. II. Comparative species genetics and evolutionary problems. *Genetics* 52:203–15
33. Hubby, J. L., Throckmorton, L. H. 1968. Protein differences in *Drosophila*. IV. A study of sibling species. *Am. Nat.* 102:193–205
34. Johnson, D. E. 1975. Genetic differentiation in two members of the *Drosophila athabasca* complex. PhD thesis. Univ. Chicago, Chicago. 118 pp.
35. Johnson, G. B. 1973. Enzyme polymorphism and biosystematics: the hypothesis of selective neutrality. *Ann. Rev. Ecol. Syst.* 4:93–116
36. Johnson, W. E., Carson, H. L., Kaneshiro, K. Y., Steiner, W. W. M., Cooper, M. M. 1975. Genetic variation in Hawaiian *Drosophila*. II. Allozymic differentiation in the *D. planitibia* subgroup. In *Isozymes*, ed. C. L. Markert, 4:563–83. New York: Academic. 964 pp.
37. Kanapi, C. G., Wheeler, M. R. 1970. Comparative isozyme patterns in three species of the *Drosophila nasuta* complex. *Texas Rep. Biol. Med.* 28:261–78
38. Kastritsis, C. D. 1967. A comparative study of the chromosomal polymorphs in the incipient species of the *Drosophila paulistorum* complex. *Chromosoma* 23:180–202
39. Kastritsis, C. D. 1969. A cytological study on some recently collected strains of *Drosophila paulistorum*. *Evolution* 23:663–75
40. Kojima, K. I., Gillespie, J., Tobari, Y. N. 1970. A profile of *Drosophila* species' enzymes assayed by electrophoresis. I. Number of alleles, heterozygosities, and linkage disequilibrium of glucose-metabolizing systems and some other enzymes. *Biochem. Genet.* 4:627–37

41. Lakovaara, S., Saura, A., Falk, C. T. 1971. Genetic distance and evolutionary relationships in the *Drosophila obscura* group. *Evolution* 26:177–84
42. Lakovaara, S., Saura, A., Lankinen, P., Pohjola, L., Lokki, J. 1976. The use of isoenzymes in tracing evolution and in classifying *Drosophila. Zool. Scr.* 5: 173–79
43. Lewontin, R. C. 1974. *The Genetic Basis of Evolutionary Change.* New York: Columbia. 346 pp.
44. Mayr, E. 1963. *Animal Species and Evolution.* Cambridge: Belknap. 797 pp.
45. McDowell, R. E., Prakash, S. 1976. Allelic heterogeneity within allozymes separated by electrophoresis in *Drosophila pseudoobscura. Proc. Natl. Acad. Sci. USA* 73:4150–53
46. Miller, D. D., Westphal, R. J. 1967. Further evidence on sexual isolation within *Drosophila athabasca. Evolution* 21:479–92
47. Nair, P. S., Brncic, D., Kojima, K. 1971. Isozyme variations and evolutionary relationships in the mesophragmatica species group of *Drosophila. Univ. Tex. Publ.* 7103:17–28
48. Nei, M. 1971. Interspecific gene differences and evolutionary time estimated from electrophoretic data on protein identity. *Am. Nat.* 105:385–98
49. Nei, M. 1972. Genetic distance between populations. *Am. Nat.* 106:283–92
50. Nei, M. 1975. *Molecular Population Genetics and Evolution.* Amsterdam: North-Holland. 288 pp.
51. Pasteur, G. 1970. A biometrical study of the semispecies of the *Drosophila paulistorum* complex. *Evolution* 24: 156–68
52. Powell, J. R. 1975. Protein variation in natural populations of animals. In *Evolutionary Biology,* ed. Th. Dobzhansky, M. K. Hecht, W. C. Steere, 8:79–119. New York: Plenum. 396 pp.
53. Prakash, S. 1969. Genic variation in a natural population of *Drosophila persimilis. Proc. Natl. Acad. Sci. USA* 62:778–84
54. Prakash, S. 1972. Origin of reproductive isolation in the absence of apparent genic differentiation in a geographic isolate of *Drosophila pseudoobscura. Genetics* 72:143–55
55. Prakash, S., Lewontin, R. C., Hubby, J. L. 1969. A molecular approach to the study of genic heterozygosity in natural populations. IV. Patterns of genic variation in central, marginal, and isolated populations of *Drosophila pseudoobscura. Genetics* 61:841–58
56. Richardson, R. H., Smouse, P. E. 1976. Patterns of molecular variation. I. Interspecific comparisons of electromorphs in the *Drosophila mulleri* complex. *Biochem. Genet.* 14:447–66
57. Richardson, R. H., Richardson, M. E., Smouse, P. E. 1975. Evolution of electrophoretic mobility in the *Drosophila mulleri* complex. In *Isozymes,* ed. C. L. Markert, 4:533–45. New York: Academic
58. Richardson, R. H., Smouse, P. E., Richardson, M. E. 1977. Patterns of molecular variation. II. Associations of electrophoretic mobility and larval substrate within species of the *Drosophila mulleri* complex. *Genetics* 85:141–54
59. Rockwood, E. S. 1969. Enzyme variation in natural populations of *Drosophila mimica. Univ. Tex. Publ.* 6918: 111–32
60. Rockwood, E. S., Kanapi, C. G., Wheeler, M. R., Stone, W. S. 1971. Allozyme changes during the evolution of Hawaiian *Drosophila. Univ. Tex. Publ.* 7103:193–212
61. Rogers, J. S. 1972. Measures of genetic similarity and genetic distance. *Univ. Tex. Publ.* 7213:145–53
62. Selander, R. K. 1976. Genic variation in natural populations. In *Molecular Evolution,* ed. F. J. Ayala, pp. 21–45. Sunderland, Mass: Sinauer. 277 pp.
63. Singh, R. S., Hubby, J. L., Throckmorton, L. H. 1975. The study of genic variation by electrophoretic and heat denaturation techniques at the octanol dehydrogenase locus in members of the *Drosophila virilis* group. *Genetics* 80: 637–50
64. Singh, R. S., Lewontin, R. C., Felton, A. A. 1976. Genetic heterogeneity within electrophoretic "alleles" of xanthine dehydrogenase in *Drosophila pseudoobscura. Genetics* 64:609–29
65. Spassky, B., Richmond, R. C., Pérez-Salas, S., Pavlovsky, O., Mourão, C. A., Hunter, A. S., Hoenigsberg, H., Dobzhansky, Th., Ayala, F. J. 1971. Geography of the sibling species related to *Drosophila willistoni* and of the semispecies of the *Drosophila paulistorum* complex. *Evolution* 25:129–43
66. Throckmorton, L. H. 1962. The problem of phylogeny in the genus *Drosophila. Univ. Tex. Publ.* 6205:207–343
67. Throckmorton, L. H. 1965. Similarity versus relationship in *Drosophila. Syst. Zool.* 14:221–36

68. Throckmorton, L. H. 1968. Biochemistry and taxonomy. *Ann. Rev. Entomol.* 13:99–114
69. Throckmorton, L. H. 1968. Concordance and discordance of taxonomic characters in *Drosophila* classification. *Syst. Zool.* 17:355–87
70. Throckmorton, L. H. 1969. Discussion. Molecular data in animal systematics. In *Systematic Biology,* pp. 360–63. Washington DC: Natl. Acad. Sci. Publ. No. 1692
71. Throckmorton, L. H. 1972. *Genetic change and speciation in the virilis group of Drosophila.* Presented at 14th Int. Congr. Entomol. Canberra
72. Throckmorton, L. H. 1972. *Abstr. 14th Int. Congr. Entomol.*, p. 16 (Abstr.)
73. Throckmorton, L. H. 1975. The phylogeny, ecology, and geography of *Drosophila.* In *Handbook of Genetics,* ed. R. C. King, 3:421–69. New York: Plenum. 874 pp.
74. Throckmorton, L. H., Hubby, J. L. 1963. Toward a modern synthesis of evolutionary thought. *Science* 140: 628–31
75. Wasserman, M. 1963. Cytology and phylogeny of *Drosophila. Am. Nat.* 97:333–52
76. Winge, H. 1965. Interspecific hybridization between the six cryptic species of *Drosophila willistoni* group. *Heredity* 20:9–19
77. Winge, H., Cordeiro, A. R. 1963. Experimental hybrids between *Drosophila willistoni* Sturtevant and *Drosophila paulistorum* Dobzhansky and Pavan from southern marginal populations. *Heredity* 18:215–22
78. Yang, S. Y., Wheeler, L. L., Bock, I. R. 1972. Isozyme variations and phylogenetic relationships in the *Drosophila bipectinata* species complex. *Univ. Tex. Publ.* 7213:213–27
79. Zouros, E. 1974. Genic differentiation associated with the early stages of speciation in the *mulleri* subgroup of *Drosophila. Evolution* 27:601–21

ENVIRONMENTAL EFFECTS OF DAMS AND IMPOUNDMENTS

❖4125

R. M. Baxter

Canada Centre for Inland Waters, Applied Research Division, Burlington, Ontario, Canada

INTRODUCTION

Most of the primary civilizations of the world emerged in or near river valleys. The construction of dams and other hydraulic structures is, therefore, one of the oldest branches of engineering (e.g. 11, 105).

The earliest dams were probably built for the purposes of irrigation, flood control, and water supply. Later, water was impounded so that its subsequent controlled release could provide a source of energy, first by the use of waterwheels and later by the use of hydroelectric generators. Other purposes include the maintenance of an adequate river flow through the year for navigation, and the provision of facilities for recreation. Most modern reservoirs are designed for two or more of these purposes. Usually "the role of water storage reservoirs ... is to impound water in periods of higher flows so that it may be released gradually during periods of lower flows" (135), but sometimes the sole purpose of the impoundment is to provide a new body of standing water for use as such; for example, for fishing or boating, or for waste-heat dissipation from a thermoelectric generating plant.

The earliest dams were probably constructed by blocking the stream with earth. Such dams are still constructed. In its simplest form, an earth-fill dam is a pile of compacted earth extending across a stream with a fairly gentle slope both upstream and downstream. Similar to earth-fill dams are rock-fill dams composed of quarried rock or natural boulders or gravel with a layer of impervious material on the upstream face.

A later development in dam construction was the invention of the masonry dam, probably in Spain (150). The earliest masonry dams were of the gravity type. Such dams are held in place by their own weight pressing against their foundations, and usually have a long sloping downstream toe to prevent overtipping.

Arched masonry dams are convex upstream with their ends abutting against the walls of the stream valley or against suitable artificial structures. Such dams are held in place by thrust against their abutments, and are steeply sloping or vertical both upstream and downstream. Dams may also be built of structural masonry, or less commonly, of steel or timber.

In almost all dams, provision must be made for the safe discharge of excess water by the construction of a spillway, and in most instances, provision is made to control the rate of discharge. In dams constructed for hydroelectric purposes, there is also a discharge of water through a penstock to the turbines, and dams for irrigation or water supply must have outlets to the irrigation canals or water supply system. Penstocks and other outlets may draw water from well below the surface of the water in the reservoir.

The water level in a reservoir is usually manipulated according to the purpose for which it was built within the constraints imposed by the prevailing climatic conditions. Reservoirs built solely for flood control are usually emptied as soon as possible after a flood. Reservoirs for other purposes, however, are usually filled at times of high flow and either drawn down gradually during the rest of the year, or maintained for several months at a fairly constant level, then drawn down rather quickly in anticipation of the next period of high flow (143).

The amount of water impounded by a dam differs greatly from one site to another. In run-of-the-river hydroelectric projects, the purpose of the dam is essentially to direct and control the flow of the stream and little water is impounded. At the other extreme, a major new lake may be formed with a retention time measurable in years.

Dams may be built on streams that leave natural lakes so that the level of these becomes subject to some degree of regulation and they may be regarded as reservoirs. The effect on large lakes is slight. The conversion of smaller lakes to reservoirs may have very considerable effects on their ecology (94).

Most commonly, impounded water is retained directly behind the dam in an on-channel or mainstream reservoir, but occasionally water may be diverted from the stream to a suitable natural or artificial basin to form an off-channel reservoir.

A special form of hydroelectrical generating station of increasing importance is the pumped storage plant (4). In such a plant, some of the generated electrical power is used to pump water to a higher level to be used to generate electricity at a later time. Essentially, this is a means of storing energy. The reservoirs used in such projects are usually small.

Man is, of course, not the only dam-building animal. The dams of beavers, although smaller than many of those made by man, can cause spectacular changes in certain areas. Streams may also become dammed temporarily or permanently by various accidents, such as clogging with masses of floating vegetation or obstruction by landslides or lava flows. Lake Tana in the Ethiopian Highlands, the source of the Blue Nile, is an important example of a lake formed by the damming of a stream by the flow of lava (19).

The most dramatic environmental effects are those associated with the very large impoundments, especially perhaps those in Africa: Lake Volta, in Ghana, the largest

DAMS AND IMPOUNDMENTS   257

in the world; Lake Nasser on the Nile, in Egypt and the Sudan; Lake Kariba on the Zambezi, between Zambia and Rhodesia; and the newest, also on the Zambezi, behind the Cobora Bassa Dam in Mozambique (56, 74). The many large reservoirs in the USSR, such as the Kuybyshev, Rybinsk, and Gorkii Reservoirs on the Volga, have received a great deal of attention. One South American reservoir, Lake Brokopondo in Surinam, has been studied by ecologists (90, 91). The largest North American reservoir is the Smallwood Reservoir in Labrador, Canada. The James Bay Hydroelectric Project in Quebec, Canada, will involve the construction of several large reservoirs (23, 48, 76, 137). In the United States, the best known is probably Lake Mead, on the Colorado River, although at least two others cover larger areas: the Garrison Reservoir and the Oahe Reservoir, both on the Missouri River. A more extensive list of large reservoirs of the world has been compiled (34).

Many studies have also been made on small reservoirs, such as those under the jurisdiction of the Tennessee Valley Authority (e.g. 172). Such countries as Rumania (40) and Spain (104) are dotted with small reservoirs. The study of reservoirs in Spain and in Czechoslovakia (64, 66, 67) has been a major activity of the limnologists of those countries in recent years.

The control of water for the benefit of man has been traditionally regarded as a particularly noble activity. According to one interpretation of Goethe's *Faust, Part Two*, it was Faust's hydrological endeavors that ultimately saved him from damnation. In our own time, such projects as the Tennessee Valley Authority have been rightly regarded as marvellous accomplishments. At the same time, the possible dangers of manipulating the flow of water have been recognized in moral and legal codes; the Negative Confession in the Egyptian Book of the Dead (16) contains the words "I have not turned back water at its springtide. I have not broken the channel of running water." The traditional Ethiopian law code, the Fetha Negast, recognizes the duty of people in the highlands to let waters flow to people in the lowlands (35). Doubtless, many more such examples could be found. Large modern projects have made provision to avoid at least the most obvious objectionable side effects; for example, the operating schedule of the TVA reservoirs includes periodic fluctuations in the water level to destroy the larvae of malaria-transmitting mosquitoes (143).

An attitude of optimism and enthusiasm still survives; it has recently been said (162) that "dams are the only solution to Man's angst." This attitude, however, is by no means universal. In recent years some very large projects have been undertaken to produce man-made lakes approaching in size all but the biggest of natural lakes. Some of these projects have had effects that were not anticipated or were ignored by their planners, and these have given rise to widespread concern. The literature on reservoir ecology is extensive, and is scattered through journals devoted to a wide range of disciplines, reports from various agencies, and books. Entry to this literature has been greatly facilitated by previous reviews (32, 41, 94, 106, 113, 138) and by the published reports of several symposia (1, 3, 33, 96, 118, 121). Translations of a number of Russian works are available (146, 147, 166, 175).

It has been pointed out more than once that a large impoundment may be regarded as a large-scale ecological experiment that may reveal or at least illustrate certain general ecological principles. The emphasis in this review is on the peculiarities of reservoirs and how they differ from natural lakes. Much excellent work on reservoirs has advanced our general knowledge of limnology rather than our specific knowledge of reservoirs. For the most part, I have not referred to this.

MORPHOLOGY AND PHYSICAL AND CHEMICAL LIMNOLOGY OF MAN-MADE LAKES

General Morphological Characteristics of Reservoirs

When a new lake is formed by damming a stream, its shape will usually be different from that of most natural lakes (68). The shoreline development (the ratio of the length of the shoreline to the circumference of a circle of the same area as the lake) is almost always higher than for natural lakes. If the river is confined between high banks, the new lake will be long and narrow; if there are tributaries, the water will back up into them, giving the new lake a dendritic form. If the river flows through flat terrain and has low banks, the water will spread over the plain forming what Russian scientists (175) call a lobed reservoir. The Manicouagan Reservoir in Quebec is ring-shaped, having been constructed by blocking the outlet of a meteorite crater with an elevated center (79, 148). It has, perhaps, the greatest insulosity of any large body of water in the world. When a reservoir is formed by damming the outlet of a lake, the change in morphology is less pronounced; but here too, the new shoreline will be more developed than the old. Off-channel reservoirs, on the other hand, being often completely artificial constructions, may be unusually regular in shape.

Reservoirs formed by damming rivers frequently differ from natural lakes in the shape of their longitudinal profiles. Whereas natural lakes are normally deepest somewhere near the middle, river reservoirs are almost always deepest just upstream from the dam. They have been referred to as "half-lakes" (32). One consequence of this is that surface currents passing down the reservoir will not dissipate themselves in shallow water as they do in natural lakes, but may be deflected downward or reflected backward at the dam.

The new shoreline, a shoreline of submergence in Johnson's (77) terminology, will not persist. It is acted upon by waves, by currents, and in cold regions, by ice; it is modified at a rate depending on the energy available, the shape of the reservoir and the surrounding terrain, and the resistance of the material. Ultimately the development of the shoreline is reduced, although in the earlier stages it may increase (159).

The extent of shoreline modification in a reservoir is likely to be greater than in a comparable natural lake because the annual drawdown exposes a larger area to the effects of shore processes.

Empirical equations exist (81, 85, 95) for calculating the available wave energy and the amount of material that will be eroded, but attempts to predict the changes that will occur have not been very successful (25).

In high latitudes where the inundated area may consist of muskeg overlying permafrost, shoreline changes may be particularly extensive and prolonged as the

DAMS AND IMPOUNDMENTS 259

permafrost melts. The time required to establish a new shoreline of moderate permanence under such conditions may be more than 30 years (115).

Although the modification of the shorelines of reservoirs is usually a fairly gradual process, on at least one occasion it has been catastrophic. In October 1963, a large landslide fell into the newly filled reservoir behind the Vaiont Dam in Italy, spilling an equal volume of water over the dam into the valley below with the loss of more than 2000 lives. This disaster was probably due in part to seismic activity, perhaps induced by the filling of the reservoir (141).

Sedimentation

Most of the material eroded from the banks of the reservoir will probably be deposited elsewhere within it. How it is deposited will depend on the morphometry of the lake and the currents in it. If the bottom has a low slope and there are strong currents along the shore, the material may be deposited near the shore, forming spits or sand bars. If the bottom slope is high, and nearshore currents are weak as, for example, in Lake Diefenbaker on the South Saskatchewan River in Canada (22), the material will be transported to the deeper parts of the lake and deposited there.

If the reservoir is on a river that carries a heavy load of sediment, this material will also contribute to the modification of the morphology of the lake. The sediment load of a stream may be produced either by sheet erosion of the surrounding land or by erosion of the banks of the stream itself or its tributaries. Of these, sheet erosion is usually the more important. Its extent depends on rainfall, slope, soil, and the vegetational cover; it may be greatly modified by land use.

Material carried by a stream may be divided into relatively coarse material pushed or rolled along the bottom (bed load) and finer, suspended sediment. When a stream carrying such material enters a body of standing water and its flow rate decreases, the material may be deposited, forming a delta. The coarsest material is dropped first, forming the foreset bed. As the delta grows, the rate of flow of water over it is further reduced, and material is deposited on top of it, forming a topset bed. Finer material may be deposited beyond the leading edge of the growing delta as a bottomset bed. Delta formation in reservoirs is complicated by the practice of drawdown. Material deposited when the water is high may be eroded away again when the level drops. Consequently, the shape and position of the delta are constantly changing, and the nature of these changes is not easy to predict.

If the gradient in the river is not very steep, the deposition of material will not be confined to the reservoir itself, but may extend upstream, sometimes for many miles. This leads to reduction in the capacity of the river channel and a consequent danger of flooding at high flows.

Physical Limnology

The retention time of water in many reservoirs is fairly short, so that its circulation is likely to be dominated by near-field effects, i.e. those associated with the inflow and outflow, rather than the far-field effects (thermal circulation and currents generated by winds) that are dominant in most natural lakes. The situation in reservoirs is rendered still more complex by the fact that the discharge is frequently effected from a point at some depth below the surface. This depth may be varied. Discharge may be effected from more than one level at the same time. If the water in the reservoir is stratified, the outgoing water will be drawn from a relatively thin layer at the level of the outlet, so that internal currents (withdrawal currents) are set up and little vertical movement occurs (174).

The water entering a reservoir frequently differs from the water already present in temperature, or in content of dissolved or suspended solids, or in some combination of these, and consequently, in density. The incoming water does not then mix immediately with the water of the reservoir, but moves downstream and laterally above, below, or within it, as an overflow, underflow, or interflow. Such flows are referred to as density currents. A convergence line often appears where the inflowing water plunges below the surface. A compensating upstream current is generated, carrying back debris which is immobilized where the two currents meet, making a convergence line visible even if its position has not already been revealed by a difference in turbidity between the inflowing and the surface water. The first density currents to be observed in a reservoir were apparently those in the Norris Reservoir, the largest of the storage reservoirs of the Tennessee Valley Authority (171).

When the inflowing water owes its greater density in whole or in part to suspended material, the resulting underflow is known as a turbidity current. The difference in density may be only a few percent; or the water may be so loaded with silt as to form a virtual slurry, so that the distinction between a turbidity current and a sliding mudflow may be almost obscured. The differences have been discussed by Kuenen (87). Turbidity currents caused by very dense suspensions are important in the sea (78) but in lakes and reservoirs, the density difference is usually rather small. Formulas exist for the calculation of the flow rates of turbidity currents (151).

The best known turbidity currents are those in Lake Mead which may extend for the whole length of the lake—about 160 km (51). Turbidity currents are important not only for their contribution to the flow patterns within a reservoir, but because they can carry silt for a long distance into it, depositing some of it on the way, and so contribute to the formation of bottomset deposits.

Slapy Reservoir on the Vltava River in Czechoslovakia provides a good example of the complexity of the flow and stratification patterns in reservoirs of relatively short retention time (65). This reservoir is 42 km long and 313 m in mean width and has a maximum depth of 53 m. Water is withdrawn from it at a depth of 35 m. The filling time at the mean flow rate of the river is 38.5 days. From February or March until August, the outflow water is cooler than the inflow, and during the rest of the year, the inflow is cooler. Through most of the year, the surface water is warmer and less dense than either the inflowing or outflowing water, so that the inflowing water plunges beneath the surface, mixing, to some degree, with hypolimnic water but not to any great extent with surface water. The consequence is that the renewal time for surface water is very considerably higher than the mean renewal time for the reservoir as a whole, the reverse of the usual situation in natural lakes. This tendency of the inflowing water to pass under the epilimnic water has been accentuated by the construction upstream of another reservoir (the Orlik Reservoir), the cool outflowing water of which becomes the inflow of Slapy.

DAMS AND IMPOUNDMENTS   261

Slapy Reservoir also displays a degree of horizontal zonation not usually found in natural lakes. The upstream region of relatively high flow has a length of 10–15 km and merges through a transition zone of 5–10 km with a relatively stagnant section of about 20 km. Finally, there is a zone of about 5 km where the reservoir is directly influenced by the turbines. The biological consequences of this horizontal zonation are discussed below.

If a reservoir is formed in a wooded area and the trees are not removed before flooding, they may have a significant effect on the circulation pattern of the reservoir. In the Nam Ngum Reservoir in Laos, standing trees almost completely abolish wind-generated turbulence and consequently reduce the extent of mixing. This gives rise to a complex pattern of stratification with a surface layer of warm water and an intermediate cooler layer from which water is drawn to the turbines, and a still cooler bottom layer apparently consisting of water that has been brought in by streams from the mountain during the coolest part of the year (W. M. Lewis, in preparation). Standing trees also interfere with mixing in the Brokopondo Lake in Surinam (90).

The exposed parts of flooded trees are quickly destroyed by wind and waves combined with biological action. The fully submerged parts, however, persist for a long time. In the Gouin Reservoir in Quebec (latitude 48 degrees N), the wood of submerged trees showed little change 55 years after the reservoir had been filled (28). In warmer climates, the process of decomposition is presumably more rapid; but even there, submerged trees may be expected to persist for many years.

Chemical Limnology

The chemical composition of the water of natural lakes and mature man-made lakes, particularly the concentrations of conservative constituents, is largely determined by the chemistry of inflows and of precipitation into it. However, in new impoundments it may be influenced by the leaching of soluble material from the flooded ground. If the amount of soluble material is small and the retention time of the water is short, this effect may be brief. If the amount of soluble material is large, as in Lake Mead (63), and especially if the retention time is long, the chemistry of the impounded water may continue to differ from that of the inflow for a long time. The presence of density currents may lead to unusual and complex patterns of chemical stratification if the chemical composition of the inflowing water is different from that of the water already present in the reservoir (152, 153, 171).

The decomposition of submerged vegetation often leads to a depletion of oxygen in the depths of the reservoir. The peculiar profile of most reservoirs, as compared with natural lakes, may permit the accumulation of a mass of stagnant water in the deepest part against the dam (165). This bottom layer can become anoxic (36), and reduced substances such as sulfide, ferrous, and manganous ions may accumulate. Nutrient substances may be leached from the underlying soil or released by the decomposition of submerged vegetation. The biological consequences of this are discussed in the next section.

Impoundment by itself usually improves the quality of water for domestic and industrial use (134); suspended solids have an opportunity to settle, the bacterial population decreases, the dissolved oxygen concentration may increase, and the color of the water may decrease [probably by photo-oxidation of the dissolved humic substances (155)]. In a new reservoir, however, many of these effects may be more than balanced by the depletion of dissolved oxygen and the leaching of humic substances from the soil (46). As the reservoir matures, the quality of the water gradually improves. To hasten improvement, it has sometimes been the practice to strip the area to be flooded (i.e. remove the vegetation and top soil) before the reservoir is filled (155). The usefulness of this practice depends among other things on the nature of the soil, climatic conditions, and the retention time of the reservoir. In a group of reservoirs in New England, water over the unstripped soil attained normal conditions in an average of six years (145). In experiments carried out in southeastern Quebec using several types of soil in small basins, the concentrations of most constituents over unstripped soils approached those over stripped soils in a year or so (18).

Often the construction of dams on a river is only one aspect of increasing industrial activity in a region; to the effects of impoundment are thus added the effects of domestic and industrial effluents. In the Saint John River in New Brunswick, Canada, the construction of dams has reduced the rate of flow of the river, and consequently, its ability to become oxygenated; and at about the same time, the establishment of industries along the river has contributed effluents of high biochemical oxygen demand (29).

The flooding of previously dry ground may lead to the release into the water of toxic substances there either naturally or as a result of human activity. The alteration of the pattern of erosion and sedimentation may lead to the release of pollutants (such as mercury) which are known to accumulate in sediments.

BIOLOGY OF RESERVOIR ECOSYSTEMS

Some General Principles

An ecosystem consists of a biological community and the habitat in which it lives. The two interact so that both change. The formation of a new lake provides a new habitat for the development of a new ecosystem. This section considers the biological factors that come into play when a new lake is created, and traces the way the biological community develops.

A number of general biological principles have been discerned by ecologists which make it possible to predict, in its broad outlines, the course of development of a new ecosystem and to provide a general conceptual framework for interpreting the observed events. The epistemological status of these principles is a matter of question (127). At least, however, they are helpful in organizing and systemizing large numbers of diverse observations. Among the most useful of these are the following:

THIENEMANN'S RULES 1. The greater the diversity of conditions in a locality, the larger the number of species in the biological community. 2. The more conditions in a locality deviate from the normal, and thus from the optimum for most species, the smaller the number of species and the greater the biomass of each. 3. The longer a locality has been in the same condition, the richer its biological community (160).

THE CONCEPT OF SUCCESSION Students of terrestrial botany have long recognized that the development of a new plant community follows a recognizable pattern (e.g. 84). This concept has been generalized and refined by Odum (119, 120) with an emphasis on certain characteristic changes that an ecosystem exhibits as it matures. Among the most important are the following: A balance is approached between gross production and community respiration (in a new ecosystem either may exceed the other); there is an increase in species diversity and spatial heterogeneity, in the complexity of food chains, and in orderliness (a decrease in entropy, an increase in information content).

THE CONCEPT OF PULSE STABILITY This concept, also Odum's (119, 120), argues for the maintenance of an ecosystem in a relatively immature state by the periodic or quasi-periodic imposition of a physical perturbation. In the present context, periodic flooding is the perturbation of interest. Many of the most dramatic ecological effects of impoundment can be interpreted as results of the interference with previously existing pulse-stabilized ecosystems, or the establishment of new ones.

THE CONCEPT OF THE ECOTONE As the concept of pulse stability concerns the ecological consequence of variability in time, so the concept of the ecotone comprehends the consequences of variability in space. An ecotone is a transition between two communities or environments. It is usually narrow and characterized by a greater diversity of species than the communities or environments on either side. The littoral zone of a lake or river may be regarded as an ecotone (102, 120).

THE THEORY OF ISLAND BIOGEOGRAPHY A lake is in some ways the aquatic equivalent of an island (83). The concepts of island ecology (98) are of value in understanding the ecology of lakes, especially the colonization of new ones. Specifically, the distinction between K-selection and r-selection first proposed by MacArthur & Wilson (98) and subsequently developed by a number of authors (e.g. 131) appears useful. The terms K and r refer to the parameters of the logistic equation: $N = K/(1+e^{a-rt})$, which generally describes the growth of populations. N is the number of individuals at time t, and the parameters K and r are respectively the maximum population possible in the given habitat and the specific growth rate of the species in question. K-selected species have a high capacity for maintaining their numbers under competitive conditions, whereas r-selected species can increase their numbers rapidly. New ecosystems are usually colonized by r-selected species.

These concepts are not independent. The high diversity in spatially variable environments to which Thienemann's first rule refers could be attributed, in part, to the large number of ecotones in such environments. Poorly diversified yet relatively productive ecosystems such as hot springs (15), highly saline waters (173), or soda lakes (158) to which the second rule applies may be considered immature ecosystems that are unable to develop because the few species able to live in them cannot bring about any modification of their physical and chemical characteristics. Thienemann's third rule calls attention to one of the most prominent features of the process of maturation of ecosystems.

DIFFERENCES BETWEEN STREAMS AND LAKES When a stream is dammed, the new lake provides a very different habitat from that provided by the stream (71). Because flow is always turbulent in a stream, thermal stratification rarely develops and oxygen depletion rarely occurs unless the stream is grossly polluted. Because floating organisms are being swept away continually, the population of plankton is low, whereas the benthic population may be high. The benthic organisms of streams often display morphological adaptations to life in flowing water.

An even more fundamental difference between the biological communities of streams and those of lakes is the source of the energy needed to maintain them. The communities of standing waters rely for the most part on photosynthesis. In streams, however, the ultimate energy source is allochthonous organic material that is heterotrophically metabolized (72). Part of this is in particulate form and is utilized by benthic organisms either directly or after being attacked by microorganisms (7); and part is in the form of dissolved organic compounds, mostly leached from the soil. These together may provide more than 99% of the utilizable energy input (38). Although a few predominately heterotrophic lakes exist, especially in the tropics (9), the concept of lakes as fundamentally autotrophic systems and streams as heterotrophic ones appears useful and generally valid.

Since in streams respiration always exceeds production, it might be thought that a stream is a permanently immature ecosystem. In fact, it is not meaningful to regard a stream as an ecosystem (72); this term can only be usefully applied to the entire watershed.

When a river is dammed and a new lake is formed, two things may therefore be expected to happen: (*a*) the lotic benthos will perish to be replaced eventually by lentic organisms; and (*b*) plankton populations will develop, so that the importance of photoautotrophic processes will increase. Ultimately, the amount of heterotrophic activity will probably decrease as the area of water increases relative to the length of shoreline providing litter. Initially, however, there will almost always be a burst of intensified heterotrophic activity as organic matter in the drowned vegetation and flooded soil is utilized.

Development of the Benthos

Among the first very large reservoirs to be constructed were those on the Volga in the USSR. The development of the benthic populations of the Gorkii and Kuybyshev reservoirs has been described in some detail (110). These reservoirs were filled in the autumn. By spring, most of the organisms originally present had perished and so the benthic fauna was very sparse. Beginning in July, larvae of the chironomid *Chironomus* (*Tendipes*) *plumosus* appeared and reached such numbers that when the midges emerged they interfered with navigation (175). Several other species of chironomids subsequently appeared in smaller numbers, along with motile aquatic arthropods and the freshwater mussel, *Dreissena*, which has motile larvae. Other molluscs, and oligochaetes such as tubificids, remained confined to the old river bed. During the following season, the chironomid population declined and the solely aquatic species continued to spread into the newly flooded area. This trend continued during the third season and the benthic population approached that of a natural lake.

Chironomids are well adapted to be the first colonizers of newly flooded areas. They may reach the area either as the winged adults or as larvae (26). Their fecundity is high; they may be regarded as typical r-selected organisms. Furthermore, some species can endure the low oxygen concentrations that are likely to occur in new impoundments. It is scarcely surprising, therefore, that chironomids have been among the first organisms to appear in new impoundments under a considerable range of conditions. Thus, a beaver dam in Algonquin Park, Ontario, was constructed in the early summer, and the insect fauna almost immediately changed (154). The numbers of Ephemeroptera, Trichoptera, and Plectoptera decreased, and the number of chironomids increased. This trend was even more pronounced in the following year. The study was carried out by trapping emerging adults. No information was obtained about other components of the benthic fauna.

In a broader study on a small impoundment in southern Ontario (123), the reservoir was filled in the spring and the major component of the benthic fauna was initially oligochaetes that invaded the flooded area from the original stream bed. However, during the summer, chironomid larvae became dominant. The chironomid population was smaller during the second season, and the species distribution showed a shift from species characteristic of a highly eutrophic lake to those favored by less eutrophic conditions. During much of the study period, oribatoid mites were the third largest group (after oligochaetes and chironomids) in the population.

In Barrier Lake, a reservoir on the Kananaskis River in Alberta, Canada, the population of chironomids developed very quickly after filling; during the next two years, chironomids made up most of the benthic biomass (116). Here, too, a succession was observed from organisms favored by eutrophic conditions to those preferring relatively oligotrophic conditions. A decade later, chironomids had decreased and the benthic fauna now included substantial numbers of pisidia and oligochaetes (117). In the reservoir formed by the Goczalkowice Dam on the Vistula River in Poland, a similar succession (chironomids to oligochaetes to molluscs) was observed (86).

It seems reasonable to suppose that in the temperate regions, small areas of newly flooded land will be colonized first by oligochaetes from the former stream; but if the area flooded is large, the first to arrive will be chironomids, and oligochaetes and other purely aquatic organisms will invade the area later. Occasionally, however, the damming of small lakes has been followed by a decrease in the population of chironomids (149).

The development of the benthic populations of Lakes Volta and Kariba has recently been reviewed (106). As in large temperate impoundments, the first colonizers were chironomid larvae; in Lake Volta one species, *Chironomus transvaalen-*

sis, was predominant. In Lake Kariba, the population after filling was more diverse, but when the water level rose again after the first drawdown, *C. transvaalensis* became dominant here too. In Lake Volta, the chironomid population became smaller but more diverse as the lake filled. This was attributed to changes in the bottom material along the shore when the shoreline was subjected to increased wave action owing to the increased wind fetch over the lake (130). In both lakes, the benthos was initially confined to a narrow strip along the shore because of the absence of oxygen in the deeper water. As the depth of oxygenation increased, the organisms moved outward and downward. In contrast to what was observed in temperate impoundments, the population of oligochaetes never became very high. This has been attributed to the greater rate of decomposition in the tropics, which prevents organic matter from accumulating below the mud–water interface, so that little food was available for burrowing organisms (106).

There is frequently considerable longitudinal variation in the benthos of reservoirs. In Barrier Lake, the largest population was found in the upper part of the reservoir where sediment rich in organic material was regularly deposited and the area of bottom laid bare at drawdown was relatively small (116). Likewise, in Lake Volta, after the benthic biomass in the lake as a whole began to decline, a higher population was maintained in the areas most directly influenced by inflows (128).

If the flooded area contains standing trees, these will provide a substrate for a variety of invertebrates. In the reservoirs of the Volga, the first invertebrates were almost exclusively chironomids (109). Later, mussels attached themselves to the trees and accounted for most of the biomass (97). In Lake Volta, the situation was entirely different; although some chironomids were present, the predominant form was the larva of the burrowing mayfly, *Povilla adusta*, which excavates holes in wood where it hides (129). In Lake Kariba, the submerged trees were of species having harder wood than those at Lake Volta, and they apparently resisted the attack of the mayfly nymphs (106). Consequently, after the lake filled, the fauna of the trees consisted largely of chironomids and oligochaetes. When the lake was drawn down the now exposed dead trees were attacked by the larvae of wood-boring beetles, which tunnelled between the bark and the wood. When the lake was filled again, substantial numbers of *Povilla* nymphs and smaller numbers of other insect species appeared, living in the tunnels that had been made by the beetles.

Organisms like *Povilla* do not use wood as food; they make tunnels only as places of refuge. Indeed, no freshwater invertebrates seem to eat wood the way teredenid molluscs do in the sea and many insects do on land. The decomposition of submerged wood is only brought about by the slow action of bacteria and fungi, which no doubt accounts for wood lasting a long time in fresh water, even if it is well oxygenated.

The conditions of gentle flow of relatively clean water that prevail in many reservoirs and their associated structures appear particularly favorable to certain sessile organisms, notably bryozoans and sponges. Thus they may occur in large numbers particularly near the outlet pipes; they have long been referred to as "pipe moss" by water engineers (111).

Development of the Plankton

Impoundment may influence the plankton population of a body of water in a number of ways. Reduction of the rate of flow of a stream will allow the sparse plankton already present to multiply before it is swept away. Often a large plankton population develops when the reservoir is filling and there is little outflow (106). Even after the reservoir is full and outflow becomes substantial, a fairly large plankton population may persist.

Various authors have attempted to define the conditions of flow under which the transformation of a stream ecosystem to a lake ecosystem may occur. Margalef (101) has suggested that the conditions under which a flowing, rather than a static, ecosystem will exist may be defined by the inequality: $V^2 > 4A'r$, where V is the rate of flow, A' is the coefficient of eddy diffusivity, and r is the rate of increase of the plankton. This relationship is based on an expression relating to the development of the plankton of the sea (139), where cells are being removed not by horizontal flow but by vertical settling. Russian scientists (164) suggest an even simpler criterion; in Russian reservoirs a lacustrine plankton population develops if the rate of flow is less than 0.2 m sec^{-1}. It seems unlikely that any simple criterion is universally applicable. Typically, many reservoirs contain regions of fairly rapid flow where the plankton populations will be small, and other regions of slack water where considerable populations may develop.

Where the development of a plankton population is mainly due to the decrease in flow rate without addition of nutrients, as in the temporary reservoirs of the Nile, a decrease in the concentration of nutrients in solution is likely (133). However, if the impoundment causes the flooding of a considerable area, the growth of phytoplankton may be increased by the nutrients made available by leaching from the soil and by decomposition of drowned vegetation. The first effect to be observed may be an increase in primary productivity rather than of standing crop. In these cases, the nutrients made available may be taken up so quickly that little or no increase in their concentrations may be observed (140).

When a lake is impounded, the earliest effect to be observed may be a decrease in phytoplankton standing crop, as a result of dilution, especially if the inflowing water is highly colored (122). Artificial lakes frequently are relatively eutrophic at the beginning. Blooms of blue-green algae are often observed, although their development may be suppressed if the water is highly turbid (105). However, under the flow conditions prevailing in many reservoirs, eutrophy is unlikely to persist. In stratified natural lakes with fairly long retention times, the behavior of nutrients is controlled to a large extent by biological processes. These processes are probably less significant in many reservoirs because of their low retention times, which permit nutrients to be flushed away fairly quickly (161). Nutrient loss will be hastened by discharge from the hypolimnion, since the concentration of nutrients is higher in the hypolimnion than in the epilimnion for much of the year.

Large variations in the water levels of reservoirs may hasten the return of nutrients from the sediments in the areas that are laid bare. This certainly happens in tropical impoundments. In temperate regions, the normal drawdown regime exposes large areas of the bottom only during the cold part of the year when the release of nutrients from the sediments will be slow. However, if an area is kept exposed for a year or more and then reflooded, significant amounts of nutrients can be released to the water (161).

The zooplankton of reservoirs, like the zoobenthos, changes in species distribution following impoundment, but the change is less dramatic. Species typical of rivers decrease in numbers or disappear, whereas characteristically lacustrine species appear or increase.

Conditions in reservoirs appear to favor the development of certain unusual zooplankters. Thus, the copepod *Arctodiaptomus ibericus* is widely distributed in Spanish reservoirs but has not yet been found in natural waters (103). Even more surprisingly, a previously undescribed species of water mite (*Piona limnetica* Biesiadka) was found to constitute a substantial part of the zooplanktonic biomass in a thirty-year-old reservoir in the Canal Zone of Panama (47). Water mites are not usually an important part of the zooplankton (69) and the presence of these predacious organisms influenced the general composition of both the zooplankton and phytoplankton communities.

If a river is regulated over a considerable range of latitude, the series of reservoirs may provide a series of stages by which northern species of zooplankton may be introduced into the more southerly regions, as in a number of rivers in the USSR (30).

A longitudinal zonation of plankton, similar to that observed with benthos, occurs in many reservoirs. In Slapy Reservoir, the area most influenced by the inflowing river has a higher population of green algae, and a lower population of blue-green algae than has the central, more lake-like zone. All forms of phytoplankton decrease in the area directly influenced by the outflow. The zooplankton population is relatively sparse in the upper region as compared with the main body of the reservoir. There is also some lateral variation with local concentrations of plankton in bays and inlets (65).

In Lake Volta, blue-green algae predominated at first in the upper and diatoms in the more lake-like lower part. Subsequently, diatoms have established themselves more widely throughout the reservoir (12).

The Littoral Region

The ecotone between land and water is normally a productive and interesting region. Its nomenclature is complex (68); for the present discussion, following Pieczynska (132), the term *eulittoral* will designate the area between the highest and lowest waterlines, together with the part of the land that is splashed by waves and the area below the lowest water level that is intermittently uncovered by waves. The terrestrial zone immediately above the eulittoral will be referred to as the *epilittoral*.

The characteristics of the plant communities of eulittorals, subjected to a variety of flood regimes, have been summarized (60). The development of reservoir eulittoral ecosystems is dominated above all by the practice of drawdown. In natural lakes, the zones between the highest and lowest waterlines are commonly subjected to a short period of flooding in the spring or during the rainy season, followed by a long period of exposure. Many organisms adapted to this regime have evolved;

diversified and productive pulse-stabilized communities usually exist in such regions. In reservoirs, the situation is commonly reversed, there being a long period of flooding and a short period of exposure. This fundamentally unnatural regime stabilizes an ecosystem of a degree of immaturity which in high latitudes amounts virtually to barrenness; indeed, Swedish ecologists have proposed a new term, *aridal,* to designate the drawdown zone of reservoirs (94). In tropical regions where growth of terrestrial vegetation is possible during the period when the drawdown zone is exposed, the situation is somewhat different. The littoral regions of pumped storage reservoirs are subject to particularly severe conditions. Here the period of fluctuation of the water level may be as short as a day, often with a superimposed weekly fluctuation, since water is usually pumped into the reserves at night and during weekends (4).

Both aquatic and terrestrial organisms adjacent to the periphery of the reservoir are adversely affected. During the period of high water, organisms enter the drawdown zone only to perish when the water level falls. The effect drastically reduces the diversity of the benthos in temperate regions (70, 117). In a tropical reservoir, Lake Kariba, chironomid larvae are stranded by the falling water, but the population in the water is maintained because adult females are always available to lay eggs below the receding water line (106). There is evidence, however, that even in the tropics drawdown reduces the diversity of the benthic population (106).

In the tropics, terrestrial grasses may grow on the drawdown zone when it is uncovered. Around Lake Kariba, this grass provides grazing for many large animals. When the water rises again, nutrients are leached from their dung, which probably increases the productivity of the aquatic ecosystem (106). In the temperate regions, however, since the drawdown zone is usually exposed at the coldest time of the year, little growth is possible. Despite attempts to plant amphibious trees or shrubs, such as willow, on the edges of reservoirs (44) or, in fairly mild regions, to seed the drawdown zone with cereals (39), these regions will remain unattractive at best. The prospect of the transformation of the eulittoral of a lake into a near-barren mud flat has often aroused very bitter opposition among biologists, as for example in Tasmania where the flooding of Lake Pedder became a major political issue (126).

When the water level of a lake or stream rises following impoundment, the existing epilittoral will be flooded and destroyed. This, like the eulittoral, constitutes an interesting and productive ecotone, providing food and shelter for many animals, including birds and mammals both large and small. Eventually a new epilittoral will establish itself above the new eulittoral, although the large variations in water level may prevent it from ever becoming as productive as the original. In warm climates but not in colder regions, the new epilittoral should establish itself fairly quickly. Since this zone is particularly important in cold regions, the destruction of the epilittoral has been a matter of particular concern in such developments as the James Bay project (48, 76).

Fish and Other Vertebrates

It has long been known that new impoundments frequently provide excellent fishing, both for sport and on a commercial scale (32). More recently, high fish yields have been obtained in the large African reservoirs (106). A factor that contributes to this plenty is undoubtedly the availability of large numbers of benthic organisms as food. In temperate regions, the high population of fish usually declines after a few years. Soviet scientists (5) have observed three phases in the development of many of their reservoirs. The first phase of high productivity, based largely on benthic organisms feeding on drowned vegetation, is followed by a phase of trophic depression when this material is exhausted or rendered unavailable by silting. This phase, which may last for several decades, gradually leads to a slow increase in productivity as the growth and settling of plankton provide a new layer of organic sediment. In Lake Volta, fish catches have declined in the 10 years since the dam was closed, but they are still good (49).

The high fish yields in new impoundments are probably not due solely to the abundant food supply, but also to the cover provided for young fish by the flooded vegetation. Increased fish yields in Lake Mead in years following unusually high water levels have been attributed at least in part to this (61).

However, impoundments may also be harmful to fish. In temperate and tropical regions many species spawn in nests in the shallow water near the shore, and these may be laid bare when the water level drops, so that the eggs or young perish (94, 144). There is also a danger of large numbers of fish being killed if the bottom water of the reservoir has become anoxic and is then mixed quickly with the surface water as a result of a change in weather conditions (32).

The impoundment of a river often leads to changes in the kinds and numbers of fish parasites; sometimes the level of infestation may become very high. A number of factors are involved in this. The number of zooplankters that serve as intermediate hosts increases, and the feeding habits of the fish may change (8, 62, 89).

The long-term effects of impoundments on populations of nonmigratory fish, therefore, depend on a large number of factors, including habits of the particular species affected, the drawdown regime of the reservoir, and the prevailing climate (which, in turn, imposes constraints on the drawdown regime). The many dams constructed by the Tennessee Valley Authority have greatly benefited the fisheries in the area involved, both in terms of numbers of fish and in terms of the species available (172). These dams were built in a region where the climate is relatively mild. During most of the summer, the water levels are kept constant (except for fluctuations to control malaria). In higher latitudes, the situation may be different. In lake reservoirs in Sweden, populations of littoral species have declined, whereas pelagic species have not been greatly affected (94). The time of spawning may be delayed if the time of autumn cooling or spring warming of the water is changed (100). Flooding of spawning grounds may discourage breeding although fish may continue to spawn in the same places after the water level rises (24).

Methods proposed for reducing the harmful effects of drawdown on the fish populations include the construction of sub-impoundments that retain water when the level in the main impoundment drops (54), the construction of floating nesting platforms (175), and a drawdown regime which avoids laying bare the littoral at the most critical times (61).

The rapid fluctuations of water level in pumped storage reservoirs probably have less direct effect on fish than the long-term fluctuations in conventional reservoirs.

Apparently the selection of sites for nest construction and spawning takes several days, so that the eggs are not likely to be laid in the drawdown zone (57).

In general, impoundment is not likely to be disastrous to the fish populations of any body of water. The species composition of the population may be changed to some degree.

Many mammals are trapped and drowned during the filling of large new reservoirs. In some instances, large operations have been mounted to rescue them. At least one such operation, at Lake Brokopondo in Surinam, has provided interesting new zoogeographical information (91). After the initial loss, the effect of an impoundment on wildlife depends on the prevailing conditions. In some areas, river valleys provide the best wildlife habitat; if one of these is flooded, there will be a permanent decrease in the animal population. Elsewhere, however, flooding may favor wildlife by providing water and opportunities for grazing on the drawdown zone.

Similar considerations apply to birds. Many nests are destroyed when a reservoir is filled during the nesting season, and the amount of suitable habitat for some species is permanently reduced. However, reservoirs may provide significant new habitat for waterfowl in areas where natural standing water is scanty, and may lead them to modify their migratory and nesting habits to some degree (6). Where vegetation has an opportunity to grow on the drawdown zone, it is used by birds as well as by mammals, e. g. around the Koka Reservoir in Ethiopia, where large numbers of Egyptian geese, spur-wing geese, and other waterfowl are often seen on the drawdown zone (personal observation). In Kariba Reservoir, birds feed on chironomid larvae stranded when the water level drops (106).

DOWNSTREAM EFFECTS OF IMPOUNDMENTS

Many of the effects produced by dams on the stream below them are the reverse of those produced on the lake above them. What is retained in the lake (heat, silt, inorganic or organic nutrients) is lost to the stream. Moreover, the large annual variation in the water level of the lake will almost certainly be associated with a decrease in the annual variation of water level in the stream. All these, in turn, may have certain biological consequences, sometimes far away.

Other effects may be caused by the mere existence of the dam as a barrier to free passage up and down stream, and by the induction of a precipitous fall of water or, alternatively, by the elimination of a precipitous fall of water when water is diverted upstream from a waterfall. This may lead to the destruction of interesting local spray-zone ecosystems (14, 82).

If sediment is deposited in the reservoir, the clear water leaving it may pick up a new load of sediment, eroding the shores and stream bed below the dam as it does so. The loss of organic detritus from the stream may reduce the level of heterotrophic activity, at least until a new source of detritus is provided by the plankton of the reservoir. The amount of primary production in the stream will be favored by the decrease in turbidity; it will tend to be decreased, however, if the growth of plankton following impoundment has led to a depletion of nutrients from the water. On the other hand, if the amounts of nutrients introduced into the water by flooding are in excess of what can be taken up by the plankton of the reservoir, primary productivity downstream may increase, at least for a time (88).

The complex flow pattern in many reservoirs may have an important influence on the downstream temperature regime. In summer, solar radiation on the reservoir will be converted to thermal energy that will heat the epilimnion but have little effect on the hypolimnion. Thus, the epilimnion serves as a trap for heat that would otherwise have served to warm the water of the stream. In winter, after stratification has broken down, some of this heat will enter the outflow. The overall effect, therefore, makes the stream below the dam cooler in summer and warmer in winter than it was before the dam was built.

This can give rise to density currents flowing upstream in tributaries entering the main stream below the dam. These will flow on the bottom of the streambed during the summer, and over the surface of the tributary streams during the colder part of the year. Among other effects, such upstream density currents can adversely affect water quality in the tributaries (20).

The biological effects of the change in the temperature regime may be quite severe. In the South Saskatchewan River below Lake Diefenbaker, the benthos has been found to be impoverished as much as 110 km downstream (93). Below the dam of a mountain reservoir in Colorado (169) the diversity was less than that in similar but unregulated streams, but the biomass was higher. (Compare Thienemann's second principle.)

The construction of a dam usually leads to a decrease in the amplitude of long-period, especially annual, variations in water level downstream. However, if the dam has been built to generate electrical power, the operation of the plant may introduce small-amplitude short-period variations as the discharge is varied in accordance with the demand for electricity. These variations can be destructive of benthic organisms and cause a considerable reduction in diversity (37, 163).

At higher trophic levels in particular, it is exceedingly difficult to predict the net effect. Decreased turbidity may increase primary productivity (thus perhaps increasing the available food supply) and will make it easier for fish to find food; but it will also make it easier for predators to find fish. Cooling of the stream may make it possible for cold-water fish, such as salmonids, to survive where they were unable to before (172). At the same time, however, it may decrease the supply of benthic food organisms whose numbers may be still further reduced by a decrease in heterotrophic metabolism and by the destructive effects of short-term variations in the water level.

Another danger for fish and other aquatic animals below dams is gas-bubble disease (58, 112, 142, 156), which resembles the "bends" in divers (27). If a fish ingests water supersaturated with gases, the excess gas may come out of solution as bubbles. These lodge in various parts of the fish's body and cause, depending on their size and location, injury or death. Water may become supersaturated with gases in two ways that are relevant to the present discussion. When air and water are mixed in a turbine, the pressure may be great enough to force gas into solution (99). When water plunges over a spillway into a deep basin, entrained air bubbles may be carried to a considerable depth where the hydrostatic pressure may be great enough to force gas into solution (10). The degree of supersaturation required to cause gas-bubble disease depends on the age and species of the fish, but as little as 18% supersaturation may be sufficient (142).

Dams may pose a serious obstacle to the upstream movement of anadromous fish, as well as to the downstream movement of the smolts. Of the two, the former is probably the more serious. Since some species of Pacific salmon spend as little as two years in the sea, blockage of streams for this short a time as, for example, during the construction of a dam, could effectively eliminate the population of these species from the stream. Even a partial blockage could obscure the olfactory and tactile clues by which the fish are guided to their spawning grounds (59). Pacific salmon do not feed on the way to their spawning grounds, and the energy reserve in their bodies is little more than sufficient to bring them to their destination (73). Hence, time lost wandering in a region of slack water above a dam could reduce their chances of reproducing.

Dams also pose an obstacle to the downstream movement of catadromous fish, of which eels are the most important. The fate of eels in the hydroelectric turbines at Cornwall on the Saint Lawrence Seaway has been gruesomely described (92). However, young eels (elvers) can negotiate obstacles that would be impossible for adult salmon, and there is no evidence that elvers seek out the streams from which their parents came, so that the danger of the depletion of eel populations as a consequence of hydroelectric development does not appear serious.

The construction of fish ladders and other devices to permit fish to pass dams is a well-developed branch of engineering (21). Doubt has been expressed, however, (41) whether such structures are adequate in all cases. Alternatively, fish may be caught below the dam and transported by truck to the undisturbed higher parts of the river.

Some of the most dramatic effects of impoundments, good or bad, result from the change of the downstream flood regime. One purpose of building dams is to reduce annual variations in water level, making the floodplain habitable throughout the year and allowing its ecosystem to become more mature; or, more commonly perhaps, causing it to be replaced by a different ecosystem maintained in a state of immaturity by the practice of agriculture.

Reduction of this variation is not an unmixed benefit; one consequence of the building of the Aswan Dam is that agriculture in the Nile Valley now requires fertilizer, whereas in the past the soil was fertilized naturally by the deposition during the flood period of silt from the Ethiopian Highlands. Elsewhere, the fertilization of the river by the floodplain may be more important than the fertilization of the floodplain by the river. This seems to be an important factor in maintaining the fish populations of many African rivers (170).

A good example of unplanned large-scale alteration of a pulse-stabilized ecosystem is provided by the effect on the Peace-Athabasca Delta in the Canadian province of Alberta of the construction of the W. A. C. Bennett Dam on the Peace River (124, 125, 136). This region was usually flooded in the spring by the water of the Athasbasca River that entered it from the south. Water from the Peace River, which passes to the north of the Delta to join the Slave River, contributed little to the actual flooding, but for flooding to occur it was necessary for the level of Peace River to be high enough to cause the water of the Athabasca River to back up into the Delta. When the flow of the Peace River was regulated, the Delta began to be transformed from marshland into meadow. This was considered undesirable, particularly because the local inhabitants subsisted on trapping and fishing in the marsh (136). A weir has now been built which will hold back the water sufficiently to permit flooding as before.

Dams on northward-flowing rivers in high latitudes are a cause for particular concern. Gill (45) has discussed the possible consequences of damming the MacKenzie River, which flows from Great Slave Lake to the Beaufort Sea through the Canadian Territory of MacKenzie. Present ecological conditions in the Delta of the MacKenzie are maintained by pulse stabilization, so regulation of the river flow would almost certainly cause considerable changes in this region. Moreover, the spring breakup of ice in the lower reaches of the river is hastened by the hydrostatic pressure generated by flood waters from the more southerly part of the river flowing under it. Consequently, a reduction in the maximum flow of the river would probably delay the beginning of spring. Somewhat similar concerns have been expressed with regard to a proposed hydroelectric project on the Ob River in Siberia (167).

If a river runs into the sea, regulating the river will probably have an effect in and around its mouth. Since the construction of the Aswan High Dam, the area of the Nile Delta has been reduced because of the disturbance of the equilibrium between erosion by the sea and deposition of sediment by the river (2). The sardine catches in the Mediterranean near the mouth of the Nile have decreased, owing either to the absence of nutrients formerly provided by the river or to the dispersal of the fish over a wider area (43).

An estuary is a complex and productive ecotone, maintained over the long term by a complex flow pattern known as a haline circulation (114). Fresh water from the river flows seaward over salt water flowing landward. Nutrient-rich seawater is entrained in the fresh water, leading to a high level of productivity in the upper layer (42). Any change in the flow pattern of the river will influence the haline circulation and the salinity gradient. This can have dramatic biological consequences. The influence of the Volta Dam on the bivalve mollusc *Egeria radiata* (9) may serve as an example. This freshwater mollusc requires a slight salinity of about 1‰ for reproduction. Construction and operation of the dam caused shifts of the spawning grounds many kilometers up and down the river.

In higher latitudes where estuarine and nearshore marine environments are more variable than in the tropics, such dramatic effects are perhaps not to be expected. Thus, the marine zooplankton in James Bay is largely composed of euryhaline and eurythermal species that will probably not be much affected by regulation of the inflowing rivers (52, 53). There are grounds for believing, however, that regulation of rivers entering such areas as the Strait of Georgia (42) and the Gulf of Saint Lawrence (114) will influence their fish populations. To what extent is difficult at present to predict.

DAMS AND IMPOUNDMENTS   275

OTHER CONSEQUENCES OF IMPOUNDMENTS

Both during its construction and afterwards, a large dam has various frequently undesirable side-effects. These consequences are common to many large industrial or technical projects. The following discussion is limited to effects peculiar to the construction of dams and the impoundment of water.

The presence of a new body of standing water produces changes in the climate in its vicinity; these are proportional to its size (17). There may be changes in the annual precipitation pattern, an increase in low stratus clouds and fog, a decrease in air temperatures in spring and an increase in the fall, all leading in high latitudes to a delay in the beginning and end of the growing season. The range of diurnal air temperature will decrease. Changes of this nature have been predicted in the vicinity of the reservoirs of the James Bay project (13).

Recently it has become apparent that large impoundments may induce seismic activity in their neighborhoods (50, 55, 80, 108). The shocks that occur are fortunately usually small, and there is considerable difficulty in attributing any given shock to the effect of impoundment. The stresses set up by the weight of water impounded even in very large reservoirs are too small by one or two orders of magnitude to exert any geophysical effect by themselves. The effect therefore is either one of increasing the pressure of groundwater in fissures in the rocks, thus enabling slippage to occur under the influence of preexisting geophysical forces, or perhaps of the addition of a critical increment to pre-existing stresses.

The incidence of certain diseases associated with water may increase in the vicinity of new impoundments. Schistosomiasis almost always increases in the vicinity of impoundments in the tropics (31, 168), partly because of the enlarged habitat made available for the snails that are its intermediate hosts. Malaria may also increase because of the enlarged breeding areas made available to the mosquitoes that transmit the various forms of this disease. The breeding of malaria-transmitting mosquitoes in the reservoirs of the Tennessee Valley Authority is prevented by periodic changes in the water level, but this is not always practicable. The malaria vectors in some regions, such as Lake Volta, would be favored by fluctuating water levels (168).

On the other hand, onchocerciasis, or river blindness, is likely to decrease as a consequence of impoundment. This disease is transmitted by a black fly that breeds in running water. Many of the rapids where it breeds will be flooded by impoundment, although new breeding areas may become available below the dam (31, 168).

Among the most distressing consequences of large reservoir construction has been the disruption of the everyday lives of the people of the region. Unlike small reservoirs for such purposes as irrigation and flood control, large hydroelectric reservoirs often contribute little in any direct way to the well-being of those most affected. This has been a particularly serious problem where the people involved have been living in closely knit tribal communities, or where their livelihood has been dependent on hunting, fishing, or trapping. The large impoundments in Africa have, unfortunately, caused a good deal of distress (168). Much of this probably resulted from inefficient and insufficiently planned resettlement schemes and it may be hoped that past experience will provide a guide to future arrangements of this sort. In other instances, people have shown considerable ingenuity in adapting, as for example, by planting crops on the drawdown zone or by changing their fishing techniques.

In northern Canada, which is becoming an increasingly important source of hydroelectric power as the potentialities of more southerly sites become exhausted, similar considerations apply. In these regions communities of indigenous peoples, largely Cree and Inuit (Eskimo), have maintained viable social and economic systems based on traditional patterns of fishing, trapping, and hunting. The disruption of the indigenous community in the vicinity of the Peace-Athabasca Delta has already been mentioned. Similar concerns have been expressed in connection with a large hydroelectric development in northern Manitoba (115, 157) and particularly with the James Bay development (76, 137). Such concerns have led to considerable research in the latter area (23) and to the signing of a detailed agreement between the governments of Canada and Quebec and the representatives of the indigenous communities (75).

SUMMARY AND CONCLUSIONS

Although reservoirs have sometimes been referred to as "embryo lakes" (106), they are probably better regarded as a distinct type of freshwater ecosystem differing from both streams and lakes. Many are characterized by a highly developed shoreline, a longitudinal profile with its maximum depth near the downstream end, a complicated flow pattern often involving discharge from the hypolimnion, and a pattern of seasonal variation in water level involving a long period of flooding and a short period of exposure. Because reservoirs are frequently built on streams carrying a heavy sediment load, the deposition and distribution of this material within the reservoir are often more important in reservoirs than they are in natural lakes. Therefore constraints on the nature of the developing biological community are imposed when a new reservoir is constructed.

The environmental changes below a dam may be no less dramatic than those above it. The two sets of events show a certain symmetry; each is to a large extent the inverse of the other. Below the dam, as above, the change in flow regime is the immediate cause of many other changes. This symmetry may be remarkably precise. For example, it has been estimated (144) that the increase in annual yield of fish from the Nasser Reservoir above the Aswan Dam may correspond closely to the decrease in annual yield from the Nile Delta and the eastern Mediterranean following the closure of the dam.

After the completion of some of the large African dams during the 1960s, many observers expressed surprise at the environmental consequences. This surprise may appear naive in retrospect, but certain of the consequences still seem remarkable—for example, the enormous populations of mayfly nymphs that developed in Lake Volta.

Large impoundments now exist under a variety of climatic and geographical conditions. Since not all the hydraulic head of the world's rivers has yet been utilized, it seems likely that more remain to be built. Certainly the rate of construction of smaller reservoirs shows no sign of diminishing (107). Will there be further ecological surprises?

The first great African impoundments were genuine novelties; nothing like them had ever existed before, and there was no basis in experience on which to predict their consequences. Subsequent tropical impoundments will not be novelties and their effects should be predictable, in their broad outlines, from earlier experience.

The development of reservoirs in the temperate regions has been more gradual. Experience with smaller reservoirs has consequently been applicable, within limits, to larger ones. Moreover, the generally lower rate of biological processes at higher latitudes has made their effects less dramatic, and perhaps has allowed more time to arrest and reverse undesirable effects before they became irreversible, as for example, the drying up of the Peace-Athabasca Delta. It seems unlikely that subsequent impoundments in the temperate regions will give rise to any large-scale surprises.

On the more detailed scale likely to be of importance to man, much remains to be learned. How the flooding of a certain area will influence the populations of furbearing animals in it, or how the damming of a certain river will influence the runs of salmon up it, are likely to be matters of intense concern to the people whose livelihood depends on these resources. Such questions can only be answered by a careful and thoughtful investigation of all possible aspects of the ecology of the region.

ACKNOWLEDGMENTS

I have been greatly helped in the preparation of this review by many colleagues throughout the world who have provided me with information and suggestions. I have also been helped by several colleagues at CCIW who have offered helpful criticism of successive drafts of the manuscript. To all these I extend my thanks.

I am grateful to Dr. G. K. Rodgers, Associate Director for Research, CCIW, for permission to publish this review.

Opinions expressed in this paper do not necessarily represent the official opinion of Environment Canada.

Literature Cited

1. Ackerman, W. C., White, G. F., Worthington, E. B., Ivens, J. L., eds. 1973. *Man-Made Lakes: Their Problems and Environmental Effects.* Washington, DC: American Geophysical Union. 847 pp.
2. Aleem, A. A. 1972. Effect of river outflow management on marine life. *Mar. Biol.* 15:200–8
3. American Fisheries Society. 1967. *Reservoir Fishery Resources Symposium.* Athens, Georgia: Univ. Georgia. 569 pp.
4. American Fisheries Society. 1976. Biological considerations of pumped storage development. *Trans. Am. Fish. Soc.* 105:155–80
5. Baranov, I. V. 1966. Biohydrochemical classification of the reservoirs in the European U.S.S.R. See Ref. 166, pp. 139–83
6. Barclay, J. S. 1976. Waterfowl use of Oklahoma reservoirs. See Ref. 121, pp. 141–51
7. Bärlocher, F., Kendrick, B. 1974. Dynamics of the fungal population on leaves in a stream. *J. Ecol.* 62:761–91
8. Bauer, O. N., Stolyarov, V. P. 1961. Formation of the parasite fauna and parasitic diseases of fishes in hydroelectric reservoirs. In *Parasitology of Fishes,* ed. V. A. Dogiel, Y. K. Petrushevski, Yu I. Polyanski, pp. 246–54. Edinburgh and London: Oliver & Boyd
9. Beadle, L. C. 1974. *The Inland Waters of Tropical Africa.* London: Longmans. 365 pp.
10. Beiningen, K. T., Ebel, W. J. 1970. Effect of John Day Dam on dissolved nitrogen concentrations and salmon in the Columbia River, 1968. *Trans. Am. Fish. Soc.* 99:664–71
11. Biswas, A. K. 1975. A short history of hydrology. In *Selected Works in Water Resources,* ed. A. K. Biswas, pp. 57–78. Champaign, Ill: Int. Water Resour. Assoc.
12. Biswas, S. 1969. The Volta Lake; some ecological observations on the phytoplankton. *Int. Ver. Theor. Angew. Limnol. Verh.* 17:259–72
13. Bondy, D. A., 1976. Prediction of climatic changes. See Ref. 23, pp. 497–98
14. Brassard, G. R., Frost, S., Laird, M., Olsen, O. A., Steele, D. H. 1971. Studies of the spray zone of Churchill Falls, Labrador. *Biol. Conserv.* 4:13–18
15. Brock, T. D., 1970. High temperature systems. *Ann. Rev. Ecol. Syst.* 1:191–220
16. Budge, E. A. W., 1967. *The Egyptian Book of the Dead.* New York: Dover. 377 pp.
17. Butorin, N. V., Vendrov, S. L., Dyakonov, K. N., Reteyum, A. Yu., Romanenko, V. I. 1973. Effect of the Rybinsk reservoir on the surrounding area. See Ref. 1, pp. 246–50
18. Campbell, P. G., Bobée, B., Caillé, A., Demalsy, M. J., Demalsy, P., Sasseville, J. L., Visser, S. A. 1975. Preimpoundment site preparation: a study of the effects of topsoil stripping on reservoir water quality. *Int. Ver. Theor. Angew. Limnol. Verh.* 19:1768–77
19. Cheesman, R. E. 1936. *Lake Tana and the Blue Nile.* Reprinted 1968. London: Cass. 400 pp.
20. Churchill, M. A., 1947. Effect of density currents upon raw water quality. *J. Am. Water Works Assoc.* 39:357–60
21. Clay, C. H. 1961. *Design of Fishways and Other Fish Facilities.* Ottawa: Dep. Fish. Canada. 301 pp.
22. Coakley, J. P., Hamblin, P. F. 1967. *Investigation of Bank Erosion and Nearshore Sedimentation in Lake Diefenbaker.* Burlington, Ontario: Canada Cent. Inland Waters. 18 pp.
23. *Compte Rendu, Environnement—Baie James, Symp.* 1976. Montreal: Soc. Energ. Baie James, Environ. Can. 883 pp.
24. Cuerrier, J.-P. 1954. The history of Lake Minnewanka with reference to the reaction of Lake Trout to artificial change in environment. *Can. Fish Cult.* 15:1–9
25. Cyberski, J. 1973. Erosion of banks of storage reservoirs in Poland. *Hydrol. Sci. Bull.* 18:317–20
26. Davies, B. R. 1976. The dispersal of chironomidae larvae: a review. *J. Entomol. Soc. South. Afr.* 39:39–62
27. D'Aoust, B. G., Smith, L. S. 1974. Bends in fish. *Comp. Biochem. Physiol.* 49A:311–21
28. Destin du bois submergé lors de la création d'un reservoir dans une région boisée. 1973. Université du Québec: INRS-Eau. Rapp. Ann. 1972–73. pp. 27–28
29. Dominy, C. L., 1973. Recent changes in Atlantic Salmon (*Salmo salar*) runs in the light of environmental changes in the Saint John River, New Brunswick, Canada. *Biol. Conserv.* 5:105–13
30. Dzyuban, N. A. 1962. Reservoirs as a

zoogeographical factor. *Tr. Zon. Sov. Tipol. Biol. Rib. Ispol. Vnut. Vod. Youzh. Zony. SSR.:* 105–10 (In Russian. Natl. Lend. Libr. Sci. Tech. translation RTS 2936)
31. Egbuniwe, H. 1976. Public health aspect of tropical water resources development. *Water Resour. Bull.* 12: 393–98
32. Ellis, M. M. 1941. Freshwater impoundments. *Trans. Am. Fish. Soc.* 71:80–93
33. Environmental impact assessment and hydroelectric projects: hindsight and foresight in Canada. 1975. *J. Fish. Res. Bd. Can.* 21:97–209
34. Fels, E., Keller, B. 1973. World register of man-made lakes. See Ref. 1, pp. 43–49
35. *The Fetha Negast. The Law of the Kings,* trans. Abba Paulos Tzadua. 1968. Addis Ababa: Faculty of Law, Haile Sellassie I University. 339 pp.
36. Fiala, L. 1966. Akinetic spaces in water supply reservoirs. *Int. Ver. Theor. Angew. Limnol. Verh.* 16:685–92
37. Fisher, S. G., LaVoy, A. 1972. Differences in littoral fauna due to fluctuating water levels below a hydroelectric dam. *J. Fish. Res. Bd. Can.* 29:1472–1476
38. Fisher, S. G., Likens, G. E. 1973. Energy flow in Bear Brook, New Hampshire: an integrative approach to stream ecosystem metabolism. *Ecol. Monogr.* 43:421–39
39. Fowler, D. K., Maddox, J. B. 1974. Habitat improvement along reservoir inundation zones by barge hydroseeding. *J. Soil Water Conserv.* 29:263–65
40. Gastescu, P., Breier, A. 1973. Artificial lakes of Rumania. See Ref. 1, pp. 50–55
41. Geen, G. H., 1974. Effects of hydroelectric development in Western Canada on aquatic ecosystems. *J. Fish. Res. Bd. Can.* 31:913–27
42. Geen, G. H. 1975. Ecological consequences of the proposed Moran Dam on the Fraser River. See Ref. 33, pp. 126–35
43. George, C. J., 1972. The role of the Aswan High Dam in changing the fisheries of the Southeastern Mediterranean. In *The Careless Technology,* ed.M. T. Farvar, J. P. Milton, pp. 159–78. Garden City, New York: Natural History Press
44. Gill, C. J., Bradshaw, A. D. 1971. Some aspects of the colonization of upland reservoir margins. *J. Inst. Water Eng.* 25:165–73
45. Gill, D., 1971. Damming the Mackenzie: A theoretical assessment of the long-term influence of river impoundment on the ecology of the Mackenzie River Delta. In *Proc. Peace Athabasca Delta Symposium,* pp. 204–222. Edmonton: Univ. Alberta
46. Gjessing, E. T., Samdal, J. E. 1968. Humic substances in water and the effect of impoundment. *J. Am. Water Works Assoc.* 60:451–54
47. Gliwicz, Z. M., Biesiadka, E. 1975. Pelagic water mites (Hydracarina) and their effect on the plankton community in a neotropical man-made lake. *Arch. Hydrobiol.* 76:65–88
48. Glooschenko, V. 1972. The James Bay power proposal. *Nat. Can. (Ottawa)* 1(1):4–10
49. Goodwin, P. 1976. Volta ten years on. *New Sci.* 71:596–97
50. Gough, D. I., Gough, W. I. 1970. Load-induced earthquakes at Lake Kariba-II. *Geophys. J. Roy. Astron. Soc.* 21:79–101
51. Gould, H. R., 1960. Turbidity currents. *Comprehensive Survey of Sedimentation in Lake Mead, 1948–49. Geological Survey Professional Paper 295,* pp. 201–7
52. Grainger, E. H., 1976. The marine plankton of James Bay. See Ref. 23, p. 111
53. Grainger, E. H., McSween, S. 1976. Marine zooplankton and some physical-chemical features of James Bay related to La Grande hydro-electric development. *Dep. Environ., Fish. Mar. Serv., Res. Dev. Dir. Tech. Rep. 650.* 94 pp.
54. Grimas, U. 1965. Inlet impoundments. An attempt to preserve littoral animals in regulated subarctic lakes. *Rep. Inst. Freshwater Res. Drottningholm,* 46: 22–30
55. Gupta, H. K., Rastogi, B. K. 1976. *Dams and Earthquakes.* Amsterdam: Elsevier. 229 pp.
56. Hall, A., Davies, B. R., Valente, I. 1976. Cabora Bassa: some preliminary physicochemical and zooplankton pre-impoundment results. *Hydrobiologia,* 50:17–25
57. Hauk, F. R., Edson, Q. A. 1976. Pumped storage: its significance as an energy source and some biological ramifications. See Ref. 4, pp. 58–64
58. Harvey, H. H. 1975. Gas diseases in fishes - a review. In *Chemistry and Physics of Aqueous Gas Solutions,* ed. W. A. Adams, G. Greer, J. E. Desnoyers, G. Atkinson, G. S. Kell, K. B. Oldham, J. Walkley, pp. 450–85. Princeton, NJ: Electrochemical Society
59. Hasler, A. D. Orientation and fish migration. In *Fish Physiology,* ed. W. S.

Hoar, D. J. Randall, 6:429–510. New York: Academic.
60. Hejny, S. 1971. The dynamic characteristic of littoral vegetation with respect to changes of water level. *Hidrobiologia* 12:71–85
61. Hoffman, D. A., Jonez, A. R. 1973. Lake Mead, a case history. See Ref. 1, pp. 220–33
62. Hoffman, G. L., Bauer, O. N. 1971. Fish parasitology in water reservoirs: a review. Reservoir Fisheries and Limnology. *Am. Fish. Soc. Spec. Publ. No. 8*, pp. 495–511
63. Howard, C. S. 1960. Chemistry of the water. See Ref. 51, pp. 115–24
64. Hrbáček, J., ed. 1966. *Hydrobiological Studies I*. Prague: Academia. 408 pp.
65. Hrbáček, J. 1969. Water passage and the distribution of plankton organisms in Slapy Reservoir. See Ref. 118, pp. 144–54
66. Hrbáček, J., Straskraba, M. 1973. *Hydrobiological Studies II*. Prague: Academia. 348 pp.
67. Hrbáček, J., Straskraba, M. 1973. *Hydrobiological Studies III*. Prague: Academia. 310 pp.
68. Hutchinson, G. E. 1957. *A Treatise on Limnology, Vol. 1*. New York: Wiley, 1015 pp.
69. Hutchinson, G. E. 1966. *A Treatise on Limnology, Vol. 2*. New York: Wiley, 1155 pp.
70. Hynes, H. B. N. 1961. The effect of water-level fluctuations on littoral fauna. *Int. Ver. Theor. Angew. Limnol. Verh.* 14:652–56
71. Hynes, H. B. N. 1969. Life in freshwater communities. See Ref. 118, pp. 25–31
72. Hynes, H. B. N. 1975. The stream and its valley. *Int. Ver. Theor. Angew. Limnol. Verh.* 19:1–15
73. Idler, D. R., Clemens, W. A. 1959. The energy expenditures of Fraser River sockeye salmon during the spawning migration to Chilko and Stuart Lakes. *Prog. Rep. Int. Pacific Salmon Fish. Comm.* 25 pp.
74. Jackson, P. B. N., Davies, B. R. 1976. Cabora Bassa in its first year: some ecological comparisons. *Rhod. Sci. News.* 10:128–33
75. *The James Bay Agreement*. 1975. Quebec City: Editeur officiel du Québec
76. *James Bay Hydro-Electric Project. Environmental Concerns*. 1975. Ottawa: Environment Canada. 45 pp.
77. Johnson, D. W. 1919. *Shore Processes and Shoreline Development*. Facsimile edition 1965. New York: Hafner. 584 pp.
78. Johnson, M. A. 1964. Turbidity currents. *Oceanogr. Mar. Biol. Ann. Rev.* 2:31–43
79. Jones, H. G., Leclerc, M., Meybeck, M., Ouellet, M., Rousseau, A. 1976. Etude limnologique préliminaire du réservoir Manicouagan, Québec. *Int. Ver. Theor. Angew. Limnol. Verh.* 19: 1758–67
80. Judd, W. R., ed. 1974. Seismic effects of reservoir impounding. *Eng. Geol. Amsterdam* 8:1–212
81. Kachugin, E. G. 1966. The destructive action of waves on the water reservoir banks. *Int. Assoc. Sci. Hydrol. Symp. Garda*. 1:511-17
82. Kallio, P. 1969. A task for ecologists around waterfalls in Labrador-Ungava. *Science* 166:1598–1601
83. Keddy, P. A. 1976. Lakes as islands: the distribution of two aquatic plants, *Lemna minor* L. and *Lemna trisulca* L. *Ecology* 57:353–59
84. Kershaw, A. K. 1973. *Quantitative and Dynamic Ecology*. London: Edward Arnold. 308 pp. 2nd ed.
85. Kondratjev, N. E. 1966. Bank formation of newly established reservoirs. *Int. Assoc. Sci. Hydrol. Symp. Garda.* 1:804–11
86. Krzyzanek, E. 1970. Formation of bottom fauna in the Goczalkowice dam reservoir. *Acta Hydrobiol.* 12:399–421
87. Kuenen, Ph. H. 1956. The difference between sliding and turbidity flow. *Deep Sea Res.* 3:134–39
88. Kujawa, M. 1974. Plankton studies on a recently impounded reservoir. *J. Environ. Health*. 37:252–55
89. Lawler, G. H. 1970. Parasites of coregonid fishes. In *Biology of Coregonid Fishes*, ed. C. C. Lindsay, C. J. Wood, pp. 279–309. Winnipeg: Univ. Manitoba Press
90. Leentvaar, P. 1966. The Brokopondo Lake in Surinam. *Int. Ver. Theor. Angew. Limnol. Verh.* 16:680–84
91. Leentvaar, P. 1973. Lake Brokopondo. See Ref. 1, pp. 186–96
92. Lefolii, K. 1970. *The St. Lawrence Valley*. Toronto: Nat. Sci. Canada Ltd. 160 pp.
93. Lehmkuhl, D. M. 1972. Changes in thermal regime as a cause of reduction of benthic fauna downstream of a reservoir. *J. Fish. Res. Bd. Can.* 29:1329–32
94. Lindstrom, T. 1973. Life in a lake reservoir. *Ambio* 2:145–53

95. Linsley, R. K., Kohler, M. A., Paulhus, J. L. H. 1949. *Applied Hydrology.* New York: McGraw-Hill. 689 pp.
96. Lowe-McConnell, R. H., ed. 1966. *Man-Made Lakes.* London: Academic 218 pp.
97. Luferov, V. P. 1969. Brief comparative description of the epifauna in the flooded forests of the Volga reservoirs. See Ref. 146, pp. 14–19
98. MacArthur, R. H., Wilson, E. O. 1967. *The Theory of Island Biogeography.* Princeton: Princeton Univ. Press. 203 pp.
99. Macdonald, J. R., Hyatt, R. A. 1973. Supersaturation of nitrogen in water during passage through hydroelectric turbines at Mactaquac Dam. *J. Fish. Res. Bd. Can.* 30:1392–94
100. Machniak, K. 1975. The effects of hydro-electric development on the biology of northern fishes (reproduction and population dynamics). I. Lake Whitefish, *Coregonus clupeaformis* (Mitchill). II. Northern Pike *Esox lucius* (Linnaeus). III. Yellow Walleye *Stizostedion vitreum vitreum* (Mitchill) IV. Lake Trout *Salvelinus namaycush* (Walbaum). *Environ. Can. Fish. Mar. Serv. Tech. Rep. 527, 528, 529, 530*
101. Margalef, R. 1960. Ideas for a synthetic approach to the ecology of running waters. *Int. Rev. Gesamten Hydrobiol. Hydrogr.* 45:133–53
102. Margalef, R. 1968. *Perspectives in Ecological Theory.* Chicago: Univ. Chicago Press. 111 pp.
103. Margalef, R. 1973. Plankton production and water quality in Spanish reservoirs. First report on a research project. Paper prepared for XI Congress, International Commission on Large Dams, Madrid.
104. Margalef, R. 1976. Typology of reservoirs. *Int. Ver. Theor. Angew. Limnol. Verh.* 19:1841–48
105. Matheny, R. T. 1976. Maya lowland hydraulic systems. *Science.* 193:639–46
106. McLachlan, A. J. 1974. Development of some lake ecosystems in tropical Africa, with special reference to the invertebrates. *Biol. Rev. Cambridge Philos. Soc.* 49:365–97
107. Mermel, T. W. 1976. International activity in dam construction. *Int. Water Power Dam Constr.* 28(4):66–69
108. Milne, W. G., ed. 1976. Proceedings of the 1st International Symposium on Induced Seismicity *Eng. Geol. Amsterdam* 10:83–338
109. Morduchai-Boltovskoi, F. D. 1955. Raspredelenie bentosa v Rybinskom vodovhranilishche. *Tr. Biol. Stan. "Borok", Akad. Nauk SSSR.* 2:36–53 (Cited in Ref. 129)
110. Morduchai-Boltovskoi, F. D. 1961. Die Entwicklung der Bodenfauna in den Stauseen der Wolga. *Int. Ver. Theor. Angew. Limnol. Verh.* 14:647–51
111. Morgan, A. H. 1930. *Field Book of Ponds and Streams.* New York and London: Putnam's. 448 pp.
112. Nebeker, A. V. 1976. Survival of *Daphnia*, crayfish, and stoneflies in air-supersaturated water. *J. Fish. Res. Bd. Can.* 33:1208–12
113. Neel, J. K. 1966. Impact of reservoirs. *Limnology in North America*, ed. D. G. Frey, pp. 575–92. Madison: Univ. Wisconsin Press. 734 pp.
114. Neu, H. J. A. 1975. Runoff regulation and its effects on the ocean environment. *Can. J. Civ. Eng.* 2:583–91
115. Newbury, R., Malaher, G. W. 1972. The destruction of Manitoba's last great river. *Nat. Can. Ottawa* 1(4):4–13
116. Nursall, J. R. 1952. The early development of a bottom fauna in a new power reservoir in the Rocky Mountains of Alberta. *Can J. Zool.* 30:387–409
117. Nursall, J. R. 1969. Faunal changes in oligotrophic manmade lakes: experience on the Kananaskis River system. See Ref. 118, pp. 163–75
118. Obeng, L. E., ed. 1969. *Man-Made Lakes: The Accra Symposium.* Accra: Ghana Univ. Press
119. Odum, E. P. 1969. The strategy of ecosystem development. *Science.* 164:262–70
120. Odum, E. P. 1971. *Fundamentals of Ecology*, Philadelphia: Saunders. 574 pp. 3rd ed.
121. Oklahoma Geological Survey. 1976. *Oklahoma Reservoir Resources.* Oklahoma Acad. Sci. Publ. No. 5. Norman, Okla: Okla. Geol. Surv. 151 pp.
122. Ostrofsky, M. L., Duthie, H. C. 1975. Primary productivity, phytoplankton, and limiting nutrient factors in Labrador lakes. *Int. Rev. Gesamten Hydrobiol. Hydrogr.* 60:145–58
123. Paterson, C. G., Fernando, C. H. 1969. Macroinvertebrate colonization of the marginal zone of a small impoundment in Eastern Canada. *Can. J. Zool.* 47:1229–38
124. *The Peace-Athabasca Delta, A Canadian Resource.* 1972. Ottawa: Information Canada. 144 pp.

125. *The Peace-Athabasca Delta Project. Technical Report.* 1973. Ottawa: Information Canada. 176 pp.
126. *Pedder Papers. Anatomy of a Decision.* 1972. Parkville, Australia: Austral. Conserv. Found. 63 pp.
127. Peters, R. H. 1976. Tautology in evolution and ecology. *Am. Nat.* 110:1–12
128. Petr, T. 1969. Development of bottom fauna in the man-made Volta lake in Ghana. *Int. Ver. Theor. Angew. Limnol. Verh.* 17:273–82
129. Petr, T. 1970. Macroinvertebrates of flooded trees in the man-made Volta lake (Ghana) with special reference to the burrowing Mayfly *Povilla adusta* Navas. *Hydrobiologia* 36:373–98
130. Petr, T. 1971. Establishment of chironomids in a large tropical manmade lake. *Can. Entomol.* 103:380–85
131. Pianka, E. R. 1972. r and K selection or b and d selection? *Am. Nat.* 106:581–88
132. Pieczynska, E. 1972. Ecology of the eulittoral zone of lakes. *Ekol. Pol.* 20:637–732
133. Prowse, G. A., Talling, J. F. 1958. The seasonal growth and succession of plankton algae in the White Nile. *Limnol. Oceanogr.* 3:222–38
134. Purcell, L. T. 1939. The aging of reservoir waters. *J. Am. Water Works Assoc.* 31:1775–1806
135. "Reservoir." *Encyclopaedia Britannica.* 1969. Chicago: William Benton
136. *The Restoration of Water Levels in the Peace-Athabasca Delta. Reports and Recommendations.* 1973. Edmonton, Alberta: Environ. Conserv. Auth. 136 pp.
137. Richardson, B. 1972. *James Bay.* San Francisco: Sierra Club. 190 pp.
138. Ridley, J. E., Steel, J. A. 1975. Ecological aspects of river impoundments. *River Ecology,* ed. B. A. Whitton, pp. 565–87. Berkeley: Univ. California Press. 725 pp.
139. Riley, G., Stommel, H., Bumpus, D. F. 1949. Quantitative ecology of the plankton of the Western North Atlantic. *Bull. Bingham Oceanogr. Collect.* 12:1–169
140. Rodhe, W. 1964. Effects of impoundment on water chemistry and plankton in Lake Ransaren (Swedish Lappland). *Int. Ver. Theor. Angew. Limnol. Verh.* 15:437–43
141. Rothé, J. P. 1973. Summary: geophysics report. See Ref. 1, pp. 441–54
142. Rucker, R. R. 1972. Gas-bubble disease of salmonids: a critical review. *US Bur. Sport Fish. Wildlife. Tech. Pap. No. 58.* 11 pp.
143. Rutter, E. J., Engstrom, L. R. 1964. Hydrology of flood control. Part III. Reservoir regulation. In *Handbook of Applied Hydrology,* ed. V. T. Chow, Sect. 25, pp. 60–97. New York: McGraw-Hill
144. Ryder, R. A., Henderson, H. F. 1975. Estimates of potential fish yield for the Nasser Reservoir, Arab Republic of Egypt. *J. Fish. Res. Bd. Can.* 32:2137–51
145. Saville, C. M. 1925. Color and other phenomena of water from an unstripped reservoir in New England. *J. N. Engl. Water Works Assoc.* 39:145–70
146. Shtegman, B. K., ed. 1969. *Plankton and Benthos of Inland Waters.* Jerusalem: Israeli Program Sci. Transl. 391 pp.
147. Shtegman, B. K., ed. 1969. *Production and circulation of organic matter in inland waters.* Jerusalem: Israeli Program Sci. Transl. 287 pp.
148. Siever, R. 1975. The earth. *Sci. Am.* 233(3):82–90
149. Sinclair, D. C. 1965. *The Effects of Water Level Changes on the Limnology of Two British Columbia Coastal Lakes, with Particular Reference to the Bottom Fauna.* MS Thesis. Univ. British Columbia, Vancouver. 84 pp.
150. Smith, N. A. F. 1969. Early Spanish dams. *Endeavour* 28:13–16
151. Snegirev, I. A. 1964. Calculation of the movement of bottom flows, saturated with suspended particles in reservoirs. *Sov. Hydrol. Sel. Pap.,* 6:621–26
152. Soltera, R. A., Gasperino, A. F., Graham, W. G. 1974. Chemical and physical characteristics of a eutrophic reservoir and its tributaries: Long Lake, Washington. *Water Res.,* 8:419–31
153. Soltera, R. A., Gasperino, A. F., Graham, W. G. 1975. Chemical and physical characteristics of a eutrophic reservoir and its tributaries: Long Lake, Washington - II. *Water Res.* 9:1059–64
154. Sprules, W. M. 1940. The effect of a beaver dam on the insect fauna of a trout stream. *Trans. Am. Fish. Soc.* 70:236–48
155. Stearns, R. H. 1916. Decolorization of water by storage. *J. N. Engl. Water Works Assoc.* 30:20–34
156. Stroud, R. K., Bouck, G. R., Nebeker, A. V. 1975. Pathology of acute and chronic exposure of salmonid fishes to super-saturated water. See Ref. 58, pp. 435–49

157. Lake Winnipeg, Churchill and Nelson Rivers Study Board. 1975. Summary Report. Winnipeg, Manitoba. 64 pp.
158. Talling, J. F., Wood, R. B., Prosser, M. V., Baxter, R. M. 1973. The upper limit of photosynthetic productivity by phytoplankton: evidence from Ethiopian soda lakes. *Freshwater Biol.* 3:53–76
159. Tarverdiyev, R. B. 1972. Changes in the morphometric characteristics of the Mingechaur reservoir since the time it was filled. *Sov. Hydrol. Sel. Pap.* 5:452–56
160. Thienemann, A. 1954. Ein drittes biozönotisches Grundprinzip. *Arch. Hydrobiol.* 49:421–42
161. Toetz, D. W. 1976. Mineral cycling in reservoirs. See Ref. 121, pp. 21–28
162. Toran, J. 1973. The consequence of building dams on the environment. *Proc. First World Congr. Water Resour., Chicago.* pp. 45–47
163. Trotzky, H. M., Gregory, R.W. 1974. The effect of water flow manipulation below a hydroelectric power dam on the bottom fauna of the Upper Kennebec River, Maine. *Trans. Am. Fish. Soc.* 103:318–24
164. Tseeb, Ya. Ya. 1962. On certain regular features associated with the formation of the hydrobiological regime in the Kakhovsk Reservoir. See Ref. 30, pp. 204–10 (Translation RTS 2937)
165. Tyler, P. A., Buckney, R. T. 1974. Stratification and biogenic meromixis in Tasmanian reservoirs. *Aust. J. Mar. Freshwater Res.* 25:299–313
166. Tyurin, P. V., ed. 1966. *The Storage Lakes of the USSR and their Importance for Fishery.* Jerusalem: Israeli Program Sci. Transl. 244 pp.
167. Vendrov, S. L. 1965. A forecast of changes in natural conditions in the northern Ob' basin in case of construction of the lower Ob' hydro project. *Izv. Akad. Nauk SSR. Ser. Geogr. No. 5, 37–49.* Transl. in *Soviet Geography: review and translation* 6(10):3–8
168. Waddy, B. B. 1975. Research into the health problems of man-made lakes, with special reference to Africa. *Trans. R. Soc. Trop. Med. Hyg.* 69:39–50
169. Ward, J. V. 1974. A temperature-stressed stream ecosystem below a hypolimnial release mountain reservoir. *Arch. Hydrobiol.* 2:247–75
170. Welcomme, R. L. 1975. The fisheries ecology of African floodplains. *CIFA Tech. Pap No. 3.* Rome: FAO
171. Wiebe, A. H. 1939. Density currents in Norris Reservoir. *Ecology,* 20:446–50
172. Wiebe, A. H. 1960. The effects of impoundments upon the biota of the Tennessee River System. *Int. Union Conserv. Nature Nat. Resour. Seventh Tech. Meet.* 4:101–17
173. Williams, W. D. 1972. The uniqueness of salt lake ecosystems. In *Productivity Problems in Freshwaters,* ed. Z. Kajak, A. Hillbricht-Ilkowska, pp. 349–61. Warsaw: PWN-Polish Scientific Publishers. 918 pp.
174. Wunderlich, W. O., Elder, R. A. 1973. Mechanics of flow through man-made lakes. See Ref. 1, pp. 300–10
175. Zhadin, V. I., Gerd, S. V. 1963. *Fauna and Flora of the Rivers, Lakes and Reservoirs of the USSR.* Jerusalem: Israeli Program Sci. Transl. 626 pp.

THE ECOLOGY
OF FISH MIGRATIONS

♦4126

William C. Leggett

Department of Biology, McGill University, Montreal, Quebec, Canada H3A 1B1

INTRODUCTION

Much of what is now known about the migrations of fish has been derived from the study of commercial catch records. Such data frequently provide more information on the distribution of fishing effort than on the distribution of fish (42, 55, 74, 127) and thus require great care in interpretation. Well designed field studies of migration are very costly and have therefore been restricted in number and confined largely to commercially important species. These studies, too, experience significant problems imposed by gear selection and by the efficiency and distribution of effort (1, 2, 29, 30, 44, 55, 95, 127, 129, 137, 158, 201). For these reasons the literature relating to field investigations of fish migrations is concentrated on a relatively small number of species and is largely descriptive and nonsynthetic in nature. Only with the very recent development of underwater telemetry and sonar techniques has it become possible continuously to observe the behavior of migrating fish under field conditions (128, 191, 192).

These inherent difficulties have restricted most experimental studies of fish migrations to the laboratory. Laboratory studies have focused on the physiological factors involved in the initiation of migratory behavior, and on the possible mechanisms of orientation (reviewed in 4, 11, 24, 40, 186, 207, 208, 215).

This review attempts to place the migrations of fish into an ecological perspective by integrating the findings of field, laboratory, and theoretical studies of the subject.

GUIDANCE MECHANISMS

The design of studies of the factors responsible for initiating and guiding fish migrations has been greatly influenced by the phenomenon of homing. During the late 1920s and 1930s several large-scale investigations of the biology of Pacific and Atlantic salmon (reviewed in 127) provided convincing evidence of their return to natal rivers following the ocean migration. The apparent precision of this return led to the early acceptance of the Parent Stream Theory (53) and to a search for additional evidence of such behavior. The most celebrated examples of homing continue to be those attributed to members of the family Salmonidae (36, 46, 73, 110, 174, 198). However, it is now known that the phenomenon is widespread among both freshwater and marine species (38, 55, 82, 113, 149, 201, 212). This finding has led to an extensive series of investigations of the biological basis of homing. These studies have been focused on two main questions: (a) by what means are fish guided during their open water movements; and (b) how is home ultimately recognized?

Orienting Cues—Open Water Migrations

The apparent precisions with which fish are able to return to their home areas (15, 36, 46, 53, 94, 196, 204) and the precise timing frequently associated with this return (108, 133, 144) have led to the assumption that precise orientation and possibly even true navigation is required during the open water phase of the migration (13, 84, 100, 101, 109). This assumption has, in turn, had a major influence on the studies undertaken in an attempt to determine the means by which fish are guided in their movements. The relatively advanced understanding of the mechanisms of orientation in other animal groups (77, 78, 139, 140, 151) has also guided these investigations.

SUN ORIENTATION The ability of fish to utilize the sun as an aid to orientation has been clearly established. In an early attempt to study this phenomenon, Hasler et al (103) observed the movements of white bass (*Roccus chrysops*) by tagging them with small floats. Under clear skies the movements of fish displaced from their spawning areas exhibited a northerly bias. Under overcast skies, or when fitted with opaque eye caps, the movement appeared to be random. Follow-up experiments revealed that white bass, pumpkinseed sunfish (*Lepomis gibbosus,*) and bluegill sunfish (*L. macrochirus*) could be trained to utilize an artificial sun to orient to a compass direction. Subsequent laboratory studies by several investigators (24, 104, 113, 131, 186–188) have demonstrated that a variety of fish species can find and hold compass directions by using information provided by the sun's azimuth, and that they can correctly compensate for daily and seasonal variation in the rate of change of the azimuth angle. This ability to compensate for the movement of the sun appears to be innate, but may be aided by information provided by daylength and the sun's altitude (24).

Conclusive evidence of the use of a sun compass by fish at liberty in their environment has been more difficult to obtain. The most convincing demonstration to date is that of Winn, Salmon & Roberts (213) who studied the orientation of parrotfish (*Scario guacamaia* and *S. coelestinus*) during their return to offshore caves following feeding excursions in shallow water. When displaced, unimpaired adults followed a SE compass course appropriate for a return to the capture area. This SE orientation was disturbed when clouds obscured the sun or when the displaced fish were blinded by the application of eyecaps before release. When unimpaired fish were released following a six hour shift in their biological clock, they followed a NNW compass course appropriate to the 165° clockwise shift in the sun's azimuth that occurs during six hours at the latitude of the study area. Sun-oriented behavior was not observed in juveniles, or in adults occupying shoreline caves, suggesting that the direction to offshore caves was learned. McCleave (152) and McCleave & Horrall (153) observed impaired homing ability in blind cutthroat trout (*Salmo clarki*) in Yellowstone Lake. However, ultrasonic tracking of blind and unimpaired fish revealed no difference in orientation between the two groups. Attempts to shift the photoperiod by six hours had no effect on orientation. The shift of photoperiod may have been unsuccessful (154). Groot (85) and Stasko et al (193, 194) have reported appropriately oriented movement in juvenile and adult sockeye and adult pink (*O. gorbuscha*) salmon under cloudy skies and at night. Dodson & Leggett (59, 60) observed similar behavior in migrating American shad (*Alosa sapidissima*).

POLARIZED LIGHT Partially polarized light is common in nature. The degree of polarization varies diurnally, being maximal at dawn and dusk when the sky may be up to 90%, and clear oceanic water near the surface up to 60%, polarized (207-209). The first evidence that animals other than man can perceive polarized light was provided in 1948 when von Frisch (77) published the results of his now famous behavioral studies with honeybees. Field studies of the orientation of yearling sockeye salmon during their seaward migration from Babine Lake led Groot (85) to propose that these fish might derive directional cues from the pattern of polarized light during the dawn and dusk peaks in their migratory activity. The ability of juvenile sockeyes to discriminate between vertical and horizontal planes of linearly polarized light was subsequently demonstrated (57) in a series of food conditioning experiments conducted in the laboratory. Waterman & Forward (209) are the only investigators to date to offer direct experimental evidence of the ability of fish to orient to linear polarized light under field conditions. In their experiments with *Zenarchopterus dispar,* the pattern of orientation was far from obvious and was demonstrated only after significant data selection, defended by the authors on the basis of several criteria. Unfortunately, the ecological significance of this behavior in *Z. dispar* could not be evaluated because of the limited knowledge of its life history and general behavior. While evidence of the ability of saltwater and freshwater fishes to detect patterns of polarized light continues to mount, no definitive evidence of its use as an orienting cue by migrating fish has yet been given (208).

GEOMAGNETIC AND GEOELECTRIC FIELDS Some groups of fish are known to possess extreme sensitivity to electric fields (155). Electric fields generated by the movement of ocean currents through the earth's magnetic field appear to be of sufficient magnitude to be detected by migrating fish such as salmon and eels (178, 179). Directional information may also be provided to migrating fish as a result of electric currents generated by the movement of individuals or schools of fish through the earth's magnetic field (26, 216). Royce et al (181) have suggested that geoelectric fields may provide orienting cues to migrating Pacific salmon whose homing migrations are made at a time when atmospheric conditions obscure the sun almost continually (191).

Evidence is rapidly accumulating in support of the hypothesis that European eels (*Anguilla anguilla*) are capable of detecting and orienting to geomagnetic fields. Five year old eels tested in the presence of natural magnetic fields at three different locations in Russia exhibited strong directional tendencies. The preferred direction was different at each study site and was eliminated when the geomagnetic field was obscured by the operation of Helmholtz coils. When tested in the presence of artificial magnetic fields the eels oriented parallel to the lines of force in a homogeneous field, and perpendicular to the lines of force in a gradient. These findings suggest the eels were attempting to minimize the influence of the field on their receptor organs (26.) Ovchinnikov, Gleizer & Galaktionov (170) reported evidence of ontogenetic and seasonal changes in the orientation of eels to magnetic fields. Three year old eels tested in an experimental maze exhibited a strong orientation in the 60°–240° direction. Elimination of the magnetic field with Helmholtz coils resulted in uniform orientation in the maze. Four year old (pigmented) eels behaved similarly but preferred the 90°–270° orientation. Tesch (201) reported silver and yellow eels tracked in the North Sea followed constant but reciprocal compass courses appropriate to their migratory phase. These headings were preferred regardless of the direction of displacement. He concluded that this compass sense was mediated directly by geomagnetic fields because the currents in the tracking areas were confused and could not provide consistant bearings via induced geoelectric fields. Zimmerman & McCleave (216) were unable to duplicate the results of Branover et al (26) and Ovchinnikov, Gleizer & Galaktionov (170) with American eels (*A. rostrata*). However, their survey of the published work led them to conclude that American and European eels can obtain orienting information from the earth's magnetic field. There is, however, no clear indication of how this is achieved.

INERTIAL GUIDANCE Barlow (18) proposed that animals may be capable of inertial navigation, perhaps utilizing the labyrinth of the inner ear as a sensor. Jones (127) criticized this suggestion, noting that the apparent threshold response of vertebrates to accelerations is three to four orders of magnitude greater than that required for a precision guidance system. Studies of turning angles in goldfish (*Carassius auratus*) by Kleerekoper et al (135) indicate that goldfish perform a sequence of left turns followed by a sequence of right turns. The turning angles in both sequences are nonrandom and result in a zero difference in cumulated radians of left and right turns. They hypothesized, as did Barlow, that the angular displacement in one direction may be recorded by the inner ear and compensated for by angular displacements in the opposite direction. Active turning would increase the acceleration involved, and overcome the threshold problem previously discussed. It is not known how long such orientation could be maintained. Pink and chum (*O. keta*) salmon fry (111) and goldfish (125) can maintain clockwise or counterclockwise orientation for considerable periods of time. Jones (127) suggested that the labyrinth may be instrumental in maintaining this behavior.

RANDOM WALKS Proponents of the hypothesis that a high degree of orientation is required in open water migrations have not been without their critics. The most vociferous of these was A. G. Huntsman, who repeatedly argued that investigators generally ignored or rejected any mechanism in fish migration other than precise orientation (115–117.) Huntsman's persistence led to a veritable storm of activity among fishery scientists, all designed to prove him wrong. His several defeats on specific points led to a somewhat uncritical acceptance of the concept of highly directed movement by migrating fish, a concept that was strengthened by early successes in the search for celestial aids to orientation. This historical development may explain, in part, the general tendency to discount the contribution of fish which depart from expected migratory routes or patterns of behavior to the overall understanding of the mechanics of fish migration (17, 36, 55).

Several authors have recently critically reassessed the importance of precise orientation in fish migration. This reassessment has been sparked, in part, by the failure to demonstrate any evidence of true navigational ability in fish, and by the limited amount of definitive evidence of the operation of proposed orienting mechanisms under natural conditions.

Saila (182) evaluated the possibility that the known homing performance of winter flounder (*Pseudopleuronectes americanus*) to Green Hill Pond on the Rhode Island coast from offshore areas could be achieved by random movement. Saila's model incorporated several biologically defensible assumptions, the most important being that the fish were at all times constrained between the coast and the 20 fathom depth contour. Calculations showed that random search would bring 75% of the population to the coast within 90 days, a time shorter than the period during which flounder are known to enter Green Hill Pond to spawn. After reaching the coast, the fish were assumed to locate the pond by a combination of pilotage and random searching close to shore where home water clues would prevail.

In a second study, Saila & Shappy (183) tested the hypothesis that observed homing ability of Pacific salmon could be achieved by random movement at sea. This analysis was based on a Monte-Carlo-type model in which salmon swam straight paths along courses whose orientation was determined at random. Step length, the distance between successive turns, was varied or biased with limits. The starting point of the homing migration was 2224 km west of the coast, and a fish was assumed to have homed successfully if it reached an area of coast extending 37 km on either side of the home river. Within this area final homing was assumed to result from a combination of pilotage, random search, and responses to home river cues. The model allowed 175 days for homing. Under conditions of random search and equal step length in all directions, no homing occurred. However, a very small bias in step length toward the general direction of home was sufficient to achieve homing success comparable to the observed percentage return of salmon tagged on the high seas. Saila & Shappy suggested that the required directional bias might be provided by sun orientation. Geoelectric fields, polarized light, and behavioral responses to a suitable environmental field (see the section below on maximization of comfort) could also provide the necessary bias.

Patten (171) criticized the Saila & Shappy model for the severity of constraints required to apply it and for its lack of relationship to salmon biology. He proposed a new model based more firmly on the known facts of salmon biology. In simplified terms, Patten proposed that salmon respond in a rational way to the various characteristics of the environment evaluated enroute. For salmon such characteristics might include food, predators, temperature, salinity, olfactory stimuli, sun position, etc. Patten's model, which demonstrated that a very low level of rational decision making (i.e. 'proper' response to the environment) yields acceptably high probabilities of successful homing, verified the earlier conclusion of Saila and Shappy that there is no necessity of assuming navigational ability, or even precise orientation, on the part of migrating salmonid fishes. Jones (127) applied Griffins' (88) radial linear scattering model to known facts of salmon biology in a further demonstration that highly oriented movement is not a prerequisite to successful homing.

A critical appraisal of the results of laboratory and field studies of orientation and migratory behavior in fish adds further support to the hypothesis that a relatively low degree of orientation may be involved in migrations. For example, tagged plaice (*Pleuronectes platessa*) are known to disperse in near random fashion after spawning. Later, the density centers of separate stocks return to the home area even though individually tagged fish are frequently recovered great distances from these density centers. De Veen notes that a small directional bias can account for the movement of the density center back to the home area. However, the near random dispersion of individual fish is generally forgotten or neglected in descriptions of plaice migrations, yielding the impression that individual fish are performing great acts of precise migration (55). The return of Atlantic salmon from ocean feeding areas to the mouths of their home rivers is also commonly depicted as a highly oriented movement (36). However, tagging studies reveal that individual salmon are frequently captured in coastal areas great distances from their home rivers (52, 69, 160, 195, 199, 210).

The migratory behavior of individual fish, as revealed by ultrasonic and sonar tracking, provides further evidence of a comparatively high frequency of seemingly inappropriate orientation during the migration (59, 60, 86, 102, 148, 153, 154, 193, 194). In each of these studies, there was clear evidence of a bias in the general direction of the home or, in the case of one study (86), toward the outlet of a lake. However, in no case was the movement precisely oriented toward the objective of the migration and in all cases the variability in the behavior of individual fish was high.

Analysis of the results of laboratory studies of the orienting ability of individual fish indicates that considerable wandering from the optimum course is to be expected. For example, Braemer (24) provided graphical examples of the direction taken by individual fish in choice experiments using the sun, or an artificial sun, as the external orientation cue. While these data clearly show a concentration of choices in the appropriate directions, the spread of critical choices by individual fish commonly exceeds 100° and frequently exceeds 180°. This indicates that, while individual fish are capable of obtaining directional information from the sun, and of correcting for hourly changes in the azimuth, the resulting orientation is only approximate. This would provide a strong bias in one direction, but not a precise course over great distances. The performance of individual fish in response to polarized light is similar (209).

MAXIMIZATION OF COMFORT Patten's (171) proposal that salmon homing could be achieved by a low frequency of appropriate behavioral responses to a host of environmental factors has recently been expanded into a generalized model of fish migration (13, 14). This appears to be the first attempt to build a quantitative ecological model of fish migrations. Balchen (13, 14) argues that fish movement, at any moment in time, is a "simpleminded process of maximizing comfort." Comfort does not refer to the conscious comfort or pain experienced by humans. Rather, maximization of comfort is seen as an unconscious seeking for optimum physiological and neurological states, the nature of which may be altered by biochemical processes under neuro-endocrine control.

The potential influence of neuro-endocrine activity on migratory behavior and orientation is great. Hormones are known to influence general excitability, olfaction, vision, taste, feeding activity, maturation, and salinity and temperature preference (reviewed in 215). Seasonal or ontogenetic changes in endocrine activity could thus profoundly influence the relative sensitivity of numerous neurological and physiological processes, thereby regulating both the timing and the route of migrations. (17)

The attainment of what Balchen terms the optimum state must involve compromise, because many competing benefits and costs are present at one time. The nature of this compromise must be governed by the instantaneous biological sensitivity associated with each of the comfort states. This hypothesis appears to be consistent with a large body of experimental and observational data relating to fish migrations. The well known temperature preferendum, a species characteristic that may vary with seasons, age, and physiological state (72, 159, 197), is an excellent example of single factor optimization. One of the finer examples of the compromise involved in the 'maximization of comfort' is provided by Fry's study (79) of the summer migrations of cisco (*Leucichthys artedi*) in Lake Nipissing. In this lake an early summer migration to deep water occurs in response to the development of unfavorable temperatures in the epilimnion. In August and September a reverse migration occurs, this time in response to declining O_2 and increasing CO_2 levels in the hypolimnion. Penetration of the thermocline occurs when the problems presented by O_2 and CO_2 levels counteract the temperature preference. The movement of adult and juvenile fish occurs at different times because of size-related temperature preferences. Feeding, which is normally active during the summer, ceases for a prolonged period after movement in both directions, indicating that the movement occurs at the expense of growth . A similar compromise regulates the movements of *Coregonus muksun* in Tiksi Bay, where, in response to low O_2 levels, schools leave the offshore feeding areas following satiation but return regularly because of limited food availability in the better oxygenated coastal zone. Bitukov (22) reports that the diel vertical migrations of herring (*Clupea harengus*) in the Baltic also appear to involve a compromise between the degree of satiation and the temperature preferendum. The principal food of herring in the Baltic is most abundant in the cold deep-water layer. Herring feed there by day, but migrate to surface waters, which are nearer the preferred temperature, at night. Collins (47) reported compromise behavior in the response of upstream migrant alewives (*Alosa pseudoharengus*) to experimentally regulated temperatures and CO_2 levels. The relative influence of these factors on the choice of migration routes depended on the magnitude of the difference between them. For example, a 2° C temperature difference overbalanced a 2.0 ppm CO_2 difference, but was balanced by a 2.4 ppm CO_2 differential.

The migrations of juvenile Pacific salmon provide additional evidence of the role of physiological and neurological optimization in migration. Byrne (34) reported a rapid ontogenetic change in the response of juvenile sockeye salmon to light. During the first nine days following emergence, activity was bimodal, with a slight dawn peak and a high peak at night. Beginning at 10–14 days after emergence and continuing until the end of the study (11 months), activity was strongly unimodal, with maximum activity occurring during the day. This changing light-response is directly related to the migratory requirements of the stock. Early dark-active behavior results in downstream drift to the lake nursery area. The subsequent light-active behavior is appropriate to feeding in the lake.

The process of optimization in juvenile sockeye is, however, more complex than these experiments suggest. This complexity is introduced by the fact that sockeye spawn in both inlet and outlet streams. Brannon's (25) outstanding study of the migrations of juvenile sockeye has shown that movement to the lake nursery areas is also governed by racially specific rheotactic responses that can be moderated by olfactory stimuli. Sockeye from both inlet and outlet streams exhibit negative rheotaxis at high current velocities and positive rheotaxis at low velocities. However, the threshold velocity at which the change from positive to negative rheotaxis occurs differs among groups. Fish spawned in outlet streams have threshold velocities above their maximum sustained swimming speed, and perhaps equal to their maximum swim speed. For these fish upstream migration to the lake is the only behavioral response possible. Fish spawned in inlet streams have lower thresholds, and thus exhibit negative rheotaxis at lower current velocities. This leads to downstream movement to the lake nursery areas. This relatively simple behavioral response to current velocity also regulates more complex migratory behavior. Fish spawned in Weaver Creek must move downstream to the Harrison River and then upstream in Harrison River to the Harrison Lake nursery area. High current velocities in Weaver Creek induce a negative rheotaxis and downstream movement. Water velocities in Harrison River are below the threshold for this stock. When fry enter Hamilton River a reversal of rheotaxis occurs and upstream migration to Hamilton Lake results. This behavior can be altered by olfactory stimuli if the fry make an inappropriate movement enroute to the nursery area. Richkus (176) reported similar current regulated reversals of rheotaxis in migrating juvenile alewives.

The timing of the seaward migration of juvenile Pacific salmon, and the rate at which this movement occurs, appear to be governed by ontogenetic changes in salinity preference. McInerney (156) studied the modal salinity preferences of five species of Pacific salmon during their first year of life. In all five the initial preference was for fresh water. This preference changed gradually, the rate differing among species, until all five exhibited a distinct preference for saltwater. The change in salinity preference was consistent, in each case, with the known migratory patterns of the species. A study of the salinity gradients in estuaries led McInerney to

ECOLOGY OF FISH MIGRATIONS   293

conclude that the seaward progress of juvenile salmon was regulated by the gradual increase in salinity preference. Hurley & Woodall (118) confirmed this pattern of salinity preference in pink salmon, but noted that early emerging fry developed their saltwater preference more gradually than those emerging later. The delayed development of saltwater preference in early emerging fry is believed to synchronize the period of seaward migration. Such changes are known to be moderated by neuroendocrine activity which is, in turn, regulated by photoperiod (8–10, 68, 110, 112).

In addition to the change in salinity preference, pink salmon fry exhibit a distinct change in temperature preference approximately 30–40 days after emergence. Before this time fry prefer temperatures in the range 11.7°–13.3° C; subsequently they prefer cooler temperatures (9.4°–10.6° C). This combination of ontogenetic changes in salinity and temperature preference is believed to regulate the initial saltwater movements of pink salmon. Immediately after leaving freshwater, juvenile pinks are found in inshore waters and bays where salinities are reduced and temperatures are high. After approximately one month they move to deeper, colder, more saline water further offshore. The timing of this movement corresponds with the downward shift in temperature preference and the development of full seawater preference.

Direct evidence of the role of optimizing behavior in the regulation of long range open water migrations is less abundant. A growing body of data appears to be in general agreement with the hypothesis.

TEMPERATURE   There is a strong relationship between oceanic temperature patterns and the movement of fish (33, 41, 45, 56, 64, 65, 71, 80, 81, 106, 121, 124, 130, 134, 138, 142, 150, 159, 161, 162, 164, 175, 185, 202). The north–south movements of American shad on the Atlantic coast of North America are regulated by the seasonal movement of waters in the 13°–18° C range (146). Movement within this narrow temperature band regulates the timing of the return to home rivers and synchronizes this return with the occurrence of optimum conditions for egg and larval survival. The seasonal north–south movements of albacore tuna (*Tunnus germo*) on the west coast of North America are known to coincide with the seasonal distribution of waters in the 14.4°–16.1° C range (2). A similar pattern has been reported on the Asian coast (44). The east–west movements of this species occur along a path bounded on the north by the 14° C isotherm and on the south by the 20° C isotherm. The width and position of this zone varies seasonally and annually, and influences the point at which tuna approach the North American coast (44). Similar temperature-related movements are reported for other tuna species, each species having a distinct temperature preference (143). The relationship between temperature and the migration of cod (*Gadus morhua*) and capelin (*Mallotus villosus*) in the Barents Sea is so direct that it is possible to predict the relative success of the Finnish and Murman fisheries four to six months in advance. In cold years these fish move further to the west with the warm water mass, and occur in greatest abundance on the Finnish coast. When the winter conditions are less severe they do not migrate as far and, as a result, the Murman fishery predominates (137).

The relationship between temperature and migrations may be indirect. Several authors have suggested a relationship between temperature and the occurrence of preferred food sources (56, 142, 146). The best documented example is the correlation between the movement of adult and subadult herring and the advance of the biological spring in the Norwegian and Greenland Seas (172). There the pattern and speed of migration vary annually, primarily in response to varying intensities of the inflow of warm Atlantic waters. This variation in warmth accelerates or retards the onset of phytoplankton and zooplankton blooms. A similar relationship among water temperature, food availability, and migrations has been reported for the anchovy (*Engraulis encrasicholus*) (20).

Temperature patterns may also influence migratory behavior directly through regulation of orientation (61, 166) or of swim speeds (23, 28).

OCEAN FRONTS   The migratory routes of fishes often follow ocean discontinuities or 'fronts' (12, 17, 19, 43, 75, 76, 80, 136, 143, 158). These fronts, characterized by steep salinity and/or temperature gradients, are common at the interface between major water masses or currents. High primary and secondary productivity is generally associated with these areas and may be the primary reason for the accumulation of fish there. (89, 143). In addition, these zones may act as barriers and/or guides to migrants at different seasons or at different stages in their development (13, 76, 136, 143).

A major difficulty in the analysis of the role of specific water masses, oceanic convergences and divergences, and frontal zones in determining the distribution and migration of fish has been the general lack of concurrent observations of fish abundance and hydrographic conditions over large areas. The utility of such information is vividly illustrated in the results of an outstanding investigation (80) of concurrent studies of oceanographic conditions and the distribution of homing sockeye salmon in western Alaskan and east Bering Sea waters during May and June, 1962–1968. Earlier studies (12) had shown that immature sockeye undertake regular north–south movements from wintering areas along a broad front ranging from longitude 155°–175° east, at 45°–50° north latitude, to summering areas predominately above 49° north latitude in the area of the Aleutian Islands and the Alaskan Peninsula. These movements showed no apparent relationship to any single oceanographic feature, and were assumed by the author to result primarily from temperature preference and food availability. This behavior pattern is altered during the fall before the attainment of sexual maturity when, instead of migrating south in the normal manner, maturing fish remain in the northern (summering) area throughout the fall and winter. Fujii's studies (80) revealed that mature and maturing sockeye concentrate in an area south of the Aleutians in May and June in preparation for the homing migration to rivers entering Bristol Bay. Their immediate entry into the Bering Sea is prevented by cold, high-salinity water in the Aleutian passes, the result of strong upwelling. This temperature/salinity barrier disappears when a rapid and continuous northbound flow of Alaskan Stream water enters the Bering Sea via the Aleutian passes as a result of spring tides and wind action. This condition normally develops in early June, and movement of sockeye into the Bering Sea commences immediately. This northward movement is largely restricted to the passes located east of longitude 175° west, where the northward flowing water has a higher temper-

ECOLOGY OF FISH MIGRATIONS   295

ature and lower salinity than is found in the western passes. Once in the Bering Sea, the sockeye migrate within a low salinity zone bounded on the north by the 33.1–33.2‰ halocline. This low salinity area is an extension of a mixed continental mass lying to the east. Therefore, if fish move west or north they encounter low-temperature/high-salinity water that acts as a reflecting barrier. Movement to the east maintains them in preferred temperature/salinity conditions. In the continental slope area, the sockeye follow a mixed water stream of high-temperature/low-salinity water flowing eastward through the center of Bristol Bay. This water mass, the principal migration path of sockeye, is made up primarily of Alaskan Stream water which, having entered the Bering Sea via the Aleutian passes, is carried to Bristol Bay by the counterclockwise currents in the area. Fujii hypothesizes that different races of sockeye respond differently to the salinity-temperature characteristics of this water mass. The tendency for larger sockeye to migrate close to the northern boundary of the zone, and for chum salmon to be more abundant than sockeye in the vertically isohaline waters at the northern boundary, may be reflections of racial and species differences in response to these variables. Such behavior could influence the relative points of contact of different races with the coast.

CURRENTS   Currents are also believed to play a significant role in fish migrations, although the mechanics of this have for the most part gone uninvestigated (127, 155). The subject has been extensively reviewed in recent years (4, 127, 141, 142). In addition to developing frontal zones and providing cues for orientation (see above), currents are known to influence migrations by transporting eggs and larvae from spawning to nursery areas, and by transporting adults. The transportation of eggs and larvae (6, 21, 37, 108, 122, 150) was once considered to be completely passive but recent studies have shown fish larvae may exert significant control over the rate, and possibly the direction, of their movements (21, 51). The transport of adults appears to be regulated, both in rate and direction, by vertical migration and by altered rheotaxis. These behavioral responses are regulated by photoperiod and/or lunar cycles (6, 126, 127; F. R. Harden Jones, personal communication).

The close relationship between hydrographic conditions and fish behavior may explain the significant shifts in centers of distributions of some species that have occurred in historical times. Petterson (173) notes that the extreme southern extension of herring movements in the twelfth and thirteenth centuries, which formed the basis of the so called Hanseatic Herring Fishery, was coincident with a major southward movement of cold water masses. Dunbar (65, 66) interprets changes in the distribution of Greenland cod and the sudden appearance of large numbers of salmon in the Greenland area in the late 1950s as the result of known changes in ocean climate. Variations in the timing and abundance of Pacific salmon migrations into specific rivers have also been attributed to large changes in oceanographic conditions (17).

Home Recognition

While homing is known to occur in a large number of freshwater and marine species (see above), the environmental cues used by migrating fish to identify the home area have been studied in only a small number of anadromous, intertidal, and freshwater species. Species making restricted movements within relatively small geographical areas appear to use both olfaction and local topographical cues to locate and identify the home area (5, 90, 113, 127). The principal mechanism of home area identification in anadromous species appears to be olfaction.

OLFACTION Buckland (31) seems to have been the first to propose that each river possesses a characteristic odor, and that detection and identification of this odor forms the basis of homing. This hypothesis remained untested for almost 80 years until A. D. Hasler and his coworkers began a long series of field and laboratory studies of the relationship between olfaction and homing. In the first of these investigations Walker & Hasler (206) demonstrated the ability of bluntnose minnows (*Hyborhynchus notatus*) to detect and discriminate between dilute rinses from a variety of aquatic plants. Subsequent experiments (96) showed that this species could discriminate between waters from two Wisconsin streams. These findings led to a series of papers (see 97–99, 101) extolling the virtues and requirements of the olfactory hypothesis of homing, but offering little new data. However, in recent years, evidence in support of the theory has accumulated rapidly. The olfactory sensitivity of fish is acute. Eels are capable of detecting and responding to β-phenylethyl alcohol at a dilution equivalent to one half teaspoonful in Lake Constance. This implies detection of 2 or 3 molecules on the olfactory epithelium (200). Similar olfactory sensitivity is known in salmon and other species (27, 91, 119).

Electrophysiological (EEG) studies of the olfactory response (48, 91, 203) have confirmed that salmon, carp (*Cyprinus carpio*), and rainbow trout (*S. gairdneri*) can distinguish between home and nonhome waters, and between different nonhome waters. These findings support the suggestion that each river has a characteristic odor that can be recognized by migrating fish. In the most rigorous test of the olfactory hypothesis to date, Cooper & Hasler (49) have recently demonstrated the ability of coho salmon to 'imprint' (109, 127) to a synthetic chemical compound (morpholine), retain this information until maturity, and use it to identify a stream scented with the chemical. Earlier criticisms of the use of morpholine in EEG response experiments (92, 93) were evaluated and effectively dismissed.

In a parallel study (50), the homing of morpholine-imprinted coho to a foreign water source into which morpholine had been released at low concentrations (3×10^{-4} to 1×10^{-5} mg liter^{-1}) was evaluated. In four experiments over two years, morpholine-imprinted adults returned to Oak Creek at a ratio of 8.8:1 (1739:197) over the controls. In the third year, when no morpholine had been released into the creek, imprinted and control fish entered the creek in equal numbers (55:51). Behavioral experiments, employing ultrasonic telemetry techniques (189), have demonstrated that morpholine-imprinted coho stop at morpholine trails laid down in their paths whereas unimprinted controls do not.

These findings are in accord with, and lend support to, a considerable body of observational data relating to the movements and homing performance of salmonid and other fishes. Imprinting of salmon in natural waters apparently occurs during a very short interval immediately before and possibly during the seaward migration

(36, 62, 123, 127, 204). There is, in addition, some evidence from EEG studies (169) that salmon do not follow an increasing gradient of a single odorant, but rather remember separately, and in order, each odor locus (stream) along the seaward migration, and react to each independently during the return migration. This hypothesis is supported by the observation that homing success is significantly reduced if fish are transported over a portion of their downstream migration (35, 205).

The nature of the olfactory cues has been variously characterized as volatile, nonvolatile, organic, and inorganic (70, 105, 119, 120, 157.) This may reflect the difference in cues among streams. Pheromones may also be important in the olfactory response to home water under natural conditions. Several authors (58, 63, 163, 168) using EEG techniques have shown that char (*Salvelinus alpinus*), chinook, and coho salmon can distinguish between waters in which they have been held, and waters in which other species or conspecifics from other populations (races?) have lived. White (211) and Soloman (190) have reported the entry of adult Atlantic salmon into streams (known to contain no spawning populations,) within one year following stocking with fry and juveniles. These authors concluded that the entering adults were attracted to the rivers by pheromones given off by the juveniles.

The importance of olfactory cues in guiding the final stages of the homing migration is indicated by the significant reduction in homing success of olfactory occluded fish (54, 59, 87, 152). The available evidence indicates that the olfactory cue serves as a sign stimulus releasing an appropriate behavioral response to other orienting cues (25, 51, 59, 60, 127).

The extent of the penetration of detectable home stream odors into lakes or the sea is not precisely known. Undoubtedly it is influenced by discharge volumes, currents, and the olfactory sensitivity of the animal. Huntsman (114) noted that the large discharge of the St. John River, N. B., produces a 'freshwater trail' detectable to Grand Manan Island, a distance of 75 km, and that drift netters fishing for salmon concentrated their efforts in this trail. The influence of Connecticut River water on the hydrographic characteristics of eastern Long Island Sound is extensive, and can be traced to Montauk Point at its eastern entrance, a distance of 65 km (177), where it is believed to initiate the final stages of homing in American shad (60). Jones (127) has calculated that Fraser River sockeye may detect Fraser River water as much as 160–320 km from the Juan de Fuca Strait.

Atmospheric conditions may alter the distribution and concentration of odorants. The largest runs of salmon into rivers on the east coast of Kamchatka, where rivers are large and discharges massive, occur when winds are offshore causing the freshwater plumes to be distributed over large areas of the Sea of Okkotsk (67). On the west coast of South Sakhalin, offshore winds disperse the freshwater plume and dilute its effect. Here, where rivers and discharge volumes are small, the largest runs occur during strong onshore winds that concentrate the freshwater plume near the river mouth. The influence of winds on the timing of fish migrations has also been reported (107, 147).

The role of olfaction in the orientation of migrations in offshore waters has not been investigated. However, in view of the major importance of this mechanism in the final stages of homing, of the known hydrological and biological differences among various water masses (12, 75, 89, 136), of the possibility of sequential use of olfactory cues in homing (35, 127, 168, 205) and of the apparent role of olfactory cues as sign stimuli triggering appropriate behavioral responses to other directional cues (25, 51, 59, 60, 127) it is highly possible that this mechanism also operates in open water orientation remote from the home area.

GENETIC FACTORS Bams (16) has recently demonstrated the importance of genetic factors in homing. The survival and homing success of an introduced donor stock of pink salmon were compared with those of a hybrid stock created by crossing females from the donor with males of the local residual stock. Survivals, from fry to returning (coastal) adult, were identical, and comparable to those of other years. Imprinting alone brought home some pure donor stock, but their numbers were significantly below those of the hybrid stock whose rate of return to the natal river was similar to that of the local population. However, the male genetic component alone was not sufficient to achieve normal accuracy of return to the natal tributary.

Ecological Significance of Homing

In the context of this discussion, homing implies a return to the natal area for reproduction. Such behavior is believed to maximize reproductive success by synchronizing the return of mature animals to the spawning grounds when conditions are optimum for egg and larval development, and by regulating the number of adults utilizing a given area, thus avoiding under- or overutilization of the habitat (165, 167, 169). Reproductive isolation achieved through homing also facilitates the development of population-specific adaptations to the particular habitat occupied. Banks (17) has suggested in this regard that the precision of homing in fish is proportional to the degree of adaptation required to complete the life history. The genetically regulated interpopulation differences in the migratory behavior of sockeye salmon fry (25) are an excellent example of such adaptation. Khalturin (132) and Schaffer & Elson (184) have noted a positive correlation between body size and the harshness of the upriver migration in stocks of salmon. These patterns appear genetically fixed and operate to balance the energy stored as fat and protein during the marine migration with the energy demands of the freshwater migration and spawning.

American shad populations native to North American Atlantic coast rivers exhibit profound differences in reproductive characteristics over the range of the species (145). Mean age at maturity and the frequency of repeat spawning increase with the latitude of the home river, while relative and absolute fecundity vary inversely with these population characteristics and with latitude. These reciprocal trends in life history characteristics are independent of growth parameters and result in near uniform lifetime egg production in all populations. The principal factors influencing this pattern of reproductive variability are the environmental cost of the spawning migration (83) and environmental stability during the critical egg and larval stages (145). Similar but less pronounced differences in reproductive characteristics also occur in populations homing to three tributaries of the St. John River,

N.B. (39). Pacific salmonids also exhibit latitudinal variations in fecundity and repeat-spawning related to the requirements of the reproductive habitat (32, 180, 214).

The evolution and maintenance of highly specialized reproductive and behavioral adaptations demand a correspondingly high degree of reproductive isolation. It is not inconceivable that in addition to the evolution of behavior patterns that enable migrants to return to their natal rivers, there might also have evolved, in some populations, patterns of behavior that would prevent straying fish from entering rivers other than home. The intensity of home stream fidelity demonstrated by sockeye salmon in Brooks and Karluk Lakes, Alaska, supports this view (94).

Additional examples of behavioral rejection of nonhome water have been reported by Sutterlin & Gray (198) and by Jensen & Duncan (123). The available data are inadequate to permit full evaluation of this hypothesis. If such behavior does exist, it would help to explain the apparent inconsistency, noted by Jones (127), between the high home-stream fidelity demonstrated by salmon, and the surprisingly low proportion of maturing salmon that enter freshwater to spawn.

SUMMARY AND CONCLUSIONS

Early assumptions that a high degree of orientation, possibly even true navigation, was required to achieve the homing success exhibited by many species have been shown to be in error. Rather, a small amount of homeward bias appears sufficient to ensure successful homing. The results of a still limited number of studies in which the migratory behavior of individual fish has been observed clearly indicate, however, that migratory movements are not random, but are oriented, with varying precision, in the general direction of home.

Several species of fish are known to be capable of obtaining directional information from the sun, polarized light, and geomagnetic fields. Some limited inertial guidance may also be involved. It is unlikely, however, that these mechanisms alone are responsible for oriented migrations in fish, if for no other reason than that minor miscalculation of heading, or the influence of varying current strengths and directions would create significant errors in a course regulated solely by such cues.

An impressive body of literature supports the hypothesis that fish migrations involve a continuous optimization of physiological and neurological states in response to a multiplicity of environmental stimuli. The nature of this optimum state varies seasonally, and with ontogeny, in response to changes in neuroendocrine activity. Racial and species differences in the timing and pattern of these responses are assumed to have developed through natural selection.

Recognition of the home area apparently involves both olfactory and local topographical cues. In long-distance migrants the former predominate. Each river, and apparently each tributary, has a characteristic odor. Young salmon, and possibly other species, imprint to this odor shortly before, or during, the seaward migration. During the homing migration this olfactory cue serves as a sign stimulus releasing appropriate behavioral responses to other orienting cues. The nature and regulation of these behavioral responses have not been extensively studied to date. The importance of olfaction in the homing migrations of exclusively marine species is unknown.

Homing may result in reproductive isolation. This isolation is essential to the development of complex behavioral, energetic, and reproductive adaptations to the reproductive habitat occupied.

The recent development of ultrasonic telemetry and sonar tracking techniques has greatly expanded the horizons of investigations into the nature of migratory behavior in fish. Early studies using these technologies have revealed that the migratory behavior of fishes is more complex, and more closely related to the immediate environment of the animal, than had previously been assumed. Properly used, these tools should greatly accelerate our understanding of the sequence of physiological and behavioral interactions between the animal and its environment that are, clearly, the ecological basis of migration.

ACKNOWLEDGMENTS

I am deeply indebted to Mrs. Rosemary Ernhofer, who conducted the literature search for this review, and to the library staffs of McGill University and the St. Andrews Biological Station, Environment Canada, for their assistance in obtaining the required materials. This review was written at Laval University, Quebec, while on sabbatical leave from McGill University. Financial support was provided by the National Research Council of Canada (Grant No. A 6513). The support and encouragement of these organizations are gratefully acknowledged.

Literature Cited

1. Ahlstrom, E. H. 1954. Distribution and abundance of egg and larval populations of the Pacific sardine. *U.S. Fish. Wildl. Serv., Fish. Bull.* 56:83–140
2. Alverson, D. L. 1961. Ocean temperatures and their relation to albacore tuna (*Thunnus germo*) distribution in waters off the coast of Oregon, Washington and British Columbia. *J. Fish. Res. Bd. Can.* 18:1145–52
3. Arnold, G. P. 1969. The reactions of the plaice (*Pleuronectes platessa* L.) to water currents. *J. Exp. Biol.* 51:681–97
4. Arnold, G. P. 1974. Rheotropism in fishes. *Biol. Rev.* 49:515–76
5. Aronson, L. R. Further studies on orientation and jumping behavior in the gobiid fish, *Bathygobius soporator. Ann. NY Acad. Sci.* 188:378–407
6. Bagenal, T. B. 1966. The ecological and geographical aspects of the fecundity of the plaice. *J. Mar. Biol. Ass. UK* 46:161–86
7. Baggerman, B. 1957. The role of external factors and hormones in migration of sticklebacks and juvenile salmon. In *Symposium on Comparative Endocrinology,* ed. A. Gorbman, 24–37. New York: Wiley
8. Baggerman, B. 1960. Salinity preference, thyroid activity and the seaward migration of four species of Pacific salmon (*Oncorhynchus*). *J. Fish Res. Bd. Can.* 17:295–322
9. Baggerman, B. 1960. Factors in the diadromous migrations of fish. *Symp. Zool. Soc. London,* 1:33–60
10. Baggerman, B. 1962. Some endocrine aspects of fish migration. *Gen. Comp. Endocrinol. Suppl. 1,* 188–205
11. Baggerman, B. 1963. The effect of TSH and antithyroid substances on salinity preference and thyroid activity in juvenile Pacific salmon. *Can. J. Zool.* 41:307–19
12. Bakkala, R. G. 1971. Distribution and migration of immature sockeye taken by U.S. research vessels with gillnets in offshore waters, 1956–67. *Bull. Int. N. Pac. Fish. Comm.* 27:1–70
13. Balchen, J. G. 1976. Principles of migration of fishes. *SINTEF:* The engi-

neering research foundation at the Technical University of Norway, Trondheim. Teknisk notat nr. 81 for NTNF/NFFR. 33 pp.
14. Balchen, J. G. 1976. Modelling of the biological state of fishes. *SINTEF: Eng. Res. Found. Tech. Univ. Norway, Trondheim. Teknisk notat nr. 62 for NTNF/NFFR.* 25 pp.
15. Ball, O. P. 1955. Some aspects of homing in cutthroat trout. *Proc. Utah Acad. Sci., Arts, Lett.* 32:75–80
16. Bams, R. A. 1976. Survival and propensity for homing as affected by presence or absence of locally adapted paternal genes in two transplanted populations of pink salmon (*Oncorhynchus gorbuscha*). *J. Fish Res. Bd. Can.* 33:2716–25
17. Banks, J. W. 1969. A review of the literature on the upstream migration of adult salmonids. *J. Fish. Biol.* 1:85–136
18. Barlow, J. S. 1964. Inertial navigation as a basis for animal navigation. *J. Theor. Biol.* 6:76–117
19. Beardsley, G. L. Jr. 1969. Distribution and apparent relative abundance of yellowfin tuna (*Thunnus albacares*) in the eastern tropical Atlantic in relation to oceanographic features. *Bull. Mar. Sci.* 19:48–56
20. Berenbeim, D. Ya., Dubrovin, I. Ya., Studenikina, E. M. 1973. Forecast of the start of the fall migration of the anchovy, *Engraulis encrasicholus* (L.) through the Kerch Strait. *J. Ichthyol.* 13:313–17
21. Bishnai, H. M. 1960. The effect of water currents on the survival and distribution of fish larvae. *J. Cons., Cons., Perma. Int. Explor. Mer* 25:134–46
22. Bityukov, E. P. 1959. A contribution to the problem of vertical migrations of *Clupea harengus membras* L. *Dokl. Akad. Nauk SSSR. Transl. Biol. Sci.* 128:766–69
23. Blaxter, J. H. S., Dickson, W. 1959. Observations on the swimming speeds of fish. *J. Cons., Cons. Perma. Int. Explor. Mer* 24:472–79
24. Braemer, W. A. 1960. A critical review of the sun-azimuth hypothesis. *Cold Spring Harbor Symp. Quant. Biol.* 25:413–28
25. Brannon, E. L. 1972. Mechanisms controlling migration of sockeye salmon fry. *Bull. Int. Pac. Salmon Fish. Comm.* 21:1–86
26. Branover, G. G., Vasil'Yev, A. S., Gleizer, S. I., Tsinober, A. B. 1971. A study of the behavior of eels in natural and artificial magnetic fields and an analysis of its reception mechanism. *J. Ichthyol.* 11:608–14
27. Brett, J. R., MacKinnon, D. 1954. Some observations on olfactory perception in migrating adult coho and spring salmon. *J. Fish. Res. Bd. Can.* 11:310–18
28. Brett, J. R., Hollands, M., Alerdice, D. F. 1958. The effect of temperature on the cruising speed of young sockeye and coho salmon. *J. Fish. Res. Bd. Canada* 15:587–605
29. Bridger, J. P. 1956. On night and day variations in catches of fish larvae. *J. Cons., Cons. Perma. Int. Explor. Mer* 22:42–57
30. Bridger, J. P. 1958. On efficiency tests made with a modified Gulf III high speed tow net. *J. Cons., Cons. Perm. Int. Explor. Mer* 23:357–65
31. Buckland, F. 1880. *Natural History of British Fishes.* London: Unwin. 420 pp.
32. Buckley, R. V. 1967. Fecundity of steelhead trout, *Salmo gairdneri* from Alsea River, Oregon. *J. Fish. Res. Bd. Can.* 24:917–26
33. Buen, F. de. 1927. Substitucion alternativa de la especies emigrantes. *Bol. Pescas* 12(135):337–42
34. Byrne, J. E. 1971. Photoperiodic activity changes in juvenile sockeye salmon (*Onchorhynchus nerka*). *Can. J. Zool.* 49:1155–58
35. Carlin, B. 1955. Tagging of salmon smolts in the River Lagan. *Rep. Inst. Freshwater Res. Drottningholm* 36:57–74
36. Carlin, B. 1968. Migration of salmon. *Atl. Salmon Ass. Spec. Publ. Montreal, Canada* pp. 14–22
37. Carruthers, J. N., Lawford, A. L., Veley, V. F. C. 1959. Fishery hydrography: Broad-strength fluctuations in various North Sea fish with suggested methods of prediction. *Kiel. Meeresforsch.,* 8(1):5–15
38. Carscadden, J. E., Leggett, W. C. 1975. Meristic differences in spawning population of American shad (*Alosa sapidissima*): Evidence for homing to tributaries in the St. John River, New Brunswick. *J. Fish. Res. Bd. Can.* 32:653–60
39. Carscadden, J. E., Leggett, W. C. 1975. Life history variations of American shad, *Alosa sapidissima* (Wilson), spawning in tributaries of the St. John River, New Brunswick. *J. Fish. Biol.* 7:595–609
40. Cheal, M. 1975. Social olfaction: a review of the ontogeny of the olfactory

influences on vertebrate behavior. *Behav. Biol.* 15(1):1-26
41. Chidambaram, K. 1950. Studies on the length frequency of the oil sardine, *Sardinella longiceps* (Cuv. and Val.) on certain factors influencing their appearance on the Calicut coast of Madras Presidency. *Proc. Indian Acad. Sci.* 3113:252-86
42. Cleaver, F. C. 1949. The Washington otter trawl fishery with reference to the petrale sole. *Wash. Dep. Fish. Biol. Rep.* No. 49A, 1-45
43. Cleaver, F. C. 1964. Origins of high seas sockeye salmon. *US Fish. Wildl. Serv. Fish. Bull.* 63(2):445-76
44. Clemens, H. B. 1963. A model of albacore migration in the North Pacific Ocean. *FAO Fish. Rep. No. 6., Vol. 3. Experience Paper* 31:1537-48
45. Clemens, W. A. 1959. Some problems in the behavior of Pacific salmon in the ocean. *Ann. Biol. Colloq.* 20:30-35
46. Clemens, W. A., Foerster, R. E., Pritchard, A. L. 1939. The migration of Pacific salmon in British Columbia waters. *Publ. Am. Ass. Advan. Sci.* (8) 51-59
47. Collins, G. B. 1952. Factors influencing the orientation of migrating anadromous fishes. *US Fish. Wildl. Serv., Fish. Bull.* 52:375-96
48. Cooper, J. C., Hasler, A. D. 1974. Electroencephalographic evidence for retention of olfactory cues in homing coho salmon. *Science* 183:336-38
49. Cooper, J. C., Hasler, A. D. 1976. Electrophysiological studies of morpholine imprinted coho salmon (*Oncorhynchus kisutch*) and rainbow trout (*Salmo gairdneri*). *J. Fish. Res. Bd. Can.* 33: 688-94
50. Cooper, J. C., Schalz, A. T., Horrall, R. M., Hasler, A. D., Madison, D. M. 1976. Experimental confirmation of the olfactory hypothesis with homing, artificially imprinted coho salmon (*Onchorhynchus kisutch*). *J. Fish. Res. Bd. Can.* 33:703-10
51. Creutzberg, F. 1963. The role of tidal streams in the navigation of migrating elvers (*Anguilla vulgaris* Turt.) *Ergeb. Biol.* 26:118-27
52. Dahl, K., Somme, S. 1938. Salmon markings in Norway 1937. *Skr. Norske. Vidensk-Akad. Oslo* 1(2):1-45
53. Davidson, F. A. 1937. Migration and homing in Pacific salmon. *Science* 86:1-4
54. DeLacy, A. C., Donaldson, L. R., Brannon, E. L. 1969. Homing behavior of chinook salmon. *Contrib. Univ. Wash. College (Sch.) Fish.* 300:59-60
55. DeVeen, J. F. 1970. On the orientation of the plaice (*Pleuronectes platessa* L.): I. Evidence for orienting factors derived from the ICES transplantation experiments in the years 1904-1909. *J. Cons., Cons. Perm. Int. Explor. Mer* 33(2): 192-227
56. Devold, F. 1967. The behavior of the Norwegian tribe of Atlanto-Scandian herring. *FAO Conf. Fish Behav. Relat. Fish. Tech. Tactics, Bergen, Norway. 19-27 Oct, 1967.* 8 pp.
57. Dill, P. A. 1971. Perception of polarized light by yearling sockeye salmon (*Oncorhynchus nerka*). *J. Fish. Res. Bd. Can.* 28:1319-22
58. Dizon, A. E., Horrall, R. M., Hasler, A. D. 1973. Olfactory electroencephalographic responses of homing coho salmon, *Oncorhynchus kisutch*, to water conditioned by conspecifics. *US Fish. Wildl. Serv., Fish Bull.* 71:893-96
59. Dodson, J. J., Leggett, W. C. 1973. Behavior of adult American shad (*Alosa sapidissima*) homing to the Connecticut River from Long Island Sound. *J. Fish. Res. Bd. Can.* 30:1847-60
60. Dodson, J. J., Leggett, W. C. 1974. Role of olfaction and vision in the behavior of American shad (*Alosa sapidissima*) homing to the Connecticut River from Long Island Sound. *J. Fish. Res. Bd. Can.* 31:1607-19
61. Dodson, J. J., Young, J. C. 1977. Temperature and photoperiod regulation of rheotropic behavior in prespawning common shiners, *Notropis cornutus* (Mitchill). *J. Fish. Res. Bd. Can.* 34:341-46
62. Donaldson, L. R., Allen, G. H. 1958. Return of the silver salmon, *Oncorhynchus kisutch* to point of release. *Trans. Am. Fish. Soc.* 87:13-22
63. Doving, K. B., Nordeng, H., Oakley, B. 1974. Single unit discrimination of fish odours released by char (*Salmo alpinus* L.) populations. *Comp. Biochem. Physiol.* 47A: 1051-63
64. Duncan, F. M. 1945. *Wonders of migration.* London: Low, Warston. 150 pp.
65. Dunbar, M. J. 1972. The nature and definition of the marine subarctic, with a note on the sealife of the Atlantic salmon. *Trans. Roy. Soc. Can.* (4) 10:249-57
66. Dunbar, M. J. 1973. On the west Greenland sea-life area of the Atlantic salmon. *Arctic* 26(1):1-6

67. Dvinin, P. A. 1952. The salmon of South Sakhalin. *Izu. Tikhookean nanchno-issled. Inst. rub Khoz. Okeanogr.* 37, 69–108. (*Fish. Res. Bd. Can. Transl.* 120, 1957)
68. Eales, J. G. 1963. A comparative study of thyroid function in migrant juvenile salmon. *Can. J. Zool.* 41:811–24
69. Edman, G. 1954. Märkning au lax utanför Hallands kust sommaren *Sv. Fish. Tidskr.* 63:166–71
70. Fagerlund, U. H. M., McBride, J. R., Smith, M., Tomlinson, N. 1963. Olfactory perception in migration salmon III. Stimulus for adult sockeye salmon in home stream waters. *J. Fish Res. Bd. Can.* 20:1457–63
71. Favorite, F., Hanavan, M. G. 1963. Oceanographic conditions and salmon distribution south of the Alaska Peninsula and Aleutian Islands, 1956. *Int. N. Pac. Fish. Comm. Bull.* 11:57–72
72. Ferguson, R. G. 1958. The preferred temperature of fish and their midsummer distribution in temperate lakes and streams. *J. Fish. Res. Bd. Can.,* 15: 607–24
73. Foerster, R. E. 1968. The sockeye salmon. *Fish. Res. Bd. Can. Bull.* 162. 422 pp.
74. Forrester, C. R. 1969. Results of English sole tagging in British Columbia waters. *Bull Pac. Mar. Fish Commn.* : (7):2–10
75. French, R. R., Bakkala, R. G. 1974. A new model of ocean migrations of Bristol Bay sockeye salmon. US Fish Wild. Serv. Fish. Bull. 72:589–614
76. French, R. R., McAlister, W. B. 1970. Winter distribution of salmon in relation to currents and water masses in the northeastern Pacific Ocean and migrations of sockeye salmon. *Trans. Am. Fish. Soc.* 99:649–63
77. Frisch, K. von. 1948. Gelöste und ungelöste Rätsel der Bienensprache. *Naturwissenschaften* 35:38–43
78. Frisch, K. von. 1965. *Tanzsprache und Orientierung der Bienen.* Berlin: Springer. 578 pp.
79. Fry, F. E. J. 1937. The summer migration of the cisco, *Leucichthys artedi* (Le Sueur), in Lake Nipissing, Ontario. *Univ. Tor. Stud. Biol. Ser. No. 44, Publ. Ont. Fish. Res. Lab., No. 55.* pp. 1–91
80. Fujii, T. 1975. On the relation between the homing migration of the western Alaska sockeye salmon, *Oncorhynchus nerka* (Walbaum) and oceanic conditions in the eastern Bering Sea. *Mem. Fac. Fish. Hokkaido Univ.* 22(2):99–192
81. Fuks, V. R. 1960. The effect of internal tidal waves on the daily vertical migration of commercial fish. *Rybn. Khoz. Okeanogr.* 46:189–96
82. Gerking, S. R. 1959. The restricted movement of fish populations. *Biol. Rev.* 34:221–42
83. Glebe, B. D., Leggett, W. C. 1976. Weight loss and energy expenditure of American shad during the freshwater migration. *Final Rep. Proj. AFC-8, US Natl. Mar. Fish Serv.* 110 pp.
84. Green, J. M. 1971. High tide movements and homing behavior of the tidepool sculpin. *J. Fish. Res. Bd. Can.* 28:383–89
85. Groot, C. 1965. On the orientation of young sockeye salmon (*Oncorhynchus nerka*) during their migration out of lakes. *Behavior (Suppl.) XIV.* 198 pp.
86. Groot, C. 1972. Migration of yearling sockeye salmon as determined by time lapse photograph of sonar observations. *J. Fish. Res. Bd. Can.* 29:1431–44
87. Groves, A. B., Collins, G. B., Trefethen, P. S. 1968. Roles of olfaction and vision in choice of spawning site of homing adult chinook salmon (*Oncorhynchus tshawytscha*). *J. Fish Res. Bd. Can.* 25:867–76
88. Griffin, D. R. 1952. Bird navigation, with an Appendix by Ernst Mayr on German experiments on the orientation of migratory birds. *Biol. Rev.* 27:359–400
89. Griffiths, R. C. 1963. Studies of oceanic fronts in the mouth of the Gulf of California, an area of tuna migrations. *FAO Fish. Rep.* 3(6):1583–1605
90. Gunning, G. E. 1959. The sensory basis for homing in longear sunfish, *Lepomis megalotis megalotis* (Rafinesque). *Invest. Indiana Lakes Streams* 5:103–30
91. Hara, T. J. 1970. An electrophysiological basis for olfactory discrimination in homing salmon: A review. *J. Fish. Res. Bd. Can.* 27:565–86
92. Hara, T. J. 1974. Is morpholine an effective olfactory stimulant in fish? *J. Fish. Res. Bd. Can.* 31:1547–50
93. Hara, T. J., MacDonald, S. 1975. Morpholine as olfactory stimulus in fish. *Science* 187:81–82
94. Hartman, W. L., Raleigh, R. F. 1964. Tributary homing of sockeye salmon at Brooks and Karluk lakes. *J. Fish. Res. Bd. Can.* 21:485–504
95. Hartt, A. C. 1966. Migrations of salmon in the North Pacific Ocean and Bering

Sea as determined by seining and tagging 1959-60. *Int. N. Pac. Fish. Comm., Bull. 19,* 141 pp.
96. Hasler, A. D. 1951. Discrimination of stream odor by fishes and its relationship to parent stream behavior. *Am. Nat.* 85:223-38
97. Hasler, A. D. 1954. Odour perception and orientation in fishes. *J. Fish. Res. Bd. Can.* 110:107-29
98. Hasler, A. D. 1956. Perception of pathways by fish in migration. *Quart. Rev. Biol.* 31(3):200-209
99. Hasler, A. D. 1967. Animal orientation and navigation. Underwater guideposts for migrating fishes. *Oregon State Univ. Biol. Colloq.* 27:1-20
100. Hasler, A. D. 1960 Homing orientation in migrating fishes. *Ergebn. Biol.* 23:94-115
101. Hasler, A. D. 1960. Guideposts of migrating fishes. *Science,* 132:785-92
102. Hasler, A. D., Gardella, E. S., Horrall, R. M., Henderson, H. F. 1969. Open-water orientation of White Bass, *Roccus chrysops,* as determined by ultrasonic tracking methods. *J. Fish. Res. Bd. Can.* 26:2173-92
103. Hasler, A. D., Horrall, R. M., Wisby, W. J., Braemer, W. 1958. Sun orientation and homing in fishes. *Limnol. Oceanogr.* 3(4):353-61
104. Hasler, A. D., Schwassmann, H. O. 1960. Sun orientation of fish at different latitudes. *Cold Spring Harbor Symp. Quant. Biol.* 25:429-41
105. Hasler, A. D., Wisby, W. J. 1951. Discrimination of stream odors by fishes and its relation to parent stream behavior. *Am. Nat.* 85:223-38
106. Hatanaka, M. 1956. Biological studies on the population of the saury, *Cololabis saira* (Brevoort). Part 2. Habits and migrations. *Tohoku J. Agric. Res.* 6:313-40
107. Hayes, F. R. 1953. Artificial freshets and other factors controlling the ascent and population of Atlantic salmon in the La Have River, Nova Scotia. *Fish. Res. Bd. Can. Bull. 99.* 114 pp.
108. Hela, I., Laevastu, T. 1962. *Fisheries hydrography.* London: Fishing News Books Ltd., 132 pp.
109. Hess, E. H. 1973. *Imprinting. Early experience and the developmental psychobiology of attachment.* New York: Van Nostrand. 472 pp.
110. Hoar, W. S. 1952. Thyroid function in some anadromous and landlocked teleosts. *Trans. Roy. Soc. Can.* (3) 45 (5):39-53
111. Hoar, W. S. 1956. The behavior of migrating pink and chum salmon fry. *J. Fish. Res. Bd. Can.* 13:309-25
112. Hoar, W. S. 1963. The endocrine regulation of migrating behavior in anadromous teleosts. *Proc. 16th Int. Congr. Zool.* 3:14-20
113. Hong Woo, K. 1974. Sensory basis of homing in the intertidal fish *Oligocottus maculosus* Girard. *Can. J. Zool.* 52:1023-29
114. Huntsman, A. G. 1934. Factors influencing return of salmon from the sea. *Trans. Am. Fish. Soc.* 64:351-55
115. Huntsman, A. G. 1939. Migration and conservation of salmon. In *The Migration and Conservation of Salmon,* ed. F. R. Moulton, 32-44. Lancaster: The Science Press. 106 pp.
116. Huntsman, A. G. 1950. Factors which may influence migration. *Salm. Trout Mag* 130:227-30
117. Huntsman, A. G. 1952. Wandering versus homing in salmon. *Salm. Trout Mag.* 136:185-91
118. Hurley, D. A., Woodall, W. L. 1968. Responses of young pink salmon to vertical temperature and salinity gradients. *Prog. Rep. Int. Pac. Salm. Fish. Comm.* 19:1-80
119. Idler, D. R., Fagerlund, U., Mayoh, H. 1956. Olfactory perception in migrating salmon. I. L-serine, a salmon repellent in mammalian skin. *J. Gen. Physiol.* 39:889-92
120. Idler, D. R., McBride, J. R., Jonas, R. E. E., Tomlinson, N. 1961. Olfactory perception in migrating salmon. II. Studies on a laboratory bioassay for home stream water and mammalian repellent. *Can. J. Biochem. Physiol.* 39:1575-84
121. Jackman, L. A. J., Steven G. H. 1955. Temperatures and mackerel movements in the inshore waters of Torbay, Devonshire. *J. Cons., Cons. Perm. Int. Explor. Mer* 33:363-85
122. Jacquaz, B., Able, K. W., Leggett, W. C. 1977. Ecology of larval capelin (*Mallotus villosus*) in the estuary and northwestern Gulf of St. Lawrence. *J. Fish. Res. Bd. Canada* 34. In press
123. Jensen, A. L., Duncan, R. N. 1971. Homing of transplanted coho salmon. *Prog. Fish. Cult.* 33:216-18
124. Jones, D., Miller, P. J. 1966. Seasonal migrations of the common goby, *Pomatoschistus microps* (Kroyer), in Morecambe Bay and elsewhere. *Hydrobiologia* 27:515-28

125. Jones, F. R. H. 1957. Rotation experiments with blind goldfish. *J. Exp. Biol.* 34:259–75
126. Jones, F. R. H. 1962. Further observations on the movements of herring (*Clupea harengus* L.) shoals in relation to tidal current. *J. Cons., Cons. Perm. Int. Explor. Mer* 27:52–76
127. Jones, F. R. H. 1968. *Fish Migration*. New York: St. Martins. 325 pp.
128. Jones, R. 1959. A method of analysis of some tagged haddock returns. *J. Cons., Cons. Perm. Int. Explor. Mer* 25:58–72
129. Jones, R. 1966. Manual of methods for fish stock assessment. *F.A.O. Fish. Biol. Tech. Pap. (51) Suppl. 1*
130. Jones, S. 1958. On the late winter and early spring migration of the Indian shad, *Hilsa ilisha* (Hamilton), in the Gangetic Delta. *Indian J. Fish* 4:304–14
131. Kalmus, H. 1964. Comparative physiology: Navigation by animals. *Ann. Rev. Physiol.* 26:109–30
132. Khaltruin, K. D. 1967. The value of angular displacement in the description of the anadromal migrations in the genus Salmo. *Proc. Acad. Sci. USSR* 177:736–38
133. Killick, S. R. 1955. The chronological order of Fraser River Sockeye salmon during migration, spawning and death. *Int. Pac. Salm. Fish. Comm. Bull. VII.* 95 pp.
134. Kinnear, B. S., Fuss, C. M. Jr. 1971. Thread herring distribution off Florida's west coast. *Commer. Fish. Rev.* 33(7–8):27–39
135. Kleerekoper, H., Timms, A. M., Westlake, G. F., Davy, F. B., Malar, T., Anderson, V. M. 1969. Inertial guidance system in the orientation of the goldfish (*Carassius auratus*). *Nature* 223:501–2
136. Kolesnikov, V. G. 1966. O vliyanii okeanologicheskikh uslovii sa puti migratsii norvezhskoi sel'di. (The effect of oceanologic conditions on the migration path of Norwegian herring). In *Materialy Sessii Uchennogo soveta PINRO po resultatam issledovanii v 1964.* (Contributions to the Session of the Academic Council of the Polar Scientific Research Institute of Marine Fisheries and Oceanography on the observations of 1964) *Murmansk* 6:130–41
137. Kondo, H., Hirano, Y., Nakayama, N., Miyake, M. 1963. Offshore distribution and migration of Pacific salmon (genus *Oncorhynchus*) based on tagging studies (1958–1961). *Int. N. Pac. Fish. Comm., Bull.* 17:1–128
138. Konstantinov, K. G. 1965. Water temperature as a factor guiding fish during their migrations. *Spec. Publ. Int. Comm. NW. Atl. Fish.* 6:221–24
139. Kramer, G. 1950. Orientierte Zugaktivität gekäfigter Singvögel. *Naturwissenschaften* 37:188
140. Kramer, G. 1952. Experiments on bird orientation. *Ibis* 94:265–85
141. Laevastu, T. 1965. Interpretation of fish distributions in respect to currents in the light of available laboratory and field observations. *Int. Comm. NW Atl. Fish. Spec. Publ.* 6:249–56
142. Laevastu, T., Hela, I. 1970. *Fisheries Oceanography: New Ocean Environmental Services.* London: Fishing New Books Ltd., 227 pp.
143. Laevastu, T., Rosa, H. Jr. 1963. Distribution and relative abundance of tuna in relation to their environment. *FAO Fish. Rep.* 3(6):1835–51
144. Leggett, W. C. 1976. The American shad (*Alosa sapidissima*) with special reference to its migration and population dynamics in the Connecticut River. In *The Connecticut River Ecological Study: The Impact of a Nuclear Power Plant,* ed. D. Merriman, L. M. Thorpe, 169–225. Amer. Fish. Soc. Monogr. No. 1
145. Leggett, W. C. 1969. Studies on the reproductive biology of the American shad (*Alosa sapidissima,* Wilson). A comparison of populations from four rivers of the Atlantic seaboard. PhD thesis. McGill University, Montreal. 125 pp.
146. Leggett, W. C., Whitney, R. R. 1972. Water temperature and the migrations of American shad. *US Fish. Wildl. Serv., Fish. Bull.* 70:659–70
147. Lorz, H. W., Northcote, T. G. 1965. Factors affecting stream location and timing and intensity of entry by spawning kokanee (*Oncorhynchus nerka*) into an inlet of Nicola Lake, British Columbia. *J. Fish. Res. Can.* 22:665–86
148. Madison, D. M., Horrall, R. M., Stasko, A. B., Hasler, A. D. 1972. Migratory movements of adult sockeye salmon (*Oncorhynchus nerka*) in coastal British Columbia as revealed by ultrasonic tracking. *J. Fish. Res. Bd. Can.* 29:1025–33
149. Malinin, L. K. 1969. Home range and homing instinct in fishes. *Zool. Zh.* 48:381–91
150. Martin, W. R., Jean, Y. 1964. Winter cod taggings off Cape Breton and on

offshore Nova Scotia banks 1959–1962. *J. Fish. Res. Bd. Can.* 21:215–38
151. Matthews, G. V. T. 1955. *Bird Navigation.* London: Cambridge. 197 pp.
152. McCleave, J. D. 1967. Homing and orientation of cutthroat trout (*Salmo clarki*) in Yellowstone Lake, with special reference to olfaction and vision. *J. Fish. Res. Bd. Can.* 24:2011–44
153. McCleave, J. D., Horrall, R. M. 1970. Ultra-sonic tracking of homing cutthroat trout, *Salmo clarki,* in Yellowstone Lake. *J. Fish. Res. Bd. Can.* 27:715–30
154. McCleave, J. D., Labar, G. W. 1972. Further ultrasonic tracking and tagging studies of homing cutthroat salmon (*Salmo clarki*) in Yellowstone Lake, *Trans. Am. Fish. Soc.* 101:44–54
155. McCleave, J. D., Rommel, S. A. Jr., Cathcart, C. L. 1971. Weak electric and magnetic fields in fish orientation. *Ann. NY Acad. Sci.* 188:270–82
156. McInerney, J. E. 1964. Salinity Preference: an orientation mechanism in salmon migration. *J. Fish. Res. Bd. Can.* 21:995–1018
157. Miles, S. G. 1968. Rheotaxis of elvers of the American eel (*Anguilla rostrata*) in the laboratory to water from different streams in N.S. *J. Fish. Res. Bd. Can.* 25:1591–1602
158. Miller, D., Cotton, J. B. Jr., Marak, R. R. 1963. A study of the vertical distribution of larval haddock. *J. Cons., Cons. Perm. Int. Explor. Mer* 28:37–49
159. Mironova, H. V. 1961. Migrations of immature codfish and reasons for their variability. *Biol. Abstr.* 43:754
160. Moriarty, C. 1968. Movements of salmon around Ireland 10: From the north Mayo coast (1962–1964). *Proc. R. Ir. Acad. Sect. B* 66:1–7
161. Nakamura, R. 1976. Temperature and the vertical distribution of two tidepool fishes (*Oligocottus maculosus* and *O. snyderi*) *Copeia* 1976 (1):143–52
162. Natarov, V. V., Noirkov, N. P. 1970. Oceanographic conditions in southeast Bering Sea and some distributional characteristics of halibut. *Biol. Abstr.* 54(7):3516
163. Nordeng, H. 1971. Is the local orientation of anadromous fishes determined by pheromones? *Nature* 233:411–13
164. Northcote T. G. 1962. Migratory behavior of juvenile rainbow trout, *Salmo gairdneri,* in outlet and inlet streams of Loon Lake, B.C. *J. Fish. Res. Bd. Can.* 19:201–10

165. Northcote, T. G. 1967. The relation of movement and migrations to production in freshwater fishes. In *The Biological Basis of Freshwater Fish Production,* ed. S. D. Gerking, 315–44. Oxford: Blackwell. 495 pp.
166. Northcote, T. G. 1969. Patterns and mechanisms in the lakeward migratory behavior of juvenile trout. In *Symposium on Salmon and Trout in Streams,* ed. T. G. Northcote, 183–203. Vancouver: Univ. Brit. Col. Press. 388 pp.
167. Northcote, T. G. 1977. Migratory strategies and production in freshwater fishes. In *Ecology of Freshwater Fish Production,* ed. S. D. Gerking. Oxford: Blackwell. In press
168. Oshima, K., Hahn, W. E., Gorbman, A. 1969. Olfactory discrimination in natural waters by salmon. *J. Fish Res. Bd. Can.* 26:2111–21
169. Oshima, K., Hahn, W. E., Gorbman, A. 1969. Electroencephalographic olfactory responses in adult salmon to waters traversed in the homing migration. *J. Fish Res. Bd. Can* 26:2123–33
170. Ovchinnikov, V. V., Gleizer, S. I., Galaktionov, G. Z. 1973. Characteristics of the orientation of the European eel, *Anguilla anguilla* (L.), at some stages of migration. *J. Ichthyol.* 13 (3):455–63
171. Patten, B. C. 1964. The rational decision process in salmon migration. *J. Cons. Cons., Perm. Int. Explor. Mer* 28:410–17
172. Pavshtiks, E. A. 1959. Seasonal changes in plankton and feeding migrations of herring. In: The herring of the North European Basin and adjacent seas. *US Fish. Wildl. Serv. Spec. Sci. Rep. Fish.* 327:104–39
173. Pettersson, O. 1926. Currents and fish migrations in the transition area. *J. Cons., Cons., Perma. Int. Explor. Mer* 1:322–26
174. Pritchard, A. L. 1948. A discussion of the mortality in pink salmon (*Oncorhynchus gorbuscha*) during their first period of marine life. *Trans. R. Soc. Can.* 42: Sect. V, 125–33
175. Radoirch, J. 1963. Effect of ocean temperature on the seaward movements of striped bass, *Roccus saxatilis,* on the Pacific coast. *Calif. Fish Game* 49:191–206
176. Richkus, W. A. 1975. The response of juvenile alewives to water currents in an experimental chamber. *Trans. Am. Fish. Soc.* 104:494–98

177. Riley, G. A. 1956. Oceanography of Long Island Sound, 1952–1954. II. Physical Oceanography. *Bull. Bingham Oceanogr. Collect. Yale Univ. New Haven, Conn.* 17:9–30
178. Rommel, S. A. Jr., McCleave, J. D. 1973. Sensitivity of American eels (*Anguilla rostrata*) and Atlantic salmon (*Salmo salar*) to weak electric and magnetic fields. *J. Fish. Res. Bd. Can.* 30: 657–63
179. Rommel, S. A. Jr., McCleave, J. D. 1973. Prediction of oceanic electric fields in relation to fish migration. *J. Cons., Cons., Perma. Int. Explor. Mer* 35:27–31
180. Rousefell, G. A. 1957. Fecundity of North American Salmonidae. *US Fish. Wildl. Serv. Fish. Bull.* 57:451–68
181. Royce, W. F., Smith, L. S., Hartt, A. 1968. Models of oceanic migrations of Pacific salmon and comments on guidance mechanisms. *US Fish. Wildl. Serv. Fish. Bull.* 66:441–6
182. Saila, S. B. 1961. A Study of winter flounder movements. *Limnol. Oceanogr.* 6:292–98
183. Saila, S. B., Shappy, R. A. 1963. Random movement and orientation in salmon migration. *J. Cons. Cons., Perm. Int. Explor. Mer* 28:153–66
184. Schaffer, W. M., Elson, P. F. 1975. The adaptive significance of variations in life history among local populations of Atlantic salmon in North America. *Ecology* 56(3):577–90
185. Schmidt-Koenig, K. 1975. Migration and homing in animals. *Zoophysiol. Ecol.* 6:31–43
186. Schwassmann, H. O. 1967. Orientation of Amazonian fishes to the equatorial sun. *Atas do Simposio sobre a Biota Amazonica 3 (Limnologia):* 201–20
187. Schwassmann, H. O., Braemer, W. 1969. The effect of experimentally changing photoperiod on the sun orientation rhythm of fish. *Physiol. Zool.* 34:273–86
188. Schwassmann, H. O., Hasler, A. D. 1964. The role of the sun's altitude in orientation of fish. *Physiol. Zool.* 37(2): 163–78
189. Schalz, A. T., Cooper, J. C., Madison, D. M., Horrall, R. M., Hasler, A. D., Dizon, A. E., Poff, R. Olfactory imprinting in coho salmon: behavioral and electrophysiological evidence. *Proc. Conf. Great Lakes Res., Huron, Ohio.* 16:143–53
190. Solomon, D. J. 1973. Evidence for pheromone-influenced homing by migrating Atlantic salmon, *Salmo salar* (L). *Nature* 244:231–32
191. Stasko, A. B. 1971. Review of field studies on fish orientation. *Ann. NY Acad. Sci.* 188:12–29
192. Stasko, A. B. 1975. Annotated bibliography of under-water biotelemetry. *Environ. Can. Fish. Mar. Serv. Res. Dev. Tech. Rep.* 534. 31 pp.
193. Stasko, A. B., Horrall, R. M., Hasler, A. D., Stasko, D. 1973. Coastal movements of mature Fraser River pink salmon as revealed by ultrasonic tracking. *J. Fish. Res. Bd. Can.* 30:1309–16
194. Stasko, A. B., Horrall, R. M., Hasler, A. D. 1976. Coastal Movements of adult Fraser River sockeye salmon (*Oncorhynchus nerks*) observed by ultrasonic tracking. *Trans. Am. Fish. Soc.* 105:64–71
195. Struthers, G. 1975. Recaptures of salmon tagged as smolts in the River Tay, Scotland from 1967 to 1973. *Cons Perm. Int. Explor. Mer* C.M. 1975/M: 14. 10 pp.
196. Stuart, T. A. 1957. The migrations and homing behavior of brown trout (*Salmo trutta L.*) *Freshwat. Salm. Fish. Res.* 18. 27 pp.
197. Sullivan, C. M., Fisher, K. C. 1953. Seasonal fluctuations in the selected temperature of speckled trout, *Salvelinus fontinalis* (Mitchell). *J. Fish. Res. Bd. Can.* 10:187–95
198. Sutterlin, A. M., Gray, R. 1973. Chemical basis for homing of Atlantic salmon (*Salmo salar*) to a hatchery. *J. Fish Res. Bd. Can.* 30:985–89
199. Swain, A., Parry, M. L. I. 1975. The migration of salmon (*Salmo salar* L.) from the River Ure, Yorkshire. *Cons. Perm. Int. Explor. Mer* CM 1975/M:9. 6 pp.
200. Teichmann, H. 1959. Über die Leistung des Geruchsinnes beim Aal (*Anguilla anguilla* (L)). *Z. Vergl. Physiol.* 42:206–54
201. Tesch, F. W. 1975. Migratory behavior of displaced yellow eels (*Anguilla anguilla*) in the North Sea. *Helgol. Wiss. Meeresunters.* 27 (2):190–98
202. Uda, M. 1952. On the relation between the variation of the important fisheries conditions and the oceanographical conditions in the adjacent waters of Japan. I. *J. Tokyo Univ. Fish.* 38:363–89
203. Ueda, K., Hara, T. J., Satou, M., Kaji, S. 1971. Electrophysiological studies of olfactory discrimination of natural waters by himé salmon, a landlocked Pacific salmon, *Oncorhynchus nerka. J.*

Fac. Sci. Univ. Tokyo (4 Zool.) 12:167–82
204. Vreeland, R. R., Wahle, R. J., Arp, A. H. 1975. Homing behavior and contribution to Columbia River fisheries of marked coho salmon released at two locations. *US Natl. Mar. Fish. Bull.* 73:717–25
205. Wagner, H. H., 1969. Effect of stocking location of juvenile steelhead trout (*Salmo gairdneri*) on adult catch. *Trans. Am. Fish Soc.* 98:27–34
206. Walker, T. J., Hasler, A. D. 1949. Detection and discrimination of odors of aquatic plants by *Hyborhynchus notatus* (Raf.). *Physiol. Zool.* 22:45–63
207. Waterman, T. H. 1972. Visual direction finding by fishes. In *Animal Orientation and Navigation,* ed. (S. R. Galler, K. Schmidt-Koenig, G. T. Jacobs, R. E. Belleville, 437–56. Washington DC: NASA
208. Waterman, T. H. 1975. Natural polarized light and e-vector discrimination by vertebrates. In *Light as an Ecological Factor, II,* ed. G. C. Evans, R. Bainbridge, O. Rockham 305–35. Oxford: Blackwell
209. Waterman, T. H., Forward, R. B. Jr. 1972. Field demonstration of polarotaxis in the fish *Zenarchopterus. J. Exp. Zool.* 180:33–54
210. Went, A. E. J. 1958. Salmon movements around Ireland. VIII. From drift nets along the coast of County Donegal. (1953–1957). *Proc. R. Ir. Acad.* 59B: 205–12
211. White, H. C. 1934. A spawning migration of salmon in E. Apple river. *Rep. Biol. Bd. Can.* 1933:41
212. Williams, G. 1957. Homing behavior of California rocky shore fishes. *Univ. Calif. Publ. Zool.* 59(7):249–84
213. Winn, H. E., Salmon, M., Roberts, N. 1964. Sun-compass orientation by parrot fishes. *Z. Tierpsychol.* 21:798–812
214. Withler, I. L., 1966. Variability in life history characteristics of steelhead trout (*Salmo gairdneri*) along the Pacific coast of North America. *J. Fish. Res. Bd. Can.* 23:365–94
215. Woodhead, A. D., 1975. Endocrine physiology of fish migration Oceanography and marine biol. *Ann. Rev.* 13:287–382
216. Zimmerman, M. A., McCleave, J. D. 1975. Orientation of elvers of American eels (*Anguilla rostrata*) in weak magnetic and electric fields. *Helgol. Wiss. Meeresunters.* 27 (2):175–89

ND
ASSESSING ELECTROPHORETIC SIMILARITY: The Problem of Hidden Heterogeneity[1]

♦4127

George B. Johnson
Department of Biology, Washington University, St. Louis, Missouri 63130

INTRODUCTION

Within the last decade zone electrophoresis has seen extensive use in population biology. It has become a fundamental tool of population geneticists, and indeed there are few areas of evolutionary biology where the technique has not been applied. The power of the technique for population biology is that it permits characterization of discrete gene products: Even within a complex extract a particular enzyme may be discriminated on a gel by assaying activity with an enzyme-specific substrate. Because it permits the direct visualization of a broad array of single gene products, electrophoresis has been used intensively by population geneticists to "count" genetic variants in nature. The levels of genetic variability within natural populations can be directly estimated by enumerating the numbers of electrophoretic variants at specific enzyme loci.

The basic assumption of such an approach is that the variation is discontinuous, so that a particular variant may be assigned to an allelic class unambiguously. The approach thus provides a minimal estimate of the number of variant classes, as it does not consider variation which may exist within an electrophoretically uniform class.

This limitation is fundamental to the application of electrophoresis to population biology: The power of electrophoretic approaches is in providing direct evidence of *difference* between two sampled variants, by demonstrating that the two variants move to different band positions when analyzed under identical conditions.

When two variants exhibit bands at the *same* position on a gel, however, one may not conclude that the variants are identical. The inference in this case is much weaker, and depends upon the amount of variation undetected by the technique.

[1]Carnegie Institute of Washington, Department of Plant Biology Publication No. 573.

THE ASSESSMENT OF HOMOLOGY

The use of electrophoresis to document *similarity* is now becoming widespread, despite the unavoidable ambiguity. This is particularly true of *comparative* electrophoretic approaches, which critically depend upon assessment of homology, and are now being made at all levels of biological organization. Numerous recent reviews are available (18, 21, 25, 42, 55).

Homology Within Species

Allele identification at a locus is used in the enumeration of allele frequencies within populations, as discussed above.

When different populations of a species are compared, identification of homologous alleles provides the basis of the comparison, which thus depends directly upon the probability of undetected difference. When populations have been separated in space or time, the degree of genetic discontinuity may strongly affect the amount of undetected difference between them.

When the degree of functional difference between allelic variants is assessed directly by examining enzyme activity, or more indirectly by examining viability differences (72), interpretation of results depends critically upon assessments of homology. Experimental evidence of functional difference between the variants derives from the demonstration of greater difference between the two classes than within them. When the classes are heterogeneous to an unknown degree, then identical activity (rate, K_m, etc) or viability of the two classes cannot be interpreted; and when difference is observed, it need not apply to all forms.

Homology Between Species

The degree of genetic difference between species may be quantitatively estimated using electrophoretic data, if a random sample of loci is examined. A variety of measures of "genetic distance" are employed (2, 5, 20, 21, 25, 48, 50, 51); in essence, each compares the probability of obtaining the same variant when one species is sampled twice, to the probability of obtaining the same variant when two species are each sampled once (6). The more similar the species, the more nearly equal are the two probabilities. Thus the estimation of genetic distance rests upon a collection of experimental determinations of homology. The argument to genetic difference is made by documentation of electrophoretic similarity. The validity of such estimates will be a sensitive function of the degree of undetected variation.

A novel approach to estimating the degree of genetic difference between species compares for a given locus the shapes of the allele frequency distributions (58). The comparison depends ultimately upon estimating the number and frequency of alleles and the average spacing of the mobility differences between them. This approach does not require the one-to-one matching of variants implicit in genetic distance estimates, but it is very sensitive to the number of alleles detected in each of the species being compared. Thus the degree of undetected variation will strongly influence the validity of such comparisons.

Taxonomy

Finally, it should be noted that it is possible to construct detailed phylogenies employing electrophoretic data (1, 5). Such phylogenies incorporate all of the ambiguities of within-species and between-species homology judgments. The greater genetic discontinuities involved in phylogenetic comparisons strongly influence the probability of undetected variation and correspondingly weaken each of the many inferences of homology upon which such phylogenies are based.

Thus, for a broad range of investigations electrophoretic analysis is employed in experimental designs in which the basic empirical statement is that two bands migrate to an identical position on a gel. Serious error may occur in all such investigations when similar R_f in a gel does not reflect genetic identity.

"HIDDEN" VARIATION IN INSECTS

Within the past few years a variety of studies of electrophoretically detected enzyme variation in insects have suggested the presence of additional variant classes not normally resolved by standard electrophoretic techniques. The earliest suggestion of such cryptic variation was the observation that when variants were analyzed on carefully standardized gels, there was more between-individual variation than predicted by experimental error (24, 52). The first clear examples of cryptic variation were not, however, electrophoretic, but rather involved differential heat stability. When individuals of a presumptively homogeneous line of OdH in *Drosophila pseudoobscura* are analyzed electrophoretically and the gels incubated at high temperature, some individuals show more stability than others (3); while such individual differences may in principle reflect post-translational events rather than allelic differences, subsequent studies strongly suggested that the variation has a genetic basis (62, 63). In the few years since these initial reports, a variety of studies have confirmed the existence of heat-stability variants: PGM in *D. melanogaster* (66), AdH in *D. melanogaster* (45), esterase in *D. melanogaster* (9, 71). A particularly careful analysis of thermal variants of AdH in *D. melanogaster* (65) confirmed in this instance that the heat-labile variant mapped to the same genetic locus as the AdH structural gene.

Cryptic variation of a very different sort was reported soon after the thermal-stability studies first appeared (27): By varying the pore size of electrophoretic gels, many new variant classes were discriminated. In the first case analyzed in detail, αGpdH in the butterfly *Colias meadii,* five variants were detected where only one is seen on a 5% acrylamide gel, and all variants appear to segregate in crosses in a Mendelian manner (29). In a subsequent study of the same butterfly, 14 loci were examined, and a total of 103 variant classes detected where only 40 were seen in standard gels (33). Similar patterns of variation have since been reported for the EST-5 locus of *Drosophila pseudoobscura* (30, 34), the EST-C locus of *D. mulleri* (35), and the RuDP carboxylase loci of C-4 plants (13).

A third sort of variation is reported in the initial pore-size variation study: variation in isoelectric point, pI. Allozymes with different pI's may have the same *net* charge at one pH, and thus migrate similarly on gels. Such differences may be detected directly by measuring pI values (19, 29, 46), or indirectly by altering the pH under which electrophoresis is carried out: The net charge of variants of different pI will be altered to different degrees by a pH shift, so that at the new pH their mobilities will differ. This second approach has recently been used to demonstrate successfully the existence of pI variants at the XdH locus in *Drosophila melanogaster* (Singh et al, in preparation).

In addition to these three classes of electrophoretically cryptic variants, thermal stability variants, gel sieving variants, and pI variants, there are a variety of reports of heterogeneity in enzyme activity the basis of which is unclear (different levels of enzyme activity may involve differential synthesis, different allelic activities, or variation of some epigenetic factor). In most studies no enzyme purification is attempted, leading to significant ambiguity: When the enzyme of interest constitutes a small fraction of the total protein used to express its specific activity, then relatively minor fluctuations in other protein constituents unrelated to that enzyme may significantly alter specific activity. Numerous recent studies of AdH in *D. melanogaster* provide an example of the degree of heterogeneity reported (10–12, 15–17, 40, 47, 67) and the ambiguities in interpretation (4, 22, 43, 44, 53, 54, 64, 68–70).

THE PROBLEM OF HIDDEN HETEROGENEITY

Although the initial reports of electrophoretically cryptic variation occurred only a few years ago, available data suggest that the result is a general one, and the universal occurrence of large amounts of previously undetected variability seems likely. The existence of such variation constitutes a serious problem in the interpretation of many of the comparative electrophoretic approaches now used in population biology and biosystematics, because, as discussed above, these approaches often rely upon assessment of *homology*. No ascertainment of "identity" may be accepted as valid if the elements being compared are sampled from heterogeneous groups! It follows that there will be significant ambiguity in *any* comparative use of electrophoresis which does not involve prior characterization of the heterogeneity.

The recent discovery of large amounts of "hidden heterogeneity" thus argues strongly that new approaches are required in comparative electrophoretic studies. At a minimum, such new approaches should satisfy two requirements: (*a*) The empirical procedure should optimally be capable of resolving all variation which is in principle detectable. (*b*) Characterizations of variants should be directly comparable from one study to another.

This paper presents a comparative approach which can in principle account for much of the previously undetected heterogeneity. It involves an experimental protocol which will resolve most amino acid substitutions altering protein structure and charge [although direct fingerprinting of purified proteins, when practical, will provide an even more powerful approach (39)], and produces data which characterize the physical *state* properties of the proteins themselves, ± explicit estimates of experimental error (rather than the single-point *rate* determinations of protein behavior which usual gel procedures provide, and which depend entirely upon the conditions of the particular analysis).

AN ANALYTIC APPROACH TO ELECTROPHORESIS

To assess potential variation at a particular enzyme locus, one needs to employ techniques affording maximal resolution, and to characterize carefully the physical properties of the protein. When practical, several parameters reflecting the catalytic behavior of allozymes may also be investigated. The following properties are of particular interest:

Physical Properties

1. The *size* of the protein, which determines the degree to which the protein's migration in electrophoresis is impeded by the fibers of the gel. Allelic changes may also involve the degree of subunit aggregation (8, 23, 38, 41).
2. The *asymmetry* or conformation of the protein, which also affects the degree of protein-gel fiber interaction.
3. The *isoelectric point,* or pH at which a protein has no net charge, which reflects the balance of positive and negative charges.
4. The *net charge,* or free mobility independent of gel fiber interactions, which for a given pH reflects the absolute excess of negative (or positive) charges.

Catalytic Properties

5. The *substrate binding affinity* ($S_{0.5}$ or K_m) of enzyme for its substrate and cofactors.
6. The *maximal reaction velocity* (V_{max}) at saturating concentrations of substrate.
7. The *functional stability,* as measured by the denaturation constant ($D_{0.5}$), which expresses the rate of decay in V_{max} due to denaturing conditions of temperature or of ionic strength (such as induced by high concentrations of urea).
8. The E_A, or *Arrhenius constant,* which expresses the dependence of V_{max} on temperature.
9. The three T values, or *thermal binding constants,* which express the dependence of $S_{0.5}$ on temperature.

Knowledge of the first four physical parameters will permit prediction of a protein's mobility, R_f, under any specified conditions of gel pore size and pH. They represent those protein characteristics which determine electrophoretic mobility, and are empirically sufficient to describe it. Characterization of the latter five catalytic properties will permit prediction of a protein's activity at any level of substrate concentration or temperature. It is of particular importance that all nine of these parameters may be estimated ± explicit experimental error.

Of the electrophoretically cryptic classes of variation discussed above: (*a*) The variation in thermal stability concerns parameter 7, the thermal denaturation rate constant. Variation in $D_{0.5}$ presumably reflects acid substitutions in the interior of the protein which alter its hydrophobic character and thus its liability to thermal perturbation. Such variation is not detected electrophoretically if it does not alter the shape or charge of the protein. (*b*) The gel sieving variation reflects parameter 2, the protein's conformation, together with its size (parameter 1) and charge (parameter 4). These three parameters determine R_f at any given pH. Variation in conformation presumably reflects amino acid changes which alter the shape of the proteins; they often involve changes in net charge as well. Such variation is often not detected electrophoretically because of the opposing influences of size and charge (29). In addition, surface polar–nonpolar amino acid substitutions at the subunit binding site of multimers might alter the degree of subunit association (parameter 1) without altering subunit MW. (c) The pH-dependent variation concerns parameter 3, the isoelectric point. Variation in pI presumably reflects amino acid substitutions which alter charge, either directly or by affecting the ionization of neighboring amino acids (26). Such variation is not detected electrophoretically when variants have the same realized net charge at the particular pH employed in the analysis.

The four physical properties may be explicitly characterized in a direct and experimentally undemanding fashion and are thus well suited for comparative approaches in population biology and biosystematics. The five catalytic properties involve more sophisticated experimental procedures, and relatively few explicit characterizations have been undertaken. While of unusual interest from an adaptive point of view, they may prove less practical as biosytematic characters.

CHARACTERIZING THE PHYSICAL PROPERTIES OF ALLOZYMES

In migrating in an electrophoretic gel, a protein interacts with the fibers of that gel. The degree of interaction may be studied by varying the pore size of the gel (gel sieving analysis). Such analysis permits direct assessment of the size and asymmetry of the protein, and the protein's extrapolated mobility at infinitely large gel pore size may be used to assess its net charge.

To determine the protein's isoelectric point, pI, one performs electrophoresis in a pH gradient: The protein migrates through the gradient until it reaches its isoelectric pH, at which point it has no net charge and ceases to migrate. Measurement of pH at this equilibrium position provides a direct measure of isoelectric point.

Gel Sieving Analysis

RATIONALE The migration of a protein in polyacrylamide gel electrophoresis may be expressed (7, 59–61) as

$$R_f = (M_o/u_f) e^{K_r T}, \qquad 1.$$

where R_f = mobility of the protein relative to the front; u_f = apparent mobility of a moving boundary in front of the resolving phase (a constant known for most common buffer systems); M_o = free electrophoretic mobility of protein; K_R = the retardation (frictional/hydrodynamic) coefficient ($K_R = K_r/2.303$); T = percent acrylamide (which determines pore size and is inversely proportional to it).

By examining R_f at several values of T, it is possible to estimate values of free electrophoretic mobility, M_o, and retardation coefficient, K_R, for any protein which can be discretely detected in a heterogeneous mixture. In practice one performs a linear regression of log R_f on T [a "Ferguson plot" (14)]: The resulting line has a

ASSESSING ELECTROPHORETIC SIMILARITY   315

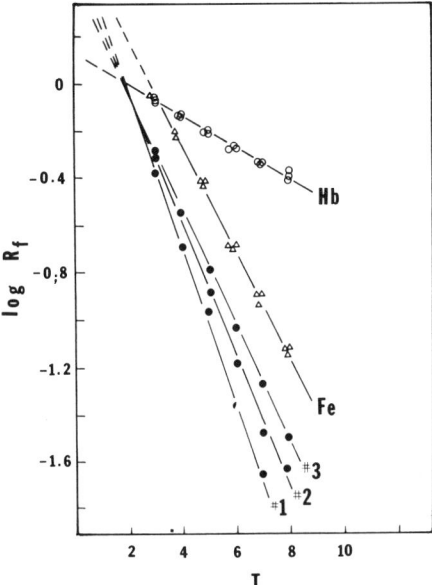

Figure 1   Gel sieving analysis of three individuals of *Colias meadii* fumarase. Hemoglobin (Hb) and ferritin (Fe) are internal standards run in the same sample gels as the butterfly sample (33).

slope of K_R and an intercept of log M_o/u_f (Figure 1). This approach has been described in detail (28, 29, 33). It provides a direct, reproducible, and very sensitive characterization of those physical properties of proteins which alter electrophoretic mobility.

STANDARDIZATION   High resolution electrophoresis requires rigorous standardization. This is particularly true in comparative studies. The simplest rigorous procedure involves running two internal standard proteins in each gel migration path (24). The standards should be chosen so that one resembles the experimental protein(s) of interest, while the other is quite different in both K_R and M_o; the two standards should thus respond quite differently to any alteration of experimental conditions, and the ratio of their mobilities provides a sensitive index of experimental error (28).

One may standardize individual determinations of K_R and M_o for an enzyme by expressing the R_f values of the enzyme relative to the corresponding R_f value of the (similar) internal standard determined from the same gels. For this point-to-point standardization, the migration of the enzyme of interest is expressed relative to that of the internal standard for each individual gel. The appropriate regression equation is thus:

$$\log\left(\frac{R_f^{enz}}{R_f^{std}}\right) = \log\left(\frac{M_o^{enz}}{M_o^{std}}\right) + \left(K_R^{enz} - K_R^{std}\right) T. \qquad 2$$

In accordance with equation 1, the error in M_o is taken as multiplicative, while the error in K_R is taken as additive (33). One may use known values of $K_R{}^{std}$ and $M_o{}^{std}$ determined from many experiments to estimate standardized values for experimental bands.

Molecular weight, ± 20%, may be determined from the value of K_R by established calibration procedures (Figure 2) (29). The vertical scatter in Figure 2 represents not errors in experimental determination, which are much lower, but rather conformational variation among proteins of similar molecular weight.

CLASSIFICATION ANALYSIS In order to demonstrate that an enzyme characterized in one individual is electrophoretically different from that characterized in another, it is necessary to demonstrate that K_R, or M_o, or both differ significantly between the two forms. In comparing estimates of K_R and of M_o, it is important to note that K_R and M_o pairs are determined from a single linear regression as slope and intercept, and error in the estimation of an intercept is not independent of error in the estimation of the corresponding slope.

One simple approach to estimating the error associated with a mobility estimate independent of that associated with a corresponding K_R estimate is to express mobility in terms of the R_f observed at the mean value of T. A linear regression may be considered to rotate around such a midpoint as its slope varies, the midpoint remaining unchanged despite great changes in the Y-intercept. The midpoint (mid-Y) is simply the mean mobility value obtained for the acrylamide concentrations employed; for $T = 3, 4, 5, 6, 7$, and 8, mid-Y corresponds to the mobility expected on a gel of 5.5% acrylamide concentration. The error in the estimate of this parameter is independent of the error in the estimate of the slope K_R.

When sampling from a natural population, there will be an error variance in K_R and in mid-Y associated with each protein type analyzed. To document the existence of multiple classes requires an independent estimate of experimental error. This estimate may be readily obtained from the internal standards run in the same gels. In practice the behavior of one internal standard is normalized to that of the other (dissimilar) internal standard (even better results would be possible using a

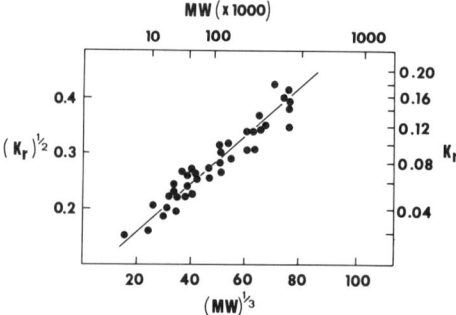

Figure 2 Retardation coefficient (K_R) as a function of molecular weight (29).

third internal standard closely resembling the first), and the error in its estimate taken as a measure of experimental error. In plotting K_R, mid-Y estimates from a natural population, points reflecting homologous proteins should have a distribution no greater than that seen for the standard. A significantly greater distribution is evidence of heterogeneity. Distributions may be compared in terms of coefficients of variation ([standard deviation/mean] × 100) associated with mid-Y and K_R values.

In Figures 1 and 3, data on enzymes of *Colias meadii* (33) are presented to illustrate the analytic, standardization, and classification procedures.

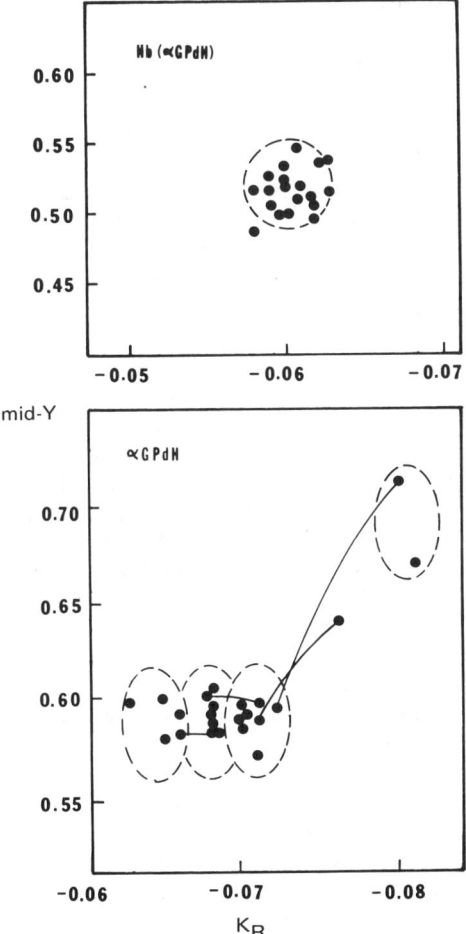

Figure 3 Gel sieving analysis of a population sample of *Colias meadii* for variation at the α-glycerophosphate dehydrogenase locus (33).

Isoelectric Focusing

Characterization of a protein's isoelectric point is carried out with standardization and classification procedures analogous to those described above. For an individual analysis, the protein sample is applied to a large-pore electrophoretic gel together with ampholine to establish the pH gradient and several standard proteins [the experimental procedure is detailed in (29)], the resulting gel sliced longitudinally, one portion stained to locate the band of interest, the corresponding unstained portion excised, and its pH determined. The experimental pH value may be standardized by reference to a similar internal standard protein whose isoelectric point is determined from the same gel. The procedure permits standardization of any error affecting both proteins. By running analyses in triplicate, error estimates on the value of pI may be obtained.

Procedures for Routine Analysis

The analytic approach described above permits direct estimation of molecular weight, K_R, M_o, and pI, and provides with each estimate an explicit chacterization of the experimental error associated with that estimate. The approach has great powers of discrimination and is capable of revealing many previously hidden patterns of variation. Its great strength lies in three facts: (a) It characterizes real physical properties of proteins. (b) It provides an empirically sufficient electrophoretic description (no other parameters are required to predict electrophoretic behavior). (c) *It provides stated estimates of error with every data statement.* This last point is of particular importance. It will permit, for the first time, direct comparison of data collected in different laboratories. I regard this as a matter of the most fundamental importance.

At an absolute minimum, the analytic approach described above requires that each individual of a sample be analyzed on five electrophoretic gels: one to determine isoelectric point and four to determine K_R and M_o. In practice, three replicates of the pI determination and six $\%T$ values in gel sieving analysis will significantly reduce experimental error. Heterogeneity among individuals of a sample is documented by comparing the variance with that of corresponding standards. To conduct such a procedure in routine analysis involves a significant increase in experimental effort—one must run at least five times as many gels. But one clear lesson of the last few years is that the detection of previously "hidden" variation is going to require considerable additional effort. An analytic approach has, in addition to its resolution, the added benefit that, for no extra expenditure of effort, data are obtained with explicit estimates of associated error.

Current procedures, which involve a single determination of R_f on a 10–12% starch gel or a 5–7% acrylamide gel, have permitted the processing of an immense amount of material in a quick and convenient fashion. With the absence a decade ago of experimental information on natural genic variation, just such a "rough and ready" technique was required and its widespread use quickly succeeded in providing a broad data base concerning patterns of enzyme polymorphism. This situation no longer obtains, for at least two reasons: (a) We are now addressing far more sophisticated questions, often of a comparative nature, where the resolution of the technique is of primary importance. (b) With the large volume of data now being produced, objective means are required for comparison of diverse data sets, particularly for studies of a comparative or biosystematic nature. While single-gel determinations of R_f continue to provide the most ready means of preliminarily assessing levels of enzyme polymorphism, it is my personal opinion that they should in future be considered only that, preliminary. More analytic empirical procedures are required to document satisfactorily the electrophoretic nature of genic variants.

CHARACTERIZING THE CATALYTIC PROPERTIES OF ALLOZYMES

Relatively little effort has been expended in characterizing the catalytic properties of allozymes, and most available data concern relatively loosely defined parameters such as "activity" or heat stability on gels. As the former is concentration-dependent and the latter nonquantitative, neither is particularly useful in comparative studies. Because of the large investment required in both instrumentation and time, it is not likely that quantitative comparisons of catalytic properties will be of widespread utility in population biology and biosystematics. I will, however, review briefly the sorts of data required for such comparisons.

$S_{0.5}$ and V_{max}

Very loosely, these two parameters describe realized and ideal kinetic behavior under standard conditions. Values of $S_{0.5}$ and V_{max} may be conveniently determined as the slope and intercept of a reciprocal plot of activity vs substrate concentration (the Lineweaver-Burke plot of the Michaelis-Menton relationship). Because the plot is a linear regression, $S_{0.5}$ and V_{max} are obtained $\pm$ known error. As in the Ferguson plots described in the determination of K_R and M_o, the errors in slope and intercept are related and this covariance must be taken into account. At a minimum, then, estimation of these parameters requires four or more determinations of activity, each at a different substrate concentration. Temperature must be known and precisely controlled. Note, however, that the preparation need not be purified (although the removal of low molecular weight endogenous metabolites is of considerable experimental benefit), as the estimation of $S_{0.5}$ and V_{max} is independent of enzyme concentration and thus of specific activity. For reactions which involve cofactors such as NAD or ATP, or allosteric effectors, their $S_{0.5}$ values should also be obtained.

$D_{0.5}$

The stability of an allozyme to denaturing conditions is most easily determined in quantitative terms by an extension of the kinetic analysis: When the value of V_{max} is known, one may characterize stability in terms of a denaturation constant, $D_{0.5}$, which expresses the rate at which V_{max} is diminished by addition of 6 M urea to the reaction mixture. Alternatively, one may choose to define it in terms of a temperature shift. Urea has the advantage that the denaturation is biochemically more straightforward and the disadvantage that one must be particularly careful to purify the urea immediately prior to each analysis.

E_A

The temperature dependence of V_{max} is log-linear [the Arrhenius relationship, $\ln V_{max} = -(E_A/RT) +$ constant E_B, where R is the gas constant]. One thus determines V_{max} at several temperatures. A plot of $\ln V_{max}$ vs $1/T$ yields a slope $-E_A \times 1/R$ and intercept E_B. As the Arrhenius plot is a linear regression, one may obtain error estimates on E_A and E_B, with the proviso, noted before for Ferguson and Lineweaver-Burke plots, that there is covariance of slope and intercept.

T

The temperature dependence of substrate binding is parabolic in form, and may be expressed as $ax^2 + bx + c$, where a, b, and c are the thermal binding constants T_1, T_2, and T_3. These parameters may be determined from the same data set used to estimate E_A, by plotting $S_{0.5}$ as a function of reaction temperature. T values and error estimates may be obtained from regression on the parabolic equation.

The importance of the five catalytic properties described above is that they are convenient estimators of fundamental enzymic properties. Together they provide an empirically sufficient description of the catalytic sensitivity of the allozyme to metabolic and thermal variation (which between them probably constitutes a major portion of adaptive stresses). All estimates are obtained with explicit associated error statements, so that any judgements of difference in function may be presented with stated confidence limits.

No complete analysis of this sort has yet been reported for allozymes. Indeed, I know of no single case where the binding affinity ($S_{0.5}$ or K_m) of an allosteric effector for an enzyme has been determined for two alleles of a locus. However, any quantitative treatment comparing functional differences between allozymes will, if carried out rigorously, require an approach such as described above.

COMPARING ELECTROPHORETIC VARIANTS

Published data on the allozymes of the α-glycerophosphate dehydrogenase locus in *Colias meadii* provide an example of analytic characterization of alleles by procedures such as described here. Determination of the physical parameters (molecular weight, K_R, M_o, and pI) are presented in Table 1 for each of five allozymes. None of these five forms may be distinguished from one another on 5% acrylamide gels (27, 29). These parameters permit prediction of mobility (R_f) under any conditions of gel pore size or pH. To calculate R_f at a given pH, the values of K_R and M_o may be substituted into equation 1. To estimate the effect of a given change in pH, one may use the value of pI to calculate (roughly) from the Henderson-Hackelbach equation the effect upon the dissociation fraction. To assess the potential resolution of two different pI alleles at a locus, one may take charge as proportional to

ASSESSING ELECTROPHORETIC SIMILARITY 321

Table 1 Functional properties of variants at the α-glycerophosphate dehydrogenase locus of *Colias meadii*[a]

αGpdH Allele	$R_f(5\%T)$	$K_R \pm \sigma$	$M_o \pm \sigma$	$pI \pm \sigma$
1	0.54	−0.088 ± .011	1.72 ± .019	5.83 ± .02
2	0.64	−0.070 ± .004	1.65 ± .005	6.17 ± .02
3	0.63	−0.067 ± .003	1.55 ± .005	6.17 ± .01
4	0.64	−0.064 ± .003	1.55 ± .003	6.11 ± .01
5	0.62	−0.059 ± .013	1.41 ± .015	6.38 ± .03

[a] $S_{0.5}$, V_{max}, and $D_{0.5}$ were determined at 20°C as described in (31). Determinations were carried out on homogenates of 20 individual progeny of a single cross, all identical in R_f at 7%T.

dissociation fraction $[M_o^1/M_o^2 = (dissociation\ fraction)^1/(dissociation\ fraction)^2]$, and calculate from the change in dissociation fraction the effect on M_o, and thus on R_f.

Data on the five catalytic properties are available for two of the variants (31, 32) and are presented in Table 2. The two allozymes do not differ in $D_{0.5}$ and are similar in maximal activity (V_{max} and E_A), but differ very significantly in substrate binding ($S_{0.5}$) and its thermal dependence (T_1, T_2, T_3).

The physical and catalytic characterizations presented in Tables 1 and 2 permit far less ambiguous assessments of homology than have been possible previously. They will differentiate any amino acid substitution having significant effects upon subunit aggregation, net charge, or protein shape, as well as more subtle structural changes altering catalytic properties and their liability to thermal perturbation. Only truly conservative substitutions, neutral in their chemical consequences, will escape detection.

It should be noted that the three parameters of Table 1, presented for the five alleles of αGpdH, constitute a 3 × 5 matrix whose eigenvalue characterizes the structural complexity of variation at this locus. Similarly, the eigenvalue of the 8 × 2 matrix of Table 2 characterizes catalytic complexity.

Table 2 Physical properties of variants at the α-glycerophosphate dehydrogenase locus of *Colias meadii* (29, 33)

αGpdH Allele	$S_{0.5} \pm \sigma$[a]	$V_{max} \pm \sigma$[b]	$D_{0.5} \pm \sigma$[c]	$E_A \pm \sigma$[d]	E_B (×10⁻³) ± σ[e]	T_1 (×10⁻²)[f]	T_2[f]	T_3[f]
2	378 ± 36	1.02 ± .10	22 ± 4	11.8 ± .3	−3.19 ± .10	3.00	−22.30	1.31
4	497 ± 22	0.98 ± .08	27 ± 2	13.4 ± 1.0	−3.47 ± .29	3.71	7.10	−0.04

[a] $S_{0.5}$, or K_m, is the substrate concentration at one half maximal velocity, V_{max}. The units are μM substrate.
[b] The units are μM substrate per mg protein per min.
[c] The time required to decrease velocity from V_{max} by 50% with the addition of 6 M urea, in sec.
[d] The intercept of a plot of V_{max} vs $1/T$, with temperature in degrees Kelvin.
[e] The slope of a plot of V_{max} vs $1/T$.
[f] Coefficients of a parabolic curve fit of a plot of $S_{0.5}$ vs temperature.

ANALYSIS OF HOMOLOGY WITHIN SPECIES

A variety of current problems in population biology depend importantly upon assessment of homology within species. One example of a situation in which hidden heterogeneity may lead to significant mistakes in homology assignment occurs at the esterase-5 locus of *Drosophila pseudoobscura* (34). Analyses of genetic variation at this locus have been used to address the question of the selective significance of allozyme variation by assessing the degree of site-to-site similarity in allele frequencies (56, 57), an estimation which depends critically upon the ability to successfully identify identical allozymes within each of the several sites. Additional arguments have involved the numbers of alleles and levels of heterozygosity at this and other loci (42), parameters which also depend critically upon the resolution of the electrophoretic approach. Seven allelic lines of this locus have been analyzed by gel sieving procedures (34), each presumptively uniform as judged by starch gel electrophoresis. Comparing K_R and M_o values among individuals sampled from each line, it was determined that most lines were in fact heterogeneous: In all, no fewer than 14 distinct variants were detected among the 7 lines. The existence of this added variation suggests that previous results indicating between-population similarity need reevaluation. Similar hidden heterogeneity among presumptively uniform allelic classes has been detected at the esterase-C locus of *D. aldrichi* (35) and the alcohol dehydrogenase locus of *C. mojavensis* (36).

An example of major changes in allele frequency within a population which are not detected by standard procedures is provided by the αGpdh locus of *Colias meadii*. Detailed characterization of αGpdh variation within one alpine population has been reported (31, 32). Analyzed in standard 5% acrylamide gels, no variation is detected (29), due to the opposing effects of charge and asymmetry (Figure 4). In fact, when pore size is altered, five different variants are detected. αGpdh variant frequencies were estimated over several years at sites along an alpine-montane transect through this population (37). Significant changes occur in the frequencies of particular variants along the transect, and indeed in the third year an entirely new variant appears in appreciable frequency! None of these differences are detectable on 5% acrylamide gels, and analysis on such gels would have led to the conclusion that the population was uniform and homogeneous in space and time. Current analyses, determining K_R, M_o, and pI values for allozymes at each of 12 loci sampled at five sites along the transect indicate that there is considerable genetic structure to the population, none of it detectable by standard methods (Johnson, in preparation).

When *difference* in mobility on gels is used as a taxonomic or genetic character, no problem arises with respect to uncharacterized heterogeneity. If a particular allozyme variant is used as a "diagnostic locus" for a species, for example, all of the variants which might be hidden within such an allozyme mobility class each individually differ from any of the allozymes of other species. Differential mobility on a single gel is sufficient evidence to establish structural difference. The problem, as we have seen, arises with the converse: Identical mobility on a single gel is not sufficient

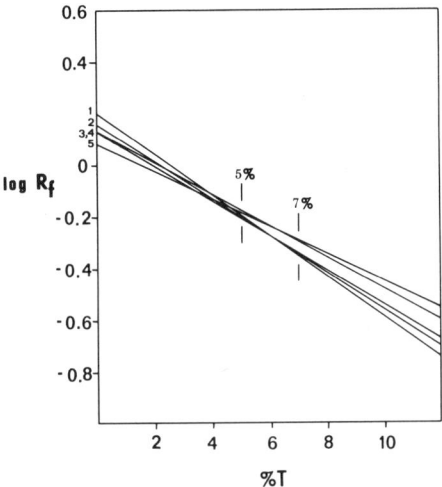

Figure 4 Gel sieving behavior of five variants of α-glycerophosphate dehydrogenase in *Colias meadii* (29).

evidence to establish structural identity. Thus the report of a "founder" population of *D. pseudoobscura* in Bogota, Colombia (56) was based upon the observation in 5% acrylamide gels of momorophism in the isolated Bogota population for polymorphic alleles common in the central North American populations. Later work characterizing previously hidden electrophoretic heterogeneity at the XdH locus has revealed that with respect to this locus the Bogota population is in fact heterogeneous, rather than monomorphic, and that none of its alleles occur in North American populations (Singh, personal communication)!

Characterization of the degree of genic difference between two populations of a species is usually summarized in terms of some frequency-weighted measure of similarity of allozyme composition (2, 5, 20, 48–50). The statistic most frequently employed is Masatoshi Nei's index of genetic distance, D, which is defined as $D = 2\ cn_T\ \lambda_a\ t$ (48), where c = proportion of amino acid substitutions detectable by electrophoresis; n_T = total number of amino acid codons; λ_a = rate of amino acid substitution per year; and t = time since genetic isolation.

In order to estimate D it is necessary to examine a large number of different proteins, assuming that $cn_t\lambda_a$ is the same for all proteins. D is then inversely proportional to the fraction of electrophoretically identical proteins between the two species.

The extensive amounts of hidden heterogeneity revealed by even the few studies reported to date suggest that the value of c will have to be reassessed; the commonly employed value of 0.63 (based upon probabilities of charge-changing amino acid substitutions) assumes an overly simplified view of the role of charge in determining electrophoretic mobility (26, 34). The value of c ought in future to be evaluated depending upon the loci actually examined, the probability of detecting an amino acid substitution at each locus being assessed independently.

In assessing the effects of hidden variability upon relative values of D, it is important to note that the value of D is frequency-weighted; differences in common alleles will have a far greater effect upon the value of D than differences in rare alleles. For the limited data now available, the frequency distribution of hidden alleles is decidedly skewed, with a few common alleles and many rare ones (33). Thus much of the hidden heterogeneity may have only a limited influence on the value of D, always providing, of course, that homology is indicated for the common variants. Should further heterogeneity be detected among the common variants, then problems may arise using D as an index of genic similarity, as it is not a good statistic for the limiting cases of almost complete incongruence (50).

In general, one would expect that the effect of large amounts of previously undetected variation will be to increase the value of c and thus increase the absolute value of D, and to lessen the utility of D in performing more dissimilar comparisons.

ANALYSIS OF HOMOLOGY BETWEEN SPECIES

As pointed out above, the existence of added heterogeneity will serve to lessen the utility of comparisons involving more dissimilar taxonomic groups. This can be seen in comparisons of allozymes of four *Colias* species, none of them closely related to one another. A cursory examination of patterns of variation as indicated on 7% acrylamide gels suggests a rough similarity between the species (23): The mobilities of the most common alleles are similar for the allozymes of the different species. When the physical characteristics of the allozymes are compared, however, a very different pattern emerges (Table 3): Most of the common allozymes of one species

Table 3 Homology among four species of the genus *Colias*

	C. meadii			C. alexandra			C. scudderi			C. philodice		
Locus[a]	$R_f(5\%T)$[b]	$-K_R$	M_o	R_f	$-K_R$	M_o	R_f	$-K_R$	M_o	R_f	$-K_R$	M_o
Adenylate kinase-1	0.71	0.027	1.11	0.66	0.044	1.25	0.67	0.035	1.35	0.64	0.033	1.10
Adenylate kinase-2	0.62	0.035	1.06	0.72	0.044	1.37	0.81	0.064	1.85	0.72	0.055	1.57
Fumarase	0.14	0.220	0.20	0.11	0.104	0.38	0.10	0.11	0.43	0.10	0.11	0.40
Glucose-6-P dehydrogenase	0.38	0.086	1.16	0.41	0.085	1.25	0.40	0.092	1.34	0.39	0.089	1.27
α-glycero-P dehydrogenase	0.62	0.068	1.57	0.63	0.069	1.61	0.63	0.068	1.58	0.61	0.069	1.61
Glutamate dehydrogenase	0.26	0.240	4.74	0.31	0.181	4.61	—	—	—	0.30	0.206	3.69
Hexokinase-1	0.78	0.052	1.64	0.79	0.044	1.52	0.81	0.043	1.54	0.80	0.044	1.50
Hexokinase-2	0.73	0.050	1.50	0.84	0.052	1.77	0.84	0.058	1.88	0.72	0.049	1.45
Malic enzyme	0.37	0.148	2.31	0.38	0.122	1.89	0.38	0.132	2.00	0.37	0.132	1.95
Malate dehydrogenase-1	0.27	0.068	0.69	0.31	0.064	0.74	0.27	0.106	1.03	0.29	0.069	0.73
Malate dehydrogenase-2	0.11	0.342	6.21	0.11	0.307	4.43	0.02	0.268	6.50	0.06	0.18	8.80
Phosphoglucomutase	0.54	0.056	1.18	0.52	0.057	1.16	0.56	0.056	1.23	0.56	0.049	1.14
Triose-P-isomerase	0.62	0.068	1.56	0.65	0.079	1.94	0.65	0.074	1.73	0.63	0.066	1.57

[a]For each of 14 loci, gel sieving data are presented for the most common allelle in *C. meadii* (33), and the most similar homolog in each of the other species occurring at a frequency ⩾ 10% (G.B. Johnson, unpublished data).

[b]Mobility on standard 5% acrylamide gels, equivalent to that seen on 10–11% starch gels.

do not occur in other species. Thus D will be of limited utility in characterizing the degree of genic relatedness of these species. It should be noted, however, that this same effect of added variation renders D a very useful statistic for comparing degrees of genic relatedness of different populations within the same species.

Average electrophoretic mobility has been suggested as an alternative taxonomic approach which is less directly dependent upon assessment of allozyme homology (58). A comprehensive survey of hidden variability in two subdivided and one disjunct population of *Colias meadii* is now underway, which should permit assessment of the effects of hidden variation upon the utility of various approaches (Johnson, in preparation).

THE EXPERIMENTAL MEASUREMENT OF GENETIC DISTANCE

The single weakest element in current approaches to estimating degree of genic similarity and genetic distance is not the metric employed. Nei's and other estimators of genic similarity seem able to satisfactorily accommodate the new patterns of variation which are being detected. The weakness of current approaches, in my opinion, is that explicit confidence intervals are not placed upon estimates of genic similarity, so that the statistical significance of a difference in D may not be directly assessed. *What is required is a way of expressing* D *that explicitly states the uncertainty of the estimate.*

The analytic procedures described in this review provide data which in principle will permit an experimentally explicit characterization of genetic distance:

1. For any individual determination, the error limits of K_r, M_0, and pI determinations provide an explicit probability that the values of these estimators are as stated.
2. For any sampled population, internal standards provide an explicit measure of sampling error.
3. Knowledge of 1 and 2 permits calculation of the conditional probability that a given individual is a member of a particular variant class.
4. Allele frequency estimates may then be expressed as a joint probability of the individual estimates.

Any individual estimate of genic similarity (assignment of an individual to a particular allelic class) thus may be presented with an explicit statement of the probability that a difference would have been detected in that data set if it were present. Any frequency-weighted group estimate, such as genetic distance, may be presented with an explicit statement of the probability of error in allele assignments and thus in allele frequency determinations.

Literature Cited

1. Avise, J. C., Ayala, F. J. 1975. Genetic change and rates of cladogenesis. *Genetics* 81:757-73
2. Balakrishnan, V., Sanghvi, L. D. 1968. Distance between populations on the basis of attribute data. *Biometrics* 24:859-65
3. Bernstein, S. C., Throckmorton, L. H., Hubby, J. L. 1973. Still more genetic variability in natural populations. *Proc. Natl. Acad. Sci. USA* 70:3928-31
4. Birley, A. J., Barnes, B. W. 1973. Genetical variation for enzyme activity in a population of *Drosophila melanogaster*. I. Extent of the variation for alcohol dehydrogenase activity. *Heredity* 31:413-16
5. Cavalli-Sforza, L. L., Edwards, A. W. F. 1967. Phylogenetic analysis: models and estimation procedures. *Am. J. Human Genet.* 19:233-57
6. Cavalli-Sforza, L. L. 1973. Some current problems of human population genetics. *Am. J. Human Genet.* 25:82-104
7. Chrambach, A., Rodbard, D. 1971. Polyacrylamide gel electrophoresis. *Science* 172:440-51
8. Cobbs, G. 1976. Polymorphism for dimerizing ability at the esterase-5 locus in *Drosophila pseudoobscura*. *Genetics* 82:53-62
9. Cochrane, B. J. 1976. Evidence for the existence of a second locus affecting electrophoretic mobility of esterase-6 in *Drosophila melanogaster*. *Genetics* 83:s16
10. Day, T. H., Hillier, P. C., Clarke, B. 1974. Properties of genetically polymorphic isozymes of alcohol dehydrogenases in *Drosophila melanogaster*. *Biochem. Genet.* 11:141-53
11. Day, T. H., Hillier, P. C., Clarke, B. 1974. The relative quantities and catalytic activities of enzymes produced by alleles at the alcohol dehydrogenase locus in *Drosophila melanogaster*. *Biochem. Genet.* 11:155-65
12. Day, T. H., Needham, L. 1974. Properties of alcohol dehydrogenase in a strain of *Drosophila melanogaster* homozygous for the *AdH*-slow allele. *Biochem. Genet.* 11:167-75
13. Enama, M. 1977. Molecular weight variation of phosphoenolpyruvate carboxylases from C_4 plants. *Carnegie Inst. Wash. Yearb. 75*. In press
14. Ferguson, K. 1964. Starch gel electrophoresis: Application to the classification of pituitary proteins and polypeptides. *Metabolism* 13:985-1002
15. Gibson, J. 1970. Enzyme flexibility in *Drosophila melanogaster*. *Nature* 227: 959-60
16. Gibson, J. B., Miklovich, R. 1971. Of variation in alcohol dehydrogenase in *Drosophila melanogaster*. *Experimentia* 27:99-101
17. Gibson, J. B. 1972. Differences in the number of molecules produced by two allelic electrophoretic enzyme variants in *Drosophila melanogaster*. *Experientia* 28:975-76
18. Harris, H., Hopkinson, D. A., Robson, E. B. 1974. The incidence of rare alleles determining electrophoretic variants: Data on 43 enzyme loci in man. *Ann. Hum. Genet. London* 37:237-53
19. Hayes, M. B., Wellner, D. 1969. Microheterogeneity of L-amino acid oxidase. *J. Biol. Chem.* 244:6636-44
20. Hedrick, P. W. 1971. A new approach to measuring genetic similarity. *Evolution* 26:276-80
21. Hedrick, P. W., Ginevan, M. E., Ewing, E. P. 1976. Genetic polymorphism in heterogeneous environments. *Ann. Rev. Ecol. Syst.* 7:1-32
22. Hewitt, N. E., Pipkin, S. B., Williams, N., Chakrabartty, P. K. 1974. Variation in ADH activity in class I and class II strains of Drosophila. *J. Hered.* 65: 141-48
23. Hopkinson, D. A., Edwards, Y. H., Harris, H. 1976. The distribution of subunit numbers and subunit sizes of enzymes: A study of the products of 100 gene loci. *Ann. Hum. Genet. London* 39:383-411
23b. Johnson, F. M., Powell, A. 1974. The alcohol dehydrogenases of *Drosophila melanogaster:* Frequency changes associated with heat and cold shock. *Proc. Natl. Acad. Sci. USA* 71:1783-84
24. Johnson, G. B. 1971. Analysis of enzyme variation in natural populations of the butterfly *Colias eurytheme*. *Proc. Natl. Acad. Sci. USA* 68:997-1001
25. Johnson, G. B. 1973. Enzyme polymorphism and biosystematics: The hypothesis of selective neutrality. *Ann. Rev. Ecol. Syst.* 4:93-116
26. Johnson, G. B. 1974. On the estimation of effective number of alleles from electrophoretic data. *Genetics* 78:771-76
27. Johnson, G. B. 1975. Enzyme polymorphism and adaptation. *Stadler Symp.* 7:91-116

28. Johnson, G. B. 1975. The use of internal standards in electrophoretic surveys of enzyme polymorphism. *Biochem. Genet.* 13:833–47
29. Johnson, G. B. 1976. Hidden alleles at the α-glycerophosphate dehydrogenase locus in *Colias* butterflies. *Genetics* 83:149–67
30. Johnson, G. B. 1976. Evaluating the charge state model of electrophoretic mobility. *Genetics* 83:s36
31. Johnson, G. B. 1976. Polymorphism and predictability at the α-glycerophosphate dehydrogenase locus in *Colias* butterflies; Gradients in allele frequencies within a single population. *Biochem. Genet.* 14:403–26
32. Johnson, G. B. 1976. Enzyme polymorphism and adaptation in alpine butterflies. *Ann. Mo. Bot. Garden* 63:248–61
33. Johnson, G. B. 1977. Characterization of electrophoretically cryptic variation in the alpine butterfly *Colias meadii*. *Biochem. Genet.* 15:665–93
34. Johnson, G. B. 1977. Evaluation of the stepwise mutation model of electrophoretic mobility: Comparison of the gel sieving behavior of alleles at the esterase-5 locus of *Drosophila pseudoobscura*. *Genetics.* In press
35. Johnson, G. B. 1977. Hidden heterogeneity among electrophoretic alleles. In *Measuring Selection in Natural Populations*, ed. F. Christiansen, T. Fenchel, pp. 223–44. Berlin: Springer
36. Johnson, G. B. 1977. Heterogeneity among the F and S allelic classes of the alcohol dehydrogenase locus of *Drosophila mojavensis*. Submitted for publication
37. Johnson, G. B. 1976. Localized patterns of enzyme polymorphism within a single population of alpine butterflies *Hereditas.* In press
38. Johnson, G. B. 1977. Isozymes, allozymes, and enzyme polymorphism: structural constraints on polymorphic variation. In *Isozymes: Current Topics in Biological and Medical Research*, Vol. 1, ed. M. Tarrazzi, J. Scandalios, G. Whitt. New York: Liss
39. Kung, S., Lee, C., Wood, D., Moscarello, M. 1977. Evolutionary conservation of chloroplast genes coding for the large subunits of fraction 1 protein. *Plant Physiol.* In press
40. Deleted in Proof
41. Koehn, R. K., Eanes, W. F. 1977. Subunit size and genetic variation of enzymes in natural populations of *Drosophila*. *Theor. Popul. Biol.* In press
42. Lewontin, R. C. 1974. *The Genetic Basis of Evolutionary Change*. New York: Columbia. 346 pp.
43. McDonald, J. F., Avise, J. C. 1976. Evidence for the adaptive significance of enzyme activity levels: Interspecific variation in α-GPdH and ADH in *Drosophila*. *Biochem. Genet.* 14:347–55
44. McKenzie, J. A., Parsons, P. A. 1974. Microdifferentiation in a natural population of *Drosophila melanogaster* to alcohol in the environment. *Genetics* 77:385–94
45. Milkman, R. 1976. Further evidence of thermostability variation within electrophoretic mobility classes of enzymes. *Biochem. Genet.* 14:383–87
46. Milkman, R., Koehler, R. 1976. Isoelectric focusing of MdH and 6PGdH from *E. coli* of diverse natural origins. *Biochem. Genet.* 14:517–22
47. Morgan, P. 1975. Selection acting directly on an enzyme polymorphism. *Heredity* 35:124–27
48. Nei, M. 1971. Interspecific gene differences and evolutionary time estimated from electrophoretic data on protein identity. *Am. Nat.* 105:385–98
49. Nei, M. 1971. Identity of genes and gentic distance between populations. *Genetics* 68:s47
50. Nei, M. 1972. Genetic distance between populations. *Am. Nat.* 106:283–92
51. Nei, M. 1975. *Molecular Population Genetics and Evolution*. Amsterdam: North-Holland. 288 pp.
52. Petrakis, P. L., Brown, C. W. 1970. A high order of heterogeneity in the serum albumin of *Ensatina eschscholtzi*, a Pacific coast salamander. *Comp. Biochem. Physiol.* 32:475–87
53. Pipkin, S. B., Rhodes, C., Williams, N. 1973. Influence of temperature on *Drosophila* alcohol dehydrogenase polymorphism. *J. Hered.* 64:181–85
54. Pipkin, S. B., Potter, J. H., Lubega, S., Springer, E. 1975. Further studies on alcohol dehydrogenase polymorphism in Mexican strains of *Drosophila melanogaster*. *Isozymes* 4:547–62
55. Powell, J. 1975. Protein variation in natural populations of animals. *Evol. Biol.* 8:79–119
56. Prakash, S., Lewontin, R. C., Hubby, J. L. 1969. A molecular approach to the study of genic heterozygosity in natural populations. IV. Patterns of genic variation in central, marginal, and isolated populations of *Drosophila pseudoobscura*. *Genetics* 61:841–58

57. Prakash, S., Lewontin, R. C., Crumpacker, D. W., Jones, J. S. 1976. A molecular approach to the study of genic heterozygosity. VI. The remarkable genetic similarity of geographical populations of *D. pseudoobscura*. In preparation
58. Richardson, R. H., Smouse, P. E. 1976. Patterns of molecular variation. I. Interspecific comparisons of electromorphs in the *Drosophila mulleri* complex. *Biochem. Genet.* 14:447–66
59. Rodbard, D., Chrambach, A. 1970. Unified theory for gel electrophoresis and gel filtration. *Proc. Natl. Acad. Sci. USA* 65:970–77
60. Rodbard, D., Chrambach, A. 1971. Estimation of molecular radius, free mobility, and valance using polyacrylamide gel electrophoresis. *Anal. Biochem.* 40:95–134
61. Rodbard, D., Chrambach, A. 1974. Quantitative polyacrylamide gel electrophoresis: Mathematical and statistical analysis of data. In *Electrophoresis and Isoelectric Focusing in Polyacrylamide Gel*, ed. R. Allen, H. Maurer, pp. 28–62. New York: deGruyter
62. Singh, R., Hubby, J. L., Lewontin, R. C. 1974. Molecular heterosis for heat sensitive enzyme alleles. *Proc. Natl. Acad. Sci. USA* 71:1808–10
63. Singh, R. S., Hubby, J. L., Throckmorton, L. H. 1975. The study of genic variation by electrophoretic and heat denaturation techniques at the octonal dehydrogenase locus in members of the *Drosophila Virilis* group. *Genetics* 80:637–50
64. Singh, R. S. 1976. Substrate-specific enzyme variation in natural populations of *Drosophila. Genetics* 82:507–26
65. Thorig, G., Schoone, A., Scharloo, W. 1975. Variation between electrophoretically identical alleles at the alcohol dehydrogenase locus in *Drosophila melanogaster. Biochem. Genet.* 13:721–30
66. Trippa, G., Loverre, A., Catamo, A. 1976. Thermostability studies for investigating non-electrophoretic polymorphic alleles in *Drosophila melanogaster. Nature* 260:42–44
67. Vigue, C. L., Johnson, F. M. 1973. Isozyme variability in species of the genus *Drosophila*. VI. Frequency-property-environment relationship of allelic alcohol dehydrogenases in *D. melanogaster. Biochem. Genet.* 9:213–27
68. Ward, R. D. 1974. Alcohol dehydrogenase in *Drosophila melanogaster*: Activity variation in natural populations. *Biochem. Genet.* 12:449–58
69. Ward, R. D., Hebert, P. D. N. 1972. Variability of alcohol dehydrogenase activity in a natural population of *Drosophila melanogaster. Nature New Biol.* 236:243–44
70. Ward, R. D. 1975. Alcohol dehydrogenase activity in *Drosophila melanogaster*: A quantitative character. *Genet. Res.* 26:81–93
71. Wright, T., MacIntyre, R. 1965. Heat-stable and heat-labile esterase-6^F enzymes in *Drosophila melanogaster* produced by different Est-6^F alleles. *J. Elisha Mitchell Sci. Soc.* 81:17–19
72. Yamazaki, T. 1971. Measurement of fitness at the esterase-5 locus in *Drosophila pseudoobscura. Genetics* 67:579–603

APHID ECOLOGY:
Life Cycles, Polymorphism, and Population Regulation

❖4128

A. F. G. Dixon
School of Biological Sciences, University of East Anglia, Norwich,
Norfolk NR4 7TJ, England

INTRODUCTION

Aphids are plant sucking bugs which occur throughout the world. The greatest number of species are in the temperate regions, where few higher plants are free from aphids. They differ from other plant sucking bugs of the Aphidoidea in that the females of at least a few generations are parthenogenetic and viviparous. Although many species are small and inconspicuous, they frequently become abundant. As many as 2000 million per acre (0.4 ha) may live on the above-ground parts of plants, and the roots may support a further 260 million (46, 92, 117, 148). Aphids, like many other insects, are capable of migrating great distances (up to 1300 km) by means of wind (25, 48, 70).

Polymorphism is characteristic of aphids. Asexual aphids of some species can either possess wings (in which case they are termed alatae), or lack wings (these morphs are called apterae). Typically there are several structurally different morphs in a species, including both sexual and asexual forms.

LIFE CYCLES AND POLYMORPHISM

Aphids which damage crops are mainly host-alternating or polyphagous non-host-alternating species, and this has obscured the fact that most aphids are monophagous. Only 10% of aphid species alternate seasonally between different species of plants (47). Aphids develop an array of different morphs which are associated with the passage of the seasons, movement from host to host, and overcrowding.

The life cycle of the sycamore aphid, *Drepanosiphum platanoidis* (Schrk.), which has only three morphs, and that of two highly polymorphic species, *Periphyllus testudinaceus* (Fernie) and *Rhopalosiphum padi* (L.), are illustrated in Figure 1.

P. testudinaceus and *D. platanoidis* spend their whole lives on sycamore, *Acer pseudoplatanus* L. When sycamore ceases to grow in summer both of these aphids produce estivating forms. *P. testudinaceus* estivates as a flattened first instar nymph, whereas *D. platanoidis* estivates as an adult. Not all aphids that are confined to woody hosts in summer estivate. Some such as the lime aphid, *Eucallipterus tiliae* L., and the oak aphid, *Tuberculoides annulatus* (Hartig), continue to reproduce, although at a lower rate than at the beginning of the year (37). How certain aphids are able to reproduce on woody hosts in summer while others, even on the same host, are not, needs study. It is possible that some species are able to convert the few amino acids in the phloem sap of woody plants during summer (158) into the total range of amino acids necessary for their growth and development.

In spring and early summer *P. testudinaceus* grows and reproduces so long as the twigs and leaves of sycamore grow. As soon as the plant ceases to grow the aphid produces estivating nymphs. An aphid can be made alternately to produce estivating and non-estivating nymphs simply by transferring it alternately between mature and growing leaves (122). Second generation individuals of *D. platanoidis* anticipate the onset of harsh conditions associated with the cessation of growth of their host plant. Even when reared under the same conditions as those experienced by first generation aphids, second generation aphids have a well developed fat body and poorly developed gonads, as if about to enter estivation (39). Whether they estivate and for how long depends on the extent of crowding they (39) and their mothers have experienced (Chambers, personal communication). High population density in summer usually is a consistent feature of *D. platanoidis* populations. By estivating as first instar nymphs, *Periphyllus* can respond directly to its environment; whereas *D. platanoidis*, which estivates as an adult, must anticipate these conditions if it is to respond effectively.

At the beginning of autumn both species of aphid resume development. The timing of this in the sycamore aphid depends on the degree of crowding experienced by the second generation during nymphal and adult stages (38, 39). Short day length can also induce reproduction (26). Estivating nymphs of *P. testudinacea* resume development either when a specific time has elapsed or when leaf tissue becomes senescent (122).

Host-alternating species like *R. padi* leave their primary host when it ceases to grow and return to it when it becomes senescent in autumn (35). Such aphids are highly polymorphic (Figure 1). By exploiting the longer growth phase of grasses in summer, *R. padi* can extend the season when conditions are favorable for reproduction (74), or by moving from plant to plant, can avoid the consequences of the increase of natural enemies (35).

The components of an aphid's environment that induce the change from the unwinged to the winged condition and from asexual to sexual reproduction have been studied in considerable detail by Bonnemaison (14) and Lees (83–85, 88, 89, 91). In some of the aphids that have winged and wingless forms the appearance of winged individuals is associated with crowding, the "group effect" described by Bonnemaison (14). The component of crowding that is most important in inducing

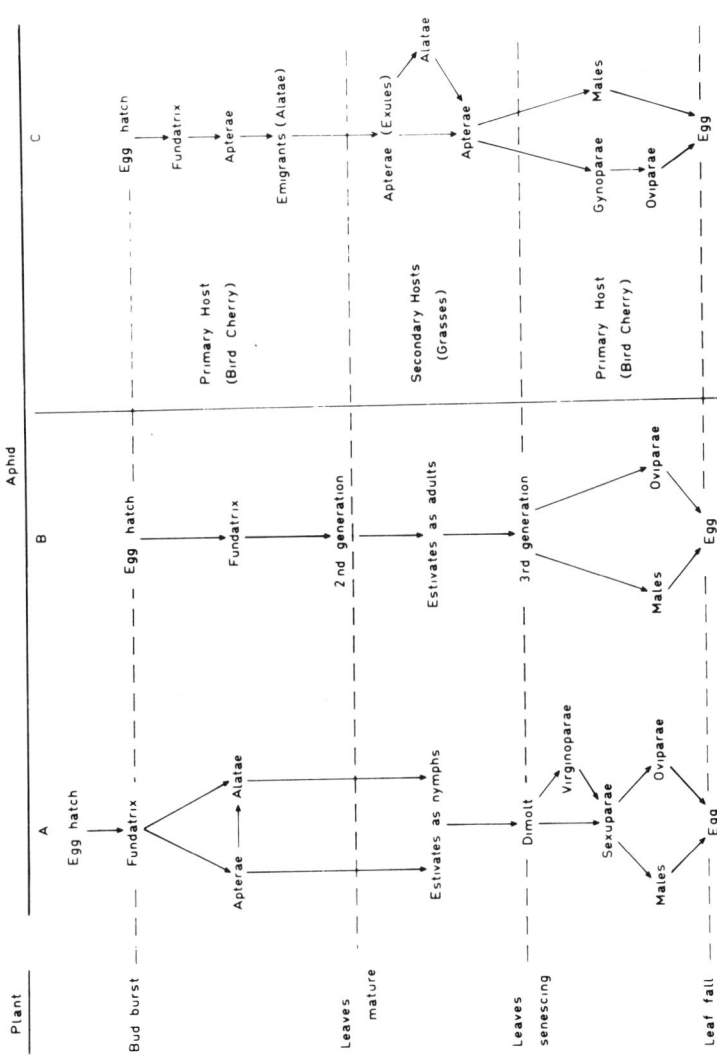

Figure 1. The life cycles of the sycamore aphids. (*a*) *Periphyllus testudinaceus;* (*b*) *Drepanosiphum platanoidis;* (*c*) the bird cherry-oat aphid *Rhopalosiphum padi.*

this change of morph is tactile stimulation (88). In other species this switch in development is associated with changes in the quality of the host plant (35, 49, 101). In these species crowding causes the switch to occur earlier, but it has not yet been determined whether this is due to direct effects of crowding or to aphid-induced changes in the quality of the plant (35).

In many species the change from asexual to sexual reproduction is induced by short day length. In other aphids low temperature or quality of nutrition determines the change (24, 49, 121). However, the response of aphids to these conditions is governed by a biological clock which blocks the appearance of sexuals in spring. The properties of this "facteur fundatrix" or biological clock have been investigated in detail by Lees, who has shown that it measures absolute time and is fairly temperature-stable (84, 85). As for other animals it is the length of the night rather than the day that is critical. Surprisingly day and night are perceived by a photoperiodic receptor located in the mid-dorsal region of the aphid's brain; light passes through the cuticle and stimulates this receptor directly (86). The physiological aspects of aphid polymorphism are well described in the papers of Bonnemaison (14) and Lees (83–88, 90, 91) and in the reviews of this work by Lees (87), Hille Ris Lambers (61), Kunkel & Kloft (81), and Schaefers (119).

Little is known about the ecological aspects of these seasonal changes in the biology of aphids. The following is an attempt to summarize the rather scanty information on how aphids optimize their exploitation of plants, food resources which change in both time and space.

Aphids lay eggs from mid-summer to autumn. The embryos complete development only after they have experienced the cold conditions of winter; they first enter diapause (14). This prevents premature hatching in late autumn. The effect of temperature on egg diapause and the relationship between the number of eggs laid and the number that hatch have been studied in detail for *Aphis fabae* Scop. and *Brevicoryne brassicae* (L.) (3, 14, 147). However, it is not known why aphid eggs hatch when they do.

Growing shoots and leaves are a rich source of food for aphids, which exploit the amino-nitrogen in these structures. However, in many species of trees (29) the period of growth can be short. It is therefore critical for aphids to synchronize their development with that of their host plant. This need may account for the high degree of host specificity in most aphids.

As an example of synchronized development, bud burst in sycamore varies from tree to tree and can occur from early March to late May. Early and late opening trees are consistently early or late each year. However, the time of bud burst of each tree depends on springtime temperatures and thus varies from year to year. The egg hatch of *D. platanoidis* is often not synchronized with the bud burst of a particular tree. Aphids that hatch before bud burst move onto a bud and feed on the bud scales. Many of these aphids, exposed on the smooth surfaces of the buds, are washed away by rain or eaten by birds. The longer the period between egg hatch and bud burst the smaller the chance that an aphid will survive. Because aphids which hatch before bud burst suffer heavy mortality, it is surprising that hatching occurs before bud burst is complete. However, aphids that infest leaves at bud burst are heavier, more fecund, and develop faster than aphids that start their development on older leaves. If hatch were delayed until after bud burst is complete on all trees, the quality of the food available to the aphids would be poor. Since sycamore aphids apparently have no way of determining whether they are on an early or late opening tree, it is interesting that peak egg hatch occurs at about the time of average bud burst for all trees in an area. The advantages of hatching early or late, should they be on early or late opening trees, appear to be sufficient to extend the period of egg hatch in both directions from the mean (40).

The response shown by tree dwelling aphids to the improved quality of the food available in autumn can be viewed in a similar way. Tamaki (130) has shown that leaf fall and weather are the most important factors determining the increase in numbers of *Myzus persicae* (Sulz.) on peach trees in autumn. In sycamore, terminal bud formation is induced by short day length whereas leaf fall is induced by low temperatures (141). As with bud burst, trees vary in the time they shed their leaves. When production of sexuals and egg laying occur late in the year, the aphids are endangered by early autumns. Sycamore aphids have partly overcome this problem. Rather than suddenly switching to the production of sexuals [as is implied in much of the literature on aphids (87)], the population changes over gradually, starting in late summer (31). Thus even in early autumns some sexuals and eggs are produced. However, the significance of the aphid's response to changing day length and temperature relative to the host plant's responses to these factors still needs to be resolved.

Some species of aphids [e.g. the lime aphid, *E. tiliae,* and walnut aphid, *Chromaphis juglandicola* (Kalt.)] produce their first sexuals in the middle of the year (20, 36). This feature is of adaptive significance because both these species show marked fluctuations in numbers over relatively short periods of time and can be rare or locally extinct in autumn (32).

Many aphids in temperate regions overwinter as eggs that can withstand a temperature of $-20°C$ (114). However, it is puzzling why sexual union is needed for the production of eggs. The energy used to produce males could be used to produce egg laying females. That many aphids, which so effectively employ parthenogenetic reproduction for most of the year, retain the sexual phase in the annual cycle, has been used by Williams to support the view that there must be a short-term advantage in sex: the production of more variable offspring (153). In host-alternating aphids the return to the primary host in autumn must result in the mixing of aphids from different populations. This would increase recombination among diverse genotypes (9). The extraordinarily fit individuals produced by recombination are likely to survive periods of population decline. However, this assumes that the nature of selection does not change but that its intensity does (94). Using a simulation of an aphid population incorporating fitness through genetic control of fecundity, Gilbert & Gutierrez have suggested that sexual reproduction would serve to optimize the relationship between fecundity of the parthenogenetic morphs and number of sexuals produced (52). Although in this particular study the model is oversimplified, it is by experimenting with models incorporating the important mortality factors and some knowledge of genetic variation in aphids that we are likely to establish the value of sex to aphids.

Some species of aphid have no sexual phase in certain parts of their range or produce sexuals only occasionally. In other species sexuals are produced by some individuals in the population but not by others (9-11). After severe winters when only eggs survive, clones having no sexual phase (i.e. the anholocyclic) develop again from the sexually reproducing (i.e. holocyclic) part of the population (10).

In several host-alternating species the proximate factors inducing the appearance of alate emigrants are well understood but the significance of the timing is not. Size, number, and chances of survival before migration are likely to be important in determining when in summer aphids will leave their primary host. If the switch to alate production were delayed, more aphids could be produced, but they would be smaller because the nutritive quality of their host would then be poorer (42). Also, the number and voracity of the aphids' natural enemies might then be greater.

The reproductive strategies of the structurally similar alate morphs of the host-alternating aphid, *R. padi*, are different (Table 1 and Figure 1). Emigrants and gynoparae are adapted to maintain a high reproductive rate in a temporarily favorable environment, whereas the alate exules, by living longer and maintaining a lower reproductive rate, are better able to survive the harsher conditions prevailing in late summer (41). In the closely related *R. insertum* (Walker), whose summer generations live on the roots rather than the tops of grasses, the alate exules have a short

Table 1 Differences between emigrants, alate exules, and gynoparae of *Rhopalosiphum padi*

	Emigrants	Alate exules	Gynoparae
Fecundity	38	25	15
Length of adult life (days)	13	17	8
Length of reproductive life (days)	10	14	6
Weight (μg)	777	330	457
Number of offspring born in the first:			
(a) two days	10	7	12
(b) four days	15	16	15
Number of well-developed embryos	19	8	13
Percentage of total fecundity present as well developed embryos in teneral adults	50	32	87
Weight of offspring (μg)	23	25	18

adult life and a high reproductive rate (24). A study of apterae is likely to reveal similar differences between morphs and species.

Each morph has a particular role to perform and is part of a sequence which ends with the egg laying forms. The most successful genotype will be that which optimizes the combined strategies of all the morphs in the sequence. There has been a tendency to treat all alates as similar, and likewise all apterae (132). However, there is now evidence to indicate that morphs differ not only in structure but in biology. Even within a morph there is variation. It is this phenotypic variation, much of which is anticipatory, that is largely responsible for the success of aphids.

POPULATION DYNAMICS

Population studies of aphids present problems rarely met in studies of univoltine insects. As noted above, many aphids are highly polymorphic with each morph adapted to a particular way of life; their relatively short development time and long reproductive life result in overlapping generations, and within a morph there is a wide range of phenotypic variation; additionally, aphids migrate.

Estimating rates of increase of aphid populations has posed serious problems. Direct observations of reproductive rates in the field have rarely been made (29, 38). Most researchers have favored an indirect method, either estimating reproductive rates from census data by Hughes' method (64–66), or recording reproductive rates in the laboratory at known temperatures and then generalizing to field conditions (112). In order to make direct observations in the field, aphids have to be caged. The effects of caging have not been explored; one would expect that the sheltered conditions and relief from crowding afforded by cages would be to the aphids' advantage.

Hughes' method has been recommended for use in population studies of *M. persicae* for the International Biological Program (11, 93). From the ratio of numbers of individuals in the first three instars it is possible to determine the aphids' rate of increase. However, the method is unlikely to give an accurate estimate of reproductive rate, especially when it is changing in time (19).

Estimating the reproductive rate of aphids in the field from results obtained under constant physical conditions in the laboratory also presents problems. The aphids are kept singly or in small groups in cages; they are often on young plants or on stem or leaf cuttings. Young plants are likely to be nutritionally superior to mature plants, and cuttings rapidly undergo changes that would make them more suitable sources of food than the plant from which they were taken. Although tedious, a direct measurement of reproductive rate gives the most accurate estimate. If laboratory results are to be used, then some attempt should be made to relate the estimate to the rate measured in the field at certain times of the year.

Laboratory measurements of the intrinsic rate of increase (r) of aphids have been useful in comparing the reproductive potentials of different morphs of one species (45) and of different species (22). Laboratory measurements have also been used in assessing the effect of temperature on the rate of increase (22, 123). However, in estimating r it is assumed that the population has achieved a stable age distribution;

therefore, estimates of this population parameter are inapplicable to the study of aphid populations in the field.

The following is an account of the more detailed population studies that have been made on aphids.

Black Bean Aphid, Aphis fabae

This aphid overwinters on spindle, *Euonymus europaeus* L.; leaves it in June to colonize beans (*Vicia faba* L.), sugar beet (*Beta vulgaris* L.), and various other herbaceous plants (71, 145); and returns to spindle in autumn. The more aphids that return to spindle in autumn, the more eggs are laid. More than 60% of these eggs hatch the following spring. Way & Banks (147, 149) indicate that intraspecific competition at high densities can result in only a few aphids surviving. This situation has been well documented for the increase in numbers of aphids on individual bean plants (147). There is an inverse relationship between the number of aphids colonizing a plant and the number of alates which leave it. An initial infestation of 2 aphids per plant gives rise to four times more alates than an infestation of 32 aphids per plant. When plants are heavily colonized by emigrants in spring, each plant produces relatively few summer migrants; yet the net effect is the production of considerably more summer migrants than in years when colonization by emigrants is poor. Therefore, even though intraspecific competition reduces the production of alates per bush or bean plant, when aphids are abundant there is still a tendency for the population to produce more alates than when aphids are rare because more units of the host plant are infested. This tendency is supported by Behrendt's 13 year study of *A. fabae,* which shows that the greater the number of aphids hatching in spring the greater the number of summer migrants (4, 5). Therefore although predators can inflict heavy mortality at these stages in the life cycle (2), they are ineffective in regulating the numbers of the aphid. Possibly this mortality is compensated by a reduction in intraspecific competition.

To this point in the life cycle of *A. fabae* more aphids leave each stage than enter it. However, there is an inverse relationship between the abundance of summer migrants and autumn migrants (143). Although there is no detailed information on how this is achieved, it has been suggested that it is due primarily to insect predators, in particular to ladybird beetles. The large numbers of predators nourished by the abundant aphid population on secondary hosts in early summer are thought to be able to reduce the numbers of aphids that colonize other plants (e.g. *Chenopodium album* L.) in late summer. These same predators can also prevent the production of large numbers of emigrants from spindle the following spring. This results in aphid outbreaks in alternate years as described by Way (143, 151), Behrendt (4–6), and Müller (102). Furthermore, weather disturbs the aphid's rate of increase, its successful colonization of plants, and the activity of its predators (143).

The population dynamics of this species—especially the fate of the offspring of the summer migrants—has not yet been fully investigated. In years when they are produced in large numbers, summer migrants must face a shortage of suitable host plants. This could lead to intense intraspecific competition and to the production of few autumn migrants. Warm years which favor the production of summer migrants reduce the abundance of secondary host plants like *C. album* (152) and must further intensify intraspecific competition. Because of their low fecundity (45), many of the small summer migrants produced when the aphid is abundant on beans (144, 148) may have difficulty in colonizing other host plants in late summer. The relative importance of these factors is unknown.

Cabbage Aphid, Brevicoryne brassicae

Hafez (57) studied changes in the numbers of this aphid over a period of two years in Europe. Hughes (65) made a similar study over three years in Australia, where the aphid has an anholocyclic life cycle.

In Europe the cabbage aphid declines in numbers in the middle of the year. This Hafez attributed to the action of insect predators. The autumnal increase in numbers is terminated by low temperature and short day length. The latter also induces the production of sexual forms (57). Hughes' more penetrating study attempted to reveal the way in which the various population processes interact. He claimed that the aphid rapidly increased in number until its food supply became inadequate. The population then diminished in size as a consequence of the interaction between emigration, a fall in reproductive rate, and the mortality inflicted by the aphid's natural enemies. This study showed the advantage of using a physiological time scale when dealing with poikilothermic animals. Unfortunately Hughes' estimates of reproductive rate may have been wrong (19). Nevertheless his study introduced a number of conceptual advances, including ideas about aphid and parasite strategies that resulted from exploring the model developed for this species of aphid (53, 68).

From a study of colonies of *B. brassicae,* Way concluded that intraspecific mechanisms are all-important in governing the size of an aggregate. With an increase in the number of aphids in an aggregate the proportion emigrating increases and the fecundity and/or rate of larviposition decreases (144, 146, 150). This study does not reveal what determines the number of aggregates. Way suggests (146) that predators, by dispersing aggregated aphids, increase the number of aggregates and release the crowding-induced restraints on reproduction.

The role of qualitative changes in the plant in determining the rate of development and increase of the aphid has been ignored. Hughes (65) implicates only those changes in food quality which arise as a consequence of overpopulation. However, it is known that cabbage aphids reproduce fastest on cabbage plants 2–2½ months old (95). The relative growth rates of these aphids are closely correlated with the amount of threonine in the host plant. Threonine levels are lowest in the youngest and oldest plants (138). Seasonal changes in the quality of the host plant could be important in determining the dynamics of this species of aphid.

Rain and temperature affect survival and rates of increase. If climatic conditions are favorable for the survival of the nymphs which hatch from overwintering eggs, then outbreaks can occur like those recorded in Finland in 1920, 1950, and 1951 (95).

Cereal Aphids

Infestation of cereals by aphids can result in considerable losses of grain, both in quality and quantity (79, 80, 116, 155). The ability to predict years of heavy infestation could avert such losses and so a great deal of work has been done on cereal aphids. However, in spite of the information that has been gained, as recently as 1974 Dean (23) stated justifiably that "little is known about the ecology of cereal aphids and of the factors controlling their populations."

Several species of aphid infest cereals. The commonest are *R. padi, Sitobion avenae* (F.), *Metopolophium dirhodum* (Wlk.), and *Sitobion fragariae* (Wlk.). They differ in their distribution, and in relative abundance from year to year (21, 82). Alates infest cereals early in southeast Scotland in years when March temperatures are above average. If early infestation is followed by above average temperatures in June, then the number of aphids is likely to reach pest proportions (126). In southern England the number of alates that infests the crop at the beginning of a year appears to be of little importance in the subsequent buildup of aphids. However, as in Scotland, warm temperatures later in the year can result in heavy infestations (113).

High levels of aphid infestation are also associated with a poor faunal diversity. A rich faunal diversity is thought to indicate a high level of aphid predation by general predators (113, 127, 140). Coccinellidae, Syrphidae, and Neuroptera—the predators usually associated with aphids—are thought to be less important than general predators in determining the level of aphid infestation (113). Coccinellids are capable of reducing the rate of increase of aphid populations (116), but they tend to appear after the aphids are numerous (23). However, Jones (72, 73) has suggested that the buildup of hyperparasites in one year could reduce the number of overwintering primary parasites. This would reduce their effect on aphid colonies the next spring and allow the aphids a more rapid proliferation. It might be a factor in the periodic outbreaks of cereal aphids in Britain.

In these studies the role of the plant in determining the rate of increase of the aphid has again been largely ignored. Wing formation and alate dispersal have, however, been linked with the stage of growth of the host plant (1, 21), and the decrease in numbers with the maturity of the crop (126).

Green Spruce Aphid, Elatobium abietinum *(Walker)*

When abundant, this aphid (which is confined to spruce) can cause defoliation of its host, especially of Sitka spruce (76, 103, 106, 107, 109). In northwestern Europe it overwinters as a viviparous female. This results not from the presence of an anholocyclic race but from the climatic conditions that prevail there. A few eggs were produced in Britain in the unusually dry autumns of 1972 and 1973, when shoot growth ceased early in the year. Sexuals develop on dormant plants kept at low temperatures and short days (8).

When reared at 0°C this aphid can survive temperatures of -15°C, an ability that has been attributed to the presence of glycerol-like substances in the aphid's body (114). Other aphids which overwinter as virginoparae contain large quantities of lipid material that could serve as an antifreeze or a nutrient reserve (128).

Outbreaks of the spruce aphid are associated with mild winters (7, 18, 115). When winter temperatures fall to -7°C or below, ice forming in the needles of the host causes the attached aphids to freeze (114). In mild winters spruce aphids survive well and may even reproduce.

Ohnesorge (104) suggests that the mortality induced by winter weather is the most important factor in determining the numbers of the spruce aphid, with density-dependent factors operating less often. The migration of alates, even though limited in non-outbreak years, may help to cause populations to decline in summer (69, 108). However, the major factor in this decline is thought to be the mortality caused by food shortage (77, 110, 111). High population densities result in severe defoliation. Aphid feeding may also cause a deterioration in the quality of food available during summer (77). Parry (110, 111) attributes the summer population collapse to the low levels of the essential amino acids iso-leucine, histidine, and methionine in the shoots of spruce in June and July. This deficiency results in the development of small adults which produce few offspring. Predator and parasite exclusion experiments carried out by Hussey (69) reveal that the aphid's natural enemies also play an important role in the summer population collapse. Exclusion of natural enemies resulted in a 29% daily increase in surviving offspring per mature female over the period 28 June to 6 August. Unprotected colonies almost vanished during the same period.

The information available does not enable a reliable assessment of the interaction of the factors affecting the spruce aphid nor does it define the relationship between the number of aphids in spring and the number present the following autumn.

Lime and Sycamore Aphids, Eucallipterus tilae *and* Drepanosiphum platanoidis

Seasonal changes in the abundance of these two species have been followed for 9 and 15 years, respectively (Dixon, unpublished results). Because these aphids are restricted to their host trees, it has been possible to follow populations on individual trees from year to year. Only a few trees were studied, but a 20 meter sycamore tree has 116,000 leaves and can support as many as 2.5×10^6 aphids. Changes in numbers within and between years in the aerial population density of the sycamore aphid over Britain are similar to the changes in numbers observed on the eight trees studied.

Neither parasites nor predators were found to be important in regulating the numbers of these aphids (30, 44, 54, 55, 58, 59, 60, 154). However, other workers (105) have claimed that parasites can keep the numbers of the lime aphid small. It is possible that in certain parts of the lime aphid's range natural enemies play an important role, but more convincing evidence is needed.

Each aphid that hatches in spring and survives, initiates a clone that reproduces parthenogenetically through summer and into autumn before giving rise to egg-laying forms (oviparae). When large numbers of aphids emerge from overwintering eggs and colonize the unfurling leaves, few oviparae develop in autumn. Conversely in years when few aphids are present in spring, large numbers of oviparae are present in autumn. The implication of this observation for both species is that the inhibiting effect on further population increase when the aphids are abundant is overcompensating. Thus following a disturbance in abundance, populations oscillate from year to year in waves of decreasing amplitude (28, 32).

Autumns with little wind allow the sycamore aphid to attain high levels of abundance (Dixon, unpublished results). Few aphids are lost by being brushed off the leaves and more of the leaves are suitable for colonization (43). For the lime aphid low temperatures in spring inhibit population increase until later in the year. As a consequence, above average numbers of oviparae develop and more eggs are laid (32). Weather thus disturbs the populations of these aphids.

The regulatory mechanisms have been revealed by studying laboratory populations and by field experiments. Even in the absence of natural enemies laboratory populations of the lime aphid suffer a sudden decline in numbers in June or July if the numbers are high at the beginning of the season (16, 32). This does not appear to be due to deterioration in the quality and quantity of food available, although aphid-induced changes in the plant do occur (33, 34). Populations of the sycamore aphid are more difficult to keep in the laboratory. However, study has revealed relationships similar to those seen in the field between numbers of aphids present in spring and those present in autumn (Chambers, unpublished data).

The causal mechanism is intraspecific (16, 38, 75). High population densities result in proportionately more migratory activity and the production of small individuals with low reproductive rates. Simulation models of lime aphid populations show that the combined effects of these two factors could account for the collapse of the population (Barlow, unpublished data). The poor quality of the individuals that are left behind after migration precludes any great increase in numbers during the remainder of the season. Predators influence the rate of recovery. In the sycamore aphid, migratory activity is less marked. It is the poor quality of the aphids present in autumn that results in low populations following springs of high aphid abundance (27, 38).

Crowding in both species results in the production of small aphids that have a low reproductive rate and a long pre-reproductive period (Figures 2 and 3). Physical contact between lime aphids is not necessary for stunting to take place; the effect can be produced through the plant. Crowding experienced by previous generations is also important in determining size. The offspring of small lime aphids develop into small adults which are only slightly larger than their parents even when reared in isolation (16). Therefore recovery in size and fecundity is likely to be slow.

The components of both the lime and sycamore aphid population systems are shown in Figure 2. Development of simulation models for these two aphids has improved our understanding of the two systems. In particular they have revealed that density-dependent effects on the quality of individual aphids are capable of regulating the populations.

Walnut Aphid, Chromaphis juglandicola

This aphid is specific to walnut, *Juglans regia* L., on which it overwinters as an egg that hatches when the buds burst in spring. The walnut is grown commercially in California and in the past has suffered heavy aphid infestations that have resulted in serious reduction in the yield and quality of the nut crop (98–100). Before the introduction of the parasite *Trioxys pallidus* (Hal.) to control the aphid, it multiplied rapidly in spring, reached a peak between late June and early July, and then declined, only to multiply again in late August or early September (124, 125). In years when the first peak reached high levels of infestation the autumn peak was low.

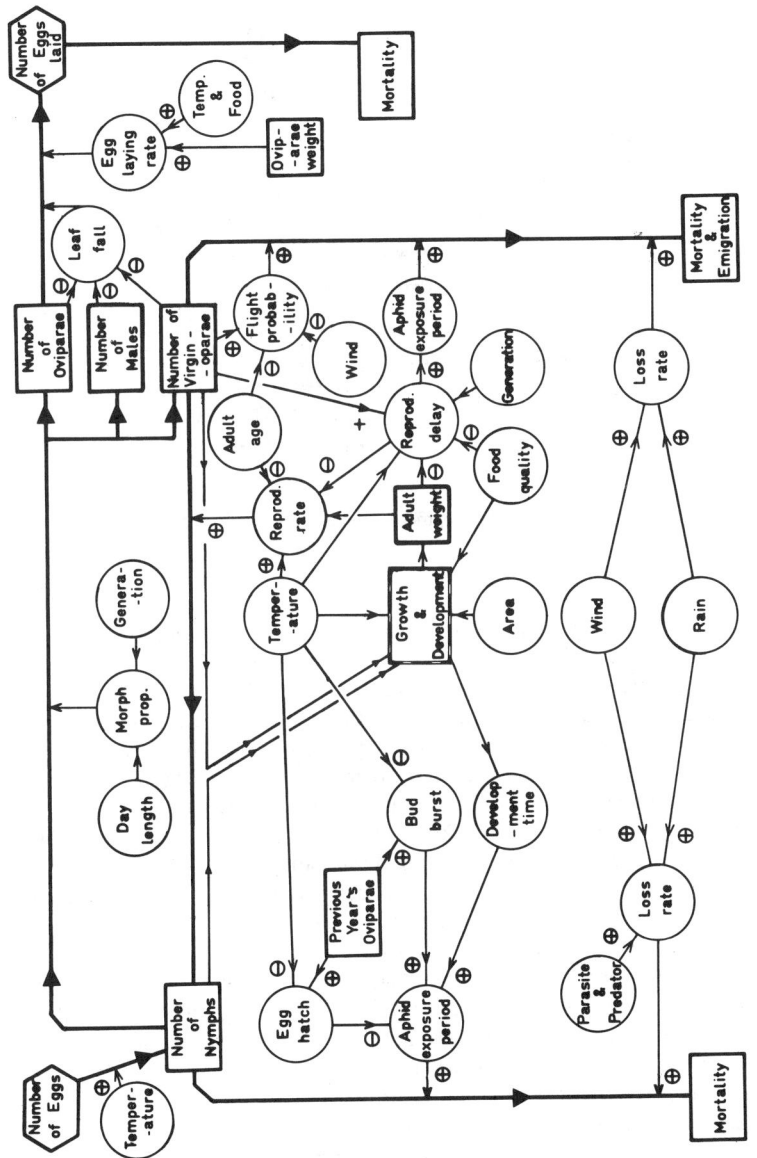

Figure 2. Components of the lime and sycamore aphid population systems.

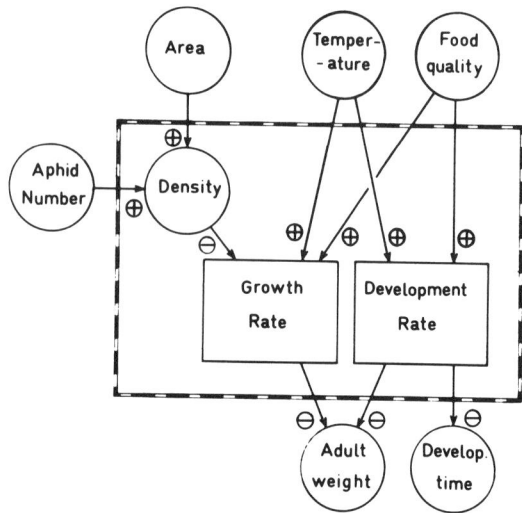

Figure 3. Details of the growth and development component of the lime and sycamore aphid population systems.

The decline in numbers in summer has been attributed to high temperatures, leaf age, the effect of aphids feeding on the leaves, and predation by ladybird beetles (124). Rapid increase in numbers can occur in summer, however. It is therefore unlikely that high temperature is a major factor in the decline (51), although it could be a contributory factor through its effect on aphid size and flight activity. Frazer & Van den Bosch (51) claim that the leaves are made unsuitable by sooty mold and debris from the spring peak of aphids. Sluss (124) and Davis (20) favor aphid-induced changes in the leaves that make them less acceptable to aphids even when washed free of the remains of previous aphid infestation.

Observations in several parts of California, each carried out for only a season or two (96, 98, 100, 124), provide evidence that high numbers in spring result in high peak numbers in June or July followed by low numbers in August and September. Low numbers in spring result in either a small peak in June or July and high numbers in August or September, or a single peak of abundance in August or September. Thus the population changes shown by the walnut aphid are similar to those shown by the lime and sycamore aphid. Davis (20) presents evidence for qualitative changes in both aphids and plants associated with high levels of aphid infestations, with changes in the quality of the aphid being more important.

Populations of this aphid kept on walnut saplings in a glasshouse maintained high levels of abundance throughout the summer (Figure 4a). However, the combined effect of high population density and aphid-induced deterioration in the nutritive quality of the host at a time of the year when it is usually a poor food source resulted in the production of aphids that had an extremely low reproductive rate (Figure 4b).

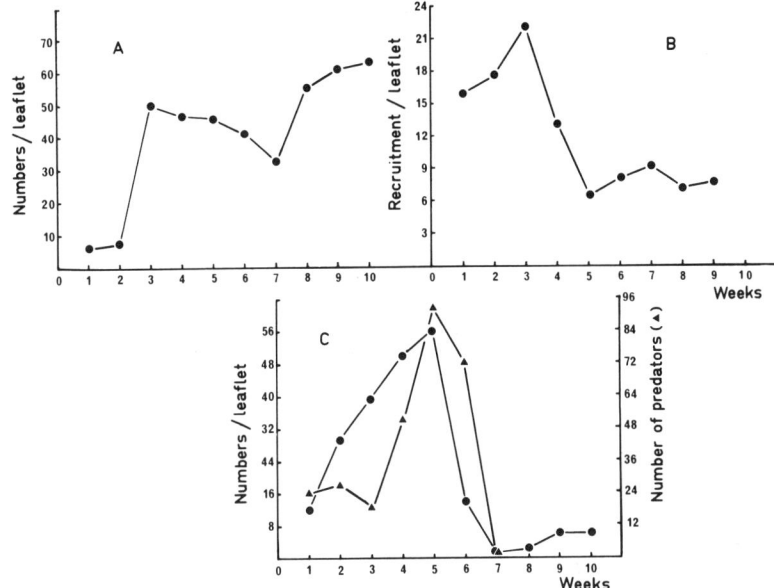

Figure 4 (*a*) The numbers of walnut aphids and (*b*) recruitment on walnut saplings over a period of 10 weeks in a glasshouse (Dixon, unpublished); (*c*) the number of aphids and the incidence of predators in the field (124).

Only in the absence of extrinsic mortality factors were the aphids able to maintain their numbers. Sluss (124) recorded the numbers of predators in the field but failed to mention the sample size. Therefore, it is not possible to assess the mortality inflicted by the predators. However, since the predators are most abundant at the time of the population collapse and would only have to consume 8 first instar nymphs per leaf to prevent recruitment, it is likely that predators played a major role in the summer decline in numbers recorded in the field (Figure 4c).

The introduction of the Iranian parasite *T. pallidus* into California in 1968 resulted in the control of the walnut aphid (96, 135). This parasite has been effective in killing a large proportion of the aphids. Because the dynamics of the aphid in the absence of the parasite is so poorly understood, it is difficult to assess the parasite's exact role. A good opportunity to further our understanding of the role of parasites in aphid population dynamics has been lost.

Other Aphids

Population studies have been made on several other aphids for shorter periods of time or in less depth. A gradual deterioration in host plant quality, which results in a drastic reduction in reproductive performance and the emigration of winged adults, combines with the effects of intraspecific competition to determine the buildup and decline in numbers of the nettle aphid, *Microlophium carnosum* (Buckt). Nettles support a rich fauna, much of which feeds on aphids. It is surprising that self-induced regulation is more important than natural enemies in determining population change (112). In years following those of heavy infestation, nettles support few aphids. This may be due either to an aphid-induced plant effect (the plant is perennial), or to the fact that it takes several generations for the aphid to recover from the effect of intense competition (112).

Aphids on nettles grown in a mixture of soil and fertilizer achieved twice and seven times the population size of aphids on nettles grown in unfertilized soil and sand, respectively. Obviously, the quality of the nettle plant is important in determining the rate of increase of the aphid (Dixon, unpublished data). It is unknown whether heavy aphid infestations in one year will affect the growth of nettles the following year.

Dispersal triggered by intraspecific competition is suggested as the main factor determining the upper size of colonies of *Pterocomma populifoliae* (Fitch) on poplar. Colonies initiated by several aphids are more likely to produce other colonists. Predators tend to prevent the development of colonies that are initiated by one or two aphids (118).

During the course of the International Biological Program studies of the population dynamics of *M. persicae* on potatoes were made simultaneously in several parts of the world. Hughes' method of developing time-specific life tables was used. [As noted earlier (19) this method is unlikely to give accurate estimates of mortality.] The species and condition of the host plant, through their effect on the aphid's multiplication rate, appear to be the most important factors determining the buildup of populations of *M. persicae* on potatoes (50, 93, 139).

The number of alates of this species caught in suction traps varies from year to year with a tendency for fewer to be caught in years following those in which the aphids were abundant. This indicates that the regulatory process overcompensates for large deviations from the equilibrium (15). In the absence of natural enemies, *M. persicae* populations on chrysanthemums remain remarkably constant in size over long periods of time (156). Therefore, intraspecific mechanisms do not appear to be involved in the overcompensated response seen in this species in the field. However, it would be rash to generalize from a single study under glasshouse conditions.

M. persicae has also been implicated in the transmission of yellowing viruses of sugar beet and of potato leaf-roll virus. The study of changes in the incidence of these diseases from year to year has revealed the importance of temperature in determining the activity and abundance of the aphid vector. In the south of England the number of frosty days during January, February, and March, and the mean temperature in April seem to be most important in determining the incidence of yellowing (142). Temperatures at these times are important in determining the size and earliness of the invasion of sugar beet crops by migrant *M. persicae*. In Scotland April temperatures are important when considering the spread of potato viruses but frost free days in the early months of the year are not (63). In those parts of the world where the aphid overwinters as an egg on its primary host, natural enemies could also be important in determining the number of emigrants available to infest secondary hosts (129).

LABORATORY STUDIES AND BIOLOGICAL CONTROL

The study of aphid/parasite systems in the laboratory has revealed that parasites are capable of affecting aphid numbers. At 21°C, *Praon exsoletum* (Nees) can kill a population of the alfalfa aphid, *Therioaphis trifolii* (Monell), whereas at 12.5°C this parasite has little effect on the aphid's rate of increase (97). Unpublished results of Way, cited in (62), show that the parasite of the cabbage aphid can induce cycles in aphid abundance; in the absence of the parasite the aphid's numbers remain relatively constant.

Parasites have been used in the biological control of aphids. Satisfactory to complete control has been reported for the following: *Acyrthosiphon pisum* (Harris) and *T. trifolii* on alfalfa; the carrot aphid, *Cavariella aegopodii* (Scop.); the walnut aphid, *C. juglandicola*; the lime aphid, *E. tiliae*; the apple aphid, *Eriosoma lanigerum* (Hausm.); and, in combination with predators, for the sugar cane aphid, *Aphis sacchari* (Zhnt.) (134, 136). Parasites have also been used with considerable success to control aphid pests in glasshouses (120, 157). However, it is difficult to ascertain from the information provided whether the parasite actually regulates the numbers of its host.

The effectiveness of a parasite depends on the conditions to which it is adapted. The French strain of *Trioxys pallidus* was not effective against the walnut aphid in the hot interior of California, but the Iranian strain of the same species was (135). The ineffectiveness of *Aphelinus mali* Hald. against the woolly apple aphid, *E. lanigerum,* in Israel is attributed to the different temperature thresholds for development in the parasite and its host (12).

Likewise in laboratory systems predators can suppress the buildup of aphid numbers. The quality of the host plant as food for the aphid is important in determining the effectiveness of the predator (62, 131). Attempts to use predators to control aphid pests have proved unsuccessful (133). This is possibly because they are not specific and are highly mobile.

SIMULATION MODELS

Recently there have been several attempts to understand the processes occurring in aphid populations by producing simulation models. Bombosch (13) used a simple model to explore the effectiveness of predators of *A. fabae*. He varied the rate of increase of the aphid, the voracity of the predators and the degree of predator synchronization. This approach has also been used by van Emden (137) to account for trends in population fluctuations of aphids in the field. This is an interesting method for comparing and assessing the potential effectiveness of predators but it assumes a constant rate of increase uninfluenced by predator activity and a predator voracity unaffected by prey density.

The model developed by Hughes for the cabbage aphid included a number of important conceptual advances (65, 66, 68). By using a physiological time scale of day degrees rather than days, Hughes neatly avoided the effect of variable temperatures on the rate of development of aphids. However, by not including the effects of food quality on rate of development, Hughes and others have introduced a significant error into their models (17, 67). Development time in *Drepanosiphum* at a constant temperature can vary from 15 to 25 days, depending on food qualty (40). After accounting for the mortality inflicted by parasites and fungal diseases, Hughes was able to estimate the maximum value for predation. However, because Hughes' method of estimating reproductive rate is unlikely to give accurate predictions, especially when the reproductive rate varies (19), the estimate of predator effectiveness is also unlikely to be accurate.

Models have also been constructed which simulate the within-season development of populations of *Masonaphis maxima* (Mason) (53) and *Aphis craccivora* Koch (56). In both these studies reproductive rate was measured in the field or in the laboratory and the results incorporated in the model. Techniques for discovering the age of aphids and for determining how long they live in the field have yet to be devised. Gilbert & Gutierrez (52) aged individuals of *M. maxima* collected in the field by determining the number of embryos they contained. By assuming that all the offspring an aphid will produce are present as embryos at the adult moult, and by knowing when the offspring will be produced, it is possible to age an aphid from the number of embryos that remain. Hopefully this technique will prove applicable to other aphid species.

These models reveal something of the causal mechanisms underlying the development of a population within a season. Their authors claim that the models closely simulate the field situation and can contribute to a high level of understanding (52). However, the models have yet to be tested against a wide range of independent data. The extension of models to incorporate changes that occur between seasons would be welcome.

We should be more critical of the ideas generated by these models especially in areas where much information already exists. Gutierrez et al (56) claimed that aphids which reproduce only parthenogenetically should maximize, not optimize, their fecundities. They supported this idea with the observation that two species of holocyclic aphids have fecundities of 40 and 54, whereas two anholocyclic species have fecundities of 98 and 105–110. However, *Megoura viciae* (Buckt.) and *A. fabae*, two other holocyclic species, have fecundities of 92–111 and 83–103, respectively (84, 132).

The theoretical study of population dynamics has greatly clarified our understanding of the interactions of population processes. However, improved theory has not been matched by an equally improved understanding of the processes that determine the numbers of animals in the field. More long term field studies and field experiments are required. Klomp's study of the pine looper has revealed that an observation period of more than 20 years may be necessary if the long term changes in the system are to be appreciated (78).

SUMMARY

It is rarely appreciated when considering aphid polymorphism and population dynamics that one is dealing with a plant–aphid interaction. Work on aphids is usually carried out by entomologists who because of their training are loathe to consider the plant component of the interaction. In order to improve our understanding, we must regard plants as an important and variable part of an aphid's environment and not merely as a source of food. Approached in this way, polymorphism makes more sense and can be seen as an adaptation to a changing but nevertheless predictable environment.

We have little detailed understanding of stability and change in field populations of aphids, but certain features of aphid population dynamics have been revealed. Extraordinary weather conditions, particularly temperatures, can have a marked effect on aphid numbers. Wind and rain have a greater effect on aphid numbers than has previously been appreciated. Qualitative changes in aphids induced by crowding appear to be the most probable regulating factor. Natural enemies can influence the rate of buildup of aphids and by interacting with other factors can shape the dynamics of certain aphid species. Whether they can regulate the numbers of an aphid species is still unknown.

Literature Cited

1. Adams, J. B., Drew, M. E. 1964. Grain aphids in New Brunswick. I. Field development on oats. *Can. J. Zool.* 42:735–40
2. Banks, C. J. 1955. An ecological study of Coccinellidae (Col.) associated with *Aphis fabae* Scop. on *Vicia faba. Bull. Entomol. Res.* 46:561–87
3. Behrendt, K. 1963. Über die Eidiapause von *Aphis fabae* Scop. (Homoptera, Aphididae). *Zool. Jahrb. Abt. Allg. Zool. Physiol. Tiere* 70:309–98
4. Behrendt, K. 1966. Population dynamics of *Aphis fabae* Scop. and the influence of Coccinellidae. In *Ecology of Aphidophagous Insects,* ed. I. Hodek, pp. 259–62. Prague: Academia
5. Behrendt, K. 1969. Über langjährige Massenwechselbeobachtungen an der Schwarzen Bohnenblattlaus, *Aphis fabae* Scopoli (Homoptera: Aphididae). *Ber. 10 Wanderversammlung Dtsch. Entomol.* 1965:335–54
6. Behrendt, K. 1971. Zur Bedeutung der Dispersion für die Abundanzdynamik von *Aphis fabae* Scop. (Homoptera: Aphididae) auf Zuckerrübenbestanden. *Zool. Jahrb. Abt. Syst. Oekol. Geogr. Tiere* 98:418–54
7. Bejer-Petersen, B. 1962. Peak years and regulation of numbers in the aphid *Neomyzaphis abietina* Walker. *Oikos* 13:155–68
8. Bevan, D., Carter, C. I. 1975. Host plant susceptibility. *For. Comm. Rep. For. Res.,* 1975
9. Blackman, R. L. 1974. Life cycle variation of *Myzus persicae* (Sulz.) (Hom., Aphididae) in different parts of the world, in relation to genotype and environment. *Bull. Entomol. Res.* 63:595–607
10. Blackman, R. L. 1975. Photoperiodic determination of the male and female sexual morphs of *Myzus persicae. J. Insect Physiol.* 21:435–53
11. Blackman, R. L. 1976. Biological approaches to the control of aphids. *Phil. Trans. R. Soc. London B* 274:473–88
12. Bodenheimer, F. S. 1947. Studies on the physical ecology of the woolly apple aphis (*Eriosoma lanigerum*) and its

parasite *Aphelinus mali* in Palestine. *Agric. Res. Stn. Rehovot Bull.* 43:1–20
13. Bombosch, S. 1963. Untersuchungen zur Vermehrung von *Aphis fabae* Scop. in Samenrübenbestanden unter besonderer Berücksichtigung der Schwebfliegen (Diptera: Syriphidae). *Z. Angew. Entomol.* 52:105–41
14. Bonnemaison, L. 1951. Contribution à l'étude des facteurs provoquant l'apparition des formes ailées et sexuées ches les Aphidinae. *Ann. Epiphyt.* C 2:1–380
15. Broadbent, L., Heathcote, G. D. 1961. Winged aphids trapped in potato fields 1942–1959. *Entomol. Exp. Appl.* 4: 226–37
16. Brown, M. I. 1975. *Intra-specific mechanisms regulating the numbers of the lime aphid.* PhD thesis. Univ. Glasgow. 104 pp.
17. Campbell, A., Frazer, B. D., Gilbert, N., Gutierrez, A. P., Mackauer, M. 1974. Temperature requirements of some aphids and their parasites. *J. Appl. Ecol.* 11:431–38
18. Carter, C. I. 1972. Winter temperatures and survival of the green spruce aphid. *For. Comm. For. Rec.* 84:1–10
19. Carter, N., Aikman, D., Dixon, A. F. G. 1978. An appraisal of the Hughes' time-specific life table analysis for determining aphid reproductive and mortality rates. To be published
20. Davis, C. S. 1957. *The biology of the walnut aphid,* Chromaphis juglandicola *(Klt) (Homoptera: Aphididae).* PhD thesis. Univ. California, Berkeley. 69 pp.
21. Dean, G. J. 1974. The four dimensions of cereal aphids. *Ann. Appl. Biol.* 77:74–78
22. Dean, G. J. 1974. Effect of temperature on the cereal aphids *Metopolophium dirhodum* (Wlk.), *Rhopalosiphum padi* (L.) and *Macrosiphum avenae* (F.) (Hem., Aphididae). *Bull. Entomol. Res.* 63:401–9
23. Dean, G. J. 1974. Effects of parasites and predators on the cereal aphids, *Metopalophim dirhodum* (Wlk.) and *Macrosiphum avenae* (F.) (Hem., Aphididae) *Bull. Entomol. Res.* 63: 411–2
24. Dewar, A. M. 1977. *Morph determination and host alternation in the apple-grass aphid* Rhopalosiphum insertum *Walk.* PhD thesis. Univ. Glasgow 120 pp.
25. Dickson, R. C. 1959. Aphid dispersal over southern California deserts. *Ann. Entomol. Soc. Am.* 52:368–72

26. Dixon, A. F. G. 1963. Reproductive activity of the sycamore aphid, *Drepanosiphum platanoides* (Schr.) (Hemiptera, Aphididae). *J. Anim. Ecol.* 32:33–48
27. Dixon, A. F. G. 1969. Population dynamics of the sycamore aphid, *Drepanosiphum platanoides* (Schr.) (Hemiptera: Aphididae): Migratory and trivial flight activity. *J. Anim. Ecol.* 38:585–606
28. Dixon, A. F. G. 1970. Stabilization of aphid populations by an aphid induced plant factor. *Nature* 227:1368–69
29. Dixon, A. F. G. 1970. Quality and availability of food for a sycamore aphid population. In *Animal Populations in Relation to their Food Resources*, ed. A. Watson, pp. 277–87. Oxford; Blackwell
30. Dixon, A. F. G. 1970. Factors limiting the effectiveness of the coccinellid beetle, *Adalia bipunctata* (L.), as a predator of the sycamore aphid, *Drepanosiphum platanoides* (Schr.). *J. Anim. Ecol.* 39:739–51
31. Dixon, A. F. G. 1971. The "interval timer" and photoperiod in the determination of parthenogenetic and sexual morphs in the aphid *Drepanosiphum platanoides*. *J. Insect Physiol.* 17: 251–60
32. Dixon, A. F. G. 1971. The role of intraspecific mechanisms and predation in regulating the numbers of the lime aphid, *Eucallipterus tiliae* L. *Oecologia Berlin* 8:179–93
33. Dixon, A. F. G. 1971. The role of aphids in wood formation. I. The effect of the sycamore aphid, *Drepanosiphum platanoides* (Schr.) (Aphididae), on the growth of sycamore, *Acer pseudoplatanus* (L.) *J. Appl. Ecol.* 8:165–79
34. Dixon, A. F. G. 1971. The role of aphids in wood formation. II. The effect of the lime aphid, *Eucallipterus tiliae* L. (Aphididae), on the growth of lime, *Tilia vulgaris* Hayne. *J. Appl. Ecol.* 8:393–99
35. Dixon, A. F. G. 1971. The life-cycle and host preferences of the bird cherry-oat aphid, *Rapalosiphum padi* L., and their bearing on the theories of host alternation in aphids. *Ann. Appl. Biol.* 68: 135–47
36. Dixon, A. F. G. 1972. The "interval timer," photoperiod and temperature in the seasonal development of parthenogenetic and sexual morphs in the lime aphid, *Eucallipterus tiliae*. *Oecologia Berlin* 9:301–10
37. Dixon, A. F. G. 1973. *Biology of Aphids*. London: Edward Arnold. 58 pp.

38. Dixon, A. F. G. 1975. Effect of population density and food quality on autumnal reproductive activity in the sycamore aphid, *Drepanosiphum platanoides* (Schr.) *J. Anim. Ecol.* 44:297–304
39. Dixon, A. F. G. 1975. Seasonal changes in fat content, form, state of gonads and length of adult life in the sycamore aphid, *Drepanosiphum platanoides* (Schr.). *Trans. R. Entomol. Soc. London* 127:87–99
40. Dixon, A. F. G. 1976. Timing of egg hatch and viability of the sycamore aphid, *Drepanosiphum platanoides* (Schr.), at bud burst of sycamore, *Acer pseudoplatanus* L. *J. Anim. Ecol.* 45:593–603
41. Dixon, A. F. G. 1976. Reproductive strategies of the alate morphs of the bird cherry-oat aphid *Rhopalosiphum padi* L. *J. Anim. Ecol.* 45:817–30
42. Dixon, A. F. G., Glen, D. M. 1971. Morph determination in the bird cherry-oat aphid, *Rhopalosiphum padi* L. *Ann. Appl. Biol.* 68:11–21
43. Dixon, A. F. G., McKay, S. 1970. Aggregation in the sycamore aphid *Drepanosiphum platanoides* (Schr.) (Hemiptera: Aphididae) and its relevance to the regulation of population growth. *J. Anim. Ecol.* 39:439–54
44. Dixon, A. F. G., Russell, R. J. 1972. The effectiveness of *Anthocoris nemorum* and *A. confuses* (Hemiptera: Anthocoridae) as predators of the sycamore aphid, *Drepanosiphum platanoides*. II. Searching behaviour and the incidence of predation in the field. *Entomol. Exp. Appl.* 15:35–50
45. Dixon, A. F. G., Wratten, S. D. 1971. Laboratory studies on aggregation, size and fecundity in the black bean aphid, *Aphis fabae* Scop. *Bull. Entomol. Res.* 61:97–111
46. Dunn, J. A. 1959. The biology of lettuce root aphid. *Ann. Appl. Biol.* 47:475–91
47. Eastop, V. F. 1973. Deductions from the present day host plants of aphids and related insects. In *Insect/Plant Relationships*, ed. H. F. van Emden, pp. 157–78. Oxford:Blackwell
48. Elton, C. S. 1925. The dispersal of insects of Spitzbergen. *Trans. Entomol. Soc. London*, pp. 289–99
49. Forrest, J. M. S. 1970. The effects of maternal and larval experience on morph determination in *Dysaphis devecta*. *J. Insect Physiol.* 16:2281–92
50. Foster, G. N., van Emden, H. F. 1976. Population dynamics on potatoes. In *Studies in Biological Control*, ed. V. Delucchi, pp. 75–86. IBP Synthesis Rep. No. 9. London: Cambridge Univ. Press
51. Frazer, B. D., van den Bosch, R. 1973. Biological control of the walnut aphid in California: The inter-relationship of the aphid and its parasite. *Environ. Entomol.* 2:561–68
52. Gilbert, N., Gutierrez, A. P. 1973. A plant–aphid–parasite relationship. *J. Anim. Ecol.* 42:323–40
53. Gilbert, N., Hughes, R. D. 1971. A model of an aphid population—three adventures. *J. Anim. Ecol.* 40:525–34
54. Glen, D. M. 1971. *The role of the blackkneed capsid* (Blepharidopterus angulatus (Fall.)) in regulating the numbers of the lime aphid (Eucallipterus tiliae (L.)). PhD thesis. Univ. Glasgow. 107 pp.
55. Glen, D. M. 1975. Searching behaviour and prey-density requirements *of Blepharidopterus angulatus* (Fall.) (Heteroptera:Miridae) as a predator of the lime aphid, *Eucallipterus tiliae* (L.), and leaf hopper *Alnetoidea alneti* (Dahlbom) *J. Anim. Ecol.* 44:115–34
56. Gutierrez, A. P., Havenstein, D. E., Nix, H. A., Moore, P. A. 1974. The ecology of *Aphis craccivora* Koch and subterranean clover stunt virus in South-East Australia. II. A model of cowpea aphid populations in temperate pastures. *J. Appl. Ecol.* 11:1–20
57. Hafez, M. 1961. Seasonal fluctuations of population density of the cabbage aphids, *Brevicoryne brassicae* (L.) in the Netherlands, and the role of its parasite, *Aphidius (Diaeretiella) rapae* (Curtis). *Tijdschr. Plantenziekten.* 67:445–548
58. Hamilton, P. A. 1969. *The role of hymenopterous parasites in the control of the sycamore aphid*. PhD thesis. Univ. Glasgow. 118 pp.
59. Hamilton, P. A. 1973. The biology of *Aphelinus flavus* (Hym.: Aphelinidae), a parasite of the sycamore aphid *Drepanosiphum platanoides* (Hemipt.: Aphididae). *Entomophaga* 18:449–62
60. Hamilton, P. A. 1974. The biology of *Monoctonus pseudoplatani, Trioxys cirsii* and *Dyscritulus planiceps*, with notes on their effectiveness as parasites of the sycamore aphid, *Drepanosiphum platanoides*. *Ann. Soc. Entomol. Fr.* (NS) 10:821–40
61. Hille Ris Lambers, D. 1966. Polymorphism in Aphididae. *Ann. Rev. Entomol.* 11:47–78
62. Hodek, I., Hagen, K. S., van Emden, H. F. 1972. Methods for studying effective-

ness of natural enemies. In *Aphid Technology*, ed. H. F. van Emden, pp. 147–88. London: Academic
63. Howell, P. J. 1973. The relationship between winter temperatures and the extent of potato leaf-roll virus in seed potatoes in Scotland. *Potato Res.* 16:30–42
64. Hughes, R. D. 1962. A method for estimating the effects of mortality on aphid populations *J. Anim. Ecol.* 31:389–96
65. Hughes, R. D. 1963. Population dynamics of the cabbage aphid, *Brevicoryne brassicae*. (L.) *J. Anim. Ecol.* 32:393–424
66. Hughes, R. D. 1972. Population dynamics. In *Aphid Technology*, ed. H. F. van Emden, pp. 275–93 London: Academic
67. Hughes, R. D. 1973. Computer simulation of aphid populations. In *Perspective in Aphid Biology, Bull. No. 2. Entomol. Soc. NZ*, ed. A. D. Lowe, pp. 85–91
68. Hughes, R. D., Gilbert, N. 1968. A model of an aphid population—a general statement. *J. Anim. Ecol.* 37:553–63
69. Hussey, N. W. 1952. A contribution to the bionomics of the green spruce aphid *Neomyzaphis abietina* Walker. *Scott. For.* 6:121–30
70. Johnson, C. G. 1969. *Migration and Dispersal of Insects by Flight.* London: Methuen 763 pp.
71. Jones, M. G. 1942. The summer hosts of *Aphis fabae*, Scop. *Bull. Entomol. Res.* 33:161–69
72. Jones, M. G. 1972. Cereal aphids, their parasites and predators caught in cages over oat and winter wheat crops. *Ann. Appl. Biol.* 72:13–25
73. Jones, M. G., Dean, G. J. 1975. Observations on cereal aphids and their natural enemies in 1972. *Entomol. Mon. Mag.* 111:69–77
74. Kennedy, J. S., Booth, C. O. 1954. Host alternation in *Aphis fabae* Scop. II. Changes in the aphids. *Ann. Appl. Biol.* 41:88–106
75. Kidd, N. A. C. 1975. *The behavioural interactions of the lime aphid Eucallipterus tiliae (L.) and their role in regulating population numbers.* PhD thesis. Univ. Glasgow. 88 pp.
76. Kloft, W., Ehrhardt, P. 1959. Zur Sitkalauskalamität in Nordwestdeutschland. *Waldhygiene* 2:47–49
77. Kloft, W., Ehrhardt, P. 1959. Untersuchungen über Saugtätigkeit und Schadwirkung der Sitkafichtenlaus, *Liosomaphis abietina* (Walk.) *Phytopathology* 35:401–10
78. Klomp, H. 1973. Population dynamics: a key to the understanding of integrated control. In *Insects: Studies in Population Management*, ed. P. W. Geier, L. R. Clark, D. J. Anderson, H. A. Nix, pp. 69–79. Canberra: Ecol. Soc. Aust.
79. Kolbe, W. 1969. Studies on the occurrence of different aphid species as the cause of cereal yield and quality losses. *Pflanzenschutz-Nachr.* 22:171–204
80. Kolbe, W. 1970. Further studies on the reduction of cereal yields by aphid infestation. *Pflanzenschutz-Nachr.* 23:144–62
81. Kunkel, H., Kolft, W. 1974. Polymorphismus bei Blattläusen. In *Sozialpolymorphismus bei Insekten: Probleme der Kastenbildung im Tierreich*, ed. G. H. Schmidt, pp. 152–201. Stuttgart: Naturwissenschaftliche Rundschau
82. Latteur, G. 1971. Evolution des populations aphidiennes sur froments d'hiver (Gembloux, 1970). *Meded. Rijksfac. Landbouwwetensch. Gent* 36:928–39
83. Lees, A. D. 1959. The role of photoperiod and temperature in the determination of parthenogenetic and sexual forms in the aphid *Megoura viciae* Buckton. I. The influence of these factors on apterous virginoparae and their progeny. *J. Insect Physiol.* 3:92–117
84. Lees, A. D. 1960. The role of photoperiod and temperature in the determination of parthenogenetic and sexual forms in the aphid *Megoura viciae* Buckton. II. The operation of the "interval timer" in young clones. *J. Insect Physiol.* 4:154–75
85. Lees, A. D. 1963. The role of photoperiod and temperature in the determination of parthenogenetic and sexual forms in the aphid *Megoura viciae* Buckton. III. Further properties of the maternal switching mechanisms in apterous aphids. *J. Insect Physiol.* 9:153–64
86. Lees, A. D. 1964. The location of the photoperiodic receptors in the aphid *Megoura viciae* Buckton. *J. Exp. Biol.* 41:119–33
87. Lees, A. D. 1966. The control of polymorphism in aphids. *Adv. Insect Physiol.* 3:207–77
88. Lees, A. D. 1967. The production of the apterous and alate forms in the aphid *Megoura viciae* Buckton, with special reference to the role of crowding. *J. Insect Physiol.* 13:289–318
89. Lees, A. D. 1967. Direct and indirect effects of day length on the aphid

Megoura viciae Buckton. *J. Insect Physiol.* 13:1781–85
90. Lees. A. D. 1972. The role of circadian rhythmicity in photoperiodic induction in animals. *Proc. Int. Symp. Circadian Rhythmicity, Wageningen, 1971*, pp. 87–110
91. Lees, A. D. 1973. Photoperiodic time measurement in the aphid *Megoura viciae*. *J. Insect Physiol.* 19:2279–2316
92. Macfadyen, A. 1953. Notes on methods for the extraction of small soil arthropods. *J. Anim. Ecol.* 22:65–77
93. Mackauer, M., Way, M. J. 1976. Population dynamics. In *Studies in Biological Control*, ed. V. Delucchi, pp. 71–75. IBP Synthesis Rep. No. 9. London: Cambridge Univ. Press
94. Manning, J. T. 1976. Is sex maintained to facilitate or minimise mutational advance? *Heredity* 36:351–57
95. Markkula, M. 1953. Biologisch-Ökologische Untersuchungen über die Kohlblattlaus, *Brevicoryne brassicae* (L.) (Hem., Aphididae). *Ann. Zool. Soc. Vanamo* 15: No. 5. 113 pp.
96. Messenger, P. S. 1975. Parasites, predators and population dynamics. In *Insects, Science and Society*, ed. D. Pimental, pp. 201–23. New York: Academic
97. Messenger, P. S., Force, D. C. 1963. An experimental host-parasite system: *Therioaphis maculata* (Buckton)–*Praon palitans* Muesebeck. *Ecology.* 44: 532–40
98. Michelbacher, A. E., Middlekauf, W. W., Wegenek, E. 1950. The walnut aphid in Northern California. *J. Econ. Entomol.* 43:448–56
99. Michelbacher, A. E., Oatman, E. 1956. Walnut aphid studies in 1955. *Calif. Agric.* March 1956: pp. 9, 10, 14
100. Michelbacher, A. E., Ortega, J. C. 1958. A technical study of insects and related pests attacking walnuts. *Calif. Agric. Exp. Stn. Bull.* 764. 86 pp.
101. Mittler, T. E. 1973. Aphid polymorphism as affected by diet. In *Perspectives in Aphid Biology, Bull. No. 2. Entomol. Soc. NZ*, ed. A. D. Lowe, pp. 65–75
102. Müller, H. J. 1966. Über Mehrjährige Coccinelliden-Fänge auf Ackerbohnen mit hohem *Aphis fabae*-Besatz. *Z. Morphol. Oekol. Tiere* 58:144–61
103. Ohnesorge, B. 1959. Die Massenvermehrung der Sitkalaus in Nordwestdeutschland. *Forstarchiv.* 30:73–78
104. Ohnesorge, B. 1963. Beziehungen zwischen Regulations-mechanismus und Massenwechselablauf bei Insekten. (Ein Beitrag zur Theorie der Populationsdynamik). *Z. Angew. Zool.* 50: 427–83
105. Olkowski, W., Olkowski, H., van den Bosch, R., Hom, R. 1976. Ecosystem management: a framework for urban pest control. *Bioscience* 26:384–89
106. Parry, W. H. 1969. A study of the relationship between defoliation of Sitka spruce and population levels of *Elatobium abietinum* (Walker). *Forestry* 42:69–82
107. Parry, W. H. 1971. Differences in the probing behaviour of *Elatobium abietinum* feeding on Sitka and Norway spruces. *Ann. Appl. Biol.* 69:177–85
108. Parry, W. H. 1973. Observations on the flight periods of aphids in a Sitka spruce plantation in north-east Scotland. *Bull. Entomol. Res.* 62:391–99
109. Parry, W. H. 1974. Damage caused by the green spruce aphid to Norway and Sitka spruce needles. *Ann. Appl. Biol.* 77:113–20
110. Parry, W. H. 1974. The effects of nitrogen levels in Sitka spruce needles on *Elatobium abietinum* (Walker) populations in north-eastern Scotland. *Oecologia Berlin* 15:305–20
111. Parry, W. H. 1976. The effect of needle age on the acceptibility of Sitka spruce needles to the aphid, *Elatobium abietinum* (Walker). *Oecologia Berlin* 23: 297–313
112. Perrin, R. M. 1976. The population dynamics of the stinging nettle aphid *Microlophium carnosum* (Bukt.) *Ecol. Entomol.* 1:31–40
113. Potts, G. R., Vickerman, G. P. 1974. Studies on the cereal ecosystem. *Adv. Ecol. Res.* 8:107–97
114. Powell, W. 1974. Supercooling and the low-temperature survival of the green spruce aphid *Elatobium abietinum*. *Ann. Appl. Biol.* 78:27–37
115. Powell, W., Parry, W. H. 1976. Effects of temperature on overwintering populations of the green spruce aphid *Elatobium abietinum*. *Ann. Appl. Biol.* 82:209–19
116. Rautapää, J. 1972. The importance of *Coccinella septempunctata* L. (Col., Coccinellidae) in controlling cereal aphids, and the effect of aphids on the yield and quality of barley. *Ann. Agric. Fenn.* 11:424–36
117. Salt, G., Hollick, F. S. J., Raw, F., Brian, M. W. 1948. The arthropod population of pasture soil. *J. Anim. Ecol.* 17:139–50

118. Sanders, C. J., Knight, F. B. 1968. Natural regulation of the aphid *Pterocomma populifoliae* on the big-tooth aspen in northern lower Michigan. *Ecology* 49:234–44
119. Schaefers, G. A. 1972. The role of nutrition in alary polymorphism among the Aphididae—an overview. *Search Cornell Univ.* 2:1–8
120. Scopes, N. E. A. 1970. Control of *Myzus persicae* on year-round chrysanthemums by introducing aphids parasitized by *Aphidius matricariae* into boxes of rooted cuttings. *Ann. Appl. Biol.* 66:323–27
121. Sethi, S. L., Swenson, K. G. 1967. Formation of sexuparae in the aphid *Eriosoma pyricola*, on pear roots. *Entomol. Exp. Appl.* 10:97–102
122. Shearer, J. W. 1976. *Polymorphism and population ecology of the European maple aphid,* Periphyllus testudinaceus *(Fernie).* PhD thesis. Glasgow Univ. 141 pp.
123. Siddiqui, W. H., Barlow, C. A., Randolph, P. A. 1973. Effects of some constant and alternating temperatures on population growth of the pea aphid, *Acyrthosiphon pisum* (Homoptera: Aphididae). *Can. Entomol.* 105:145–56
124. Sluss, R. R. 1967. Population dynamics of the walnut aphid *Chromaphis juglandicola* (Kalt.) in northern California. *Ecology* 48:41–58
125. Sluss, R. R., Hagen, K. S. 1966. Factors influencing the dynamics of walnut aphid populations in northern California. In *Ecology of Aphidophagous Insects,* ed. I. Hodek. pp. 243–48. Prague: Academia
126. Sparrow, L. A. D. 1974. Observations on aphid populations on spring-sown cereals and their epidemiology in southeast Scotland. *Ann. Appl. Biol.* 77:79–84
127. Sunderland, K. D. 1975. The diet of some predatory arthropods in cereal crops. *J. Appl. Ecol.* 12:507–15
128. Sutherland, O. W. 1968. Dormancy and lipid storage in the Pemphigine aphid *Thecabius affinis. Entomol. Exp. Appl.* 11:348–54
129. Tamaki, G. 1973. Spring populations of the green peach aphid on peach trees and the role of natural enemies in their control. *Environ. Entomol.* 2:186–91
130. Tamaki, G. 1974. Life system analysis of the autumn population of *Myzus persicae* on peach trees. *Environ. Entomol.* 3:221–26
131. Tamaki, G., Weeks, R. E. 1968. *Anthocoris melanocerus* as a predator of the green peach aphid on sugar beets and broccoli. *Ann. Entomol. Soc. Am.* 61:579–84
132. Taylor, L. R. 1975. Longevity, fecundity and size; control of reproductive potential in a polymorphic migrant, *Aphis fabae* Scop. *J. Anim. Ecol.* 44:135–59
133. van den Bosch, R. 1971. Biological control of insects. *Ann. Rev. Ecol. Syst.* 2:45–66
134. van den Bosch, R., Messenger, P. S. 1973. *Biological Control.* New York: Intertext. 180 pp.
135. van den Bosch, R., Schlinger, E. I., Hagen, K. S. 1962. Initial field observations on *Trioxys pallidus* (Haliday), a recently introduced parasite of the walnut aphid. *J. Econ. Entomol.* 58:857–62
136. van den Bosch, R., Schlinger, E. I., Hall, J. C., Puttler, B. 1964. Studies on succession, distribution and phenology of imported parasites of *Therioaphis trifolii* (Monell) in southern California. *Ecology* 45:602–21
137. van Emden, H. F. 1966. The effectiveness of aphidophagous insects in reducing aphid populations. In *Ecology of Aphidophagous Insects,* ed. I. Hodek, pp. 227–35. Prague: Academia
138. van Emden, H. F., Bashford, M. A. 1971. The performance of *Brevicoryne brassicae* and *Myzus pesicae* in relation to plant age and leaf amino acids. *Entomol. Exp. Appl.* 14:349–60
139. van Emden, H. F., Way, M. J. 1972. Host plants in the population dynamics of insects. In *Insect/Plant Relationships, Symp. 6. Roy. Entomol. Soc., London,* ed. H. F. van Emden, pp. 181–99
140. Vickerman, G. P., Sunderland, K. D. 1975. Arthropods in cereal crops: nocturnal activity, vertical distribution and aphid predation. *J. Appl. Ecol.* 12: 755–65
141. Wareing, P. F. 1954. Growth studies in woody species. VI. The locus of photoperiodic perception in relation to dormancy. *Physiol. Plant.* 7:261–77
142. Watson, M. A., Heathcote, G. D., Lauckner, F. B., Sowray, P. A. 1975. The use of weather data and counts of aphids in the field to predict the incidence of yellowing viruses of sugar-beet crop in England in relation to the use of insecticides. *Ann. Appl. Biol.* 81:181–98
143. Way, M. J. 1967. The nature and causes of annual fluctuations in numbers of *Aphis fabae* Scop. on field beans (*Vicia faba*). *Ann. Appl. Biol.* 59:175–88

144. Way, M. J. 1968. Intra-specific mechanisms with special reference to aphid populations. In *Insect Abundance*, ed. T. R. E. Southwood, pp. 18–36. Oxford: Blackwell
145. Way, M. J. 1971. A prospect of pest control. *Inaugural Lecture, Imp. Coll. Sci. Technol.*, pp. 127–62
146. Way, M. J. 1973. Population structure in aphid colonies. In *Perspectives in Aphid Biology, Bull. No. 2. Entomol. Soc. NZ*, ed. A. D. Lowe, pp. 76–84
147. Way, M. J., Banks, C. J. 1964. Natural mortality of eggs of the black bean aphid *Aphis fabae* Scop., on the spindle tree, *Euonymus europaeus* L. *Ann. Appl. Biol.* 54:255–67
148. Way, M. J., Banks, C. J. 1967. Intra-specific mechanisms in relation to the natural regulation of numbers of *Aphis fabae* Scop. *Ann. Appl. Biol.* 59:189–205
149. Way, M. J., Banks, C. J. 1968. Population studies on the active stages of the black bean aphid *Aphis fabae* Scop., on its winter host *Euonymus europaeus* L. *Ann. Appl. Biol.* 62:177–97
150. Way, M. J., and Cammell, M. E. 1970. Aggregation behaviour in relation to food utilization by aphids. In *Animal Populations in Relation to Their Food Resources*, ed. A. Watson, pp. 229–47. Oxford: Blackwell
151. Way, M. J., Cammell, M. E. 1973. The problem of pest and disease forecasting —possibilities and limitations as exemplified by work on the bean aphid, *Aphis fabae. Proc. 7th Br. Insecticides and Fungicides Conf.*, pp. 933–54
152. Weismann, L. 1967. Die Populationsdynamik der schwarzen Rübenblattlaus *Aphis fabae* Scop. an der Zuckerrübe als Grundlage der Schadensprongnose. *Z. Angew. Entomol.* 59:1–15
153. Williams, G. A. 1975. *Sex and Evolution.* Princeton: Princeton Univ. Press. 200 pp.
154. Wratten, S. D. 1973. The effectiveness of the coccinellid beetle, *Adalia bipunctata* (L.), as a predator of the lime aphid, *Eucallipterus tiliae* (L.) *J. Anim. Ecol.* 42:785–802
155. Wratten, S. D. 1975. The nature of the effects of the aphids *Sitobion avenae* and *Metopolophium dirhodum* on the growth of wheat. *Ann. Appl. Biol.* 79:27–34
156. Wyatt, I. J. 1965. The distribution of *Myzus persicae* (Sulz.) on year-round chrysanthemums. *Ann. Appl. Biol.* 56:439–59
157. Wyatt, I. J. 1970. The distribution of *Myzus persicae* (Sulz.) on year-round chrysanthemums. II. Winter season: the effect of parasitism by *Aphidius matricariae* Hal. *Ann. Appl. Biol.* 65:31–41
158. Ziegler, H. 1976. Nature of transported substances. In *Transport in Plants.* I. Phloem Transport, ed. M. H. Zimmermann, J. A. Milburn, pp. 59–100. Berlin: Springer

A HISTORY OF SAVANNA VERTEBRATES IN THE NEW WORLD. Part I: North America

✤4129

S. David Webb

Florida State Museum, University of Florida, Gainesville, Florida 32611

INTRODUCTION

Early in the Cenozoic Era North America was covered almost entirely by forest, predominantly of a mixed mesophytic nature. During the mid-Cenozoic, however, an increasing proportion of the land opened up, forests giving way to woodland savanna, thorn forest, and thorn scrub. By the Late Cenozoic forested areas had decreased still further and much of the savanna was being replaced by grassland steppe and even desert. Today over 25% of the natural vegetation of North America consists of such nonforest biomes. Evolution of an open-country fauna naturally followed these growing opportunities, with the locus of its expansion generally near the center of the continent in the zone of prevailing westerlies.

A remarkably similar series of changes affected the fauna of temperate South America during about the same 40-million-year interval. Yet these two American faunas remained totally separated during most of the Cenozoic. Not only were the two continents physically separate, but also the centers of open-country evolution were located deep within each continent.

The dramatic climax of this history came about three million years ago, when North and South America were connected by way of the isthmian land bridge. Surprisingly, the biota that then began to interchange through the American tropics included a major component of savanna taxa.

The purpose of this review is to elucidate the evolutionary history of open-country biota, and particularly the vertebrates in that biota, in North and South America. A fundamental assumption here is that the New World record of fossil vertebrates is sufficient to reveal much of this history during successive ages of the Cenozoic. It is believed that it can indicate major adaptive tendencies, relative abundance of various adaptive types, approximate phylogenetic relationships, and even geographic patterns of distribution among Cenozoic American vertebrates.

The evolution of American savanna vertebrates provides a particularly rich historical background for studies of Recent evolution and biogeography. Although that record will never be complete, a remarkably full history has been assembled from fossil documents scattered throughout the New World. The work was well begun over a century ago by such pioneers as Leidy, Cope, and Marsh in North America and the brothers Ameghino in South America, and it has continued to grow and be refined ever since. Surely these nonforest biotopes are much better preserved than forest biotopes. Thus, a relatively complete history can be told.

A recurrent question in the study of evolution is how to distinguish the selective effects of past environments from those of the present. Likewise biogeographers attempt to partition the distributional results of past environments from those of the present. For example, Diamond (21), in his review of Haffer's book *Avian Speciation in Tropical South America* (55), asks "What is the relative importance of historical factors and of continuing processes?" Seemingly troubled by the possible importance of the past, he alleges that "Haffer tacitly assumes that present conditions either fail to provide explanations or are to be invoked only as a last resort."

In selected examples paleontology provides the best available answers to such questions by adding historic perspective. If previous distributions and previous environments can be described, then the importance of their effects relative to those of the Recent may be deduced. An instructive example is Simpson's (134) comparison of present areas of Peruvian paramo "islands" with their probable areas during the last glacial interval as an explanation for the present floral diversity on each. Her conclusion that "glacial episodes thus seem to have played a major role in determining the present diversity of high paramo plant taxa . . ." underlines the importance of history. The present is surely the key to the past, but one may assert also with good reason, that the past is the key to the present.

THE MEANING OF SAVANNA

The word *savanna* (from the West Indian *zabana*) refers to subtropical grassy or shrubby plains with scattered trees (127). It is construed here in its broadest sense to include all open-country formations with at least a few trees; only open steppe and grasslands are excluded. Thorn scrub and open deciduous forest are included. Such usage was employed by Gregory (53) in his important essay on the history of North American savanna vertebrates. It is difficult to refine the application of *savanna* to fossil vertebrate samples, unless they happen to be associated with floral samples. But it appears from Duellman's analysis (25) of faunal relationships in Central America and from Müller's studies (102) throughout the Neotropical Region that the vertebrates from all savanna formations have much closer faunal affinity among themselves than with any forest formations. Indeed, recent vertebrate faunas of the savannas have closer affinities with grassland and desert faunas, and also with nonforest highland faunas, than with forest faunas. Likewise, plant ecologists often comment upon the importance of the break between woodland (with closed canopy) and woodland savanna (with broken canopy) (16, 162). Thus, a broad construction of the term savanna seems to have a natural basis.

SAVANNA VERTEBRATE HISTORY 357

Two general aspects of savanna evolution seem worthy of special note. First is the rapid and innovative change that often characterizes taxa in savanna environments. Among mammals, for example, the groups that have developed hypsodont (high-crowned) dentition, clearly an adaptation to feeding in open plant formations, have experienced very rapid evolution and, in most cases, considerable diversification. Perhaps the best examples are horses and ruminants among large herbivores and voles among small herbivores. Likewise, Stebbins (138, 139) cites a number of plant families, including Gramineae, Leguminosae, and Compositae, that have risen rapidly to importance during the mid- to Late Cenozoic cycle of increasing aridity. He emphasizes the importance of savanna-like formations as a "species pump" where evolving phyla could shift first to more xeric and then back to more mesic situations. Savannas were central to an expanding complex of open-country environments that provided a major theater for rapid evolution.

A second notable feature of most savanna vertebrates is their mobility. Their locomotor systems generally show specialization of one sort or another for foraging widely, although exceptions, especially wholly fossorial groups, may be noted. Scansorial, cursorial, ambulatory, and volant modes are characteristic in open country (59). Most taxa with such locomotor styles have large home ranges and cruising ranges; many attain large size (11, 12, 69, 137). Such mobility tends to produce high genetic continuity between adjacent populations. Savanna vertebrates do not recognize barriers to their dispersal as readily as do forest dwellers (102, 105). Presumably this explains the broad continuity throughout the neotropical region between savanna faunas and other open country faunas that Müller contrasts with the more insular nature of forest centers. Sears (127) summarizes the Late Cenozoic significance of increased vagility in open country animals as follows: "Although living organisms had accomplished marvels of diffusion during the preceding multimillions of years, migration during the Late Tertiary took on new dimensions."

With these generalizations in mind, it is appropriate to examine what is known of the history of New World savanna vertebrates. In this first part we consider the North American fauna. The second part will trace the history of the South American vertebrate fauna and the great interamerican interchange.

North American Protosavanna (Paleocene-Eocene)

There are hints that an archaic open-country biota began to develop in the Rocky Mountain region early in the Tertiary. Although paleobotany documents a subtropical forest of broad-leafed angiosperms and no grasses across most of North America, there were surely many local breaks in that cover. Even if these breaks were mainly edaphic and erosional features of temporary significance, they might have formed a mobile mosaic available to a protosavanna biota.

The evidence for a possible protosavanna biota in the Early Tertiary consists of an association of red-banded sediments and certain large vertebrates that may have been open-country opportunists. In his study of Late Paleocene mammals from the Crazy Mountain Field, Simpson (135) drew attention to the vastly increased proportion of terrestrial periptychids, phenacodontids, and arctocyonids in certain localities in contrast with the more diverse samples of smaller, mainly arboreal mammals from the principal quarry sites. Similarly, Wilson (165) found that such ferungulates constituted 91% of the floodplain facies in his Angel's Peak local fauna. Van Houten (146) further documented the two recurrent associations of large ungulates with red-banded (open-country) sediments and of small arboreal mammals with drab (forest) sediments in Late Paleocene and Early Eocene sites throughout the Rocky Mountain region. "These observations," he concluded, "strongly suggest that the grey layers accumulated in swampy woodlands, while the red layers were deposited on flood plains in savannas that displaced the wooded areas from time to time."

When one scrutinizes the presumed protosavanna vertebrates for indications of savanna adaptations, he finds that they are generally quite subtle; certainly they do not include the obvious extremes of hypsodonty, cursoriality, or fossoriality that one finds later in the Tertiary. Nonetheless, the teeth of several larger-bodied taxa appear adapted for chopping coarse fodder, and a few show some modest specializations in their limbs for more efficient open-country locomotion. Perhaps the best example is *Meniscotherium* of the Late Paleocene and Early Eocene, a hare-sized animal with precociously developed molar crescents, molarized premolars, and incipiently cursorial limb elongation (46). The crescentic teeth of the pantodonts suggest a general tendency within that order toward ingesting coarse vegetation, and this seems to be corroborated in *Titanoides* of the Late Paleocene by its digging forelimbs. *Coryphodon* of the Early Eocene has generally been supposed to be an amphibious grazer, much like a *Hippopotamus,* and Simons (133) has noted striated grooves on the lingual base of the large lower canines that may indicate rooting food habits. The multihorned graviportal uintatheres of the Late Paleocene and Early Eocene have crested molars and molariform premolars that probably are adaptations to chopping moderately coarse fibers; their hornlike protuberances presumably imply herding behavior characteristic of open-country ungulates but not of forest dwellers (158). The taeniodonts developed truly hypsodont teeth, enormous canines, and large compressed claws, possibly for ant eating or root grubbing (107). While none of these are proven savanna dwellers, they do form a recurrent association of probable open-country taxa. The opportunities toward which they were groping would soon be seized by the early savanna biota.

North American Woodland Savanna

LATE EOCENE TO EARLY OLIGOCENE North America's first true savanna regions, maintained by secular climatic conditions, emerged in the Middle and Late Eocene. Botanical and sedimentological evidence from the Rocky Mountain Province clearly indicate a marked increase in seasonal aridity during Middle Eocene time. In the geologic record of that time, the vast Lake Gosiute and the other structurally controlled Green River Lakes retreat, the organically rich Green River shales show regular seasonal varving, evaporites increase markedly (including some unique saline minerals), and deeply oxidized red-beds accumulate extensively (13). The first records of grass pollen appear, and nearly half the leaves in the Green River Flora (88) are small and compound. The families Leguminosae, Sapindaceae, and Anacardiaceae predominate and presumably represent a savanna woodland that dominated the lower slopes and was maintained there by the winter dry season.

MacGinitie called the climate of the Green River "Orizaban-subtropical" after the closely comparable woodland savanna flora now living on the slopes of snowcapped Mt. Orizaba near Veracruz, Mexico.

These same drying trends continued through the Late Eocene and Early Oligocene. The Green River Lakes had completely vanished by the Late Eocene, and the Uinta Basin sediments indicate an increasingly arid regime from the Bridgerian through the Duchesnean (latest Eocene) stages (9). Persistent volcanic activity in the Yellowstone and Absaroka centers may have contributed to further disruption of persistent forest patches. Leaf and pollen floras of the Late Eocene to Middle Oligocene, including the fabulous Florissant Flora, have been well studied. Woodland savanna, including such characteristic genera as *Ephedra, Mahonia,* and *Vauquelinia,* came to resemble that now living in the Chihuahuan region. Substantial amounts of gramineaceous pollen occurred, with a major peak of abundance in the Early Oligocene. By this time the subtropical forests had retreated "off the south toe of the Rocky Mountains" (74).

The appearance of an early savanna biota in North America produced a schism between the New and Old World. Along with tropical floras, the rich prosimian primate fauna of the Bridgerian (Middle Eocene) was the last in North America with clear Old World affinities (50, 152). By the Late Eocene, primates are notably scarce in the Rocky Mountain area. Their center of diversity presumably had shifted south into Middle America, and they were wholly cut off from their Old World cousins.

Besides decimation of arboreal species, the Late Eocene fauna of the Rocky Mountain region evinces two more positive responses to the expanding savanna flora: First, a number of native North American groups developed special adaptations to savanna living; and second, a number of new groups immigrated from Asia where they had already attained a degree of savanna adaption. In *Hyopsodus,* one of the most abundant native mammals in most Eocene deposits, Gazin (47) noted in later Eocene materials, "a distinct trend toward a more nearly lophodont [crest-toothed] condition in the molars . . . presumably in response to increasing aridity and coarser food." Among protrogomorphous rodents, several groups seem to have moved rapidly into the new adaptive zones available in the savanna. The genus *Manitsha,* which became half again as large as a modern beaver, may have been the "ground squirrel" of the Eo-Oligocene as its massive cheek teeth and heavy limbs indicate. The protoptychids were the "kangaroo rats" of that period and showed elongate hind limbs and progressive hystricomorph elaboration of their masticatory muscles (151, 173). The cylindrodontids had dentitions precociously specialized for coarse fibrous foods or sandy roots (9, 42, 173). It is noteworthy, however, that none of these native groups is thought to be closely related to later savanna-adapted taxa. Rather, they seem to have been replaced by them.

A major contingent of savanna-adapted vertebrates entered North America from Asia during the Late Eocene. The marked shift in faunal resemblance from Europe in the Early Eocene to Asia in the Late Eocene can be accounted for by final opening of the North Atlantic Ocean and reorganization of Asian plates (95, 123). The most notable immigrants are the selenodont artiodactyls, including the families Camelidae, Hypertragulidae, Leptomerycidae, Agriochoeridae, and possibly also the ancestors of Oromerycidae, Protoceratidae, and Merycoidodontidae. The first rabbits (*Mytonolagus*) entered during the Uintan (Late Eocene). It is noteworthy that the tortoises had entered the New World from Europe (3) two stages earlier than the hares, thus anticipating Aesop's account. Likewise, the first myomorph rodents appeared: *Protadjidaumo* and *Namatomys* representing the eomyids, and *Simimys* representing the zapodids or the cricetids or both (77, 78, 168). The hyaena-like Hyaenodontidae and the hippo-like Amynodontidae also entered North America at this time. The major source of early woodland savanna vertebrates thus appears to have been Asiatic.

It may be more than coincidence that such possible protosavanna inhabitants as the taeniodonts, uintatheres, and condylarths made their last appearances during the Uintan (158). The Late Eocene and Early Oligocene have long been recognized as a time of major vertebrate faunal turnover, especially in North America (9, 75). Possibly the rapid expansion of the early savanna biota in the midcontinent had much to do with these major faunal changes.

It would be of great interest to know the geographic extent of the Late Eocene and Early Oligocene savanna in North America. The principle body of floral and faunal evidence reviewed here accumulated in depositional basins in the Rocky Mountain Region, but there is also a substantial record from the Pacific Coastal Region. Floral evidence from the Pacific Northwest clearly shows that seasonal aridity and savanna vegetation had not extended into that area (74, 171, 172). On the other hand, the Late Eocene fauna from southern California includes several of the same presumed savanna taxa that occur in the Rocky Mountain Region; notable among a similar diversity of selenodont artiodactyls are the genera *Protoreodon, Protylopus, Poebrodon*, and *Mytonolagus* (9, 14, 48, 76). During the Eo-Oligocene, temperatures dropped and the seasonally arid *Madro-Tertiary Geoflora* began to invade California at low latitudes and low elevations fostering southern connections between the Rocky Mountain Region and parts of southern California (1, 4). The Eo-Oligocene Vieja faunas in West Texas and south through Chihuahua and Puebla, now being studied intensively by Wilson and associates (10, 35, 38, 40, 56, 57, 94, 140, 164, 174), contain a large number of the same presumed savanna vertebrates that occur in the Rocky Mountain Region, thus strengthening the assumption of southward continuity of savannas from the Rocky Mountain Region onto the Mexican Plateau. The Gulf Coast and eastern North America are not affected by these midcontinental developments and retain a moist subtropical vegetation (49).

OLIGOCENE   The scenic badlands of southwestern South Dakota and northwestern Nebraska, formed predominantly from the Oligocene White River Group, are the richest fossil vertebrate deposits in the world. It is possible to document the progress of the savanna biota during the critical phases of the Oligocene with some confidence. In the Orellan (Middle Oligocene) the first considerable number of taxa had developed hypsodont (high-crowned), or at least mesodont cheek teeth (53, 141, 160; see Figure 1). At the same time, many of these same taxa underwent diversification and adaptive radiation, suggesting that opportunities in savanna habitats were expanding.

SAVANNA VERTEBRATE HISTORY 361

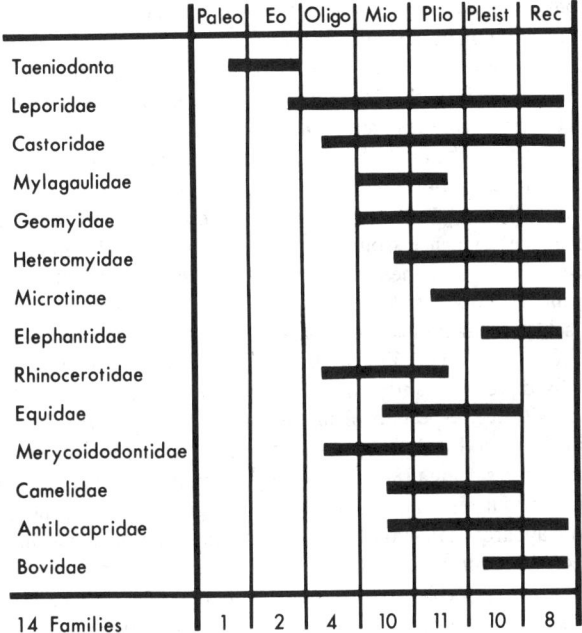

Figure 1 Time ranges of hypsodont mammals in North America. [After (53, 160)]

It is fortunate that in the White River Group different depositional facies can be distinguished, thus permitting an objective separation of different sets of fossil vertebrates. In the simplest model, silts and channel sandstones are taken to represent a stream-border community (the stream community itself being very limited and quite obvious), and the finer clayey sediments are taken to sample the open-country community (17, 93, 126, 136, 169). Despite the fact that the channels surely do not preserve biocenoses, and that even the clays have a biased sample (149), there is remarkable agreement among these studies as to the dominant members of the two major terrestrial communities. For example, the six commonest taxa in Wilson's (169) stream-border fauna from Whitneyan (Late Oligocene) deposits are, in order, *Protoceras* (a horned camel), *Elomeryx* (a crescent-toothed anthracothere), *Miohippus* (an early horse), *Subhydracodon* (a rhinoceros), *Agnotocastor* (the earliest beaver), and the piglike entelodonts; and these are likewise all prominent members of Matthew's (93) "forest fauna." Such facies analysis documents at least the principle taxa of the Oligocene forest and nonforest associations. *Palaeolagus* (a rabbit) and *Leptomeryx* (a ruminant) are both found to be common open-country inhabitants. There is also general agreement that the less common Leporidae (e.g. *Megalagus*) and other ruminants such as the Hypertragulidae and the Camelidae were also predominantly nonforest and nonstream-border dwellers. Clearly the little hypertragulid *Hypisodus* has all the marks of a plains grazer, including markedly hypsodont teeth, enlarged bullae, and elongate cursorial limbs. The very abundant merycoidodonts, especially *Leptauchenia,* give ambiguous results; some workers find them more common in the open-country facies (93, 169), while others find them more frequently in the channel facies (17, 126). *Mesohippus* (a horse) and *Perchoerus* (a peccary) also occur in both. One can more easily imagine savanna taxa ranging to the stream borders than the reverse. Other possibilities are that some postmortem transport has distorted the analysis or that differently adapted sibling species have not been distinguished (148). In any event, several leporids and ruminants are determined by such independent evidence to represent an early savanna biome.

This independent evidence from facies analysis may shed light on an enigma regarding the dentitions of these early lophodont (crest-toothed) and selenodont (crescent-toothed) herbivores. As Gregory (53) observes, "the dentitions of Uintan and Orellan selenodonts are more comparable to those of browsing deer and chevrotains than to any grazers." Yet the facies analysis seems to demand that many such animals lived predominantly away from the stream border (gallery) forest. And the botanical evidence, especially from the rich contemporary Florissant Flora, makes it clear that woodland savanna was widespread, and that grasses were quite abundant within the savanna (86). A majority of ruminants living in such habitats today have hypsodont teeth and many are largely grazers. Yet in the Oligocene, a majority of the apparent inhabitants of the midcontinental savannas had teeth that must be called mesodont, or at best subhypsodont. Possibly their feeding strategies were not so narrowly specialized as in many modern savanna herbivores. More probably the absence of markedly hypsodont contemporaries may have permitted the mesodont masticators to utilize foods that at present would be preempted by more efficient feeders on coarse fiber. Thus, even though an Oligocene leptomerycid and a modern *Hyemoschus* (of equatorial African rainforests) are comparable in size and overall dental character, the leptomerycid's feeding strategy may have differed drastically because no better grazers then existed. Prairie dogs (*Cynomys*) bite off grasses and herbs close to the ground, yet their teeth are mesodont and only moderately lophate. The evidence thus suggests that feeding preferences and behavioral patterns were adapted toward open-country living first; and that dental and other morphological changes followed gradually over a period of several million years.

Because of their relative scarcity, no rodents other than *Ischyromys* have been sorted out by such facies analysis. Nevertheless, the rodent fauna tends to show the same general evolutionary response as the ruminants and rabbits, with markedly increased diversification of hypsodont and otherwise savanna-adapted groups. Many of the old protrogomorph groups were apparently replaced by modern families. The first beavers, *Agnotocastor,* appear in the Early Oligocene (27). The first North American heteromyid, *Meliakrouniomys,* is recognized in the Early Oligocene (Chadronian) with progressive zygomasseteric structure and lophate, four-cusped cheek teeth; by Middle Oligocene (Orellan) the group diversifies considerably with such distinct genera as *Heliscomys, Akmaiomys,* and *Aptetomeus* (28, 57, 115). Likewise, the first Cricetidae in North America appear "at the beginning of the Oligocene" (168). This depends in part on the definition of that family and in part on what the earlier *Simimys* is. Still, the appearance of *Nonomys* (30,

77) and a new eumyine (J. H. Wahlert and E. Lindsay, unpublished data) in the Chadronian seems to indicate the modest beginning of the series of New World radiations of the Cricetidae. As Wilson (168) puts it: "The abundance of typical cricetids in the Middle Oligocene and their rarity, amounting almost to absence, in the Early Oligocene also may be a response to increasing amounts of subhumid savanna."

Wilson suggests that two fundamental structural advances triggered the revolution among modern rodent families at that time. The first, in most instances, was incisor strengthening by the change from uniserial to pauciserial enamel structure; the second, achieved by parallel evolution in many groups, was attaining greater masticatory power by shifting the several masseter muscle origins forward for greater purchase. Altogether these cranial and dental changes seem to be adaptations for securing and chopping coarser fodder. Evidently the Oligocene rodent revolution was a response to the new evolutionary opportunities in the savanna.

One of the characteristic elements of savanna and other open-country situations is a diversity of fossorial vertebrates, including both insectivores and rhizophores. In the White River one seems to find few, if any, rodents clearly specialized for a fully fossorial life, but highly specialized fossorial adaptations are evident in several insectivores. These include *Proterix* and *Cryptoryctes* among the true insectivores, and *Glyptosaurus* among amphisbaenid lizards (8, 44, 112, 113).

The commonest White River carnivore, the foxlike *Hesperocyon,* has been determined to belong to the open-plains facies by Clark, Beerbower, & Kietzke (17). *Hyaenodon* may well belong to that community also. Most other carnivores are either too rare to be assigned with certainty or else are truly eurytopic.

The possibility that faunal provinces may be distinguished within the Oligocene vertebrate samples of North America has intrigued several able paleontologists (63, 126, 159). Unfortunately many of the apparent differences between the "Plains Province" of the Big Badlands and the "Mountain Province" sampled at Canyon Ferry and Pipestone Springs in Montana have broken down either because they were based on sampling errors or because such central Wyoming sites as Bate's Hole, Flagstaff Rim, and Beaver Divide have blurred the boundaries. Furthermore, the distinctions of higher elevation and more intense volcanic activity that were thought to distinguish the "Mountain Province" in the Oligocene are removed now that the Wiggins Formation (volcanic conglomerates) has been shown to be Eocene (29, 81). Nonetheless, as Robinson (122) notes, the possibility of developing provincial concepts still seems better in the Oligocene than earlier. The work of Wilson and associates in the Big Bend of Texas now provides a number of excellent samples to compare with the northern "plains" and "mountain" samples.

MIOCENE According to the classic story, grasslands first blanketed the Great Plains during the Miocene. The burden of this story was usually borne by the horse, in particular by the Hemingfordian transition from *Parahippus* and its browsing antecedents to *Merychippus* and its grazing descendants. While modern advances in the field of geochronology have provided improved estimates of the age and duration of these changes, and while improved sampling of the fossil faunas, especially by the Frick Laboratory, has refined and complicated the known phylogeny of Equidae (85), the well-known history of the grazing horse is still, as it were, engraved in tablets of stone.

The story becomes more interesting when additional vertebrate groups are considered. In the first place, it gains credence if more than one taxon seems to have responded to the supposed spread of grasslands. If such changes as increased crown height, enamel folding, and cement deposition on the cheek teeth were confined only to the horse, one might well argue that some *Parahippus* populations had merely crossed a genetic threshold. Instead, a large number of unrelated taxa show comparable trends accelerating during an adjacent age or two. This point has always been implicit in the best discussions of the subject. For example, in his early studies of Middle and Late Miocene faunas of the Great Basin and Mohave Desert, J. C. Merriam (97, 98) recognized the diversification and dominance not only of the merychippine, protohippine, and hipparionine horses, but also of merycodontine pronghorns and hypsodont camelids. In his great book *A History of Land Mammals of the Western Hemisphere,* Scott (125) summarized the situation as follows: "... an especially characteristic feature of later Miocene faunas was the large number of species, belonging to several different orders, which had the hypsodont, or persistently growing type of grinding teeth; many of the horses, camels, ruminants and rodents displayed this structure and as was first pointed out by Kowalevsky, the explanation is probably to be found in the spread of grasslands at the expense of the forest...."

The diversity of ungulates in North America reached its apogee in the Miocene, with an array of taxa fully comparable to that in African savannas today. Besides hypsodont horses, camels, and pronghorns, one finds an extraordinary diversity of hypsodont oreodonts, diceratherine, teleoceratine and other advanced rhinos, and finally at least four kinds of gomphotheriid proboscideans. There is a notable correlation between those lineages that are most hypsodont and those that undergo the greatest diversification. For example, gomphotheriid proboscideans outstrip mammutids, protolabine camels diversify more than giraffe camels, and hypsodont horses surpass brachydont horses.

Functional studies of these Miocene ungulates usually have featured their dental and masticatory adaptations to chopping coarse fodder. For both phylogenetic and functional reasons, the postcranial skeletons deserve equal consideration, and a number of recent studies have begun to right that balance. It is impressive to find that in many ungulate groups, modifications for open-country locomotion appear at least as early, and at least as markedly, as do their masticatory modifications. Thus, Camp & Smith showed that the hypsodont lineages of horses acquired their digital springing ligaments and other progressive skeletal features about as early as they did their high-crowned teeth (15, 65). Similarly, the first hypsodont camelids also developed the pacing gait and digitigrade padded feet for efficient travel in open landscapes (155). The general coincidence of masticatory and locomotor modifications in each of these ungulate groups further implies that the underlying cause of these changes is environmental in nature.

Fully as impressive as the open-country adaptations of Miocene ungulates are those of Miocene rodents. The outlines of their history are only now coming into sharp focus, thanks to the Hibbard Method of collecting microvertebrates and to a new generation of rodent specialists. As in the ungulates, one notices a marked tendency in many Miocene rodents to develop hypsodont dentitions presumably for masticating coarse fibers, sandy roots, or both (118). Here too one often finds simultaneous change in the locomotor system; in most hypsodont rodents the change involves adaptations for burrowing. The most spectacular Miocene adaptations for burrowing and fodder chopping occur in the now extinct mylagaulid rodents, beginning with *Mesogaulus* in the Hemingfordian and continuing with *Mylagaulus* and the horned genera *Epigaulus* and *Ceratogaulus* in the Barstovian and younger (32). Ochotonid rabbit genera multiplied rapidly (52). The greatest diversification among Miocene savanna rodents has been documented for the geomyoids (pocket gophers, etc) in a number of recent papers (45, 79, 116, 117, 119, 131, 142). *Entoptychus* in the Upper John Day Formation of Oregon exhibits the most rapid evolutionary change documented in any Tertiary land mammal lineage, with "at least eight successive stages of cheek tooth hypsodonty during its existence" of about two million years (36, 116). Heteromyidae continued to diversify through the Miocene, especially during the Barstovian and Clarendonian (79). Kangaroo rats (*Eodipodomys*), with fully specialized hind limbs, hypsodont, dentine-exposed teeth, and enlarged auditory bullae, appeared in the Late Miocene (Late Clarendonian) (150). Even among beavers, one finds root-eating burrowers, as the repeated discovery of *Palaeocastor* (of Arikareean age) in the magnificent burrows known as "Devil's Corkscrews" clearly demonstrates (109). This striking diversity of burrowing rodents in the Miocene contrasts with the limited number of known burrowing forms (few of which are rodents) in the Oligocene. Among burrowing nonmammals, rattlesnakes and owls appear to be new plains-adapted groups (62, 100).

Since its brilliant enunciation by Kowalevsky (71), paleontologists and botanists generally have accepted the view that the Miocene outburst of hypsodont mammals in north temperate regions was an adaptation to widespread grasslands or steppe vegetation. This view seemed to be confirmed by Elias's documentation (26) in the High Plains of abundant fossil hulls of *Stipidium* and *Berriochloa*, representing the needlegrass (*Stipa*) group, in the same Miocene deposits as the world's most diverse assemblage of hypsodont horses. Yet, as it now appears, this classic view was only about half true.

These supposed Miocene grasslands were considerably abridged by MacGinitie's study (87) of the Kilgore Flora from Late Miocene deposits in Nebraska. The Kilgore leaf and pollen flora represents a savanna "with mesic forests along the streams and open grassy forests on the interfluves . . . [and] no treeless prairies." It consisted of grasses, along with trees such as "small live oaks, pines, blackberry, and persimmon, with shrubs of *Mahonia,* currant, hawthorne, sagebrush, and relatively abundant species of composites."

In his excellent review of this problem, Gregory (53) points out three kinds of vertebrate evidence that convincingly corroborate the floral evidence, and permit recognition of woodland savanna biota across much of the midcontinent. First, he notes the continued occasional presence of arboreal rodents, and one may add, the last North American primates and abundant insectivores as well (66, 83, 84). Second, among ungulates a considerable diversity of brachydont browsers and mixed feeders persist, including anchitheriine horses, chalicotheres, bunodont peccaries, moschid and dromomerycid ruminants, protoceratids, miolabine camels and, most notably, aepycameline (giraffe) camels over 12 feet tall. Third, the high total diversity of large and medium-sized herbivores in sample after sample equals that in the present African savannas and is much higher than that in treeless prairies (144). Thus Gregory's analysis of the vertebrate faunas corroborates and amplifies MacGinitie's analysis of the Kilgore Flora: Both clearly tell us that through the Late Miocene the Great Plains supported a woodland savanna with broad interdigitations of riparian forest.

The Miocene lasted nearly as long as the Eocene, spanning some 20 million years, and therefore may be expected to include major changes within its span. Thanks to a rich North American Miocene record and to a comprehensive review of its biochronology by Tedford et al (in press), patterns of faunal change can be clearly discerned within that epoch. Tedford et al recognized seven successive biostratigraphic units or subunits with durations of between two and three million years each. Fortunately, major faunal shifts occur much less frequently than these units could discriminate; thus Tedford et al recognize only three vertebrate chronofaunas. These chronofaunas maintain essentially unified generic composition while evolving through long spans of time, thus implying extended periods of ecosystem stability. The First Miocene Chronofauna in North America spans the Late Arikareean and Early Hemingfordian intervals, the second spans the Late Hemingfordian and Early Barstovian, and the third, the "Clarendonian Chronofauna" (153), spans the Late Barstovian, Clarendonian, and Early Hemphillian. The first and second of these Miocene chronofaunas each lasted about five million years, and the third endured about eight million years. By definition, each chronofauna, roughly corresponding to Early, Middle, and Late Miocene, ends with a major faunal breakdown and the next begins with a fundamental faunal reorganization. Yet in this case, these breaks do not terminate, but merely punctuate, the overall trends of adaptation and diversification in the principal savanna lineages. For example, from *Parahippus*-like stock in the First Miocene Chronofauna come diverse genera of protohippine and merychippine horses in the Second Miocene Chronofauna, and from them emerge about eight genera of pliohippine and hipparionine horses in the Third Miocene Chronofauna. Gregory's (53) usage unites these three Miocene chronofaunas along with two earlier Oligocene chronofaunas (namely the Whitneyan-Arikareean and the Orellan) in a greater North American mid-Tertiary chronofauna. This broader usage seems to exceed the original essentially generic-level definition of Olsen's and might be termed a chain of chronofaunas or a "zeugochronofauna." In any event, the overall continuity of successive autochthonous radiations in North American savanna lineages from Oligocene through Miocene included five successive chronofaunas.

The degree of continuity in North American Miocene savanna lineages can be crudely measured by asking what percentage are autochthonous products of this continent. The relatively high percentage that one indeed finds tends to corroborate Gregory's (53) emphasis on the "repeated association in successive deposits . . . of the same lineages from Orellan through Barstovian." Indeed that continuity can be followed into the Hemphillian. Horses, camels, and oreodonts are clearly native stocks, and account for most of the Miocene diversity of hypsodont ungulates (not to mention much of the diversity of brachydont ungulates). Likewise, among hypsodont rodents the heteromyids, geomyids, mylagaulids, and most of the castorids have their Oligocene ancestors in North America.

Of immigrant stock, only the merycodontine pronghorns may be counted among the first rank of savanna-adapted vertebrates. Appearing in the Early Hemingfordian, their herds diversified and spread abundantly over western North America during the Second Miocene Chronofauna. And their descendants, the antilocaprine pronghorns, were equally important members of the Third Miocene Chronofauna and thereafter (156).

This argument becomes blurred somewhat if one considers immigrant groups that do not fully exemplify open-country adaptations (Figure 2). About a dozen of the 34 Miocene mammalian immigrant genera tabulated by Tedford et al might be regarded as forest-edge or mixed-feeding types, and thus in the broad sense savanna-adapted. These include the amphicyonids (64), the chalicotheres, the cervoid ruminants, the proboscideans, the ochotonid rabbits, the zapodid rodents, and the cricetid rodents that diversified considerably in the Barstovian. Presumably they were able to enter the more densely wooded margins of the major North American savannas and to penetrate the broader riparian forest corridors (166, 167), thus edging into the savanna fauna. This pattern further supports the view that the midcontinent supported woodland savannas, not pure grasslands.

It is interesting to note the steady increase in Old World immigrants through the Early Miocene (Gering to Marsland). Their numbers increased markedly through a period of five million years and then declined in the mid-Miocene (Figure 2). Surely this is not a stochastic pattern; rather it appears as if the doors were opening and then closing at a remarkably steady rate. Presumably, this reflects a gradual opening and reclosing of the forested habitats connecting temperate latitudes of North America and Asia. An even more marked increase in immigration rate occurred in the Late Pliocene and Pleistocene (120, 154).

The new view of woodland savanna rather than treeless prairie blanketing the interfluves during the Miocene is reasonably well documented by floral and faunal evidence in the Great Plains. One may ask whether, or to what degree, the same pattern extended to other parts of North America. The answer is that the pattern was by no means uniform, and the the Late Miocene seems to be a time of maximum regional differentiation in North America (Tedford et al, in press).

In the Pacific Northwest a sequence of rich Oligocene and Miocene floras has permitted Wolfe & Hopkins (171, 172) to document the vegetational history of this region in some detail. In the prevailing warm-temperate forest, the frequency of trees with entire-margined leaves dropped notably in the mid-Oligocene (about 31 million years ago); several lesser fluctuations followed. The percentage then moved dramatically downward again in the Middle Miocene, about 12 million years ago. While there has been some discussion both of how well leaf margin percentages track climatic changes and of the regional significance of a local flora [with its own edaphic, orographic, and altitudinal peculiarities (6)], the large array of floras

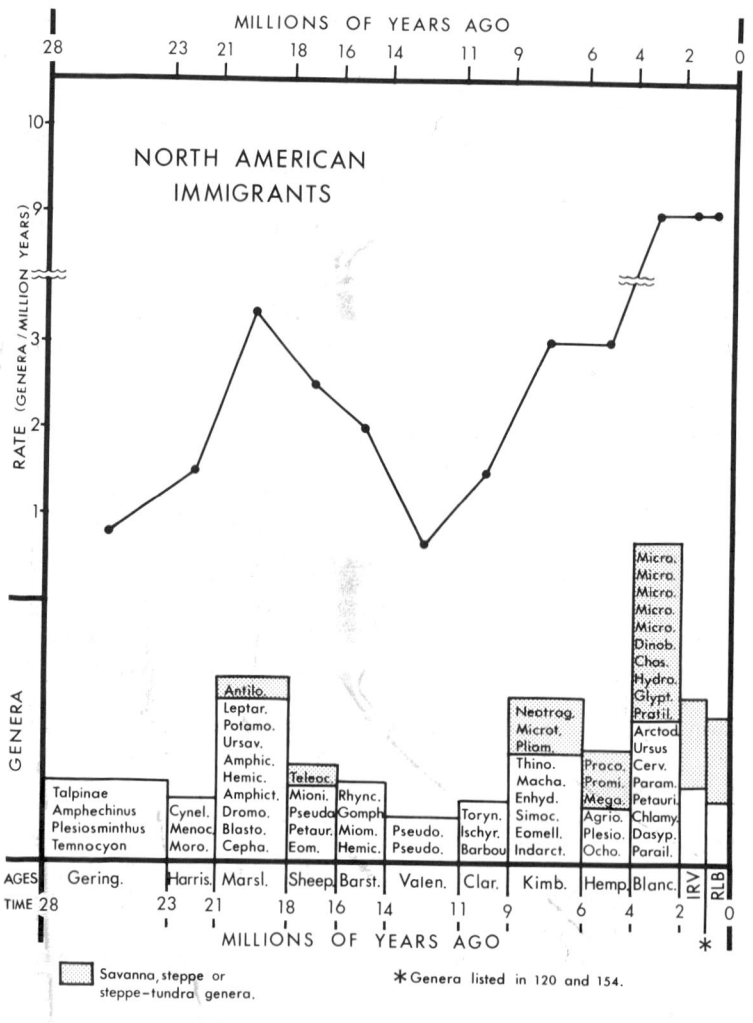

Figure 2 Immigrant mammal genera in the Middle and Late Cenozoic of North America. Note two distinct peaks, one in the mid-Miocene, one in the Plio-Pleistocene. [After (120, 154) and Tedford et al, in press]

sampled allays most serious doubts (22). Thus it seems clear that during the Middle Miocene in the Pacific Northwest mesophytic forests continued to predominate, although they moved markedly toward a cool-temperate composition. Evidence from vertebrates of coastal California gives much the same impression (67, 80).

At approximately this same time vertebrate faunas are also well represented in the Columbia Plateau. Indeed some Miocene faunas occur in direct association with floras, the most notable case being the Mascall flora and fauna (24). Shotwell (132) drew an interesting facies distinction between his Quartz Basin fauna and the contemporaneous fauna from the Red Basin and the other Oregon sites. The "Quartz Basin facies" produces diverse heteromyids, geomyids, cricetids, and zapodids, but few sciuromorph rodents; the widespread *Monosaulax* beavers and *Merychippus* horses are each represented in the Quartz Basin by a species with more complex enamel; and such probable browsers as *Hypohippus, Ticholeptus,* and *Rakomeryx* are conspicuous by their absence. This "Quartz Basin facies" is rare and suggests that within the predominant deciduous forests of the Columbia Plateau there were patches of more open (perhaps shrubby) woodland. This facies becomes more evident in Barstovian faunas as one moves southward into Nevada and the Mohave Desert; and it incorporates a greater diversity of merycodontine pronghorns and protohippine horses (130). The Tonopah local fauna of southern Nevada (58) also represents the "Quartz Basin facies."

In the Stewart Valley of west-central Nevada one finds both the "Quartz Basin facies" and the "Red Basin facies" (a deciduous forest biota) in separate but nearly correlative deposits. These local faunas are bracketed by local floras: the Fingerrock Flora below, the Stewart Valley Flora stratigraphically above. In agreement with the mammalian evidence, these floras show a marked facies shift in probable Barstovian time (5, 170): The older flora, despite its domination by live oaks, resembles the mesophytic floras of Oregon; whereas the younger flora has "few species in common with floras to the north," but has a distinctive spruce-cedar–live oak association, including some early records of uniquely Great Basin species, conspicuous pine and grass contributions, and a generally less humid character. Thus during the early part of the Late Miocene one can plot the boundary between the mesic biota of Oregon and northern California and the emerging semiarid biota of Nevada and southern California.

By the latest Miocene a scrub or steppe-adapted Great Basin fauna had expanded to nearly its present range. It was characterized by the appearance of abundant arvicolines (*Promimomys* and *Microtoscoptes*), endemic antilocaprines (*Ilingoceras* and *Sphenophalos*), badgers (*Pliotaxidea*), and by the spread of steppe-adapted horses (*Pliohippus*) at the expense of forest-adapted horses (*Hipparion*) (130).

It is more difficult at present to discuss the Late Miocene biota of the Colorado Plateau. Several samples, notably from the Flint Creek beds of western Montana and the Lemhi Valley in Idaho resembling the Mascall and other mesic-adapted faunas of the Columbia Plateau, have been found (68, 72, 103, 111, 122, 143). While there is clear evidence here of continuity between the plains and the far west, the mammalian faunas include predominantly mesic forest types and few, if any, species adapted to less humid conditions.

The Middle and Late Miocene record provides numerous suggestions that the evolution of less humid savanna biota was centered about the Mexican Plateau. The history of the flora has been treated under the rubric "Madro-Tertiary Geoflora." One of its earliest records is the Tehachapi Local Flora associated with the Phillips Ranch Local Fauna of Hemingfordian age in southern California, and the surviving elements now reside in the arid and semiarid southwest (6, 175). A number of savanna-adapted Miocene vertebrates seem to follow a southern distribution between the Plains and the Southwest. For example, the precociously hypsodont gazelle camels (*Stenomylus*) have been discovered in Middle Miocene sites in southern California, Arizona, and New Mexico (45, 176). And the important Zia Sand Fauna from New Mexico produces a progressive aridity-adapted fauna, including stenomylines, Cynarctoides, mylagaulids, geomyids, and heteromyids, during successive phases of the Middle Miocene (43, 45). These faunas agree with the floras of the Rio Grande Trench in indicating a downdropped southern savanna corridor connecting the Plains faunas with those of the Southwest (7).

The accepted evolutionary history of the Miocene peccaries *Cynorca* and *Dyseohyus*, revised by Woodburn (175), emphasizes the same pattern. In the earlier Miocene *Cynorca* was widely distributed in mesic situations from Oregon and California to the Texas Gulf Coast and High Plains and a little later in Florida and Maryland. The mid-Miocene range of its descendant, *Dyseohyus*, however, shows evidence of adaptation to subhumid conditions and a more southerly distribution, which appears to have its center in the Mexican Plateau, with one northern arm (*D. fricki*) reaching southern California and southern Nevada, and another (*D. stirtoni*) extending into the High Plains via the Rio Grande Trench (in Texas). Confirmation of this pattern comes from Dalquest & Mooser's (20) recognition of *D. stirtoni* in the Zoyatal Local Fauna in central Mexico.

The four known Miocene faunules of central Mexico have received special attention, partly because of their possible focal role in the history of the semiarid biota of North America. While the results are still tantalizing, these faunules do show remarkable continuity with the Great Plains savanna fauna at least in the grazing horses, camels, and oreodonts (20, 33, 34). Ferrusquia (34) suggests these samples represent both "una sabana, ambiente apropiado para *Merychippus, Oxydactylus* y probablamenta el protoceratido traguloide, y un bosque lluvioso tropical, ambiente adecuado para *Merychyus* y *Gomphotherium*."

One notes that in the course of the Miocene an interesting shift occurred in the affinities of the vertebrate fauna around the Gulf of Mexico. During the Early Miocene the Garvin Gully Fauna of Texas and the Thomas Farm Local Fauna of Florida exhibit considerable endemism in the herpetofauna and among the browsing artiodactyls where the Florida tragulids, nothokematids, and peculiar protoceratids seem largely to supplant the moschids and dromomerycids so characteristic of the Plains (2, 39, 108, 163). The Gaillard Cut Local Fauna of Panama (161) samples virtually the same Gulf Coast Fauna that Tedford et al have appropriately distinguished as the Miocene Gulf Coast Chronofauna.

By mid-Miocene (Barstovian) time the peculiarities of the Gulf Coast Fauna have nearly disappeared and closer faunal continuity is reestablished with the Great

Plains Fauna. This is most clearly indicated by the Trinity River Local Fauna, in which the diverse sample of artiodactyls wholly resembles plains samples (37, 108; Tedford et al, in press). And the same carries up through the later Miocene Cold Spring and Lapara Creek faunas.

The faunal nexus between the Plains and the subhumid region of the Mexican Plateau, established during the Middle Miocene, persisted into Late Miocene. But evidently major barriers were beginning to separate these faunas from the Great Basin and Mohave Desert faunas to their west and northwest. The beginnings of such a separation may be perceived earlier in the specific distinction between *Dyseohyus fricki* (in the far west) and *D. stirtoni* (from Mexico to the High Plains). It is more evident, however, by the Late Miocene (Clarendonian) when a major suite of hypsodont horses, including *Calippus, Astrohippus,* and *Pseudhipparion,* and the shovel-tusked gomphothere *Amebelodon* range from the Plains and the Gulf Coast into Arizona and Chihuahua, but not into the far west (73, 153). One may speculate that the region of the present Sonoran and Mohave Deserts may already have begun to form a barrier to savanna-adapted taxa in the Great Plains and Chihuahua regions.

Presumably, the increased Late Miocene continuity between the Gulf Coast Fauna and the Plains Fauna represents an eastward expansion of the savanna corridor that earlier had been largely confined to the Rio Grande Trench. Presumably, the corresponding reduction in the area of subtropical forest enclaves led to the extinction of most of the peculiar Miocene browsing fauna of the Gulf Coast, and to the final break between the mesic forests of the southeast and those of the tropics. A probable broadening of the Gulf Coastal savanna corridor does not contradict the more general view of Graham (49) that warm-temperate forests of essentially modern aspect prevailed over most the southeastern United States. But Graham's (51) study of a Late Miocene pollen flora in Veracruz shows no tropical rainforest elements, up to 45% grass pollen, and oak and pine-oak woodland predominating, suggesting that the temperate character of the northern Gulf Coast woodland savanna extended far to the south as well.

The Late Miocene extension of woodland savanna around much of the Gulf Coast also produced a major disjunction between the mesic forests of eastern United States and those of Central America. This disjunction is very clearly documented in fossil Amphibia and Reptilia of North America (31). Until about Late Miocene time, north-temperate records of many groups of mesic herptiles are relatively abundant (2, 31, 61, 62, 70, 82, 99, 110, 145). Subsequently, as Estes' review (31) clearly shows, many of these taxa disappear from the fossil record (mainly in temperate latitudes) only to reappear as Recent members of Central American mesic faunas (124). Jaçanas (106) provided a similar example among birds. The Late Miocene establishment of a broad Gulf Coastal woodland savanna produced the final disjunction between forest biotas of Middle America and the eastern United States (90, 91).

North American Steppe (Pliocene and Pleistocene)

The last step in dismantling the once continuous forests of North America came in the Pliocene, about five million years ago. Woodland savanna changed to steppe in the High Plains and over major parts of the arid southwest. The evidence, as Gregory (53) clearly analyzed it, consists of four points: absence of arboreal species, absence of browsing (especially giraffoid) species, restricted diversity even of grazing species, and (as a composite of these) lower overall faunal diversity. Gregory went on to show that in the Great Plains these features only partly characterize the Early Hemphillian (Latest Miocene), but are fully applicable to the Late Hemphillian and Blancan (Late Pliocene) faunas. The final shift from woodland savanna to treeless prairie (except narrow belts of riparian forest) accounts for the final demise of the Third Miocene Chronofauna of Tedford et al in the Late Hemphillian. Thus the midcontinental steppe came into existence no more than five or six million years ago.

The change from woodland savanna to steppe had a dramatic effect on the vertebrate fauna. In the High Plains, the browsing horses, the tapirs, most of the peccaries, most of the proboscideans, and all of the moschid and dromomerycid ruminants disappeared—presumably in response to the nearly total disappearance of their browse. What might seem more surprising is the loss (without apparent replacement) of predominantly grazing types, including four genera of hypsodont horses, two genera of hypsodont camels, and three genera of antilocaprines. Furthermore, at the end of the Hemphillian less than two million years later, another chronofaunal breakdown occurred with the further loss from the former savanna fauna of three more genera of hypsodont horses, the last mylagaulid rodents, the last New World rhinocerotids, and three more genera of antilocaprines. Among rodents, the cricetids experienced an important radiation, especially the more hypsodont groups such as sigmodontines and phyllotines (Baskin, in press). Wilson (166) observed that "there is an increase in hypsodonty in Middle Pliocene forms ... in response to the increasing aridity ... a response which culminated in the upper Pliocene appearance of many of our Recent genera with long-crowned teeth."

The spread of steppe and desert conditions in the Pliocene has long been suggested by students of paleobotany. In the Great Plains grass hulls are common (26, 41), as are caliche deposits, suggesting increased evaporation, but no major flora directly confirms the current view of forests narrowly restricted to riparian sites. The most direct Pliocene evidence of steppe vegetation comes from southern California where the Mt. Eden Flora produces desert border chaparral (4).

The relative recency (five million years) of steppe conditions in the Great Plains is reflected in the relative paucity of endemic steppe-adapted ("pure grassland") taxa in the modern biota. Most of the grasses themselves range more widely and more diversely into surrounding woodland biomes, particularly to the southeast; the grasshoppers also have greater diversity outside the Central Plains with major centers of diversity to the west and southwest. Similar patterns are evident within the mammals (23). And to most birds that range through the region, as Mengel (96) puts it, the steppe per se represents "a hostile Sea."

During the Pliocene the expanding steppe of central North America did not connect broadly with the central Asiatic steppe. Instead the two steppes were at first separated in Beringia and in boreal latitudes by broad bands of forest, mesophytic during most of the Pliocene and earliest Pleistocene and coniferous taiga during much of the Pleistocene, which continued to filter the steppe elements. Not until the

Late Pleistocene did continuous tundra and steppe-tundra corridors connect the Old and New World steppe biota (54, 60, 128). Instead, forest-adapted vertebrate immigrants, such as *Petaurista*, a flying squirrel, *Parailurus*, a panda, and *Ursus*, bears enter during Hemphillian and Blancan time (121, 144a). Nonetheless, Hemphillian deposits of North America yield an increasingly important contingent of Old World steppe elements that filtered through the northern forests. Most notable are the arvicoline rodents, which began to enter North America in increasing numbers and with characteristically explosive diversification upon arrival (89). Zapodine jumping mice may have come from the same source at this time, or they may actually have gone in the opposite direction. *Neotragocerus* of Hemphillian deposits is the first of a number of Old World bovids to disperse to the New World; and *Procoileus* is the first of a number of progressive Cervidae adapted to some degree of open-country living. The rate of immigration from the Old World continued to accelerate through the Blancan and younger stages (120). Even so, a majority of the surviving plains and desert vertebrates of North America have North American ancestors from the Miocene or older; the dominance of native stock merely became less clear in the Late Cenozoic than it had been in the Miocene.

A renewed pattern of provincialism developed within temperate North America during the Pliocene, principally because certain forest enclaves were not greatly affected by the onset of drier climatic conditions. In the Pacific Northwest and northern California many elements of the older mesic biota persisted. The aplodontids were represented then by *Liodontia*, as they are now by *Aplodontia* (129); their poor urine concentration (104) partly explains their consistent history of restriction to mesic forest.

The biota around the Gulf of Mexico was not so deeply affected by the Pliocene onset of drier conditions. There presumably the prevailing tendency toward desiccation was countered by the effects of summer rains from the Gulf, and these same effects seem to have extended west onto the Mexican Plateau into Chihuahua. The evidence for this is the remarkable resemblance between the savanna vertebrates of the Yepomera Local Fauna in Chihuahua and the Bone Valley Fauna in Florida (73, 157). Furthermore, the El Ocote Local Fauna near Aguascalientes samples much the same fauna in southern Mexico (101). Thus, although the Late Miocene savannas gave way to Pliocene grasslands in the Great Plains, they persisted on through the Pliocene in a broad belt southward from about latitude 30 degrees north.

This subtropical savanna biota that persisted around the Gulf of Mexico and into Central America played an extremely important role in the interamerican faunal interchange which began in the latest Pliocene. We shall take up this subject in Part II of this review.

Extinctions

As much of North America's savanna vegetation gave way to steppe, the diversity of the savanna biota declined markedly (18, 19, 53). A majority of the large ungulates disappeared (53, 154), whereas most of the surviving steppe- and desert-adapted taxa were small to medium-sized.

The resemblances between the Late Pliocene (Hemphillian and Blancan) extinctions and the Late Pleistocene (Rancholabrean) extinctions in North America are striking. While there are differences between the two sets of circumstances (notably the presence of man in the Late Pleistocene), the similarities should not be dismissed lightly. In both instances, a cumulative set of large mammal extinctions coincided with climatic deterioration, and a corresponding increase in small mammal taxa. The Late Pliocene extinctions involved about 30 mammalian genera in North America, while the Late Pleistocene extinctions involved about 20 genera (92, 147, 154). Thus, even though the former extinctions may have been spread over a longer period of time (up to three million years), the changes seem to have been more devastating. Furthermore, several of the large vertebrates that are counted as Late Pleistocene extinctions are still living in Central America and/or South America; these include giant tortoises, capybaras, jaguars, spectacled bears, tapirs, and llamas. The fact that the same genera (if not species) survived in subtropical savannas but became extinct in north temperate steppes suggests that environmental deterioration had something to do with their north temperate demise. Reed (114) notes a similar pattern of reduced extinction in subtropical latitudes in the Old World (e.g. elephants and rhinoceroses became extinct in northern Eurasia, but live on at lower latitudes). Thus, the reduction in savanna area may be causally related to the loss of large vertebrates in the Late Cenozoic of North America.

ACKNOWLEDGMENTS

I thank Richard Tedford, J. W. Hardy, and Jon Baskin for helpful criticism of this manuscript. Much of the author's work in Floridian vertebrate paleontology was supported by NSF Grants GB 3862 and 33500.

Literature Cited

1. Addicott, W. 1970. Tertiary palaeoclimatic trends in the San Joaquin Valley, California. *USGS Prof. Pap.* 644, D:1–19
2. Auffenberg, W. 1963. The fossil snakes of Florida. *Tulane Stud. Zool.* 10:131–216
3. Auffenberg, W. 1974. Checklist of fossil land Tortoises. *Bull. Fla. State Mus., Biol. Sci.* 18:121–251
4. Axelrod, D. I. 1950. Evolution of desert vegetation in western North America. *Carnegie Inst. Wash. Publ.* 590:215–306
5. Axelrod, D. I. 1956. Mio-Pliocene floras from west-central Nevada. *Univ. Calif. Publ. Geol. Sci.* 33:1–322
6. Axelrod, D. I., Bailey, H. P. 1969. Paleotemperature analysis of Tertiary floras. *Palaeogeogr. Palaeoclimatol. Palaeoecol.* 6:163–95
7. Axelrod, D. I., Bailey, H. P. 1976. Tertiary vegetation, climate, and altitude of the Rio Grande depression, New Mexico-Colorado. *Paleobiology* 2:235–55
8. Bjork, P. R. 1975. Observations on the morphology of the hedgehog genus *Proterix. Univ. Mich. Pap. Paleontol. (Hibbard Mem.)* 12:81–88
9. Black, C. C., Dawson, M. R. 1966. A review of Late Eocene mammalian faunas from North America. *Am. J. Sci.* 264:321–49
10. Black, C. C., Stephens, J. J. 1973. Rodents from the Paleogene of Guanajuato, Mexico. *Occas. Pap. Mus. Texas Tech. Univ.* 14:1–10
11. Bourlière, F. 1963. Observations on the ecology of some large African mammals. In *African Ecology and Human Evolution*, ed. F. C. Howell, F. Bourlière, pp. 43–54. Chicago: Univ. Chicago Press
12. Bourlière, F. 1975. Mammals, small and large: the ecological implications of size. In *Small Mammals: Their Produc-*

tivity and Population Dynamics, ed. F. B. Golley et al, pp. 1–8. London: Cambridge
13. Bradley, W. H. 1947. Limnology and the Eocene lakes of the Rocky Mountain Region. *Bull. Geol. Soc. Am.* 59:635–48
14. Brattstrom, B. H. 1955. New snakes and lizards from the Eocene of California. *J. Palenotol.* 29:145–49
15. Camp, C. L., Smith, N. 1942. Phylogeny and function of the digital ligaments of the horse. *Univ. Calif. Mem.* 13(2):69–124
16. Carpenter, J. R. 1940. The grassland biome. *Ecol. Monogr.* 10:617–84
17. Clark, J., Beerbower, J. R., Kietzke, K. K. 1967. Oligocene sedimentation, stratigraphy, paleoecology and paleoclimatology in the Big Badlands of South Dakota. *Fieldiana Geol. Mem.* 5:158
18. Clements, F. E. 1936. The origin of the desert climax and climate. In *Essays in Geobotany in honor of William Albert Setchell,* ed. T. H. Goodspeed, pp. 87–140. Berkeley, Calif: Univ. Calif.
19. Cloudsley-Thompson, J. L. 1964. Wild animals in arid zones. In *Life in Deserts,* ed. J. L. Cloudsley-Thompson, M. J. Chadwick, pp. 29–43. London: Foulis
20. Dalquest, W. W., Mooser, O. 1974. Miocene vertebrates from Aguascalientes, Mexico. *Pearce-Sellards Ser. Tex. Mem. Mus.* 21:1–10
21. Diamond, J. M. 1975. Neotropical Biogeography (Rev.). *Science* 187:830–31
22. Dorf, E. 1969. Paleobotanical evidence of Mesozoic and Cenozoic climatic changes. *Proc. N. Am. Paleol. Conv., Vol. 1 pt. D: Paleoclimatol.* 323–47 Lawrence, Kans.: Allen
23. Dort, W. Jr., Jones, J. K. Jr. 1970. *Pleistocene and Recent Environments of the Central Great Plains.* Lawrence, Kans.: Univ. Kansas Press. 433 pp.
24. Downs, T. 1956. The Mascall Fauna from the Miocene of Oregon. *Univ. Calif. Publ. Geol. Sci.* 31:199–354
25. Duellman, W. E. 1966. The Central American herpetofauna: an ecological perspective. *Copeia* 1966(4):700–19
26. Elias, M. K. 1942. Tertiary prairie grasses and other herbs from the High Plains. *Geol. Soc. Am. Spec. Pap.* 41. 176 pp.
27. Emry, R. J. 1972. A new species of *Agnotocastor* (Rodentia, Castoridae) from the Early Oligocene of Wyoming. *Am. Mus. Novit.* 2485:1–7
28. Emry, R. J. 1972. A new heteromyid rodent from the Early Oligocene of Natrona County, Wyoming. *Proc. Biol. Soc. Wash.* 85:179–90
29. Emry, R. J. 1975. Revised Tertiary stratigraphy and paleontology of the western Beaver Divide, Fremont County, Wyoming. *Smithson. Contrib. Paleobiol.* 25:1–20
30. Emry, R. J., Dawson, M. R. 1972. A unique cricetid (Rodentia, Mammalia) from the Early Oligocene of Natrona County, Wyoming. *Am. Mus. Novit.* 2508:1–14
31. Estes, R. 1970. Origin of the Recent North American lower vertebrate fauna: an inquiry into the fossil record. *Forma Functio.* 3:139–63
32. Fagan, S. R. 1960. Osteology of *Mylagaulus laevis,* a fossorial rodent from the Upper Miocene of Colorado. *Univ. Kans. Paleontol. Contrib. Vertebrata* 9:1–32
33. Ferrusquia-Villafranca, I. 1974. Tres edades radiometricas Oligocenicas y Miocenicas de Rocas Volcanicas de las regiones Mixteca Alta y Valle de Oaxaca, Estado de Oaxaca (1). *Bol. Asoc. Mex. Geol. Pet.* 26:249–62
34. Ferrusquia-Villafranca, I. 1975. Mamiferos Miocenicos de Mexico: Contribucion al conocimiento de la paleozoogeografia del continente. *Rev. Inst. Geol., Univ. Nac. Mex.* 75:12–18
35. Ferrusquia-Villafranca, I., Wood, A. E. 1969. New fossil rodents from the Early Oligocene Rancho Gaitan local fauna, northeastern Chihuahua, Mexico. *Pearce-Sellards Ser. Tex. Mem. Mus.* 16:1–13
36. Fisher, R. V., Rensberger, J. M. 1972. Physical stratigraphy of the John Day Formation, central Oregon. *Univ. Calif. Publ. Geol. Sci.* 101:1–33
37. Forstén, A. 1970. The late Miocene Trail Creek mammalian fauna. *Contrib. Geol.* 9:39–51
38. Forstén, A. 1971. Early Tertiary vertebrate faunas, Vieja Group, Trans-Pecos Texas: Equidae. *Pearce-Sellards Ser. Tex. Mem. Mus.* 18:1–16
39. Forstén, A. 1975. The fossil horses of the Texas Gulf Coastal Plain: a revision. *Pearce-Sellards Ser. Tex. Mem. Mus.* 22:1–86
40. Fries, C. H., Hibbard, C. W., Dunkle, D. W. 1955. Early Cenozoic vertebrates in the red conglomerate at Guanajuato, Mexico. *Smithson. Misc. Coll.* 123(7):1–25

41. Frye, J. C., Leonard, A. B. 1959. Correlation of the Ogallala Formation (Neogene) in western Texas with type localities in Nebraska. *Bur. Econ. Geol. Univ. Tex. Rept. Invest.* 39:4–46
42. Galbreath, E. C. 1969. Cylinodrodont rodents from the lower Oligocene of northeastern Colorado. *Trans. Illinois State Acad. Sci.* 62:94–97
43. Galusha, T. 1966. The Zia Sand Formation, New Early to Medial Miocene Beds in New Mexico. *Am Mus. Novit.* 2271:1–12
44. Gawne, C. E. 1968. The genus *Proterix* of the upper Oligocene of North America. *Am. Mus. Novit.* 2315:26
45. Gawne, C. E. 1975. Rodents from the Zia Sand Miocene of New Mexico. *Am. Mus. Novit.* 2586:1–25
46. Gazin, C. L. 1965. A study of the Early Tertiary condylarthran mammal *Meniscotherium*. *Smithson. Misc. Coll.* 149(2):98
47. Gazin, C. L. 1968. A study of the Eocene condylarthran mammal, *Hyopsodus*. *Smithson. Misc. Coll.* 153:1–90
48. Golz, D. J. 1976. Eocene Artiodactyla of Southern California. *Bull. Los Angeles Cty. Mus. Nat. Hist.* 26:1–85
49. Graham, A. 1965. Origin and evolution of the biota of southeastern North America: evidence from the fossil plant record. *Evolution* 18:571–85
50. Graham, A. 1972. Outline of the origin and historical recognition of floristic affinities between Asia and eastern North America. In *Floristics and Paleofloristics of Asia and Eastern North America*, ed. A. Graham, pp. 1–18. Amsterdam: Elsevier
51. Graham, A. 1975. Late Cenozoic evolution of tropical lowland vegetation in Veracruz, Mexico. *Evolution* 29:723–35
52. Green, M. 1972. Lagomorpha from the Rosebud Formation, South Dakota. *J. Paleontol.* 46:377–85
53. Gregory, J. T. 1971. Speculations on the significance of fossil Vertebrates for the Antiquity of the Great Plains of North America. *Abh. Hess. Landesamtes Bodenforsch. (Tobien Fest.)* 60:64–72
54. Guthrie, R. D. 1968. Paleoecology of the large-mammal community in interior Alaska during the Late Pleistocene. *Am. Midl. Nat.* 79:346–63
55. Haffer, J. 1974. Avian Speciation in tropical South America. *Publ. Nuttall Ornithol. Club* 14:1–390
56. Harris, J. M. 1967. *Toxotherium* (Mammalia:Rhinocerotoidea) from western Jeff Davis County, Texas. *Pearce-Sellards Ser. Tex. Mem. Mus.* 9:3–7
57. Harris, J. M., Wood, A. E. 1969. A new genus of eomyid rodent from the Oligocene Ash Spring local fauna of Trans-Pecos, Texas. *Pearce-Sellards Ser. Tex. Mem. Mus.* 14:1–7
58. Henshaw, P. C. 1942. A Tertiary mammalian fauna from the San Antonio Mountains near Tonopah, Nevada. *Carnegie Inst. Wash. Contrib. Paleontol.* 530:79–168
59. Hildebrand, M. 1976. Analysis of tetrapod gaits: General considerations and symmetrical gaits. In *Neural Control of Locomotion*, ed. R. M. Herman et al, pp. 203–236. New York: Plenum
60. Hoffmann, R. S. 1973. An ecological and zoogeographical analysis of animal migration across the Bering Land Bridge during the Quaternary Period. In *The Bering Land Bridge and its Role for the History of Holarctic Floras and Faunas in the Late Cenozoic*, ed. A. I. Kozlovsky et al, pp. 10–11. Khabarovsk, USSR: Acad. Sci. USSR, Far East. Sci. Cent.
61. Holman, J. A. 1968. Lower Oligocene amphibians from Saskatchewan. *Quart. J. Fla. Acad. Sci.* 31:273–89
62. Holman, J. A. 1976. Snakes from the Rosebud Formation (Middle Miocene) of South Dakota. *Herpetologica* 32:41–48
63. Hough, J., Alf, R. 1956. A Chadron mammalian fauna from Nebraska. *J. Paleontol.* 30:132–40
64. Hunt, R. M. Jr. 1972. Miocene amphicyonids (Mammalia, Carnivora) from the Agate Spring Quarries, Sioux County, Nebraska. *Am. Mus. Novit.* 2506:1–39
65. Hussain, T. 1975. Evolutionary and Functional Anatomy of the Pelvic Limb in Fossil and Recent Equidae. *Anat. Histol. Embryol.* 4:179–222
66. Hutchison, J. H. 1972. Review of the Insectivora from the Early Miocene Sharps Formation of South Dakota. *Los Angeles Cty. Mus. Contrib. Sci.* 235:1–16
67. Hutchison, J. H., Lindsay, E. H. 1974. The Hemingfordian mammal fauna of the Vedder Locality, Branch Canyon Formation, Santa Barbara County, California. Part I: Insectivora, Chiroptera, Lagomorpha, and Rodentia (Sciuridae). *PaleoBios* 15:1–19
68. Izett, G. A. 1975. Late Cenozoic sedimentation and deformation in northern

Colorado and adjoining areas. *Geol. Soc. Am. Mem.* 144:179–209
69. Kingdon, J. 1971. *East African Mammals, Vol. 1.* London: Academic. 446 pp.
70. Kluge, A. 1966. A new pelobatine frog from the lower Miocene of South Dakota with a discussion of the evolution of the *Scaphiopus-Spea* complex. *Los Angeles Cty. Mus. Contrib. Sci.* 113: 1–26
71. Kowalevsky, W. 1873–1874. Monographie der Gattung *Anthracotherium* Cuv. und Versuch einer natürlichen Classification der fossilen Hufthiere. *Palaeontographica* 22:131–385
72. Kuenzi, W. D., Fields, R. W. 1971. Tertiary stratigraphy, structure, and geological history, Jefferson Basin, Montana. *Bull. Geol. Soc. Am.* 82:3373–94
73. Lance, J. F. 1950. Paleontologia y estratigraphia del Plioceno de Yepomera, estado de Chihuahua la parte: Équidos, excepto *Neohipparion*. *Univ. Nac. Aut. Mex. Bull.* No. 54:i-viii, 1–81
74. Leopold, E., MacGinitie, H. D. 1972. Development and affinities of Tertiary floras in the Rocky Mountains. In *Floristics and Paleofloristics of Asia and Eastern North America*, ed. A. Graham, pp. 147–200. Amsterdam: Elsevier
75. Lillegraven, J. A. 1972. Ordinal and familial diversity of Cenozoic mammals. *Taxon* 21:261–74
76. Lillegraven, J. A. 1976. Didelphids (Marsupialia) and *Uintaso* (? Primates) from later Eocene sediments of San Diego County, California. *Trans. San Diego Soc. Nat. Hist.* 18(5):85–112
77. Lillegraven, J. A., Wilson, R. W. 1975. Analysis of *Simimys* simplex, an Eocene rodent (? Zapodidae). *J. Paleontol.* 49:856–74
78. Lindsay, E. 1968. Rodents from the Hartman Ranch Local Fauna, California. *PaleoBios* 6:1–22
79. Lindsay, E. 1972. Small mammal fossils from the Barstow Formation, California. *Univ. Calif. Publ. Geol. Sci.* 93:1–104
80. Lindsay, E. 1974. The Hemingfordian mammal fauna of the Vedder Locality, Branch Canyon Formation, Santa Barbara County, California. Part II: Rodentia (Eomyidae and Heteromyidae). *PaleoBios* 16:1–20
81. Love, J. D., McKenna, M. C., Dawson, M. R. 1976. Eocene, Oligocene, and Miocene rocks and vertebrate fossils at the Emerald Lake Locality, 3 miles south of Yellowstone National Park, Wyoming. *Geol. Surv. Prof. Pap.* 932:A:1–28
82. Lynch, J. D. 1971. Evolutionary relationships, osteology, and zoogeography of leptodactyloid frogs. *Misc. Publ. Univ. Kans. Mus. Nat. Hist.* 53:1–238
83. MacDonald, J. R. 1970. Review of the Miocene Wounded Knee faunas of southwestern South Dakota. *Bull. Los Angeles Cty. Mus. Nat. Hist.* 8:1–82
84. MacDonald, L. J. 1972. Monroe Creek (Early Miocene) microfossils from the Wounded Knee Area, South Dakota. *Rep. Invest. So. Dakota Geol. Surv.* 105:1–43
85. MacFadden, B. J. 1976. Cladistic analysis of primitive equids, with notes on other perissodactyls. *Syst. Zool.* 25: 1–14
86. MacGinitie, H. D. 1953. Fossil Plants of the Florissant beds, Colorado. *Carnegie Inst. Wash. Publ.* 599:198
87. MacGinitie, H. D. 1962. The Kilgore flora. A Late Miocene flora from northern Nebraska. *Univ. Calif. Publ. Geol. Sci.* 35(2):67–158
88. MacGinitie, H. D. 1969. The Eocene Green River flora of northwestern Colorado and northeastern Utah. *Univ. Calif. Publ. Geol. Sci.* 83:1–203
89. Martin, L. D. 1975. Microtine Rodents from the Ogallala Pliocene of Nebraska and the early evolution of the Microtinae in North America. *Univ. Mich. Mus. Paleontol. Pap.* 12:101–10
90. Martin, P. S. 1958. A biogeography of reptiles and amphibians in the Gomez Farias Region, Tamaulipas, Mexico. *Misc. Publ. Mus. Zool. Univ. Mich.* 101:1–102
91. Martin, P. S., Harrell, B. E. 1957. The Pleistocene history of temperate biotas in Mexico and the United States. *Ecology* 38:468–80
92. Martin, P. S., Wright, H. E. Jr. 1967. *Pleistocene Extinctions: The Search for a Cause.* New Haven, Conn.: Yale Univ. Press. 453 pp.
93. Matthew, W. D. 1901. Fossil mammals of the Tertiary of northeastern Colorado. *Mem. Am. Mus. Nat. Hist.* 1(2): 353–447
94. McGrew, P. O. 1971. Early Tertiary vertebrate faunas, Vieja Group Trans-Pecos, Texas, Equidae. *Pearce-Sellards Ser. Tex. Mem. Mus.* 18:1–15
95. McKenna, M. C. 1972. Possible biological consequences of plate tectonics. *BioScience* 22:519–525
96. Mengel, R. M. 1970. The North American Central Plains as an isolating agent

in bird speciation. In *Pleistocene and Recent Environments of the Central Great Plains,* ed. W. Dort Jr., J. K. Jones Jr., pp. 279–340. Lawrence, Kans.: Univ. Kansas Press
97. Merriam, J. C. 1916. Tertiary vertebrate fauna from the Cedar Mountain region of western Nevada. *Univ. Calif. Publ. Geol. Sci.* 9:161–98
98. Merriam, J. C. 1919. Tertiary mammalian faunas of the Mohave Desert. *Univ. Calif. Publ. Geol. Sci.* 11:437–586
99. Meszoely, C. A. M. 1970. North American fossil anguid lizards. *Bull. Mus. Comp. Zool.* 139:87–150
100. Miller, A. H. 1944. An avifauna from the lower Miocene of South Dakota. *Univ. Calif. Publ. Geol. Sci.* 27:85–100
101. Mooser, O. 1973. Pliocene horses of the Ocote local fauna, Central Plateau of Mexico. *Southwest. Nat.* 18:257–68
102. Müller, P. 1973. The dispersal centres of terrestrial vertebrates in the neotropical realm. *Biogeographica (The Hague, Netherlands)* 2:1–244
103. Nichols, R. 1976. Early Miocene mammals from the Lemhi Valley of Idaho. *Tebiwa* 18:9–48
104. Nungesser, W. C., Pfeiffer, E. W. 1965. Water balance and maximum concentrating ability in the primitive rodent, *Aplodontia rufa. Comp. Biochem. Physiol.* 14:289–97
105. Olrog, C. C. 1969. Birds of South America. *Monogr. Biol.* 19:849–78
106. Olson, S. L. 1976. A Jaçana from the Pliocene of Florida (Aves: Jacanidae). *Proc. Biol. Soc. Wash.* 89:259–64
107. Patterson, B. 1949. Rates of evolution in taeniodonts. In *Genetics, Paleontology, and Evolution* ed. G. L. Jepsen, E. Mayr, G. G. Simpson, pp. 243–278. Princeton, N. J.: Princeton Univ. Press
108. Patton, T. H., Taylor, B. 1971. The Synthetoceratinae (Mammalia, Tylopoda, Protoceratidae). *Bull. Am. Mus. Nat. Hist.* 145:119–218
109. Peterson, O. A. 1904. Description of new rodents and discussion of the origin of *Daemonelix. Mem. Carnegie Mus.* II(4):139–202
110. Rabb, G. B., Marx, H. 1973. Major ecological and geographic patterns in the evolution of colubroid snakes. *Evolution* 27:69–83
111. Rasmussen, D. L. 1969. *Late Cenozoic geology of the Cabbage Patch Area, Granite and Powell Counties, Montana.* PhD Thesis. Univ. Montana, Missoula, 188 pp.
112. Reed, C. A. 1954. Some fossorial mammals from the Tertiary of western North America. *J. Paleontol.* 28:104–11
113. Reed, C. A. 1956. A new species of the fossorial mammal *Arctoryctes* from the Oligocene of Colorado. *Fieldiana Geol. Mem.* 10:305–311
114. Reed, C. A. 1970. Extinction of mammalian megafauna in the Old World Late Quaternary. *BioScience* 20:284–288
115. Reeder, W. G. 1960. Two new rodent genera from the Oligocene White River Formation (family Heteromyidae). *Fieldiana Geol. Mem.* 10:511–23
116. Rensberger, J. M. 1971. Entoptychine pocket gophers (Mammalia, Geomyoidea) of the Early Miocene John Day Formation, Oregon. *Univ. Calif. Publ. Geol. Sci.* 90:1–163
117. Rensberger, J. M. 1973. Pleurolicine rodents (Geomyoidae) of the John Day Formation, Oregon and their relationships to taxa from the Early and Middle Miocene, South Dakota. *Univ. Calif. Publ. Geol. Sci.* 102:1–95
118. Rensberger, J. M. 1973. An occlusal model for mastication and dental wear in herbivorous mammals. *J. Paleontol.* 47:515–28
119. Rensberger, J. M. 1973. *Sanctimus* (Mammalia, Rodentia) and the phyletic relationships of the large Arikareean geomyoids. *J. Paleontol.* 47:835–53
120. Repenning, C. A. 1967. Palearctic-Nearctic mammalian dispersal in the Late Cenozoic. In *The Bering Land Bridge,* ed. D. M. Hopkins, pp. 288–311. Stanford, Calif.: Stanford Univ. Press
121. Robertson, J. R. 1976. Latest Pliocene mammals from Haile XV A, Alachua County, Florida. *Bull. Fla. State Mus. Biol. Sci.* 20(3):111–86
122. Robinson, P. 1970. The Tertiary deposits of the Rocky Mountains. A summary and discussion of unsolved problems. *Univ. Montana Contrib. Geol.* 9:86–96
123. Savage, D. E. 1971. The Sparnacian-Wasatchian mammalian fauna; Early Eocene of Europe and North America. *Abh. Hess. Landesamtes Bodenforsch.* 60:154–58
124. Savage, J. M. 1966. The origins and history of the Central American herpetofauna. *Copeia* 1966(4):719–66
125. Scott, W. B. 1937. *A History of Land Mammals in the Western Hemisphere.* New York: Macmillan. 786 pp.

126. Scott, W. B., Jepsen, G. L., Wood, A. E. 1941. The mammalian fauna of the White River Oligocene. *Trans. Am. Philos. Soc. (N.S.)* 28:1–980
127. Sears, P. B. 1969. *Lands Beyond the Forest.* New York: Prentice-Hall. 206 pp.
128. Sher, A. V. 1974. Pleistocene Mammals of the far northeast USSR and North America. *Int. Geol. Rev.* 16:1–89
129. Shotwell, J. A. 1958. Inter-community relationships in Hemphillian (Mid-Pliocene) mammals. *Ecology* 39:271–82
130. Shotwell, J. A. 1961. Late Tertiary biogeography of horses in the northern Great Basin. *J. Paleontol.* 35:203–17
131. Shotwell, J. A. 1967. Late Tertiary geomyoid rodents of Oregon. *Bull. Mus. Nat. Hist. Univ. Oregon* 9:1–51
132. Shotwell, J. A. 1968. Miocene mammals of southeast Oregon. *Bull. Mus. Nat. Hist. Univ. Oregon* 14:1–67
133. Simons, E. L. 1960. The Paleocene Pantodonta. *Trans. Am. Philos. Soc.* 50:1–99
134. Simpson, B. B. 1975. Pleistocene changes in the flora of the high Tropical Andes. *Paleobiol.* 1:273–94
135. Simpson, G. G. 1937. The Fort Union Crazy Mountain Field, Montana, and its mammalian faunas. *Bull. U.S. Natl. Mus.* 169:1–287
136. Stanley, K. O. 1976. Sandstone petrofacies in the Cenozoic High Plains sequence, eastern Wyoming and Nebraska. *Bull. Geol. Soc. Am.* 87:297–309
137. Stanley, S. M. 1973. An explanation for Cope's rule. *Evolution* 27:1–26
138. Stebbins, G. L. 1952. Aridity as a stimulus to plant evolution. *Am. Nat.* 86:33–44
139. Stebbins, G. L. 1974. *Flowering plants: Evolution above the Species Level.* Cambridge, Mass.: Harvard Univ. Press. 399 pp.
140. Stevens, M. S., Stevens, J. B., Dawson, M. R. 1969. New Early Miocene Formation and vertebrate local fauna, Big Bend National Park, Brewster County, Texas. *Pearce-Sellards Ser. Tex. Mem. Mus.* 15:53
141. Stirton, R. A. 1948. Observations on evolutionary rates in hypsodonty. *Evolution* 1:32–41
142. Storer, J. E. 1973. The entoptychine geomyid *Lignimus* (Mammalia: Rodentia) from Kansas and Nebraska. *Can. J. Earth Sci.* 10:72–83
143. Sutton, J. F., Black, C. C. 1972. Oligocene and Miocene deposits of Jackson Hole, Wyoming. In *Guidebook of the Field Conference on the Tertiary Biostratigraphy of Southwestern Wyoming,* coord. R. M. West, pp. 73–79. Garden City, N.Y.: West
144. Talbot, L. M., Talbot, M. H. 1963. The high biomass of wild ungulates on East African savanna. *Trans. 28th N. Am. Wildl. Conf.,* pp. 465–476
144a. Tedford, R. H., Gustafson, E. P. 1977. First North American record of the extinct panda *Parailurus. Nature* 265:621–23
145. Tihen, J. A. 1964. Tertiary changes in the herpetofaunas of temperate North America. *Senckenbergiana Biol.* 45:265–79
146. Van Houten, F. B. 1945. Review of latest Paleocene and Early Eocene mammalian faunas. *J. Paleontol.* 19:421–61
147. Van Valen, L. 1969. Evolution of communities and Late Pleistocene extinctions. *Proc. N. Am. Paleontol. Conv.* E:469–85
148. Voorhies, M. R. 1969. Taphonomy and population dynamics of an Early Pliocene vertebrate fauna, Knox County, Nebraska. *Univ. Wyom. Contrib. Geol. Spec. Pap.* 1:69
149. Voorhies, M. R. 1969. Sampling difficulties in reconstructing Late Tertiary mammalian communities. *Proc. N. Am. Paleontol. Conv.* E:454–68
150. Voorhies, M. R. 1975. A new genus and species of fossil kangaroo rat and its burrow. *J. Mammal.* 56:160–76
151. Wahlert, J. H. 1973. *Protoptychus,* a hystricomorphous rodent from the Late Eocene of North America. *Mus. Comp. Zool., Breviora* 1419:1–14
152. Walker, A. 1972. Dissemination and segregation of early primates in relation to continental configuration. In *Calibration of Hominoid Evolution,* ed. W. W. Bishop, J. A. Miller, pp. 195–218. Edinburgh: Scottish Academic Press
153. Webb, S. D. 1969. The Burge and Minnechaduza Clarendonian mammalian faunas of north-central Nebraska. *Univ. Calif. Publ. Geol. Sci.* 78:191
154. Webb, S. D. 1969. Extinction-origination equilibrium in Late Cenozoic land mammals of North America. *Evolution* 23:688–702
155. Webb, S. D. 1972. Locomotor evolution in camels. *Forma Functio.* 5:99–112
156. Webb, S. D. 1973. Pliocene pronghorns of Florida. *J. Mammal.* 54:203–21
157. Webb, S. D., Tessman, N. 1968. A Pliocene vertebrate fauna from low eleva-

tion in Manatee County, Florida. *Am. J. Sci.* 266:777–811
158. Wheeler, W. H. 1961. Revision of the uintatheres. *Bull. Peabody Mus. Nat. Hist.* 14:1–93
159. White, T. E. 1954. Preliminary analysis of the fossil vertebrates of the Canyon Ferry Reservoir Area. *Proc. U.S. Natl. Mus.* 103(3326):395–438
160. White, T. E. 1959. The endocrine glands and evolution, No. 3: Os cementum, hypsodonty, and diet. *Contrib. Mus. Paleontol. Univ. Mich.* 13:211–65
161. Whitmore, F. C., Stewart, R. H. 1965. Miocene mammals and Central American seaways. *Science* 148:180–85
162. Whittaker, R. H., Niering, W. A. 1975. Vegetation of the Sta. Catalina Mountains, Arizona. V. Biomass, production and diversity along the elevation gradient. *Ecology* 56:771–90
163. Wilson, J. A. 1956. Miocene formations and vertebrate biostratigraphic units, Texas Coastal Plain. *Bull. Am. Assoc. Pet. Geol.* 40(9):2233–46
164. Wilson, J. A. 1974. Early Tertiary vertebrate faunas, Vieja Group and Buck Hill Group, Trans-Pecos Texas: Protoceratidae, Camelidae, Hypertragulidae. *Tex. Mem. Mus. Bull.* 23:1–34
165. Wilson, R. W. 1951. Preliminary Survey of a Paleocene faunule from the Angels Peak Area, New Mexico. *Univ. Kans. Publ. Mus. Nat. Hist.* 5:1–11
166. Wilson, R. W. 1960. Early Miocene rodents and insectivores from northeastern Colorado. *Univ. Kans. Paleontol. Contrib., Vertebrata, Art.* 7:1–92
167. Wilson, R. W. 1967. Fossil mammals in Tertiary correlations. *Univ. Kans. Geol. Spec. Publ.* (*Moore Volume*) 2:590–606
168. Wilson, R. W. 1972. Evolution and extinction in Early Teritary rodents. *Int. Geol. Congr.* 24:217–24
169. Wilson, R. W. 1975. The National Geographic Society-South Dakota School of Mines and Technology Expedition into the Big Badlands of South Dakota, 1940. *Natl. Geog. Soc. Rep. 1890–1954 Proj.*, pp. 79–85
170. Wolfe, J. A. 1964. Miocene floras from Fingerrock Wash, southwestern Nevada. *Geol. Surv. Prof. Pap.* 454-N:1–36
171. Wolfe, J. A. 1975. Some aspects of plant geography of the northern hemisphere during the Late Cretaceous and Tertiary. *Ann. Mo. Bot. Gard.* 62:264–79
172. Wolfe, J. A. Hopkins, D. M. 1967. Climatic changes recorded by Tertiary land floras in northwestern North America. In *Tertiary Correlations and Climatic Changes in the Pacific*, ed. K. Hatai, pp. 67–76. Sendai, Japan: Sasaki
173. Wood, A. E. 1962. The Early Tertiary rodents of the family Paramyidae. *Trans. Am. Philos. Soc.* 52:1–261
174. Wood, A. E. 1973. Eocene rodents, Pruett Formation, Southwest Texas; their pertinence to the origin of the South American Caviamorpha. *Pearce-Sellards Ser. Tex. Mem. Mus.* 20:1–40
175. Woodburne, M. O. 1969. Systematics, biogeography, and evolution of *Cynorca* and *Dyseohyus* (Tayassuidae). *Bull. Am. Mus. Nat. Hist.* 141(2):273–355
176. Woodburne, M. O., Tedford, R. H., Stevens, M. S. Taylor, B. E. 1974. Early Miocene mammalian faunas, Mojave Desert, California. *J. Paleontol.* 48:6–26

Ann. Rev. Ecol. Syst. 1977. 8:381–405
Copyright © 1977 by Annual Reviews Inc. All rights reserved

CIRCANNUAL RHYTHMS IN BIRD MIGRATION ❖4130

Eberhard Gwinner

Max-Planck-Institut für Verhaltensphysiologie, D 8131 Erling-Andechs, Federal Republic of Germany

INTRODUCTION

It had been suspected for a long time that the annual cycles of many physiological and behavioral activities in birds are controlled not only by environmental variables such as photoperiod and temperature but also by an endogenous annual rhythmicity [summaries in (2, 5, 6, 13, 38, 42, 60, 61, 66–69)]. The participation of internal periodic time-measuring processes was postulated particularly for bird species that live, at least for part of the year, in environments deficient in reliable seasonal variations. Many permanent inhabitants of the tropics, and migratory species that breed in the temperate zones and winter in regions close to the equator, fall into this category (e.g. 22–24, 59–62, 66–70, 72, 73, 83). Suggestive evidence in support of this hypothesis had been available for a long time; yet it was only after Pengelley & Fisher (77) first demonstrated the existence of internal annual clocks in a hibernating rodent (the golden-mantled ground squirrel *Citellus lateralis*) that intense investigations into this problem were initiated with birds. These studies have revealed endogenous annual rhythms in a variety of bird species. They have contributed a great deal to our understanding of the mechanisms that guarantee proper adjustment of biological activities to external seasonal changes [summaries in (12, 43, 44, 48, 56, 76)].

In the first two sections of the present review some of the experimental data on endogenous annual rhythms in birds are summarized, including a discussion of some general features of these rhythms. The third section concentrates on one especially well-investigated aspect of the adaptive value of these rhythms, namely on their significance for the control of migrations in long-distance-migrating birds.

DISTRIBUTION AND PROPERTIES OF CIRCANNUAL RHYTHMS

An early experiment demonstrating circannual rhythms in birds was carried out on a Palaearctic long distance migrant, the willow warbler (*Phylloscopus trochilus*), a species that winters in tropical and southern Africa (37, 38, 42–44). Figure 1 illustrates some of the problems that confront those individuals of this species that

381

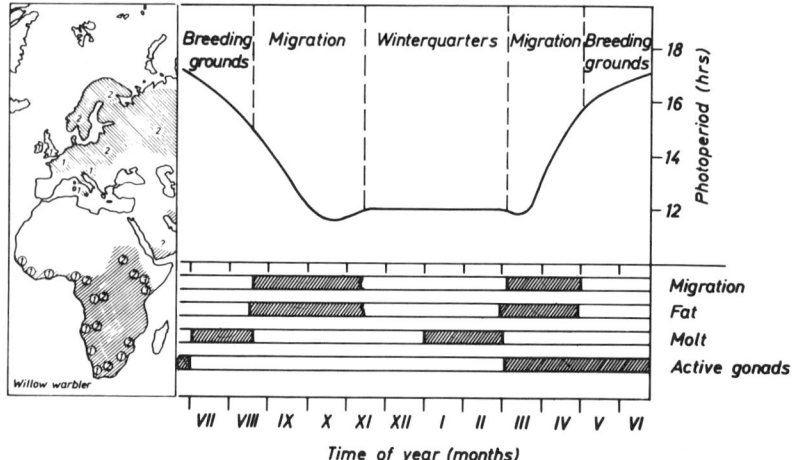

Figure 1 (*left*) Annual distribution of the willow warbler, *Phylloscopus trochilus*. Numbers refer to the breeding and wintering areas of different races. (*right, upper figure*) Photoperiod experienced in the course of a year by individuals that winter close to the equator. (*right, lower figure*) Schematic representation of the occurrence of various annual events. [Distribution map after (28); from (49)]

winter close to the equator. The upper part of the figure shows that these birds are exposed to pronounced photoperiodic changes only while they are in their temperate zone breeding grounds and during part of their migration, but that they are exposed to a constant photoperiod of about 12 hr from the beginning of October to the end of March. Moreover, in many of these tropical regions other environmental factors such as temperature or precipitation vary quite unpredictably from year to year. The factors known or suspected as external annual timing cues for birds are then of little use during six months of the year. Nevertheless, a variety of activities are carried out at precise times during that period, as shown schematically in the lower section of Figure 1. A complete molt takes place in January and February; gonadal growth is initiated in the winter quarters or shortly after the beginning of homeward migration; and, most importantly, spring migration and the fattening associated with it commence year after year at precisely the same times (38).

Figure 2 summarizes the results of an experiment which shows that the timing of this sequence is basically independent of external information. Willow warblers

---

Figure 2 (*opposite*) Seasonal variation in migratory restlessness, body weight, and molt in willow warblers, *Phylloscopus trochilus*. Birds were kept until late September under the natural photoperiod of their breeding grounds (Eni = Erling, Germany, 48° N, 11°11' E); seven birds were then displaced to within the wintering grounds of the species (B = Bukavu, Congo, 2°14' S, 28°39' E) and kept there under natural photoperiods (*upper seven diagrams*). Nine birds were transferred at the same time to a constant 12-hr photoperiod (LD 12:12, 200:0.02 lux). ↓: date of hatching. Shaded area: nocturnal migratory restlessness. Solid bars: molt of the large wing and tail feathers. Shaded bars: molt of body feathers. ——body weight. †: bird died. [After (43)]

CIRCANNUAL RHYTHMS IN BIRD MIGRATION 383

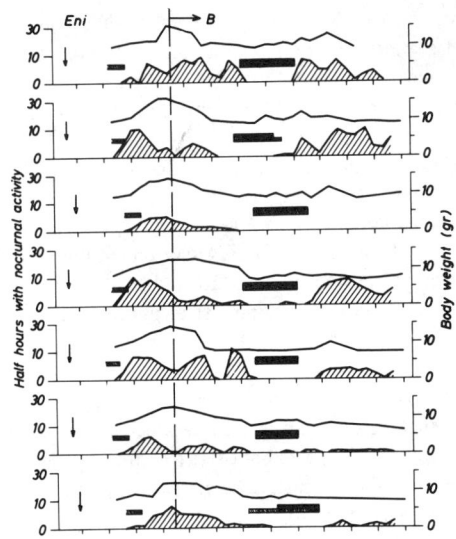

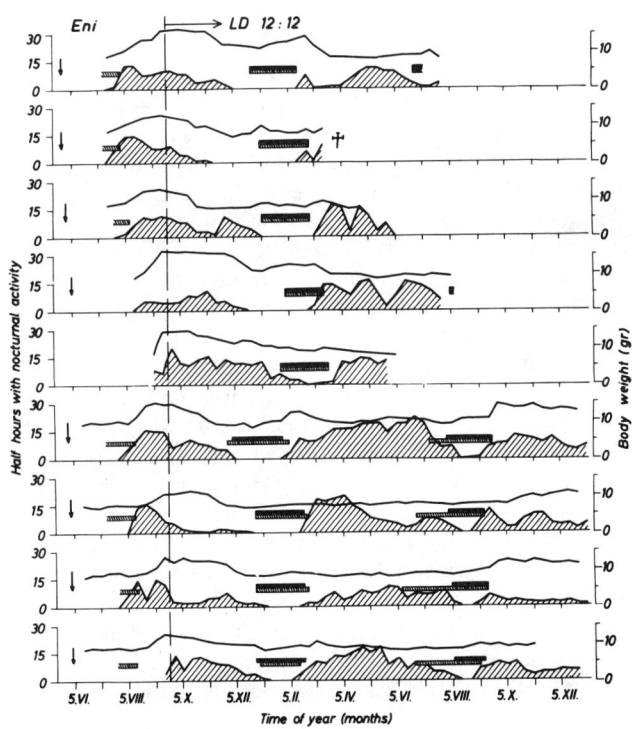

that were maintained in temperature-controlled chambers with a constant 12-hr photoperiod from the end of September, molted and came into spring migratory condition at the same time as control birds kept in cages on their wintering grounds. Moreover, birds living for more than one year under constant experimental conditions also carried out the following summer molt and began to show autumnal migratory restlessness at the appropriate time (38).

The suggestion derived from these results—that the annual cycles of migratory restlessness, fattening, and molt are controlled by an endogenous annual clock—has been verified in experiments on birds of various species maintained under seasonally constant conditions for an extended period of time (10–12, 18, 19, 37, 38, 43). As an example, Figure 3 shows how these rhythms persist in garden warblers (*Sylvia borin*) kept under three different constant photoperiods for more than 2½ yr. That these rhythms are truly endogenous is proven by the fact that their period deviates from exactly 1 yr, as can be seen more clearly in Figure 4. In both garden warblers and blackcaps (*S. atricapilla*) successive postnuptial molts occur earlier each year than the previous year, suggesting that the period of this "freerunning" molt rhythm is shorter than 12 months. In other words, under seasonally constant conditions the subjective calendar of these birds deviates from the true objective calendar, an observation that excludes uncontrolled external seasonal time cues as possible causes of the observed rhythms. Hence they have been called *circannual* rhythms.

Circannual rhythms persisting for at least two cycles under constant environmental conditions with periods different from 1 yr have been demonstrated in at least nine species of birds and in functions as diverse as gonadal size (Figure 5), migratory restlessness, migratory fattening, molt, and food preference. In addition, there are data suggesting the existence of circannual rhythms in at least ten other species. Most of the evidence is compiled in tabular form in (13) and (48). More recent results may be found in (50, 55, 56). Clear circannual rhythms are known to occur in at least 15 species of animals other than birds, including molluscs, arthropods, reptiles, and mammals. Summaries and critical evaluations of these data are to be found in (13) and (43).

Despite the rapidly increasing number of studies demonstrating circannual rhythms in animals, it is still not clear how widespread they are. Experiments exist in which circannual rhythms could not be found, but these should not be overrated. The range of conditions in which circannual rhythms can express themselves may be limited and may vary among species. For instance, starlings (*Sturnus vulgaris*) showed a well-defined annual rhythm of testicular size when kept in a constant 12-hr photoperiod, whereas no periodicity was observed when the photoperiods were less than 11 hr or more than 13 hr (91). Conversely, sika deer (*Cervus nippon*) replaced antlers on the basis of a circannual rhythmicity when the animals were kept in an 18- or 6-hr photoperiod, but not in a 12-hr photoperiod (36). Similar cases are discussed in (43). These results indicate that lack of circannual rhythmicity in experiments where animals were tested in a restricted number of conditions provides no unambiguous answer to the general question of their distribution among animals. Even so, studies of closely related species differing in their general biology and ecology suggest that there can be considerable differences in the involvement of

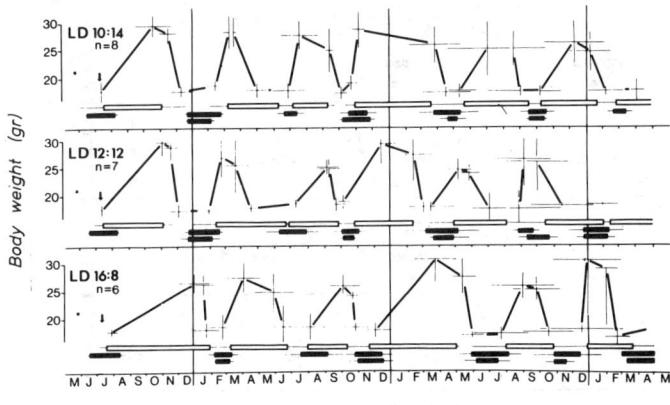

Figure 3 Circannual rhythms of migratory restlessness, body weight, and molt in three groups of garden warblers (*Sylvia borin*) kept for 33 months under three different constant photoperiods. The curves representing changes in body weight connect: the means of the last minima before the beginning of fattening; the means of the maximal body weights; the means of the body weights preceding the reduction of fat; and the means of the first minima after obesity (with standard errors of the weights and time). Open bars: time during which the birds showed migratory restlessness. Solid bars: molt (*upper row:* molt of body feathers; *lower row:* molt of the large wing and tail feathers). [After (19)]

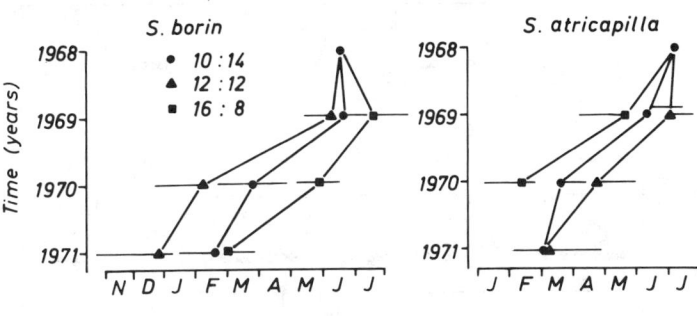

Figure 4 Free-running circannual rhythms of postnuptial molt in garden warblers (*left*) and blackcaps (*right*) kept for 2½ yr under a constant 10-hr (●), 12-hr (▲), and 16-hr (■) photoperiod. The symbols indicate the date at which the four to eight individuals of each group started to molt in successive years. Horizontal lines: standard deviations. [After (19)]

Testicular width (mm)
10
6
2
1
2
3
4
Time (years)

Photoperiod (hrs)
15
12
9
Testicular width (mm)
1/year
1½/year
2/year
3/year
4/year
5/year
180 270 360 90 180 270 360 90 180 270 360 90 180 270 360 90 180 270 360
Phase of the photoperiodic cycle (°)

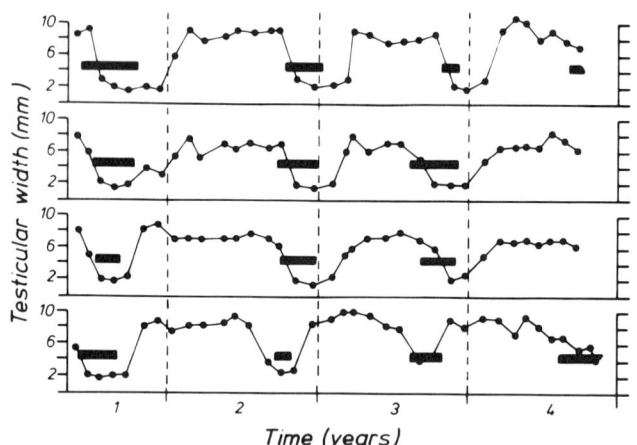

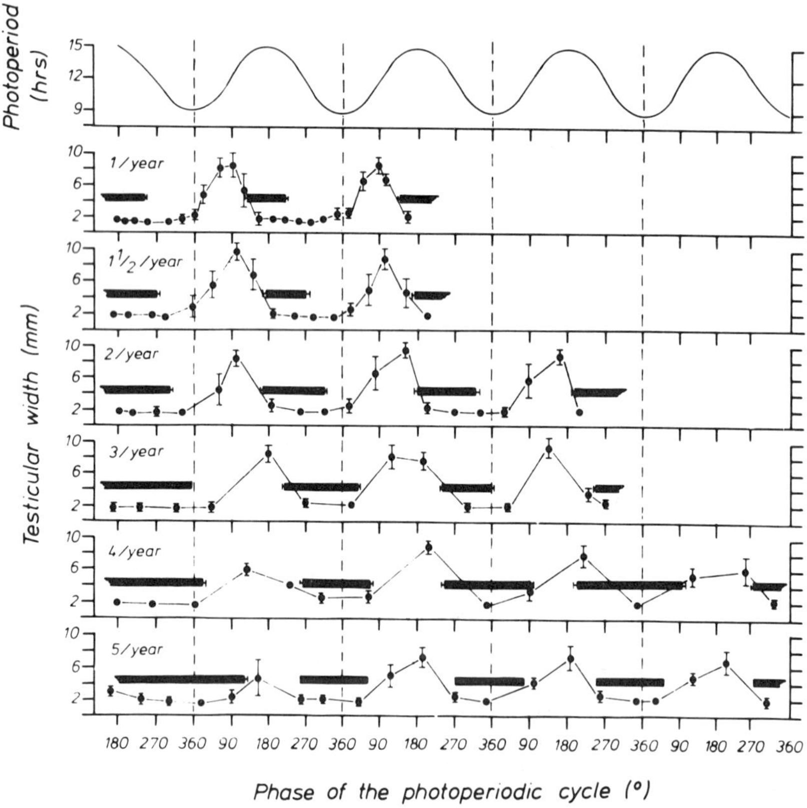

circannual rhythms. For instance, the endogenous control of migratory restlessness, body weight, and molt is quite rigid in the willow warbler, a long-distance migrant. Its rhythms persist for at least three cycles in a constant 12-hr photoperiod (37, 38, 43). On the other hand, the endogenous control of these activities is much less rigid in the closely related chiffchaff (*Phylloscopus collybita*), a short-distance migrant (43). Its rhythms damp out within the first cycle. A similar relationship between migratory distance and the extent to which a circannual rhythmicity is involved was also found in several warblers of the genus *Sylvia* (11, 12). These data are evidence for the assumption that a rigid endogenous control of annual biological rhythms may have developed as a result of the need for precise seasonal timing in relatively constant or unpredictable environments.

However, while this may be true in warblers, there must be other selection pressures favoring the development of circannual rhythms, since pronounced circannual rhythms also occur in short-distance migrants such as the starlings (46, 50, 91), and even in resident birds that permanently inhabit temperate or northern latitudes (10, 12). The question remains unanswered of why these animals have developed circannual rhythms when they could easily have depended on external seasonal cues for the timing of their annual activities.

SYNCHRONIZATION OF CIRCANNUAL RHYTHMS

Although the periods of circannual rhythms deviate from 1 yr in constant conditions, they are, of course, exactly 1 yr under natural conditions. This indicates that in the natural environments of birds there must be seasonally changing variables that synchronize circannual rhythms with the natural year. Exogenous rhythms, capable of synchronizing (entraining) endogenous rhythms are generally called synchronizers, or *Zeitgeber* (1).

A suitable way of testing whether an external rhythm is a *Zeitgeber* entails altering its period. If it is a *Zeitgeber*, the internal rhythm should follow the altered period, at least within certain limits (e.g. 3). Figure 5 shows the results of an experiment of this type, which demonstrates that the annual cycle of photoperiod (i.e. the light fraction of the 24-hr day, which changes as a function of season) is

◄─────────────

Figure 5 (*opposite, upper diagram*) Rhythms of testicular width (curves) and molt (bars) in four European starlings (*Sturnus vulgaris*) kept for 3½ yr under constant photoperiodic conditions (*upper three diagrams:* LD 11:11; *fourth diagram:* LD 12:12). (*lower diagram*) Rhythms of testicular width and molt in six groups of starlings exposed to sinusoidal changes of photoperiod as shown in the upper panel. The amplitude and the general shape of these photoperiodic cycles were the same in all groups but their duration varied from one cycle per year (duration: 365 days, second panel) to five cycles per year (duration: 73 days, lowest panel). Each group numbered 8–12 birds. Sizes of testes were monitored at 2–4 week intervals. Duration of photoperiodic cycles is normalized to 360° and data are plotted relative to the phase of the photoperiodic cycle (0° = phase of shortest photoperiod; 180° = phase of longest photoperiod). Vertical lines at the curve points and horizontal lines at the bars: standard deviations. [After (52) and unpublished data]

a potent *Zeitgeber* of circannual rhythms in the European starling. The rhythms of gonadal size and molt, which are controlled by an endogenous annual rhythmicity in this species (upper graph), can be synchronized with photoperiodic cycles drastically shorter than 1 yr (lower graph). Even a photoperiodic cycle with a period of only 2.4 months (5 cycles per year) is capable of synchronizing both the testicular and the molt rhythm, suggesting a wide range of entrainment of these biological periodicies.

Photoperiod was established long ago as the most significant environmental variable for the control of annual rhythms in birds [summaries in (32–34)]. Until quite recently it had been believed that photoperiodic alterations acted as an obligatory causal stimulus, eliciting the overt annual rhythms rather than entraining a preprogrammed endogenous rhythmicity. However, photoperiodic effects have now been described even in species that are known to be equipped with a circannual rhythmicity. It is therefore likely that the annual photoperiodic cycle acts as a *Zeitgeber* not only in the starling but in other species as well. The specific way in which photoperiod exerts its effect in many species is compatible with this idea. [For discussion of these problems see (2, 43, 46, 51, 52).] To what extent *Zeitgeber* other than photoperiod play a role in birds is unknown.

The annual rhythm of photoperiod is also the only known synchronizer of circannual rhythms in other classes of animals. Its effectiveness as a *Zeitgeber* has been shown most convincingly in the sika deer, where the circannual rhythm of antler growth and shedding was synchronized with up to four photoperiodic cycles per year (35).

TEMPORAL CONTROL OF MIGRATION

Circannual rhythms may have great significance in the initiation of migration at the appropriate times of the year for long-distance migrants that winter in unpredictable tropical environments. The onset of spring migration has now been shown to be under circannual control in at least seven species of birds [*Phylloscopus trochilus:* (37, 38, 43); *Sylvia borin, S. atricapilla:* (18, 19); *Sylvia cantillans, S. melanocephala, S. sarda, S. undata:* (10, 11, 12)]. In at least five others this type of endogenous control is very likely [*Phylloscopus collybita:* (43); *Puffinus tenuirostris:* (70); *Spiza americana:* (99); *Fringilla montifringilla:* (81); *Lanius collurio:* (55)].

The initiation of migration is not the only function of circannual rhythms in the control of migratory behavior. There is now a substantial body of data suggesting that an endogenous time-program, controlled by a circannual rhythmicity, can also determine the temporal course and possibly the distance of migration, at least in some long-distance migrants on their first fall migration. The following paragraphs focus on the discussion of this function.

Hypothesis of an Endogenous Time-Program Controlling Migration

The idea of an endogenous time-program for the temporal control of autumn migration was first suggested when the temporal course of fall migratory restlessness (*Zugunruhe*) of caged, first-year willow warblers was compared with the temporal course of actual migration in freeliving conspecifics (38, 39, 42). In these studies, the total number of half-hour intervals with activity per night was taken as a measure of migratory restlessness; this parameter is supposed to represent the periods during which a bird is in the migratory mood during a particular night. Migratory speed was estimated from the dates at which willow warblers pass through various places along their migratory route. The comparison revealed that the *Zugunruhe* of the caged birds and the calculated migratory speed of freeliving birds show parallel changes: There is a rapid increase in migratory restlessness at the beginning of the migratory season just like the increase in migratory speed of freeliving conspecifics. The period of most intense migratory restlessness coincides with the period of highest migratory speed, during which freeliving willow warblers cross the Sahara desert. Then migratory restlessness slowly decreases as freeliving birds also reduce their migratory speed.

A similar relationship between the pattern of autumnal migratory restlessness of caged birds under constant photoperiodic conditions and that of actual migration in freeliving conspecifics has been found in other species as well (9, 20, 39, 64). On the basis of these results the following hypothesis was proposed (38):

1. Nocturnal activity measured as *Zugunruhe* in caged birds is the expression of an endogenous temporal program controlled by a circannual rhythmicity that determines duration and temporal variation in migratory activity. If this is true then the nocturnal activity displayed by caged birds should be equivalent to a given amount of actual migratory activity, i.e. a given distance travelled by free individuals during the same time.
2. Furthermore, it is proposed that this temporal program is organized to produce just enough migratory activity (i.e. migratory time) during the migratory season to reach the specific winter quarters. Thus, the bird will have reached the vicinity of its wintering area when its program has run down.

If the mechanism suggested here exists, a previously unexplained phenomenon could be better understood. Evidence suggests that birds travelling to their winter quarters for at least the second time in their lives find their wintering grounds through goal-orientation; i.e. they know the geographic coordinates of their wintering area. On the other hand, inexperienced first-year birds of several species are apparently only capable of direction-orientation; i.e. they migrate in a given compass direction (30, 79, 80, 88–90). This leads to the question of what factors are responsible for the termination of migration in birds capable only of maintaining a given direction. Several mechanisms have been suggested, but none has yet received convincing experimental support [see (31, 41, 87, 94, 95) for discussion]. The present hypothesis proposes that the distance of migration, and hence the area in which migration ends, is determined in these first-year migrants by an endogenous program that controls the time a bird spends in its migratory flight. Such an orientation mechanism would fall into the category of vector-orientation (86, 96).

In the following sections this hypothesis is tested and investigated in terms of (*a*) whether the program depends, in fact, on endogenous time; (*b*) whether it is species-specific; and (*c*) whether it is organized in such way as to assist a bird in finding its winter quarters.

Test of the Hypothesis

A TIME PROGRAM? The hypothesis assumes that the intensity of migratory activity developed at any time during the fall migratory season depends essentially on the phase of the endogenous circannual oscillator and hence on the bird's subjective time. It should therefore be independent of energy expenditure. This prediction has been verified in an experiment on the garden warbler, which made use of the fact that the expression of its nocturnal migratory restlessness depends on a minimum light intensity at night. Figure 6 shows that temporary exposure of a group of garden warblers to complete darkness at night results in a drastic reduction of migratory

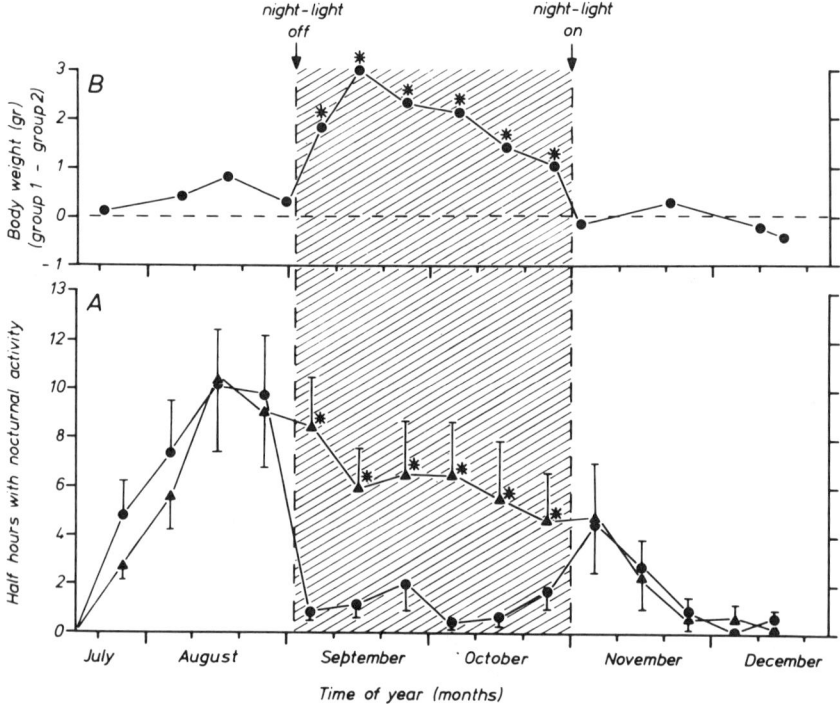

Figure 6 (*A*) Temporal course of migratory restlessness in two groups of garden warblers kept under a constant 12-hr photoperiod throughout the fall migratory season. ▲ : Control birds (group 1)—exposed to a dim light at night (0.01 lux) throughout the experiment. ● : Experimental birds (group 2)—also exposed to the same dim night light except for the period between 1 September and 30 October (shaded area) when lights were switched off at night. Ordinate: half-hour intervals during which a bird was active each night; mean values are plotted for successive thirds of a month and for all eight individuals of each group. Vertical bars: standard errors. (*B*) Difference in the average body weight (measured early in the morning) between birds of group 2 and those of group 1. ✳ : significant differences between the values of the two groups. [After (47)]

restlessness and an increase in body weight compared to the control birds that were permanently exposed to a dim night light. The fact that body weight was increased in the experimental birds during the period of reduced nocturnal activity suggests that less energy was expended during that time. However, the experimental treatment had no discernible aftereffect on the intensity of migratory restlessness. Except during the time when the lights were off, the temporal pattern of the migratory restlessness of the two groups was indistinguishable (47).

That the gross pattern of nocturnal migratory activity is largely independent of the energetic state of the birds is also shown by the experiments summarized in Figures 7 and 8 (14, 15). Reduction of body weight by controlled starvation to a level just above the lean weight does not affect nocturnal activity (Figure 7). A drastic reduction of migratory restlessness occurs only when starvation results in weight levels indicating the complete exhaustion of fat reserves (Figure 8). When starvation is discontinued and body weight has recovered, *Zugunruhe* intensity

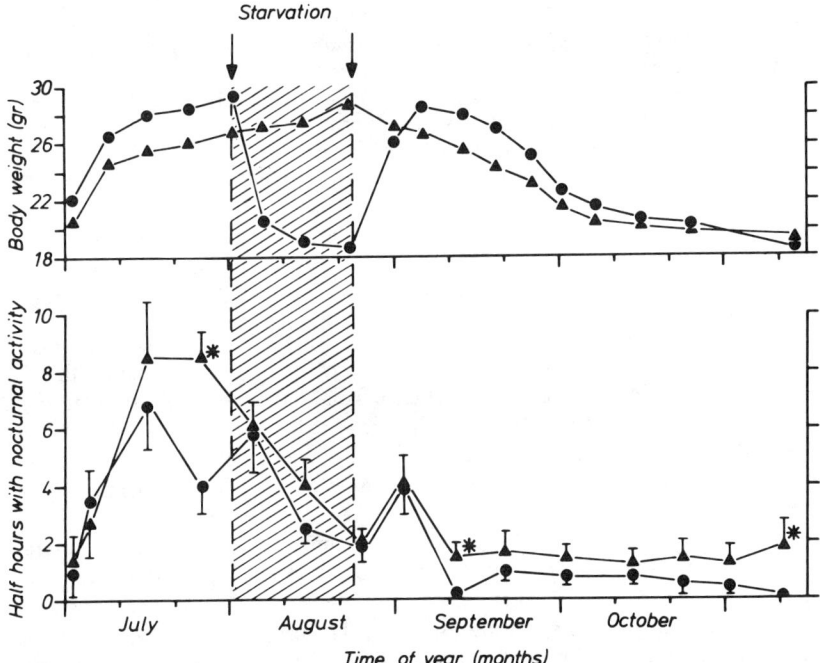

Figure 7 Temporal course of body weight (*upper diagram*) and migratory restlessness (*lower diagram*) in two groups of ten garden warblers each kept under a constant 10-hr photoperiod (LD 10:14) throughout the fall migratory season. ▲ : Control birds. ● : Experimental birds in which the body weight was reduced to a level just above fat-free weight by feeding a low-calory diet during the time indicated by shading. For other explanations see Figure 6. [After (14)]

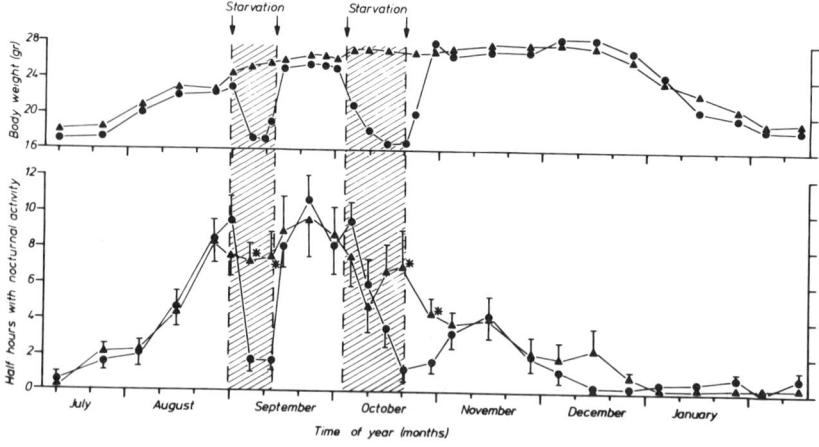

Figure 8 Temporal course of body weight (*upper diagram*) and migratory restlessness (*lower diagram*) in two groups of ten garden warblers each kept under the natural light conditions of their southwestern German breeding grounds throughout the fall migratory season. ▲ : Control birds. ● : Experimental birds in which the body weight was reduced to a level at or slightly below lean body weight by feeding a low-calory diet from 30 August to 12 September and from 3 to 19 October (shaded areas). For other explanations see Figure 6. [After (15)]

returns to the level of the control birds. The pattern of body weight shown by the experimental birds in comparison with the control birds suggests that the temporal changes in body weight are also endogenously programmed, as has been shown before for other species of birds (e.g. 63) and mammals (e.g. 58, 74).

SPECIES-SPECIFIC? The hypothesis assumes that the temporal program that is supposed to control migratory distance by controlling the time spent for migration is species-specific. If this program is, in fact, reflected in the pattern of *Zugunruhe* of caged birds, closely related species travelling over different distances in fall migration should develop different total quantities of *Zugunruhe*. This prediction was verified in comparative studies with various warbler species of the genera *Phylloscopus* (39) and *Sylvia* (9, 20), which cover different distances on their migrations. For species kept under identical conditions, the total amount of time spent in autumnal migratory restlessness was found to be longer the greater the distance between breeding grounds and winter quarters. The proportionality between these two measures is evident from Figure 9, in which the results obtained with eight different warbler species are summarized. Figure 10 illustrates an example of the difference in the pattern of migratory restlessness between two closely related species; the willow warbler is a long-distance migrant wintering in central and southern Africa, and the chiffchaff is a short-distance migrant, wintering in the Mediterranean area and in northern Africa (see Figure 11).

Differences in intensity of migratory restlessness have also been found in various populations or races of the same species migrating over different distances, e.g. in

CIRCANNUAL RHYTHMS IN BIRD MIGRATION   393

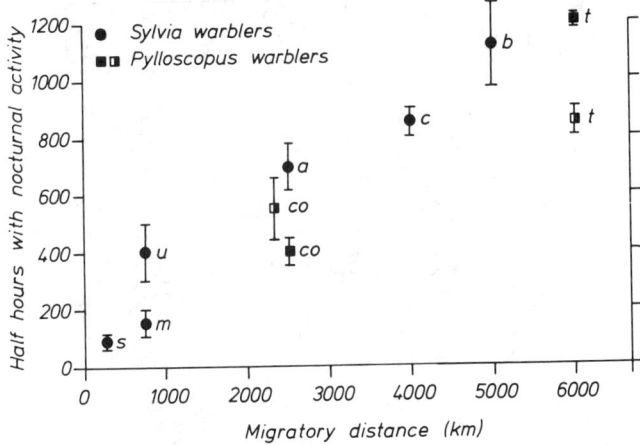

Figure 9 Relationship between the distance covered during migration by freeliving birds and the total time spent in fall migratory restlessness by caged conspecifics in eight warbler species. a = *Sylvia atricapilla*; b = *S. borin*; c = *S. cantillans*; m = *S. melanocephala*; s = *S. sarda*; u = *S. undata*; co = *Phylloscopus collybita*; t = *P. trochilus*. Each symbol shows the mean duration (with standard errors) of migratory restlessness (expressed as the total number of half-hour intervals with nocturnal activity) of a group of six to nine caged birds. ●, ■: Results from birds kept under photoperiodic conditions of their breeding grounds. ◻: Results from birds transferred by the end of September from the photoperiodic conditions of their breeding grounds to a constant 12-hr photoperiod. [After (9, 20, 39)]

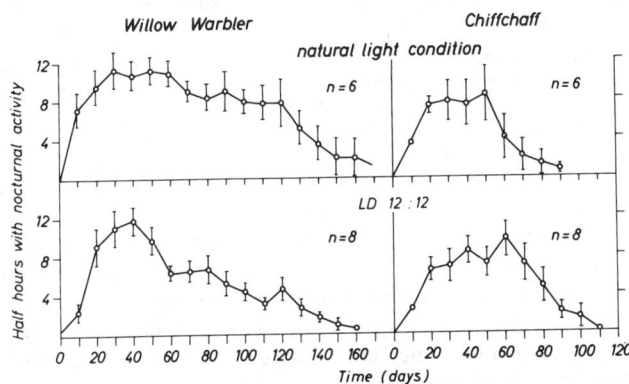

Figure 10 Variation in fall migratory restlessness in two groups of willow warblers and chiffchaffs. Birds were either kept in natural photoperiodic conditions of their breeding grounds (*upper panel*) or they were transferred from such conditions to a 12-hr photoperiod (*lower panel*). Time of transfer for willow warblers was between days 50 and 80; for chiffchaffs between days 0 and 30. Vertical bars: standard errors. For further explanations see Figure 6. [After (39)]

Figure 11 Breeding areas and winter quarters of the chiffchaff and the willow warbler. Numbers refer to breeding and wintering areas of different races. Large solid circles are calculated endpoints of migration (with standard errors) for two groups of chiffchaffs and three groups of willow warblers kept in different experimental conditions. [After (42)]

two populations of the blackcap [(64); Berthold, personal communication]. Comparable differences are suggested by results obtained from three races of the white-crowned sparrow, *Zonotrichia leucophrys* (42). Yet, in two other studies the expected differences between populations could not be detected [(55); Berthold, in preparation].

CONTROL OF DISTANCE? The hypothesis assumes that the temporal program is organized in such a way as to motivate the bird to spend that amount of time for its migratory flight which is necessary to reach the species-specific winter quarters. This most important implication of the hypothesis is, at the same time, the most difficult one to test. It still lacks confirmation. Because the migratory time-program is supposed to be reflected in the pattern of *Zugunruhe* of caged birds, the most direct way of testing this prediction would be to multiply the number of hours during which a bird shows migratory restlessness by the average speed at which the individuals of a species fly during migration. If the hypothesis is correct, the distance so calculated and the actual distance covered by freeliving birds should be identical.

Unfortunately this approach is difficult for at least two reasons: (*a*) the average speed of the migratory flight is not known for any of the species studied; and (*b*) the actual distance covered by freeliving individuals of the species studied is not known. Although we can calculate the direct distance between the breeding and wintering areas, it is known from radar investigations and from direct observations of migrating birds that the actual distance migrated is considerably longer. Migration does not proceed along straight lines but rather in a zig-zag pattern. How much this adds, on the average, to the direct distance, cannot even be reasonably estimated.

As a result of these difficulties, the prediction in question can only be tested in an indirect way. Perhaps the most successful approach is based on the comparison between known migratory performances of freeliving birds with the intensity of *Zugunruhe* shown by caged birds at corresponding times (38). The hypothesis assumes that the amount of *Zugunruhe* displayed by a caged bird within any given time-interval (U_i) is equivalent to the distance D_i traveled during this interval by freeliving conspecifics. The ratio U_i/D_i should then equal the ratio U_g/D_g, where U_g is the total amount of *Zugunruhe* of the caged bird and D_g the total distance travelled by freeliving conspecifics. U_i and U_g are known. If in addition we knew D_i we could calculate the distance equivalent to the total amount of *Zugunruhe* of the caged birds using the formula $D_g = U_g \cdot D_i/U_i$. If the hypothesis proposed here is correct, this theoretical distance should be approximately that normally covered by the species during migration.

Fortunately information about D_i is available from recoveries of banded birds. To test the hypothesis, data from 24 willow warblers and 20 chiffchaffs that had been banded during fall migration in northern Europe and recovered during the same migratory season in southern Europe were used to determine 24 and 20 D_i values, respectively. Using the above formula, average individual D_g values were calculated from these D_i values for a total of 21 caged willow warblers and 14 caged chiffchaffs with known U_i and U_g. These mean D_g values were averaged again for the six to eight individuals of three experimental groups of willow warblers and two experimental groups of chiffchaffs to which these birds belonged. Then the endpoints of the route along which each experimental group would have migrated were plotted on a map under the assumption that the birds had travelled along the normal flyway. Figure 11 shows that the birds would, in fact, have ended up in their respective winter quarters.

While the hypothesis of a temporal control of migratory distance is obviously difficult to test in laboratory studies, it suggests a variety of field experiments with banded birds. For instance, the hypothesis implies predictable shifts in the wintering area of birds that either have been prevented from migration during part of the fall migratory season or have been displaced forward or backward along the migratory route. Unfortunately the recovery rates of banded birds of all those species that have been studied extensively in the laboratory are much too low (mainly because of the small size of the birds) to encourage promising experiments. There are, however, results from experiments with other, larger species that can be used to evaluate the hypothesis. For instance, Rüppell & Schüz (84) caught first year carrion crows (*Corvus corone*) in fall, halfway along their migratory route, and displaced them to a location beyond their wintering area. Recoveries of banded birds in winter revealed that the displaced individuals had continued to migrate along the original direction and over distances similar to that which had separated them from their actual winter quarters at the point of capture. Comparable results suggesting the participation of endogenous time factors in the control of fall migration of first year European starlings were obtained by Perdeck (79, 80).

The results of a detention experiment carried out by Bellrose (7) are also in agreement with the hypothesis proposed here. Blue-winged teal (*Anas discors*) were caught in Illinois on their first fall migration and detained until all the adults had passed through. When released, they continued to migrate in the appropriate south-southeasterly direction. An analysis of the data published by Bellrose revealed that recoveries of detained birds were, on the average, at shorter distances than recoveries of birds that had been caught and immediately released after banding. While about 50% of the latter birds were recovered at distances greater than 2000 km from the releasing site, only about 10% of the former were found that far away.

Photoperiodic Modification of the Program

The notion that onset, end, and temporal course of migratory activity are controlled by an endogenous circannual rhythmicity does not, of course, imply that these parameters are independent of environmental influences. On the contrary, since it has been shown that photoperiod acts as a *Zeitgeber* on the endogenous rhythms (see the section above on the synchronization of circannual rhythms) it is to be expected that this variable affects the migratory program as part of this rhythm as well.

In most of the experiments discussed until now, the birds were either kept in constant photoperiodic conditions or in photoperiodic conditions similar to those they would have experienced during migration; almost no attention was given to the possible modifying effects of photoperiod. The little information available suggests, however, that photoperiod may affect the temporal pattern of migration quite drastically, and in a biologically significant way. For example, Figure 12 shows that fall migratory restlessness begins much earlier and the preceding plumage development proceeds much faster in willow warblers kept from an early age in a 12-hr photoperiod than in birds kept in an 18-hr photoperiod (53). The accelerating effect of shorter photoperiods on the timing of developmental processes and on the onset of fall *Zugunruhe* is probably advantageous for those individuals that hatch late in the season and thus grow up under decreasing daylengths. To avoid unfavorable environmental conditions in fall these birds initiate migration at an earlier age than birds hatched earlier in the year. Consequently, they have to pass more quickly through the postjuvenile processes preceding migration (4, 8, 17, 29, 40, 57, 82). The results shown in Figure 12, as well as similar results obtained from garden warblers and blackcaps (17), suggest that the differences between birds hatched early and late are at least partly due to the different photoperiodic conditions present during their development.

Whereas in spring and summer it is the shorter photoperiod which has an accelerating effect on seasonal activities, the opposite holds true in fall, as can be seen in Figure 13. The onset and end of molt and the onset of spring migratory restlessness occur earlier in birds transferred in mid-September to an 18-hr photoperiod than in birds transferred to a 12-hr photoperiod (43, 44). This accelerating effect of long photoperiods in winter may be advantageous for those willow warblers that cross

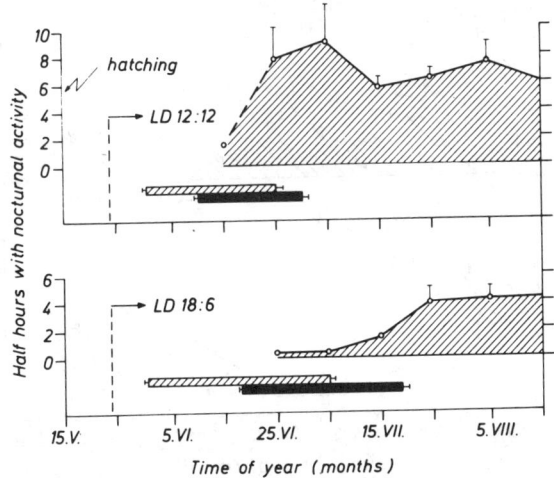

Figure 12 Migratory restlessness (shaded area), second period of plumage development (shaded bars), and postjuvenile molt (black bars) of two groups of eight and nine willow warblers transferred at an average age of nine days (24 May) from the natural photoperiodic conditions of their breeding grounds to a constant 12-hr (LD 12:12) and 18-hr (LD 18:6) photoperiod, respectively. Vertical lines at the symbols and horizontal lines at the bars: standard errors. For further explanation see Figure 6. [After (54)]

the equator on their fall migration. Since in these regions daylength increases with latitude during the northern winter, birds migrating far south are exposed to longer photoperiods than their conspecifics further north. As a result, fall migratory restlessness may be curtailed and the processes preceding spring migration may be speeded up. This in turn may enable an earlier onset of homeward migration. An earlier onset of migration in birds wintering further south is known for a variety of species (e.g. 25).

Photoperiodic effects similar to those described above for the willow warbler have also been shown in a few other species, mainly warblers (17, 19). Still, the exact way in which photoperiod modifies the endogenous pattern has not as yet been worked out. It is clear, however, that the final evaluation of the hypothesis of an endogenous timing of migration will depend on detailed information about the modifying effects of photoperiod.

Endogenously Controlled Changes in Migratory Direction

The results presented so far support the idea of an endogenous control of the temporal course of migration. There is now evidence suggesting that in addition, the spatial course of migration may depend to some extent on an endogenous circannual rhythmicity as well.

Like many other long-distance migrants, garden warblers breeding in western and central Europe reach their African winter quarters by first flying in a southwesterly direction and later changing their migration direction to the south or southeast,

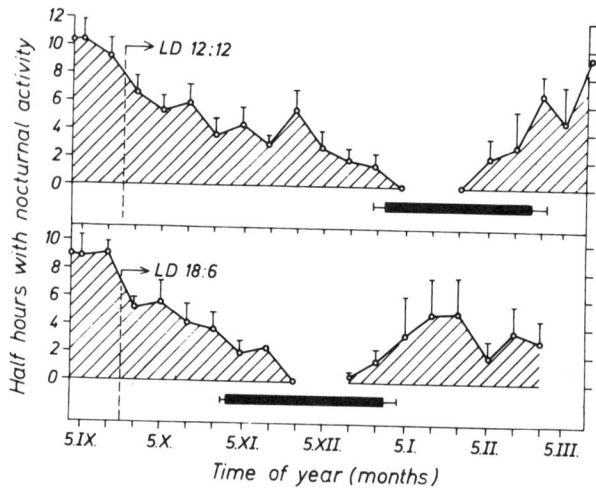

Figure 13 Migratory restlessness (shaded area) and prenuptial molt (black bars) of two groups of six and seven willow warblers transferred at an average age of 110 days (20 September) from the natural photoperiodic conditions of their breeding grounds to a constant 12-hr (LD 12:12) and 18-hr (LD 18:6) photoperiod, respectively. For further explanation see Figure 12. [From data in (43)]

instead of flying directly south from their breeding grounds (65, 100). This change in direction at about the end of September has been found to occur in captive garden warblers tested in orientation cages as well. In these experiments a total of 59 garden warblers were hand-raised and subsequently kept under a constant 12-hr photoperiod throughout their first fall migratory season. At regular intervals each bird was transferred to a round cage in which the mean direction of its nocturnal perch-hopping activity was established. Under test conditions the birds were deprived of celestial visual cues, but were exposed to the normal magnetic field of the earth, which can be used by garden warblers for direction-orientation (98). The results, presented in Figure 14, show a significant counterclockwise shift of the mean direction preferred by the birds in August and September compared to that preferred later in the season. The direction and the magnitude of this shift were similar to those in freeliving conspecifics (Gwinner & Wiltschko, in preparation). Because the experimental birds were kept throughout the autumnal testing period under constant photoperiodic conditions, these results suggest that this directional shift depends on the same internal timing mechanism as the other activities discussed above.

Interplay with Other Seasonal Activities

So far the discussion has only been concerned with the significance of circannual rhythms in the control of migration. However, migration is not an isolated phenomenon but rather only one of a great variety of seasonal activities. Some of these

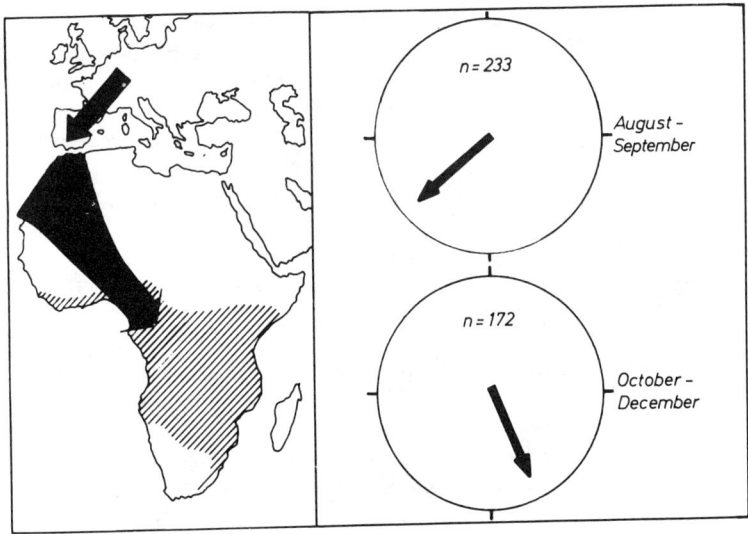

Figure 14 (*left*) Schematic of changes in migratory direction of garden warblers on fall migration. (*right*) Average direction of migratory restlessness of garden warblers tested in circular orientation cages in August and September (*upper diagram*) and in October through December (*lower diagram*). Numbers in the circular diagrams refer to the number of individual test nights during which the birds exhibited migratory restlessness. The mean direction of each of these nights was used as the statistical unit for further analysis. The mean direction preferred in August/September and that preferred in October/December are significantly different ($p < 0.001$, Watson & William test). (After Gwinner & Wiltschko, in preparation)

activities are in conflict with the demands of migration (e.g. reproduction and molt); others are prerequisites for its success (e.g. migratory fattening). It is not surprising, therefore, that the processes preceding, accompanying and following migratory activity are also often endogenously programmed and timed relative to migration in a biologically significant manner. This has become clear from investigations of closely related species with different migratory habits. Comparative investigations of seasonal phenomena in first-year birds of various species of *Phylloscopus*—warblers that migrate over different distances and begin fall migration at different times— have revealed a variety of adaptations to migration (40). For instance, it has been shown that species starting autumnal migration early differ from later migrants in that (*a*) their postjuvenile molt begins and ends earlier and is shorter; (*b*) during molt they replace fewer feathers; (*c*) they initiate migratory fattening and migratory restlessness at an earlier age; (*d*) migratory restlessness starts earlier relative to the end of postjuvenile molt; (*e*) they deposit more fat. These and other differences between species, initially observed in birds kept under natural photoperiodic conditions, have been shown even for individuals maintained from an early age in constant and identical environmental conditions (53). Hence, these species differences must be preprogrammed into the endogenous temporal organization of the birds.

Similar conclusions have been drawn from differences observed between other species (17, 45), races (54), and even populations (16, 55) kept under the same constant environmental conditions. A discussion of these results is beyond the scope of the present paper; most of them are summarized in (16) and (44).

Critique of the Hypothesis

In the preceding paragraphs much of the evidence supporting the hypothesis of an endogenous timing of bird migration has been presented. There are, however, observations that, at least at first sight, are hard to reconcile with it, and that show the limitations of the proposed mechanism. The following emphasizes objections to the most important prediction of this hypothesis, which states that an endogenous time-program determines the distance covered by birds on their first fall migration. Similar objections are valid for the hypothesis of an endogenous control of changes in migratory direction.

The principal weakness of the proposed mechanism would be its inaccuracy. If migratory distance is merely a passive consequence of the time spent on the migratory flight, the overall distance should depend drastically on environmental conditions, particularly on the weather situation. For instance, a bird flying with tail winds covers a much greater distance than a bird flying against head winds. Even if one assumes that the proposed migratory program is adjusted to the average meteorological situation prevailing along the migratory route, considerable seasonal and year-to-year variations might be expected in migratory distance and, correspondingly, in the location of the wintering grounds. These considerations do, in fact, argue against an exclusively temporal control of migratory distance in species with remote and highly localized wintering areas. It remains to be seen, however, whether these objections are valid generally. In most of the species investigated, including for instance the willow warbler and the garden warbler, the wintering grounds extend over a very wide range of latitudes; it is quite possible that even birds of the same population spend the winter hundreds or even thousands of kilometers apart. If this conjecture is confirmed by recoveries of banded birds, the low precision of the proposed mechanism could perhaps be tolerated for these species.

In addition, the possibility should be considered that the precision is improved by built-in compensatory processes that reduce the effects of external disturbances. Recent findings of Czeschlik (27) suggest such a compensatory mechanism in the garden warbler. In his experiments, caged birds were exposed alternately to a few nights of almost complete darkness (0.001 lux) and then a few nights of dim light (0.1 lux). During the dark nights migratory restlessness was drastically reduced (as in the experiment shown in Figure 7). When the birds were exposed to the dim night light again, they tended in the first few nights to show more migratory restlessness than was usual for that time of the year. These preliminary results may reflect the action of a mechanism that guarantees that, having been prevented from migration by adverse weather conditions, birds make up for the delay by additional flight time during the following nights.

Radar observations of migrating birds have revealed results that might also be interpreted in terms of an internal compensatory mechanism. It was found that migrating birds show a tendency to keep their ground speed constant by increasing air speed under head wind conditions and decreasing it under tail wind conditions (e.g. 21). In the latter situation, air speed was reduced by as much as 30%, to a rate below the optimal range speed (78). Alternative explanations are possible, yet it is tempting to interpret these wind-dependent changes in air speed as the expression of a mechanism that keeps the bird in synchrony with its endogenous migratory program. To achieve this, even the opportunity of saving energy might be neglected.

Even if compensatory processes for improving precision do exist, however, it is not likely that the mechanism proposed here is the only one to determine migration distance in first-year migrants of all species. The wintering areas of many species cover quite narrow latitudinal ranges (e.g. a band about 300 km wide in the orphean warbler, *Sylvia hortensis*). In these instances it seems necessary that any internal mechanism be assisted by external factors, at least during the last part of migration. Moreover, there is direct experimental evidence for the participation of external stimuli in many short-distance migrants, even when their wintering areas are large. For instance, it was found in several of these species that the fall migratory restlessness of caged birds not only exceeded the time normally spent for migration (42, 71, 97), but also showed more intra-individual variability than in long-distance migrants of the same genus (20, 39). Migratory restlessness in short-distance migrants appears to be more susceptible to exogenous variables than in long-distance migrants (27, 75, 85, 92, 93). The relative significance of internal and external mechanisms controlling migratory activity may then differ between long- and short-distance migrants. In the former, internal mechanisms may be of prime significance; in the latter, external factors may dominate. Hence migratory restlessness in many short-distance migrants may express only vaguely the general readiness of a bird to migrate; the actual time course and the distance of migration would then be largely determined by external factors.

Finally, it should be emphasized that even in long-distance migrants there is still no convincing evidence in support of the hypothesis. All the evidence that is available is of an indirect nature, based on comparisons between the migratory restlessness of caged birds and the performances of freeliving conspecifics. This approach is impeded by a variety of methodological difficulties that are hard to overcome. Three of them deserve mentioning.

1. It is not yet clear which parameter of migratory restlessness in caged birds is the best measure of migratory readiness. Most studies mentioned in this review used the total number of half-hour intervals during which a bird showed perch-hopping activity each night, because this parameter is thought to be a rough measure of the time a bird is in the migratory mood. Recent results of Czeschlik (27) suggest that hopping activity may not be as good a measure as "whirring," the wing-fluttering behavior shown by warblers and other migrants during the migratory seasons. When both activities were measured simultaneously in garden warblers and blackcaps, it turned out that under all test conditions the birds spent more time whirring than hopping (26, 27). It is not clear whether this is true during the entire fall migratory season. But if it is so, then the theoretical flight distance (calculated from the flight speed of a bird and the time during which it shows migratory restlessness) would be different depending on whether "hopping time" or "whirring time" were taken as a basis for the calculation.

2. It is not yet clear to what extent the conditions under which the caged birds were tested are comparable to those experienced by freeliving birds. It is well known that both migratory restlessness and actual migration are affected by a variety of external variables, particularly weather conditions. In the experiments mentioned, temperature and nocturnal light intensity were in the range where migratory restlessness is little affected (27). There are, however, other weather variables, the effects of which are as yet barely known. Hence, the possibility cannot be excluded that even a slight change in any of these factors would drastically alter the pattern and/or duration of migratory restlessness. The consequences for the testing of the hypothesis would be similar to those mentioned in the preceding paragraph.

3. Finally, any rigorous test of the hypothesis on the basis of comparisons between migratory restlessness and true migratory flight requires a detailed knowledge of the flight course of birds during their migrations. As mentioned above, this information is not available and there is little hope that it will become available in the near future for any of the small long-distance migrants used in most investigations of this subject.

In view of these difficulties, it is doubtful whether laboratory studies of migratory restlessness in caged birds—although very helpful in the investigations carried out so far—will ever allow a rigorous testing of the proposed mechanisms. At the moment, field experiments of the type described above, in the section on control of distance, appear to be the more promising approach.

Literature Cited

1. Aschoff, J. 1954. Zeitgeber der tierischen Tagesperiodik. *Naturwissenschaften* 41:49–56
2. Aschoff, J. 1955. Jahresperiodik der Fortpflanzung bei Warmblütern. *Stud. Gen.* 8:742–76
3. Aschoff, J. 1960. Exogenous and endogenous components in circadian rhythms. *Cold Spring Harbor Symp. Quant. Biol.* 25:11–28
4. Baggott, G. K. 1970. The timing of the moults in the pied wagtail. *Bird Study* 17:45–46
5. Baker, J. R. 1950. The seasons in a tropical rain-forest. Part 7. Summary and general conclusions. *J. Linn. Soc. London, Zool.* 41:248–58
6. Baker, J. R., Ranson, R. M. 1938. The breeding seasons of southern hemisphere birds in the northern hemisphere. *Proc. Zool. Soc. London Ser. A* 108:101–41
7. Bellrose, F. C. 1958. The orientation of displaced waterfowl. *Wilson Bull.* 70:20–40
8. Berthold, P. 1970. Zur Jahresperiodik von Staren (*Sturnus vulgaris*) aus Früh- und Spätbruten. *Vogelwelt* 91:88–95
9. Berthold, P. 1973. Relationships between migratory restlessness and migratory distance in six *Sylvia* species. *Ibis* 155:594–99
10. Berthold, P. 1973. Circannuale Periodik bei Teilziehern und Standvögeln. *Naturwissenschaften* 60:522–23
11. Berthold, P. 1974. Circannuale Periodik bei Grasmücken (*Sylvia*) III. Periodik der Mauser, der Nachtunruhe und des Körpergewichtes bei mediterranen Arten mit unterschiedlichem Zugverhalten. *J. Ornithol.* 115:251–72
12. Berthold, P. 1974. Circannual rhythms in birds with different migratory habits. In *Circannual Clocks,* ed. E. T. Pengelly, pp. 55–94. New York, San Francisco, London: Academic. 523 pp.
13. Berthold, P. 1974. *Endogene Jahresperiodik. Innere Jahreskalender als Grundlage der jahreszeitlichen Orientierung bei Tieren und Pflanzen.* Konstanz: Universitätsverlag. 46 pp.
14. Berthold, P. 1975. Migratory fattening. Endogenous control and interaction with migratory activity. *Naturwissenschaften* 62:399
15. Berthold, P. 1976. Über den Einfluß der Fettdeposition auf die Zugunruhe

bei der Gartengrasmücke *Sylvia borin*. *Vogelwarte* 28:263-66
16. Berthold, P. 1977. Über die Steuerung der Jugendentwicklung bei verschiedenen Populationen derselben Art. Untersuchungen an südfinnischen und südwestdeutschen Gartengrasmücken *Sylvia borin*. *Vogelwarte* 29: In press
17. Berthold, P., Gwinner, E., Klein, H. 1970. Vergleichende Untersuchungen der Jugendentwicklung eines ausgeprägten Zugvogels, *Sylvia borin*, und eines weniger ausgeprägten Zugvogels, *S. atricapilla*. *Vogelwarte* 25:297-331
18. Berthold, P., Gwinner, E., Klein, H. 1971. Circannuale Periodik bei Grasmücken (*Sylvia*). *Experientia* 27:399
19. Berthold, P., Gwinner, E., Klein, H. 1972. Circannuale Periodik bei Grasmücken I. Periodik des Körpergewichtes, der Mauser und der Nachtunruhe bei *Sylvia atricapilla* und *S. borin* unter verschiedenen konstanten Bedingungen. *J. Ornithol.* 113:170-90
20. Berthold, P., Gwinner, E., Klein, H., Westrich, P. 1972. Beziehungen zwischen Zugunruhe und Zugablauf bei Gartengrasmücken und Mönchsgrasmücken. *Z. Tierpsychol.* 30:26-35
21. Bruderer, B. 1971. Radarbeobachtungen über den Frühlingszug im Schweizerischen Mittelland. *Ornithol. Beob.* 68:89-158
22. Chapin, J. P. 1932. The birds of the Belgian Congo, Vol. 1. *Bull. Am. Mus. Nat. Hist.* 65:1-75B
23. Chapin, J. P. 1954. The calendar of the wideawake fair. *Auk* 71:1-15
24. Chapin, J. P., Wing, L. W. 1959. The wideawake calendar, 1953 to 1958. *Auk* 76:153-58
25. Curry-Lindahl, K. 1963. Molt, body weights, gonadal development and migration in *Motacilla flava*. *Proc. XIIth Int. Ornithol. Congr.*, pp. 960-73
26. Czeschlik, D. 1974. A new method of recording migratory restlessness in caged birds. *Experientia* 30:1490
27. Czeschlik, D. 1976. *Der Einfluß des Wetters auf die Zugunruhe von Garten- und Mönchsgrasmücken (Sylvia borin und S. atricapilla)*. PhD thesis. Univ. Innsbruck, 80 pp.
28. Dementiev, G. P., Gladkow, N. A. 1954. *The birds of the Sovjet Union.*, Vol. 6. Moscow: Sovietskaya Nauka. (In Russian). 792 pp.
29. Dolnik, V. R., Blyumental, T. I. 1967. Autumnal premigratory and migratory periods in the chaffinch (*Fringilla coelebs coelebs*) and some other temperate-zone passerine birds. *Condor* 69:435-68
30. Drost, R. 1938. Über den Einfluß von Verfrachtungen zur Herbstzugzeit auf den Sperber, *Accipiter nisus*. *Proc. 9th Int. Ornithol. Congr.*, pp. 503-21
31. Emlen, S. T. 1975. Migration: Orientation and navigation. In *Avian Biology*, ed. D. S. Farner, J. R. King, 5:129-219. New York, San Francisco, London: Academic. 523 pp.
32. Farner, D. S., Follett, B. K. 1966. Light and other environmental factors affecting avian reproduction. *J. Anim. Sci.* 25 (Suppl.):90-105
33. Farner, D. S., Lewis, R. A. 1971. Photoperiodism and reproductive cycles in birds. In *Photophysiology*, ed. A. C. Giese, 6:325-70. New York, San Francisco, London: Academic. 388 pp.
34. Farner, D. S., Lewis, R. A., Darden, T. R. 1973. Photoperiodic control mechanisms: homoiothermic animals. In *Biology Data Book*, ed. P. L. Altman, D. S. Dittmer, 2:1047-52. Bethesda, Md: Fed. Am. Soc. Exp. Biol.
35. Goss, R. J. 1969. Photoperiodic control of antler cycles in deer. I. Phase shift and frequency changes. *J. Exp. Zool.* 170:311-24
36. Goss, R. J. 1969. Photoperiodic control of antler cycles in deer. II. Alterations in amplitude. *J. Exp. Zool.* 171:223-34
37. Gwinner, E. 1967. Circannuale Periodik der Mauser und der Zugunruhe bei einem Vogel. *Naturwissenschaften* 54:447
38. Gwinner, E. 1968. Circannuale Periodik als Grundlage des jahreszeitlichen Funktionswandels bei Zugvögeln. Untersuchungen am Fitis (*Phylloscopus trochilus*) und am Waldlaubsänger (*P. sibilatrix*). *J. Ornithol.* 109:70-95
39. Gwinner, E. 1968. Artspezifische Muster der Zugunruhe bei Laubsängern und ihre mögliche Bedeutung für die Beendigung des Zuges im Winterquartier. *Z. Tierpsychol.* 25:843-53
40. Gwinner, E. 1969. Untersuchungen zur Jahresperiodik von Laubsängern. Die Entwicklung des Gefieders, des Gewichts und der Zugunruhe bei Jungvögeln der Arten *Phylloscopus bonelli*, *P. sibilatrix*, *P. trochilus* und *P. collybita*. *J. Ornithol.* 110:1-21
41. Gwinner, E. 1971. Orientierung. In E. Schüz, *Grundriss der Vogelzugskunde*, pp. 299-348. Berlin, Hamburg: Parey. 390 pp.
42. Gwinner, E. 1971. Endogenous timing factors in bird migration. In *Animal*

Orientation and Navigation, ed. S. R. Galler et al, pp. 321–38. Washington DC: NASA. 606 pp.
43. Gwinner, E. 1971. A comparative study of circannual rhythms in warblers. In *Biochronometry,* ed. M. Menaker, pp. 405–27. Washington DC: Natl. Acad. Sci. 662 pp.
44. Gwinner, E. 1972. Adaptive functions of circannual rhythms in warblers. *Proc. 15th Int. Ornithol. Congr.:* 218–36
45. Gwinner, E. 1973. Die Dauer der Grossgefiedermauser beim Fitis (*Phylloscopus trochilus*) und beim Zilpzalp (*P. collybita*) unter verschiedenen photoperiodischen Bedingungen. *J. Ornithol.* 114:507–10
46. Gwinner, E. 1973. Circannual rhythms in birds: their interaction with circadian rhythms and environmental photoperiod. *J. Reprod. Fert.* 19(Suppl.): 51–65
47. Gwinner, E. 1974. Endogenous temporal control of migratory restlessness in warblers. *Naturwissenschaften* 61: 405
48. Gwinner, E. 1975. Circadian and circannual rhythms in birds. See Ref. 31, pp. 221–85
49. Gwinner, E. 1975. Adaptive significance of circannual rhythms in birds. In *Physiological Adaptation to the Environment,* ed. F. J. Vernberg. pp. 417–33. New York: Intext Educational Publ. 576 pp.
50. Gwinner, E. 1975. Die circannuale Periodik der Fortpflanzungsaktivität beim Star (*Sturnus vulgaris*) unter dem Einfluss gleich- und andersgeschlechtiger Artgenossen. *Z. Tierpsychol.* 38: 34–43
51. Gwinner, E. 1977. Über die Synchronisation circannualer Rhythmen bei Vögeln. *Vogelwarte* 29: In press
52. Gwinner, E. 1977. Photoperiodic synchronization of circannual rhythms in the European starling (*Sturnus vulgaris*) *Naturwissenschaften* 64:44
53. Gwinner, E., Berthold, P., Klein, H. 1971. Untersuchungen zur Jahresperiodik von Laubsängern. II. Einfluss der Tageslichtdauer auf die Entwicklung des Gefieders, des Gewichts und der Zugunruhe bei *Phylloscopus trochilus* und *P. collybita. J. Ornithol.* 112: 253–65
54. Gwinner, E., Berthold, P., Klein, H. 1972. Untersuchungen zur Jahresperiodik von Laubsängern. III. Die Entwicklung des Gefieders, des Gewichts und der Zugunruhe südwestdeutscher und skandinavischer Fitisse (*Phylloscopus trochilus trochilus* und *P. t. acredula*). *J. Ornithol.* 113:1–8
55. Gwinner, E., Biebach, H. 1977. Endogene Kontrolle der Mauser und der Zugdisposition bei südfinnischen und südfranzösischen Neuntötern (*Lanius collurio*). *Vogelwarte* 29: In press
56. Gwinner, E., Dorka, V. 1976. Endogenous control of annual reproductive rhythms in birds. *Proc. 16th Int. Ornithol. Congr.,* pp. 223–34
57. Hankioja, E. 1969. Weights of reed buntings (*Emberiza schoeniclus*) during summer. *Ornis Fenn.* 46:13–21
58. Heller, H. C., Poulson, T. L. 1969. Circannian rhythms. II. Endogenous and exogenous factors controlling reproduction and hibernation in chipmunks (*Eutamias*) and ground squirrels (*Spermophilus*). *Comp. Biochem. Physiol.* 33:357–83
59. Homeyer, E. F. von. 1881. *Die Wanderungen der Vögel.* Leipzig: Grieben's. 415 pp.
60. Immelmann, K. 1963. Tierische Jahresperiodik in ökologischer Sicht. *Zool. Jahrb. Abt. Syst. Oekol. Geogr. Tiere* 91:91–200
61. Immelmann, K. 1967. Periodische Vorgänge in der Fortpflanzung tierischer Organismen. *Stud. Gen.* 20:15–33
62. Immelmann, K. 1971. Ecological aspects of periodic reproduction. See Ref. 31, 1:341–89
63. King, J. R. 1963. Autumnal migratory-fat deposition in the white-crowned sparrow. *Proc. 13th Int. Ornithol. Congr.,* pp. 940–49
64. Klein, H. 1974. The adaptational value of internal annual clocks in birds. See Ref. 12, pp. 347–91
65. Klein, H., Berthold, P., Gwinner, E. 1973. Der Zug europäischer Garten- und Mönchsgrasmücken (*Sylvia borin* und *S. atricapilla*). *Vogelwarte* 27:73–134
66. Marshall, A. J. 1959. Internal and environmental control of breeding. *Ibis* 101:456–78
67. Marshall, A. J. 1960. Annual periodicity in the migration and reproduction of birds. *Cold Spring Harbor Symp. Quant. Biol.* 25:499–505
68. Marshall, A. J. 1960. The role of the internal rhythm of reproduction in the timing of avian breeding seasons, including migration. *Proc. XIIth Int. Ornithol. Congr.,* pp. 475–82
69. Marshall, A. J. 1961. Breeding seasons and migration. In *Biology and Compar-*

ative *Physiology of Birds*, ed. A. J. Marshall. 307–39. London: Academic. 518 pp.
70. Marshall, A. J., Serventy, J. B. 1959. Experimental demonstration of an internal rhythm of reproduction in a transequatorial migrant (the short-tailed shearwater, *Puffinus tenuirostris*). *Nature* 184:1704–5
71. Mewaldt, L. R., Morton, M. L., Brown, J. L. 1964. Orientation of migratory restlessness in *Zonotrichia*. *Condor* 66: 377–417
72. Miller, A. H. 1959. Reproductive cycles in an equatorial sparrow. *Proc. Natl. Acad. Sci.* 45:1095–1100
73. Miller, A. H. 1962. Bimodal occurrence of breeding in an equatorial sparrow. *Proc. Natl. Acad. Sci.* 48:396–400
74. Mrosovsky, N., Fisher, K. C. 1970. Sliding set points for body weight in ground squirrels during the hibernation season. *Can. J. Zool.* 48:241–47
75. Palmgren, P. 1937. Auslösung der Frühlingszugunruhe durch Wärme bei gekäfigten Rotkehlchen, *Erithacus rubecula* (L.) *J. Ornithol.* 14:71–73
76. Pengelley, E. T. 1971. Annual biological clocks. *Sci. Am.* 224:72–78
77. Pengelley, E. T., Fisher, K. C. 1963. The effect of temperature and photoperiod on the yearly hibernating behavior of captive golden-mantled ground squirrels (*Citellus lateralis tescorum*). *Can. J. Zool.* 41:1103–20
78. Pennycuick, C. J. 1969. The mechanics of bird migration. *Ibis* 111:525–56
79. Perdeck, A. C. 1958. Two types of orientation in migrating starlings, *Sturnus vulgaris*, and chaffinches, *Fringilla coelebs* L. as revealed by displacement experiments. *Ardea* 46:1–37
80. Perdeck, A. C. 1964. An experiment on the ending of autumn migration in starlings. *Ardea* 52:133–39
81. Pohl, H. 1971. Circannuale Periodik beim Bergfinken. *Naturwissenschaften* 58:572–73
82. Promptov, A. N. 1949. Sozonnye migratsii ptits kak biofiziologicheskaia problema. *Izv. Akad. Nauk, SSSR Ser. Biol.*, pp. 30–39
83. Rowan, W. 1926. On photoperiodism, reproductive periodicity, and the annual migrations of birds and certain fishes. *Proc. Boston Soc. Nat. Hist.* 38: 147–89
84. Rüppell, W., Schüz, E. 1948. Ergebnis der Verfrachtung von Nebelkrähen (*Corvus corone cornix*) während des Wegzuges. *Vogelwarte* 15:30–36

85. Schildmacher, H. 1938. Zur Auslösung der Frühlings-Zugunruhe durch Wärme bei gekäfigten Rotkehlchen, *Erithacus r. rubecula* (L.). *Vogelzug* 9: 7–14
86. Schmidt-Koenig, K. 1970. Ein Versuch, theoretisch mögliche Navigationsverfahren zu klassifizieren und relevante sinnesphysiologische Probleme zu umreissen. *Verh. Dtsch. Zool. Ges.* 64: 243–45
87. Schmidt-Koenig, K. 1975. *Migration and Homing in Animals*. Berlin, Heidelberg, New York: Springer. 99 pp.
88. Schüz, E. 1949. Die Spät-Auflassung ostpreussischer Zugstörche in Westdeutschland 1933. *Vogelwarte* 15:63–78
89. Schüz, E. 1950. Zur Frage der angeborenen Zugwege. *Vogelwarte* 15:219–26
90. Schüz, E. 1951. Überblick über die Orientierungsversuche der Vogelwarte Rossitten (jetzt: Vogelwarte Radolfzell) *Proc. Xth Int. Ornithol. Congr. Uppsala*, pp. 249–68
91. Schwab, R. G. 1971. Circannian testicular periodicity in the European starling in the absence of photoperiodic change. See Ref. 43, pp. 428–47
92. Wagner, H. O. 1957. Vogelzug, Umweltreize und Hormone. *Verh. Dtsch. Zool. Ges.*, pp.289–98
93. Wagner, H. O., Schildmacher, H. 1937. Über die Abhängigkeit des Einsetzens der nächtlichen Zugunruhe von der geographischen Breite. *Vogelzug* 8: 18–19
94. Wallraff, H. G. 1960. Können Grasmücken mit Hilfe des Sternenhimmels navigieren? *Z. Tierpsychol.* 17:165–77
95. Wallraff, H. G. 1960. Does celestial navigation exist in animals? *Cold Spring Harbor Symp. Quant. Biol.* 25: 451–61
96. Wallraff, H. G. 1971. Fernorientierung der Vögel. *Verh. Dtsch. Zool. Ges.* 65: 201–14
97. Weise, C. M. 1963. Annual physiological cycles in captive birds of differing migratory habits. *Proc. 13th Int. Ornithol. Congr., Ithaca.*, pp. 983–93
98. Wiltschko, W. 1974. Der Magnetkompass der Gartengrasmücke (*Sylvia borin*) *J. Ornithol.* 115:1–7
99. Zimmermann, J. L. 1966. Effects of extended tropical photoperiod and temperature on the dickcissel. *Condor* 68:377–87
100. Zink, G. 1973. *Der Zug europäischer Singvögel*. Möggingen: Vogelzug-Verlag. 123 pp.

EXTRAFLORAL NECTARIES AND PROTECTION BY PUGNACIOUS BODYGUARDS

❖4131

Barbara L. Bentley

Department of Ecology and Evolution, State University of New York, Stony Brook, NY 11794

> Nectar glands . . . are of indirect use by attracting suitable pollinators to flowers, by luring prey to the digestive apparatus of some carnivorous plants, or by maintaining upon the plant a bodyguard of pugnacious insects which more or less efficiently protect it against certain of its enemies. . . . "Nuptial" nectary glands are those which attract pollinators . . . and "extra-nuptial" nectary glands are those which have no value in this respect. . . . The subject of extra-nuptial nectar and its relation to ants, is deserving of much fuller discussion. Broadly speaking this class of extra-nuptial nectar glands, by their secretion, attracts to the plants which bear them, hordes of ants (rarely wasps) which constitute a temporary and changing bodyguard, disputing the presence of all other insects with the exception of their proteges the sugar-secreting aphids, coccids, etc, and resisting often furiously and effectively, the onslaught of ruminants and other large animals. That this is a true explanation of the reason for the existence of these structures is generally admitted.
>
> Delpino, 1874

> While floral nectaries obviously serve the purpose of pollination and act as an indirect bait for visitors like insects, birds, and bats, the function of extra-floral nectaries is related not to the environment but to plant metabolism. . . . The question of whether these nectaries play an ecological role in the biocenosis of the forest (as food for Hymenoptera, for example) cannot be answered, likewise the true purpose of these nectaries is still a mystery.
>
> Schremmer, 1970

INTRODUCTION

Extrafloral nectaries are sugar producing glands found outside the flower. They most commonly occur on the leaf blade, rachis, petiole, stem, or in proximity to the reproductive parts of a plant, but they may also be found on bracts, stipules, or cotyledons of some species (179). Although Vergil (*Georgica,* Book IV) describes nectaries as sugar secreting organs in flowers, it was not until 1762 that the distinction between floral and extrafloral nectaries was recognized (81).

Extrafloral nectaries occur in a wide variety of plant taxa, particularly among the angiosperms (25). In some families, such as the Passifloraceae and Leguminosae, many species have extrafloral nectaries, yet in other families, such as the Graminae, the presence of nectaries may be restricted to one or two species (20, 22). Their occurrence may be highly variable within a single species [e.g. *Gossypium* spp. (141)] or may vary with the season or the maturity of the plant [e.g. *Populus glandulifera* (160)]. An abridged list of families with some representative genera is given in Table 1. A more thorough treatment is given by Zimmerman (179) and Schnell et al (148).

Extrafloral nectaries show enormous variability in structure, from simple glandular surfaces (e.g. *Malpighia glabra*) to elaborate cups, flaps, or stalks with hairs and contrasting pigmentation (148, 179). Some examples are shown in Figure 1.

Cytologically, extrafloral nectaries resemble floral nectaries. As active secretory cells, they have very large numbers of mitochondria, dense protoplasts, and large nuclei (62, 64, 75, 76, 78, 114, 115, 148). Vascular tissue often extends to the underlying parenchyma of the gland and the sugar is primarily derived from the phloem (1, 61, 66, 69, 161, 168, 169) but may be synthesized in the region of the nectary (6).

The mechanisms for the secretion of nectar are not well understood, perhaps because of the wide morphological variation of nectaries (61, 78, 83, 114, 136). Many authors believe that nectar secretion is in fact "excretion" of surplus sugars in the plant as the associated organs shift from a "sink" to a "source" of carbohydrates during development (46, 67, 69, 80, 141, 156), especially if growth is limited by available nitrogen (84, 151), or during water stress when sugar concentrations in the associated tissues increase (56, 84), or with increased light intensity (80, 171). Others suggest that sugar "excretion" from extrafloral nectaries is an incidental event associated with the excretion of water and "superabundant salts" under conditions of high humidity (43, 44, 155). Most authors agree, however, that nectar secretion is the result of active metabolic processes (84); the cellular structure is typical of active metabolic tissue (61, 77, 114), oxygen consumption in the gland increases during secretion (68, 114), and phosphomonoesterase is specifically associated with the cytoplasm of extrafloral nectaries (63).

All studies on the composition of extrafloral nectar have shown that sucrose, glucose, and fructose are by far the most abundant solutes, but other sugars, amino acids, and miscellaneous organic compounds may be present in some species (see Table 1). The concentration of the nectar solution depends upon a variety of factors including the proximity of phloem vessels (61, 114), the proportion of xylem in the vascular trace (66, 114), and the photosynthetic rates of the associated organs (80, 141). The data are difficult to interpret, however, because they all are related to the volume of nectar collected. Volume is strongly affected by evaporation under low relative humidity (84, 164), hygroscopic absorption of water by the sugars at high relative humidity (28), and by local edaphic factors (113).

The activity patterns of nectaries are frequently quite clear and show some interesting ecological correlates. First, in contrast to floral nectaries, extrafloral nectaries usually do not have an easily recognized diurnal pattern of secretion: Nectar is often produced by the plant throughout the day and night (13, 28, 137,

EXTRAFLORAL NECTARIES 409

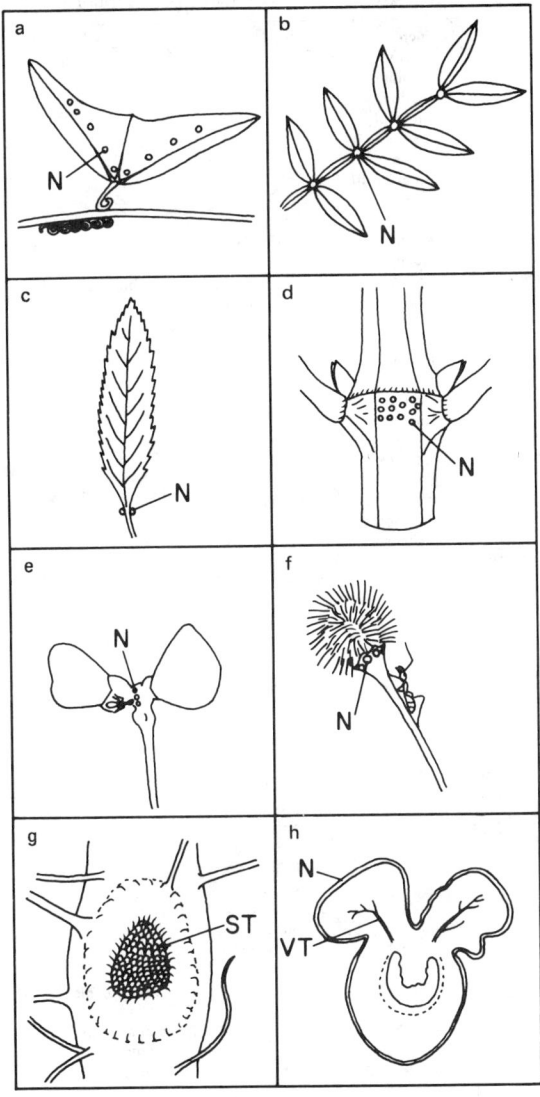

Figure 1 Examples of extrafloral nectaries. (*a*) *Passiflora platyloba* (from a photograph); (*b*) *Inga* sp. (from a photograph); (*c*) *Turnera trioniflora* (from a live specimen); (*d*) *Pseudocalymma* sp. (redrawn from Seibert); (*e*) *Vigna* sp. with a *Camponatus planatus* forager (from a photograph); (*f*) *Bixa orellana* with an *Ectotomma tuberculatum* forager (from a photograph); (*g*) *Gossypium hirsutum* (redrawn from a scanning electron micrograph in Wergin et al); (*h*) *Prunus cerasus,* cross section through the petiole at the level of the nectary (redrawn from Schnell et al). N, extrafloral nectary; ST, secretory tissue; VT, vascular trace. All photographs were taken by the author.

Table 1 Composition of extrafloral nectar

Taxon	Sugars	Amino acids	Other	Reference
Andropogon gayanum	sucrose glucose fructose maltose raffinose arabinose xylose 3 unidentified sugars			(21)
Campsis radicans	sucrose glucose fructose			(59)
Gossypium barbadense	sucrose glucose fructose	glutamic and β alanine	uronic acid?	(123)
Gossypium barbadense	glucose fructose sucrose			(164)
Gossypium hirsutum	glucose fructose sucrose			(28)
Gossypium tomentosum	sucrose glucose ribose rhamnose raffinose	none		(37)
Helianthella quinquenervis	sucrose	alanine arginine aspartic acid cystine glutamic acid glutamine glycine histidine isoleucine leucine lysine methionine proline serine threonine tryptophan valine tyrosine phenylalanine	amino butyric acid	(D. W. Inouye and R. S. Inouye, unpublished data)
Orchidaceae	fructose glucose sucrose	"protein"	mucilage organic acids water soluble vitamins	(7)
Orchidaceae	fructose sucrose glucose raffinose maltose gentiobiose stachyose melibiose lactose melezitose		a number of unidentifed oligosaccharides	(97, 98)
Prunus persica	glucose fructose sucrose			(135)
Turnera ulmifolia	sucrose glucose fructose			(60)

139, 140, 162). Again, because of the compounding effects of evaporation and hygroscopic absorption, volumetric data are equivocal. Second, active nectaries are those found on younger portions of the plant (13, 30, 42, 46, 139, 143, 160) usually above the first 5–10 nodes below an actively growing meristem. Frequently the most active nectaries are those associated with developing reproductive portions of the plant (13, 89, 98, 123, 129). While nectaries are usually mature and active long before the associated organ is fully expanded (42, 135), nectary "death" (cessation of secretion, lignification, withering) occurs soon after the organ is fully mature (13, 34–36, 46, 47). Third, the greatest secretory activity is usually found at the height of the growing season in temperate zone plants (8, 123), but continues throughout the year in moist tropical regions (B. L. Bentley, unpublished field notes) and may be independent of the phenological stage of the plant (30). In many *Populus* species, however, only the first leaves produced in the spring bear nectaries (160), except for root sprouts which have nectaries on all leaves produced throughout the growing season (B. L. Bentley, unpublished field notes). (An interesting correlate is that new growth on Poplar *trees* ceases with the production of a winter bud, while growth of a root sprout continues until it is killed by frost.) Fourth, attack by sucking insects may increase the rate of secretion by nectaries (123) and this increase is positively correlated with increased infestation levels.

These patterns may not exclude the "excretory" role of nectaries, but at least two conflict with the "source/sink" hypothesis: (*a*) developing buds, flowers, and fruits never become a carbohydrate source, and yet nectaries frequently remain active on these organs until maturity; and (*b*) sucking insects reduce the "source pressure" of an organ (and may even cause it to become a sink), and yet nectar secretion increases with increasing infestation.

The genetic control for the development of extrafloral nectaries has only been investigated in one genus, *Gossypium* (cotton), where it was found that the inheritance of "nectariless" was determined by two pairs of recessive genes (122, 141). The lack of extrafloral nectaries was linked to male sterility (142), but was independent of the inheritance of floral nectaries (85, 142). However, there is a great deal of phenotypic variation in the expression of the genes controlling nectary development in cotton. For example, a short growing season or "poor growing conditions" (presumably low water availability) will produce plants with few or no extrafloral nectaries, even though their selfed progeny will have nectaries (122, 141). Phenotypic plasticity, however, is not characteristic of all taxa containing nectary-bearing plants. Indeed, in many groups, such as the Caesalpinioideae (147) and the Bignoniaceae (150), the occurrence of nectaries is constant enough for use as a taxonomic guide (25, 118, 162). But in other groups, variability within a population renders the presence of nectaries useless for taxonomic purposes (84).

If extrafloral nectaries in most plants have a similar, simple genetic basis, then the evolution of nectaries may be a relatively "easy" process. That nectaries occur in an extremely wide variety of taxa and yet may not be present in closely related taxa, is further evidence for the "easy" evolution of nectaries (Table 2). Possible evolutionary pathways will be discussed by T. Elias in a future symposium on nectaries (AIBS 1977). But the "ease" of evolution does not explain why so many plant taxa have evolved extrafloral nectaries.

Table 2 Taxonomic distribution of extrafloral nectaries

Family[a]	Genus[b]	Reference[c]
Acanthaceae	*Acanthus*	L. Durkee, pers. comm.
Agavaceae	*Dracaena*	(70)
	Sansevieria	(70)
Balsaminaceae	*Impatiens*	(17,43,44,80,164)
Bignoniaceae (150)	*Bignonia (Spathodea)*	(88, 137)
	Campsis	(59)
	Cydista	(53)
	Tecoma	(129)
Bixaceae	*Bixa*	(13)
Cactaceae (110)	*Cereus*	T. Elias (pers. comm.)
	Phyllo-Cactus	(172)
Capparidaceae	*Capparis*	(133)
Caprifoliaceae	*Sambucus*	(17)
	Viburnum	(17)
Caryophyllaceae	*Drynaria*	(15, 17)
	Silene	(17)
Compositae (102)	*Calycadenia*	(29)
	Helianthella	(89)
Convolvulaceae	*Batatas (Ipomoea)*	(100)
Costaceae	*Costus*	(138)
Cucurbitaceae (33)		
Dioscoreaceae	*Dioscorea*	(127)
Euphorbiaceae	*Hevea*	(67)
	Macaranga	(143, 153)
	Mercurialis	(63)
	Pedilanthus	(47)
	Poinsettia[d]	(157, 166, 167)
	Ricinus	(17, 74, 75, 140)
Gesneriaceae		(S. Kleinfelt, pers. comm.)
Goodeniaceae	*Scaevola*	(39)
Graminae (Poaceae)	*Andropogon*	(21, 22)
Gurneraceae (14)		
Guttiferae	*Hypericum*	(38)
Leguminosae[e]	*Acacia*	(9, 15, 17, 30, 82, 92, 93, 144)
	Apuleia	(17)
	Bauhinia	(15)
	Caesalpinia	(15)
	Cassia	(15, 71)
	Dialium	(17)
	Erythrina	(119)
	Humbolttia	(17)

Table 2 *(Continued)*

	Heterostemon	(17)
	Inga	(Bentley, unpubl. field notes)
	Macrolobium	(17)
	Pithecellobium	(17, 58)
	Pueraria	(116)
	Robinia	(15, 91)
	Tachigalia	(17)
	Vicia	(17, 64)
	Vigna	(Bentley, unpubl. field notes)
Malpigheaceae (152)	*Banisteria*	(17)
	Clonodia	(17)
	Hemsleyna	(17)
	Thryallis	(17)
Malvaceae (90)	*Gossypium*	(17, 28, 108, 123, 139, 162)
	Hibiscus	(103, 108)
Marantaceae	*Calathea*	(H. Kennedy, pers. comm.)
Oleaceae	*Ligustrum*	(17)
	Olea	(130)
	Osmanthus	(121)
	Phillyrea	(130)
	Syringa	(131)
Orchidaceae[e] (154)	*Acropera*	(45)
	Cattleya	(7, 70, 98)
	Coelogyne	(98)
	Coryanthes	(45, 98)
	Cymbidium	(70, 98 105)
	Dendrobium	(97, 98)
	Epidendrum	(7, 98)
	Gongora	(45)
	Laelia	(98)
	Notylia	(45)
	Oncidium	(98)
	Penera	(17)
	Schomburgkia	(98)
	Vanda	(7, 98)
	Vanilla	(45, 98)
Passifloraceae	*Passiflora*[e]	(18, 101, 175)
	Tetrastylis	(Bentley, unpubl. field notes)
Polygonaceae	*Polygonum*	(145)
Polypodiaceae	*Pteris (Pteridium)*	(46, 103, 104, 109, 149)
Rosaceae	*Malus*	(178)
	Prunus	(42, 54, 134, 135, 178)
	Pyrus	(178)
Rubiaceae	*Coprosoma*	(73, 79)
	Hamelia[f]	(Bentley, unpubl. field notes)

Table 2 *(Continued)*

Family[a]	Genus[b]	Reference[c]
Salicaceae	*Populus*	(160)
	Salix	(87, 149)
Simaroubaceae	*Ailanthus*	(48, 132)
Smilacaceae	*Smilax*	(Bentley, unpubl. field notes)
Sterculiaceae	*Ayenia*	(41)
	Byttneria	(2, 40)
Ternstroemiaceae	*Marcgravia*	(15, 17, 86)
Turneraceae	*Turnera*	(60; Bentley, unpubl. field notes)
Verbenaceae	*Avicennia*	(Bentley, unpubl. field notes)
	Clerodendrum	(35, 16, 17)
	Duranta	(15, 115)
	Gmelina	(34)
	Holmskioldia	(17)
	Petrea	(36)
Vitaceae	*Vitis*	(Bentley, unpubl. field notes)

[a] Reference numbers after a family name refer to studies on a major portion of the family.
[b] All species within a genus do not necessarily have extrafloral nectaries.
[c] Most references were chosen for anatomical, physiological, or ecological information rather than taxonomic surveys.
[d] The nectaries of *Poinsettia* are extrafloral, but in fact ecologically function as floral nectaries by attracting pollinators.
[e] Extrafloral nectaries occur in many more species than can be given here.
[f] The nectaries of *Hamelia* are anatomically floral nectaries, but continue to secrete for at least 21 days after the corolla falls (B. L. Bentley, unpublished field notes). Ants are attracted to the nectar and may protect the developing fruits. Thus these nectaries may be considered ecologically "extrafloral" nectaries.

THE ADAPTIVE SIGNIFICANCE OF EXTRAFLORAL NECTARIES

The adaptive significance of extrafloral nectaries has been disputed for over 100 years (3, 9, 12, 13, 16, 17, 19, 45, 46, 50–52, 104, 111, 117, 123, 126, 134, 135, 143, 146, 149, 159, 160, 162, 163, 173, 174). The controversy has been between two groups (26), the "protectionists" (who have supported the idea that the ants visiting the nectaries protect the plant from herbivorous animals), and the "exploitationists" [who have felt that the plants "have no more use of their ants than dogs do their fleas" (173), and that the secretion of nectar has some purely physiological function]. Early work supporting the protectionists' interpretation was based more on armchair biology than on well documented facts (16). Many early evolutionists had difficulty accepting the possibility that the evolution of some plant characteristics could be the result of selective pressures from animal activities and vice versa (57). What the protectionists have needed to settle the issue is (*a*) experimental evidence to establish a solidly based set of correlative observations, and (*b*) general acceptance of the concepts of coevolution between plants and animals. Since coevolution is now widely accepted by biologists (57, 72), I devote the remainder of this paper to a presentation of evidence in support of the protectionists' hypothesis that plants bearing extrafloral nectaries are protected by "a bodyguard of ants, maintained by the foliar glands . . . rendering the plant unsafe for lepidopterous larvae or other herbivores to frequent the plant" (160).

Correlative Data in Support of the Protectionists' Hypothesis

For the beneficial interaction between ants and plants bearing extrafloral nectaries to occur, ants must be present on the plant, they must show aggressive behavior towards potential herbivores, and (to be most effective) they should be potential predators on the herbivore species. In addition, the plant must be vulnerable to, and actually subject to, herbivore attack at least during some stage of life. To be most efficient, nectar flow should vary directly with herbivore activities.

LOCAL AND GEOGRAPHIC DISTRIBUTION OF ANTS AND EXTRAFLORAL NECTARIES   If the protectionist position is valid, then the global and local distribution of nectary plants must correlate with areas of high ant abundance. The distribution of ant species is fairly well documented (128, 173, 174, 176). By far the greatest diversity and abundance are in tropical regions. And, although plants bearing extrafloral nectaries occur in the temperate zone (59, 89), the greatest diversity and abundance of nectary plants are in the tropics of both the Old and New Worlds (148, 179). An exception which "proves" this correlation is the distribution of ants and nectary plants at high elevations in these two latitudinal regions. Nectary plants and ants are rare in high-altitude forests in the tropics (31), but occur at similar elevations in temperate zone mountains (89). At high elevations in the tropics, almost continuous cloud cover is present (producing typical "cloud forest" vegetation). Air temperatures remain low and soil moisture remains high. Ants are extremely rare under these conditions: Ground nesting ants cannot exist in very wet soils and arboreal species are extremely sensitive to air temperatures. In the temperate zone, montane soils are considerably drier and diurnal air temperatures during the summer can be sufficiently high for survival of many poikilothermous organisms.

The local distribution of nectary plants also correlates with high abundance of ants. In a study of the distribution of ants and nectary plants in northeast Costa Rica, I found that both ants and plants bearing extrafloral nectaries were most abundant at forest edges or clearings (12). In these areas experimental nectary plants showed less insect damage than elsewhere. Ridley (143) found a similar pattern in the distribution of *Macaranga hypoleuca* in Malaya. On the other hand, extrafloral nectaries are not found on hydrophytes or extreme xerophytes (84), plants adapted to habitats where ants are extremely rare.

The abundance of ant foragers on individual nectary plants can be quite variable (13, 89) due, in part, to the patchy distribution of ant nests in most areas. This patchiness is a result of a variety of interacting factors, such as the distribution of appropriate nest sites, the dispersal patterns of founding queens, the abundance of prey species, and the proximity of other ant colonies (32). Both Inouye & Taylor (89) and I (13) have used these "natural experimental" conditions to show that those plants closer to ant nests (or nest units) have both a greater number of ant foragers visiting the nectaries and a reduced level of insect damage.

For some tropical ant species, extrafloral nectar may be a major source of sugar for the colony (other species, of course, may tend membracids or coccids). And so the proximity of nectary plants to the nest may be critical for the survival of the ant colony. Some pseudomyrmecines, in fact, will move the nest (or nest unit) closer to active nectaries (D. H. Janzen, personal communication). Although ants are by far the most frequent visitors to extrafloral nectaries, other insects could play a role in protection of the plant. For example, *Trigona* bees are extremely aggressive at the nectaries of *Cassia* growing on the Osa peninsula in Costa Rica, and nectar can be extremely important in the nutrition and local distribution of parasitic Hymenoptera which often parasitize larval herbivores (49, 55, 106, 107, 135, 158, 160).

AGGRESSIVE BEHAVIOR OF ANTS ON NECTARY PLANTS Many ant species are notoriously aggressive (173, 174, 177) and this is certainly true of those species common on extrafloral nectaries (13, 160). This aggression in ants can be related to three major factors: "ownership behavior" (170) around the nest sites and food sources, predatory behavior, and pheromone-mediated attack responses.

"Ownership behavior" is defined by Way (170) as aggression towards all organisms intruding on an area monopolized by an ant colony. It is most pronounced near the nest and can be elicited at a food source, especially if the location of the food source is relatively permanent. I have seen some ants (notably *Ectotomma tuberculatum*) stand poised with mandibles open ready to attack any object approaching a nectary or other food (13).

Many woody species of plants with extrafloral nectaries have hollow internodes (e.g. *Macaranga*), thorns (e.g. *Acacia*), or extremely soft pith (e.g. *Bixa*) which can be quickly discovered and utilized as nest sites (domatia) by ants (4). Ridley (143) feels that this is an important component in the protective interaction because potential domatia may "induce [the ants] to remain as permanent guardians of the plant." And, since ownership behavior is more pronounced near the nest, the propensity of ants to nest in or near nectary plants will increase their effectiveness as herbivore deterrents.

The swarming behavior of some ant species, especially in response to disturbance, may also play a role in defense of the plant. This behavior, more characteristic of smaller-sized species, can be elicited through pheromones by threat to an individual ant or by discovery of an especially rich food source. Nectar per se usually does not trigger swarming, but foragers visiting nectaries may discover potential prey in the area and release the appropriate swarming behavior. I have seen this occur when a small species of *Crematogaster* attacked and carried off a lepidopteran larva which had been feeding near the nectaries of *Cassia biloba*.

Aggressive behavior is more characteristic of predatory ant species (176) than, say, of scavengers. But the ants need not kill or even catch the herbivore. They need only force it to leave the plant. Mere disruption of herbivorous activity can constitute protection.

FORAGING PATTERNS OF ANTS ON NECTARY PLANTS   Most ant foragers at nectaries are in search of sugar for adult nutrition (125, 135, 165) and protein foods to feed to the larvae (165). Since extrafloral nectar is primarily sucrose, glucose, and fructose, it is obviously an appropriate source of sugars. Although the role of amino acids in nectar has yet to be determined, it may be critical in adult ant nutrition, especially among those species restricted to liquid foods during adult life (5). (The variability in amino acid composition and concentrations among various species of nectary plants suggests the possibility of competition for ants among the plant species: Those plants providing a richer food may have a competitive advantage over those whose nectar is low in amino acids.)

Since insect prey, however, probably constitutes the major protein source for the colony (165), ants should forage on the plant at times when prey is abundant. This is exactly when the plant requires protection (assuming, of course, that at least some of the prey individuals are potential herbivores on the plant). In tropical regions, insects are present on vegetation throughout the day and night (94–96). Ants are also present at active nectaries both night and day (12, 13). Not all ant species forage during all 24 hours, of course. Most ant species have fairly distinct periods of activity (B. L. Bentley, unpublished field notes), but the combination of species on any individual plant usually results in a steady visitation frequency throughout the day and night (13).

The predatory activities of ant foragers can have a significant effect, reducing the number of herbivores attacking a plant. Negm & Hensley (124) found, for example, a correlation between the number of ants and the percent of pest larvae (*Diatraea saccaralis*) destroyed in experimental plots of sugar cane. The ants, *Solenopsis* spp., fed on the eggs, larvae, and pupae of the prey, and it appeared that the discovery of a good source stimulated further searching in the immediate area.

Vanderplank (165) found that the ant *Oecophylla longinoda* can have a significant effect on the yield from coconut plantations. When he experimentally removed *Oecophylla* from the trees, yield dropped from 61.2 nuts per tree to 18.1. Although he had no data on actual predation by the ants, he observed that palm canopies with more than ten nest units were devoid of other living insects.

Foster & Dagg (65) have some evidence which indicates that browsing giraffes in East Africa are deterred by ants on the swollen thorn *Acacia*. After a giraffe has browsed for less than two minutes, its head is covered with ants, and the giraffe moves on to another plant. Those plants which have a few or no ants are often heavily browsed.

VULNERABILITY OF THE PLANT TO HERBIVORE ATTACK   For the protectionists' position to be tenable some part(s) of the plant must be vulnerable to herbivore attack. If other anti-herbivore mechanisms, such as toxic compounds or dense hairs, are sufficient to deter all herbivores, "protection" by ants is meaningless. But as Ehrlich & Raven (57) point out, very few plants are totally immune to herbivore attack. This does not mean that extrafloral nectaries cannot function in conjunction with other mechanisms. On the contrary, extrafloral nectaries often occur on plants noted for their toxicity (e.g. *Passiflora* and *Ricinus*), thorns (e.g. *Acacia*), or coriaceous leaves (e.g. *Avicennea* and *Macaranga*).

The most vulnerable plant parts are succulent tissues rich in proteins, carbohydrates, or oils—in other words, new vegetative growth and developing buds, flowers, or fruits.

In his observations on *Macaranga*, Ridley (143) concluded that defense is only required by young leaves during the early stages of development. Unprotected young leaves, though they may not be entirely destroyed, are damaged to such an extent that "the tree has a weak and wretched appearance and does not develop to [full] size or robustness." This pattern holds true for other nectary plants as well: Young growth on *Passiflora* is preferred by *Heliconius* butterfly larvae (11); damage by European red mites is greater on young foliage of *Prunus* (138), and young cotton plants are more subject to attack by the cotton leaf perforator, *Bucculatrix turberiella* (10).

The reproductive parts of a nectary plant may also be subject to herbivore attack. The buds of *Bixa orellana* in Costa Rica, for example, can be damaged by chewing insects such as chrysomelid beetles and orthopterans. This damage can significantly reduce the proportion of buds which mature into fruits (12). Inouye & Taylor (89) found that tephritid fly larvae could consume more than 85% of the ovules and developing seeds of *Helianthella quinquenervis*, the aspen sunflower. As a final example, Bradley & Hinks (23) found that sawfly damage on Jack pine in Manitoba was lower on trees with aphids, as long as the aphids were tended by ants. In this case, the aphids might be considered analogues of extrafloral nectaries.

[In the absence of ants, nectar itself may be an important food for herbivorous pests. For example, the longevity and the fecundity of several lepidopterous pests are greater on cotton with extrafloral nectaries than on a nectariless strain (10, 27, 37, 99, 112, 120). As a result of these studies, many agriculturalists support the development of nectariless strains of cotton (141). This, of course, may be necessary if cotton continues to be treated with pesticides toxic to predatory ant species. However, as Trelease pointed out in 1879 (159), when beneficial parasites and predators are introduced into cotton fields, extrafloral nectar may be necessary for their survival.]

PRESENCE OF HERBIVORES Even if the plant is vulnerable to herbivore attack, "protection" by ants is unnecessary if herbivores are not present. Under most natural conditions, the distribution of the herbivore species closely coincides in time and space with the distribution of its host plant. Natural selection favors plants which grow outside the area or season of high herbivore frequency, but there is an equivalent selective pressure on the herbivores to "track" these evolutionary changes in their host plants (57).

The presence of herbivorous insects on plants has, of course, been extremely well documented. And among those nectary plants whose ecology is known, herbivore activity is almost always noted. Ridley (143) for example, noted tortricid and bombycid larvae on the foliage of ant-free *Macaranga hypoleuca* (143); *Heliconius* is frequently host-specific on *Passiflora* (11); and the insect pests on cotton are infamous. Other herbivores, notably mammals, may feed on nectary plants (and be deterred by the ants). Of this there is no quantitative documentation, but only anecdotal data on giraffes feeding on *Acacia* in Africa (65).

The absence of herbivores in some areas suggests the possibility of some interesting patterns in ant–plant relationships. For example, Inouye & Taylor (89) noted that *Helianthella* extends to higher elevations than its seed predators, and thus predation is less at high elevations (82% at 2734 m vs 25% at 3505 m). They present no data on the ant–plant interaction at higher elevations, but one might expect reduced protective activities above 3500 m.

The absence of a large browser community in Australia is associated with the absence of extrafloral nectaries among many *Acacia* species (a genus otherwise noted for consistent presence of nectaries). Brown (26) suggests this is not due to any differences in the ants, but to the long-continued absence of large, tree-browsing herbivores in the evolutionary history of Australia.

CONTROL OF NECTAR FLOW In the evolution of extrafloral nectaries, selection should favor those attributes which maximize benefit to the plant (e.g. protection from herbivores) and minimize the cost (e.g. nectar production). Since ants are attracted in greater abundance to nectaries producing greater quantities of nectar (see Figure 2), increased protection might be gained by increased nectar production. But this of course would increase cost: Nectar secretion requires metabolic energy, and nectar itself is composed of metabolically useful compounds. If a plant could control the time and/or location of nectar production, thus maintaining a high ant visitation rate to its vulnerable parts at times when the herbivores are active, then it could reduce the cost of protection.

The specific location of active nectaries on young foliage, on developing buds, etc, suggests that selection has favored specific genetic control of the location of extrafloral nectaries. Secretion of nectar is restricted to specific regions of the plant (143), and these regions are usually those of greatest vulnerability.

The role ants might play in this selective process may be illustrated by the floral characteristics of *Bixa orellana*, a plant widely distributed throughout the New World tropics (13). At low elevations in Costa Rica, *Bixa* has large nectaries encircling the peduncle [see Figure 1, (*a–f*)]. Ants visit these nectaries and can reduce herbivore activities on the bud. At San Vito de Java, however, *Bixa orellana* does not have extrafloral nectaries, and the bud is protected by thick, leathery sepals. San Vito is a mid-elevation tropical area where ant species are neither abundant nor diverse.

The active life of a nectary is limited in most plants. Many authors have noted that nectaries on young vegetative growth begin secretion soon after the new leaves have differentiated, then cease production when the leaves have expanded fully (42,

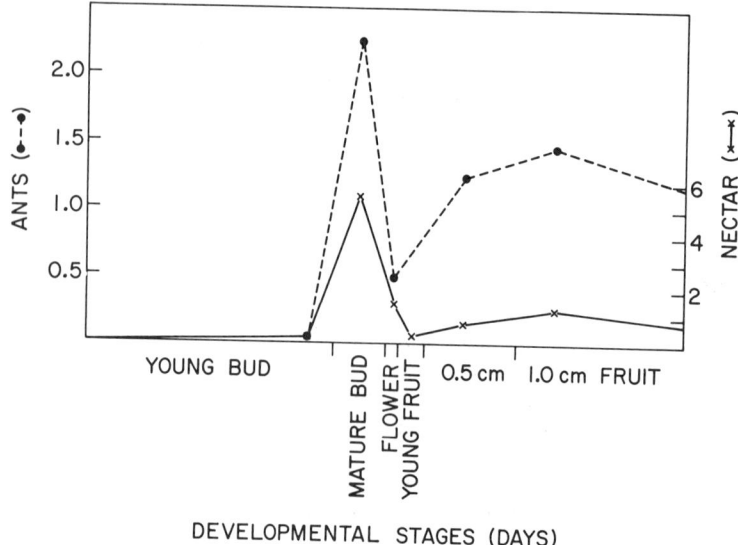

Figure 2 Ant activity and nectar flow rates from six developmental stages of the buds, flowers, and fruits of *Bixa orellana* [from Bentley (13)].

46, 47, 134). As can be seen in Figure 2, secretion of nectar in *Bixa* is greatest at the "mature bud stage," drops off when the flower is open, then again increases during the early stages of fruit maturation. Herbivores feed on the expanding buds and immature seeds, but usually do not damage the ephemeral flower (13).

Phenotypic plasticity in the secretory controls by the plant may restrict nectar flow either to times when it can "afford" nectar secretion (e.g. not during periods of physiological stress) or to times when the probability of herbivore damage is high. The "poor growing conditions" described by Rhyne (141) may be the same environmental conditions that suppress insect activity, or may simply require the plant to allocate its energy reserves to the immediate physiological demands of stress.

The loss of extrafloral nectaries in some populations suggests that there is a cost to nectar production. For example, D. B. McKey (personal communication) has recently studied two populations of *Leonardoxa africana* in Cameroon. A population growing in the Duala-Edéa Reserve in southern Cameroon has a very close relationship with an as yet unidentified ant species. In this population, 99% of the plants have at least two extrafloral nectaries per leaflet. In the Korup Reserve in northern Cameroon the ants are absent and only 0–64% of the plants have nectaries. McKey suggests that both the reduced frequency and the increased variance in the occurrence of extrafloral nectaries is the result of reduced selective pressures for nectaries in the northern population. And thus the "cost" of nectar production when unaccompanied by ants, puts plants with nectaries in the north at a selective disadvantage.

The above discussion emphasizes that our understanding of the control of nectar flow is really quite incomplete. We need more studies to determine not only the proximal physiological mechanisms controlling nectar secretion (24), but also the exact role of control in the ant-plant-herbivore interaction.

Experimental Studies Demonstrating Protection

Many authors have asserted that ants attracted to extrafloral nectar protect the plant (26, 45, 46, 50–52, 59, 60, 98, 143, 159, 160), but have failed to provide experimental evidence in support of their assertions. In the last 10 years, however, the first experimental data on protection were provided by Janzen (92, 93), who demonstrated the obligate relationship between bull's horn acacia (*Acacia cornigera*) and *Pseudomyrmex ferruginea* in Central America. The plant provides domatia (hollow thorns), a protein food (Beltian bodies), and sugar (extrafloral nectar) in exchange for protection against herbivores, competing plant species, and possibly fire. Although extrafloral nectar is exploited by the ants, the role of nectaries per se cannot be separated from the other resources offered by the plant. Thus, in 1970 I set up experiments in which I provided ordinary bean plants (*Phaseolus vulgaris*) with "nectar" (25% sugar/water). In habitats where ants were abundant, the experimental plants had significantly more ants and less herbivore damage than the control plants. In habitats where ants were rare, there was no difference between the treatment and control plants (12). In the same year I found that individual *Bixa orellana* plants which had higher ant visitation rates set more seeds than those with fewer or no ants at the nectaries (13). And, in 1973 Inouye & Taylor (89) excluded ants from *Helianthella* and found that these plants suffered significantly higher rates of pre-dispersal seed predation than control plants.

Thus we can conclude that at least some species of plants with extrafloral nectaries in both the temperate zone and the tropics are protected from herbivore damage by the ants visiting the nectaries. And the enormous body of correlative data strongly suggests that most plants with nectaries gain similar benefits from their ants.

ACKNOWLEDGMENTS

My appreciation goes to C. R. Carroll and D. B. McKey for critical reviews of the manuscript, to M. B. Wolstoff for bibliographic help, to J. Roe for the artwork, and to S. C. Carroll for diversion.

This work was supported in part by a grant from the Organization for Tropical Studies (PG 71-1), and is publication No. 198 from the Department of Ecology and Evolution, State University of New York at Stony Brook.

Literature Cited

1. Agthe, C. 1951. Über die physiologische Herkunft des Pflanzennektars. *Ber. Schweiz. Bot. Ges.* 61:240–73
2. Arbo, M. M. 1972. Structure and ontogeny of the leaf nectaries in the genus *Byttneria* (Sterculiaceae). *Darwiniana* 17:104–58
3. Aufrecht, S. 1891. *Beitrag zur Kenntnis extrafloraler Nektarien.* PhD thesis. Univ. Zurich. 44 pp.
4. Bailey, I. W. 1924. Notes on neotropical ant-plants, III, *Cordia nodosa,* Lam. *Bot. Gaz.* 77:32–49
5. Baker, H. G., Baker, I. 1973. Amino acids in nectar and their evolutionary significance. *Nature* 241:543–45
6. Bargoni, N. 1972. Synthesis of sucrose in the nectary of *Convolvulus sepium. Bull. Soc. Ital. Biol. Sper.* 48:1159–60
7. Baskin, S. I., Bliss, C. A. 1969. Sugar occurring in the extrafloral exudates of the Orchidaceae. *Phytochemistry* 8:1139–45
8. Bedford, H. W. 1923. The asal of cotton and its causes in Sudan. *Wellcome Trop. Res. Lab. Khartoum. Entomol. Sect. Bull.* 19. 38 pp.
9. Belt, T. 1874. *The Naturalist in Nicaragua.* London: Murray. 403 pp.
10. Benschoter, C. A., Leal, M. P. 1974. Relation of cotton plant nectar to longevity and reproduction of the cotton leaf-perforator in the laboratory. *J. Econ. Entomol.* 67:217–18
11. Benson, W. W., Brown, K., Gilbert, L. 1976. Coevolution of plants and herbivores: passion flower butterflies. *Evolution* 29:659–80
12. Bentley, B. L. 1976. Plants bearing extrafloral nectaries and the associated ant community: interhabitat differences in the reduction of herbivore damage. *Ecology* 54:815–20
13. Bentley, B. L. 1977. The protective function of ants visiting the extrafloral nectaries of *Bixa orellana* L. (Bixaceae). *J. Ecol.* 65:27–38
14. Berger, M.-G. 1919. *Étude organographique, anatomique, et pharmacologique de la famille des Gurneracees.* Paris: Bigot Frères. 270 pp.
15. Bhattacharyya, B., Maheshwart, J. K. 1971. Studies on extrafloral nectaries of the Leguminales I. Papilionaceae, with a discussion on the systematics of the Leguminales. *Proc. Indian Natl. Sci. Acad.* 37:11–30
16. Blatter, E. 1928. Myrmecosymbiosis in the Indo-Malayan flora. *J. Indian Bot. Soc.* 7:176–185
17. Bohmker, H. 1917. Beitrage zur Kenntnis der floralen und extrafloralen Nektarien. *Beih. Bot. Zentralbl.* 33:169–247
18. Bohmker, H. 1917. Der Bau der extrafloralen und floralen Nektarien von *Passiflora coerulea* L. *Beih. Bot. Zentralbl.* 33:247–50
19. Bonnier, G. 1879. Les nectaires. Étude critique, anatomique, et physiologique. *Ann. Sci. Natur., Bot. Ser. VI.* 8:5–212
20. Bowden, B. 1963. Studies on *Andropogon gayanus* var *bisquamulatus. Bot. J. Linn. Soc.* 58:509–19
21. Bowden, B. 1970. The sugars in the extrafloral nectar of *Andropogon gayanus* var *bisquamulatus. Phytochemistry* 9:2315–18
22. Bowden, B. 1971. Studies on *Andropogon gayanus* Knuth, VI, The leaf nectaries of *Andropogon gayanus* var *bisquamulatus* (Hochst) Hack, (Graminae). *Bot. J. Linn. Soc.* 64:77–80
23. Bradley, G. A., Hinks, J. D. 1968. Ants, aphids, and Jack pine in Manitoba. *Can. Entomol.* 100:40–50
24. Brown, H. D. 1960. Conductivity as a measure of electrolyte content in nectar. *Bot. Gaz.* 122:314–15
25. Brown, W. H. 1938. The bearing of nectaries on the phylogeny of flowering plants. *Proc. Am. Philos. Soc.* 79:549–96
26. Brown, W. L. 1960. Ants, acacias, and browsing mammals. *Ecology* 41:587–92
27. Butler, G. D. 1968. Sugar for the survival of *Lygus hesperus* on alfalfa. *J. Econ. Entomol.* 61:854–5
28. Butler, G. D., Loper, G. M., McGregor, S. E., Webster, J. L., Margolis, H. 1972. Amounts and kinds of sugars in the nectars of cotton (*Gossypium* spp.) and the time of their secretion. *Agron. J.* 64:364–68
29. Carlquist, S. 1959. The leaves of *Calycadenia* and its glandular appendages. *Am. J. Bot.* 46:70–80
30. Carne, W. M. 1913. The secretion of nectar by extrafloral glands in the genus *Acacia* (wattles). *Aust. Nat.* 2:198–99
31. Carroll, C. R. 1972. Baiting for ants from the Cerro to San Isidro. In *Tropical Biology: an Ecological Approach,* 72-2:334–36. San José, Costa Rica: Organization for Tropical Studies
32. Carroll, C. R. 1974. *The Structure of Tropical Arboreal Ant Communities.* PhD thesis. Univ. Chicago. 150 pp.

33. Chakravarty, H. L. 1948. Extrafloral glands of Cucurbitaceae. *Nature* 162:576–77
34. Chavan, A. R., Deshmukh, Y. S. 1960. The ontogeny of extrafloral nectaries in the genus *Gmelina*. *J. Indian Bot. Soc.* 39:410–14
35. Chavan, A. R., Deshmukh, Y. S. 1964. Studies in organogeny and ontogeny, IV, a study of the floral organogeny and ontogeny of extrafloral nectaries in the genus *Clerodendron*. *J. Maharaja Sayagirao Univ. Baroda* 13:27–29
36. Chavan, A. R., Deshmukh, Y. S. 1964. Studies in organogeny and ontogeny, V, a study of the floral organogeny and ontogeny of the extrafloral nectaries in *Petrea bolubilis*. *J. Maharaja Sayagirao Univ. Baroda* 13:63–67
37. Clark, E. W., Lukefahr, M. J. 1956. A partial analysis of cotton extrafloral nectar and its approximation as a nutritional medium for adult pink bollworms (*Pectinophora gossypiella*). *J. Econ. Entomol.* 49:875–76
38. Clos, D. 1868. Des glandes dans le genre *Hypericum*. *Mem. Acad. Sci. Toulouse* 6:257–66
39. Collins, M. I. 1918. On the leaf anatomy of *Scaevola crassifolia*, with special reference to the epidermal secretion. *Proc. Linn. Soc. NSW* 43:247–59
40. Cristobal, C. L. 1971. Mirmecofilia en *Byttneria* (Sterculiaceae). *Kurtziana* 6:271–74
41. Cristobal, C. L. 1972. Sobres las species de *Ayenia* (Sterculiaceae) con nectarios foliares. *Darwiniana* 16:603–12
42. Cusset, G., Schnell, R. 1963. Glandularisation et foliarisation. *Bull. Jard. Bot. L'État* 33:530–35
43. Dahlgren, K. V. O. 1940. On the secretion of sugar in *Impatiens*. *Sv. bot. Tidskr.* 34:53–55
44. Dahlgren, K. V. O. 1945. On the excretion of sugar in *Impatiens*. *Ark. Bot.* (Stockh.) 32:53–5
45. Darwin, C. 1884. *The Various Contrivances by which Orchids are Fertilized by Insects*. London: Murray. 300 pp.
46. Darwin, F. 1877. On the glandular bodies on *Acacia sphaerocephala* and *Cecropia peltata* serving as food for ants, with an appendix on the nectar-glands of the common brake fern, *Pteris aquilina*. *J. Linn. Soc. London Bot.* 15:398–409
47. Dave, Y. S., Patel, N. D. 1975. A developmental study of extrafloral nectaries in spurge (*Pedilanthus tithymaloides*, Euphorbiaceae). *Am. J. Bot.* 62:808–12

48. Davies, P. A. 1945. Leaf glands on *Ailanthus altissima*. *Trans. Ky. Acad. Sci.* 12:31–33
49. DeBach, P. 1964. *Biological Control of Insect Pests and Weeds*. New York: Reinhold. 844 pp.
50. Delpino, F. 1874–1875. Ulteriori osservazioni e considerazioni sulla dichogamia nel regno vegetale. *Soc. Ital. Sci. Nat.* 16:151–349; 17:266–407
51. Delpino, F. 1886. Raporto tra insetti e netterii estranuziale in alcune piante. *Mem. Acad. Sci. Bologna* 7:215–323
52. Delpino, F. 1886. Funzione mirmecofila nel regno vegetale. *Mem. Acad. Sci. Bologna* 8:601–650
53. Dop, P. 1927. Les glandes florales externes des Bignoniacees. *Bull. Soc. Hist. Nat. Toulouse.* 56:189–98
54. Dorsey, M. J., Weiss, E. 1920. Petiolar glands in the plum. *Bot. Gaz.* 69:391–406
55. Doutt, R. L. 1959. The biology of parasitic hymenoptera. *Ann. Rev. Entomol.* 4:161–82
56. Eaton, F. M., Ergle, D. R. 1948. Carbohydrate accumulation in the cotton plant at low moisture levels. *Plant Physiol.* 23:169–87
57. Ehrlich, P. R., Raven, P. H. 1964. Butterflies and plants: a study in coevolution. *Evolution* 18:586–608
58. Elias, T. S. 1972. Morphology and anatomy of foliar nectaries of *Pithecellobium macrodenium* (Leguminosae). *Bot. Gaz.* 133:38–42
59. Elias, T. S., Gelband, H. 1975. Nectar: its production and function in trumpet creeper. *Science* 189:289–91
60. Elias, T. S., Rozich, W. R., Newcombe, L. 1975. The foliar and floral nectaries of *Turnera ulmifolia*. *Am. J. Bot.* 62:570–76
61. Esau, K. 1965. *Plant Anatomy*. New York: Wiley. 767 pp.
62. Eyme, J. 1966. Infrastructure des constituants cellulaires des tissues excreteurs de nectaires floraux. *J. Microsc. Paris* 5:46
63. Figier, J. 1969. Incorporation de glycine-^{3}H chez les glandes petiolaire de *Mercurialis annua*. *Planta* 87:275–89
64. Figier, J. 1971. Étude infrastructurale de la stipule de *Vicia faba* L. au niveau du nectaire. *Planta* 98:31–49
65. Foster, J. B., Dagg, A. I. 1972. Notes on the biology of the giraffe. *E. Afr. Wildl. J.* 10:1–16
66. Frei, E. 1955. Die Innervierung der floralen Nektarien dikotyler Pflanzen-

familien. *Ber. Schweiz. Bot. Ges.* 65:60–114
67. Frey-Wyssling, A. 1933. Über die physiologische Bedeutung der extrafloralen Nektarien von *Hevea brasiliensis*. *Ber. Schweiz. Bot. Ges.* 42:109–122
68. Frey-Wyssling, A., Zimmermann, M., Maurizio, A. 1954. Über den enzymatischen Zuckerumbau in Nektarien. *Experimenta* 10:490–92
69. Frey-Wyssling, A. 1955. The phloem supply to the nectaries. *Acta Bot. Neerl.* 4:358–69
70. Frey-Wyssling, A., Hausermann, E. 1960. Deutung der gestaltlosen Nectarien. *Schweiz. Bot. Ges.* 70:150–162
71. Gaillochet, J. 1967. Les mouvements foliolaires chez le *Cassia fasciculata* Mich.: influence de la glande petiolaire sur le retour a l'état normal après excitation mecanique. *C.R. Seances Soc. Biol. Paris* 161:661–67
72. Gilbert, L. E., Raven, P. H., eds. 1975. *Coevolution in Animals and Plants*. Austin, Texas: Univ. Texas Press. 246 pp.
73. Gonçalves da Cunha, A. 1931. Le développement des cavites nectariferes de la feuille de *Coprosma Baueri*. *C.R. Seances Soc. Biol. Paris* 108:206–7
74. Gonçalves da Cunha, A. 1931. L'anatomie des nectaries du petiole de la feuille de *Ricinus communis* L. *C. R. Seances Soc. Biol. Paris*. 108:90
75. Gonçalves da Cunha, A. 1931. La Mechanisme de l'excretion du nectar dans les nectaries extranuptiaux de *Ricinus communis*. *C.R. Seances Soc. Biol. Paris* 108:205
76. Gonçalves da Cunha, A. 1936. Quelques observations cytologiques sur les nectaries du petiole de la feuilles de *Ricinus communis* L. *Bol. Soc. Portugesa Sci. Nat.* 112:12
77. Gonçalves da Cunha, A. 1937. Sur l'origine mitochondriale de la secretion nectarifere chez *Ricinus communis* L. *C.R. Seances Soc. Biol. Paris* 125:563–64
78. Gonçalves da Cunha, A. 1938. Études cytophysiologigues sur les nectaries du petiole de la feuille de *Ricinus communis* L. *Bol. Soc. Broterianas* 13:1–28
79. Gonçalves da Cunha, A., Sobrinho, G. 1931. Les pores de la feuille de *Coprosma*. *Bol. Soc. Portugesa Sci. Nat.* 11:131–33
80. Groner, M. G. 1937. Sugar excretion in *Impatiens sultani*. *Am. J. Bot.* 26:464–67
81. Hall, B. M. 1762. *Dissertatio Botanica sistens Nectaria florum*. Upsala: Dissertationes academica (Linné) No. 122. 16 pp.
82. Hardy, A. D. 1912. The distribution of leaf glands in some Victorian Acacias. *Vict. Nat.* 29:26–32
83. Haupt, H. 1902. Zur Sekretionsmechanik der extrafloralen Nektarien. *Flora* 90:1–41
84. Helder, R. J. 1958. The excretion of carbohydrates (nectaries). In *Handbuch der Pflanzenphysiologie*, ed. W. Ruhland, 6:978–90. Heidelburg: Springer
85. Holder, D. G., Jenkins, J. N., Maxwell, F. G. 1968. Duplicate linkages of glandless and nectariless genes in upland cotton, *Gossypium hirsutum* L. *Crop Sci.* 8:577–80
86. Howard, R. A. 1970. The ecology of an elfin forest in Puerto Rico, 10. Notes on two species of *Marcgravia*. *J. Arn. Arb.* 51:41–55
87. Ianishevskii, D. E. 1941. The extrafloral nectar-glands of *Salix*. *Tr. Bot. Inst. Akad. Nauk. SSSR, Ser. 4* 5:258–94
88. Inamdar, J. A. 1969. Structure and ontogeny of foliar nectaries and stomata in *Bignonia chamberlaynii* Sims. *Proc. Indian Natl. Acad. Sci.* 70:232–40
89. Inouye, D. W., Taylor, O. R. 1977. An experimental investigation of a plant-ant-seed predator system from a high altitude temperate region. *Ecology.* In press
90. Janda, C. 1931. Die extranuptialen Nektarien der Malvaceen. *Oesterr. Bot. Z.* 86:81–130
91. Janota, D. 1956. Erste Erfahrungen mit dem Nectarbetrag der *Robinia pseudoacacia* und *Phacelia tanacetifolia*. *Vys. Sk. Zemed. Praze Fac. Provozne Ekon. Cesk. Budejovicich Sb. Ref. Czech.* 1956:225–34
92. Janzen, D. H. 1966. The interaction of the bull's-horn acacia (*Acacia cornigera* L.) with one of its ant inhabitants (*Pseudomyrmex ferruginea* F. Smith) in eastern Mexico. *Univ. Kansas Sci. Bull.* 47:315–558
93. Janzen, D. H. 1966. Coevolution of mutualism between ants and acacias in Central America. *Evolution* 20:249–75
94. Janzen, D. H. 1973. Sweep samples of tropical foliage insects: description of study sites, with data on species abundance and size distributions. *Ecology* 54:664–87
95. Janzen, D. H. 1973. Sweep samples of tropical foliage insects: effects of seasons, vegetation types, elevation, time of

day, and insularity. *Ecology* 54:687-708
96. Janzen, D. H., Schoener, T. W. 1968. Differences in insect abundance and diversity between wetter and drier sites during a tropical dry season. *Ecology* 49:96-110
97. Jeffery, D., Arditti, J. 1969. The separation of sugars in orchid nectars by thin layer chromatography. *Bull. Am. Orchid. Soc.* 38:751-60
98. Jeffery, D. C., Arditti, J., Koopowitz, H. 1970. Sugar content in floral and extrafloral exudates of orchids: pollination, myrmecology and chemotaxonomic implications. *New Phytol.* 69:187-95
99. Jenkins, J. N., Maxwell, F. G., Parrott, W. L. 1967. Field evaluation of glanded and glandless cotton (*Gossypium hirsutum*) lines for boll weevil susceptibility. *Crop Sci.* 7:437-40
100. Kerner, A. 1877. Das extraflorale Nectarium bei *Batatas edulis. Bot. Z.* 27:780
101. Killip, E. P. 1938. The American species of Passifloraceae. *Field Mus. Nat. Hist. Publ. Bot. Ser.* 19:Parts I & II
102. Klug, J. 1926. Über die Sekretdrusen bei den Labiaten und Compositem. PhD thesis. Univ. Frankfurt. 74 pp.
103. Koernicke, M. 1918. Über die extrafloralen Nektarien auf den Laubblättern einiger Hibisceen. *Flora* 111/12:526-40
104. Lawton, J. H. 1976. The structure of the arthropod community on bracken. *Bot. J. Linn. Soc.* 73:187-216
105. Laychock, S. 1971. *The extrafloral nectaries of a Cymbidium hybrid.* Senior thesis. Brooklyn Coll., NY. 21 pp.
106. Leius, K. 1960. Attractiveness of different foods and flowers to the adults of some hymenopterous parasites. *Can. Entomol.* 92:369-76
107. Leius, K. 1967. Influence of wild flowers on parasitism of tent caterpillar and codling moth. *Can. Entomol.* 99:444-46
108. Lewton, F. L. 1925. The value of certain anatomical characters in classifying the Hybisceae. *J. Wash. Acad. Sci.* 15:165-72
109. Lloyd, F. E. 1901. The extranuptial nectaries in the common brake, *Pteridium aquilinum. Science* 13:885
110. Lloyd, F. E., Ridgway, C. S. 1912. The behavior of the nectar glands in the cacti with a note on the development of the trichomes and areolar corl. *Plant World* 15:145-56
111. Lubbock, J. 1878. *On certain relations between Plants and Insects: a lecture.* London & Glasgow: Longman. 24 pp.
112. Lukefahr, M. J. 1960. Effects of nectariless cottons on populations of three lepidopterous insects. *J. Econ. Entomol.* 53:242-44
113. Lukefahr, M. J., Griffin, J. A. 1956. The effects of food on the longevity and fecundity of pink bollworm moths. *J. Econ. Entomol.* 49:876
114. Lüttge, U. 1971. Structure and function of plant glands. *Ann. Rev. Plant Physiol.* 22:23-44
115. Maheshwari, J. K. 1954. The structure and development of extrafloral nectaries in *Duranta plumieri* Jacq. *Phytomorphology* 4:208-11
116. Maheshwari, P. 1931. Contributions to the morphology of *Albizzia lebbek. J. Indian Bot. Soc.* 10:241-64
117. Martinet, J. 1872. Organes de secretion des vegetaux. *Ann. Sci. Nat. Ser. 5.* 14:91-232
118. Masters, M. T. 1870. Contributions to the natural history of the Passifloraceae. *Trans. Linn. Soc. London* 27:593-646
119. Mattei, G. E. 1925. Una *Erythrina* a necttari estranuziali dell' Eritrea. *Boll. Soc. Sci. Nat. Econ. Palermo* 7:19-21
120. Meredith, W. R., Ranney, C. D., Laster, M. L., Bridge, R. R. 1973. Agronomic potential of nectariless cotton. *J. Environ. Qual.* 2:141-44
121. Metcalf, C. R. 1938. Extrafloral nectaries in *Osmanthus* leaves. *Kew Bull.* 6:254-56
122. Meyer, C. R., Meyer, V. G. 1961. Origin and inheritance in nectariless cotton. *Crop Sci.* 1:167-70
123. Mound, L. A. 1962. Extrafloral nectaries of cotton and their secretions. *Emp. Cotton Grow. Rev.* 39:254-61
124. Negm, A. A., Hensley, S. D. 1969. Evaluation of certain biological control agents of the sugarcane borer in Louisiana. *J. Econ. Entomol.* 62:1008-16
125. Nishida, T. 1958. Extrafloral glandular secretion as a food source for certain insects. *Proc. Hawaii Entomol. Soc.* 16:379-86
126. Ono, K. 1907. Studies on some extranuptial nectaries. *J. Coll. Sci. Imp. Univ. Tokyo* 23:1-28
127. Orr, M. 1926. On the secretory organs of the Dioscoreaceae. *Notes Royal Bot. Gard. Edinburgh* 15:133-46
128. Owen, D. F. 1971. *Tropical Butterflies, the ecology and behaviour of butterflies in the tropics with special reference to African species.* Oxford: Clarendon. 214 pp.

129. Parija, P., Samal, K. 1936. Extra-floral nectaries in *Tecoma capensis* Lindl. *J. Indian Bot. Soc.* 15:241–46
130. Patel, R. C., Inamdar, J. A. 1972. Studies in the trichomes and nectaries of some Gentianales. *Biol. Land Plants* 1972:328–40
131. Patel, R. C., Inamdar, J. A. 1973. Studies on nectaries in the Oleaceae. *Biol. Land Plants* 1973:271–77
132. Petaj, V. 1916. Die extrafloralen Nektarien auf den Blättern von *Ailathus glandulosa*. *Rad. Jugosl. Akad. Znan. Kuj.* 215:59–81
133. Poulsen, M. 1879. Det extraflorale nectarium hos *Caparis cynophallophorus*. Kjobenhavn: Vidensk. Medd. Nat. Hist. For. Kjobenhavn
134. Putman, W. L. 1958. Mortality of the European red mite (*Acarina*, Tetranychidae) from secretion of peach leaf nectaries. *Can. Entomol.* 90:720–21
135. Putman, W. L. 1963. Nectar of peach leaf glands as insect food. *Can. Entomol.* 95:108–9
136. Rachmilevitz, T., Fahn, A. 1973. Ultrastructure of nectaries of *Vinca rosea* L., *V. major* L. and *Citrus sinensis*, Osbeckev. Valencia and its relation to the mechanism of nectar secretion. *Ann. Bot. London* 37:1–19
137. Rao, V. S. 1926. A short note on the extrafloral nectaries in *Spathodea stipulata*. *J. Indian Bot. Soc.* 5:113–16
138. Rao, V. S. 1963. The epigynous glands of Zingiberaceae. *New Phytol.* 62:342–49
139. Reed, E. L. 1917. Leaf nectaries of *Gossypium*. *Bot. Gaz.* 63:229–31
140. Reed, E. L. 1923. Extra-floral nectar glands of *Ricinus communis*. *Bot. Gaz.* 76:102–106
141. Rhyne, C. L. 1965. Inheritance of extrafloral nectaries in cotton. *Advan. Front. Plant Sci.* 13:121–35
142. Rhyne, C. L. 1972. Linkage of indehiscent anthers and lack of leaf nectaries in *Gossypium hirsutum* L. *Cotton Grow. Rev.* 49:57–60
143. Ridley, H. N. 1910. Symbiosis of ants and plants. *Ann. Bot. London* 24:457–83
144. Rudd, V. 1964. Nectaries in some Leguminosae. *Aust. Exp. Stn. Bull.* 3:17–21
145. Salisbury, E. J. 1909. The extra-floral nectaries of the genus *Polygonum*. *Ann. Bot. London* 23:229–41
146. Schimper, A. F. W. 1888. *Die Wechselbeziehungen zwischen Pflanzen und Ameisen im tropischen Amerika*. Botanische Mitteilungen aus den Tropen, pp. 1–94. Jena: Fischer
147. Schnell, R., Cusset, G. 1963. Glandularisation et Foliarisation. *Bull. Jard. Bot. Brux.* 33:525–30
148. Schnell, R., Cusset, G., Quenum, M. 1963. Contribution a l'étude des glandes extra-florales chez quelques groupes de plantes tropicales. *Rev. Gen. Bot.* 70:269–342
149. Schremmer, F. 1970. Extranuptiale Nektarien: Beobachtungen an *Salix eleagnos* und *Pteridium aquilinum*. *Oesterr. Bot. Z.* 117:205–22
150. Seibert, R. J. 1948. The use of glands in a taxonomic consideration of the family Bignoniaceae. *Ann. Mo. Bot. Gard.* 35:123–37
151. Shuel, R. W. 1954. Nectar secretion in relation to nitrogen supply, nutritional status, and growth of the plant. *Can. J. Agric. Sci.* 35:124–38
152. Small, J. K. 1910. Malpighiaceae. *North Am. Flora* 25:117–71
153. Smith, W. 1903. *Macaranga triloba*, a new myrmecophilous plant. *New Phytol.* 2:79–82
154. Soysa, S. W. 1940. Orchids and ants. *Orchidol. Zeylan.* 7:88–91
155. Stahl, E. 1920. Zur Physiologie und Biologie der Exkrete. *Flora* 113:1–132
156. Stocking, C. R. 1956. Excretion by glandular organs. In *Handbuch der Pflanzenphysiologie*, ed. W. Ruhland, 3:489–502. Heidelberg: Springer
157. Stone, W. E. 1892. The chemical composition of the nectar of *Poinsettia*. *Bot. Gaz.* 17:192–93
158. Townes, H. 1958. Some biological characteristics of the Ichneumonidae (Hymenoptera) in relation to biological control. *J. Econ. Entomol.* 51:650–52
159. Trelease, W. 1879. Nectar and its uses. In Comstock, J. H., *Report upon Cotton Insects*, Part 3:317–43. Washington DC: USDA
160. Trelease, W. 1881. The foliar nectary glands of *Populus*. *Bot. Gaz.* 6:284–90
161. Trelease, W. 1920. Physiology of nectar secretion. *Am. Bee J.* 60:7–9
162. Tyler, F. J. 1908. The nectaries of cotton. *USDA Bur. Plant Ind. Bull.* 131:45–56
163. Uxkull-Guldenbrandt, Nieuwenhuis von. 1907. Extrafloralen Zuckerabscheidungen und Ameisenschutz. *Ann. Gard. Bot. Buitenzorg.* 21:195–328
164. Valbusa, U. 1936. Sui Nettari extranuziale della *Impatiens sultani*

Hook. *Nuovo G. Bot. Ital.* (NS) 43: 754–56
165. Vanderplank, F. L. 1960. The bionomics and ecology of the red tree ant *Oecophylla* and its relationship to the coconut bug. *J. Anim. Ecol.* 29:15–33
166. Vansell, G. H. 1940. Nectar secretion in *Poinsettia* blossoms. *J. Econ. Entomol.* 33:409–10
167. Vansell, G. H. 1944. Cotton nectar in relation to bee activity and honey production. *J. Econ. Entomol.* 37:528–30
168. Vasil'ev, A. E. 1969. Submikroskopichoskaya morfologiya kletok nektarnikow. *Bot. Z. Leningrad* 54:1015–31
169. Watari, S. 1959. Anatomical studies on some leguminous leaves with special references to the vascular system in petioles and rachises. *Tokyo Imp. Univ. Fac. Sci. J.* 21:7–16
170. Way, M. J. 1963. Mutualism between ants and honeydew-producing Homoptera. *Ann. Rev. Entomol.* 8:307–44
171. Weber, F. 1951. *Impatiens*-Nektar. *Phyton* 3:110–11
172. Weingart, W. 1920. Extranuptiale Nektarien an einen *Phyllo-Cactus. Kakteen. Berlin* 30:136–38
172a. Wergin, W. P., Elmore, C. D., Hanny, B. W., Ingber, B. F. 1975. Ultrastructure of the subglandular cells from the foliar nectaries of cotton in relation to the distribution of plasmodesmata and the symplastic transport of nectar. *Am. J. Bot.* 62:842–49
173. Wheeler, W. M. 1910. *Ants, Their Structure, Development and Behavior*. New York: Columbia Univ. Press. 663 pp.
174. Wheeler, W. M. 1942. Studies of Neotropical ant-plants and their ants. *Bull. Mus. Comp. Zool. Harvard. Univ.* 90:1–262
175. Wilde, W. J. J. O. de. 1972. The indigenous Old World *Passiflora. Blumea* 20:227–50
176. Wilson, E. O. 1959. Some ecological characteristics of ants in New Guinea rain forests. *Ecology* 40:437–47
177. Wilson, E. O. 1971. *The Insect Societies*. Cambridge, Mass: Belknap. 548 pp.
178. Woodcock, E. F., Tullis, E. C. 1928. Extrafloral glands of *Malus malus* and *Pyrus communis. Pap. Mich. Acad. Sci. Arts Lett.* 8:239–43
179. Zimmerman, J. 1932. Über die extrafloralen Nektarien der Angiospermen. *Beih. Bot. Zentralbl.* 49:99–196

RELATIVE BRAIN SIZE AND BEHAVIOR IN ARCHOSAURIAN REPTILES

❖4132

James A. Hopson
Department of Anatomy, University of Chicago, Chicago, Illinois 60637

INTRODUCTION

The dominant terrestrial vertebrates of Mesozoic time, 225 to 65 million years ago, were reptiles of the subclass Archosauria, a name appropriately meaning "ruling reptiles." During Triassic time, the order Thecodontia, the basal group of archosaurs, radiated to produce four orders: the Saurischia and Ornithischia, which together we call dinosaurs; the Pterosauria, flyers and gliders with wings of skin, each stretched across a single elongated finger; and the Crocodylia, the familiar crocodiles that have been the dominant freshwater predators for their 200 million years of existence. Also part of this great radiation are the birds—offshoots of small carnivorous dinosaurs of the Late Jurassic (97), but so modified from their reptilian forebears that they have been removed to their own class, the Aves.

The pterosaurs, like the birds, possessed an insulating cover; it consisted of hair-like structures (121) rather than feathers. Thus the pterosaurs were probably endothermic, maintaining a high body temperature using metabolically produced heat. The other Mesozoic archosaurs lacked a surface insulation and have been thought to be ectothermic, using the sun and other external heat sources to warm their bodies, as crocodilians and all other reptiles must do today.

The traditional image of dinosaurs (28, 114) is of sluggish creatures with tiny brains capable of little more than reflexive responses to environmental stimuli. Such a low order of behavioral complexity would have been sufficient for meeting the modest demands imposed by low, reptilian metabolic rates. Although according to this view they were ectothermic, dinosaurs maintained relatively constant body temperatures due to the thermal inertia of their great bulk.

This concept was challenged by L. S. Russell (112), who pointed out that the skeletons of dinosaurs were more like those of birds than those of crocodiles (which appear typically reptilian), suggesting that dinosaurs were more like birds in soft anatomy and physiology. The theory of dinosaur endothermy has been more extensively developed by Ostrom (94), Ricqlès (105–109), and especially Bakker (5, 7–11). Dinosaurs, they argue, had levels of metabolism and heat production as high as those of living mammals and birds and, like these modern endotherms, were capable of sustaining high levels of activity for long periods. Their theory is based on three main arguments: (*a*) The similarity of dinosaur limb proportions and joint morphology to these features in mammals and birds indicates comparably high running speeds and therefore comparably high levels of aerobic exercise metabolism (5, 7–12). (*b*) Similar ratios of predator to prey biomass in dinosaurian and mammalian communities indicate comparable energy requirements for the predators (8–10). (*c*) Similar bone histology in dinosaurs and living endotherms indicates rapid, endothermic-level, growth rates (105–109). Counterarguments (17, 27, 41, 44, 45, 84, 113, 116, 117) and further statements in support (9, 33, 34, 109) have been made.

The traditional idea that dinosaurs possessed brains capable of only the simplest reflexive behaviors has been challenged on several fronts. Jerison (72, 73) showed that once the negatively allometric relationship of brain to body weight is taken into account, dinosaurs had the proper amount of brain for reptiles of their huge size. The bizarre horns, crests, and other "ornaments" of dinosaurs have been interpreted by Davitashvili (32) as weapons and display organs for use in intraspecific social interactions. More recent studies of such structures suggest behaviors similar to those seen in living reptiles. New studies of the behavior of living crocodilians indicate that the behavior of modern archosaurian reptiles can be quite complex.

In this review I reexamine the evidence on relative brain size in dinosaurs, pterosaurs, and *Archaeopteryx* (the earliest known bird), and review the social behavior of living crocodilians and of dinosaurs (as inferred from the study of footprints, fossil eggs, and skeletal morphology). In the final part I examine the implications of archosaur brains and behavior for the theory of dinosaur endothermy.

THE BRAINS OF FOSSIL ARCHOSAURS

Introduction

The external morphology and relative size of the brain in extinct vertebrates can be studied by means of endocasts—casts of the cranial cavity of the skull. Endocasts are known for all orders of archosaurs (67). Because brain size scales allometrically to body size, comparative statements about brain size must take body size into account. The empirically determined relationship between the size (weight or volume) of the brain and that of the body in vertebrates is represented by the power function:

$$E = k\, P^{0.67}, \qquad 1.$$

where E is brain size, P is body size, and k is a constant equal to the y intercept on a log–log plot (72, 73). The higher the value of k, the larger the brain at any given body size. Plots of brain size against body size on a log–log scale for a variety of living vertebrates yield two parallel clusters of points with principal axes showing slopes of approximately 0.67, but with y intercepts showing an approximate ten-fold difference between the lower clusters (bony fishes, amphibians, reptiles) and the upper one (birds, mammals) (73). For ease of comparison the data points representing a taxonomic group may be enclosed in minimum convex polygons (73; Figure 1).

Comparison of relative brain size (level of encephalization) among members of a group may be made by comparing the measured brain size of an animal with the "expected" or "predicted" brain size for a reference animal of its body size. Jerison (73) calculates an "Encephalization Quotient" (EQ) for a species from the equation $EQ_i = E_i/E_e$, where E_i is the measured brain size and E_e is the expected brain size determined from equation 1. EQ values below 1.0 indicate a relatively small-brained animal, above 1.0 a large-brained animal. Jerison (73) uses a k-value of .007 for determining expected brain sizes of lower vertebrates from equation 1. In order to make comparisons within the Archosauria, I use a k-value of .005, which is the y intercept of the axis determined by passing a line of slope 2/3 through the approximate center of a cluster of points for three living crocodilians (Figure 1).

The volume of fossil endocasts can be determined by water displacement or graphic double integration (72, 73). However, because the brain of lower vertebrates fills the cranial cavity to a varying degree, it is uncertain what fraction of the endocast volume represents the volume of the brain in a given case. Jerison (72, 73) uses a value of 50% for dinosaurs but considers it likely that the fraction of the cranial cavity occupied by the brain was a negative function of body size (73). This means that with increasing body size the brain would fill a decreasing fraction of the cranial cavity. My studies (66, 67) of the endocasts of living and fossil reptiles suggest that the relationship of brain to endocast volume as a function of body size is not regular across a broad taxonomic spectrum. The endocasts of dinosaurs of widely varying body sizes reflect details of brain morphology (i.e. are "brainlike") to varying degrees independent of absolute size. It is doubtful that the entire brain ever occupies less than 60% of the cranial cavity in larger crocodilians (66). Because their endocasts are usually as "brainlike" as those of large crocodilians, I consider this to be a reasonable (though undemonstrable) minimum limit for fossil archosaurs as well. In the analyses that follow, however, I use Jerison's estimate of 50%, except where noted.

Estimates of body size (volume or weight) in fossil vertebrates are usually determined from scale models (29) or from equations, empirically derived from living species, relating body weight to body length (73, 104). In this paper I use Colbert's (29) weight estimates for most of the dinosaurs, Bakker's (10) for *Euoplocephalus,* and Russell's (110) for *Stenonychosaurus.* That of the stegosaur *Kentrosaurus* was extrapolated from Colbert's weight for the larger *Stegosaurus.* The weight for *Iguanodon* is from Hopson (67). The body weights for pterosaurs are from Jerison (73) and that for *Archaeopteryx* is from Hopson (in preparation).

Dinosaurs

It has become a truism that dinosaurs had among the smallest brains for their body size of any vertebrates and that their behavior consisted of relatively simple, rigidly stereotyped actions (28, 114). Jerison (72, 73), however, has analyzed the relationship between brain size and body size in dinosaurs and has shown that when size-related allometry is taken into account, the brains of dinosaurs fall within the expected range for reptiles of their body size.

Because the aim of Jerison's study of dinosaur brains was to test whether they conformed to the pattern he had determined for other "lower" vertebrates, he did not attempt to compare relative brain size *among* dinosaurs. In order to investigate the patterns of variation in brain size in the subgroups of dinosaurs I have extended Jerison's analysis by adding several genera to his sample, and have modified some of his data on the basis of new information (67). I follow Jerison in using half the volume of the endocast to represent brain volume, except for the sauropods. The heads of these enormous animals are relatively small and the endocast indicates that the brain was anteroposteriorly compressed and transversely widened, perhaps as a consequence of "packaging" of the brain in a restricted space. The shape of the endocast suggests that the brain was closely confined within the walls of the endocranial cavity; therefore, I have used the entire endocast to represent brain size. Brain size for the coelurosaur *Stenonychosaurus* is based on a reconstruction of the entire brain from a partial endocast, the unrepresented portion of the brain being modeled on that of *Caiman,* with the braincase of the related coelurosaur *Saurornithoides* (13) providing control on overall proportions. The endocast shows clear evidence that the forebrain filled the cranial cavity (110).

The results of my analysis are displayed in Figure 1. All suborders of saurischians and ornithischians are represented. The upper polygons enclose all points for living birds and mammals, the lower ones for living reptiles (heavy lines) and extinct reptiles (light lines). All of the polygons have been modified from those published by Jerison by the addition of new data (99, 101). The line of slope 2/3 and y intercept of .005 is taken to represent the expected brain size of a "typical" archosaur ($EQ = 1.0$) against which all species are compared. Note that the brain size of *Stenonychosaurus* falls within the size range for birds, as was independently determined by Russell (110). I have used Russell's body weight estimate of 45 kg in determining its relative brain size; Bakker (10) gives a weight of about 10 kg for the same individual, which would yield a much higher EQ.

In order to compare relative brain sizes within the dinosaurs, EQs for all genera have been determined and are displayed by suborder (by infraorder for the Theropoda) in Figure 2. Note that except for the lowest and the highest groups, EQ shows a poor correlation with body size; therefore, low EQs are not attributable merely to large body size. I shall argue below for a close correlation between EQ and certain behavioral parameters (as determined from functional interpretations of skeletal structure), as Platel (101) has done for lizards.

In the groups of dinosaurs represented in Figure 2, both EQ and inferred speed and degree of agility during locomotion increase from lower left to upper right. Sauropods are graviportally adapted quadrupeds comparable in limb proportions to elephants but generally of much larger size. It is reasonable to suppose that their great body weights imposed on them a pattern of slow locomotor speeds (1, 2). Despite the fact that the entire endocast volume is used to represent brain size,

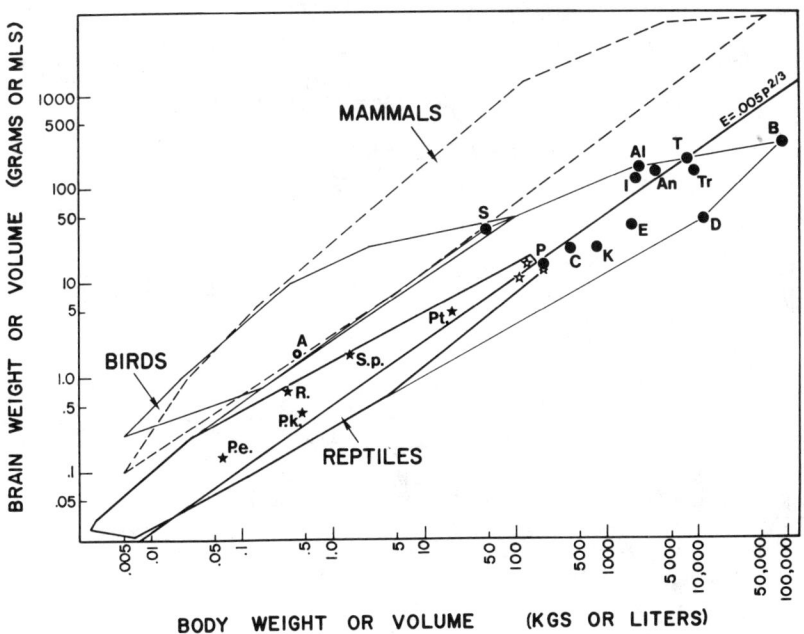

Figure 1 Brain:body size relations in archosaurs superimposed on minimum convex polygons of living reptiles, birds, and mammals. Scale is log-log. Line with 2/3 slope passes through approximate center of cluster of points for living crocodilians (open stars).

PTEROSAURS (solid stars): P.e., *Pterodactylus elegans*; P.k., *Pterodactylus kochi*; Pt., *Pteranodon* sp.; R., *Rhamphorhynchus*; S.p., *Scaphognathus purdoni.*

DINOSAURS (closed circles): Carnosaurs: Al, *Allosaurus*; T, *Tyrannosaurus.* Coelurosaur: S, *Stenonychosaurus.* Sauropods: B, *Brachiosaurus*; D, *Diplodocus.* Ornithopods: An, *Anatosaurus*, C, *Camptosaurus*, I, *Iguanodon.* Ankylosaur: E, *Euoplocephalus.* Stegosaur: K, *Kentrosaurus.* Ceratopsians: P, *Protoceratops*; T, *Triceratops.*

BIRD (open circle): A, *Archaeopteryx.*

sauropods had unusually small brains. Brain size may be influenced by the adaptive requirements of head size (as in large- and small-headed sea snakes; P. S. Ulinski, personal communication); therefore, I hesitate to draw further conclusions about additional behavioral correlates of the very low *EQ*s of sauropods.

The armored ankylosaurs and plated stegosaurs share several features that may be correlated with their low (0.52–0.56) *EQ*s. Both are quadrupeds whose limb structure suggests a lack of agility and speed and whose tails served as important defensive weapons. The armor of ankylosaurs was an adaptation for passive defense and the plates of stegosaurs may have been utilized in threat displays (32). Neither speed nor extreme wariness would seem to have been essential for the avoidance of predators in either group. An interesting parallel is seen in those acanthopterygian

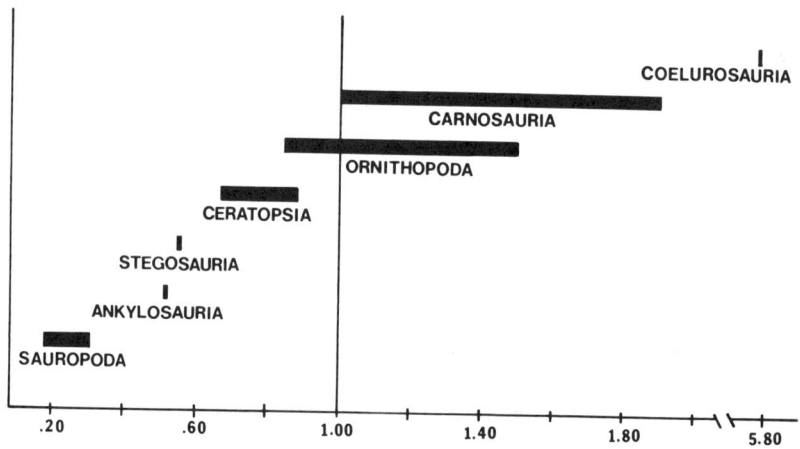

Figure 2 Encephalization Quotients (EQs) for the suborders of dinosaurs (infraorders Carnosauria and Coelurosauria for the suborder Theropoda) calculated with reference to line with the equation $E = .005\ P^{2/3}$ in Figure 1. Living crocodilians lie at 1.00. Note the break in the scale between 2.00 and 5.80.

fishes that rely on passive defenses (spines, poisons, camouflage) against predators and that have low EQs relative to related fishes that depend upon evasive action to escape from predators (14).

The larger ceratopsians, with their great horned heads, relied on active defensive strategies and presumably required somewhat greater agility than the tail-weaponed forms, both in fending off predators and in intraspecific combat bouts (42). The smaller ceratopsians, lacking true horns, would have relied on sensory acuity and speed to escape from predators. *Protoceratops*, possibly a facultative biped, had a higher EQ than the obligate quadruped *Triceratops*.

Locomotion in ornithopod ornithischians and theropod saurischians was normally bipedal, although ornithopods could walk and stand quadrupedally. Among the ornithopods, the EQ of *Camptosaurus*, an early, medium-sized form, is below 1.0, close to that of *Protoceratops*. The EQs of the much larger *Iguanodon* (1.5) and of the hadrosaur *Anatosaurus* (1.3) fall within the range of the carnosaurian theropods. Unlike other ornithischians, ornithopods lacked special defensive weapons for use against predators; they presumably relied on acute senses (91) and relatively fast speeds (46) to evade the large carnosaurs. The interpretation of the cranial crests of hadrosaurs as visual and acoustic display structures (35, 64) also suggests the existence of complex intraspecific behaviors in this group.

Theropods were the only carnivores among the dinosaurs, and they were also obligate bipeds. They possessed the relatively largest brains among the dinosaurs, although large ornithopods fall within the theropod range. Active predators tend to have relatively large brains in lizards (100, 101), fishes (15), and mammals (73). Theropods also differ from large ornithopods in possessing grasping forelimbs. The

BRAIN SIZE AND BEHAVIOR IN ARCHOSAURS   435

smaller theropods, coelurosaurs of most classifications (97), had by far the relatively largest brains among reptiles. They were lightly built and undoubtedly extremely agile (93, 110). Russell (110) notes that the presence of grasping hands in *Stenonychosaurus* was associated with orbits that are directed anterolaterally, indicating the possession of a large binocular visual field; he suggests that such coelurosaurs utilized their forelimbs in the capture of small active prey such as mammals. Relatively large endocasts or cranial cavities have been noted in other coelurosaurian theropods, e.g. in ornithomimids (111), *Oviraptor* (89), and *Dromaeosaurus* (30). Therefore, the extraordinary high *EQ* of *Stenonychosaurus* probably reflects the general condition among advanced coelurosaurs.

The pattern of association of *EQ* with locomotor habits in dinosaurs parallels in a general way that described in lizards by Platel (101). Also of significance in relation to *EQ* in dinosaurs seems to be carnivory vs herbivory, increased use of the forelimbs as grasping organs, and possibly increased sociality.

Pterosaurs

The endocast of pterosaurs is extraordinarily brainlike in details of its morphology, indicating that the brain essentially filled the cranial cavity (38, 73). Like that of living birds, the pterosaur brain possessed reduced olfactory bulbs and very large optic lobes, which were displaced laterally (away from the dorsal midline) by an enlarged cerebellum (38, 67). The floccular lobe of the cerebellum was greatly enlarged. These features indicate that, as in modern birds, olfaction was reduced in importance, vision was emphasized, and the equilibrium sense and motor coordination were increased. Despite the reflection of a birdlike habitus, the pterosaur brain was still reptilian in relative size (73), as indicated in Figure 1. The brainlike appearance of the endocast may have been a consequence of weight-reduction specialization of the skull, so that the cranial bones became very thin and closely fitted to the brain (73).

The Earliest Bird, Archaeopteryx

The British Museum specimen of *Archaeopteryx* includes an impression of the right cerebral hemisphere and optic lobe which indicates that the brain filled the cranial cavity. Early studies emphasized the reptilian nature of the brain (37), but Jerison (71, 73) has determined that when the correct midline of the cast is identified, the width of the brain is revealed to be greater than previously thought. He determined a brain volume of 0.92 ml and a maximum possible body weight of 500 g, yielding a relative brain size for *Archaeopteryx* that lies above the reptile polygon but still well below that for modern birds. With the addition of new reptile data (101), Jerison's estimate of relative brain size for *Archaeopteryx* would now fall within the expanded polygon for living reptiles (Figure 1).

Aided by the figures and description of the recently described Eichstatt specimen of *Archaeopteryx* (120), I have made a new determination of brain size in the British Museum *Archaeopteryx* specimen (Hopson, in preparation). My volume determination is 1.76 ml, nearly twice Jerison's value. I have determined a *maximum* body weight of 400 g for this specimen, which value is used in Figure 1 (although I believe the true weight was closer to 300 g). The revised estimates of brain and body size in the British Museum *Archaeopteryx* give a relative brain size that falls within the lower limits of the bird and mammal polygons in Figure 1. Thus, the brain of *Archaeopteryx* lay well outside the range for living reptiles, although it is comparable in relative size to that of the coelurosaurian theropod *Stenonychosaurus*.

When compared with the brains of modern birds, that of *Archaeopteryx* had a relatively unexpanded forebrain and its cerebellum was small (120). The optic lobes were not displaced laterally from the dorsal midline, as in pterosaurs and modern birds, and were wider than the forebrain in both known endocasts (120; Hopson, in preparation). The smallness of its cerebellum indicates that the brain of *Archaeopteryx* was less adapted to the aerial niche than was that of pterosaurs.

The nature of the adaptive zone in which avian brain enlargement occurred is uncertain. Jerison (73) assumes, as do I, that *Archaeopteryx* was capable of powered flight over short distances (19, 26, 123, 124) and that it was at least partially arboreal. However, because pterosaurs did not undergo comparable increase in brain size, Jerison doubts that avian brain enlargement was due to the selective pressures of the aerial adaptive zone. He suggests instead that Mesozoic birds inhabited woodlands (unlike pterosaurs, which, he suggests, were associated with open expanses of water). In woodlands, the confusingly variegated background provided strong selection pressures for enlargement of the brain as a processing center for visual information. Such an enlargement occurred much later in primates.

I believe another hypothesis is suggested by the evidence of a close phylogenetic link between *Archaeopteryx* and large-brained coelurosaurs (97). Ostrom (96) suggests that birds originated as ground-living, cursorial predators that, like coelurosaurs, used their forelimbs as well as their jaws in the capture of small, active prey. In such a niche, selection would be expected to increase the coordination of rapid but controlled movement of the hands with visual information about small, swiftly moving prey; it would therefore act largely on the visual centers of the brain, as in Jerison's hypothesis. However, the enlargement of the brain into the lower end of the bird range probably occurred in the ground-living coelurosaurian ancestor of birds, long before the evolution of *Archaeopteryx* and of flight.

THE BEHAVIOR OF ARCHOSAURIAN REPTILES

The Behavior of Living Crocodilians

The brains of most extinct archosaurs fell within the relative size range of living reptiles. As the only surviving reptilian-level archosaurs, the Crocodylia are the best living analogs on which to base inferences about the possible level of behavioral complexity in Mesozoic archosaurs. Their principal contribution, I believe, will be to caution us against underestimating the capacities of the reptilian-level nervous systems of dinosaurs and other extinct archosaurs (65).

The complexity of crocodilian behavior has only lately become apparent as the result of long-term observations in the wild and under semi-captive conditions. It is now clear that several species show territorial behavior, dominance hierarchies, elaborate courtship behaviors, cooperative food-getting behaviors, and intensive parental care of the eggs and young. Mediating many aspects of their social interactions is an extensive repertoire of vocalizations (25, 50, 61, 63, 82). Unfortunately, only a few species have been studied intensively, principally the Nile crocodile, *Crocodylus niloticus* (31, 59, 86, 103); the American crocodile, *C. acutus* (77, 78); and the American alligator, *Alligator mississippiensis* (51, 52, 87). Except where noted, most of what follows is based on these species.

During the breeding season, males of the Nile and American crocodiles and of the American alligator become territorial and establish a dominance hierarchy. The dominant males patrol their territories, excluding subdominant males but allowing females to move freely through them. Fighting occurs among male crocodiles (31, 86); however, Pooley & Gans (103) and Lang (77) describe ritualized fighting in *C. niloticus* and *C. acutus* in which stereotyped noninjurious biting behaviors are observed. Subdominants in both species show a submissive snout-lifting behavior (77, 86), which serves to expose the throat to the dominant (103). In American alligators, fighting appears to be more aggressively pursued (51), and the submissive snout-lifting gesture has not been observed (78).

Dominance has been described in female alligators (51); dominant females exclude other females from their territory, which includes a nesting site. Dominant males of all three species prevent subdominant males from engaging in courtship displays (51, 77, 86). Modha (86) saw courtship displays only in the territorial dominant males of *C. niloticus,* and Lang (77) states that females of *C. acutus* court primarily with territorial males. In all species, the territorial males elicit the approach of females by slapping the head against the surface of the water (51, 77, 86); alligators also bellow, as do spectacled caimans (3), but the two species of *Crocodylus* are much less vocal (77, 86).

Courtship is a mutual activity in which both sexes go through elaborate ritualized behaviors that may last for hours and may be repeated over several days before copulation occurs. Both sexes appear capable of initiating courtship (52, 77). The Nile crocodile appears to be monogamous, at least during the sexually active part of the annual breeding cycle (103). The American alligator, on the other hand, appears to be polygamous (51).

The female of all species constructs a nest, which may be a hole in sand or a mound of vegetation (24, 57). The number of eggs is dependent on body size, varying from about 15 to 80 or 90 in *C. niloticus* and *A. mississippiensis* (58, 103). The female of most species remains near the nest and guards it against predators. In the Nile crocodile, she lies by the nest for 84–90 days, apparently without feeding (103); Cott (31) reports that during much of this time she is in a comatose state. The male remains nearby as well. In a pair of captive *Caiman crocodilus* both the male and the female guarded the nest (3).

When the young are ready to hatch they call from within the egg. Lee (82) provides evidence suggesting that vocal communication in American alligators serves to synchronize hatching. The nearby female responds to the calls of the unhatched young and digs open the nest. This behavior has been observed in *C. niloticus* (86, 103), *C. acutus* (77), *C. moreleti* (68), *A. mississippiensis* (62), and *Caiman crocodilus* (in which a male opened the nest) (3). Opening of the nest by the parent is essential to release of the young in *C. niloticus* (31, 86) and it may also be important in the American alligator (82). Both the male and the female of the

Nile crocodile and spectacled caiman have been observed to take unhatched eggs in their mouths and gently crack them until the young are freed (3, 103). It is evident that synchrony of hatching of a clutch is extremely important because of the necessity for the parent to release the hatchlings from the nest and because their immediate post-hatching association affords essential protection to the young.

It is now reliably documented that the young are carried from the nest to water in the mouth of the mother in *C. niloticus* (102, 103), *C. acutus* (88), and *C. moreleti* (68). Carrying of older juveniles in the mouth has also been observed in the American alligator (75).

For an extended period (6–8 weeks in *C. niloticus,* about a month in *Caiman crocodilus*) (3) the young of most species remain in a group in the shallows, with the mother or both parents nearby. Frequent vocal communication occurs among the young and between young and parents. During this period, the hatchlings' distress call will bring all adults in the vicinity to their defense (103). In the solitary dwarf caiman, the young remain together in small groups even though the mother is not present (85).

After a period of time, the young of the Nile crocodile (103) disperse from the home rivers and lakes, entering streams and pools not inhabited by adults and subadults [animals five feet long or longer (31)]. Cott (31) reports that "between ages of about two and five years crocodiles go into retreat." Pooley has observed that the young remain in groups and dig communal tunnels for protection and warmth and that during this period they avoid larger crocodiles (103). Dispersal of the hatchling aggregations has also been observed in captive *Caiman crocodilus* (3) and *Crocodylus moreleti* (69).

Observations and experiments by Hunt (69) on Morelet's crocodile in the Atlanta Zoo indicate that the behavior of adults toward young undergoes a change when the latter reach a total length of about 40 cm. Smaller individuals are protected, but larger ones are threatened and pursued. If caught they are held in the mouth for a period, then released unharmed. The larger juveniles become secretive and in the spatially restricted zoo enclosure take to hiding among rocks, feeding only when adults are not close by. Hunt observed that such yearlings behaved aggressively toward new hatchlings, chasing them and taking food from them; he saw one juvenile eat a hatchling. Disturbance of the hatchlings by older juveniles elicited distress calls, which brought aggressive responses from the female toward the yearlings. Hunt's observations suggest an adaptive basis for the age-related behaviors of crocodilians.

Aggregation and group vocalizations by hatchlings ensure protection by adults during the stage of life at which crocodilians are most vulnerable to predators. Since older juveniles would threaten to eat the very young or would compete with them for food, their departure from the habitat of the latter serves to increase hatchling survival. The change in behavior of the adults toward the larger juveniles may be why the latter disperse from areas of adult and hatchling habitation. These juveniles become secretive and adopt group behaviors that render them less vulnerable to predation. When they reach a certain size, they are accepted back into the adult habitat, presumably because they no longer pose a threat to the hatchlings (103).

Inferences about the Behavior of Dinosaurs

GROUP BEHAVIORS The best evidence for the existence of coordinated group behavior in dinosaurs is provided by multiple trackways that show the parallel movement of several or many individuals in the same direction. Ostrom (95) describes three sites at which 19 to 25 individuals were traveling essentially parallel to one another. The taxa include Late Triassic carnosaurs and Early Cretaceous iguanodontid ornithopods and sauropods. Evidence against the argument that the animals were constrained to follow a certain path by the existence of a physical barrier is provided by tracks that cross the main series of footprints at high angles (and that frequently belong to species other than those that made the main series). Alexander's (2) determination of dinosaur speeds from the study of trackways supports Ostrom's conclusion that the animals moved as a group; he reports that a large and small sauropod in Ostrom's series "seem to have been walking at the same speed, consistent with the suggestion that they were traveling as a herd."

Group foraging behaviors are reported for several living lizards (40), and cooperative fishing and prey-carrying behaviors occur in the Nile crocodile (103). Coordinated group behaviors have been described in hatchling and juvenile crocodiles (103) and green iguanas (23). The available evidence suggests that some dinosaurs, like some living reptiles, were gregarious, and traveled in groups showing some degree of organization.

INTRASPECIFIC COMBAT AND DISPLAY BEHAVIORS Among dinosaurs are found numerous examples of bizarre, seemingly nonadaptive structures that have until recently defied satisfactory interpretation. The Soviet paleontologist L. S. Davitashvili (32) has dealt comprehensively with these crests, spines, frills, and other features, which "cannot be put into the framework of a cliché formula: feeding, locomotion, and protection." He terms these features "perigametic," and interprets them as products of sexual selection; they increase an individual's contribution to the next generation by enhancing its success in obtaining breeding partners. Perigametic structures are associated with combat and ritual display activities directed toward conspecific rivals, and with advertisement and courtship displays directed toward potential mates.

Studies of living vertebrates have yielded generalizations about intraspecific combat and display organs that can be applied to the interpretation of the morphology of extinct species (16, 39, 53–56, 60, 118). Such structures often seem designed to minimize injury in equally armed combatants, and frequently serve as holding surfaces in head-to-head pushing or wrestling matches that test the strength of the combatants. Closely related species often show wide variation in size and shape of the structures that correlates with species-specific variation in fighting style. The size of the structure may also be correlated with body size, strength, and experience, so that its appearance alone may serve as a basis for dominance relationships. In fact, the deployment of combat weapons in threat displays toward rivals and in advertisement displays toward potential mates may take on great importance. Because larger animals generally dominate smaller ones, such displays often involve stereotyped postures that result in the greatest possible exposure of body surface to the rival;

features such as dewlaps, manes, crests, and "sails" add to the appearance of increased size. Display features provide information about the sex and species of individuals and thus are important mechanisms promoting the choice of appropriate mates; as such they function as isolating mechanisms, preventing hybridization of closely related species.

Theropods Because their teeth are such lethal weapons, extant carnivorous tetrapods (e.g. *Crocodylus*) often restrict their use against similarly armed conspecifics (77, 103); threat displays assume great importance. The carnosaurian theropods *Spinosaurus, Altispinax,* and *Acrocanthosaurus* possessed elongate neural spines, which probably supported a web of skin (32). These dorsal "sails" were probably used in lateral displays to increase apparent body size. The thin nasal horn of *Ceratosaurus* (32), the rugose ridges on the nasal and lacrimal bones of some carnosaurs (32), and the flaring facial flanges of *Dilophosaurus* (119) all presumably served to increase the formidable appearance of the head in threat displays.

Sauropodomorphs The herbivorous prosauropods and sauropods possessed few features that are unambiguously interpretable as combat or display organs. The long whip-like tails may have been utilized as weapons against predators (6) and, as in many lizards, against conspecific rivals. All possessed an enlarged sickle-like claw on the first digit of the forefoot, which may have been used as a weapon (49). Wilson (122), following a suggestion of Bakker, illustrates rearing males of the sauropod *Diplodocus* clawing at each other with this claw, and neck-wrestling like modern giraffes.

Primitive Ornithischians Few small ornithopods possessed recognizable combat or display structures. The Late Triassic heterodontosaurids show dimorphism in the premaxillary and anterior dentary teeth. Enlarged, serrated, caniniform teeth in presumed males are thought to have been used against conspecific rivals (115).

Pachycephalosaurs Several families of advanced ornithischians specialized the head for intraspecific combat. In some pachycephalosaurs the skull roof forms a rounded dome of dense, greatly thickened bone above the cranial cavity. Galton (47, 48) suggests that these domes served as battering rams in intraspecific fighting. He notes that the occiput is rotated forward beneath the skull roof so that the line of action of the postulated impact force would pass close to the occipital condyle; this would minimize the chance of violent, possibly injurious, rotation of the head. Presumed females of the dome-headed *Stegoceras* are identified by the possession of smaller domes than presumed males (47). The Mongolian genus *Homalocephale* (83) differs from other pachycephalosaurs in possessing a broader and thinner skull roof that appears to be more appropriate to pushing matches such as occur in marine iguanas (16).

Hadrosaurs All members of the Late Cretaceous ornithopod family Hadrosauridae show unusual specializations of the facial region (32, 64, 90). In the lambeosaurines, the nasal passages loop upward into prominent supracranial crests formed by the nasal and premaxillary bones. These crests show species-specific and sexually dimorphic variation and have been interpreted as visual signal structures used in displays (32, 35, 64). The elongated nasal tubes most likely served as vocal resonators (64). Primitive hadrosaurs (kritosaurs) possessed a small nasal horn used as a butting weapon in intraspecific combat. The circumnarial depression on the side of the face is interpreted as having lodged an inflatable diverticulum of the nasal passage, which served as a display to draw attention to the adjacent weapon during broadside displays. More derived groups (saurolophines, edmontosaurs, lambeosaurines) show an enhancement of the display function of the diverticulum and a regression of the weapon function of the nasal region.

Ceratopsians The great variability of horn and frill morphology in ceratopsians is interpreted by Davitashvili (32) and by Farlow & Dodson (42) as a reflection of differences in species-specific agonistic and ritual display behaviors. The low nasal horn of small protoceratopsids was used in flank butting (42). The large ceratopsids fall into two groups with differing horn patterns. I follow Langston (81) in recognizing a long-faced group (long-frilled forms plus *Triceratops*) and a short-faced group (*Centrosaurus, Monoclonius, Styracosaurus,* and *Pachyrhinosaurus*). Long-faced ceratopsids possessed long, paired brow horns and a small, median nasal horn; their probable mode of combat involved head-pushing and head-wrestling with interlocked brow horns (42). Most short-faced forms possessed a long median nasal horn and very small brow horns; because of the difficulty of countering blows from a median horn, Farlow & Dodson (41) believe that injuries may have been more frequent in this group. *Pachyrhinosaurus* lacked horns but the skull roof was covered by massive bony bosses, which were probably used in frontal ramming or pushing.

The frills of ceratopsians were probably used in frontal displays and served to increase the apparent height of the animal, especially in the long-frilled forms (42). Frills were sexually dimorphic in *Protoceratops* (36) and probably in the large ceratopsids. The secondarily shortened and unfenestrated frill of *Triceratops* served as a shield against an opponent's horns (42).

Stegosaurs The double row of large dermal plates along the back of *Stegosaurus* is interpreted by Davitashvili (32) as functioning to increase the apparent size of the animal in threat displays. Farlow et al (43) suggest that the plates functioned as forced convection "fins" for the dissipation of excess body heat. Supporting evidence comes from experiments on models and from the morphology, arrangement, and internal structure of the plates. Inasmuch as the double row of alternating dorsal plates increased the surface area of the animal, both hypotheses seem plausible; the plates may have served a dual function.

Discussion Combat and visual display structures inferred for dinosaurs do not differ in any substantial way from those possessed by living reptiles (principally lizards). Where the sexes can be differentiated (pachycephalosaurs, lambeosaurine hadrosaurs, *Protoceratops*), females possess the same combat weapons or display organs seen in males, although in slightly less developed form. This indicates the existence of agonistic behaviors between females, perhaps over nesting territories (as occurs in crocodilians and marine iguanas). Brattstrom (20) suggests that most dominance hierarchies in dinosaurs were nonaggressive. The evidence discussed here, however, indicates that aggressive interactions were common in most groups.

PARENTAL BEHAVIORS The existence of parental care of the eggs and hatchlings in crocodilians and birds suggests that similar behaviors existed in dinosaurs. In most respects, the fossil evidence suggests parental behavior patterns closer to those of crocodilians. Dinosaur eggs were laid in holes in sand (28, 74) and therefore the heat for incubation was derived from the environment. Clutch sizes are uncertain; the largest nest of *Protoceratops* contains 18 eggs arranged in three incomplete concentric circles, but a full complement may have been approximately 34 eggs (28). This clutch size is within the range of crocodilians (58) but is large for birds (76), although the male of the greater rhea (*Rhea americana*) incubates clutches of 20–50 eggs [which, however, are laid by more than one female (21, 22)]. The eggs of *Protoceratops* are very large compared with those of crocodilians; Bakker (7) estimates the hatchling may have weighed five times as much as that of a Nile crocodile of larger adult body size. Dinosaur hatchlings, like those of other reptiles, would have been precocial and capable of finding food soon after hatching.

Dinosaurs probably protected their eggs from predators just as crocodilians (103) and large ground-nesting birds such as rheas (22) do. It is also likely that hatching was synchronous, and that the hatchlings could leave the nesting area as a group in company with the parent—again as in crocodilians and rheas.

Little evidence of association of juveniles with adults is available for dinosaurs. The best evidence is provided by the Early Cretaceous Davenport Ranch sauropod trackways (5, 18, 80, 95), in which the smaller tracks are toward the center of the group and the largest are at the periphery. This suggests that the largest adults sheltered the vulnerable juveniles at the center of the herd (5, 80). The smallest footprints are about half the size of the largest, indicating that the youngest juveniles were perhaps half the height of the adults. No evidence is available about the possible association with adults of very young juveniles for any dinosaur species.

Desmond (33) suggests that sauropods may have been viviparous, giving birth to young large enough to travel immediately with the herd. However, oviparity in sauropods is indicated by large eggs attributed to the sauropod *Hypselosaurus* (28) from the Late Cretaceous of France. Packard et al (98) suggest that the evolution of viviparity was impossible for crocodilians and birds because their use of the eggshell as a calcium reservoir for the ossification of the developing skeleton would preclude the required thinning and subsequent loss of the shell in the early evolutionary stages of intrauterine development. This argument can be extended to include fossil archosaurs, which, therefore, were probably all oviparous.

Vocal communication between young and adults is well-developed in both crocodilians and birds and is likely to have been so in dinosaurs as well. Evidence for acuity of hearing into the high frequency ranges expected for the vocalizations of juveniles (63) is provided by the elongate birdlike cochlea of dinosaurs (70, 79).

CONCLUSIONS

In a review of the evidence for endothermy in dinosaurs (65), I concluded that dinosaurs could not be viewed as typical ectotherms as are all living reptiles. The evidence of mammalian-type bone histology strongly suggests that dinosaur growth rates, and therefore metabolic levels, were significantly greater than those of living ectotherms (108). On the other hand, I considered the anatomical evidence for mammalian and avian metabolic levels to be much less convincing, at least for some groups of dinosaurs. The results of the present review of relative brain size and behavior in archosaurs reinforces my earlier conclusion that dinosaurs showed a spectrum of activity levels intermediate between those of living ectotherms at the lower end and living endotherms at the upper end. This conclusion is based on the assumption that the neural requirements of a vertebrate are, to a significant degree, a function of its total level of activity and therefore of its total energy budget.

Recent studies of relative brain size in living ectothermic vertebrates (14, 100, 101) suggest a positive relationship between degree of encephalization (EQ) and levels of activity (and perhaps degree of awareness of the environment), although data on total energy budgets are not available. Especially intriguing is a study by Avery (4) of latitudinal variation in total annual activity time in four species of lizards. The total time available in a year in each species is inversely correlated with latitude; more northerly species are dormant for more of the year, and when active must spend a greater percentage of their time in basking. The amount and complexity of interactive behavior during active periods correlate with the total annual time available for activity. Furthermore, Avery observes that the behavioral plasticity or what one might term the "intelligence" of a species shows the same correlation with total annual activity time. His explanation for a relationship between complexity of interactive (including social) behavior and time available for all activity is that more northerly species must spend most of their limited active time foraging for food, leaving little time free for other activities. EQs are available for three of the four species (101); they show a direct correlation with amount and complexity of interactive behavior, "intelligence" (as subjectively determined by Avery), and total amount of time available for activity. Furthermore, annual metabolic expenditure shows the same correlation. These observations suggest that behavioral plasticity, or "intelligence," is related directly to *total* amount of activity and that it will be reflected in the relative size of the brain.

Returning to activity levels and EQs of dinosaurs, my perception of the amount and complexity of activity in the major groups of dinosaurs (inferred from functional interpretation of skeletal structure) correlates well with EQ in Figure 2. I would argue, as does Feduccia (44), that the mammalian/avian levels of activity claimed by Bakker for dinosaurs should be correlated with a great increase in motor and sensory control and this should be reflected in increased brain size. Such an increase is *not* indicated by most dinosaur endocasts. Arguments that some dinosaurs (e.g. coelurosaurs) *did* have bird-sized brains (9, 11, 34) do not remove this objection, nor does the argument that a large brain is not a necessary prerequisite for the regulation of body temperature by physiological means (34).

I conclude from the evidence of relative brain size that most dinosaurs were less active than modern endotherms. This in turn suggests that the metabolic rates of dinosaurs were also below those of living mammals and birds, inasmuch as comparable high energy budgets (in endotherms) should be reflected in comparable high levels of (food-getting) activity (7). Energy requirements of dinosaurs may perhaps be indirectly estimated in herbivorous groups from the degree of specialization of the dentition for processing vegetation [but see (7)]. The teeth of sauropods, ankylosaurs, and stegosaurs functioned only in cropping; those of ceratopsians were specialized for shearing (92); and those of hadrosaurs formed a complex grinding mill (90). Thus, the amount of food-processing taking place in the mouth correlates well with increasing EQ. This suggests that metabolic rates, like activity levels, varied greatly among dinosaurs, so that it is an oversimplification to view dinosaurs as physiologically uniform.

The very large brains of coelurosaurs, taken with skeletal evidence for high speed running and agility, lead to the conclusion that only these dinosaurs were possibly as fully endothermic as living mammals and birds.

Pterosaurs and *Archaeopteryx*, as active flyers at least for short periods, would have had periods of high aerobic metabolic activity, although their total energy requirements may not have been as great as those of modern birds. The presence of an insulating cover ("hair," feathers) indicates that metabolic heat conservation was important during periods of activity. The batlike construction of the limbs suggests that pterosaurs did little active locomoting on the ground; therefore they may have had significantly lower total energy budgets than the cursorially adapted *Archaeopteryx*. This factor may in part account for the lower EQs of pterosaurs.

Except for coelurosaurs, the range of behaviors that existed in dinosaurs, as inferred from trackways and skeletal morphology, may not have lain much outside the range observed in living ectothermic crocodilians.

ACKNOWLEDGMENTS

Thanks are due Drs. E. S. Gaffney, R. G. Northcutt, and D. A. Russell for the loan of endocasts, and Drs. G. M. Burghardt, H. W. Greene, H. A. Herzog, and R. H. Hunt for providing access to unpublished manuscripts. Drs. H. R. Barghusen and L. B. Radinsky provided helpful comments on portions of the manuscript. The research reported here was supported in part by National Science Foundation Research Grant BMS75-01159.

Literature Cited

1. Alexander, R. M. 1971. *Size and Shape, Studies in Biology No. 29.* London: Arnold. 59 pp.
2. Alexander, R. M. 1976. Estimates of speeds of dinosaurs. *Nature* 261:129–30
3. Alvarez Del Toro, M. 1969. Breeding the spectacled caiman, *Caiman crocodylus,* at Tuxtla Gutierrez Zoo. *Int. Zoo Yearb.* 9:35–36
4. Avery, R. A. 1976. Thermoregulation, metabolism and social behavior in Lacertidae. In *Morphology and Biology of Reptiles,* ed. A. d'A. Bellairs, C. B. Cox, pp. 245–59. London: Academic
5. Bakker, R. T. 1968. The superiority of dinosaurs. *Discovery, New Haven* 3(2):11–22
6. Bakker, R. T. 1971. Ecology of the brontosaurs. *Nature* 229:172–74
7. Bakker, R. T. 1971. Dinosaur physiology and the origin of mammals. *Evolution* 25:636–58
8. Bakker, R. T. 1972. Anatomical and ecological evidence of endothermy in dinosaurs. *Nature* 238:81–85
9. Bakker, R. T. 1974. Dinosaur bioenergetics—a reply to Bennett and Dalzell, and Feduccia. *Evolution* 28:497–503
10. Bakker, R. T. 1975. Experimental and fossil evidence for the evolution of tetrapod bioenergetics. In *Perspectives of Biophysical Ecology,* ed. D. Gates, R. Schmerl, pp. 365–399. New York: Springer
11. Bakker, R. T. 1975. Dinosaur renaissance. *Sci. Am.* 232(4):58–78
12. Bakker, R. T., Galton, P. M. 1974. Dinosaur monophyly and a new class of vertebrates. *Nature* 248:168–72
13. Barsbold, R. 1974. Saurornithoididae, a new family of small theropod dinosaurs from Central Asia and North America. *Palaeontol. Pol.* 30:5–22
14. Bauchot, R., Bauchot, M. L., Platel, R., Ridet, J. M. 1977. Brains of Hawaiian tropical fishes: brain size and evolution. *Copeia* 1977:42–46
15. Bauchot, R., Platel, R., Ridet, J. M. 1976. Brain-body weight relationships in Selachii. *Copeia* 1976:305–10
16. Bellairs, A. d'A. 1970. *The Life of Reptiles,* Vols. 1, 2. New York: Universe. 590 pp.
17. Bennett, A. F., Dalzell, B. 1973. Dinosaur physiology: a critique. *Evolution* 27:170–74
18. Bird, R. T. 1944. Did *Brontosaurus* ever walk on land? *Nat. Hist.* 53:61–67
19. Bramwell, C. D. 1971. Flying ability of *Archaeopteryx. Nature* 231:128
20. Brattstrom, B. H. 1974. The evolution of reptilian social behavior. *Am. Zool.* 14:35–49
21. Bruning, D. F. 1973. The greater rhea chick and egg delivery route. *Nat. Hist.* 82:68–75
22. Bruning, D. F. 1975. Social structure and reproductive behavior in the greater rhea. *Living Bird* 13:251–94
23. Burghardt, G. M., Greene, H. W., Rand, A. S. 1977. Social behavior in hatchling green iguanas: life at a reptile rookery. *Science* 195:689–91
24. Campbell, H. W. 1972. Ecological or phylogenetic interpretations of crocodilian nesting habits. *Nature* 238:404–5
25. Campbell, H. W. 1973. Observations on the acoustic behavior of crocodilians. *Zoologica* 58:1–11
26. Carey, D. J. 1972. Flying ability of *Archaeopteryx. Nature* 239:535
27. Charig, A. J. 1976. "Dinosaur monophyly and a new class of vertebrates": a critical review. See Ref. 4, 65–104
28. Colbert, E. H. 1961. *Dinosaurs: Their Discovery and Their World,* New York: Dutton. 300 pp.
29. Colbert, E. H. 1962. The weights of dinosaurs. *Am. Mus. Nov.* 2076:1–16
30. Colbert, E. H., Russell, D. A. 1969. The small Cretaceous dinosaur *Dromaeosaurus. Am. Mus. Nov.* 2380:1–49
31. Cott, H. B. 1961. Scientific results of an inquiry into the ecology and economic status of the Nile crocodile (*Crocodylus niloticus*) in Uganda and Northern Rhodesia. *Trans. Zool. Soc. London* 29:211–358
32. Davitashvili, L. S. 1961. *Teoriya Polovogo Otbora (The Theory of Sexual Selection).* Moscow: Izdatel'stvo Akad. Nauk (Acad. Sci.). 538 pp.
33. Desmond, A. J. 1976. *The Hot-Blooded Dinosaurs: A Revolution in Palaeontology.* New York: Dial/James Wade. 238 pp.
34. Dodson, P. 1974. Dinosaurs as dinosaurs. *Evolution* 28:494–97
35. Dodson, P. 1975. Taxonomic implications of relative growth in lambeosaurine hadrosaurs. *Syst. Zool.* 24:37–54
36. Dodson, P. 1976. Quantitative aspects of relative growth and sexual dimorphism in *Protoceratops. J. Paleontol.* 50:929–40
37. Edinger, T. 1926. The brain of *Archae-*

opteryx. Ann. Mag. Nat. Hist. (9)18: 151–56
38. Edinger, T. 1941. The brain of *Pterodactylus. Am. J. Sci.* 239:665–82
39. Ewer, R. F. 1968. *Ethology of Mammals*, New York: Plenum. 418 pp.
40. Farlow, J. O. 1976. Speculations about the diet and foraging behavior of large carnivorous dinosaurs. *Am. Midl. Nat.* 95:186–91
41. Farlow, J. O. 1976. A consideration of the trophic dynamics of a late Cretaceous large-dinosaur community (Oldman Formation). *Ecology* 57:841–57
42. Farlow, J. O., Dodson, P. 1975. The behavioral significance of frill and horn morphology in ceratopsian dinosaurs. *Evolution* 29:353–61
43. Farlow, J. O., Thompson, C. V., Rosner, D. E. 1976. Plates of the dinosaur *Stegosaurus:* forced convection heat loss fins? *Science* 192:1123–25
44. Feduccia, A. 1973. Dinosaurs as reptiles. *Evolution* 27:166–69
45. Feduccia, A. 1974. Endothermy, dinosaurs, and *Archaeopteryx. Evolution* 28:503–4
46. Galton, P. M. 1970. The posture of hadrosaurian dinosaurs. *J. Paleontol.* 44:464–73
47. Galton, P. M. 1970. Pachycephalosaurids—dinosaurian battering rams. *Discovery, New Haven* 6(1):23–32
48. Galton, P. M. 1971. A primitive domeheaded dinosaur (Ornithischia: Pachycephalosauridae) from the Lower Cretaceous of England and the function of the dome of pachycephalosaurids. *J. Paleontol.* 45:40–47
49. Galton, P. M. 1976. Prosauropod dinosaurs (Reptilia: Saurischia) of North America. *Postilla* 169:1–98
50. Garrick, L. D. 1975. Structure and pattern of the roars of Chinese alligators (*Alligator sinensis* Fauvel). *Herpetologica* 31:20–28
51. Garrick, L. D. 1975. Love among the alligators. *Animal Kingdom* 78(2):2–8
52. Garrick, L. D., Lang, J. W. 1975. Alligator courtship. *Am. Zool.* 15:813
53. Geist, V. 1966. The evolution of hornlike organs. *Behavior* 27:175–214
54. Geist, V. 1971. The relation of social evolution and dispersal in ungulates during the Pleistocene, with emphasis on the Old World deer and the genus *Bison. Quat. Res.* 1:283–315
55. Geist, V. 1974. On fighting strategies in animal combat. *Nature* 250:354
56. Geist, V., Walther, F., eds. 1974. *The Behaviour of Ungulates and Its Relation to Management,* Vols. 1, 2. Morges, Switzerland: IUCN. 940 pp.
57. Greer, A. E. Jr. 1971. Crocodilian nesting habits and evolution. *Fauna* 2:20–28
58. Greer, A. E. 1975. Clutch size in crocodilians. *J. Herpetol.* 9:319–22
59. Guggisberg, C. A. W. 1972. *Crocodiles: Their Natural History, Folklore, and Conservation,* Harrisburg, Pa.: Stackpole. 203 pp.
60. Harris, V. A. 1964. *The Life of the Rainbow Lizard.* London: Hutchinson. 174 pp.
61. Herzog, H. A. Jr. 1974. *The vocal communication system and related behaviors of the American alligator (Alligator mississippiensis) and other crocodilians.* MA thesis. Univ. Tennessee, Knoxville. 83 pp.
62. Herzog, H. A. Jr. 1975. An observation of nest opening by an American alligator *Alligator mississippiensis. Herpetologica* 31:446–47
63. Herzog, H. A. Jr., Burghardt, G. M. 1977. Vocal communication signals in juvenile crocodilians. *Z. Tierpsychol.* In press
64. Hopson, J. A. 1975. The evolution of cranial display structures in hadrosaurian dinosaurs. *Paleobiology* 1:21–43
65. Hopson, J. A. 1976. Hot-, cold-, or lukewarm-blooded dinosaurs? Review of *The Hot-Blooded Dinosaurs: A Revolution in Paleontology,* by Adrian J. Desmond. *Paleobiology* 2:271–75
66. Hopson, J. A. 1977. Brain size and behavior of dinosaurs. *J. Paleontol.* 51(2, Suppl.):15
67. Hopson, J. A. 1978. Paleoneurology. In *Biology of the Reptilia, Neurology A,* Vol. 10, ed. C. Gans, R. G. Northcutt, P. S. Ulinski. London: Academic. In press
68. Hunt, R. H. 1975. Maternal behavior in the Morelet's crocodile, *Crocodylus moreleti. Copeia* 1975:763–64
69. Hunt, R. H. 1977. Aggressive behavior by adult Morelet's crocodile, *Crocodylus moreleti* toward young. *Herpetologica* 33:195–201
70. Janensch, W. 1935/1936. Die Schädel der Sauropoden *Brachiosaurus, Barosaurus* und *Dicraeosaurus* aus der Tendaguru-Schichten Deutsch-Ostafrikas. *Palaeontographica* 7 (Suppl.):145–298
71. Jerison, H. J. 1968. Brain evolution and *Archaeopteryx. Nature* 219:1381–82
72. Jerison, H. J. 1969. Brain evolution and dinosaur brains. *Am. Nat.* 103:575–88

73. Jerison, H. J. 1973. *Evolution of the Brain and Intelligence.* New York: Academic. 482 pp.
74. Kurtén, B. 1968. *The Age of the Dinosaurs.* New York: McGraw-Hill. 255 pp.
75. Kushlan, J. A. 1973. Observations on maternal behavior in the American alligator, *Alligator mississippiensis. Herpetologica* 29:256–57
76. Lack, D. 1968. *Ecological Adaptations for Breeding in Birds.* London: Methuen. 409 pp.
77. Lang, J. W. 1975. The Florida crocodile: will it survive? *Field Mus. Nat. Hist. Bull.* 46(8):4–9
78. Lang, J. W. 1975. American crocodile courtship. *Am. Zool.* 16:197
79. Langston, W. Jr. 1960. The vertebrate fauna of the Selma Formation of Alabama, Pt. 6: The dinosaurs. *Fieldiana Geol. Mem.* 3:313–61
80. Langston, W. Jr. 1974. Nonmammalian Comanchean tetrapods. *Geosci. Man* 8:77–102
81. Langston, W. Jr. 1975. The ceratopsian dinosaurs and associated lower vertebrates from the St. Mary River Formation (Maestrichtian) at Scabby Butte, Southern Alberta. *Can. J. Earth Sci.* 12:1576–1608
82. Lee, D. S. 1968. Possible communication between eggs of the American alligator. *Herpetologica* 24:88
83. Maryánska, T., Osmólska, H. 1974. Pachycephalosauria, a new suborder of ornithischian dinosaurs. *Palaeontol. Pol.* 30:45–102
84. McNab, B. K., Auffenberg, W. 1976. The effect of large body size on the temperature regulation of the Komodo dragon, *Varanus komodoensis. Comp. Biochem. Physiol.* 55A:345–50
85. Medem, F. 1971. The reproduction of the dwarf caiman *Paleosuchus palpebrosus. IUCN Publ., New Ser.,* 32 (Suppl.): 159–65
86. Modha, M. L. 1967. The ecology of the Nile crocodile (*Crocodylus niloticus* Laurenti) on Central Island, Lake Rudolf. *E. Afr. Wildl. J.* 5:74–95
87. Neill, W. T. 1971. *The Last of the Ruling Reptiles: Alligators, Crocodiles, and Their Kin.* New York: Columbia Univ. 486 pp.
88. Ogden, J., Singletary, C. 1973. Night of the crocodile. *Audubon* 75(3):32–37
89. Osmólska, H. 1976. New light on the skull anatomy and systematic position of *Oviraptor. Nature* 262:683–84
90. Ostrom, J. H. 1961. Cranial morphology of the hadrosaurian dinosaurs of North America. *Am. Mus. Nat. Hist.* 122:37–186
91. Ostrom, J. H. 1964. A reconsideration of the paleoecology of hadrosaurian dinosaurs. *Am. J. Sci.* 262:975–97
92. Ostrom, J. H. 1966. Functional morphology and evolution of the ceratopsian dinosaurs. *Evolution* 20:290–308
93. Ostrom, J. H. 1969. Osteology of *Deinonychus antirrhopus,* an unusual theropod from the Lower Cretaceous of Montana. *Peabody Mus. Nat. Hist. Bull.* 30:1–165
94. Ostrom, J. H. 1969. Terrestrial vertebrates as indicators of Mesozoic climates. *Proc. N. Am. Paleontol. Conv., Chicago,* pp. 347–76
95. Ostrom, J. H. 1972. Were some dinosaurs gregarious? *Palaeogeogr. Palaeoclimatol. Palaeoecol.* 17:287–301
96. Ostrom, J. H. 1974. *Archaeopteryx* and the origin of flight. *Quart. Rev. Biol.* 49:27–47
97. Ostrom, J. H. 1976. *Archaeopteryx* and the origin of birds. *Biol. J. Linn. Soc.* 8:91–182
98. Packard, G. C., Tracy, C. R., Roth, J. J. 1977. The physiological ecology of reptilian eggs and embryos, and the evolution of viviparity within the class Reptilia. *Biol. Rev.* 52:71–105
99. Platel, R. 1972. Les relations pondérale encéphalo-somatiques chez les reptiles sauriens. *C. R. Acad. Sci. Ser. D* 274:2181–84
100. Platel, R. 1974. Poids encéphalique et indice d'encéphalisation chez les reptiles sauriens. *Zool. Anz.* 192:332–82
101. Platel, R. 1975. Nouvelles données sur l'encéphalisation des reptiles squamates. *Z. Zool. Syst. Evolutionsforsch.* 13:161–84
102. Pooley, A. C. 1974. How does a baby crocodile get to water? *Afr. Wildl.* 28:8–11
103. Pooley, A. C., Gans, C. 1976. The Nile crocodile. *Sci. Am.* 234:114–24
104. Radinsky, L. B. 1977. Evolution of brain size in carnivores and ungulates. *Am. Nat.* In press
105. Ricqlès, A. 1969. L'histologie osseuse envisagée comme indicateur de la physiologie thermique chez les tétrapodes fossiles. *C. R. Acad. Sci., Ser. D.* 268:782–85
106. Ricqlès, A. 1972. Vers une histoire de la physiologie thermique. L'apparition de l'endothermie et le concept de reptile. *C. R. Acad. Sci., Ser. D* 275:1875–78

107. Ricqlès, A. 1972. Les dinosaurs, "reptiles" à sang chaud. *La Recherche* 3(28):992–94
108. Ricqlès, A. 1974. Evolution of endothermy: histological evidence. *Evol. Theory* 1:51–80
109. Ricqlès, A. 1976. On the bone histology of fossil and living reptiles, with comments on its functional and evolutionary significance. See Ref. 4, pp. 121–50
110. Russell, D. A. 1969. A new specimen of *Stenonychosaurus* from the Oldman Formation (Cretaceous) of Alberta. *Can. J. Earth Sci.* 6:595–612
111. Russell, D. A. 1972. Ostrich dinosaurs from the Late Cretaceous of Western Canada. *Can. J. Earth Sci.* 9:375–402
112. Russell, L. S. 1965. Body temperature of dinosaurs and its relationships to their extinction. *J. Paleontol.* 39:497–501
113. Spotila, J. R., Lommen, P. W., Bakken, G. S., Gates, D. M. 1973. A mathematical model for body temperatures of large reptiles: implications for dinosaur ecology. *Am. Nat.* 107:391–404
114. Swinton, W. S. 1970. *The Dinosaurs.* New York: Wiley. 331 pp.
115. Thulborn, R. A. 1974. A new heterodontosaurid dinosaur (Reptilia: Ornithischia) from the Upper Triassic Red Beds of Lesotho. *Zool. J. Linn. Soc.* 55:151–75
116. Thulborn, R. A. 1973. Thermoregulation in dinosaurs. *Nature* 245:51–52
117. Thulborn, R. A. 1975. Dinosaur polyphyly and the classification of archosaurs and birds. *Aust. J. Zool.* 23:249–70
118. Walther, F. 1958. Zum Kampf- und Paarungsverhalten einiger Antilopen. *Z. Tierpsychol.* 15:340–80
119. Welles, S. P. 1970. *Dilophosaurus* (Reptilia: Saurischia), a new name for a dinosaur. *J. Paleontol.* 44:989
120. Wellnhofer, P. 1974. Das fünfte Skelettexamplar von *Archaeopteryx*. *Palaeontographica, Abt. A* 147:169–216
121. Wellnhofer, P. 1977. Die Pterosaurier. *Naturwissenschaften* 64:23–29
122. Wilson, E. O. 1975. *Sociobiology: The New Synthesis.* Cambridge, Mass: Harvard Univ. 697 pp.
123. Yalden, D. W. 1971. Flying ability in *Archaeopteryx*. *Nature* 231:127
124. Yalden, D. W. 1971. The flying ability in *Archaeopteryx*. *Ibis* 113:349–56

AUTHOR INDEX

A

Able, K. W., 295
Abrahamson, W. G., 166, 167
Abramsky, Z., 184
Ackerman, W. C., 257
Ackermann, H. W., 38, 41
Acquaye, D. K., 94, 96, 97
Adams, J. B., 338
Addicott, W., 360
Agthe, C., 408
Ahlstrom, E. H., 285
Ahrens, E., 53
Aikman, D., 335, 337, 344, 346
Aleem, A. A., 274
Alerdice, D. F., 294
Alexander, R. M., 432, 439
Alf, R., 363
Alfani, A., 66, 68
Ali, S. A., 8, 11
Allen, G. H., 297
Allen, J. A., 124, 127
Allen, J. S., 54
Allen, T. T., 158
Allison, F. E., 55, 57-59, 71
Alm, G., 164
Alvarez Del Toro, M., 437, 438
Alverson, D. L., 285, 293
Amadon, D., 201, 202
Ambroes, P., 53
Anderson, A., 19
Anderson, G. R., 43
Anderson, J. M., 64, 68, 70, 71
Anderson, S., 193
Anderson, V. M., 288
André, J., 110, 112, 119
Andreano, R. L., 230
Andrewes, C. H., 30, 42, 44
Andrews, R., 162
Ansari, N., 230
Appert, O., 9
Appleyard, A., 62
Arbo, M. M., 414
Arditti, J., 410, 411, 413, 421
Arnold, G. P., 285, 295
Arnold, R. W., 110, 117, 119, 120, 126, 129, 131, 134
Arnold, G. W., 179, 185
Aronson, L. R., 296
Arp, A. H., 286, 297
Arthur, W., 117, 118, 125, 127, 128
Aschoff, J., 381, 387, 388
Ashmole, N. P., 163, 164
Askew, R. R., 1, 19
Aubertin, D., 110, 111
Auerbach, S. I., 175, 180
Auffenberg, W., 360, 370, 371, 430
Aufrecht, S., 414

Auslander, D. M., 158
Ausmus, B. S., 182, 183
Avery, R. A., 443
Avise, J. C., 242, 311, 312
Axelrod, D. I., 360, 368-70, 372
Ayala, F. J., 111, 123, 125, 132, 237, 242, 243, 249, 250, 311
Azrin, N. H., 196
Azzi, G., 174

B

Bagenal, T. B., 295
Baggerman, B., 285, 293
Baggott, G. K., 396
Bailey, H. P., 368, 370
Bailey, I. W., 416
Bailey, W. A., 95
Baker, C. M. A., 113
Baker, E. C. S., 2, 7-9, 13, 16, 21
Baker, H. G., 129, 417
Baker, I., 417
Baker, J. R., 381
Baker, K. F., 175
Bakkala, R. G., 294, 298
Bakken, G. S., 430
Bakker, R. T., 430-32, 440, 442-44
Balakrishnan, V., 310, 323
Balchen, J. G., 286, 291, 294
Baldamus, E., 8
Baldwin, R. E., 230
Ball, D. F., 59
Ball, O. P., 286
Ballard, T. M., 64
Ballinger, R. E., 162
Bams, R. A., 298
Banks, C. J., 329, 332, 336, 337
Banks, J. W., 289, 291, 294, 295, 298
Bantock, C. R., 110, 111, 113, 114, 119, 120, 124, 126, 133
Baranov, I. V., 270
Barasch, C., 197
Barber, S. A., 93
Barclay, J. S., 271
Bargoni, N., 408
Barkalow, F. S. Jr., 164
Bärlocher, F., 264
Barlow, C. A., 335
Barlow, J. S., 288
Barnes, B. W., 312
Barnette, R. M., 53
Barr, L. G., 242, 250
Barrette, C., 195
Barsbold, R., 432
Bashford, M. A., 337

Baskin, S. I., 410, 413
Bauchot, M. L., 434, 443
Bauchot, R., 434, 443
Bauer, O. N., 270
BAXTER, R. M., 255-83; 264
Bayley, J. A., 124
Baynton, H. W., 90
Bazilevich, N. I., 52
Bazzaz, F. A., 63
Beadle, L. C., 264, 274
Beardsell, M. F., 95
Beardsley, G. L. Jr., 294
Bedford, H. W., 411
Beerbower, J. R., 361-63
Behling, L., 94
Behrendt, K., 332, 336
Beiningen, K. T., 273
Bejer-Petersen, B., 338
Bellairs, A. d'A., 439, 440
Bellrose, F. C., 396
Belt, T., 412, 414
Bengtson, S.-A., 133
Benicourt, C., 39
Bennett, A. F., 430
Bennett, D., 179, 185
Bennett, K. J., 94
Benschoter, C. A., 418
Benson, C. W., 2, 7
Benson, W. W., 418, 419
Bent, A. C., 13, 14
BENTLEY, B. L., 407-27; 408, 411, 412, 414-21
Berenbeim, D. Ya., 294
Berger, M.-G., 412
Bernhard, F., 101
Bernhard-Reversat, F., 85
Bernstein, S. C., 246, 248, 311
Berrie, G. K., 91
Berry, K., 33
Berthelot, N., 41
Berthold, P., 381, 384, 385, 387-89, 391-93, 396-401
Betts, F. N., 20
Bevan, D., 338
Bhattacharyya, B., 412-14
Bhuiya, M. R. H., 72
Biddulph, O., 94
Biebach, H., 384, 388, 394, 400
Biebricher, C. K., 40
Biederbeck, V. O., 71
Biesiadka, E., 268
Bigger, J. W., 180
Billes, G., 69, 70
Billings, W. D., 66, 69
Bingham, F. T., 59-61
Birch, H. F., 54, 61, 71
Birch, L. C., 165, 166
Bird, R. T., 442
Birley, A. J., 312
Birley, N., 124

449

Bishnai, H. M., 295
Bishop, J. M., 38, 42
Biswas, A. K., 255
Biswas, S., 268
Bityukov, E. P., 291
Bjork, P. R., 363
Black, C. C., 359, 360, 369
Black, F. L., 44
Blackman, R. L., 333-35
Blackstock, T. H., 124
Blair, W. F., 162
Blaise, M., 13-15
Blasco, F., 83, 86
Blatter, E., 414
Blatti, S., 40
Blaxter, J. H. S., 294
Bleak, A. T., 65
Bliss, C. A., 410, 413
Bliss, L. C., 54
Blumenthal, T., 39
Blythe, G. M., 111
Blyumental, T. I., 396
Bobée, B., 262
Bock, I. R., 238
Bodenheimer, F. S., 345
Boettger, C. R., 118
Bogatyrev, L. G., 55
Bohmker, H., 412-14
Bohn, H. L., 52, 57
Bolen, E. G., 14, 19
Bolin, B., 52, 57
Bombosch, S., 345
Bondy, D. A., 275
Bonnemaison, L., 330, 332
Bonnet, J. A., 85, 89
Bonnier, G., 414
Booth, C. O., 330
Bormann, F. H., 53, 59-61, 63, 65, 71, 72
Botstein, D., 43
Bötticher, R., 94
Bouck, G. R., 272
Bourlière, F., 54, 357
Bowden, B., 408, 410, 412
Bowling, D. J. F., 95
Bradbury, S., 117
Bradley, D. J., 216, 223-26, 229
Bradley, G. A., 418
Bradley, W. H., 358
Bradshaw, A. D., 269
Braemer, W. A., 285, 286, 290
Bramwell, C. D., 436
Brannon, E. L., 292, 297, 298
Branover, G. G., 287, 288
Brassard, G. R., 271
Brattstrom, B. H., 360, 442
Bray, J. R., 59, 62, 66, 70, 87
Breder, C. M., 1
Breier, A., 257
Bremner, J. M., 65
Brereton, J. L., 200
Brett, J. R., 294, 296

Brian, M. W., 329
Bridge, R. R., 418
Bridger, J. P., 285
Briscoe, C. B., 85
Brncic, D., 239, 250
Broadbent, F. E., 59
Broadbent, L., 344
Brock, T. D., 263
Bromley, P. T., 201
Brooke, R. K., 7
Brown, A., 69
Brown, A. W. A., 174, 186
Brown, C. W., 311
Brown, H. D., 421
Brown, J. L., 7, 15, 194, 201, 401
Brown, K., 418, 419
Brown, M. I., 340
Brown, P. E., 62, 64
Brown, S., 39
Brown, S. M., 94, 96, 97
Brown, W. H., 85, 408, 411
Brown, W. L., 414, 419, 421
Bruce, R. C., 161
Bruck, D. L., 92, 99
Bruderer, B., 401
Bruning, D. F., 442
Bruning, E., 53
Brussard, P. F., 113, 127, 129
Buckland, F., 296
Buckley, R. V., 299
Buckney, R. T., 261
Budge, E. A. W., 257
Bumpus, D. F., 267
Bundgaard, J., 123
Bunnell, F. L., 177
Buol, S. W., 62
Burford, J. R., 65
Burghardt, G. M., 437, 439, 442
Burnet, F. M., 29, 30, 37
Buschbom, R. L., 65, 69, 70
Bush, G. L., 17
Butler, G. D., 408, 410, 413, 418
Butorin, N. V., 275
Butterfield, P. A., 200
Byer, M. D., 92, 99
Byerly, T. C., 174, 186
Byrne, J. E., 292

C

Caillé, A., 262
Cain, A. J., 110, 111, 113-18, 120-23, 125-28, 130-34
Caldwell, J., 166, 167
Caldwell, M. M., 69
Cameron, R. A. D., 110, 111, 116, 119, 120, 122, 125-28, 133, 134

Cammell, M. E., 336, 337
Camp, C. L., 364
Camp, L. B., 69
Campbell, A., 346
Campbell, C. A., 58, 71, 123, 125
Campbell, H. W., 437
Campbell, P. G., 262
Capek, V., 8, 9, 13-16
Carey, D. J., 436
Carlin, B., 286, 289, 290, 297, 298
Carlquist, S., 412
Carmichael, G. G., 39
Carmody, G., 242
Carne, W. M., 411, 412
Carpenter, J. R., 356
Carre, D., 39
Carroll, C. R., 415, 416
Carruthers, J. N., 295
Carscadden, J. E., 286, 299
Carson, H. L., 239-43, 249, 250
Carter, C. I., 338
Carter, M. A., 110, 111, 115-18, 120, 122, 124-26, 128, 133, 134
Carter, N., 335, 337, 344, 346
Caspar, D. L. D., 43
Catamo, A., 311
Cathcart, C. L., 287, 295
Catsky, J., 87
Caughley, G., 155, 158
Cavalli-Sforza, L. L., 310, 311, 323
Cernuda, C. F., 61
Cesnola, A. P. D., 128
Chakrabartty, P. K., 312
Chakravarty, H. L., 412
Chance, E., 3, 7, 8, 11, 13-15
Chandler, R. F. Jr., 62
Chapin, J. P., 381
Chapman, H. D., 94, 96, 97
Chapman, S. B., 54, 61
Charig, A. J., 430
Charlesworth, B., 147
Chavan, A. R., 411, 414
Cheal, M., 285
Cheesman, R. E., 256
Cheever, A. W., 224, 226
Cheng, T., 231
Child, G. I., 100
Chrambach, A., 314
Christiansen, F. B., 123
Chuong, P. H., 101
Churchill, M. A., 272
Cintrón, G., 64, 65, 67
Clark, E. W., 410, 418
Clark, J., 361-63
Clarke, B., 312
Clarke, B. C., 117, 118, 120, 123-28, 132, 133
Clatworthy, J. N., 167

AUTHOR INDEX 451

Clausen, J., 166
Clay, C. H., 273
Clayton, F. E., 239
Cleaver, F. C., 285, 294
Clemens, H. B., 285, 293
Clemens, W. A., 273, 286
Clements, F. E., 373
Clements, R. G., 100
Clinning, C. F., 2, 3, 8, 10, 11, 13, 16, 21
Clos, D;, 412
Cloudsley-Thompson, J. L., 373
Coakley, J. P., 259
Cobbs, G., 313
Cochrane, B. J., 247, 248, 311
Cody, M. L., 163, 168
Cohen, D., 153
COHEN, J. E., 209-33; 156, 209, 212, 214, 215, 230
Colbert, E. H., 429, 431, 435, 442
Cole, D. W., 52, 53
Cole, L. C., 153
Cole, P. E., 31
Coleman, D. C., 65, 66, 69
Collazo, D. A., 242
Collier, B. D., 177, 189
Collins, G. B., 291, 297
Collins, J. P., 153
Collins, M. I., 412
Commoner, B., 183
Conaway, C. H., 164
Conway, G. R., 153
Cook, A. W., 3, 10, 11
Cook, L. M., 111, 113, 114, 119, 120, 124, 126, 128
Cooke, F., 19
Coons, L. W., 162
Cooper, J. C., 296
Cooper, M. M., 240, 242, 249, 250
Cordeiro, A. R., 242
Corey, R. B., 180
Cornfield, A. H., 72
Cornforth, I. S., 101
Cortez, J., 69, 70
Cott, H. B., 437, 438
Cottam, C., 14, 19
Cotton, J. B. Jr., 285, 294
Coulson, J. C., 204
Courtney, J., 10
Coutinho, A. B., 220, 221
Coutinho, F. A. B., 220, 221
Cox, G. S., 68
Coyne, J. A., 247, 248
Coyne, P. I., 66, 69
Cram, W. J., 94
Creed, E. R., 117
Creutzberg, F., 295, 297, 298
Cristobal, C. L., 414
Crocker, R. L., 53
Crofton, H. D., 223

Cronin, E. W., 15
Crook, J. H., 200, 201
Crow, J. F., 128, 131
Crumpacker, D. W., 322
Cuerrier, J.-P., 270
Currey, J. D., 110, 111, 114-17, 120, 122, 123, 125-27, 130, 132, 133
Curry-Lindahl, K., 397
Cusset, G., 408, 411, 413, 415, 419
Cyberski, J., 258
Czeschlik, D., 400-2

D

Dagg, A. I., 417, 419
Dahl, K., 290
Dahlgren, K. V. O., 408, 412
Dahlman, R. C., 58, 60
Dale, W. L., 90
Dalquest, W. W., 370
Dalton, G. E., 173, 181, 184
Daly, G. T., 55
Dalzell, B., 430
D'Aoust, B. G., 272
Darbishire, F. V., 62
Darden, T. R., 388
Darley, J. A., 15
Darnell, J. E., 30, 38
Darwin, C., 12, 18, 20, 413, 414, 421
Darwin, F., 408, 411, 413, 414, 420, 421
Dave, Y. S., 411, 412, 420
Davidson, F. A., 285, 286
Davies, B. R., 257, 265
Davies, C. R., 94
Davies, P. A., 414
Davies, P. W., 116
Davis, C. S., 333, 342
Davis, D. E., 9, 13, 19
Davitashvili, L. S., 430, 433, 439-41
Davy, F. B., 288
Dawkins, H. C., 87
Dawson, M. R., 359, 360, 362, 363
Dawson, P. S., 165, 166
Day, J. C. L., 120, 123
Day, T. H., 312
Dayhoff, M. O., 43
Dean, G. J., 335, 338
Dearman, R. S., 111, 126
DeBach, P., 416
De Boois, H. M., 65, 68, 70
de Chavigny, J., 8
De Jong, C., 64, 65, 69-72
Delacour, J., 196, 201, 202, 204
DeLacy, A. C., 297
Delnicki, D., 14, 19

Delpino, F., 414, 421
Delwiche, C. C., 52, 57
Demalsy, M. J., 262
Demalsy, P., 262
Dementiew, G. P., 382
Denaeyer-DeSmet, S., 53
Deng, C., 38, 42
Denman, D., 174, 184, 188
DePuit, E. J., 69
de Rham, P., 101
Deshmukh, Y. S., 411, 414
Desmond, A. J., 430, 442
Deuver, M. J., 100
DeVeen, J. F., 285, 286, 289, 290
Devold, F., 294
Dewar, A. M., 332, 335
Diamond, J. M., 356
Dickinson, C. H., 59
Dickson, B. A., 53
Dickson, R. C., 329
Dickson, W., 294
Diener, T. O., 38, 39
Diesselhorst, G., 4
Dietz, K., 221, 223, 224
Dill, P. A., 287
Dinger, B. E., 53, 59
Diver, C., 110, 111, 113, 120, 123, 128
DIXON, A. F. G., 329-53; 330, 332-35, 337, 339, 340, 344, 346
Dizon, A. E., 296, 297
Dobzhansky, Th., 128, 136, 165, 166, 237, 242, 243
Dodson, J. J., 287, 290, 294, 297, 298
Dodson, P., 430, 434, 441, 443
Dolnik, V. R., 396
Dolotov, V. A., 53, 56
Dominy, C. L., 262
Donaldson, L. R., 297
Donselaar, J. Van, 54
Dop, P., 412
Dorf, E., 369
Dorka, V., 381, 384
Dorsey, M. J., 413
Dort, W. Jr., 372
Douglas, L. A., 54, 69, 70
Doutt, R. L., 416
Dove, W. F., 31
Doving, K. B., 297
Dow, D. D., 3
Dowdeswell, W. H., 120, 123
Downs, T., 369
Doyle, M., 40
Draycot, W. M., 129
Drew, M. E., 338
Drost, R., 389
Dubrovin, I. Ya., 294
Duellman, W. E., 356
Dunbar, M. J., 295

Duncan, R. N., 297, 299
Dundee, D. S., 129
Dunkle, D. W., 360
Dunmire, W. W., 164
Dunn, J. A., 329
Dunnet, G. M., 19
Duthie, H. C., 267
Duvigneaud, P., 53
Dvinin, P. A., 297
Dyakonov, K. N., 275
Dykeman, W. R., 68
Dyrness, C. T., 52, 53
Dyson-Hudson, R., 174, 184, 188
Dzieciolowski, R., 201
Dzyuban, N. A., 268

E

Eales, J. G., 293
Eanes, W. F., 313
Eastop, V. F., 329
Eaton, F. M., 408
Ebel, W. J., 273
Edinger, T., 435
Edman, G., 290
Edmonds, R. L., 54
Edson, Q. A., 271
Edwards, A. W. F., 310, 311, 323
Edwards, G., 7
Edwards, N. T., 53, 59, 63-65, 67, 70, 71, 189
Edwards, P. J., 85-88, 90, 93, 97, 98, 101, 102
Edwards, Y. H., 313, 324
Egan, M. E., 162
Egbert, A. C., 185
Egbuniwe, H., 275
Ehrenfeld, J. G., 242, 250
Ehrhardt, P., 338, 339
Ehrlich, P. R., 415, 418
Ehrman, L., 242
Eisenberg, J. F., 194
Elder, R. A., 260
Elias, M. K., 365, 372
Elias, T. S., 410, 412-15, 421
Elkington, T. T., 87
Elliot, H. F. I., 20
Elliott, P. F., 5, 14
Elliott, P. O., 165, 166
Ellis, M. M., 257, 258, 269, 270
Ellis, R. C., 70
Elson, P. F., 298
Elton, C. S., 329
Elton, R. A., 39, 123
Emberton, L. R. B., 117
Emlen, S. T., 389
Emry, R. J., 362, 363
Enama, M., 311

Endler, J. A., 132
Engstrom, L. R., 256, 257
Epstein, E. H., 230
Ergle, D. R., 408
Esau, K., 408
Essex, M., 36
Estacio, F., 185
Estes, R., 371
Estes, R. D., 194, 197
Evans, C. E., 56
Evans, H. J., 113
Everett, K. R., 54
Ewel, J. J., 101
Ewer, R. F., 439
Ewing, E. P., 310
Eyme, J., 408
Eze, J. M. O., 91

F

Fagan, S. R., 365
Fagerlund, U. H. M., 296, 297
Fahn, A., 408
Falk, C. T., 237
Fankhauser, D. P., 15
Fareed, G. C., 33
Farlow, J. O., 430, 434, 439, 441
Farner, D. S., 388
Farooq, M., 230
Farris, J. S., 243, 250
Faust, W. R., 4, 5
Favre, J., 110
Feduccia, A., 430, 443
Fels, E., 257
Felton, A. A., 247
Fenner, F., 29, 30, 36, 44, 46
Fenner, M., 95
Ferguson, K., 314
Ferguson, R. G., 291
Ferguson-Lees, I. J., 3, 11
Fernandez, O. A., 69
Fernando, C. H., 265
Ferrusquia-Villafranca, I., 360, 370
Fiala, L., 261
Fields, R. W., 369
Fiering, M. B., 188, 189
Figier, J., 408, 412, 413
Fine, P. E. M., 209, 214
Fisher, K. C., 291, 381, 392
Fisher, R. A., 113
Fisher, R. V., 365
Fisher, S. G., 264, 272
Fitch, H. S., 162
Fittkau, E. J., 53
Flerov, K. K., 204
Florence, R. G., 101
Flower-Ellis, J. G. K., 54
Foerster, R. E., 286
Follett, B. K., 388

Force, D. C., 345
Ford, E. B., 11, 16
Ford, M. A., 95
Forrest, J. M. S., 332
Forrester, C. R., 285
Forstén, A., 360, 370, 371
Forward, R. B. Jr., 287, 290
Foster, G. N., 344
Foster, J. B., 417, 419
Foster, N. W., 53
Fowler, D. K., 269
Frädrich, H., 195
Frank, M. L., 64, 65, 68
Franklin, W. L., 194, 195
Franz, G., 62
Frazer, B. D., 342, 346
Fredga, K., 113
Fredriksen, R. L., 52, 53
Frei, E., 408
French, R. R., 294, 298
Frey-Wyssling, A., 408, 412, 413
Friedmann, H., 1-3, 5, 7, 9-11, 13-16, 18-22
Friend, M. T., 54, 61
Fries, C. H., 360
Frisch, K. von, 286, 287
Frith, C. B., 11
Froment, A., 65, 68, 70
Fromming, E., 111
Frost, S., 271
Fry, C. H., 10
Fry, F. E. J., 291
Frye, J. C., 372
Fuggles-Couchman, N. R., 20
Fujii, T., 294

G

Gadgil, M., 166-68
Gaillochet, J., 412
Gaines, M. S., 166, 167
Gainey, P. L., 62
Galaktionov, G. Z., 288
Galbreath, E. C., 359
Gallo, R. C., 33, 37, 38, 42
Galton, P. M., 430, 434, 440
Galusha, T., 370
Gans, C., 437-40, 442
Ganzhara, N. F., 58
Gardella, E. S., 290
Garland, T. R., 65, 69, 70
Garrett, H. E., 68
Garrick, L. D., 437
Gasperino, A. F., 261
Gastescu, P., 257
Gaston, A. J., 3-5, 7, 10, 11, 13, 14, 22
Gates, D. M., 93, 430
Gawne, C. E., 363, 365, 370
Gazin, C. L., 358, 359

AUTHOR INDEX

Geen, G. H., 257, 273, 274
GEIST, V., 193-207; 194-97, 199-202, 204, 205, 439
Gelband, H., 410, 412, 415, 421
George, C. J., 274
Gerd, S. V., 257, 258, 264, 270
Gerking, S. R., 286
Gersper, P. L., 54
Gessel, S. P., 59-61
Gibbs, M., 174, 186
Gibson, J., 312
Gibson, J. B., 312
Gilbert, L. E., 415, 418, 419
Gilbert, N., 333, 337, 346
Gilkeson, R. A., 55
Gill, C. J., 269
Gill, D., 274
Gillespie, D., 37, 38, 42
Gillespie, D. H., 39
Gillespie, J., 238
Ginevan, M. E., 310
Giunchedi, L., 38
Gjessing, E. T., 262
Gladkow, N. A., 382
Glebe, B. D., 298
Gleizer, S. I., 287, 288
Glen, D. M., 334, 339
Gliwicz, Z. M., 268
Glooschenko, V., 257, 269
Godlin, M. M., 54
Goffman, W., 230
Golley, F. B., 69, 100
Golz, D. J., 360
Gonçalves da Cunha, A., 408, 412, 413
Goodall, D. W., 96, 177
Goodhart, C. B., 111, 115, 116, 120, 123, 125, 129-33
Goodwin, D., 10, 11
Goodwin, P., 270
Gorbman, A., 297, 298
Gore, A. J. P., 55, 59
Gorham, E., 59, 62, 66, 70, 87
Gosnell, H. T., 14
Gosper, D., 11
Goss, R. J., 384, 388
Gosz, J. R., 53, 59, 60, 63, 65, 71, 72
Gough, D. I., 275
Gough, W. I., 275
Gould, H. R., 260
Graham, A., 359, 360, 371
Graham, W. G., 261
Grainger, E. H., 274
Gramet, P., 10
Gray, R., 286, 299
Green, J. M., 286
Green, M., 365
Green, R., 153
Greene, H. W., 439

Greenham, C. G., 96
Greenland, D. J., 100
Greenwood, J. J. D., 110, 117, 120, 123, 125-27
Greer, A. E., 437, 442
Greer, A. E. Jr., 437
Gregory, F. G., 96
Gregory, J. T., 356, 360-62, 365-67, 372, 373
Gregory, R. W., 272
Grier, C. C., 52-54
Griffin, D., 40
Griffin, D. R., 290
Griffin, J. A., 408
Griffiths, R. C., 294, 298
Grimas, U., 270
Grime, J. P., 92, 111, 126
Grishina, L. A., 55
Groner, M. G., 408, 412
Groot, C., 287, 290
Gross, R. A., 54
Groves, A. B., 297
Grubb, E. A. A., 88
GRUBB, P. J., 83-107; 83-93, 95-99
Guerrucci, M.-A., 133
Guerrucci-Henrion, M.-A., 111, 114, 119, 120, 133
Guggisberg, C. A. W., 437
Guha, M. M., 96
Gullstrand, C., 54
Gunning, G. E., 296
Gupta, H. K., 275
Gurney, J. H., 3
Gustafson, E. P., 373
Guthrie, R. D., 373
Gutierrez, A. P., 333, 346
Guyton, A. C., 196
GWINNER, E., 381-405; 381, 382, 384, 385, 387-401

H

Haas, H. J., 56
Haase, A. T., 40
Haber, W., 64
Hadidi, A., 39
Hadley, M., 54
Hadley, N. F., 162
Haenni, A. L., 39
Hafez, M., 337
Haffer, J., 356
Hagen, K. S., 340, 343, 345
Haggstrom, A., 113
Hahn, W. E., 297, 298
Hairston, N. G., 147, 209, 212-14, 216, 217, 221, 224-26, 230
Haldane, J. B. S., 136
Halkka, O., 133
Hall, A., 257

Hall, B. M., 407
Hall, J. C., 345
Hamblin, P. F., 259
Hamilton, M. E., 11
Hamilton, P. A., 339
Hamilton, W. D., 15, 19
Hamilton, W. J., 19, 20
Hamilton, W. J. III, 11
Hamrick, J. L., 166, 167
Hamzah, A., 93
Hanawalt, R. B., 53-55, 62
Hankioja, E., 396
Hanshaw, B. B., 60, 63
Hanson, H. C., 194
Hara, T. J., 296
Hardcastle, A., 11
Hardy, A. D., 412
Hardy, D. E., 239
Hardy, R. J., 215, 216
Harley, J. L., 63, 93, 94
Harper, J. L., 166, 167
Harper, T. L., 180
Harrell, B. E., 371
Harris, H., 310, 313, 324
Harris, J. M., 360, 362
Harris, V. A., 439
Harris, W. F., 53, 59
Harrison, C. J. O., 16
Hartman, W. L., 286, 299
Hartt, A., 287
Hartt, A. C., 285
Harvey, H. H., 272
Harvey, P. H., 110, 111, 119, 120, 124, 127
Hasler, A. D., 273, 286, 287, 290, 296, 297
Hassell, M. P., 153
Hatrick, A. A., 95
Hauk, F. R., 271
Haupt, H., 408
Hausermann, E., 412, 413
Havas, P., 65, 68, 70
Havens, Y. H., 62
Havenstein, D. E., 346
Haverschmidt, F., 9
Hawley, D. A., 39
Hayes, F. R., 297
Hayes, M. B., 312
Haynes, F. N., 120, 134
Hayward, G. S., 41
Hazelwood, D. H., 164
Healy, W. R., 161
Heath, D. J., 117
Heathcote, G. D., 344
Hebert, P. D. N., 312
Hedgecock, D., 242, 249
Hedrick, P. W., 310, 323
Heed, W., 322
Hejny, S., 268
Hela, I., 286, 294, 295
Helder, R. J., 408, 411, 415
Heller, H. C., 392

Henderson, H. F., 270, 276, 290
Henshaw, P. C., 369
Hensley, S. D., 417
Herrick, F. H., 9
Herskowitz, I., 43
Herzog, H. A. Jr., 437, 442
Hess, E. H., 286, 296
Hester, J. B., 53
Hewitt, N. E., 312
Heydecker, W., 95
Heyligers, P. C., 54
Hibbard, C. W., 360
Hickman, J. C., 167
Hiesey, W. M., 166
Hildebrand, M., 357
Hilger, F., 67
Hill, W. G., 131
Hille Ris Lambers, D., 332
Hillier, P. C., 312
Hines, A. H. Jr., 156
Hinks, J. D., 418
Hirano, Y., 285, 293
Hirsch, W. M., 215, 221-25, 227-29
Hirshfield, M. F., 153, 158
Hnatiuk, R. J., 90
Hoar, W. S., 286, 288, 293
Hoddenbach, G. A., 162
Hodek, I., 345
Hoenigsberg, H., 242, 243
Hoffman, D. A., 270
Hoffman, G. L., 270
Hoffman, R. S., 373
Holcomb, L. S., 9
Holder, D. G., 411
Holford, T. R., 215, 216
Hollands, M., 294
Hollick, F. S. J., 329
Holling, C. S., 127, 176, 188, 189
Holman, J. A., 365, 371
Holmes, R. T., 196
Holmquist, K. A., 54
Hom, R., 339
Homeyer, E. F. von, 381
Hong Woo, K., 286, 296
Hoogland, J. L., 8, 19
Hope, G. S., 85
Hopkins, D. M., 360, 367
Hopkinson, D. A., 310, 313, 324
HOPSON, J. A., 429-48; 430-32, 434-36, 440, 441, 443
Horejsi, B. L., 205
Horrall, R. M., 286, 287, 290, 296, 297
Horsley, D. T., 117, 118, 125, 127, 128
Hosking, E., 7
Hougen, V. H., 54

Hough, J., 363
Howard, C. S., 261
Howard, H. E., 14, 15
Howard, P. J. A., 59
Howard, R. A., 87, 414
Howell, P. J., 344
Howells, W. W., 10
Hoy, G., 21
Hozumi, K., 64
Hrbáček, J., 257, 260, 268
Hubby, J. L., 235-37, 240, 243, 245-50, 311, 322, 323
Hückler, U., 17
Hudson, W. H., 14
Huey, R. B., 162
Hughes, R. D., 335, 337, 346
Hunt, R., 92
Hunt, R. H., 437, 438
Hunt, R. M. Jr., 367
Hunt, W. G., 132
Hunter, A. S., 237, 242, 243
Hunter, H. C., 8, 16
Huntsman, A. G., 289, 297
Hurley, D. A., 286
Hussain, T., 364
Hussell, D. J. T., 163
Hussey, N. W., 339
Hutchinson, G. E., 258, 268
Hutchison, J. H., 366, 369
Huttel, C., 85
Huxley, J., 128, 136
Hyatt, R. A., 272
Hylmö, B., 93
Hynes, H. B. N., 264, 269

I

Ianishevskii, D. E., 414
Idler, D. R., 273, 296, 297
Immelmann, K., 381
Inamdar, J. A., 412, 413
Innis, G. S., 177
Ino, Y., 64, 67, 69, 70, 72
Inouye, D. W., 411, 412, 415, 416, 418, 419, 421
Irving, A. J., 120, 134, 135
Irwin, M. P. S., 2
Isaacson, A. J., 200
Ivanov, V. V., 55
Ivens, J. L., 257
Iwaki, H., 182
Izett, G. A., 369

J

Jackson, P. B. N., 257
Jackson, W. B., 164
Jacob, R. J., 41
Jacquaz, B., 295
Jaffrey, I. S., 242

Jagnow, G., 62
Jameson, D. A., 185
Janda, C., 413
Janensch, W., 442
Janota, D., 413
Janzen, D. H., 85, 412, 417, 421
Jarman, P. J., 194
Jarvis, P., 87
Jaspers, E. M. J., 34, 39
Jean, Y., 295
Jeffery, D. C., 410, 411, 413, 421
Jeffrey, D. W., 55
Jenkins, J. N., 411, 418
Jenkinson, D. S., 71
Jenner, E., 3, 11, 12
Jenni, D. A., 194
Jenny, H., 53, 56, 59-62, 72
Jensen, A. L., 297, 299
Jensen, M. K., 2-4, 11, 16, 21
Jensen, R. A. C., 2-4, 8, 10, 11, 13, 16, 21
Jensen, V., 57, 66
Jeppesen, L. L., 125
Jepsen, G. L., 361-63
Jerison, H. J., 430, 431, 434-36
Jobin, W. R., 219, 230
Johansson, L. G., 54
Johnson, A. W., 14
Johnson, C. G., 329
Johnson, D. E., 238
Johnson, D. W., 258
Johnson, F. L., 54
Johnson, F. M., 312
JOHNSON, G. B., 309-28; 235, 310-18, 320-24
Johnson, M. A., 260
Johnson, M. S., 113, 132
Johnson, W. E., 240-42, 249, 250
Joklik, W. K., 30, 38, 39
Jonas, R. E. E., 297
Jones, B. M. G., 87
Jones, D., 41
Jones, D. M., 39
Jones, E. W., 53
Jones, F. R. H., 285, 288, 290, 295-99
Jones, H. G., 258
Jones, J. K., 193
Jones, J. K. Jr., 372
JONES, J. S., 109-43; 110, 114, 117-20, 122, 123, 126, 127, 129, 133-35, 322
Jones, M. G., 336, 338
Jones, M. J., 54
Jones, P. J., 161
Jones, R., 285
Jones, R. C., 194
Jonez, A. R., 270

AUTHOR INDEX 455

Jonsson, S., 54
Jordan, C. A., 127
Jordan, C. F., 59, 62, 64, 65, 67
Jordan, M. J., 72
Jordan, P., 209, 210, 212
Jorgensen, J. R., 72
Jourdain, F. C. R., 3, 8-11, 13, 14, 16, 19
Joyner, D. E., 6, 18, 20
Judd, W. R., 275
Junker, J., 230

K

Kachugin, E. G., 258
Kado, C. I., 42
Kagan, I. G., 217
Kaji, S., 296
Kallio, P., 271
Kalmus, H., 286
Kanapi, C. G., 239, 240
Kanemasu, E. T., 64
Kaneshiro, K. Y., 239, 240, 242, 249, 250
Karsten, P., 202
Kastritsis, C. D., 243
Katznelson, B. H., 72
Kayama, R., 54
Kearsey, M. J., 120, 128
Keck, D. D., 166
Keddy, P. A., 263
Keeling, C. D., 73
Keith, L. B., 163, 164
Keller, B., 257
Kelley, A. C., 230
Kelley, J. J., 66, 69
Kempthorne, O., 147
Kendrick, B., 264
Kennedy, C. R., 226
Kennedy, J. S., 330
Kercher, J. R., 189
Kerner, A., 412
Kershaw, A. K., 263
Kestemont, P., 53
Key, B. A., 95
Khaltruin, K. D., 298
Khoury, G., 33
Kidd, N. A. C., 340
Kietzke, K. K., 361-63
Kiff, L. F., 2, 5
Killick, S. R., 286
Killip, E. P., 413
Kimball, B. A., 64
Kimball, S., 242
Kimura, M., 128, 131
King, C. E., 165, 166
King, J. M. B., 113, 127, 131, 134
King, J. R., 4, 7, 11, 392

King, M.-C., 148
Kingdon, J., 357
Kira, T., 53, 54, 67, 70
Király, Z., 94
Kirita, H., 64, 67, 70
Kirkham, D. R., 64-66, 69, 70
Kjellstrand, A. M., 54
Klaas, E. E., 4-6
Kleerekoper, H., 288
Klein, C. J., 71
Klein, H., 384, 385, 388, 389, 392-94, 396-401
Klemmedson, J. O., 53
Klinge, H., 53, 86, 98, 100, 101
Kloft, W., 338, 339
Klomp, H., 346
Klueter, H. H., 95
Klug, J., 412
Kluge, A., 371
Knight, C. A., 42
Knight, F. B., 344
Knudson, D. L., 31, 42
Koehler, R., 312
Koehn, R. K., 313
Koernicke, M., 413
Kohler, M. A., 258
Kojima, K., 239, 250
Kojima, K. I., 120, 128, 238
Kolbe, W., 337
Kolesnikov, V. G., 294, 298
Kolft, W., 332
Kondo, H., 285, 293
Kondratjev, N. E., 258
Kononova, M. M., 53-62, 70-73
Koo, R. C. J., 96, 97
Koopowitz, H., 410, 411, 413, 421
Köster, F., 11, 19
Kothbauer, H., 113, 120
Kovda, V. A., 174, 184
Kowal, J. M. L., 100
Kowalevsky, W., 365
Kozyreva, M. G., 59
Kramer, F. R., 31
Kramer, G., 286
Krebs, C. J., 127
Kreulen, D. A., 111, 114, 123, 126
Krizek, D. T., 95
Krug, A., 43
Krumholz, L. A., 164, 168
Krzyzanek, E., 265
Kucera, C. L., 58, 60, 64-66, 69, 70
Kuenen, Ph. H., 260
Kuenzi, W. D., 369
Kujawa, M., 272
Kung, S., 312
Kunkel, H., 332
Kuroda, N., 7, 8

Kurstak, E., 30, 36
Kurtén, B., 442
Kushlan, J. A., 438
Kvassman, S., 113
Kylin, A., 93

L

Labar, G. W., 287, 290
Lack, D., 1-4, 6, 7, 12-14, 146, 163, 193, 442
Laevastu, T., 286, 293-95
Laird, M., 271
Lake, J. R., 31
Lakovaara, S., 237
Lamba, B. S., 10, 11
Lamotte, M., 110, 111, 113, 117-20, 124, 125, 128, 130, 133, 134
Lance, J. F., 371, 373
Landers, T. A., 39
Lang, A., 110, 114
Lang, G. E., 60
Lang, J. W., 437, 440
Langston, W. Jr., 441, 442
Lankinen, P., 237
Lannom, J. R., 162
Larcher, W., 87, 92, 95
Laskey, A. R., 14, 15
Laster, M. L., 418
Latteur, O. H., 9, 13
Latteur, G., 338
Lauckner, F. B., 344
Laura, R. D., 71
Laver, W. G., 33
LaVoy, A., 272
Lawford, A. L., 295
Lawler, G. H., 270
Lawton, J. H., 413, 414
Laychock, S., 413
Leal, M. P., 418
Lechevalier, M. P., 72
Leclerc, M., 258
Le Dû, R., 8
Lee, C., 312
Lee, D. S., 437
Lee, K., 216, 225, 230
Lee, T. N. H., 33
Leentvaar, P., 257, 261, 271
Lees, A. D., 330, 332, 333, 346
Lefolii, K., 273
LEGGETT, W. C., 285-308; 286, 287, 290, 293-95, 297-99
Lehmkuhl, D. M., 272
Leigh, E. G. Jr., 86, 87, 90, 92
LEITH, B. H., 109-43
Leius, K., 416
Lemon, E. R., 64
Leonard, A. B., 372
Leopold, E., 359, 360

Levan, G., 113
Levins, R., 187
Lewis, E. R., 216, 225, 230
Lewis, R. A., 388
Lewis, T., 214-17, 221, 224, 225, 229
Lewontin, R. C., 131, 136, 148, 149, 153, 155, 165, 166, 236-38, 240, 247, 310, 311, 322, 323
Leyton, F. L., 413
Leyton, M. K., 222-24
Lichtenberg, F. von, 221
Lieff, B. C., 4, 20
Lieth, H., 62, 64
Likens, G. E., 52, 53, 57, 59-61, 63, 65, 71-73, 264
Lillegraven, J. A., 360, 363
Lindeman, R. L., 175
Lindsay, E. H., 360, 365, 369
Lindstrom, T., 256, 257, 269, 270
Linhart, H., 217, 218, 221, 224
Link, J. A., 3
Linsley, R. K., 258
Littauer, U. Z., 39
Liversidge, R., 3, 4, 7, 8, 10, 11
Lloyd, F. E., 412, 413
Lloyd, J. R., 83
Loeb, L. A., 40
Lofts, B., 6
Lokki, J., 237
Lommen, P. W., 430
Loper, G. M., 408, 410, 413
Lord, R. D. Jr., 164
Lorz, H. W., 297
Lossaint, P., 54, 69, 70
Lotka, A. J., 147
LOUCKS, O. L., 173-92; 91, 174, 180, 184
Lousier, J. D., 59
Love, J. D., 363
Loveless, A. R., 97
Loverre, A., 311
Lowe, C. H., 162
Lowe-McConnell, R. H., 257
Lubbock, J., 414
Lubega, S., 312
Luferov, V. P., 266
Lugo, A., 64, 65, 67
Lugo, A. E., 69
Lukefahr, M. J., 408, 410, 418
Lundegårdh, H., 62, 63
Lunt, H. A., 59
Luria, S. E., 30, 38
Lüttge, U., 408
Lutz, H. J., 62
Lynch, J. D., 371
Lynes, H., 20

M

MacArthur, R. H., 145, 147, 153, 263
Macdonald, G., 221-26, 228, 230
Macdonald, J. R., 272, 366
MacDonald, K. B., 69-72
MacDonald, L. J., 366
MacDonald, S., 296
MacFadden, B. J., 364
MacFadyen, A., 69, 329
MacGinitie, H. D., 358-60, 362, 365
MacGinnis, J. T., 164
Machniak, K., 270
MacInnes, C. D., 4, 20
MacIntyre, R., 311
Mackauer, M., 335, 344, 346
MacKinnon, D., 296
MacLean, S. A., 196
MacPherson-Stewart, S. F., 111, 126
Maddox, J. B., 269
Madison, D. M., 290, 296
Mäenpää, E., 65, 68, 70
Maheshwari, J. K., 408, 414
Maheshwari, P., 413
Maheshwart, J. K., 412-14
Maiorana, V. C., 161
Makatsch, W., 1-3, 7-9, 11, 13, 16
Malaher, G. W., 259, 276
Malar, T., 288
Malchevsky, A. S., 1, 16
Maldague, M. E., 67
Malek, E. A., 209
Malinin, L. K., 286
Manners, I. R., 183, 184
Manning, J. T., 333
Manwell, C., 113
Marak, R. R., 285, 294
Maramorosch, K., 30, 36
Margalef, R., 205, 257, 263, 267, 268
Margolis, H., 408, 410, 413
Margulis, L., 38
Markkula, M., 337
Markov, M. V., 181
Marks, R. H., 54, 62
Marples, T. G., 54
Marshall, A. J., 381, 388
Marshall, E. R., 197
Marshall, J. T., 197
Marshall, W. G., 55
Martin, L. D., 373
Martin, M. A., 33
Martin, P. S., 371, 374
Martin, W. R., 295
Martinet, J., 414

Marx, H., 371
Maryánska, T., 440
Masters, M. T., 411
Matheny, R. T., 255, 267
Mattei, G. E., 412
Matthew, W. D., 361, 362
Matthews, G. V. T., 286
Matthews, R. E. F., 29, 31, 34-37
Maurizio, A., 408
Maxson, L. R., 148
Maxwell, F. G., 411, 418
May, R. M., 153, 223-25, 229
Mayfield, H., 2-5, 10, 14, 15, 21
Mayoh, H., 296, 297
Mayr, E., 20, 128, 136, 147, 235, 243, 247
McAlister, W. B., 294
McAuslan, B. R., 30, 44
McBride, J. R., 297
McCallum, K. J., 58
McCleave, J. D., 287, 288, 290, 295, 297
McClelland, J. E., 54
McColl, J. G., 71
McCracken, G. F., 113
McDonald, J. F., 312
McDowell, R. E., 247
McFee, W. W., 53, 60
McGeen, D. S., 2, 13
McGeen, J. J., 2, 13
McGinnis, J. T., 100
McGregor, S. E., 408, 410, 413
McGrew, P. O., 360
McInerney, J. E., 292
McInnes, D., 20
McKay, S., 340
McKenna, M. C., 359, 363
McKenzie, J. A., 312
McLachlan, A. J., 257, 265-67, 269-71, 276
McNab, B. K., 430
McNaughton, S. J., 167
McNeill, W. H., 44
McSween, S., 274
McVean, D. N., 90
Mech, L. D., 194
Medem, F., 438
Medica, P. A., 162
Mengel, R. M., 372
Meredith, W. R., 418
Mermel, T. W., 277
Merriam, J. C., 364
Mertz, D. B., 165, 166
Meslow, E. C., 163, 164
Messenger, P. S., 342, 343, 345
Meszoely, C. A. M., 371
Metcalf, R. L., 413
Mewaldt, L. R., 401
Meybeck, M., 258

AUTHOR INDEX 457

Meyer, C. R., 411
Meyer, V. G., 411
Michelbacher, A. E., 340, 342
Michelson, E. H., 219
Middlekauf, W. W., 340, 342
Miklovich, R., 312
Miles, E. F., 56
Miles, J., 96
Miles, S. G., 297
Milkman, R., 311, 312
Miller, A. H., 365, 381
Miller, D., 285, 294
Miller, D. D., 238
Miller, H., 40
Miller, M. J., 39
Miller, P. C., 177
Miller, R. B., 96
Mills, D. R., 31
Milne, W. G., 275
Mims, C. A., 30, 44
Minderman, G., 58-60, 64
Minkowski, K., 197
Mironova, H. V., 291
Mirsky, P. J., 19
Mitchell, K. J., 95
Mittler, T. E., 332
Mitzutani, S., 40
Miyake, M., 285, 293
Miyata, I., 88
Modha, M. L., 437
Mogen, C. A., 54
Monsi, M., 64, 67, 69, 70, 72
Montieth, J. L., 91
Moore, J. C., 164
Moore, P. A., 346
Mooser, O., 370, 373
Morduchai-Boltovskoi, F. D., 264, 266
Moreau, R. E., 2
Morel, G., 7
Morel, M.-Y., 2-7, 10, 11, 13, 14, 21, 22
Morgan, A. H., 266
Morgan, P., 312
Morgensen, J., 3, 9
Moriarty, C., 290
Mörnsjö, B., 72
Morris, D., 116
Morrison, I. K., 53
Morton, M. L., 401
Moscarello, M., 312
Moss, R., 204
Mound, L. A., 410, 411, 413, 414
Mountford, M. D., 153
Mountfort, G., 3, 11
Mourao, C. A., 242, 243
Moynihan, M., 200, 201
Mrosovsky, N., 392
Muench, H., 212-14, 230
Müller, H. J., 336

Muller, K. A., 9
Müller, P., 356, 357
Mullins, M. G., 94
Mundy, P. J., 3, 10, 11
Murphy, G. I., 145, 147, 153, 155, 164
Murphy, R. M., 71
Murray, J. J., 110, 111, 113, 114, 117, 118, 120, 123, 128, 129, 132, 134
Murton, R. K., 6, 200
Myers, J. H., 127

N

NAHMIAS, A. J., 29-49; 30, 35, 40, 41, 43
Nair, P. S., 239, 241, 250
Nakane, K., 53, 62, 66, 67
Nakayama, N., 285, 293
Narayan, O., 40
Narayanan, R., 96
Narokova, R. P., 54
Nasell, I., 215, 216, 221-25, 227-29
Nathans, D., 33
Nathanson, N., 45
Nebeker, A. V., 272
Needham, L., 312
Neel, J. K., 257
Negm, A. A., 417
Nei, M., 236, 238, 310, 323, 324
Neill, W. T., 437
Neu, H. J. A., 274
Neufeldt, I., 10, 11, 16
Nevling, L. I., 90
Newbury, R., 259, 276
Newcombe, L., 410, 414, 421
Newill, V. A., 209
Newman, G. A., 4
Newton, A., 8, 16
Ngugen, T. T. H., 64, 66, 67, 69, 70
Nice, M. M., 4, 7, 11, 13, 15
Nichols, R., 369
Nicolai, J., 1, 2, 10, 12, 14, 17, 18
Niering, W. A., 62, 356
Niethammer, G., 7, 11, 15
Nihlgard, B., 53
Nilsson, A., 133
Nilsson, B., 72
Nishida, T., 417
Nishihara, T., 31
Niveleau, A., 39
Nix, H. A., 346
Nobel, P. S., 91
Noble, K., 114, 133
Nolan, V., 19, 20

Nordeng, H., 297
Nordstrom, S., 133
Norman, R. F., 7
Norris, R. T., 2, 11
Northcote, T. G., 294, 297, 298
Nungesser, W. C., 373
Nursall, J. R., 265, 266, 269
Nye, P. H., 94, 101

O

Oakley, B., 297
Oatman, E., 340
Obeng, L. E., 257
Obreebski, S., 242
O'Connor, R. J., 10
O'Donald, P., 111, 116, 126, 128
Odum, E. P., 61, 175, 263
Odum, H. T., 64, 65, 67, 69, 85, 90, 92, 93, 95, 99
Ogawa, H., 53, 54
Ogden, J., 438
Ohmart, R. D., 11
Ohnesorge, B., 338, 339
Oka, H.-I., 167
Olkowski, H., 339
Olkowski, W., 339
Ollagnier, M., 96
Olrog, C. C., 357
Olsen, O. A., 271
Olson, J. S., 55, 59, 60, 62, 100, 175, 176
Olson, S. L., 371
O'Neill, R. V., 177
Ono, K., 414
Oppenheimer, H. R., 91
Organ, J. A., 161
Orgel, L. E., 40
Orians, G. H., 19, 20, 194, 197
Orr, M., 412
Ortega, J. C., 340, 342
Oshima, K., 297, 298
Osmólska, H., 435, 440
Oster, G. F., 150, 155, 158
Ostrofsky, M. L., 267
Ostrom, J. H., 429, 430, 434-36, 439, 440, 442, 444
Ottow, H., 21
Ottow, J., 4
Ouellet, M., 258
Ouellette, R., 62, 64
Ovchinnikov, S. M., 54
Ovchinnikov, V. V., 288
Overton, W. S., 177
Ovington, J. D., 100
Owen, D. F., 133, 415
Owen, J. H., 4, 11, 13, 14
Oxford, G. S., 113, 114

P

Pachikina, L. I., 54
Packard, G. C., 442
Padgett, T., 38, 42
Padilla-Saravia, B., 61
Painter, P. R., 153
Palmgren, P., 401
Pareek, O. P., 95
Parija, P., 411, 412
Park, T., 165, 166
Parker, W. S., 156
Parkhurst, D. F., 91
Parkin, D. T., 110, 117-20, 125, 127, 128, 133
Parkinson, D., 59
Parrott, W. L., 418
Parry, M. L. I., 290
Parry, W. H., 338, 339
Parsons, P. A., 312
Pasteur, G., 242
Patel, N. D., 411, 412, 420
Patel, R. C., 413
Paterson, C. G., 265
Patten, B. C., 63, 176, 289, 291
Patterson, B., 358
Patton, T. H., 370, 371
Paul, E. A., 58
Paulhus, J. L. H., 258
Paulini, E., 230
Pavlovsky, O., 237, 242, 243
Pavshtiks, E. A., 294
Payne, K., 2, 4, 12, 14-16
PAYNE, R. B., 1-28; 1-4, 6-8, 10, 12-18, 20-22
Pecan, E. V., 51
Pengelley, E. T., 381
Penhoet, E., 40
Pennington, T. D., 83
Pennycuick, C. J., 401
Perdeck, A. C., 389, 396
Pérez-Salas, S., 242, 243
Perrin, R. M., 335, 344
Perrins, C. M., 161
Perrot, J.-L., 109, 110
Perrot, M., 109, 110
Petaj, V., 414
Peters, R. H., 262
Petersen, L. J., 38
Peterson, K. M., 66, 69
Peterson, O. A., 365
Petr, T., 266
Petrakis, P. L., 311
Petrusewicz, K., 165, 166
Pettersson, O., 295
Pfeiffer, E. W., 373
Phillips, W. W. A., 3
Phillipson, J., 60, 63, 68, 70
Pianka, E. R., 145, 156, 162, 263
Pieczynska, E., 268

Pigott, C. D., 89
Pimentel, D., 174, 184, 188
Pinck, L. A., 71
Pipkin, S. B., 312
Pitelka, F., 196
Pitman, M. G., 94
Plass, G. N., 73
Platel, R., 432, 434, 435, 443
Poff, R., 296
Pohjola, L., 237
Pohl, H., 388
Pollak, E., 147
Pooley, A. C., 437-40, 442
Portmann, A., 198
Pospelova, E. B., 55
Post, W., 5
Potter, J. H., 312
Potts, D. C., 123
Potts, G. R., 338
Poulsen, M., 412
Poulson, T. L., 392
Pound, G. S., 187
Powell, J., 310
Powell, J. R., 235, 237, 242
Powell, W., 333, 338
Powers, W. L., 64
Pozsár, B. I., 94
Prakash, S., 237, 247, 322, 323
Preble, N. A., 9, 11
Prevett, J. P., 4, 20
Prevot, P., 96
Price, D., 40
Price, D. J., 110, 111, 113, 119, 120, 126
Price, G. R., 151
Price, P. W., 165, 166
Priston, R. A. J., 31
Pritchard, A. L., 286
Prochiantz, A., 39
Promptov, A. N., 396
Prosser, M. V., 264
Prout, T., 123
Prowse, G. A., 267
Pugh, G. J. F., 59
Purcell, L. T., 261
Putman, R. J., 60, 63, 68, 70
Putman, W. L., 410, 411, 413, 414, 416, 417, 420
Puttler, B., 345

Q

Quenum, M., 408, 415

R

Rabb, G. B., 371
Rachmilevitz, T., 408
Radinsky, L. B., 431
Raethel, S., 196
Raleigh, R. F., 286, 299

Ralls, K., 197
Ralph, C. P., 9, 11
Ramakrishnan, P. S., 167
Rand, A. S., 162, 439
Randall, P. J., 96
Randolph, P. A., 335
Ranger, G. A., 7, 10, 11, 15
Ranney, C. D., 418
Ranson, R. M., 381
Rao, V. S., 408, 409, 412, 418
Rapp, M., 54, 87, 96, 101
Rasmussen, D. L., 369
Rastogi, B. K., 275
Ratcliffe, F. N., 36, 46
Ratsey, M., 114
Rauner, Y. L., 182, 183
Raunkiaer, C., 84, 92
Rautapää, J., 337, 338
Raven, P. H., 415, 418
Raw, F., 329
RAWLINGS, P., 109-43
Read, J. W., 59
Reader, R. J., 55
REANNEY, D. C., 29-49; 33, 37, 39, 41, 43
Reddingius, J., 154
Reed, C. A., 363, 374
Reed, C. F., 110, 129
Reed, E. L., 411-13
Reed, R. A., 4, 10, 14, 16
Reeder, W. G., 362
Rees, W., 128
Reichle, D. E., 53, 59, 175, 180, 189
Reifschneider, D., 174, 184, 188
Reiners, N. M., 55, 71
Reiners, W. A., 52, 54, 55, 57, 58, 62, 64-66, 68-73
Rennie, D. A., 58
Rensberger, J. M., 365
Rensch, B., 8, 22
Repenning, C. A., 367, 368, 373
Repnevskaja, M. A., 66, 68, 72
Reteyum, A. Yu., 275
Rey, E., 8, 13
Rhodes, C., 312
Rhyne, C. L., 408, 411, 418, 420
Richards, A. V., 118, 129, 134
Richards, P. W., 53, 84, 85, 90-92
Richardson, A. M. M., 111, 116, 118, 126
Richardson, B., 257, 276
Richardson, M. E., 244, 250
Richardson, R. H., 244, 250, 310, 325
Richkus, W. A., 292
Richmond, R. C., 242, 243, 249

AUTHOR INDEX

Ricklefs, R. E., 21, 22
Ricqlès, A., 430, 443
Ridet, J. M., 434, 443
Ridgell, R. H., 59
Ridgway, C. S., 412
Ridley, H. N., 411, 412, 414-16, 418, 419, 421
Ridley, J. E., 257
Rightmire, C. T., 60, 63
Riley, G., 267
Riley, G. A., 297
Ripley, S. D., 11
Risser, R. G., 54
Robbins, K. C., 43
Roberts, N., 286
Robertson, A., 131
Robertson, J. R., 373
Robertson, R. J., 7
Robinson, P., 363, 369
Robson, E. B., 310
Rochereau, S., 174, 184, 188
Rochow, J. J., 87
Rocklin, S., 150
Rockwood, E. S., 240
Rodbard, D., 314
Rodhe, W., 267
Rodin, L. E., 52
Rodrigues, W. A., 53, 86, 101
Rogers, J. S., 238
Roizman, B., 41
Romanenko, V. I., 275
Romell, L. G., 52-54, 57, 62, 65, 66
Rommel, S. A. Jr., 287, 295
Rongstad, O. J., 163, 164
Rosa, H. Jr., 293, 294
Rosen, D. E., 1
Rosenfield, P. L., 215, 230
Rosenzweig, M. L., 153, 156
Rosner, D. E., 441
Rosswall, T., 54
Rotarides, M., 117
Roth, J. J., 442
Rothé, J. P., 259
Rothstein, S. I., 2, 5, 8, 9
Roughgarden, J., 147
Rousefell, G. A., 299
Rousseau, A., 258
Rowan, W., 381
Royce, W. F., 287
Rozich, W. R., 410, 414, 421
Rubilin, Y. V., 53, 56, 59
Rucker, R. R., 272, 273
Rudd, V., 412
Rühling, A., 72
Ruinen, J., 102
Ruiz-Reyes, J., 90
Rundgren, S., 133
Rüppell, W., 395
Russell, D. A., 431, 432, 435
Russell, E. J., 62
Russell, G. J., 39

Russell, L. S., 429
Russell, R. J., 339
Russell, R. S., 93
Rutter, E. J., 256, 257
Rydén, B. E., 54
Ryder, R. A., 270, 276
Ryszkowski, L., 175, 181-84

S

Sacchi, C. F., 120
Sadler, K. C., 164
Saila, S. B., 289
Salem, A. E., 56, 60, 62, 72
Salisbury, E. J., 413
Salmon, M., 286
Salomon, R., 39
Salt, G., 329
Samal, K., 411, 412
Sambrook, J., 30, 44
Samdal, J. E., 262
Samis, R., 174, 184, 188
Samuels, G., 61
Sanders, C. J., 344
Sanft, K., 202
Sanghvi, L. D., 310, 323
San Pietro, A., 174, 186
Santosa, P. D. N., 64, 66, 67, 69, 70
Sapio, S., 66, 68
Sarich, V. M., 148
Sasseville, J. L., 262
Satchell, J. E., 53
Satou, M., 296
Saura, A., 237
Savage, D. E., 359
Savage, J. M., 371
Saville, C. M., 262
Schaefers, G. A., 332
Schaffer, W. M., 145, 153, 156, 298
Schalz, A. T., 296
Schappert, H. J. V., 64, 65, 69-72
Scharloo, W., 311
Scharpenseel, H. W., 60
Schiermann, G., 4, 13
Schifferli, A., 17
Schilder, F. A., 120, 123, 131, 133
Schilder, M., 120, 123, 131, 133
Schildmacher, H., 401
Schimper, A. F. W., 414
SCHLESINGER, W. H., 51-81; 55
Schlinger, E. I., 343, 345
Schmidt-Koenig, K., 389
Schnell, R., 408, 411, 413, 415, 419
Schnetter, M., 117

Schnitzler, S., 113, 120
Schoener, T. W., 417
Scholey, G. J., 13
Schönwetter, M., 13
Schoone, A., 311
Schremmer, F., 413, 414
Schroer, F. W., 54
Schuitema, K. A., 111, 126
Schulz, J. P., 53
Schulze, E. D., 65-67, 69
Schüz, E., 389, 395
Schwab, R. G., 384, 387
Schwartz, D. M., 63
Schwassmann, H. O., 285, 286
Scopes, N. E. A., 345
Scott, W. B., 361-64
Sears, P. B., 356, 357
Sedlmair, H., 118
Seibert, R. J., 411, 412
Selander, R. K., 132, 235
Sene, F. M., 241
Seppä, J., 7
Serventy, J. B., 381, 388
Sestak, Z., 87
Sethi, S. L., 332
Shackleton, D. M., 205
Shah, K., 45
Shalaeva, N. M., 55
Shalla, T. A., 38
Shappy, R. A., 289
Sharoshkina, N. B., 54
Shearer, J. W., 330
Sheldrick, P., 41
Shelford, R., 9
Shepard, M., 174, 184, 188
Sheppard, P. M., 111, 113, 115-18, 120, 126-29, 131, 134
Sher, A. V., 373
Sherman, M. S., 71
Sherman, P. W., 8, 15, 19
Shorrocks, V. M., 93, 94, 96, 97
Shotwell, J. A., 365, 369, 373
Shtegman, B. K., 257
Shuel, R. W., 408
Shugart, H. H. Jr., 189
Siccama, T. G., 61
Sick, H., 4
Siddiqui, W. H., 335
Siever, S., 258
Sij, J. W., 64
Silagi, S., 242
Silvester, W. B., 94
Simons, E. L., 358
Simpson, B. B., 166, 356
Simpson, G. G., 20, 357
Sinclair, D. C., 265
Singer, M. J., 53
Singh, H., 311
Singh, R. S., 247, 311, 312
Singletary, C., 438
Sites, J. W., 96, 97

Sivanayagam, T., 95
Skead, D. M., 10, 14
Skutch, A. F., 9, 11, 14, 196
Slade, W. R., 31
Slobin, L. I., 39
Sluss, R. R., 340, 342, 343
Small, E., 100
Small, J. K., 413
Smith, E. A., 174, 184, 188
Smith, F. B., 62, 64
Smith, F. E., 176
Smith, H. W., 55
Smith, J. M., 151
Smith, J. M. B., 90
Smith, L. S., 272, 287
Smith, M., 297
Smith, M. H., 164
Smith, N., 364
Smith, N. A. F., 255
Smith, N. G., 4, 22
Smith, P. F., 96
Smith, R. A., 215, 230
Smith, R. M., 61
Smith, S., 7
Smith, W., 412
Smith, W. K., 91
Smouse, P. E., 244, 250, 310, 325
Sneath, P. H. A., 41
Snedaker, S. C., 69
Snegirev, I. A., 260
Snow, D. W., 116
Snyder, W. C., 175
Soemarwoto, O., 182
Soerohaldoko, S., 64, 66, 67, 69, 70
Sokolova, T. A., 54
Solbrig, O. T., 166-68
Sollins, P., 53, 59, 63-65, 67
Solomon, D. J., 297
Soltera, R. A., 261
Somme, S., 290
Sonesson, M., 54
Son'ko, M. P., 54
Sorensen, L. H., 71
Soulides, D. A., 71
Southern, H. N., 2, 3, 5, 7, 8, 16, 17
Southern, W. E., 4
Southwood, T. R. E., 153
Sowls, L. K., 195
Sowray, P. A., 344
Soysa, S. W., 413
Sparrow, L. A. D., 338
Spassky, B., 237, 242, 243
Spector, D. H., 38, 42
Spedding, C. R. W., 178-80
Spencer, O. R., 11
Spiegelman, S., 31, 46
Spieth, H. T., 239
Spotila, J. R., 430
Sprague, D., 66, 70, 72

Springer, E., 312
Sprules, W. M., 265
Srebnova, L. V., 55
Stahl, E., 408
Stalker, H. D., 239
Stanhill, G., 185
Stanley, K. O., 361
Stanley, S. M., 357
Starkey, R. L., 62, 71, 72
Stasko, A. B., 285, 287, 290
Stasko, D., 287, 290
Stearns, R. H., 262
STEARNS, S. C., 145-71; 145, 164
Stebbins, G. L., 357
Steel, J., 60, 63, 68, 70
Steel, J. A., 257
Steele, D. H., 271
Steenis, C. G. G. J. van, 93
Stehelin, D., 38, 42
Steiner, W. W. M., 240, 242, 249, 250
Stent, G. S., 148
Stephens, J. J., 360
Stevens, J. B., 360
Stevens, K. R., 59
Stevens, M. S., 360, 370
Stevens, P. F., 83, 84, 89, 90, 92
Stevenson, I. L., 71, 72
Stewart, J. M., 55
Stewart, R. H., 370
Steyn, P., 3, 10
Stimming, R., 8
Stirton, R. A., 360
Stirzacker, D., 230
Stocking, C. R., 408
Stolyarov, V. P., 270
Stolzy, L. H., 72
Stommel, H., 267
Stone, E. L., 53, 60
Stone, W. E., 412
Stone, W. S., 239, 240
Storer, J. E., 365
Stotzky, G., 72
Stowring, L., 40
Straskraba, M., 257
Stresemann, E., 10
Stroud, R. K., 272
Struthers, G., 290
Stuart, T. A., 286
Stubblefield, E., 38, 42
Studenikina, E. M., 294
Sturrock, B. M., 220
Sturrock, R. F., 214, 215, 220
Subak-Sharpe, J. H., 39
Sullivan, C. M., 291
Sunderland, K. D., 338
Surorov, A. K., 54
Sutcliffe, J. F., 94
Sutherland, O. W., 338
Sutterlin, A. M., 286, 299

Sutton, J. F., 369
Svardson, G., 165
Swain, A., 290
Swenson, K. G., 332
Swinton, W. S., 429, 431
Swynnerton, C. F. M., 8

T

Takahashi, Y., 155
Talbot, L. M., 183, 366
Talbot, M. H., 366
Talling, J. F., 264, 267
Tallis, G. M., 222
Tamaki, G., 333, 345
Tamm, C. O., 96
Tanner, E. V. J., 83, 85-88, 90, 93, 95-102
Targul'yan, V. O., 54
Tartof, K. D., 40
Tarverdiyev, R. B., 258
Tassy, B., 86
Taylor, B., 370, 371
Taylor, B. E., 370
Taylor, L. R., 335, 346
Taylor, O. R., 411, 412, 415, 416, 418, 419, 421
Tedford, R. H., 370, 373
Tedrow, J. C. F., 54, 69, 70
Tegelstrom, H., 113
Teh, C. K., 41
Teichmann, H., 296
Temin, H. M., 33, 37, 38, 40
Tenney, F. G., 59
Terhune, E. C., 174, 184, 188
Tesch, F. W., 285, 286, 288
Tessman, N., 373
Thienemann, A., 263
Thomas, R. G., 95
Thompson, C. F., 19, 20
Thompson, C. V., 441
Thomsson, D. A. W., 199
Thorig, G., 311
Thorne, G. N., 95
THROCKMORTON, L. H., 235-54; 235, 236, 240, 242, 243, 245-50, 311
Thulborn, R. A., 430, 440
Tidwell, T., 242
Tihen, J. A., 371
Tikhonenko, T. I., 30, 37
Tilley, S. G., 161, 162, 168
Timin, M. E., 189
Timms, A. M., 288
Tinker, P. B., 94
Tinkle, D. W., 147, 153, 158, 162, 168
Tischler, W., 175
Tobari, Y. N., 238
Todaro, G. J., 37, 38, 42
Toetz, D. W., 267, 268

AUTHOR INDEX 461

Toha, M., 93
Tomlinson, N., 297
Toran, J., 257
Townes, H., 416
Tracey, M. L., 242, 249, 250
Tracy, C. R., 442
Travaglini, E. C., 40
Trefethen, P. S., 297
Trelease, W., 408, 411, 414-16, 418, 421
Trippa, G., 311
Trivers, R. L., 198
Trotzky, H. M., 272
Truog, E., 62
Tseeb, Ya. Ya., 267
Tsinober, A. B., 287, 288
Tsutsumi, T., 53
Tullis, E. C., 413
Turner, F. B., 162
Turner, J., 53
Tutt, H. R., 7
Tyler, F. J., 411, 413, 414
Tyler, G., 54, 72
Tyler, P. A., 261
Tyurin, P. V., 257

U

Ueda, K., 296
Ulrich, B., 53
Ulrich, M., 53
Ulrich, R., 242
Uxkull-Guldenbrandt, N. von, 414

V

Valbusa, U., 408, 410, 412
Valente, I., 257
Valli, G., 120
Valovirta, I., 133
Valverde, J. A., 11
Van Cleve, K., 66, 70, 72
van den Bosch, R., 339, 342, 343, 345
Van Der Drift, J., 72
Vanderplank, F. L., 417
Van Dyne, G. M., 184
van Emden, H. F., 337, 344, 345
Van Gundy, S. D., 72
Van Houten, F. B., 358
Van Schreven, D. A., 71
Vansell, G. H., 412
van Someren, G. R. C., 1, 10, 14, 21
Van Valen, L., 374
Varmus, H. E., 38, 42
Vasey, E. H., 93
Vasil'ev, A. E., 408
Vasil'Yev, M. A. S., 287, 288

Vassiljevskaja, V. D., 55
Veley, V. F. C., 295
Vendrov, S. L., 274, 275
Vernon, C. J., 1, 7, 21
Vickerman, G. P., 338
Victoria, J. K., 19
Vigue, C. L., 312
Vilks, K., 8
Vincent, J., 3
Vinegar, M. B., 155, 162
Virzo de Santo, A., 66, 68
Visser, S. A., 262
Vitousek, P. M., 54, 62
Vogt, K. J., 166, 167
Voipio, P., 7, 8
Volkovintser, V. I., 55
von Frisch, H., 10
von Frisch, O., 10, 11
von Lucanus, F., 8
Voorhies, M. R., 361, 362, 365
Vreeland, R. R., 286, 297
Vulto, J. C., 64

W

Waddington, C. H., 205
Waddy, B. B., 275
Wadsworth, F. H., 85, 89
Wadsworth, S. C., 41
Wagner, H. H., 297, 298
Wagner, H. O., 401
Wahba, A. J., 39
Wahle, R. J., 286, 297
Wahlert, J. H., 359
Waksman, S. A., 59, 62, 71, 72
Walden, H. W., 110
Walker, A., 359
Walker, J. M., 93
Walker, T. J., 296
Walkinshaw, L. H., 2, 4, 5, 13
Wallace, B., 237
Wallis, J. R., 56, 60, 62, 72
Wallraff, H. G., 389
Walter, H., 64
Walters, D., 189
Walther, F., 202, 439
Wanner, H., 64, 66, 67, 69, 70
Ward, J. V., 272
Ward, M. M., 96
Ward, P., 200
Ward, R. D., 312
Wardle, P., 92
Wareing, P. F., 94, 333
Warren, K. S., 209, 216, 230
Wasenius, E., 1, 16
Wasserman, M., 244
Watari, S., 408
Waterman, T. H., 285, 287, 290
Watson, A., 204

Watson, M. A., 344
Watt, K. E. F., 176
Watt, W. B., 117
Way, M. J., 329, 332, 335-37, 344, 416
Weaver, J. E., 54
Weaver, P. L., 92, 99
Webb, L. J., 84
WEBB, S. D., 355-80; 364, 366-68, 371, 373, 374
Webbe, G., 209, 210, 212, 214, 215
Weber, F., 408
Weber, K., 39
Webster, J. L., 408, 410, 413
Webster, R. G., 33
Weeden, R. B., 201
Weeks, R. E., 345
Wegenek, E., 340, 342
Weingart, W., 412
Weisbrod, B. A., 230
Weise, C. M., 401
Weismann, L., 336
Weiss, E., 413
Weiss, R., 37
Welcomme, R. L., 273
Weldon, M. D., 54
Weller, M. W., 1, 4, 6, 13, 18-21
Welles, S. P., 440
Wellner, D., 312
Wellnhofer, P., 429, 435, 436
Wells, C. G., 72
Went, A. E. J., 290
Westlake, G. F., 288
Westphal, R. J., 238
Westrich, P., 389, 392, 393, 401
Westwood, N.J., 200
Wheeler, L. L., 238
Wheeler, M. R., 239, 240
Wheeler, W. H., 358, 360
Wheeler, W. M., 414-16
White, D. O., 30, 44
White, G. F., 257
White, H. C., 297
White, T. E., 360, 361, 363
Whitmore, F. C., 370
Whitmore, T. C., 83-85, 88-90, 92, 93
Whitney, R. R., 293, 294
Whittaker, R. H., 29, 52-58, 61, 62, 68, 73, 356
Wiant, H. V. Jr., 63, 70
Wiebe, A. H., 257, 260, 261, 270, 272
Wiebe, H. H., 69
Wiens, J. A., 15
Wilbur, H. M., 147, 153, 162, 167, 168
Wilde, W. J. J. O. de., 413

Wildung, R. E., 65, 69, 70
Wiley, J. W., 5
Wiley, R. H., 196, 198, 201
Williams, G., 286
Williams, G. A., 333
Williams, N., 312
Williams, R. F., 101
Williams, W. D., 263
Williamson, M. H., 110
Williamson, P., 111, 116, 125-28
Wilson, A., 7
Wilson, A. C., 148
Wilson, E. O., 1, 145, 153, 200, 263, 415-17, 440
Wilson, J. A., 360, 370
Wilson, R. F., 69
Wilson, R. W., 358, 360-63, 367, 372
Wiltschko, W., 398
Wing, L. W., 381
Winge, H., 242
Winn, H. E., 286
Winneberger, J. H., 95
Wisby, W. J., 286, 297
Withler, I. L., 299
Witkamp, M., 57, 63-68, 70-72, 182, 183

Wolda, H., 111, 113, 114, 116, 120, 123, 125, 126, 128, 130, 134
Wolfe, J. A., 360, 367, 369
Wolman, M. G., 215, 230
Wong-Staal, F., 39
Wood, A. E., 359-63
Wood, D., 312
Wood, R. B., 264
Woodall, W. L., 286
Woodburne, M. O., 370
Woodcock, E. F., 413
Woodell, S. R. J., 60, 63, 68, 70
Woodhead, A. D., 285, 291
Woodward, F. I., 89
Woodwell, G. M., 51, 54, 61, 68
Woolhouse, H. W., 94
Worthington, E. B., 257
Wratten, S. D., 335, 337, 339
Wright, H. E. Jr., 374
Wright, S., 132, 147, 153
Wright, T., 311
Wu, A. M., 33
Wunderlich, W. O., 260
Wyatt, I. J., 344, 345
Wyllie, I., 3, 4, 7, 11, 13-15, 21

Y

Yalden, D. W., 436
Yamazaki, T., 310
Yang, S. Y., 132, 238
Yingchoi, P., 64, 66, 67, 69, 70
Yoda, K., 53, 54, 67, 70
Yom-Tov, Y., 19
Young, H., 15
Young, J. C., 294

Z

Zaragoza, L. J., 91
Zaydel'man, F. R., 54
Zhadin, V. I., 257, 258, 264, 270
Zhdanov, V. M., 30, 37
Zich, J., 189
Ziegler, H., 330
Zimmerman, J., 407, 408, 415
Zimmerman, M. A., 287, 288
Zimmermann, J. L., 388
Zimmermann, M., 408
Zink, G., 398
Zouros, E., 243
Zweep, A., 111, 126
Zweifel, R. G., 162

SUBJECT INDEX

A

Abies forests
 annual carbon release from, 68
Aburria
 social adaptations of, 201
Acacia
 cornigera
  ant protection of, 421
 extrafloral nectaries in, 412, 416-21
Acanthaceae
 extrafloral nectaries in, 412
Acanthopterygian fishes
 defenses of, 433-34
Acer
 forests
  annual carbon release from, 67
 pseudoplatanus
  as aphid host, 329-32, 339-42
Acrocanthosaurus
 display organs of, 440
Acrocephalus schoenobaenus
 parasitism of, 4-5
Acyrthosiphon pisum
 parasitic control of, 345
Adalia
 absorption of solar energy by, 117
Adaptations
 of extrafloral nectaries
  significance of, 414-21
 reproductive
  in plants, 167
 social
  in birds and ungulates, 193-205
Adenoviruses
 and oncogenesis, 37
Adenylate kinase
 in Colias homology, 324
AdH
 electrophoretic studies of, 311-12
Africa
 birds and ungulates in
  ecological evolution of, 202
 brood parasitism in, 4, 13
 dams in
  environmental effects of, 256-57, 270, 273-75
 forest distribution studies in, 83, 86, 92
 giraffes and ants in, 417, 419
 as migration destination

of birds, 381, 392, 397
Agavaceae
 extrafloral nectaries in, 412
Age
 distribution
  in life history trait evolution, 154-55, 160
 infection function of
  in schistosomiasis, 211-12
 at maturity
  and life history trait evolution, 146-48, 152, 165
 -related behaviors
  of crocodilians, 438
Agelaius xanthomus
 parasitism of, 4-5
Aggression
 in ants
  and extrafloral nectaries, 416-17
 in birds
  in host/parasite interactions, 7
 and territoriality
  among birds and ungulates, 196, 197, 200, 202-4
Agnotocastor
 and vertebrate history, 361-62
Agricultural
 ecology
  and agro-ecosystem analysis, 174-75
 economics
  simulation models for, 184
 systems analysis of, 178-79
Agriocharis ocellate
 social adaptations of, 197
Agriochoeridae
 and vertebrate history, 360
Agro-ecosystem research, 173-91
 agricultural ecology
  contribution of, 174-75
 byproduct waste problems, 183-84
 ecosystem analysis, 175-77
 focus on, 178-86
  definition of, 178-81
  as integrative science, 181-84
  simulation models, 184-86
 opportunity for, 186-90
 origins of, 175-76
 universal components,

176-77
 temperate zone
  production and cycling in, 181-82
 tropical zone
  production and cycling in, 182
Akmaiomys
 and vertebrate history, 362
Alarm calls
 of birds
  in host/parasite interactions, 7, 15
Alaska
 waters of
  fish migrations in, 294, 299
Alatae
 in aphid life cycle, 329-47
Alchornea latifolia
 mineral cycling in, 101
Alcohol dehydrogenase
 electrophoretic studies of, 322
Alectoris
 social adaptations of, 200
Alewives
 migration of, 291-92
Algae
 blue-green
  in reservoir ecosystems, 267-68
Allele identity
 in electrophoresis, 310-25
Alligator mississippiensis
 behavior of, 437
Allosaurus
 brain:body ratio in, 433
Allozyme
 catalytic properties of, 319-20
 physical properties of, 314-19
 see also Proteins, evolution
Alosa
 pseudoharengus
  migration of, 291-92
 sapidissima
  migration of, 287
Alphaviruses
 transmission of
  and temperature effects, 36
 in viral evolution, 34
Alps
 Cepaea distribution in, 110
Altispinax
 display organs of, 440
Altitude
 and ecological evolution,

463

464 SUBJECT INDEX

193, 198
and forest growth, 83-85, 88
and morph frequency distribution in Cepaea, 135
Amaranthus viridis
fertilizer effects on, 95
Amebleodon
and vertebrate history, 371
American shad
migration of, 287, 293, 297-98
Amino acids
in aphid ecology, 339
electrophoresis of
and heterogeneity, 309-25
in extrafloral nectaries, 408, 417
list of, 410
Anacardiaceae
in savanna evolution, 358
Anas
cyanoptera
breeding success of, 6
discors
migration of, 396
Anatosaurus
brain:body ratio in, 433-34
Anchovy
migration of, 294
Andes
birds of
ecological evolution of, 204
drosophila in, 237, 239, 242
forest distribution studies in, 83
Andropogon gayanum
nectar of, 410, 412
Angiosperms
extrafloral nectaries of, 408
of Tertiary period
in North American savanna, 357
and viral transmission, 35, 42
Anguilla
migration of, 288
Anhimidae
social adaptations of, 197
Animal
processes
systems analysis of, 177
Animalia
viruses occurring in, 30-33, 41
Ankylosaur
brain:body ratios in, 433-34
teeth of, 444
Anolis
life history of
reliability of, 162
Anomalospiza
brood parasitism in, 1, 13
imberbis

clutch size of, 14
Antilocapra
social adaptations of, 201-2
Antlers
casting of
as social adaptation, 201
Ants
and extrafloral nectaries
aggressive behavior, 416-17
distribution of, 415-16
foraging patterns of, 417
plant vulnerability to, 417-18
presence of herbivores, 418-19
Aphelinus mali
as aphid parasite, 345
Aphid
and ant interaction
on nectary plants, 418
as viral vector, 42
Aphid ecology, 329-47
laboratory studies and biological control, 345
life cycles and polymorphism, 329-35
population dynamics, 335-45
black bean aphid, 336-37
cabbage aphid, 337
cereal aphids, 337-38
green spruce aphid, 338-39
lime and sycamore aphids, 339-40
others, 343-45
walnut aphid, 340-43
simulation models, 345-46
Aphis
craccivora
life cycle of, 346
fabae
life cycle of, 332, 336-37, 345-46
sacchari
parasitic control of, 345
Aplodontia
and vertebrate history, 373
Apterae
in aphid life cycle, 329-47
Aptetomeus
and vertebrate history, 362
Arabinose
as nectary secretion, 410
Arboreal mammals
in New World savanna, 357, 59, 366, 372
Archaeopteryx
brain size and behavior in, 430-31, 433, 435-36, 444
Archeovirus
and viral evolution, 38
Archosaurian reptiles, 429-44
behavior in, 436-42

combat and display, 439-42
dinosaur groups, 439
living crocodilians, 436-38
parental behaviors, 442
brain size in, 429-36
archaeopteryx, 435-36
dinosaurs, 431-35
fossil brain size, 430-31
pterosaurs, 435
Arctic habitats
and social adaptations, 202-5
Arctic jaeger
as lemming predator, 189
Arctodiaptomus ibericus
in reservoirs, 268
Argentina
drosophila in, 242
Argusianinae
social adaptations of, 197-204
Argusianus argus
social adaptations of, 198-200, 202-3
Arianta arbustorum
polymorphism among, 125, 128
Arizona
drosophila in, 244
Arrhenius constant
in electrophoretic studies, 313, 320
Arthropods
circannual rhythms in, 384
Artiodactyls
and vertebrate evolution, 360, 370
Asclepias
life history of
reliability of, 167
Asia
coast of
fish migrations off, 293
Southeast
birds and ungulates in, 202
drosophila in, 238-39
vertebrate migrations from during Eocene, 359-60
Astrohippus
and vertebrate history, 371
Australia
aphids in, 337
cuckoos in
parasitism among, 2, 17
drosphila in, 239
myxomatosis in, 46
nectary plants in, 419
Avian sarcoma virus
and viral evolution, 38
Avicennea
extrafloral nectaries on, 414, 418
Aythya americana
parasitism in

SUBJECT INDEX 465

and breeding success, 6
evolution of, 18

B

Babblers
  brood parasitism in, 19
Bacillus
  and viral systematics, 41
Bacteria
  cell evolution in
    and viral agents, 37
  and fungi
    in reservoir ecosystems, 266
  growth in tropical mountain forests, 90
  lysogenic
    and viral transmission, 35
  nitrogen-fixing
    in tropical mountain forests, 94, 102
  as viral hosts
    in viral systematics, 32-33
Bacteriophage
  ϕX 174, 30
Baculoviruses
  in insect control, 46
Balsaminaceae
  extrafloral nectaries in, 412
Baltic
  fish migrations in, 291
Barbary partridge
  social adaptations of, 200
Barents Sea
  fish migrations in, 293
Barnacles
  life histories of, 156-58
Bass
  white
    migration of, 286
Batrachoseps attenuatus
  life history of
    reliability of, 161
Beavers
  dams of, 256, 265
Beetles
  life histories of
    reliability of, 165-66
  in reservoir ecosystems, 266
Behavior
  aggressive
    of ants, 416-17
  of archaeosaurian reptiles
    and brain size, 429-44
  of brood parasites, 6-12
    adaptations of young, 10-11
    breeding seasons, 6-7
    host discrimination and egg mimicry, 8-9
    host discrimination and mimicry of young, 9-10
    interactions of host and parasite, 7-8
    vocal mimicry, 12

Belonopterus cayennensis
  social adaptations of, 197
Berriochloa
  and savanna evolution, 365
Beta vulgaris
  as aphid host, 336
Bignoniaceae
  extrafloral nectaries in, 411, 412
Bilharzia
  see Schistosomiasis
Biochemical evolution
  of drosophila
    and systematics, 235-51
Biocides
  use of in agriculture, 184
Biological clock
  in aphid life cycle, 332
Biomass
  of Cepaea, 111
  distribution of
    by ecosystem types, 56
  and productivity
    in tropical rain forests, 85-88
Biome programs
  and systems analysis, 176
Biomphalaria glabrata
  as schistosome host, 210, 219-20
Bipectinata complex
  of drosophila, 238-39
Bird calls
  and mating systems, 14-15
Bird migrations
  circannual rhythms in, 381-402
  critique of hypothesis, 400-2
  direction change, 397-98
  distribution and properties of, 381-87
  endogenous time-program, 388-89
  other seasonal activities, 398-400
  photoperiodic modification, 396-97
  synchronization of, 387-88
  temporal migration control, 388-402
  test of hypothesis, 390-96
Birds
  archaeopteryx
    brain size and behavior of, 430-31, 433, 435-36, 444
  brain:body size ratios in, 433-34
  brood parasitism in, 1-22
    behavior and coevolutionary aspects of, 6-12
    evolution of, 18-22
    host breeding success, 3-6
    host selection and specialization, 1-3

mating systems and social behavior, 14-15
  population structure, 15-16
  reproductive strategies, 12-14
evolution of
  and archaeosaurian reptiles, 429, 435-36
gallinaceous
  social adaptations of, 193-205
life histories of
  reliability of, 163
  reservoir ecosystems and, 271
Bison
  social adaptations of, 202
Bixaceae
  extrafloral nectaries in, 412
Bixa orellana
  extrafloral nectaries in, 409, 416, 418-21
Blackbird
  parasitism of, 5
Body size
  of birds and ungulates
    and ecological evolution, 202-5
  and egg number
    in crocodilians, 437
  and migration
    in salmon, 298
Body weight
  in birds
    and migration, 381-402
Boreal forest
  carbon balance in, 54, 56-57, 68
Borneo
  tropical mountain forests in, 93
Bovids
  sexual monomorphism among, 200-1
Brachiosaurus
  brain:body ratio in, 433
Brain:body ratio
  in vertebrates, 430-33
Brassica pekinensis
  fertilizer effects on, 95-96
Brazil
  tropical rain forests in, 86, 97-98, 100-1
Breeding
  biology
    and brood parasitism, 12-14
  of Cepaea, 111, 123
  cycle
    of crocodilians, 437
  seasons
    of brood parasites, 6-7
  success of hosts
    effects of brood parasitism on, 3

466 SUBJECT INDEX

Brevicornye brassicae
  life cycle of, 332, 337, 345-46
Britain
  Cepaea distribution in, 111-36
  morph frequency and climate, 119-20
  see also England
Bromo mosaic virus (BMV)
  and viral systematics, 42
Brood parasitism, 1-22
  behavior and coevolutionary aspects of, 6-12
  adaptations of young parasites, 10-11
  breeding seasons, 6-7
  host discrimination and egg mimicry, 8-9
  host discrimination and mimicry of young, 9-10
  interactions of host and parasite, 7-8
  vocal mimicry, 12
  evolution of, 18-22
  host breeding success effects on, 3-6
  host selection and specialization, 1-3
  mating systems and social behavior, 14-15
  population structure, 15-16
  and gens concept, 16-17
  host specificity and speciation, 17-18
  reproductive strategies, 12-14
Brookhaven model
  of carbon balance, 73
Bryophytes
  in tropical forests, 90
Bryozoans
  and sponges
  in reservoir ecosystems, 266
Bucculatrix turberiella
  and ant protection, 418
Buceros
  social adaptations of
  and dispersal theory, 202
Bucerotidae
  social adaptations of, 202
Bulinus
  as schistosome host
  globosus, 220

C

Cacicus cela
  fledging success of
  and parasitism, 4
Cacomantis pyrrhophanus
  distribution of, 17
Cactaceae
  extrafloral nectaries in, 412
Caesalpiniodeae
  extrafloral nectaries in, 411

Caiman
  brain size and behavior of, 432
  crocodilus
  behavior of, 437-38
Calandra oryzae
  life history of
  reliability of, 166
Calcarius
  life history of
  reliability of, 163
Calcium
  and soil respiration, 72
  in tropical forest growth, 96-101
California
  aphid infestations in, 340, 342-43, 345
  barnacles of, 156-58
  bird breeding seasons in, 6
  cowbirds in, 13
  drosophila in, 244
  mountain rain forests in, 89
  soils in
  and carbon balance, 60, 62
Calippus
  and vertebrate history, 371
Calluna heath
  annual carbon release from, 69
Camels
  and vertebrate evolution, 359-61, 364, 366, 370, 372
Cameroon
  nectary plants in, 420
Campsis radicans
  nectar of, 410
Camptosaurus
  brain size and behavior of, 433-34
Canada
  dams and reservoirs in
  environmental effects of, 257-59, 261-62, 265, 272-77
Canis lupus
  pairbonding among, 194
Cantabrians
  Cepaea distribution in, 122
Capelin
  migration of, 293
Capercallie
  social adaptations of, 201
Capparidaceae
  extrafloral nectaries in, 412
Capra ibex
  and dispersal theory, 203
Capreolus
  social adaptations of, 201
Caprifoliaceae
  extrafloral nectaries in, 412

Carassius auratus
  inner-ear studies of, 288
Carbon
  and nutrient cycling
  in agro-ecosystems, 182
  Carbon balance in detritus, 51-73
  accumulation, 51-62
  future investigation, 62
  identity and turnover of compounds, 57-59
  steady-state detritus models, 59-62
  carbon cycle synthesis, 73
  carbon release, 62-72
  environmental factors in soil respiration, 70-72
  soil respiration methodology, 63-65
  soil respiration patterns, 65-70
  soil system complexities, 63
Carbon dioxide
  release of
  and carbon balance, 62-73
Carbon-14 dating
  and soil decomposition, 59-60
Carcinogenesis
  of viruses, 36-37
Caribbean
  forest distribution studies in, 83, 86
  schistosomiasis in, 230
Caribou
  social adaptations of, 198-99, 201, 203
Carnivores
  pairbonding among, 194
Carp
  olfaction studies of, 296
Carya forest
  annual carbon release from, 68
Caryophyllaceae
  extrafloral nectaries in, 412
Cassia biloba
  extrafloral nectaries of, 412, 416
Cassowaries
  defensive adaptations of, 196-97
Castanea forests
  annual carbon release from, 68
Catbirds
  parasitism of, 9
Catreus wallichii
  social adaptations of, 204
Cattails
  life histories of
  reliability of, 167
Cavariella aegopodii
  parasitic control of, 345
Cayenne lapwing

SUBJECT INDEX 467

social adaptations of, 197
Cenozoic era
  savanna vertebrates of, 355-74
Central Africa
  agro-ecosystems in, 183
Central America
  nectaries and ants in, 421
Centropodinae
  coloring and down of, 9
Centropus
  incubation and nestling periods in, 11
Centrosaurus
  display organs of, 441
Cepaea
  polymorphism in, 109-36
  climatic selection, 117-23
  density-dependent selection, 126-27
  disruptive selection, 125-26
  distribution in Neolithic, 121
  evolution of, 114-35
  feeding rates, 111
  frequency-dependent selection, 123-25
  gene frequency patterns, 114
  heterozygote advantage, 127-28
  linkage disequilibrium and coadaptation, 130-35
  mobility, 111, 125
  random processes, 128-30
  seasonal predation on, 116
  shell polymorphism, 111-14
  visual selection, 116-17
Cephalophus
  social adaptations of, 197
Ceratogaulus
  and vertebrate history, 365
Ceratopsians
  display organs of, 441
  teeth of, 444
Ceratosaurus
  display organs of, 440
Cercariae
  of schistosomes, 210, 216, 218, 222
Cervus
  elaphus
    social adaptations of, 201-2
  nippon
    circannual rhythms of, 384, 388
Chachalacas
  social adaptations of, 201-2
Chalicotheres
  arboreal browsing of, 195
Chamaepetes
  social adaptations of, 201

Char
  migration of, 297
Chemotherapy
  problems of
    and viral adaptation, 45-46
Chenopodium album
  as aphid hosts, 336
Chernozem
  and world carbon balance, 58, 61, 71
Chiffchaff
  migrations of
    and cirannual rhythms, 387, 392-95
Chimpanzees
  and life history evolution, 147
Chironomids
  in reservoirs, 264-66, 269
Chironomus
  (Tendipes) plumosus
    as benthic fauna, 264
  transvaalensis
    in Lake Volta, 265-66
Chondestes grammacus
  fledging success of
    and parasitism, 4
Chromaphis juglandicola
  life cycles of, 333, 340-43, 345
Chromatography
  in soil respiration studies, 65
Chrysanthemums
  as aphid hosts, 344
Chrysococcyx
  brood parasitism in, 8, 10, 14
  caprius
    and egg types, 16
    incubation and nestling periods in, 11
    mimicry of, 8
  klaas
    and egg types, 16
    incubation and nestling periods in, 11
Chrysomelid beetles
  and ant protection, 418
Chthamalus fissus
  life history of, 156-58
Circannual rhythms
  in bird migrations, 381-402
  critique of hypothesis, 400-2
  direction change, 397-98
  distribution and properties of, 381-87
  endogenous time-program, 388-89
  other seasonal activities, 398-400
  photoperiodic modification, 396-97
  synchronization of, 387-88

temporal migration control, 388-402
  test of hypothesis, 390-96
Cisco
  migrations of, 291
Citellus lateralis
  internal clock in, 381
Citrus aurantium
  mineral nutrients of, 96-97
Clamator
  glandarius
    incubation and nestling periods in, 11
    parasitism of, 9
  jacobinus
    brood parasitism among, 4-5
    incubation and nestling periods in, 11
    interactions with host, 7-8
  levaillantii
    incubation and nestling periods in, 11
Classification
  of rain forest types, 84
  of viruses, 30-31, 40-44
Clethra occidentalis
  mineral cycling in, 101
Climate
  changes in
    as dam effects, 275
    effects on Cepaea, 111, 117-23, 127, 134
Clocks
  see Biological clocks; Circannual rhythms; Molecular clocks
Cloud formation
  effects on tropical mountain forests, 88-89
Clupea harengus
  migration of, 291
Clutch size
  of birds
    and life history traits, 163
Cnemidophorus
  life history of
    reliability of, 162-63
Coccinellidae
  as aphid predators, 338
Coccyzus
  distribution of, 17
  incubation and nestling periods in, 11
Coconuts
  and ant protection, 417
Cod
  migration of, 293, 295
Coelodonta
  and dispersal theory, 202
Coelurosaurs
  brain size and behavior of, 433-36
Coevolution
  between plants and animals
    and extrafloral nectaries,

415
of brood parasitism
 in birds, 6-12
Colias
 absorption of solar energy
  by, 117
 homology studies of, 324
 meadii
 electrophoretic studies of,
  311, 315-17, 320-25
Collared peccary
 kin selection among, 195-96,
  201
Colombia
 drosophila in, 237, 239
Colonization
 in life history models, 145
Color
 as polymorphism in Cepaea,
  111-36
Colorado
 grazing ecosystem in
  model of, 184
Colostrum
 viral transmission via, 35
Columbidae
 lactation among males, 194
Combat behaviors
 among dinosaurs, 439-42
Competition
 among Cepaea, 110
 and aphid life cycles, 336-
  37, 344
 in ungulate evolution, 194-
  95
Compositae
 evolution of, 357
 extrafloral nectaries in,
  412
Computer
 and ecosystem analysis
  techniques, 175-91
Conifers
 growth and distribution
  in rain forests, 85-87
Convolvulaceae
 extrafloral nectaries in, 412
Cooperative food-getting
 in crocodilians, 436-38
Coots
 social adaptations of, 204
Copper
 and soil respiration, 72
 in tropical forest growth,
  96, 98-99, 101
Coregonus muksun
 migration of, 290
Corvus
 corone
 migrations of, 395-96
 parasitism of, 3
Coryphodon
 and vertebrate evolution,
  358
Costaceae
 extrafloral nectaries in,
  412

Costa Rica
 distribution of ants in
  and nectary plants, 415-16,
   418-19
 soil respiration in, 65
Cotton
 growth reduction in
  and high humidity, 95
Couinae
 coloring and down of, 9
Courtship behavior
 among Cepaea, 110, 123
 in crocodilians, 436-38
 in dinosaurs, 439
Cowbirds
 brood parasitism in, 1-2,
  8, 11, 13-15, 19-21
Cracidae
 social adaptations of, 197,
  201-2
Crax
 social adaptations of, 197
Cree Indians
 disruption of
  and dam building, 276
Crematogaster
 and ant aggression, 416
Cricetidae
 and vertebrate evolution,
  362-63
Crocodilians
 behavior of
  and archosaurian reptiles,
   436-38
Crocodylus
 behavior of, 437-38
Crop
 management
  and agricultural ecology,
   175, 186-88
 rotation
  and cycling wastes, 183-
   84
Crossoptilon
 social adaptations among,
  200, 203
Crotophaga
 incubation and nestling periods in, 11
Crowding
 effects on aphid life cycle,
  330, 332, 335, 339-40,
  347
Crows
 carrion
  migrations of, 395-96
 as cuckoo hosts, 10
Cryptoryctes
 and vertebrate evolution,
  363
Cuckoo
 brood parasitism in, 1-22
Cuculus
 canorus
 brood parasitism among,
  1-22
 dispersal distances of, 17

incubation and nestling
 periods in, 11
clamosus
 incubation and nestling
  periods in, 11
micropterus
 and egg types, 16
 incubation and nestling
  periods in, 11
saturatus, 7
solitarius
 and egg types, 16
 incubation and nestling
  periods in, 11
Cucurbitaceae
 extrafloral nectaries in,
  412
 and viral systematics, 42
Cultivated field
 as ecosystem, 187-88
Curassows
 social adaptations of, 197,
  201-2
Currents
 role in fish migrations, 295
Cynomys
 and vertebrate history, 362
Cynorca
 and vertebrate history, 370
Cyprinus carpio
 olfaction studies of, 296
Cytomegalovirus
 similarities to HSV, 41
 transmission of, 45

D

Dacrycarpus
 leaf production in
  in rain forests, 87
Dairying
 simulation models of, 184
Dams and impoundments
 environmental effects of,
  255-77
 benthos development, 264-
  67
 chemical limnology, 261-
  62
 downstream effects, 271-
  74
 fish and other vertebrates,
  269-71
 lake morphology and lim-
  nology, 258-62
 littoral region, 268-69
 other consequences, 275-
  76
 physical limnology, 259-
  61
 plankton, 267-68
 reservoir ecosystems, 262-
  71
 reservoir morphology, 258-
  59
 sedimentation, 259
Dandelions

SUBJECT INDEX 469

life history data on
  reliability of, 166-67
Data processing
  see Systems analysis
Davenport Ranch
  sauropod trackways
  and dinosaur behavior, 442
Decay
  in tropical mountain forests, 87-88, 92-93
Deer
  social adaptations of, 201-4
Denaturation constant
  in electrophoresis, 313, 319-20
Dendroica kirtlandii
  fledging success of
  and parasitism, 4-5
Deoxyribonucleic acid (DNA)
  and evolution of viruses, 29-46
Desert
  carbon balance in, 55-56, 69
  salt
  as irrigation result, 180
Desmognathus
  life history data reliability, 161
Detritus
  carbon balance in, 51-73
  accumulation, 51-62
  carbon cycle synthesis, 73
  carbon release, 62-72
  and total biomass
  in rain forests, 86
  worldwide estimation of, 52-55
Developmental time
  among Cepaea, 111, 113, 122-23
Diardigallus diardi
  social adaptions of, 203
Diatraea saccaralis
  ant predation on, 417
Dicerorhinids
  and dispersal theory, 202
Dicotyles tajacu
  kin selection among, 195-96
Dicrurus
  parasitism of, 8
Digestion
  of ungulates
  evolution of, 194-96
Dilophosaurus
  display organs of, 440
Dinosaurs
  see Archosaurian reptiles
Dioscoreaceae
  extrafloral nectaries in, 412
Diplodocus
  brain:body ratio in, 433, 440
Diploid genetics
  and ontogeny
  in life history trait evolution, 147-51, 160

Disease
  viral
  and human intervention in, 44-46
Dispersal theory
  in social adaptation and evolution, 199, 202-5
Dispersion
  of schistosomes, 223-24, 229
Display behaviors
  among dinosaurs, 439-42
Display organs
  among birds and ungulates
  evolution of, 193, 196-205
Distemper
  interspecies spread of, 44-45
Dominance hierarchies
  in crocodilians, 436-38
Drawdown
  of lakes and reservoirs
  environmental effects of, 268-71
Dreissena
  in reservoirs, 265
Drepanosiphum platanoidis
  life cycle of, 329-32, 339-42, 346
Dromaeosaurus
  brain size in, 435
Drosophila
  albomicans
  protein evolution of, 239
  aldrichi
  electrophoretic studies of, 322
  protein evolution of, 243-44
  a. americana
  protein evolution of, 236
  a. texana
  protein evolution of, 236
  arizonensis
  protein evolution of, 243-44
  athabasca
  protein evolution of, 238
  brncici
  protein evolution of, 239
  gascici
  protein evolution of, 239
  gaucha
  protein evolution of, 239
  heteroneura
  protein evolution of, 242
  kambysellisi
  protein evolution of, 239
  lacicola
  protein evolution of, 245
  life history of
  reliability of, 165-66
  major groups, 235-47
  Hawaiian, 239-42
  melanogaster, 238
  mesophragmaticas, 239
  mulleri, 243-44

nasuta complex, 239
  obscura, 237-38
  sibling species study, 236-37
  virilis, 235-36, 245-46
  willistoni, 242-43
melanogaster
  age at eclosion in, 155
  electrophoretic studies of, 311-12
  protein evolution of, 237-38, 240, 247-48
mesophragmatica
  protein evolution of, 239
mimica
  protein evolution of, 239
mojavensis
  electrophoretic studies of, 322
  protein evolution of, 243-44
montana
  protein evolution of, 245
mulleri
  electrophoretic studies of, 311
  protein evolution of, 243-44
ochrobasis
  protein evolution of, 241
pallidifrons
  protein evolution of, 239
paulistorum
  protein evolution of, 242-43
pavani
  protein evolution of, 239
persimilis
  protein evolution of, 237, 240, 247-48
pseudoobscura
  electrophoretic studies of, 311, 322-23
  protein evolution of, 237, 240, 247-48
pseudoobscura bogotana
  protein evolution of, 237, 247
pseudoobscura pseudoobscura
  protein evolution of, 247
serrata
  life history reliability of, 166
setosimentum
  protein evolution of, 241
silvestris
  protein evolution of, 242
simulans
  protein evolution of, 237, 240
sulfurigaster
  protein evolution of, 239
systematics and biochemical evolution, 235-51
  genetic variability, 246-47
  major investigations, 235-47

470 SUBJECT INDEX

molecular clocks, 249
species and speciation, 247-49
thermal physiology
and mating activity, 123
willistoni
protein evolution of, 242
Drought
and tropical forest growth, 91
Duck
parasitic, 1, 3-4, 6, 13, 18-19, 21
Duikers
social adaptations of, 197
Dumetella carolinensis
parasitism of, 9
Dyseohyus
and vertebrate evolution, 370-71

E

Earthquakes
as dam effects, 275
East Africa
agro-ecosystems in, 183
Echinochloa
life history of
reliability of, 167
Echinoderms
water vascular system of
and life history traits, 151
Ecological evolution
of birds and ungulates, 193-205
see also Life history traits
Economics
of schistosomiasis control, 230
Ecosystem analysis
and agro-ecosystem research, 175-77
Ecosystems
agro-ecosystems research, 173-91
annual release of carbon in
table of, 67-69
cultivated fields as, 187-88
detritus in soil profiles of
table of, 53-55
hot springs as, 263
reservoir, 262-71
benthos development, 264-67
fish and other vertebrates, 269-71
littoral region, 268-69
plankton, 267-68
spray-zone
and impoundment effects, 271-72
tide zones
and barnacle life histories, 156-58
Ecotone
in reservoir ecosystems,

263, 268-70, 272, 274
Ectothermy
in dinosaurs
and behavior, 429-44
Ectotomma tuberculatum
behavior of
and extrafloral nectaries, 416
Edentates
arboreal browsing among, 195
Eels
migrations of
and electric fields, 287-88
and olfaction, 296
Egeria radiata
effects of dams on, 274
Egypt
Lake Nasser in
environmental effects of, 257, 267, 273-74, 276
schistosomiasis in, 225
Elatobium abietinum
life cycle of, 338-39
Electrophoresis
assessment of, 309-25
allozyme properties, 314-20
analytic approach to, 313-14
comparing variants, 320-21
genetic distance measurement, 325
hidden heterogeneity, 312
homology assessment, 310-11
in insects, 311-12
in drosophila systematics, 244, 246-49
Elomeryx
and vertebrate evolution, 361
Empidonax virescens
fledging success of
and parasitism, 4
Encephalization quotient (EQ)
and comparative brain size
in archosaurian reptiles, 431-35, 443
Endocasts
of fossil archosaurs, 430-32, 435, 443-44
Endothermy
in dinosaurs
and behavior, 429-44
Energy budgets
and experiment difficulties
in life history trait evolution, 158-59, 162
England
aphid life cycles in, 338, 344
brood parasitism in, 4, 16-17
forest growth in, 94, 96

and drought, 91
see also Britain
Engraulis encrasicholus
anchovy
migration of, 294
Enteroviruses
transmission of, 36
Entoptychus
and vertebrate history, 365
Environment
in North America
evolution of, 355-74
Environmental effects
of dams and impoundments, 255-77
Environmental standards
effectiveness of
measurement of, 188
Enzymes
in electrophoresis, 309-25
see also Protein evolution
Eocene
woodland savanna in
in North America, 358-60
Eodipodomys
and vertebrate history, 365
Ephedra
in woodland savanna, 359
Ephemeroptera
in reservoirs, 265
Epigaulus
and vertebrate history, 365
Epstein-Barr herpesvirus
human intervention with, 45-46
Ericaceae
in tropical rain forests, 99
Eriosoma lanigerum
parasitic control of, 345
Erithacus rubecula
parasitism of, 4, 9
Escherichia coli
and viral classification, 37, 40
Esterase-6 locus (EST)
in drosophila studies, 247
electrophoretic studies of, 311, 322
Estrildids
and parasites, 10, 12-14
Estuary
effects of dams on, 274
Ethiopia
reservoirs in
environmental effects of, 256, 271, 273
Eucallipterus tiliae
life cycles of, 330, 333, 339-42, 345
Euceratherium
social adaptations of, 203
Eudynamis scolopacea
brood parasitism of, 3, 8-9
incubation and nestling periods in, 11

SUBJECT INDEX 471

Euonymus europaeus
  as aphid host, 336
Euoplocephalus
  brain size of, 431, 433
Euphorbia
  life history of
    reliability of, 167
Euphoribiaceae
  extrafloral nectaries in, 412
Europe
  aphids in, 337-38
  Cepaea distribution in, 110, 119-20, 122-23
  cuckoos in
    brood parasitism among, 3, 9
  mountain rain forests in, 89
  soils in
    carbon release from, 65
Evicting behavior
  of brood parasites, 11
  evolution of, 22
Evolution
  biochemical
    of drosophila, 235-51
  of brood parasitism in birds, 18-22
  ecological
    of birds and ungulates, 193-205
    see also Social adaptations
  of extrafloral nectaries, 411, 414-21
  of life history traits, 145-68
    age distribution assumptions, 154-55
    ambiguity in theory, 146-58
    design constraints, 151-53
    diploid genetics and ontogeny, 147-51
    evidence for diversity, 159-68
    multiple causation, 153-54
    observation and experiment difficulties, 158-59
    post-reproductive survival, 156
    summary of ambiguity sources, 159-60
    time scale choice, 156-58
  protein
    and drosophila systematics, 235-51
  of shell polymorphism
    in Cepaea, 114-35
  social
    of birds and ungulates, 193-205
  vertebrate
    in North America, 355-74
  of viruses, 29-46
    diversity among, 30-31
    genetic strategies of, 31-34
    human intervention in, 44-46
    origins and phylogeny of, 37-44
    transmission strategies of, 34-36
    viral-host interactions, 36-37
Evolutionarily stable strategy (ESS)
  and life history traits, 150-51
Evovirology, 30
  see Evolution of viruses
Extinctions
  of savanna vertebrates
    in North America, 373-74

F

Fagus forests
  annual carbon release from, 68
Farming systems
  agro-ecosystem analysis of, 178-79
Fecundity
  of American shad
    and migration, 298
  of aphids, 333-34, 337, 346
  of Biomphalaria glabrata
    and schistosomiasis, 219-21
  of Cepaea, 111, 113
  of Chironomids, 265
  and early maturity
    as genetic traits, 147, 165-66
Feedback
  and ecosystem response patterns, 189-90
Ferritin
  in gel sieving analysis, 315
Fertility
  and life history trait evolution, 148, 166
Fertilizer
  and agro-ecosystems analysis, 174, 180, 187-88
  effects of
    and mineral nutrients, 95-96
Fibrosis
  in humans
    and schistosomes, 211
Fighting
  ritualized
    in crocodilians, 437
Finches
  brood parasitism in, 1-2, 10-11, 13, 19, 21
  host specificity in, 17-18
Finland
  aphid outbreaks in, 337
Firebacks
  social adaptations of
    and dispersal theory, 203
Fish
  acanthopterygian
    defenses of, 433-34
  life histories of
    reliability of, 164-65
  in reservoir ecosystems, 269-74
Fisheries
  ecosystem analysis of, 173
  simulation models for, 184
Fish migrations, 285-300
  ecological significance of
    homing, 298-99
  guidance mechanisms, 385-95
    currents, 295
    geomagnetic and geoelectric fields, 287
    inertial, 288
    maximization of comfort, 291-93
    ocean fronts, 294-95
    polarized light, 287
    random walks, 288-91
    sun orientation, 286-87
    temperature, 293-94
  home recognition, 295-98
    genetic factors, 298
    olfaction, 296-98
Flaviviruses
  transmission of
    and temperature effects, 36
Fledging periods
  of parasitic birds, 10-12
Flood control
  use of reservoirs for, 256
Florida
  carbon experiments in, 63
  drosophila in, 242
Flounder
  homing in, 289
Flowers
  with extrafloral nectaries
    and ant protectors, 407-21
Flukes
  blood
    see Schistosomiasis
Flycatchers
  fledging success of
    and parasitism, 5
Foliar analysis
  of tropical mountain forests, 96-100
  mineral element concentrations, 98
Food shortages
  world
    and agro-ecosystem analysis, 174, 186, 190
Food webs
  energy flow in
    and systems analysis, 176, 183
Foraging

472 SUBJECT INDEX

behaviors
  of crocodilians, 439
patterns
  of ants on nectaries, 417
Forestry
  ecosystem analysis of, 173
  simulation models for, 184
Forests
  and ant distribution, 415
  boreal
    carbon balance in, 54, 56-57, 68
    detritus in
      carbon balance in, 51-73
    hardwood
      annual carbon release from, 67
      decay constants for, 60
    in North America
      and vertebrate history, 358-60, 370-73
    rainforest
      annual release of carbon from, 67
    soils of
      annual carbon release from, 67-69
    temperate
      carbon balance in, 53, 56-57, 65-68
      as drosophila habitat, 245
    tropical
      carbon balance in, 53, 56-57, 65-67
    tropical highland
      drosophila in, 241
    tropical mountain, 83-103
      biomass analysis, 100
      biomass and productivity, 85-87
      decay, 87-88
      fertilizer effects, 95-96
      foliar analysis, 96-100
      growth and distribution of, 83-103
      leaf form, 91-92
      leaf life, 90-91
      leaf structure, 88
      limitations on distribution, 88-90
      limitations on growth, 90
      mineral cycling studies, 100-2
      mineral nutrient supply, 92-102
      soil analysis, 96
      transpiration rate and mineral supply, 93-95
Fossil energy
  and agro-ecosystem analysis, 174, 186
Fossilization
  of ungulates
    and evolution studies, 205
Fossils
  of Cepaea, 123
Founder effect

in drosophila, 243
in D. pseudoobscura
  and electrophoretic studies, 323
France
  Cepaea distribution in, 119, 129
Freeze-thaw cycles
  and carbon release, 71
Frequency-dependent selection
  in Cepaea evolution, 123-25
Fringilla montifringilla
  migrations of
    circannual rhythms of, 388
Fructose
  as nectary secretion, 408, 410
Fulica
  cornuta
    social adaptations of, 204
  gigantea
    social adaptations of, 204
  rufifrons
    fledging success of, 4
Fumarase
  of Colias meadii
    electrophoresis of, 315, 324
Fungi
  growth in tropical mountain forests, 90-91, 102
  mycorrhizal
    role in carbon balance, 63, 66
  viruses occurring in, 30-31

G

Gadus morhua
  migration of, 293
Gallus
  gallus
    social adaptations of, 201
  varius
    social adaptations of, 197, 203
Gambusia
  life history data on
    reliability of, 164
Gas-bubble disease
  among fish
    as effect of dams, 272
Gazelles
  social adaptations of, 202
Geese
  and reservoir ecosystems, 271
Gel sieving analysis
  in electrophoresis, 314-17, 323
    classification analysis, 316-17
    rationale, 314-15

standardization, 315-16
Gene frequency
  distribution patterns of
    in Cepaea, 114, 117
    of drosophila groups, 237
Gene pool
  see Drosophila, systematics and biochemical evolution
Generation alternation
  of schistosomes, 209-11
Genetic
  distance
    measurement of, 310, 323, 325
  drift
    in Cepaea evolution, 128
  factors
    in fish migrations, 298
Genetics
  of egg polymorphism
    in cuckoos, 16-17
  in extrafloral nectary development, 411, 419
  and life history trait evolution, 147-51
  of shell polymorphism
    in Cepaea, 109-36
  viral strategies, 31-34
  see also Protein evolution
Gennaeus
  social adaptations of, 203-4
Genomes
  in viral evolution, 32-34
Gens concept
  and cuckoo population structure, 16-17
Gentiobiose
  as nectary secretion, 410
Geococcyx californianus
  incubation and nestling periods in, 11
Geoelectric fields
  and fish migrations, 287-88
Geomagnetic field
  and bird migrations, 398
  and fish migrations, 287-88, 299
Germany
  Cepaea linkage disequilibria in, 131
  cuckoo distribution in, 16-17
Gesneriaceae
  extrafloral nectaries in, 412
Ghana
  Lake Volta in
    environmental effects of, 256-57
  tropical mountain forests in, 86, 100
Giraffe
  browsing
    and ant behavior, 417, 419
  and ungulate evolution, 195

SUBJECT INDEX 473

Giraffe-camels
  arboreal browsing among, 195
  in vertebrate evolution, 364, 366
Glucose
  as nectary secretion, 408, 410
Glucose-6-P dehydrogenase
  in Colias homology studies, 324
Glutamate dehydrogenase
  in Colias homology studies, 324
Glycerophosphate dehydrogenase (GPdH)
  electrophoretic studies of, 311, 317, 320-24
Glyptosaurus
  and vertebrate evolution, 363
Goldfish
  inertial guidance of, 288
Goodeniaceae
  extrafloral nectaries in, 412
Gossypium
  extrafloral nectaries in, 408-11
Gramineae
  evolution of, 357
  extrafloral nectaries in, 412
  as viral hosts, 42
Granuloma
  in humans
  and schistosomes, 211
Grasslands
  carbon balance in, 54, 56-57
  soils of, 60, 65-66
Grazing
  simulation models for, 184
Great Basin
  Miocene faunas of
  and vertebrate evolution, 364-65, 369, 371
Great Plains
  and savanna vertebrate history, 360-74
Greenland
  waters of
  fish migrations in, 394-96
Gregariousness
  of birds and ungulates
  social adaptations of, 195-96, 200
Grouse
  red
  ecological specialization of, 204
Guidance mechanisms
  of migrating fish, 285-95
  currents, 295
  geomagnetic and geoelectric fields, 287
  inertial, 288
  maximization of comfort,

291-93
  ocean fronts, 294-95
  polarized light, 287
  random walks, 288-91
  sun orientation, 286-87
  temperature, 293-94
Guinea pigs
  color of
  genetic basis for, 147
Gurneraceae
  extrafloral nectaries in, 412
Gyrinophilus
  life histories of
  reliability of, 161

H

Haber method
  of measuring carbon dioxide flux, 64
Habitat
  of Cepaea, 110-11, 115, 120, 125, 129-30
  and fish migrations, 298
  of Hawaiian drosophila, 240-41
  and social evolution
  in birds and ungulates, 193-205
Hadrosaurs
  display organs of, 440-41
  teeth of, 444
Haline circulation
  effects of dams on, 274
Hawaii
  drosophila in, 239-42
  reservoir fish in
  reproductive traits of, 145
Hawaiian drosophila
  biochemical evolution of and systematics, 239-42, 249
Hawks
  as parasitic hosts, 7-8
Heat
  balance
  of cultivated fields, 181
  denaturation
  in drosophila protein studies, 246-47
  stability
  in electrophoresis, 311, 319
Helianthella quinquenervis
  ant protection of, 418-19, 421
  nectar of, 410
Helianthus
  life history of
  reliability of, 167
Heliconius
  on nectary plants, 418-19
Heliscomys
  and vertebrate evolution, 362
Helix aspersa

natality in, 123
Hemoglobin
  in gel sieving analysis, 315
Henderson-Hackelbach equation
  in electrophoresis, 320
Hepatitis virus
  human intervention with, 45
Herbaceous plants
  as aphid hosts, 336
  flowering
  life histories of, 166-67
Herbivores
  and ant protection
  of nectary plants, 414-19
Hermaphroditism
  in Cepaea, 109, 128
Herpes simplex virus (HSV)
  DNA structure of, 41
  ocular
  and chemotherapy problems, 45-46
Herpesvirus
  B
  transmission to man, 36
  human intervention with, 44-46
  and oncogenesis, 37
  probability in plants, 31
  transmission of, 35
  and viral systematics, 40-41, 43-44
Herring
  migration of, 291, 294-95
Hesperocyon
  and vertebrate evolution, 363
Heterogeneity
  electrophoretic assessment of, 309-25
  allozyme properties, 314-20
  analytic approach to, 313-14
  comparing variants, 320-21
  genetic distance measurement, 325
  hidden, 312
  homology assessment, 310-11
  in insects, 311-12
Heteronetta atricapilla
  brood parasitism in, 1, 3-4, 6
Heterozygote advantage
  and stabilizing selection
  in Cepaea, 127
Hevea brasiliensis
  foliar analysis of, 96-97
Hexokinase
  in Colias homology studies, 324
Hibernation
  of Citellus lateralis
  and internal clock, 381

SUBJECT INDEX

Hierophasis
  social adaptations of, 203
Himalayas
  birds of
    and ecological adaptation, 204
Hippopotamus
  and vertebrate evolution, 358
Histidine
  in aphid ecology, 339
  in nectar, 410
Holcus lanatus
  fertilizer effects on, 95
Holland
  Cepaea in, 125-26
Homaladotheres
  arboreal browsing among, 195
Homalocephale
  combat organs in, 440
Homing
  in fish migrations, 295-98
  genetic factors in, 298
  olfaction, 296-98
Homology
  electrophoretic assessment of
    between species, 310, 324-25
    taxonomy, 311
    within species, 310, 322-24
Honeybees
  orientation of
  and polarized light, 287
Honey guides
  brood parasitism in, 1-2, 10-11, 13-14, 21
Hoplopterus spinosus
  social adaptations of, 197
Hornbills
  social adaptations of, 202
Horses
  evolution of
  and New World savanna, 357-74
Host
  alternation
    among aphids, 329-47
  discrimination
    and brood parasitism, 8-10
  interactions
    with viruses, 36-37
  selection
    in brood parasitism, 1-3
Houppifer
  erythrophthalmus
    social adaptations of, 197, 203
  inornatus
    social adaptations of, 203
Humble-bee
  brood parasitism among, 19
Humidity

and leaf life
  in tropical mountain forests 90
  in nectary ecology, 408
  and plant growth reduction, 95, 102
Humus
  carbon in, 57-62, 65, 72-73
  mineralization of, 88, 93-93, 103
Hyaenodon
  and vertebrate evolution, 363
Hyborhynchus notatus
  migration of, 296
Hydrophytes
  and ant distribution, 415
Hyemoschus
  and vertebrate evolution, 362
Hylochoerus
  pairbonding among, 195
Hymenoptera
  and extrafloral nectaries, 416
Hyopsodus
  and vertebrate evolution, 361
Hypohippus
  and vertebrate evolution, 369
Hypselosaurus
  parental behavior of, 442
Hypsodonts
  evolution of, 357-64
  time range of, 361

I

Iberian peninsula
  Cepaea distribution in, 120
Ichneumonids
  life histories of, 165-66
Icteridae
  brood parasitism in, 13-14
Iguanodon
  brain size and behavior of, 431, 433-34
Ilingoceras
  and vertebrate history, 369
Immunity
  to schistosomiasis, 216-18
Immunotherapy
  and viral-host relationships, 45
Imprinting
  among bird parasites, 3, 17-18
  in fish olfaction studies, 296-97
India
  cuckoos in
    brood parasitism among, 2
Indicator

indicator
  social behavior of, 14
minor
  incubation of, 10, 14
xanthonotus
  social behavior of, 15
Indicatoridae
  brood parasitism in, 1-2, 10-11, 14
Indigobirds
  and host imprinting, 18
Indonesia
  rural ecology of, 182-83
Indricotheres
  arboreal browsing among, 195
Inertial guidance
  in fish migrations, 288
Influenza virus
  A
    and viral adaptations, 44
    genetic reassortment of, 33
Inga
  extrafloral nectaries, 409
Insect control
  by baculoviruses, 46
Insectivorous plants
  in tropical mountain forests, 93
Insects
  hidden variation in
    and electrophoresis, 311-12
  life histories of, 151, 165-66
  as viral vectors, 42
Intertidal zones
  and barnacle life histories, 156-58
Inuit (Eskimo)
  effects of dams on, 276
Iododeoxyuridine (IDU)
  and treatment of HSV, 45-46
Ireland
  Cepaea distribution in, 122
Irish elk
  and dispersal theory, 202
Iron
  and soil respiration, 72
  in tropical forest growth, 96-99, 101
Irrigation
  and habitat deterioration, 180
  simulation models for, 184
Ischyromys
  and vertebrate history, 362
Island biogeography
  in reservoir ecology, 263
Isoelectric point (pI)
  in electrophoresis, 311-14
Isoleucine
  in aphid ecology, 339

SUBJECT INDEX 475

in nectar, 410
Israel
  agriculture in, 185-86
  aphid control in, 345
Italy
  dams and reservoirs of, 259
Ivory Coast
  mineral cycling studies in, 101

J

Jamaica
  rainforest growth in, 85-88, 91, 95-103
Japan
  cuckoos in, 3
  soil respiration in, 66
  solar-energy utilization in, 182
Jarman-Bell principle
  and ungulate evolution, 195
Jays
  social adaptations of, 201
Juglans regia
  as aphid host, 340-43
Jungle fowls
  social adaptations of, 197, 201, 203-4
Jurassic
  dinosaurs of
    brain size and behavior of, 429
Juveniles
  of crocodilians
    behavior of, 438

K

Kentrosaurus
  brain size and behavior of, 431, 433
Kin selection
  among ungulates, 195-96
Kirchoff Law
  in climatic selection of Cepaea, 117-18
Kudu
  social adaptations of, 204

L

Lactation
  and social evolution, 194
Lactose
  in nectar, 410
Ladybird beetles
  absorption of solar energy by, 117
  as aphid predators, 342
Lagonosticta senegala
  fledging success of, 4, 6
  nest predation of, 18, 22
Lagopus
  social adaptations of, 201
Lake Kariba

dam of
  environmental effects of, 257, 266, 271
Lake Mead
  environmental effects of, 257, 260-61, 270
Lake Nasser
  environmental effects of, 257, 276
Lakes
  see Dams and impoundments
Lake Tana
  formation of, 256
Lake Volta
  environmental effects of, 256-57, 265-66, 268, 270, 274-76
Lama vicugna
  pairbonding among, 194
Lampyris noctiluca
  as Cepaea predator, 116
Lanius collurio
  migrations of, 388
  parasitism of, 9
Laos
  dams and reservoirs in, 261
Lassa fever virus
  human intervention with, 45
Latitude
  and ant distribution, 415
  and ecological evolution, 193-94, 202
  and photoperiod
    in bird migrations, 382-402
  of various ecosystems and annual carbon release, 67-70
Lauraceae
  in tropical rain forests, 99
Lava
  as stream dam, 256
Leaching
  in reservoirs, 261, 264, 267, 269
Leaf life
  and form
    in tropical mountain forests, 88-92
Leguminosae
  evolution of, 357-58
  extrafloral nectaries of, 408, 412
Lemmings
  reaction to stress, 189
Lemna
  life history of, 167
Leonardoxa africana
  and ant relationship, 420
Lepidoptera
  frequency-dependent selection in, 125
  larvae
    and extrafloral nectaries, 415-16, 418
Lepomis

migration of, 286
Leporidae
  and vertebrate evolution, 361
Leptauchenia
  and vertebrate evolution, 362
Leptomerycidae
  and vertebrate evolution, 360-62
Lepus
  life history of, 164
Leslie matrix models
  and life history trait evolution, 158
Leucichtys artedi
  migration of, 291
Leukemia
  feline
    retrovirus adaptations, 36
Leviviridae
  mutations in, 31
Lichens
  in tropical mountain forests, 90
Life cycle
  of aphids, 329-47
    laboratory studies and biological control, 345
    polymorphism, 329-35
    population dynamics, 335-45
    simulation models of, 345-46
  of schistosomes, 209-11, 225-30
Life history traits
  evolution of, 145-68
    age distribution assumptions, 154-55
    ambiguity sources, 159-60
    ambiguity in theory, 146-58
    design constraints, 151-53
    diploid genetics and ontogeny, 147-51
    evidence for diversity, 159-68
    multiple causation, 153-54
    observation and experiment problems, 158-59
    post-reproductive survival, 156
    time scale choice, 156-58
  reliability tables for
    birds, 163
    fish, 164-65
    herbaceous flowering plants, 166-68
    insects, 165-66
    lizards, 162
    mammals, 163-64
    salamanders, 161
Lignin

476 SUBJECT INDEX

in humus formation and
  decay, 58
Limnion
  and fish migrations, 291
Limnology
  physical and chemical
    of man-made lakes, 258-62
Linkage disequilibrium
  in Cepaea, 130-35
Liodontia
  and vertebrate evolution, 373
Liriodendron forest
  annual carbon release from, 67
Littoral regions
  in reservoir ecosystems, 268-70
Livestock grazing systems
  of East Africa
    agro-ecosystem analysis of, 183
Lizards
  life histories of, 162-63
Lobiophasis bulweri
  social adaptations of, 203
Lophophorus impeganus
  social adaptations of, 201
Lophura ignita
  social adaptations of, 203
Lotka-Volterra models
  and life history trait evolution, 148, 158
Lunar cycles
  in fish migrations, 295
Lycaon
  pairbonding among, 194

M

Macaques
  herpesvirus of, 36, 41
Macaranga hypoleuca
  and extrafloral nectaries, 412, 415-16, 418-19
Magnesium
  in tropical forest growth, 96, 98-101
Mahonia
  in woodland savanna, 359
Malaria
  control of, 257, 275
  models of
    and schistosomiasis, 226
Malate dehydrogenase
  in Colias homology studies, 324
Malaya
  and distribution in, 415
  tropical mountain forests in, 86, 89-91, 96, 98-99, 102-3
Malaysia
  drosophila in, 239
Malesia
  forest distribution in, 83, 86

Malic enzyme
  in Colias homology studies, 324
Mallotus villosus
  migration of, 293
Malpigheaceae
  extrafloral nectaries in, 412
Malpighia glabra
  extrafloral nectaries of, 408
Maltose
  as nectary secretion, 410
Malvaceae
  extrafloral nectaries in, 412
Mammals
  brain:body size in, 433-34
  circannual rhythms in, 384
  extinctions of
    and savanna evolution, 373-74
  hormones in
    and reversed sex roles, 151
  life histories of, 163-64
  of New World savanna, 357-74
  placental
    diversity in, 148
  and reservoir ecosystems, 271
Mammoth
  and dispersal theory, 202
Manacus manacus
  nest predation of, 22
Manganese
  in tropical forest growth, 96, 98-99, 101
Mangrove swamp
  annual carbon release from, 69
Manitsha
  and vertebrate evolution, 359
Marantaceae
  extrafloral nectaries in, 412
Marek's disease virus
  in chickens, 37
  and viral classification, 41
Masonaphis maxima
  life cycle of, 346
Massenerhebung effect
  in tropical mountains, 84-85, 88
Mate selection
  and vocal mimicry
    in vidua finches, 12
Mating
  of schistosomes, 221-25
  and social behavior
    of brood parasites, 14-15
Mayfly
  in reservoir ecosystems, 266, 276

Measles
  interspecies spread of, 44
Mediterranean
  sardine catch in
    and effects of dams, 274, 276
Megaloceros
  and dispersal theory, 202
Megoura viciae
  life cycle of, 346
Melanogaster group
  of drosophila
    protein evolution of, 238-39
Meleagris gallopavo
  social adaptations of, 197
Melezitose
  in nectar, 410
Meliakrouniomys
  and vertebrate evolution, 362
Melibiose
  in nectar, 410
Melospiza melodia
  fledging success of, 4
Mengovirus
  and viral evolution, 39
Meniscotherium
  of New World savanna, 358
Merychippus
  and vertebrate evolution, 363-65, 366, 369-70
Merychyus
  and savanna evolution, 370
Merycoidodontidae
  and vertebrate evolution, 365
Mesogaulus
  and vertebrate evolution, 365
Mesohippus
  and vertebrate evolution, 362
Mesozoic
  archosaurian reptiles of
    brain size and behavior of, 429-44
Metabolism
  of Cepaea, 111
  in dinosaurs, 430-44
Methionine
  in aphid ecology, 339
  in nectar, 410
Metopolophium dirhodum
  life cycles of, 338
Mexico
  drosophila in, 242, 244
  savanna vertebrate history in, 359-71, 374
Microbes
  in soil
    and role in carbon balance, 57-58, 63, 66, 71
Microlophium carnosum
  life cycle of, 344
Microorganisms
  role in mineralization, 182
Microtoscoptes

SUBJECT INDEX 477

and vertebrate evolution, 369
Migration
  of aphids, 329
  of birds
    circannual rhythms in, 381-402
  of fish, 285-300
    ecological significance of homing, 298-99
    guidance mechanisms, 285-95
    home recognition, 295-98
  and opportunism
    in ungulate and bird evolution, 199
Mimicry
  of host
    by bird parasites, 8-10
  of males
    evolution of, 197, 200
Mineral cycling studies
  and forest growth, 100-2
Mineral element concentrations
  in tropical rain forests
    table of, 98
Mineral nutrition
  of tropical mountain forests, 92-102
    biomass analysis, 100
    fertilizer effects, 95-96
    foliar analysis, 96-100
    mineral cycling studies, 100-2
    soil analysis, 96
    transpiration rate and mineral supply, 93-95
Minnows
  migration of, 296
Miocene
  savanna vertebrates of, 363-71
Miohippus
  and vertebrate evolution, 361
Miracidium
  and schistosomiasis, 211, 216, 226
Missouri River
  dams on, 257
Mobility
  of proteins
    in electrophoresis, 313-14, 320, 322, 324-25
Models
  in agro-ecosystem analysis, 184-86
  of carbon balance, 73
  mathematical
    of schistosomiasis, 209-31
  simulation
    of aphid ecology, 333-34, 345-46
  steady-state
    of detritus, 59-62
  stochastic and deterministic

of life history traits, 145-68
Molecular clock
  in drosophila evolution, 246, 249-51
Molluscicide
  in schistosomiasis control, 230
Molluscs
  circannual rhythms in, 384
  helicid
    frequency-dependent selection of, 124, 128
  in reservoirs, 265-66, 274
Molothrus
  ater
    breeding season of, 6
    brood parasitism among, 2, 4-5
    interactions with host, 7, 9-14
    population, 15-16
  badius
    as parasite host, 19-20
  bonariensis
    brood parasitism among, 4-5, 7
  rufoaxillaris
    brood parasitism among, 10-20
Molybdenum
  in tropical forest growth, 96
Monera
  viruses occurring in, 30-33
Monocistronic mRNA
  in viral evolution, 34
Monoclonius
  display organs of, 441
Monosaulax
  and vertebrate evolution, 369
Montane
  tropical forests
    growth and distribution of, 83-103
  Montane habitats
    of birds and ungulates
    and adaptations to, 198, 201-5
Moose
  opportunism and evolution of, 199, 203
Mora excelsa
  mineral cycling in, 101
Morph frequency distribution
  in Cepaea, 115
Morpholine
  in fish olfaction studies, 296
Morphology
  of man-made lakes, 258-59
Mortality
  in aphids, 332, 336-37, 339, 341, 343-44, 346
  of Cepaea, 111, 123
  in life history models, 145,

153-54, 161, 163-64, 166
  from schistosomiasis, 215
snail
  and schistosomiasis infection, 220-21
Mosquitoes
  control of
    and dams, 257, 275
  as vectors
    in viral transmission, 36
Motacilla alba
  parasitism of, 9
Mountain sheep
  social adaptations of, 201-3
Mozambique
  Cobora Bassa Dam in, 257
Mutation
  rate in Cepaea evolution, 128
  in viral evolution, 31, 43
Mycoplasma
  and viral classification, 37
Mycorrhiza
  in tropical rain forests, 93-94
Mylagaulus
  and vertebrate evolution, 365
Myrtaceae
  in tropical rain forests, 99
Mytonolagus
  and vertebrate evolution, 360
Myxomatosis
  and rabbit control, 46
  and viral transmission, 36
Myzus persicae
  life cycle of, 333, 335, 344
  as viral vector, 42

N

Namatomys
  and vertebrate evolution, 360
Nectaries
  extrafloral, 407-21
    adaptive significance of, 414-21
    ant protection, 415-21
    cell structure of, 408
    nectar composition, 410
    taxonomic distribution of, 412-14
Neoheterandria
  life history of, 164
Neomorphinae
  and brood parasitism, 9
Neoteny
  and paedomorphism
    as social adaptations, 201
Neotragocerus
  and vertebrate evolution, 373
Nepenthes

in tropical mountain forests, 93
Neuro-endocrine
influence on fish migration, 291, 293, 299
Neuroptera
as aphid predators, 338
New England
dams and reservoirs in, 262
New Guinea
drosophila in, 239
drought in
and forest growth, 91
forest distribution in, 83-103
New Hampshire forest
carbon experiments in, 63
New Zealand
cuckoos in
brood parasitism in, 3
tropical mountain forests in, 92
Nile river
dams on, 256-57, 267, 273-74, 276
Nitrogen
in aphid nutrition, 332
cycling of
agro-ecosystem analysis of, 182, 187
leaching and byproduct wastes, 183-84
in nectar secretion, 408
in soils
and carbon balance, 61-62, 71-72
in tropical mountain forests, 93-103
Nocturnal activity
in migrating birds, 383, 389, 391
Nonomys
and vertebrate evolution, 362-63
North America
brood parasitism in, 5, 9, 17, 20
Cepaea distribution in, 110
drosophila in, 237
fish migration in, 293-99
palaeoenvironment of, 355-74
Nothocrax
social adaptations of, 201
Nothofagus truncata
mineral nutrients of, 96
Notophthalmus
life history of, 161
Nutrient processes
systems analysis of, 177
Nutrition
mineral
see Mineral nutrition
Nylad
social adaptations of, 204

O

Ocellate pheasant
social adaptations of, 198-200, 203
Ocellate-turkey
social adaptations of, 197
Octanol dehydrogenase (ODH)
in drosophila studies, 247-48
OdH
electrophoretic studies of, 311
Odocoileus
social adaptations of, 201
Oecophylla longinoda
and plant protection, 417
Oleaceae
extrafloral nectaries in, 412
Olfaction
and fish migration, 291, 296-99
Oligocene
woodland savanna in
in North America, 358-63
Oligochaetes
in reservoirs, 265-66
Onchocerciasis
river blindness
and effects of damming, 275
Oncogenesis
of viruses, 36-37
Oncomelania
as schistosome host, 210
Oncorhynchus (salmon)
migration of, 287-88
Ontogeny
and diploid genetics
in life history trait evolution, 147-51, 160
Opportunism
role in social evolution, 199-200
Orchidaceae
extrafloral nectaries of, 410, 412
Oreophasis
social adaptations of, 201
Oribatoid mites
in reservoirs, 265
Orientation
of migrating birds, 389
Ornithischia
brain size and behavior of, 429-44
Oromerycidae
and vertebrate evolution, 360
Ortalis
social adaptations of, 201-2
Oryza
life history of, 167
Ovibos moschatus
social adaptations of, 203

Oviraptor
brain size in, 435
Ovis ammon
social adaptations of, 202-3
Ovulation
of birds
and parasitic adaptations, 6-7
Owls
plains adaptation of, 365
Oxychilus cellarius
as Cepaea predator, 116
Oxydactylus
and savanna evolution, 370
Oxygen depletion
in reservoirs, 261-62
Oxyura jamaicensis
breeding success of, 6
evolution of parasitism in, 18

P

Pachycephalosaurs
combat organs of, 440
Pachyphylls
and forest growth, 91-92, 102
Pachyrhinosaurus
display organs of, 441
Pacific Northwest
palaeoenvironment of, 360, 367, 369
Paedomorphism
as social adaptation, 201
Pairbonding
of birds and ungulates
and ecological evolution, 193-96
Palaeocastor
and vertebrate evolution, 365
Palaeoclimatology
in North America
and vertebrate evolution, 355-74
Palaeolagus
and vertebrate evolution, 361
Panama
Miocene fauna of, 370
reservoirs of, 268
tropical mountain forests in, 86, 90, 100
Papovavirus
and oncogenesis, 37
SV40
adaptation of, 45
recombination of, 33
Papuacedrus
leaf production in, 87
Parahippus
and vertebrate evolution, 363-64, 366
Parailurus
and vertebrate evolution,

SUBJECT INDEX 479

373
Paramyxoviruses
 in viral evolution, 34, 40
Parasites
 and aphids, 337, 339-41, 343, 345-46
 of fish
  in reservoirs, 270
Parasitism, brood
 see Brood parasitism
Parental behavior
 in crocodilians, 436-38
 among dinosaurs, 442
Parental care
 by males
  and pairbonding, 194, 196, 198
Parent Stream Theory
 and fish migrations, 285
Parrotfish
 migration of, 286
Parrots
 monomorphism among, 200
Parthenogenesis
 in aphids, 329-47
Partridges
 social adaptations of, 200-1
Partula
 distribution of, 132
Passer eminibey
 nesting among, 20
Passifloraceae
 extrafloral nectaries of, 408-9, 413, 418-19
Passiflora platyloba
 extrafloral nectaries in, 409, 413
Pathogen ecology
 insect and plant, 175
Pavo
 cristatus
  social adaptations of, 200
 muticus
  social adaptations of, 200, 203
Peach trees
 as aphid hosts, 333
Peacocks
 social adaptations of, 197-204
Peccaries
 and savanna vertebrates, 370-72
Penelope
 social adaptations of, 201
Penelopina
 social adaptations of, 201
Perchoerus
 and vertebrate evolution, 362
Periphyllus testudinaceus
 life cycle of, 329-31
Permafrost
 and reservoir morphology, 258-59
Peromyscus

life history of, 164
Pest control
 and agricultural ecology, 175
 simulation models of, 184
Petaurista
 and vertebrate evolution, 373
PGM
 electrophoretic studies of, 311
pH
 of soil
  in tropical mountain forests, 93
Phaenicophaeinae
 brood parasitism, 9
Phage
 specificity of hosts, 36
 see also Virus
Phaseolus vulgaris
 and ant protection, 421
Phasianus
 social adaptations of, 204
Pheasants
 social adaptations of, 198-204
Pheromones
 and ant behavior, 416
Phillippines
 rain forest growth in, 85
 schistosomiasis in, 213-14, 225
Phoenicurus phoenicurus
 parasitism of, 9
Phosphoglucomutase
 in Colias homology studies, 324
Phosphomonoesterase
 in extrafloral nectaries, 408
Phosphorus
 cycling of
  agro-ecosystem analysis of, 182
 and soil respiration, 72
 in tropical mountain forests, 93-103
Photoperiod
 in aphid life cycle, 332-33
 in bird migrations, 381-402
 in fish migrations, 293, 295
Photoperiodism
 and breeding seasons
  of brood parasites, 6-7
Photosynthesis
 efficiency studies
  in Russia, 182-83
  and humidity, 95
 in lakes and reservoirs, 264
Phyllocladus
 leaf production in, 87
Phylloscopus
 collybita
  circannual rhythms of, 387-88, 392-95

sibilatrix
 parasitism of, 9
trochilus
 migrations of, 381-84, 387-89, 393-97, 399-400
Picea forest
 annual carbon release from, 68
Picornaviruses
 poliovirus
  human intervention with, 44-45
 RNA recombination in, 31
 transmission of, 36
 in viral evolution, 34
Pigs
 pairbonding among, 195
Pine
 Jack
  ant protection of, 418
Pine looper
 and aphid life cycles, 346
Pinus forests
 annual carbon release from, 67-68
Piona limnetica Besiadka
 in reservoirs, 268
Pisidia
 in reservoirs, 265
Plaice
 migration of, 290
Plains habitats
 and social adaptations, 200-1
Plankton
 and barnacle life histories, 158
 in reservoirs, 264, 267-68
Plantae
 viruses occurring in, 30-33, 42
Plants
 annuals and perennials compared, 167
 growth of
  simulation models for, 184
 herbaceous flowering
  life history data on, 166-67
 processes of
  systems analysis of, 177
 reproductive adaptations in
  and life history models, 167
Plectoptera
 in reservoirs, 265
Plectrophenax
 life history of, 163
Plectropterus gambienses
 social adaptations of, 197
Pleistocene
 in North America
  savanna vertebrates of, 371-74
Pleuronectes platessa
 migration of, 290

480 SUBJECT INDEX

Pliocene
  in North America
  savanna vertebrates of,
    371-74
Pliohippus
  and vertebrate history,
    369
Pliotaxidea
  and vertebrate history, 369
Ploceus cucullatus
  brood parasitism in, 19
Podocarpus urbanii
  leaf life in, 87
Poebrodon
  and vertebrate evolution,
    360
Poikilothermic animals
  and ant distribution, 415
  and aphid life cycles, 337
Poisson model
  in schistosomiasis, 224-25,
    229
Poland
  agro-ecosystem study in,
    181-82
  Vistula River in, 265
Polarized light
  role in fish migration, 287,
    290, 299
Polioviruses
  human intervention with, 44-45
  in viral evolution, 34
Pollen flora
  Miocene
    and savanna evolution, 371
Polycistronic RNA
  in viral evolution, 34, 38
Polygonaceae
  extrafloral nectaries in,
    412
Polygonum
  life history of, 167
Polymerase
  in viral evolution, 40
Polymorphism
  in aphids, 329-35
  in Cepaea
    see Cepaea
  in drosophila, 246
  egg
    in cuckoos, 16-17
  in enzymes
    and electrophoresis, 318
Polynesia
  Partula distribution in, 132
Polyphenols
  in humus formation and decay,
    58
Polyplectron
  social adaptation of, 197,
    203
Polypodiaceae
  extrafloral nectaries in,
    412
Poplar
  as aphid host, 344

Population
  biology
    role of electrophoresis in,
      309-25
  density
    in life history models, 145,
      153-55, 164, 166
  dynamics
    of aphids, 335-45
  genetics
    and evolutionary view,
      149
  size
    of Cepaea, 110-11, 127
  structure
    and brood parasitism, 15-18
Populations
  and exploitation of
    systems analysis of, 176
Populus glandulifera
  extrafloral nectaries of,
    408
Potassium
  in tropical mountain forests,
    94-103
Potatoes
  as aphid hosts, 344
Potato spindle viroid
  and viral evolution, 39
Povilla adusta
  in reservoirs, 266
Poxvirus
  myxomatosis
    introduction in Australia,
      46
  and viral classification, 37-38
Prairie
  and savanna evolution, 360-67
  soils of
    and carbon balance, 61
Praon exsoletum
  in aphid control, 345
Predation
  anti-predator strategy
    of birds and ungulates,
      196, 198, 200
  and evolution of brood parasitism, 21-22
Predator-prey interactions
  systems analysis of, 176
Predators
  of aphids, 336-46
  brain size of, 433-34
  of Cepaea, 111, 116, 123-24,
    127, 136
  of fish
    and effects of dams, 272
Primates
  in New World
    evolution of, 359, 366
Proboscideans
  and vertebrate evolution,
    367, 372
Procoileus

and vertebrates evolution,
  373
Promimomys
  and vertebrate evolution,
    369
Pronghorns
  merycodontine
    and vertebrate evolution,
      364, 367, 369
Protadjidaumo
  and vertebrate evolution,
    360
Proteins
  electrophoresis of
    and heterogeneity, 309-25
  evolution
    and drosophila systematics,
      235-51
Proterix
  and vertebrate evolution,
    363
Protista
  viruses occurring in, 30-33
Protoceras
  and vertebrate evolution,
    361
Protoceratidae
  and vertebrate evolution,
    360
Protoceratops
  brain size and behavior of,
    433-34
  combat and display organs
    of, 441-42
Protoreodon
  and vertebrate evolution,
    360
Protylopus
  and vertebrate evolution,
    360
Prunus
  cerasus
    extrafloral nectaries in,
      409, 418
  persica
    nectar of, 410, 413
Przewalskium albirostris
  social adaptations of, 204
Pseudhipparion
  and vertebrate evolution,
    371
Pseudocalymma
  extrafloral nectaries in,
    409
Pseudomyrmex ferruginea
  and nectary plant protection,
    421
Pseudopleuronectes americanus
  homing of, 289
Psittacosis
  and viral classification, 37
Ptarmigan
  social adaptations of, 201
Pteranodon
  brain:body ratio in, 433
Pterocomma populifoliae
  life cycles of, 344

SUBJECT INDEX 481

Pterodactylus
 brain:body ratio in
  elegans, 33
  kochi, 433
Pterosauria
 brain size and behavior of,
  429-44
Puerto Rico
 brood parasitism in, 5
 rain forest growth in, 85-87,
  89-90, 98, 100, 102-3
 soil respiration in, 65
Puffinus tenuirostris
 migrations of
  and circannual rhythms,
   388
Pulse stability
 in reservoir ecosystems,
  263, 269, 274
Pycnonotus
 parasitism of, 8
Pyrenees
 Cepaea distribution in, 110,
  122, 130-31, 134

Q

Quail
 social adaptations of, 201
Quebec
 dams and reservoirs in,
  257-58, 276
Quelea quelea
 social adaptations of, 200
Quercus forests
 annual carbon release from,
  67-69
Quercus ilex
 foliar analysis of, 96
 growth reduction in, 95
 mineral cycling in, 101

R

Radiation
 and Cepaea evolution, 117-
  18, 133
 as growth factor
  in tropical mountain forests, 90, 102
 see also Solar radiation
Raffinose
 as nectary secretion, 410
Rain forests
 total biomass in, 86
 types of formations, 84
 see also Forests, tropical
  mountain
Rakomeryx
 and vertebrate evolution,
  369
Rangelands
 ecosystem analysis of, 173
Rangifer
 social adaptations of, 198-
  99, 201, 203
Rattlesnakes

and savanna vertebrate history, 365
Rattus
 life history of, 164
Reassortment
 in viral genetics, 33
Recombination
 in viral evolution, 31
Reindeer
 social adaptations of, 200
Reoviruses
 and viral evolution, 34, 40
Replication
 in viral transmission, 34-
  36
Reproduction
 of aphids, 329-47
 among Cepaea, 111
 of fish
  and effects of dams, 273
  and migrations, 298-99
Reproductive effort
 and life history trait evolution, 156, 160, 166-67
Reproductive strategies
 and brood parasitism
  in birds, 12-14
Reptiles
 archosaurian
  brain size and behavior in,
   429-44
 circannual rhythms in, 384
Reservoir ecosystems, 262-
  71
 benthos development, 264-
  67
 fish and other vertebrates,
  269-71
 littoral region, 268-69
 plankton, 267-68
Reservoirs
 morphology and limnology
  of, 258-62
 chemical limnology of,
  258-62
 physical limnology, 259-
  61
 sedimentation, 259
Respiration
 of soil
  and carbon release, 62-72
Respiratory viruses
 evolution of
  and human intervention,
   44
Retroviruses
 C-type
  and viral genetics, 43
  evolution of, 33, 35, 38-39,
   42-43, 46
Rhabdoviruses
 diversity of hosts, 36
 in plants, 31
Rhamphorhynchus
 brain:body ratio in, 433
Rhea americana
 parental behaviors of, 442

Rheinartia ocellata
 social adaptations of, 198-
  200, 203
Rheotaxis
 in fish migrations, 292, 295
Rhinoplax
 social adaptations of
  and dispersal theory, 202
Rhinoviruses
 transmission of, 36
Rhizopertha dominica
 life history of, 165-66
Rhopalosiphum
 insertum
  life cycle of, 334
 padi
  life cycle of, 329-31, 334,
   338
Ribonucleic acid (RNA)
 and evolution of viruses,
  29-46
Ribophage
 Leviviridae
  mutations in, 31
  and viral evolution, 39
Rice
 life histories of, 167
Ricinus
 extrafloral nectaries of,
  412, 418
Rinderpest
 interspecies spread of, 44
Roccus chrysops
 migration of, 286
Rocky Mountains
 as drosophila range, 237
 in savanna vertebrate evolution, 357-58, 360
Rodents
 in savanna vertebrate history, 360-74
Rosaceae
 extrafloral nectaries of,
  412
Rubiaceae
 extrafloral nectaries in,
  412
RuDP carboxylase loci
 electrophoretic studies of,
  311
Rumania
 dams and reservoirs in,
  257
Russia
 Cepaea distribution in, 110
 dams and reservoirs of
  environmental effects of,
   257, 264, 266-68
 photosynthesis efficiency
  studies in, 182-83
 soil studies in
  and carbon balance, 58-59,
   61, 71

S

Salamanders

482 SUBJECT INDEX

life histories of, 161
Salicaceae
 extrafloral nectaries in, 412
Saline waters
 as ecosystems, 263-64, 274
Salinity
 in arid regions, 184
 preference
 and fish migrations, 291-95
Salmo
 clarki
  homing ability of, 287
 gairdneri
  olfaction studies of, 296
  life history data on, 164
Salmon
 and dam effects, 272-73
 migrations of, 285-300
Salts
 and nectar secretion, 408
Salvelinus alpinus
 migration of, 297
Salvinia
 life history of, 167
Sandpipers
 social systems of, 196
Sanitation
 and schistosomiasis, 226
Sapindaceae
 in savanna evolution, 358
Sardines
 effects of dams on, 274
Saurischia
 brain size and behavior of, 429-44
Sauropods
 brain size and behavior of, 432-44
Saurornithoides
 brain size and behavior of, 432
Savanna vertebrates
 in North America, 355-74
 extinctions, 373-74
 meaning of savanna, 356
 Miocene, 363-71
 Oligocene, 360-63
 protosavanna, 357-58
 steppe, 371-73
 woodland savanna, 358-60
Sayornis phoebe
 fledging success of
 breeding season of, 6
 and parasitism, 4
Scaphidura oryzivora
 brood parasitism among, 3-4
Scaphognathus purdoni
 brain:body ratio in, 433
Scario (parrot fish)
 migration of, 286
Sceloporus graciosus
 life history of, 162-63
Schistosoma

haematobium, 210-11, 215-16, 220, 225, 230
japonicum, 210-11, 213-14, 217, 224-25
mansoni
 life cycle of, 209-11
 and schistosomiasis, 209-31
 sexual pairing among, 210, 217-18, 221-25
Schistosomiasis
 as effect of dams, 275
 mathematical models of, 209-31
 current infection, 212-19
 decision-making and economics, 230-31
 infection as age function, 211-12
 life cycle, 209-11, 225-30
 sex among schistosomes, 221-25
 snail population dynamics, 219-21
Sclerophylls
 and forest growth, 91-92
Scotland
 aphid infestations in, 338, 344
Cepaea distribution in, 122, 129
Seasons
 in aphid life cycles, 329-47
 effects on bird migrations and circannual rhythms, 381-402
 effects on nectar secretions, 411
 of schistosomiasis infection, 230
Sedge warblers
 parasitism of, 4-5
Sedimentation
 of reservoirs, 259, 276
Seismic activity
 as dam effect, 275
Selection
 in Cepaea evolution, 114-28
 climatic, 117-23
 density-dependent, 126-27
 disruptive, 125-26
 frequency-dependent, 123-25
 heterozygote advantage, 127-28
 visual, 116-17
 in life history trait evolution, 158
 in viral evolution, 31-34
Senecio
 in tropical mountain forests, 92
Serotypes
 in viral groups, 44
Sex roles

among birds and mammals
 social evolution of, 193-94, 196
Sexual dimorphism
 evolution of, 193, 196-202
Sexual transmission
 of viruses, 36
Sheep
 production
 agro-ecosystem analysis of, 178-79
Shells
Cepaea
 polymorphism in, 111-36
Siberia
 effects of dams in, 274
Sika deer
 circannual rhythms in, 384, 388
Silicone
 in tropical forest growth, 96
Simaroubaceae
 extrafloral nectaries in, 412
Simian papovavirus
SV40
 and polio vaccines, 45
Simimys
 and vertebrate evolution, 360, 362
Simulation models
 of aphid ecology, 333-34, 340, 345-46
Sitobion
 life cycles of, 338
Smallpox
 effects on human civilization, 44-45
Smilacaceae
 extrafloral nectaries in, 412
Smolts
 and effects of dams, 273
Snail
Cepaea
 polymorphism in, 109-36
 predatory
 Oxychilus cellarius, 116
 in schistosome life cycle, 209-31
 population dynamics, 219-21
Snowcock
 social adaptations of, 200, 204
Snowshoe hare
 life history of, 163-64
Social adaptations
 in birds and ungulates, 193-205
 dispersal theory, 202-6
 pairbonding, 193-96
 sexual mono- or dimorphism, 197-202
 weapons, 196
Soda lakes

SUBJECT INDEX   483

as ecosystems, 263-64
Sodium
  in tropical forest growth,
    96, 101
Soil
  carbon compounds in
    and world carbon balance,
    57-73
  decay and mineralization
    in tropical mountain forests, 88
  erosion control
    and agro-ecosystems analysis, 187-88
  management
    and agricultural ecology,
    175
  of tropical mountain forests,
    93-96
Solar radiation
  and reservoir ecosystems,
    272
  utilization studies in Japan,
    182
Solenopsis
  and plant protection, 417
Solidago
  life history of, 166-67
South Africa
  brood parasitism in, 2, 8
South America
  agro-ecosystems in, 183
  birds and ungulates in, 202
  brood parasitism in, 7-8,
    13, 18
  Lake Brokopondo in
    environmental effects of,
    257, 261
  tropical mountain forests in,
    92, 97
  vertebrate evolution in, 355-57
South Dakota
  badlands of
    fossil deposits in, 360-63
Southeast Asia
  drosophila in, 238-39
Soybeans
  growth reduction in
    and high humidity, 95
Spain
  Cepaea distribution in, 119
  dams and reservoirs in, 255,
    257, 268
Sparrow
  migration of
    and circannual rhythms,
    394
  nesting among, 20
Spawning
  of fish
    and effects of dams, 273
Species
  and speciation
    in drosophila evolution,
    247-49, 251
Species specific

endogenous time programs
  in birds, 392-93
Speed
  of migrating birds
    and compensatory mechanisms, 400-1
  of migrating fish
    and temperature, 294
Sphenophalos
  and vertebrate evolution,
    369
Spinosaurus
  display organs of, 440
Spiza americana
  migration of
    and circannual rhythms,
    388
Spleen necrosis virus
  and viral evolution, 40
Spreo bicolor
  parasitism of, 2
Spruce
  as aphid host, 338-39
Spur-winged goose
  social adaptations of, 197
Spur-winged plover
  social adaptations of, 197
Squirrels
  life histories of, 164
Sri Lanka
  brood parasitism in, 3
Stachyose
  in nectar, 410
Starlings
  migrations of, 384, 386-88,
    396
Stegoceras
  combat organs of, 440
Stegosaurus
  brain size and behavior of,
    431, 433-34
  display organs of, 441
  teeth of, 444
Stenomylus
  and vertebrate evolution,
    370
Stenonychosaurus
  brain size of
    and behavior, 431-33, 435-36
Steppe
  North American
    in Pliocene and Pleistocene,
    371-73
Sterculiaceae
  extrafloral nectaries in,
    412
Stipidium
  and savanna evolution, 365
Stress
  in agricultural systems,
    188-89
Sturnus vulgaris
  circannual rhythms of, 384,
    386-88
Styloviridae
  mutation rate in, 31

Styracosaurus
  display organs of, 441
Subhyracodon
  and vertebrate evolution,
    361
Sucrose
  as nectary secretion, 408,
    410
Sugar cane
  and ant predation, 417
  as aphid host, 345
Suidae
  and ungulate evolution, 195-96
Sumatra
  firebacks of
    and dispersal theory, 203
Sunfish
  migration of, 286
Sunflower
  ant protection of, 418-19,
    421
Sun orientation
  in migrating fish, 286-87,
    290, 299
Surinam
  dams and impoundments in,
    257, 261, 271
Surniculus lugubris
  brood parasitism among,
    8
Swamp and marsh
  carbon balance in, 54, 56-57, 69
Sweden
  cuckoo distribution in, 16
  reservoirs in, 270
Switzerland
  cuckoo distribution in, 16
Sycamore
  as aphid host, 329-32, 339-42
Sylvia
  migrations and circannual
    rhythms of
    atricapilla, 384-88, 393,
      396, 401
    borin, 384-85, 387-88,
      390-93, 396-401
    cantillans, 388, 393
    hortensis, 401
    melanocephala, 388, 393
    sarda, 388, 393
    undata, 388, 393
Sylvilagus
  life history of, 164
Syrmaticus
  social adaptations of, 204
Syrphidae
  as aphid predators, 338
Systematics
  Drosophila
    and biochemical evolution,
      235-51
  virus classification, 30-31,
    40-44
Systems analysis

484 SUBJECT INDEX

and agro-ecosystem
  research, 175-91

T

Taeniodonts
  and vertebrate evolution,
    358, 360-61
Tanzania
  soil and mineral shortage,
    95
Tapera naevia
  and brood parasitism, 9
Taraxacum
  life history of, 167
Tasmania
  reservoirs of, 269
Taxonomic distribution
  of extrafloral nectaries,
    412-14
Taxonomy
  and electrophoresis, 311
Tayassuidae
  and ungulate evolution, 195-
    96, 201
Teal
  migrations of
    and circannual rhythms,
      396
Teeth
  and vertebrate evolution,
    356-74
Temperate zone
  as aphid habitat, 329-47
  bird migrations in, 382-402
  drosophila habitats in, 245
  forest
    carbon balance in, 53, 56-
      57, 65-68
  grassland
    carbon balance in, 54, 56-
      57, 65-66, 69
  plants in
    extrafloral nectaries of,
      411, 421
  production and cycling in
    agro-ecosystems analysis,
      181-82
  reservoir ecosystems of,
    265, 267, 270, 277
Temperature effects
  on aphid life cycles, 332-33,
    335, 337-38, 341-42, 344-
    47
  on bird migrations, 281
  on fish life histories, 165
  on fish migrations, 291, 293-
    95
  in reservoirs
    environmental effects of,
      260, 272
  on selection
    in Cepaea, 117-23
  and soil respiration
    in carbon balance, 70-72
  of substrate binding
    in electrophoresis, 320

in tropical mountain forests
  and growth and distribution,
    87-90, 102
  on viral transmission, 36
Tennessee
  pine forests in
    and carbon release, 64,
      66
Tennessee Valley Authority
  dams and reservoirs of
    environmental effects of,
      257, 260, 270, 275
Tephritid fly
  and ant protection, 418
Territoriality
  in crocodilians, 436-38
  among ungulates, 195-97,
    202
Tertiary
  savanna vertebrates of, 357-
    74
Tetraclita squamosa
  life history of, 156-58
Tetraogallus
  social adaptations of, 200,
    204
Tetraonidae
  social adaptations of, 198,
    201
Tetrao urogallus
  social adaptations of, 201
Theaceae
  in tropical rain forests, 99
Thecodontia
  orders of
    brain size and behavior of,
      429
Theobroma cacao
  mineral nutrients in, 96
Therioaphis trifolii
  life cycle of, 345
Thienemann's rules
  in reservoir ecosystems,
    262-63
Threonine
  and aphid life cycle, 337
Thrush
  as Cepaea predator, 116,
    124
Thuja swamp
  annual carbon release from,
    69
Ticholeptus
  and vertebrate evolution,
    369
Ticks
  as vectors
    in viral transmission, 36
Titanoides
  and vertebrate evolution,
    358
Tobacco mosaic virus (TMV)
  spreading of, 36
Tobacco necrosis virus
  and viral evolution, 32-33
Tockus
  social adaptations of, 202

Togaviridae
  in viral evolution, 34
Tragelaphus
  social adaptations of, 204
Tragulidae
  social adaptations of, 197
Transpiration rate
  in tropical mountain forests,
    92-95
Triassic
  reptiles of
    brain size and behavior in,
      429-44
Tribolium
  life history of, 165-66
Triceratops
  brain:body ratio in, 433-34
  display organs of, 441
Trichoptera
  in reservoirs, 265
Trigona bees
  and extrafloral nectaries,
    416
Trinidad
  mineral cycling studies in,
    101
Triose-P-isomerase
  in Colias homology studies,
    324
Trioxys pallidus
  as aphid parasite, 340, 343,
    345
Troglyodytes aedon
  nest predation of, 22
Tropical forests
  carbon balance in, 53, 56-
    57, 65-67
  ungulate evolution in, 195,
    197
Tropical mountain forest
  growth and distribution of,
    83-103
Tropical zone
  plants in
    extrafloral nectaries of,
      411, 415, 419, 421
  production and cycling
    agro-ecosystems analysis,
      182-83
  savanna in
    carbon balance in, 54, 56,
      66, 69
Tropics
  bird migrations to, 381-402
  mountain forests in
    growth and distribution of,
      83-103
  origins of brood parasitism
    in, 22
  reservoir ecosystems in,
    270, 276
Trout
  homing ability of, 287
Tuberculoides annulatus
  life cycles of, 330
Tubificids
  in reservoirs, 265

SUBJECT INDEX 485

Tuna
  migration of, 293
Tundra
  and alpine
    carbon balance in, 54, 56-57, 66, 69
  and savanna vertebrate history, 373
  stress studies on, 189
Tunnus germo
  migration of, 293
Turdoides
  caudatus
    and parasitism, 4-5
  squamiceps
    brood parasitism in, 19
  striatus
    fledging success of, 4-5
Turdus ericetorum
  as Cepaea predator, 116
Turnera
  extrafloral nectaries in
    trioniflora, 409
    ulmifolia, 410
Turneraceae
  extrafloral nectaries in, 412
Typha
  life histories of, 167
Typhlosaurus
  life history of, 162
Tyrannosaurus
  brain:body ratio in, 433

U

Uinta Basin sediments
  in savanna evolution, 359-60, 362
Ungulates
  in savanna vertebrate evolution, 364-74
  social adaptations
    and ecology of, 193-205
Uta stansburiana
  life history of, 162-63

V

Vauquelinia
  in woodland savanna, 359
Vectors
  in viral systematics, 42
Verbenaceae
  extrafloral nectaries in, 412
Vertebrates
  savanna
    New World history of, 355-74
    see also Savanna vertebrates
    viral transmission in, 35
Viability
  of Cepaea, 111, 113
Vicia faba
  as aphid host, 336

Vicuna
  pairbonding among, 194
Vidua
  brood parasitism in, 1-7, 10-15
  chalybeata, 4-6, 10, 12, 14-16, 22
  host specificity in, 17-18
Vigna
  extrafloral nectaries in, 409
Vireo olivaceus
  parasitism in, 4
Viroid
  and virus evolution, 39-40
Viruses
  diversity among, 30-31
  evolution of, 29-46
    human intervention in, 44-46
    origins and phylogeny of, 37-44
    genetic strategies of, 31-34
    transmission strategies of, 34-36
    viral-host interactions, 36-37
    accelerated cell evolution, 37
    oncogenesis, 36-37
Visna
  DNA plasmids in, 40
Visual selection
  in Cepaea polymorphism, 116-18, 123
Vitaceae
  extrafloral nectaries in, 412
Vitamins
  production of
    and ungulate digestive system, 194
Vocalization
  in crocodilians, 437-38
  in dinosaurs, 442
Vocal mimicry
  in Vidua finches
    and parasitism, 12
Volcanoes
  Hawaiian
    as drosophila habitat, 240
Volga River
  dams on
    environmental effects of, 257, 264, 266

W

Wales
  Cepaea distribution in, 121, 132
Warblers
  migrations of, 381-402
Wastes
  from agricultural byproducts, 183-84
Water

dams and impoundments
  environmental effects of, 255-77
  as growth factor
    in tropical mountain forests, 90-91
  processes
    systems analysis of, 177
  stress
    and nectar secretion, 408
  utilization
    and agro-ecosystems analysis, 186
Waterfowl
  and reservoir ecosystems, 271
Weapons
  among birds and ungulates
    evolution of, 193, 196
Weather
  effects on aphid life cycle, 336, 338-39, 347
  effects on bird migrations, 400-1
Weaver birds
  social adaptations of, 200
Wildlife
  simulation models for, 184
Wind
  as aphid carrier, 329
  and fish migrations, 297
  as growth factor
    in tropical mountain forests, 90
Wolf
  pairbonding in, 194
Woodland
  savanna evolution
    in North America, 356, 358-60
  and shrubland
    carbon balance in, 54, 56, 68-69
Wood pigeons
  social adaptations of, 200
Wooly rhino
  and dispersal theory, 202

X

Xanthine dehydrogenase (XDH)
  in drosophila studies, 246, 248
Xerophytes
  and ant distribution, 415
Xylose
  as nectary secretion, 410

Y

Yellow fever virus
  adaptability of, 45

Z

Zambezi River
  dams on

environmental effects of, 257, 266
Zapodine mice
  in vertebrate evolution, 373
Zarhynchus wagleri
  fledging success of, 4
Zeitgeber
  in circannual rhythms for birds, 387-88, 396
Zenarchopterus dispar
  orientation of, 287
Zinc
  in tropical forest growth, 96, 98-99, 101
Zonotrichia
  capensis
    fledging success of, 4
  leucophrys
    migration of, 394
Zugunruhe
  migratory restlessness in birds, 388-97

CUMULATIVE INDEXES

CONTRIBUTING AUTHORS VOLUMES 4-8

A

Alexander, R. D., 5:325-83
Ashlock, P. D., 5:81-99
Ayala, F. J., 5:115-38

B

Baker, H. G., 5:1-24
Baxter, R. M., 8:255-83
Bentley, B. L., 8:407-27
Berlin, B., 4:259-71
Bliss, L. C., 4:359-99
Burns, C. W., 7:177-208
Bush, G. L., 6:339-64

C

Campbell, C. A., 5:115-38
Carroll, C. R., 4:231-57
Carson, H. L., 7:311-45
Chapman, A. R. O., 5:65-80
Cody, M. L., 4:189-211
Cohen, J. E., 8:209-33
Colwell, R. K., 6:281-310
Colwell, R. R., 4:273-300
Courtin, G. M., 4:359-99
Covich, A. P., 7:235-57
Cracraft, J., 5:215-61
Crowley, P. H., 7:177-208

D

Denman, K. L., 6:189-210
Dickeman, M., 6:107-37
Dixon, A. F. G., 8:329-53

E

Ehrlich, P. R., 6:211-47
Ewing, E. P., 7:1-32

F

Fox, L. R., 6:87-106
Fretwell, S. D., 6:1-13
Fuentes, E. R., 6:281-310

G

Gallucci, V. F., 4:329-57
Geist, V., 8:193-207
Giesel, J. T., 7:57-79
Gilbert, L. E., 6:365-97
Ginevan, M. E., 7:1-32

Grubb, P. J., 8:83-107
Gwinner, E., 8:381-405

H

Hall, D. J., 7:177-208
Harper, J. L., 5:419-63
Hedrick, P. W., 7:1-32
Heinrich, B., 6:139-70
Hespenheide, H. A., 4:213-29
Holling, C. S., 4:1-23
Hopson, J. A., 8:429-48
Horn, H. S., 5:25-37
Hungate, R. E., 6:39-66

I

Immelmann, K., 6:15-37

J

Jackson, R. C., 7:209-34
Jain, S. K., 7:469-95
Jander, R., 6:171-88
Janzen, D. H., 4:231-57; 7:347-91
Johnson, G. B., 4:93-116; 8:309-28
Johnson, W. E., 4:75-91
Jones, J. S., 8:109-43

K

Kaneshiro, K. Y., 7:311-45
King, J. A., 4:117-38

L

Leggett, W. C., 8:285-308
Leigh, E. G. Jr., 6:67-86
Leith, B. H., 8:109-43
Levin, D. A., 7:121-59
Levin, S. A., 7:287-310
Lillegraven, J. A., 5:263-83
Livingstone, D. A., 6:249-80
Loucks, O. L., 8:173-92
Lugo, A. E., 5:39-64

M

MacIntyre, R. J., 7:421-68
Monsi, M., 4:301-27

Müller, K., 5:309-23

N

Nahmias, A. J., 8:29-49
Noy-Meir, L., 4:25-51; 5:195-214

O

Oikawa, T., 4:301-27
Otte, D., 5:385-417

P

Pattie, D. L., 4:359-99
Payne, R. B., 8:1-28
Peet, R. K., 5:285-307
Pianka, E. R., 4:53-74
Platt, T., 6:189-210

R

Rawlings, P., 8:109-43
Reanney, D. C., 8:29-49
Riewe, R. R., 4:359-99
Rohlf, F. J., 5:101-13
Rowlands, I. W., 4:139-63
Rozeboom, L. E., 7:393-420

S

Schad, G. A., 7:393-420
Schlesinger, W. H., 8:51-81
Selander, R. K., 4:75-91
Sieburth, J. McN., 7:259-85
Simberloff, D. S., 5:161-82
Singer, M. C., 6:365-97
Smith, R. L., 7:33-55
Snedaker, S. C., 5:39-64
Soulé, M., 4:165-87
Staley, T. E., 4:273-300
Stearns, S. C., 8:145-71

T

Taub, F. B., 5:139-60
Threlkeld, S. T., 7:177-208
Throckmorton, L. H., 8:235-54

U

Uchijima, Z., 4:301-27

487

V

Vayda, A. P., 5:183-93

W

Wangersky, P. J., 7:161-76

Webb, S. D., 8:355-80
Weir, B. J., 4:139-63
White, J., 5:419-63
Whitfield, D. W. A., 4:359-99

Widden, P., 4:359-99
Wiegert, R. G., 6:311-38
Wiens, J. A., 7:81-120

CHAPTER TITLES VOLUMES 4-8

VOLUME 4 (1973)

Title	Author	Pages
Resilience and Stability of Ecological Systems	C. S. Holling	1-23
Desert Ecosystems: Environment and Producers	I. Noy-Meir	25-51
The Structure of Lizard Communities	E. R. Pianka	53-74
Genetic Variation Among Vertebrate Species	R. K. Selander, W. E. Johnson	75-91
Enzyme Polymorphism and Biosystematics: The Hypothesis of Selective Neutrality	G. B. Johnson	93-116
The Ecology of Aggressive Behavior	J. A. King	117-38
Reproductive Strategies of Mammals	B. J. Weir, I. W. Rowlands	139-63
The Epistasis Cycle: A Theory of Marginal Populations	M. Soulé	165-87
Character Convergence	M. L. Cody	189-211
Ecological Inferences from Morphological Data	H. A. Hespenheide	213-29
Ecology of Foraging by Ants	C. R. Carroll, D. H. Janzen	231-57
Folk Systematics in Relation to Biological Classification and Nomenclature	B. Berlin	259-71
Application of Molecular Genetics and Numerical Taxonomy to the Classification of Bacteria	T. E. Staley, R. R. Colwell	273-300
Structure of Foliage Canopies and Photosynthesis	M. Monsi, Z. Uchijima, T. Oikawa	301-27
On the Principles of Termodynamics in Ecology	V. F. Gallucci	329-57
Arctic Tundra Ecosystems	L. C. Bliss, G. M. Courtin, D. L. Pattie, R. R. Riewe, D. W. A. Whitfield, P. Widden	359-99

VOLUME 5 (1974)

Title	Author	Pages
The Evolution of Weeds	H. G. Baker	1-24
The Ecology of Secondary Succession	H. S. Horn	25-37
The Ecology of Mangroves	A. E. Lugo, S. C. Snedaker	39-64
The Ecology of Macroscopic Marine Algae	A. R. O. Chapman	65-80
The Uses of Cladistics	P. D. Ashlock	81-99
Methods of Comparing Classifications	F. J. Rohlf	101-13
Frequency-Dependent Selection	F. J. Ayala, C. A. Campbell	115-38
Closed Ecological Systems	F. B. Taub	139-60
Equilibrium Theory of Island Biogeography and Ecology	D. S. Simberloff	161-82
Warfare in Ecological Perspective	A. P. Vayda	183-93
Desert Ecosystems: Higher Trophic Levels	I. Noy-Meir	195-214
Continental Drift and Vertebrate Distribution	J. Cracraft	215-61
Biogeographical Considerations of the Marsupial-Placental Dichotomy	J. A. Lillegraven	263-83

The Measurement of Species Diversity	R. K. Peet	285-307
Stream Drift as a Chronobiological Phenomenon in Running Water Ecosystems	K. Müller	309-23
The Evolution of Social Behavior	R. D. Alexander	325-83
Effects and Functions in the Evolution of Signaling Systems	D. Otte	385-417
The Demography of Plants	J. L. Harper, J. White	419-63

VOLUME 6 (1975)

The Impact of Robert MacArthur on Ecology	S. D. Fretwell	1-13
Ecological Significance of Imprinting and Early Learning	K. Immelmann	15-37
The Rumen Microbial Ecosystem	R. E. Hungate	39-66
Structure and Climate in Tropical Rain Forest	E. G. Leigh Jr.	67-86
Cannibalism in Natural Populations	L. R. Fox	87-106
Demographic Consequences of Infanticide in Man	M. Dickeman	107-37
Energetics of Pollination	B. Heinrich	139-70
Ecological Aspects of Spatial Orientation	R. Jander	171-88
Spectral Analysis in Ecology	T. Platt, K. L. Denman	189-210
The Population Biology of Coral Reef Fishes	P. R. Ehrlich	211-47
Late Quaternary Climatic Change in Africa	D. A. Livingstone	249-80
Experimental Studies of the Niche	R. K. Colwell, E. R. Fuentes	281-310
Simulation Models of Ecosystems	R. G. Wiegert	311-38
Modes of Animal Speciation	G. L. Bush	339-64
Butterfly Ecology	L. E. Gilbert, M. C. Singer	365-97

VOLUME 7 (1976)

Genetic Polymorphism in Heterogeneous Environments	P. W. Hedrick, M. E. Ginevan, E. P. Ewing	1-32
Ecological Genesis of Endangered Species: The Philosophy of Preservation	R. L. Smith	33-55
Reproductive Strategies as Adaptations to Life in Temporally Heterogeneous Environments	J. T. Giesel	57-79
Population Responses to Patchy Environments	J. A. Wiens	81-120
The Chemical Defenses of Plants to Pathogens and Herbivores	D. A. Levin	121-59
The Surface Film as a Physical Environment	P. J. Wangersky	161-76
The Size-Efficiency Hypothesis and the Size Structure of Zooplankton Communities	D. J. Hall, S. T. Threlkeld, C. W. Burns, P. H. Crowley	177-208
Evolution and Systematic Significance of Polyploidy	R. C. Jackson	209-34
Analyzing Shapes of Foraging Areas: Some Ecological and Economic Theories	A. P. Covich	235-57
Bacterial Substrates and Productivity in Marine Ecosystems	J. McN. Sieburth	259-85
Population Dynamic Models in Heterogeneous Environments	S. A. Levin	287-310
Drosophila of Hawaii: Systematics and Ecological Genetics	H. L. Carson, K. Y. Kaneshiro	311-45
Why Bamboos Wait So Long to Flower	D. H. Janzen	347-91
Integrated Control of Helminths in Human Populations	G. A. Schad, L. E. Rozeboom	393-420
Evolution and Ecological Value of Duplicate Genes	R. J. MacIntyre	421-68
The Evolution of Inbreeding in Plants	S. K. Jain	469-95

VOLUME 8 (1977)

The Ecology of Brood Parasitism in Birds	R. B. Payne	1-28
The Evolution of Viruses	A. J. Nahmias, D. C. Reanney	29-49
Carbon Balance in Terrestrial Detritus	W. H. Schlesinger	51-81
Control of Forest Growth and Distribution on Wet Tropical Mountains: with Special Reference to Mineral Nutrition	P. J. Grubb	83-107
Polymorphism in Cepaea: A Problem with Too Many Solutions?	J. S. Jones, B. H. Leith, P. Rawlings	109-43
The Evolution of Life History Traits: A		

Critique of the Theory and a Review of the Data	S. C. Stearns	145-71
Emergence of Research on Agro-Ecosystems	O. L. Loucks	173-92
A Comparison of Social Adaptations in Relation to Ecology in Gallinaceous Bird and Ungulate Societies	V. Geist	193-207
Mathematical Models of Schistosomiasis	J. E. Cohen	209-33
Drosophila Systematics and Biochemical Evolution	L. H. Throckmorton	235-54
Environmental Effects of Dams and Impoundments	R. M. Baxter	255-83
The Ecology of Fish Migrations	W. C. Leggett	285-308
Assessing Electrophoretic Similarity: The Problem of Hidden Heterogeneity	G. B. Johnson	309-28
Aphid Ecology: Life Cycles, Polymorphism, and Population Regulation	A. F. G. Dixon	329-53
A History of Savanna Vertebrates in the New World. Part 1: North America	S. D. Webb	355-80
Circannual Rhythms in Bird Migration	E. Gwinner	381-405
Extrafloral Nectaries and Protection by Pugnacious Bodyguards	B. L. Bentley	407-27
Relative Brain Size and Behavior in Archosaurian Reptiles	J. A. Hopson	429-48